# Biology

## Science for Life

### WITH PHYSIOLOGY

FIFTH EDITION

## Colleen Belk
University of Minnesota–Duluth

## Virginia Borden Maier
St. John Fisher College

PEARSON

Senior Acquisitions Editor: Star MacKenzie
Project Manager: Mae Lum
Program Manager: Leata Holloway
Development Editor: Leata Holloway
Editorial Assistant: Maja Sidzinska
Development Director: Ginnie Simione Jutson
Program Management Team Lead: Michael Early
Project Management Team Lead: David Zielonka
Production Management: Lumina Datamatics, Inc.
Copyeditor: Lumina Datamatics, Inc.
Compositor: Lumina Datamatics, Inc.
Design Manager: Mark Ong
Interior Designer: Integra
Cover Designer: Richard Leeds, BigWig Design
Illustrators: Imagineering

Rights & Permissions Project Manager:
Donna Kalal
Rights & Permissions Management: Lumina
Datamatics, Inc.
Photo Researcher: Lumina Datamatics, Inc.
Manufacturing Buyer: Stacey Weinberger
Executive Marketing Manager: Lauren Harp
Cover Photo Credit: Skull: Gianluca Fabrizio/
Moment Open/Getty Images; Genetic research:
Pgiam/E+/Getty Images; Acrospaera radiolarian:
Steve Gschmeissner/Science Photo Library;
TeguhSantosa/Moment/Getty Images; Jeff Rotman/
Stockbyte/Getty Images; MedicalRF/Corbis; Matt
Walford/Cultura/Corbis; Kennan Ward/Corbis

**Library of Congress Cataloging-in-Publication Data**

Belk, Colleen M.
   Biology : science for life, with physiology / Colleen Belk, University of Minnesota-Duluth, Virginia Borden Maier, St. John Fisher College. — Fifth edition.
      pages cm
   Includes index.
   ISBN 978-0-321-92221-2 — ISBN 0-321-92221-2
   1. Biology. 2. Physiology. I. Maier, Virginia Borden. II. Title.
   QH307.2.B43 2014b
   570—dc23
                                                                                    2014038059

2  3  4  5  6  7  8  9  10—V011—18  17  16  15

                                        Student Edition ISBN-13: 978-0-321-92221-2
                                        Student Edition ISBN-10:      0-321-92221-2
                                        A la Carte Edition ISBN-13: 978-0-133-92276-9
                                        A la Carte Edition ISBN-10:     0-133-92276-6

                                        **www.pearsonhighered.com**

# About the Authors

**Colleen Belk** and **Virginia Borden Maier** collaborated on teaching biology to non-majors for over a decade at the University of Minnesota–Duluth. This collaboration has continued for an additional decade through Virginia's move to St. John Fisher College in Rochester, New York, and has been enhanced by their differing but complementary areas of expertise. In addition to the non-majors course, Colleen Belk teaches general biology for majors, genetics, cell biology, and molecular biology courses. Virginia Borden Maier teaches general biology for majors, evolutionary biology, zoology, plant biology, ecology, and conservation biology courses.

After several somewhat painful attempts at teaching the breadth of biology to non-majors in a single semester, the two authors came to the conclusion that they needed to find a better way. They realized that their students were more engaged when they understood how biology directly affected their lives. Colleen and Virginia began to structure their lectures around stories they knew would interest students. When they began letting the story drive the science, they immediately noticed a difference in student engagement and willingness to work harder at learning biology. Not only has this approach increased student understanding, but it has also increased the authors' enjoyment in teaching the course—presenting students with fascinating stories infused with biological concepts is simply a lot more fun.

# Preface

## To the Student

Is it acceptable to clone humans? When does human life begin? What should be done about our warming planet? Who owns living organisms? What are our responsibilities toward endangered species? Having taught this course for nearly 40 combined years, we understand that no amount of knowledge alone will provide satisfactory answers to these questions. Addressing them requires the development of a scientific literacy that surpasses the rote memorization of facts. To make decisions that are individually, socially, and ecologically responsible, you must not only understand some fundamental principles of biology but also be able to use this knowledge as a tool to help you analyze ethical and moral issues involving biology. This is the aim of this textbook.

To help you understand biology and apply your knowledge to an ever-expanding suite of issues, we have structured each chapter of *Biology: Science for Life* around a compelling story in which biology plays an integral role. Through the story you not only will learn the relevant biological principles but also will see how science can be used to help answer complex questions. As you learn to apply the strategies modeled by the text, you will also be developing your critical thinking skills.

Even though you may not be planning to be a practicing biologist, well-developed critical thinking skills will enable you to make better decisions about issues that affect your own life and form well-reasoned, fact-based opinions about personal, social, and ecological issues.

## To the Instructor

You are probably all too aware that teaching non-majors students is very different from teaching biology majors. You know that most of these students will never take another formal science course; therefore, your course may be the last chance for these students to appreciate how biology is woven throughout the fabric of their lives and to develop a deep understanding of the process of science. You recognize the importance of engaging non-majors because you know that these students will one day be voting on issues of scientific importance, holding positions of power in the community, serving on juries, and making health care decisions for themselves and their families. This text is designed to help you reach your goals.

By now, most non-majors biology instructors are aware that this book differs from other books in that we use a compelling storyline woven throughout the entire chapter to garner student interest. Once we draw students in, we keep them engaged by returning to the storyline again and again until the end of the chapter, when students should be able to form their own data-driven opinions about each topic. Storylines are skillfully crafted to allow the same depth and breadth of coverage as any non-majors biology text.

Our experience has taught us that students will not remember as many facts as we hope they will, but they can and do remember how to apply the scientific method to novel questions involving biology, and they can retain a strong appreciation for how science differs from other methods of understanding the world. To ensure our students leave our course with the ability to critically evaluate information they may come across, this text focuses heavily on process of science, providing opportunities for students to practice applying the scientific method and analyze data at every opportunity.

### New to the Fifth Edition

The positive feedback obtained in previous editions assured us that presenting science alongside a story works for students and instructors alike. In the fifth edition, we have added two new features and several reorganized chapters. We also updated storylines and continued to improve popular features from previous editions as well as our supplements.

### New Features: Working with Data and Sounds Right, But Is It?

In this edition, we have added new **Working with Data** questions to select figures within each chapter. Students are asked questions that guide them in how to carefully and critically analyze and interpret data in graphical, tabular, or written form. Each chapter contains at least one of these critical data analysis questions. In Chapter 6, for example, students are asked to evaluate a graph showing the cancer risks associated with smoking.

A new end-of-chapter feature, **Sounds Right, But Is It?**, addresses common misconceptions we know that our own students often have. In Chapter 4, for example, the misconception deals with whether use of laxatives can cause permanent weight loss. To help students identify and discard such misconceptions, the description of the misconception is followed by guided inquiry questions, which lead students through a careful analysis of the reliability of the misconception, using the biological concepts from the chapter covered.

## Updated Physiology Coverage and New Chapter

Content in physiology chapters has been significantly reorganized to address concerns from instructors that too much material was covered in too few chapters; what was once covered in two chapters is now spread over three. **Chapter 17** now focuses only on tissues and organs, while the new **Chapter 18** uses a discussion of the biology of the digestive and urinary systems as a way to help students understand the biological and safety consequences of binge drinking.

## Revised Unit One Coverage

Because we have found that our students need more practice analyzing pseudoscientific information they come across, we are using **Chapter 2** of the book to build on Chapter 1's introduction to the scientific method. There, students will use their newly acquired skills to learn about life and evolution in analyzing whether zombies as they are portrayed in popular culture are "alive" and whether humans are evolutionarily progressing to become higher beings. They will learn basic biochemistry while determining whether the Bermuda Triangle is a site of massive ship and plane disappearances, whether ingesting sugar causes hyperactivity, and whether tryptophan in turkey does make people tired.

## Updated Storylines

Our chapter on cellular respiration and body weight (**Chapter 4**) incorporates new meta-data showing that being underweight is less healthy than being overweight and the health consequences of being overweight start at higher weights than once thought. Our chapter on global warming and photosynthesis (**Chapter 5**) is updated to reflect the continued global changes resulting from this process. The cell division chapter (**Chapter 6**) helps students understand the biology of differently acquired cancers using the examples of the very public battles fought by celebrities like Angelina Jolie. The protein synthesis chapter (**Chapter 9**) has been updated to reflect current developments in pet and human cloning as well as so-called genetic pharming practices.

Our review of biological diversity (**Chapter 13**) now examines the question of humanity's supposed superiority over other species. The skeletal, endocrine, and muscular system coverage (**Chapter 21**) revolves around the

2014 inclusion of women's ski jumping as an Olympic sport for the first time, and the chapter on the nervous system has been revised to focus on the phenomenon of students sharing non-prescribed ADD meds with each other (**Chapter 23**).

## Improved Pedagogy

With the previous editions, we focused on improving flexibility for instructors via **A Closer Look** chapter subsections; these are now streamlined and better identified within the text. Our popular **Roots to Remember** feature that helps students build their scientific vocabulary is now integrated into the chapter itself; students can find definitions for these terms as they occur. Features that help students assess their understanding within the chapter—**Stop and Stretch** and **Visualize This** questions—have been expanded and updated in nearly every chapter. Many **Savvy Reader** essays, found in every chapter and meant to develop students as better consumers of popular media, have been updated as well.

# Supplements and Media

For the fifth edition, we've undertaken a significant revision and updating of the complete supplements package. Judi Roux EdD, a talented college instructor with years of classroom experience in non-majors biology and colleague of Colleen Belk at the University of Minnesota, Duluth, has undertaken authoring these innovative new items. We think you will find that the supplements she developed are brimming with ideas for how to reach this particular population of students. In addition to a completely revamped Instructor's Manual (for use in traditional lectures as well as flipped classrooms) and a test bank, we also provide slides, animation, and videos to enrich instruction efforts. Available online, the *Biology: Science for Life with Physiology* resources are easy to navigate and support a variety of learning and teaching styles. Judi Roux authored not only the Instructor Guide, MasteringBiology Quiz and Test Items, but the PowerPoint lectures as well.

New features in MasteringBiology include interactive concept maps and Working with Data exercises for each chapter. And our **Learning Outcomes** continue to provide support to students and instructors by organizing the chapter summary and tagging questions and activities within MasteringBiology and other ancillary material.

We believe you will find that the design and format of this text and its supplements will help you meet the challenge of helping students both succeed in your course and develop science skills—for life.

We look forward to learning about your experience with *Biology: Science for Life with Physiology, Fifth Edition.*

# Compelling Stories Highlight the Relevance of Biology to Everyday Life

Each chapter weaves a compelling story based on a current issue or hot topic that presents, explains, and demystifies biological concepts, examples, and applications.

CHAPTER **18** | **Binge Drinking**

A student is turning 21.

18.1 The Digestive
System 416

Mechanical and Chemical
Breakdown of Food
Absorption of Digested Food
Regulation of Digestive Secretions

18.2 Removing Toxins from
the Body: The Urinary
System 420

Kidney Structure and Function
Engaging Safely with Alcohol

*savvy reader*
Sexual Assault on College
Campuses 425

SOUNDS RIGHT, **BUT IS IT?** 426

## MasteringBiology®

**NEW! Storyline PPTs** help instructors incorporate the stories into their lectures with videos and pre-made lectures.

## ▲ UPDATED!

**Six thoroughly revised storylines** have been added to the Fifth Edition to highlight the relevance of biology concepts to everyday life, along with one entirely new storyline:

- **Chapter 2:** Science Fiction, Bad Science, and Pseudoscience
- **Chapter 4:** Body Weight and Health
- **Chapter 13:** The Greatest Species on Earth?
- **NEW CHAPTER! Chapter 18:** Binge Drinking*
- **Chapter 19:** Clearing the Air*
- **Chapter 21:** Human Sex Differences*
- **Chapter 23:** Study Drugs*

* Chapters 17–25 are included in the expanded version of the text that includes coverage of animal and plant anatomy and physiology.

# The Digestive and Urinary Systems

It's Saturday night and Malik is hosting a surprise party to celebrate the 21st birthday of his friend Lin. Lin is several years younger than Malik. She lived with Malik's family, sharing a room with his younger sister, for 2 years when she was a high school exchange student. Now an international student attending college in the United States, Lin has had almost no experience with alcohol. Malik knows that Lin is eagerly anticipating this birthday and that she is planning to drink at least a little alcohol. Because he feels as protective of Lin as he does his little sister, Malik wants to help Lin learn how to enjoy the benefits of alcohol consumption while limiting the negative consequences that can also occur.

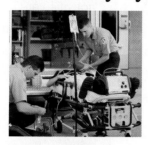

**He does not want her alcohol consumption to place her at risk of overdose …**

Some of Malik's concerns about negative consequences are based on situations he has witnessed and others on information he came across while writing a paper on alcohol abuse for a health class he took last semester.

A student who lived on Malik's dorm floor freshman year broke his ankle when he tripped while running from the police to avoid an underage consumption ticket. His chemistry lab partner broke her nose when she was riding with an intoxicated driver whose car hit a tree on a snow-covered road. While working on the paper for his health class, he came across a government website that indicated over 30,000 students required medical treatment for alcohol

poisoning last year and he does not want this to happen to Lin. He also worries about the high rate of sexual assault on college campuses. He does not want Lin to become one of the 20% of female students who will be sexually assaulted while in college.

Malik wants to develop a plan for convincing Lin, a pre-med biology major, that drinking too much is bad for her body, an argument he thinks she may find credible. Because he has heard that eating food before drinking might help absorb some of the alcohol and that alcohol consumption causes dehydration, he plans to focus his efforts on the effects of drinking on the digestive and urinary systems.

**Malik is worried about his friend Lin.**

**… or jeopardize her safety.**

◀ **NEW!**

**Chapter 18\*** covers the digestive and urinary systems, which were previously part of the chapters on the cardiovascular and respiratory systems. This new chapter presents this material in a more manageable format for instructors and students.

and alcohol is broken down and absorbed across the intestinal wall and into the bloodstream. When alcohol relaxes muscles involved with peristalsis, food spends more time in the digestive tract than normal and this increased exposure to digestive enzymes can cause diarrhea.

Malik has heard that it is good to eat a large meal before drinking. This is because the presence of food in the stomach causes the pyloric sphincter to remain closed. Since the stomach does not absorb alcohol as readily as the small intestine, preventing the alcohol from reaching the small intestine can slow the rate at which it reaches the blood stream. Therefore, Malik plans to take Lin out to eat before the birthday party.

Many of the digestive enzymes used in the small intestine are produced by an organ called the **pancreas.** Secretions from the pancreas neutralize stomach

wastes including urea and various ions. During urine **excretion,** urine leaves the kidneys and flows to the bladder.

Alcohol is a diuretic, which means that it promotes the formation of urine and increases the volume of urine that is released from the bladder, a process called **micturation.** Coupling the increased volume of urine produced with the deadening of awareness of the need to urinate that goes with intoxication can result in a very full bladder. Even though micturition is typically under conscious control, an intoxicated person that passes out before emptying the bladder may end up urinating on himself. In this case, the body overrides the conscious control of micturition to prevent a potentially lethal bladder rupture.

Alcohol is a depressant, slowing down brain function and altering perceptions, reflexes and balance, and causing slurred speech. In an attempt to prevent the depressant effects of intoxication, some of Malik's friends mix alcohol with energy drinks. Malik will recommend to Lin that she does not do this because Lin should develop an awareness of when to stop drinking. This is harder to do if the depressant effects of intoxication are, in part, masked by the stimulant effects of the energy drink.

In addition to managing wastes, the urinary system also plays an important role in regulating blood volume, acidity, and salt balance. The kidneys regulate

**The story is revisited** throughout the chapter. ▶ In these examples, the story narrative provides an opportunity for students to learn about the digestive system as they follow a student's experience with binge drinking.

# Learn and Practice Science and Literacy Skills

Building upon the popular features of previous editions, the Fifth Edition of *Biology: Science for Life* helps students develop scientific thinking skills for a lifetime of critically evaluating scientific—and pseudoscientific—information.

## NEW!

**Sounds Right, But Is It?** activities are located at the end of each chapter and challenge students to answer a series of questions that address common biology-related misconceptions.

## SOUNDS RIGHT **BUT IS IT?**

A couple with two boys is considering having another child. While they are grateful to be fertile and to have had two healthy boys, they do think it would be fun to have a girl. In investigating their odds, they came across some data on sibships, or groups of siblings with the same parents. The study they saw showed that three-fourths of sibships of three contain members of both genders versus containing all boys or all girls. The couple now believes that their next child will very likely be a girl.

**If a couple has two boys, the odds are higher than normal that their next child will be a girl.**

Sounds right, but it isn't.

1. The probabilities of independent events, that is, events not affected by previous events, are multiplied to determine their combined probability. For example, when you flip a coin twice, the outcome of the second flip is independent of the outcome of the first flip. Therefore, the likelihood of flipping heads twice is one-half times one-half or one-fourth. What is the likelihood of flipping heads three times in a row?

2. Each fertilization of an egg by a sperm is an independent event. What is the probability that a couple will have three boys in a row?

3. The total probabilities of related independent events equal one. This means that the probability that a couple with three kids will have some outcome aside from three boys is seven-eighths. Describe, in terms of gender, what those sibships could contain.

4. While it is true that in sibships with three children seven-eighths would have at least one girl, how is looking at all the possible sibships that could be produced when there are three children different than the scenario outlined above?

5. Consider your answers to questions 1–4 and explain why the original statement bolded above sounds right, but isn't.

**Sounds Right, But Is It?** misconceptions include:

- "If a product is clinically proven to do what it advertises, that means it will work for you." —Chapter 1
- "The use of tanning beds is not only safe, it improves health." —Chapter 6
- "If a couple has two boys, the odds are higher than normal that their next child will be a girl." —Chapter 8
- "The human eye is too complex to have evolved by chance from nothing." —Chapter 11
- "There is always a chance that a brain-dead person will make a full recovery." —Chapter 17*
- "The number of vaccinations given to modern children is too much for the average immune system to handle." —Chapter 20*

...and more!

**NEW! Sounds Right, But Is It?** questions in the text are also available for in-class activities using Learning Catalytics.

* Chapters 17–25 are included in the expanded version of the text that includes coverage of animal and plant anatomy and physiology.

## savvy reader

### Labeling GMOs

The following was excerpted from the GMO FREE NY (http://gmofreeny.net/thecaseforgmolabeling.html) website dated 2014 that makes the case for labeling genetically modified foods. No author of this essay is named.

"Genetic modification (GM; also called genetic engineering or GE) is biotechnology used to create new varieties of plants and animals that exhibit traits found in unrelated species, such as bacteria and viruses."

The author then points out that many countries do have laws requiring the labeling of GMO foods, and goes on to say, "Americans have been eating GMOs without their knowledge or consent since 1996. We are the ultimate guinea pigs."

1. The author chose to focus on modifications involving the transfer of genes from bacteria and viruses to crop foods instead of the transfer of a gene from one crop food to another. Why do you think the author focused on bacteria and viruses instead of other kinds of organisms?

2. Assume a food was modified with bacterial or viral DNA. What happens to the DNA of that, or any, food after it is ingested: Is it used to produce bacterial and viral proteins or is it broken down?

3. If you read the entire essay, you would find that the author focuses only on the potential negative outcomes of producing and ingesting GMOs. For example, consider the suggestion above that humans ingesting unlabeled GMOs are being treated like guinea pigs. If the author had softened his or her argument by stating that crops engineered to be more nutritious have not lived up to the initial hype surrounding them, or that the benefits of these crops may not outweigh the risks, would his or her argument for labeling foods seem more credible to you? Why or why not?

4. An opinion piece does not have the same requirements for presenting substantiating evidence as a conventional news story. In this piece, the author uses strong language, often punctuates sentences with exclamation points, does not present alternate opinions, and possibly overstates conclusions in an attempt to convince you that GM foods should be labeled as such. Did this strategy work on you, or would a more balanced approach have been more likely to get you to agree with his or her opinion?

Source: http://gmofreeny.net/thecaseforgmolabeling.html

---

**Savvy Reader** activities in each chapter investigate a short excerpt from a variety of current news sources relating to discussions in the main chapter narrative. Critical thinking questions help students evaluate scientific information and data presented in the media.

---

### Part E - Evaluation

Finally, how can you use your assessment of the authority, motivation, and reliability of the information to evaluate this web site relative to other sources? Use the scales below to assign a numerical score to this source.

| Authority | Motivation | Reliability |
|---|---|---|
| **1** — Source is a recognized authority (e.g., .gov or .edu). | **1** — Content is balanced and informational. | **1** — • Primary sources cited for all claims. • Consistent with authoritative sources. |
| **0** — Source is a nonexpert with some relevant credentials. | **0** — Motivation is unclear. | **0** — • Primary sources cited for some claims. • Not as comprehensive as other sources. |
| **-1** — Source is a nonexpert with no relevant credentials. | **-1** — Content promotes an agenda. | **-1** — • No primary sources cited for any claims. • Contradicts authoritative sources. |

Assign a numerical score for each category. Then add up the total score. (The highest possible score is 3; the lowest is −3.) In what range does this source fall?

- ○ −3 to −2
- ○ −1 to 1
- ○ 2 to 3

Submit    My Answers   <u>Give Up</u>

---

## MasteringBiology®

**NEW!** **Savvy Reader: Evaluating Sources** activities ask students to examine a website, article, or video with a critical eye on the sources and methods used to convey information.

# Engage with Data and Visual Information

The hallmark illustration style of previous editions has been enhanced in the Fifth Edition with new pedagogy to help students interpret data and other visual information.

## Working with Data ▼

Is the cancer risk associated with smoking and drinking additive or multiplicative? Explain your answer.

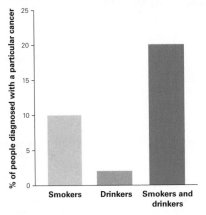

**FIGURE 6.2 Alcohol and tobacco are synergists.** Smoking cigarettes while drinking is an unhealthy practice.

## Working with Data ▶

The line looks relatively flat from 8000 B.C.E. to 1500 B.C.E., though the population doubled four times in that period. Was the population growing exponentially at this time?

2010: 7.2 billion — 7
1999: 6 billion — 6
1970: 4 billion — 5
1930: 2 billion — 4
1800: 1 billion — 3
Black Death: 14th Century — 2
Dawn of Christianity: 250 million
Egyptian empire: 100 million — 1

Agricultural era: 5 million

Human population (billions)

Year
10,000 B.C.E   8000   6000   4000   2000   0   2000 C.E

**FIGURE 14.3 Exponential growth.** The number of people on Earth grew relatively slowly until the eighteenth century. The rapid growth since then has occurred in proportion to the total, causing a J-shaped curve.

◀ ▲ NEW!

**Working with Data** questions have been added to the figure legends of selected graphs, tables, or figures, and challenge students to closely interpret the data.

## MasteringBiology®

**NEW!** **Working with Data** assignments are available for each chapter and ask students to analyze and apply their knowledge of biology to a graph or a set of data.

## Visualize This ▶

Evaporation occurs when molecules at the surface of a liquid "escape" into a gaseous phase. Where would most of these escaped molecules appear on this figure and why?

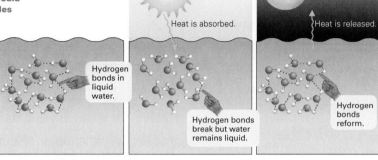

Heat is absorbed.

Heat is released.

Hydrogen bonds in liquid water.

Hydrogen bonds break but water remains liquid.

Hydrogen bonds reform.

**FIGURE 5.3 Hydrogen bonding in water.** Hydrogen bonds break as they absorb heat and reform as water releases heat.

## Visualize This ▼

Based only on structures shown in this figure, can you guess which parts of the virus are most likely to help it attach to a cell?

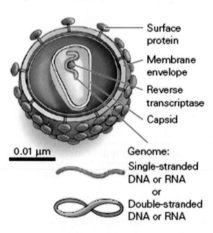

Surface protein

Membrane envelope

Reverse transcriptase

Capsid

0.01 µm

Genome:
Single-stranded DNA or RNA
or
Double-stranded DNA or RNA

**FIGURE 20.4 Viral structure.** Viruses are composed of genetic material surrounded by a protein coat. Some viruses, including the one shown, are also surrounded by an envelope.

## ◀ EXPANDED!

**Visualize This** questions within selected figure legends encourage students to look more closely at figures to more fully understand their content.

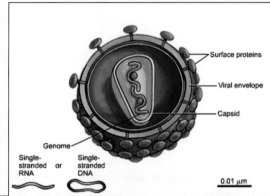

A virus does not contain cytoplasm or organelles. The viral genome can be linear or circular and can consist of single or double strands of RNA or DNA as well as other combinations. The genome is enclosed in a protein coat called the capsid; there is often an additional layer, the viral envelope, around the capsid. Many surface proteins also stick out of the outer layer of the viral coat.

Surface proteins

Viral envelope

Capsid

Genome

Single-stranded or RNA

Single-stranded DNA

0.01 µm

STOP | PLAY

BACK TO INTRO

# MasteringBiology®

**Narrated Animations** of selected figures from the text can be assigned in MasteringBiology as activities with assessment questions that include answer-specific feedback and hints.

# Tools to Learn and Visualize Key Concepts

Colleen Belk and Virginia Borden Maier incorporate many classroom-tested teaching techniques into the Fifth Edition prose and illustrations, making it easier for students to learn and remember unfamiliar biology concepts.

## The Process of Evolution

Generally, the word *evolution* means "change," and the process of evolution reflects this definition as it applies to populations of organisms. A **biological population** is a group of individuals of the same species that is somewhat independent of other groups, often isolated from them by geography. **Biological evolution**, then, is a change in the characteristics of a biological population that occurs over the course of generations. The changes in populations that are considered evolutionary are those that are passed from parent to offspring via genes.

evol- means to unroll.

### NEW! ▲

**Roots to Remember** references have been added in context within chapter discussions to help students learn the language of biology using word roots.

**A Roots to Remember summary** is also provided at the end of each chapter for quick reference.

## Roots to Remember

**These roots come from Greek or Latin and will help you decode the meaning of words:**

| | |
|---|---|
| evol- | means to unroll. Chapter term: *evolution* |
| homini- | means human-like. Chapter terms: *hominid, hominin* |
| homolog- | indicates similar or shared origin; from a word meaning "in agreement." Chapter terms: *homologous, homology* |
| macro- | means large scale. Chapter term: *macroevolution* |
| -metric | means to measure. Chapter term: *radiometric* |
| micro- | means extremely small. Chapter term: *microevolution* |
| radio- | is the combining form of radiation. Chapter term: *radiometric* |

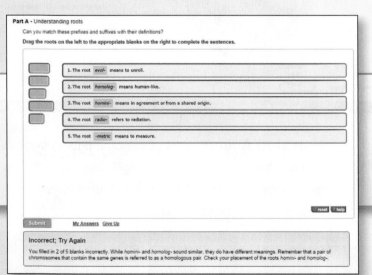

**Part A - Understanding roots**

Can you match these prefixes and suffixes with their definitions?

Drag the roots on the left to the appropriate blanks on the right to complete the sentences.

1. The root *evol-* means to unroll.
2. The root *homolog-* means human-like.
3. The root *homini-* means in agreement or from a shared origin.
4. The root *radio-* refers to radiation.
5. The root *-metric* means to measure.

↻ reset ? help

Submit    My Answers   Give Up

**Incorrect; Try Again**

You filled in 2 of 5 blanks incorrectly. While *homin-* and *homolog-* sound similar, they do have different meanings. Remember that a pair of chromosomes that contain the same genes is referred to as a homologous pair. Check your placement of the roots *homin-* and *homolog-*.

## MasteringBiology®

**Roots to Remember** coaching activities provide a fun, interactive way to learn word roots.

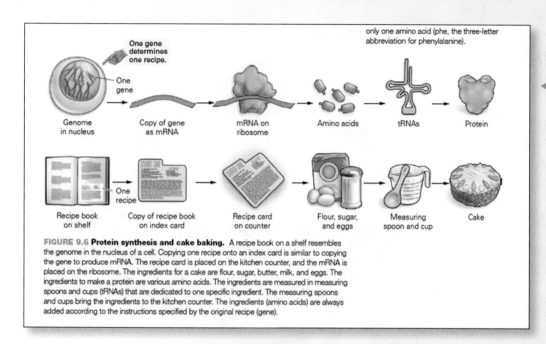

only one amino acid (phe, the three-letter abbreviation for phenylalanine).

One gene determines one recipe.

Genome in nucleus → Copy of gene as mRNA → mRNA on ribosome → Amino acids → tRNAs → Protein

Recipe book on shelf → One recipe → Copy of recipe book on index card → Recipe card on counter → Flour, sugar, and eggs → Measuring spoon and cup → Cake

**FIGURE 9.6 Protein synthesis and cake baking.** A recipe book on a shelf resembles the genome in the nucleus of a cell. Copying one recipe onto an index card is similar to copying the gene to produce mRNA. The recipe card is placed on the kitchen counter, and the mRNA is placed on the ribosome. The ingredients for a cake are flour, sugar, butter, milk, and eggs. The ingredients to make a protein are various amino acids. The ingredients are measured in measuring spoons and cups (tRNAs) that are dedicated to one specific ingredient. The measuring spoons and cups bring the ingredients to the kitchen counter. The ingredients (amino acids) are always added according to the instructions specified by the original recipe (gene).

◀ **Unique Visual Analogies** compare abstract science with familiar objects and experiences to help students grasp complex biology concepts.

**Illustrated Tables** ▶ organize information in one place and provide easy visual references to compare and contrast.

**TABLE 7.2 To what extent is IQ heritable?** A summary of various estimates of IQ heritability, their shortcomings, and the problems with using them to understand the role of genes in determining an individual's potential intelligence.

| Method of Measurement | Estimated Percentage of Phenotype Determined by Genes | Warnings When Interpreting This Result | Warnings That Apply to All Measurements of Heritability |
|---|---|---|---|
| Correlation between parents' IQ and children's IQ in a population | 42% | When parents and children live together, a correlation can't rule out environmental influence. | • Heritability values are specific to the populations for which they were measured. • High heritability for a trait does not mean that the trait will not respond to a change in the environment. • Heritability is a measure of a population, not an individual. |
| Natural experiment comparing IQ in pairs of identical twins versus nonidentical twins | 52% | Because identical twins are treated as more alike than nonidentical twins the heritability value could be an overestimate. | |
| Natural experiment comparing IQ of identical twins raised apart versus nonidentical twins raised apart | 72% | Small sample size may skew results. | |

# Support For Your Students Anytime, Anywhere

**MasteringBiology®** is an online homework, tutorial, and assessment program that helps you quickly master biology concepts and skills. Self-paced tutorials provide immediate wrong-answer feedback and hints to help keep you on track to succeed in the course.

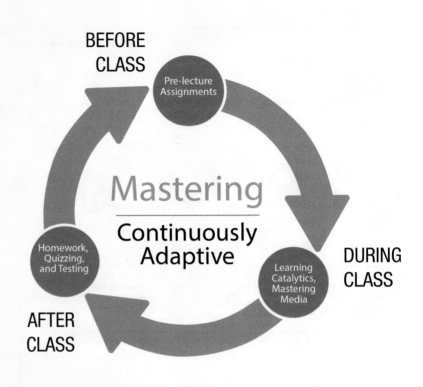

BEFORE CLASS — Pre-lecture Assignments

DURING CLASS — Learning Catalytics, Mastering Media

AFTER CLASS — Homework, Quizzing, and Testing

## Mastering
### Continuously Adaptive

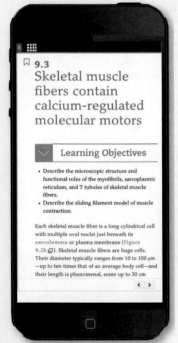

**eText 2.0**

**Dynamic Study Modules**

## BEFORE CLASS

### NEW!

◀ **eText 2.0** Allow your students to access their text anytime, anywhere.

- Now available on Smartphones and Tablets.
- Seamlessly integrated digital and media resources.
- Fully accessible (screen-reader ready).
- Configurable reading settings, including resizable type and night reading mode.
- Instructor and student note-taking, highlighting, bookmarking and search.

### NEW!

◀ **Dynamic Study Modules** help students acquire, retain, and recall information faster and more efficiently than ever before. These convenient practice questions and detailed review explanations can be accessed using a smartphone, tablet, or computer.

# DURING CLASS

## NEW!

**Learning Catalytics** is an assessment and classroom activity system that works with any web-enabled device and facilitates collaboration with your classmates. Your MasteringBiology subscription with eText includes access to Learning Catalytics.

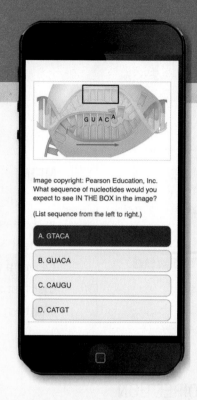

## NEW!

**Everyday Biology Videos** briefly explore interesting and relevant biology topics that relate to concepts in the course. These 20 videos, produced by the BBC, can be shown in class or assigned as homework in MasteringBiology.

# AFTER CLASS

**A wide range of question types and activities** are available for homework assignments, including the following **NEW** assignment options for the Fifth Edition:

- **Interactive Storyline Activities** tie the storyline of the chapter to key science concepts.
- **Working with Data questions** require you to analyze and apply your knowledge of biology to a graph or set of data.
- **Savvy Reader Evaluating Media activities** challenge you to evaluate various information from websites, articles, and videos.

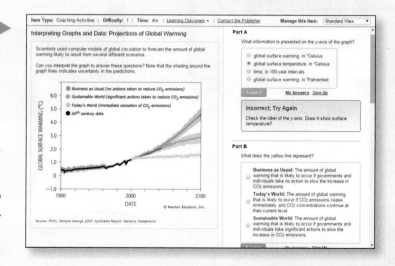

# New Resources for Flipped Classrooms and More

**New resources save valuable time both during course prep and during class.**

## NEW!

**Learning Catalytics** is a "bring your own device" assessment and classroom activity system that expands the possibilities for student engagement. Using Learning Catalytics, instructors can deliver a wide range of auto-gradable or open-ended questions that test content knowledge and build critical thinking skills. Eighteen different answer types provide great flexibility, including:

### SKETCH/DIRECTION

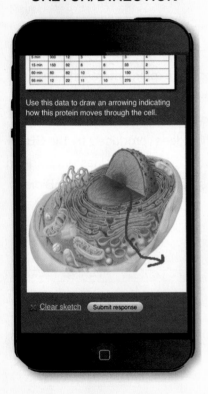

Use this data to draw an arrowing indicating how this protein moves through the cell.

✕ Clear sketch   Submit response

### MANY CHOICE

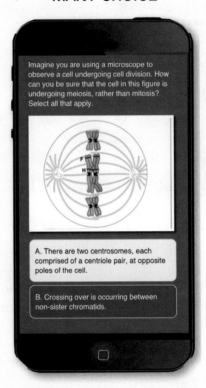

Imagine you are using a microscope to observe a cell undergoing cell division. How can you be sure that the cell in this figure is undergoing meiosis, rather than mitosis? Select all that apply.

A. There are two centrosomes, each comprised of a centriole pair, at opposite poles of the cell.

B. Crossing over is occurring between non-sister chromatids.

### REGION

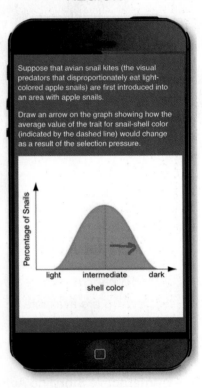

Suppose that avian snail kites (the visual predators that disproportionately eat light-colored apple snails) are first introduced into an area with apple snails.

Draw an arrow on the graph showing how the average value of the trait for snail-shell color (indicated by the dashed line) would change as a result of the selection pressure.

**MasteringBiology**®

MasteringBiology users may select from Pearson's library of Learning Catalytics questions, including two **NEW** types of questions developed from the **Stop and Stretch** and **Sounds Right, But Is It?** questions in the Fifth Edition of *Biology: Science for Life.*

# NEW!

**"Flipped Classroom" Instructor's Manual** includes many activities that have been tested by Colleen Belk, Virginia Borden Maier, and their colleagues in their own classes. Each text chapter is supplemented with a selection of in-class activities, suggestions for student "pre-work" outside of class, media references, and more. In addition, teaching tip videos by the authors are available in MasteringBiology.

---

**Lecture Activity 6.5: Meiosis Walk**

**Estimated Time to Complete:** 15–20 minutes

**Introduction:** This activity will engage students in acting out the events of meiosis. Each student will play the role of a sister chromatid. The students will act out the motions of the chromosomes during both meiotic divisions, ultimately producing four daughter cells with unique collections of chromosomes. This activity reinforces the mechanics of meiotic division.

**Material**
- Arm bands or bandanas. You will need 16 total, four each of four different colors.

**Procedures**

You will need 16 students to simulate meiosis in a cell having four pairs of chromosomes. If you wish (and if you have the space), you can modify this activity to accommodate a larger number of students, but it doesn't work well with fewer than 16. (If you have fewer than 16 students you can use pop bead chromosomes and have the students use these to simulate meiosis in small groups.) Students who are watching should be able to see the process (they can encircle the area in which the "chromosomes" will be moving), and they typically enjoy the simulation.

1. Give each participant an arm band or a bandana; those having the same color should find each other and pair up. Members of a pair will link arms to represent sister chromatids linked by a centromere (the linked arms). Ideally, each foursome will include two men and two women: The two men would link arms to represent a paternal chromosome, and the two women to represent the maternal chromosome.
2. Once you have eight chromosomes (four pairs of homologues), begin the simulation as follows: Have the linked pairs cluster in the middle of the room, representing the nucleus. They can wander around, with homologous pairs not spending any more time near each other than near other chromosomes.
3. Designate a line to serve as the equator of the cell, and two points to serve as poles.

---

## 2.1 A Definition of Life

| Living Humans | Zombies |
|---|---|
| • Grow | • do not grow from child to adult |
| • Move | • can move; hindered by injuries |
| • Reproduce and pass genetic information to offspring | • do not produce offspring; do not pass genetic information |
| • Respond to external stimuli | • respond to limited stimuli |
| • Metabolize | • do not metabolize human flesh for nourishment |
| • Maintain homeostasis | • limited homeostatic abilities do not promote healing |

# NEW!

**Storyline PPTs** for instructors allow easy integration of the stories into lecture. The PPT presentations include integrated story examples and video launcher segments to engage students.

---

# MasteringBiology®

These valuable resources are available to adopting instructors and can be downloaded from the Instructor Resources area of MasteringBiology.

# Acknowledgments

## Reviewers

Each chapter of this book was thoroughly reviewed several times as it moved through the development process. Reviewers were chosen on the basis of their demonstrated talent and dedication in the classroom. Many of these reviewers are already trying various approaches to actively engage students in lectures and to raise the scientific literacy and critical thinking skills among their students. Their passion for teaching and commitment to their students were evident throughout this process. These devoted individuals scrupulously checked each chapter for scientific accuracy, readability, and coverage level.

All of these reviewers provided thoughtful, insightful feedback, which improved the text significantly. Their efforts reflect their deep commitment to teaching non-majors and improving the scientific literacy of all students. We are very thankful for their contributions.

### Reviewers of the Fifth Edition

Joseph Ahlander  *Northeastern State University*
Josephine Arogyasami  *Southern Virginia University*
Veronica Barr  *Heartland Community College*
Kelly Barry  *Southern Illinois State University*
Katrinka Bartush  *University of North Texas*
Drew Benson  *Georgia Gwinnett College*
Wendy Birky  *California State University, Northridge*
Anne Bower  *Philadelphia University*
Jamie Burchill  *California State University, Northridge*
Rebecca Burton  *Alverno College*
David Byres  *Florida State College, Jacksonville*
Cassandra Cantrell  *Western Kentucky University*
Michelle Cawthorn  *Georgia Southern University*
Reggie Cobb  *Nash Community College*
Angela Costanzo  *Hawaii Pacific University*
James B. Courtright  *Marquette University*
Richard Cowart  *Coastal Bend Community College*
Melissa Deadmond  *Truckee Meadows Community College*
Tcherina Duncombe  *Palm Beach Community College*
Donna Ewing  *McLellan Community College*
Michele Finn  *Monroe Community College*
Barbara Frank  *Idaho State University*
Janet Gaston  *Troy University*
Richard Gill  *Brigham Young University*
Rebekka Gougis  *Illinois State University*
Tamar Goulet  *University of Mississippi*
Eileen Gregory  *Rollins College*

Jay Hatch  *University of Minnesota*
Jay Hodgson  *Armstrong Atlantic State University*
Staci Johnson  *Southern Wesleyan University*
Trey Kidd  *University of Missouri, St. Louis*
Sarah Krajewski  *Grand Rapids Community College*
Lorraine Leiser  *Southeast Community College*
Linda Moore  *Georgia Military College*
Alex Olvido  *University of North Georgia*
Jennifer O'Malley  *Saint Charles Community College*
Brent Palmer  *University of Kentucky*
Murali Panen  *Luzerne County Community College*
Monica Parker  *Florida State College*
Shelly Penrod  *Lonestar College*
Krista Peppers  *University of Central Arkansas*
Anne-Marie Prouty  *Sam Houston State University*
Yelena Rudayeva  *Palm Beach Community College*
Bill Simcik  *Lonestar College*
Indrani Sindhuvalli  *Florida State College, Jacksonville*
Jack Shurley  *Idaho State University*
Marialana Spiedel  *Jefferson College*
Brooke Stabler  *University of Central Oklahoma*
Jennifer Stovall  *Southcentral Kentucky Community & Technical College*
Sue Trammell  *John A Logan College*
Kimberly Turk  *Caldwell Community College*
Sandra Walsh  *The Citadel*
Mark Walvoord  *University of Oklahoma*
Michael Wenzel  *Folsom Lake College*
Heather Wilson-Ashworth  *Utah Valley University*

### Reviewers of Previous Editions

Daryl Adams, *Minnesota State University, Mankato*
Karen Aguirre, *Clarkson University*
Marcia Anglin, *Miami-Dade College*
Susan Aronica, *Canisius College*
Mary Ashley, *University of Chicago*
James S. Backer, *Daytona Beach Community College*
Ellen Baker, *Santa Monica College*
Gail F. Baker, *LaGuardia Community College*
Neil R. Baker, *The Ohio State University*
Andrew Baldwin, *Mesa Community College*
Thomas Balgooyen, *San Jose State University*
Tamatha R. Barbeau, *Francis Marion University*
Sarah Barlow, *Middle Tennessee State University*
Andrew M. Barton, *University of Maine, Farmington*
Vernon Bauer, *Francis Marion University*
Paul Beardsley, *Idaho State University*

Donna Becker, *Northern Michigan University*
Tania Beliz, *College of San Mateo*
David Belt, *Penn Valley Community College*
Steve Berg, *Winona State University*
Carl T. Bergstrom, *University of Washington*
Janet Bester-Meredith, *Seattle Pacific University*
Barry Beutler, *College of Eastern Utah*
Donna H. Bivans, *Pitt Community College*
Lesley Blair, *Oregon State University*
John Blamire, *City University of New York, Brooklyn College*
Barbara Blonder, *Flagler College*
Susan Bornstein-Forst, *Marian College*
Bruno Borsari, *Winona State University*
James Botsford, *New Mexico State University*
Robert S. Boyd, *Auburn University*
Bryan Brendley, *Gannon University*
Eric Brenner, *New York University*
Peggy Brickman, *University of Georgia*
Carol Britson, *University of Mississippi*
Carole Browne, *Wake Forest University*
Neil Buckley, *State University of New York, Plattsburgh*
Stephanie Burdett, *Brigham Young University*
Warren Burggren, *University of North Texas*
Nancy Butler, *Kutztown University*
Suzanne Butler, *Miami-Dade Community College*
Wilbert Butler, *Tallahassee Community College*
Tom Campbell, *Pierce College, Los Angeles*
Merri Casem, *California State University, Fullerton*
Anne Casper, *Eastern Michigan University*
Deborah Cato, *Wheaton College*
Peter Chabora, *Queens College*
Bruce Chase, *University of Nebraska, Omaha*
Thomas F. Chubb, *Villanova University*
Gregory Clark, *University of Texas, Austin*
Kimberly Cline-Brown, *University of Northern Iowa*
Mary Colavito, *Santa Monica College*
William H. Coleman, *University of Hartford*
William F. Collins III, *Stony Brook University*
Walter Conley, *State University of New York, Potsdam*
Jerry L. Cook, *Sam Houston State University*
Melanie Cook, *Tyler Junior College*
Scott Cooper, *University of Wisconsin, La Crosse*
Erica Corbett, *Southeastern Oklahoma State University*
George Cornwall, *University of Colorado*
Charles Cottingham, *Frederick Community College*
Angela Cunningham, *Baylor University*
Judy Dacus, *Cedar Valley College*
Judith D'Aleo, *Plymouth State University*
Deborah Dardis, *Southeastern Louisiana University*
Juville Dario-Becker, *Central Virginia Community College*
Garry Davies, *University of Alaska, Anchorage*
Miriam del Campo, *Miami-Dade Community College*
Judith D'Aleo, *Plymouth State University*
Edward A. DeGrauw, *Portland Community College*
Heather DeHart, *Western Kentucky University*

Miriam del Campo, *Miami-Dade Community College*
Veronique Delesalle, *Gettysburg College*
Lisa Delissio, *Salem State College*
Beth De Stasio, *Lawrence University*
Elizabeth Desy, *Southwest Minnesota State University*
Donald Deters, *Bowling Green State University*
Gregg Dieringer, *Northwest Missouri State*
Diane Dixon, *Southeastern Oklahoma State University*
Christopher Dobson, *Grand Valley State University*
Cecile Dolan, *New Hampshire Community Technical College, Manchester*
Matthew Douglas, *Grand Rapids Community College*
Lee C. Drickamer, *Northern Arizona University*
Dani DuCharme, *Waubonsee Community College*
Susan Dunford, *University of Cincinnati*
Stephen Ebbs, *Southern Illinois University*
Douglas Eder, *Southern Illinois University, Edwardsville*
Patrick J. Enderle, *East Carolina University*
William Epperly, *Robert Morris College*
Ana Escandon, *Los Angeles Harbor College*
Dan Eshel, *City University of New York, Brooklyn College*
Steve Eisenberg, *Elizabethtown Community and Technical College*
Marirose Ethington, *Genessee Community College*
Deborah Fahey, *Wheaton College*
Chris Farrell, *Trevecca Nazarene University*
Richard Firenze, *Broome Community College*
Lynn Firestone, *Brigham Young University*
Susan Fisher, *Ohio State University*
Brandon L. Foster, *Wake Technical Community College*
Richard A. Fralick, *Plymouth State University*
Barbara Frank, *Idaho State University*
Stewart Frankel, *University of Hartford*
Lori Frear, *Wake Technical Community College*
Jennifer Fritz, *The University of Texas at Austin*
David Froelich, *Austin Community College*
Suzanne Frucht, *Northwest Missouri State University*
Edward Gabriel, *Lycoming College*
Anne Galbraith, *University of Wisconsin, La Crosse*
Patrick Galliart, *North Iowa Area Community College*
Wendy Garrison, *University of Mississippi*
Anthony Gaudin, *Ivy Tech Community College of Indiana—Columbus/Franklin*
Alexandros Georgakilas, *East Carolina University*
Robert George, *University of North Carolina, Wilmington*
Tammy Gillespie, *Eastern Arizona College*
Sharon Gilman, *Coastal Carolina University*
Mac F. Given, *Neumann College*
Bruce Goldman, *University of Connecticut, Storrs*
Andrew Goliszek, *North Carolina Agricultural and Technical State University*
Beatriz Gonzalez, *Sante Fe Community College*
Eugene Goodman, *University of Wisconsin, Parkside*
Lara Gossage, *Hutchinson Community College*
Tamar Goulet, *University of Mississippi*

Becky Graham, *University of West Alabama*
Mary Rose Grant, *University of Missouri, St. Louis*
John Green, *Nicholls State University*
Robert S. Greene, *Niagara University*
Tony J. Greenfield, *Southwest Minnesota State University*
Bruce Griffis, *Kentucky State University*
Mark Grobner, *California State University, Stanislaus*
Michael Groesbeck, *Brigham Young University, Idaho*
Stanley Guffey, *University of Tennessee*
Mark Hammer, *Wayne State University*
Blanche Haning, *North Carolina State University*
Robert Harms, *St. Louis Community College*
Craig M. Hart, *Louisiana State University*
Patricia Hauslein, *St. Cloud State University*
Stephen Hedman, *University of Minnesota, Duluth*
Bethany Henderson-Dean, *University of Findlay*
Julie Hens, *University of Maryland University College*
Peter Heywood, *Brown University*
Julia Hinton, *McNeese State University*
Phyllis C. Hirsh, *East Los Angeles College*
Elizabeth Hodgson, *York College of Pennsylvania*
Leland Holland, *Pasco-Hernando Community College*
Jane Horlings, *Saddleback Community College*
Margaret Horton, *University of North Carolina, Greensboro*
Laurie Host, *Harford Community College*
David Howard, *University of Wisconsin, La Crosse*
Michael Hudecki, *State University of New York, Buffalo*
Michael E. S. Hudspeth, *Northern Illinois University*
Laura Huenneke, *New Mexico State University*
Pamela D. Huggins, *Fairmont State University*
Sue Hum-Musser, *Western Illinois University*
Carol Hurney, *James Madison University*
James Hutcheon, *Georgia Southern University*
Anthony Ippolito, *DePaul University*
Richard Jacobson, *Laredo Community College*
Malcolm Jenness, *New Mexico Institute of Technology*
Carl Johansson, *Fresno City College*
Ron Johnston, *Blinn College*
Thomas Jordan, *Pima Community College*
Jann Joseph, *Grand Valley State University*
Mary K. Kananen, *Penn State University, Altoona*
Arnold Karpoff, *University of Louisville*
Judy Kaufman, *Monroe Community College*
Michael Keas, *Oklahoma Baptist University*
Judith Kelly, *Henry Ford Community College*
Karen Kendall-Fite, *Columbia State Community College*
Andrew Keth, *Clarion University*
David Kirby, *American University*
Stacey Kiser, *Lane Community College*
Dennis Kitz, *Southern Illinois University, Edwardsville*
Carl Kloock, *California State, Bakersfield*
Jennifer Knapp, *Nashville State Technical Community College*
Loren Knapp, *University of South Carolina*
Michael A. Kotarski, *Niagara University*
Michelle LaBonte, *Framingham State College*

Phyllis Laine, *Xavier University*
Dale Lambert, *Tarrant County College*
Tom Langen, *Clarkson University*
Michael L' Annunziata, *Pima College*
Lynn Larsen, *Portland Community College*
Mark Lavery, *Oregon State University*
Brenda Leady, *University of Toledo*
Mary Lehman, *Longwood University*
Doug Levey, *University of Florida*
Lee Likins, *University of Missouri, Kansas City*
Abigail Littlefield, *Landmark College*
Andrew D. Lloyd, *Delaware State University*
Jayson Lloyd, *College of Southern Idaho*
Suzanne Long, *Monroe Community College*
Judy Lonsdale, *Boise State University*
Kate Lormand, *Arapahoe Community College*
Paul Lurquin, *Washington State University*
Kimberly Lyle-Ippolito, *Anderson University*
Douglas Lyng, *Indiana University/Purdue University*
Michelle Mabry, *Davis and Elkins College*
Stephen E. MacAvoy, *American University*
Molly MacLean, *University of Maine*
Charles Mallery, *University of Miami*
Cindy Malone, *California State University, Northridge*
Mark Manteuffel, *St. Louis Community College, Flo Valley*
Ken Marr Green, *River Community College*
Kathleen Marrs, *Indiana University/Purdue University*
Roger Martin, *Brigham Young University, Salt Lake Center*
Matthew J. Maurer, *University of Virginia's College at Wise*
Geri Mayer, *Florida Atlantic University*
T. D. Maze, *Lander University*
Steve McCommas, *Southern Illinois University, Edwardsville*
Colleen McNamara, *Albuquerque Technical Vocational Institute*
Mary McNamara, *Albuquerque Technical Vocational Institute*
John McWilliams, *Oklahoma Baptist University*
Susan T. Meiers, *Western Illinois University*
Diane Melroy, *University of North Carolina, Wilmington*
Joseph Mendelson, *Utah State University*
Paige A. Mettler-Cherry, *Lindenwood University*
Debra Meuler, *Cardinal Stritch University*
James E. Mickle, *North Carolina State University*
Craig Milgrim, *Grossmont College*
Hugh Miller, *East Tennessee State University*
Jennifer Miskowski, *University of Wisconsin, La Crosse*
Ali Mohamed, *Virginia State University*
Stephen Molnar, *Washington University*
James Mone, *Millersville University*
Daniela Monk, *Washington State University*
David Mork, *Yakima Valley Community College*
Bertram Murray, *Rutgers University*
Ken Nadler, *Michigan State University*
John J. Natalini, *Quincy University*
Alissa A. Neill, *University of Rhode Island*
Dawn Nelson, *Community College of Southern Nevada*

Joseph Newhouse, *California University of Pennsylvania*
Jeffrey Newman, *Lycoming College*
David L.G. Noakes, *University of Guelph*
Shawn Nordell, *St. Louis University*
Tonye E. Numbere, *University of Missouri, Rolla*
Lori Nicholas, *New York University*
Jorge Obeso, *Miami-Dade College, North Campus*
Erin O'Brien, *Dixie College*
Igor Oksov, *Union County College*
Kevin Padian, *University of California, Berkeley*
Arnas Palaima, *University of Mississippi*
Anthony Palombella, *Longwood University*
Marilee Benore Parsons, *University of Michigan, Dearborn*
Steven L. Peck, *Brigham Young University*
Javier Penalosa, *Buffalo State College*
Murray Paton Pendarvis, *Southeastern Louisiana University*
Rhoda Perozzi, *Virginia Commonwealth University*
John Peters, *College of Charleston*
Patricia Phelps, *Austin Community College*
Polly Phillips, *Florida International University*
Indiren Pillay, *Culver-Stockton College*
Francis J. Pitocchelli, *Saint Anselm College*
Nancy Platt, *Pima College*
Roberta L. Pohlman, *Wright State University*
Calvin Porter, *Xavier University*
Linda Potts, *University of North Carolina, Wilmington*
Robert Pozos, *San Diego State University*
Marion Preest, *The Claremont Colleges*
Gregory Pryor, *Francis Marion University*
Rongsun Pu, *Kean University*
Narayanan Rajendran, *Kentucky State University*
Anne E. Reilly, *Florida Atlantic University*
Michael H. Renfroe, *James Madison University*
Laura Rhoads, *State University of New York, Potsdam*
Ashley Rhodes, *Kansas State University*
Gwynne S. Rife, *University of Findlay*
Todd Rimkus, *Marymount University*
Laurel Roberts, *University of Pittsburgh*
Wilma Robertson, *Boise State University*
Bill Rogers, *Ball State University*
William E. Rogers, *Texas A&M University*
Troy Rohn, *Boise State University*
Deborah Ross, *Indiana University/Purdue University*
Christel Rowe, *Hibbing Community College*
Joanne Russell, *Manchester Community College*
Michael Rutledge, *Middle Tennessee State University*
Wendy Ryan, *Kutztown University*
Christopher Sacchi, *Kutztown University*
Kim Sadler, *Middle Tennessee State University*
Brian Sailer, *Albuquerque Technical Vocational Institute*
Jasmine Saros, *University of Wisconsin, La Crosse*
Ken Saville, *Albion College*
Michael Sawey, *Texas Christian University*
Louis Scala, *Kutztown University*
Debbie Scheidemantel, *Pima College*

Daniel C. Scheirer, *Northeastern University*
Beverly Schieltz, *Wright State University*
Nancy Schmidt, *Pima Community College*
Robert Schoch, *Boston University*
Julie Schroer, *Bismarck State College*
Fayla Schwartz, *Everett Community College*
Steven Scott, *Merritt College*
Gray Scrimgeour, *University of Toronto*
Roger Seeber, *West Liberty State College*
Mary Severinghaus, *Parkland College*
Allison Shearer, *Grossmont College*
Robert Shetlar, *Georgia Southern University*
Cara Shillington, *Eastern Michigan University*
Beatrice Sirakaya, *Pennsylvania State University*
Cynthia Sirna, *Gadsden State Community College*
Lynnda Skidmore, *Wayne County Community College*
Thomas Sluss, *Fort Lewis College*
Brian Smith Black, *Hills State University*
Douglas Smith, *Clarion University of Pennsylvania*
Mark Smith, *Chaffey College*
Gregory Smutzer, *Temple University*
Sally Sommers, *Smith Boston University*
Anna Bess Sorin, *University of Memphis*
Bryan Spohn Florida, *Community College at Jacksonville, Kent Campus*
Carol St. Angelo, *Hofstra University*
Amanda Starnes, *Emory University*
Susan L. Steen, *Idaho State University*
Timothy Stewart, *Longwood College*
Shawn Stover, *Davis and Elkins College*
Bradley J. Swanson, *Central Michigan University*
Joyce Tamashiro, *University of Puget Sound*
Jeffrey Taylor, *Slippery Rock University*
Martha Taylor, *Cornell University*
Glena Temple, *Viterbo*
Tania Thalkar, *Clarion University of Pennsylvania*
Jeff Thomas, *California State University, Northridge*
Alice Templet, *Nicholls State University*
Jeffrey Thomas, *University of California, Los Angeles*
Janis Thompson, *Lorain County Community College*
Nina Thumser, *California University of Pennsylvania*
Alana Tibbets, *Southern Illinois University, Edwardsville*
Martin Tracey, *Florida International University*
Jeffrey Travis, *State University of New York, Albany*
Michael Troyan, *Pennsylvania State University*
Robert Turgeon, *Cornell University*
Michael Tveten, *Pima Community College, Northwest Campus*
James Urban, *Kansas State University*
Brandi Van Roo, *Framingham State College*
John Vaughan, *St. Petersburg Junior College*
Martin Vaughan, *Indiana State University*
Mark Venable, *Appalachian State University*
Paul Verrell, *Washington State University*
Tanya Vickers, *University of Utah*

Janet Vigna, *Grand Valley State University*
Sean Walker, *California State University, Fullerton*
Don Waller, *University of Wisconsin, Madison*
Tracy Ware, *Salem State College*
Jennifer Warner, *University of North Carolina, Charlotte*
Derek Weber, *Raritan Valley Community College*
Carol Weaver, *Union University*
Frances Weaver, *Widener University*
Elizabeth Welnhofer, *Canisius College*
Marcia Wendeln, *Wright State University*
Shauna Weyrauch, *Ohio State University, Newark*
Wayne Whaley, *Utah Valley State College*
Howard Whiteman, *Murray State University*
Jennifer Wiatrowski, *Pasco-Hernando Community College*
Vernon Wiersema, *Houston Community College*
Gerald Wilcox, *Potomac State College*
Peter J. Wilkin, *Purdue University North Central*
Robert R. Wise, *University of Wisconsin, Oshkosh*
Michelle Withers, *Louisiana State University*
Art Woods, *University of Texas, Austin*
Elton Woodward, *Daytona Beach Community College*
Kenneth Wunch, *Sam Houston State University*

Donna Young, *University of Winnipeg*
Michelle L. Zjhra, *Georgia Southern University*
John Zook, *Ohio University*
Michelle Zurawski, *Moraine Valley Community College*

## The Book Team

We feel blessed to be able to work with Star MacKenzie, our editor for the last three editions, and our new development editor Leata Holloway. Both of these women are insightful, funny, kind, and generous with their time. Their commitment to producing an excellent book that meets the needs of students and instructors is unrivaled in the industry.

This book is dedicated to our families, friends, and colleagues who have supported us over the years. Having loving families, great friends, and a supportive work environment has enabled us to make this heartfelt contribution to non-majors biology education.

COLLEEN BELK AND
VIRGINIA BORDEN MAIER

*"Because science, told as a story, can intrigue and inform the non-scientific minds among us, it has the potential to bridge the two cultures into which civilization is split— the sciences and the humanities. For educators, stories are an exciting way to draw young minds into the scientific culture."*

—E.O. WILSON

# Brief Contents

# Contents

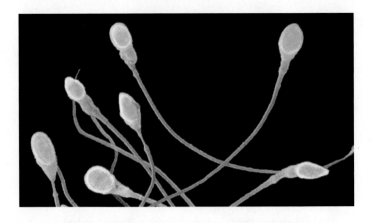

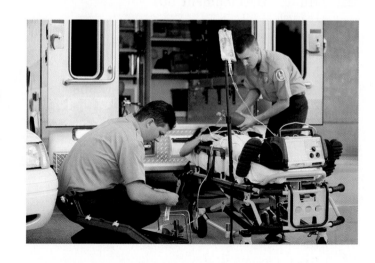

CHAPTER **1**

# Can Science Cure the Common Cold?

**Another cold! What can I do?**

# Introduction to the Scientific Method

We have all been there—you just recover from one bad head cold and on a morning soon after you notice that scratchy feeling in your throat that signals a new one is about to begin. It is always at the worst time, too, when you have an important exam coming up, a term paper due, and a packed social calendar. Why are you sick yet again? What can you do about it?

If you ask your friends and relatives, you will hear the usual advice on how to prevent and treat colds: Take massive doses of vitamin C. Suck on zinc lozenges. Drink plenty of echinacea tea. Meditate. Get more rest. Exercise vigorously every day. Put that hat on when you go outside! You are left with an overwhelming list of options, often contradictory and some contrary to common sense.

**Take massive doses of vitamin C?**

**Drink echinacea tea?**

If you keep up with health news, you may be even more confused. One website reports that a popular over-the-counter cold treatment is effective, while a local TV news story details the risks of using this remedy and highlights its ineffectiveness. How do you decide what to do?

Faced with this bewildering situation, most people follow the advice that makes the most sense to them, and if they find they still feel terrible, they try another remedy. Testing ideas and discarding ones that don't work is a kind of "everyday science." This technique has its limitations—for example, even if you feel better after trying a new cold treatment, you can't know if your recovery occurred because the treatment was effective or because the cold was ending anyway.

What professional scientists do is a more refined version of this everyday science—using strategies that help eliminate other possible explanations for a result. And while some fields of science may use unintelligible words or complicated and expensive equipment, the basic process for testing ideas is simple and universal to all areas of science. An understanding of this process can help you evaluate information about many issues that may concern and intrigue you—from health issues, to global warming, to the origin of life and the universe—with more confidence. In this chapter, we introduce you to the powerful process scientists use by asking the question we've considered here: Is there a cure for the common cold?

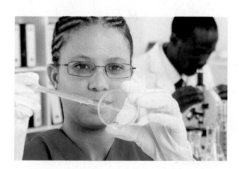

**How would a scientist determine which advice is best?**

# 1.1 The Process of Science

The term *science* can refer to a body of knowledge—for example, the science of **biology** is the study of living organisms. You may believe that science requires near-perfect recall of specific sets of facts about the world. In reality, this goal is impossible and unnecessary—we do have reference books, after all. The real action in science is not memorizing what is already known but using the process of science to discover something new and unknown.

This process—making observations of the world, proposing ideas about how something works, testing those ideas, and discarding (or modifying) our ideas in response to the test results—is the essence of the **scientific method.** The scientific method allows us to solve problems and answer questions efficiently and effectively. Can we use the scientific method to solve the complicated problem of preventing and treating colds?

**bio-** means life.

**-ology** means the study of or branch of knowledge about.

## The Nature of Hypotheses

The statements our friends and family make about which actions will help us remain healthy (for example, the advice to wear a hat) are in some part based on the advice giver's understanding of how our bodies resist colds. Ideas about "how things work" are called **hypotheses.** Or, more formally, a hypothesis is a proposed explanation for one or more observations.

Hypotheses in biology come from knowledge about how the body and other biological systems work, experiences in similar situations, our understanding of other scientific research, and logical reasoning; they are also shaped by our creative mind (**FIGURE 1.1**). When your mom tells you to dress warmly to avoid colds, she is basing her advice on the following hypothesis: Becoming chilled makes you more susceptible to illness.

**hypo-** means under, below, or basis.

The hallmark of science is that hypotheses are subject to rigorous testing. Therefore, scientific hypotheses must be **testable**—it must be possible to evaluate a hypothesis through observations of the measurable universe. Not all hypotheses are testable. For instance, the statement that "colds are generated by disturbances in psychic energy" is not a scientific hypothesis because psychic energy has not been demonstrated to exist and thus cannot be measured in a test.

**FIGURE 1.1 Hypothesis generation.**
All of us generate hypotheses. Many different factors, both logical and creative, influence the development of a hypothesis. Scientific hypotheses are both testable and falsifiable.

## Visualize This ▶

**Most colleges require students who are science majors to take courses in the humanities and social sciences, just as they require students in these majors to take science courses. What skills do scientists gain from humanities and social sciences that can help them be better scientists?**

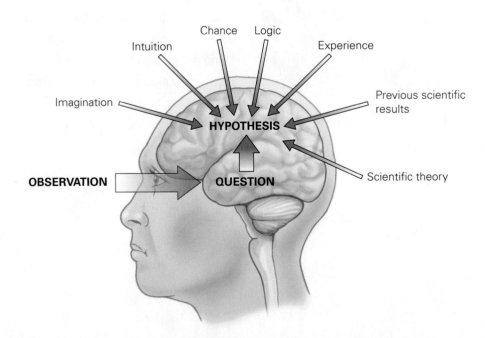

In addition, hypotheses that require the intervention of a supernatural force cannot be tested scientifically. If something is **supernatural**, it is not constrained by any laws of nature, and its behavior cannot be predicted using our current understanding of the natural world. A scientific hypothesis must also be **falsifiable**; that is, an observation or set of observations could potentially prove it false. The hypothesis that exposure to cold temperatures increases your susceptibility to colds is falsifiable; we can imagine an observation that would cause us to reject this hypothesis (for instance, the observation that people exposed to cold temperatures do not catch more colds than people protected from chills). Of course, not all hypotheses are proved false, but it is essential in science that incorrect ideas be discarded, which can occur only if it is *possible* to prove those ideas false. Lack of falsifiability is another reason supernatural hypotheses cannot be scientific. Because a supernatural force can cause any possible result, hypotheses that rely on supernatural forces cannot be falsified.

Finally, statements that are value judgments, such as, "It is wrong to cheat on an exam," are not scientific because different people have different ideas about right and wrong. It is impossible to falsify these types of statements. To find answers to questions of morality, ethics, or justice, we turn to other methods of gaining understanding—such as philosophy and religion.

## Scientific Theories

Most hypotheses fit into a larger picture of scientific understanding. We can see this relationship when examining how research upended a commonly held belief about diet and health—that chronic stomach and intestinal inflammation is caused by eating too much spicy food. This belief directed the standard medical practice for ulcer treatment for decades. Patients with ulcers were prescribed drugs that reduced stomach acid levels and advised to avoid eating acidic or highly spiced foods. These treatments were rarely successful, and ulcers were considered chronic problems.

In 1982, Australian scientists Robin Warren and Barry Marshall discovered that the bacterium *Helicobacter pylori* was present in nearly all samples of ulcer tissue that they examined (**FIGURE 1.2**). From this observation, Warren and Marshall reasoned that *H. pylori* infection—invasion of the stomach wall by the bacteria—was the cause of most ulcers. Barry Marshall even tested this hypothesis on himself by consuming live *H. pylori*. He subsequently suffered from acute stomach pain.

Warren and Marshall's colleagues were at first unconvinced that ulcers could have such a simple cause. Today, the hypothesis that *H. pylori* infection is responsible for most ulcers is accepted as fact. Why is this the case? First, no reasonable alternative hypotheses about the causes of ulcers (for instance, consumption of spicy foods) has been consistently supported by hypothesis tests; and second, the hypothesis has not been rejected—that is, there have been no carefully designed experiments that show that *H. pylori* removal fails to cure most ulcers.

The third reason that the relationship between *H. pylori* and ulcers is considered fact is that it conforms to a well-accepted scientific principle, namely, the germ theory of disease. A **scientific theory** is an explanation for a set of related observations that is based on well-supported hypotheses from several different, independent lines of research. The basic premise of germ theory is that microorganisms (that is, organisms too small to be seen with the naked eye) are the cause of some or all human diseases.

The biologist Louis Pasteur first observed that bacteria cause milk to become sour. From this observation, he reasoned that these same types of organisms could injure humans. Later, Robert Koch demonstrated a link between anthrax bacteria and a specific set of fatal symptoms in mice, providing additional evidence for the theory. Germ theory is further supported by the observation that

**(a)**

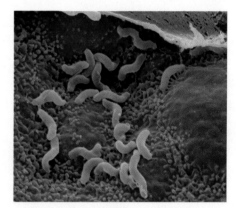

**(b)**

**FIGURE 1.2 A scientific breakthrough.** (a) *Helicobacter pylori* on stomach lining (image from electron microscope). (b) Robin Warren and Barry Marshall won the 2005 Nobel Prize in Medicine for their discovery of the link between *H. pylori* and ulcers.

**If you ask your friends and relatives, you will hear the usual advice on how to prevent and treat colds.**

---

**induc-** means to rely on reason to derive principles (also, to cause to happen).

---

---

**deduc-** means to reason out, working from facts.

---

antibiotic treatment that targets particular microorganisms can cure certain illnesses—as is the case with bacteria-caused ulcers.

In everyday speech, the word *theory* is synonymous with untested ideas based on little information. In contrast, scientists use the term when referring to well-supported ideas of how the natural world works. The supporting foundation of all scientific theories is multiple hypothesis tests.

## The Logic of Hypothesis Tests

One common hypothesis about cold prevention is that taking vitamin C supplements keeps you healthy. This hypothesis is very appealing, especially given the following generally accepted facts:

1. Fruits and vegetables contain a lot of vitamin C.
2. People with diets rich in fruits and vegetables are generally healthier than people who skimp on these food items.
3. Vitamin C is known to be an anti-inflammatory agent, reducing throat and nose irritation.

With these facts in mind, we can state the following testable and falsifiable hypothesis: Consuming vitamin C decreases the risk of catching a cold. This hypothesis makes sense given the statements just listed and the experiences of the many people who insist that vitamin C keeps them healthy.

The process used to construct this hypothesis is called **inductive reasoning**—combining a series of specific observations (here, statements 1–3) to discern a general principle. Inductive reasoning is an essential tool for understanding the world. However, a word of caution is in order: Just because the inductive reasoning that led to a hypothesis seems to make sense does not mean that the hypothesis is necessarily true. The example that follows demonstrates this point.

Consider the ancient hypothesis that the sun revolves around Earth. This hypothesis was induced based on the observations that the sun rose in the east every morning, traveled across the sky, and set in the west every night. For almost all of history, this hypothesis was considered to be a "fact" by nearly all of Western society. It wasn't until the early seventeenth century that this hypothesis was overturned—as the result of Galileo Galilei's observations of Venus. His observations proved false the hypothesis that the sun revolved around Earth. Galileo's work helped to confirm the more modern hypothesis, proposed by Nicolaus Copernicus, that Earth revolves around the sun.

So, even though the hypothesis about vitamin C is sensible, it needs to be tested to see if it can be proved false. Hypothesis testing is based on **deductive reasoning** or deduction. Deduction involves using a general principle to predict an expected observation. This **prediction** concerns the outcome of an action, test, or investigation. In other words, the prediction is the result we expect from a hypothesis test.

Deductive reasoning takes the form of "if/then" statements. That is, if our idea is correct, then we expect to observe a specific outcome from a hypothesis test. A prediction based on the vitamin C hypothesis could be: *If* vitamin C decreases the risk of catching a cold, *then* people who take vitamin C supplements with their regular diets will experience fewer colds than will people who do not take supplements.

**STOP & STRETCH** Consider the following scenario: Your coworker, Homer, is a notorious doughnut lover who has a nose for free food. You walk into the break room one morning to discover a box from the doughnut shop that is already completely empty. According to this information, what most likely happened to the doughnuts? Is this inductive or deductive reasoning?

Deductive reasoning, with its resulting predictions, is a powerful method for testing hypotheses. However, the structure of such a statement means that hypotheses can be clearly rejected if untrue but impossible to prove if true. This shortcoming is illustrated using the if/then statement concerning vitamin C and colds (**FIGURE 1.3**).

Consider the possible outcomes of a comparison between people who supplement with vitamin C and those who do not. People who take vitamin C supplements may suffer through more colds than people who do not; they may have the same number of colds as the people who do not supplement; or supplementers may in fact experience fewer colds. What does each of these results tell us about the hypothesis?

If, in a well-designed test, people who take vitamin C have more colds or the same number of colds as those who do not supplement, then the hypothesis that vitamin C provides protection against colds can be clearly rejected. But what if people who supplement with vitamin C do experience fewer colds? If this is the case, then we can only say that the hypothesis has been supported and not disproven.

Why is it impossible to say from this possible experimental result that the hypothesis that vitamin C prevents colds is true? Because there are **alternative hypotheses** that explain why people with different vitamin-taking habits vary in their cold susceptibility. In other words, demonstrating the truth of the *then* portion of a deductive statement does not prove that the *if* portion is true.

> **STOP & STRETCH**  Consider this if/then statement: If your coworker Homer ate all the doughnuts, then there won't be any doughnuts left in the box that was placed in the break room 30 minutes ago. Imagine you find that the doughnuts are all gone—did you just prove that Homer ate them? Why or why not?

Consider the alternative hypothesis that frequent exercise reduces susceptibility to catching a cold. And suppose that people who take vitamin C supplements are more likely to engage in regular exercise. If both of these hypotheses are true, then the prediction that vitamin C supplementers experience fewer colds than people who do not supplement would be true but not because the original hypothesis (vitamin C reduces the risk of colds) is true. Instead, people who take vitamin C supplements experience fewer colds because they are also more likely to exercise, and it is exercise that reduces cold susceptibility.

A hypothesis that seems to be true because it has not been rejected by an initial test may be rejected later because of a different test. This is what happened to the hypothesis that vitamin C consumption reduces susceptibility to colds. The argument for the power of vitamin C was popularized in 1970 by Nobel Prize–winning chemist Linus Pauling. Pauling based his assertion—that large doses of vitamin C reduce the incidence of colds by as much as 45%—on the results of a few studies that had been published between the 1930s and 1970s. However, repeated, careful tests of this hypothesis have since failed to support it. In many of the studies Pauling cited, it appears that alternative hypotheses explain the difference in cold incidence between vitamin C supplementers and nonsupplementers. Today, most health scientists agree that the hypothesis that vitamin C prevents colds has been convincingly falsified.

## Visualize This ▶

**According to this flowchart, scientists should consider alternative hypotheses even if their hypothesis is supported by their research. Explain why this is the case.**

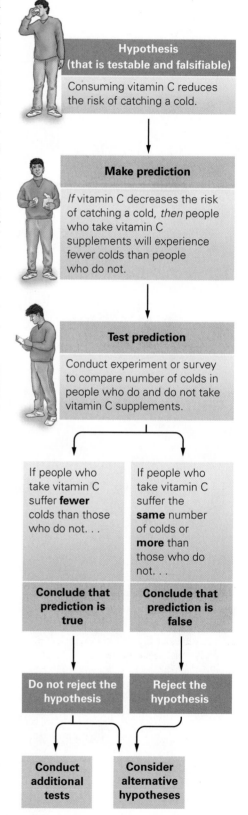

**FIGURE 1.3 The scientific method.** Tests of hypotheses follow a logical path. This flowchart illustrates the process of deduction as practiced by scientists.

The example of the vitamin C hypothesis also highlights a challenge of communicating scientific information. You can see why the belief that vitamin C prevents colds is so widespread. If you don't know that scientific knowledge relies on rejecting incorrect ideas, a book by a Nobel Prize–winning scientist may seem like the last word on the benefits of vitamin C. It took many years of careful research to show that this "last word" was, in fact, wrong.

# 1.2 Hypothesis Testing

The previous discussion may seem discouraging: How can scientists determine the truth of any hypothesis when there is always a chance that the hypothesis could be falsified? Even if one of the hypotheses about cold prevention is supported, does the difficulty of eliminating alternative hypotheses mean that we will never know which approach is truly best? The answer is yes—and no.

Hypotheses cannot be proven absolutely true; it is always possible that the true cause of a phenomenon may be found in a hypothesis that has not yet been tested. However, in a practical sense, a hypothesis can be proven beyond a reasonable doubt. That is, when one hypothesis has not been disproven through repeated testing and all reasonable alternative hypotheses have been eliminated, scientists accept that the well-supported hypothesis is, in a practical sense, true. *Truth* in science can therefore be defined as *what we know and understand based on all currently available information*. But scientists always leave open the possibility that what seems true now may someday be proven false.

An effective way to test many hypotheses is through rigorous scientific experiments. Experimentation has enabled scientists to prove beyond a reasonable doubt that the common cold is caused by a virus. A virus is a microscopic entity with a simple structure—it typically contains a short strand of genetic material and a few proteins encased in a relatively tough protein shell and sometimes surrounded by a membrane. A virus must infect a cell to reproduce. Of the over 200 types of viruses that are known to cause the common cold, most infect the cells in our noses and throats. The sneezing, coughing, congestion, and sore throat of a cold appear to result from the body's protective response—established by our immune system—to a viral invasion (**FIGURE 1.4**).

As you may know, if we survive a viral infection, we are unlikely to experience a recurrence of the disease the virus causes. For example, it is extremely rare to suffer from chicken pox twice because one exposure to the chicken pox virus (through either infection or vaccination) usually provides lifelong immunity to future infection. The huge variety of cold viruses makes immunity to the common cold—and the development of a vaccine to prevent it—very improbable. Scientists thus focus their experimental research about common colds on methods of prevention and treatment.

## The Experimental Method

**Experiments** are sets of actions or observations designed to test specific hypotheses. Generally, an experiment allows a scientist to control the conditions that may affect the subject of study. Manipulating the environment allows a scientist to eliminate some alternative hypotheses that may explain the result.

Experimentation in science is analogous to what a mechanic does when diagnosing a car problem. There are many reasons why a car engine might not start. If a mechanic begins by tinkering with numerous parts to apply all possible fixes before restarting the car, she will not know what exactly caused the problem (and will have an unhappy customer who is charged for unnecessary parts and labor). Instead, a mechanic begins by testing the battery for

# Visualize This ▼

Find two points in this process where intervention by drugs or other treatment could disrupt either the virus or the immune response and therefore lead to fewer cold symptoms.

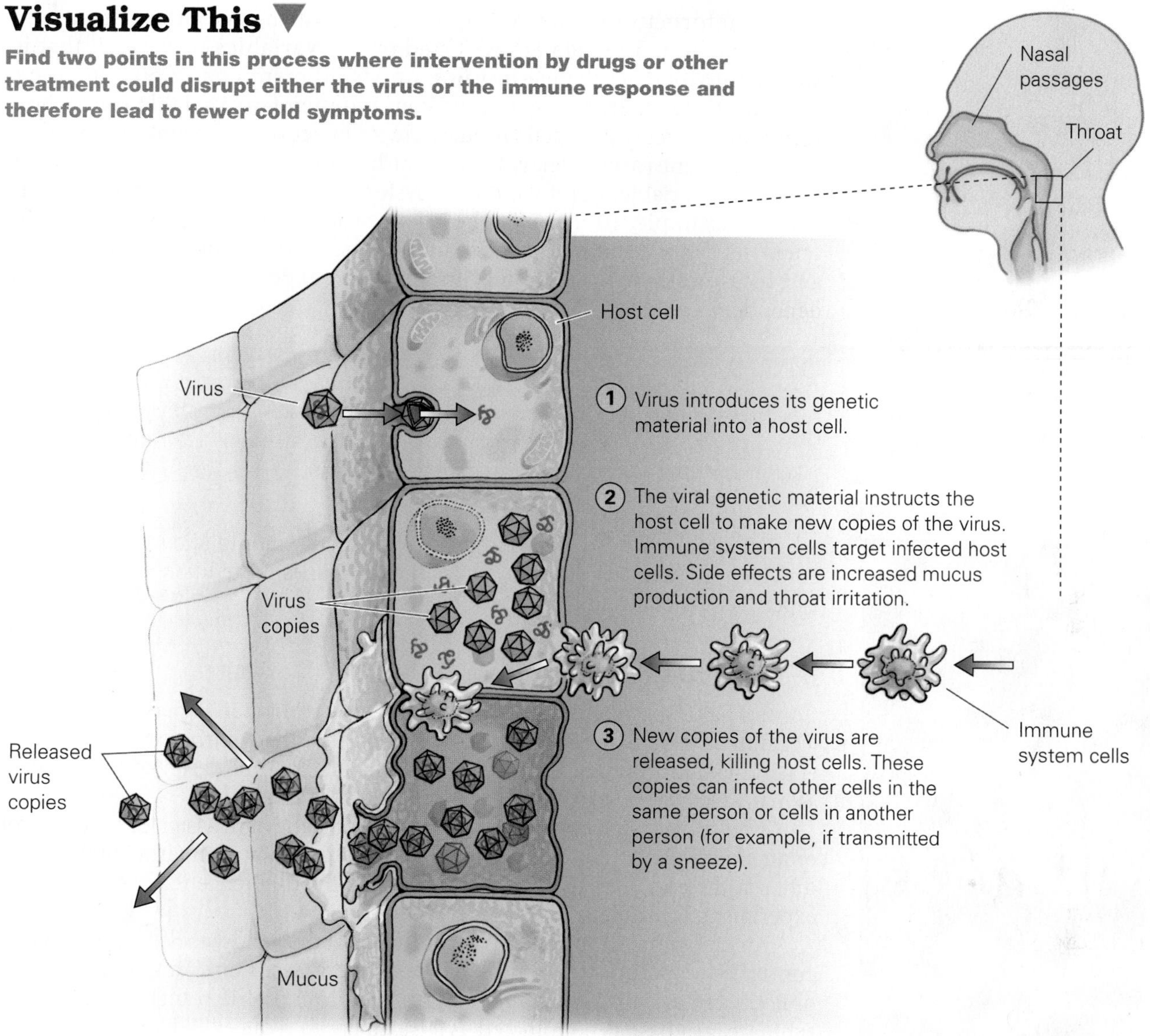

Nasal passages

Throat

Host cell

Virus

**①** Virus introduces its genetic material into a host cell.

**②** The viral genetic material instructs the host cell to make new copies of the virus. Immune system cells target infected host cells. Side effects are increased mucus production and throat irritation.

Virus copies

**③** New copies of the virus are released, killing host cells. These copies can infect other cells in the same person or cells in another person (for example, if transmitted by a sneeze).

Immune system cells

Released virus copies

Mucus

**FIGURE 1.4 A cold-causing virus.** A rhinovirus causes illness by invading nose and throat cells and using them as "factories" to make virus copies. Cold symptoms result from immune system attempts to eliminate the virus.

**FIGURE 1.5 Testing hypotheses through observation.** The fossil record provides a source of data to test hypotheses about evolutionary history.

power; if the battery is charged, then she checks the starter motor; if the car still doesn't start, she looks over the fuel pump; and she continues in this manner until identifying the problem. Likewise, a scientist systematically attempts to eliminate hypotheses that do not explain a particular phenomenon.

Not all scientific hypotheses can be tested through experimentation. For instance, hypotheses about how life on Earth originated or the cause of dinosaur extinction are usually not testable in this way. These hypotheses are instead tested using careful observation of the natural world. For instance, the examination of fossils and other geological evidence allows scientists to test hypotheses regarding the extinction of the dinosaurs (**FIGURE 1.5**).

# Even if you feel better after trying a new cold treatment, you can't know if your recovery occurred because the treatment was effective.

**FIGURE 1.6 *Echinacea purpurea,* an American coneflower.** Extracts from the leaves and roots of this plant are among the most popular herbal remedies sold in the United States.

The information collected by scientists during hypothesis testing is known as **data.** The data are collected on the **variables** of the test, that is, any factor that can change in value under different conditions. In an experimental test, scientists manipulate an **independent variable** (one whose value can be freely changed) to measure its effect on a **dependent variable.** The dependent variable may or may not be influenced by changes in the independent variable, but it cannot be systematically changed by the researchers. For example, to measure the effect of vitamin C on cold prevention, scientists can vary individuals' vitamin C intake (the independent variable) and measure their susceptibility to illness on exposure to a cold virus (the dependent variable).

**STOP & STRETCH** Some of the experiments that established the link between *H. pylori* and stomach ulcer consisted of feeding gerbils *H. pylori* and measuring the rate of ulcer development compared to untreated animals. What are the independent and dependent variables in these experiments?

Data obtained from well-designed experiments should allow researchers to convincingly reject or support a hypothesis. This is more likely to occur if the experiment is controlled.

## Controlled Experiments

Control has a specific meaning in science. A **control** for an experiment is a subject similar to an experimental subject except that the control is not exposed to experimental treatment. Controlled experiments are thus designed to eliminate as many alternative hypotheses as possible.

Once subjects are enlisted in an experiment, they are assigned to a control or an experimental group. If members of the control and experimental groups differ at the end of a well-designed test, then the difference may be due to the experimental treatment.

Our question about effective cold treatments lends itself to a variety of controlled experiments on possible drug therapies. For example, an extract of *Echinacea purpurea* (a common North American prairie plant) in the form of echinacea tea has been promoted as a treatment to reduce the likelihood as well as the severity and duration of colds (**FIGURE 1.6**). A scientific experiment on the efficacy of *Echinacea* involved asking individuals suffering from colds to rate the effectiveness of a tea in relieving their symptoms. In this study, people who used echinacea tea felt that it was 33% more effective. The "33% more effective" is in comparison to the rated effectiveness of a tea that did not contain *Echinacea* extract—that is, the results from the control group.

Control groups and experimental groups must be as similar as possible to each other to eliminate alternative hypotheses that could explain the results. In the case of cold treatments, the groups should not systematically differ in age, diet, stress level, or other factors that might affect cold susceptibility. One effective way to minimize differences between groups is the **random assignment** of individuals. For example, a researcher might put all of the volunteers' names in a hat, draw out half, and designate the people drawn as the experimental group and the remaining people as the control group. As a result, each group should be a rough cross-section of the population in the study. In the echinacea tea experiment just described, members of both the experimental and control groups were female employees of a nursing home who sought relief from their colds at their employer's clinic. The volunteers were randomly assigned to either the experimental or control group as they came into the clinic.

**STOP** & **STRETCH** In the echinacea tea experiment, a nonrandom assignment scheme might have put the first 25 visitors to the clinic in the control group and the next 25 in the experimental group. Imagine that in this version of the experiment, the experimental group did recover in fewer days than the control group. Describe an alternative hypothesis related to the nonrandom assignment of subjects that was not eliminated by this experimental design and that could explain these results.

The second step in designing a well-controlled experiment is to treat all subjects identically during the course of the experiment. In this study, all participants, whether in the control or experimental group, received the same information about the supposed benefits of echinacea tea, and during the course of the experiment, all participants were given tea to drink five to six times daily until their symptoms subsided. However, individuals in the control group received "sham tea" that did not contain *Echinacea* extract. Treating all participants the same ensures that no factor related to the interaction between subject and researcher influences the results.

The sham tea in this experiment would be equivalent to the sugar pills that are given to the control group during drug trials. Like other intentionally ineffective medical treatments, sham tea is a **placebo.** Employing a placebo generates only one consistent difference between individuals in the two groups—in this case, the type of tea they consumed.

In the echinacea tea study, the data indicated that cold severity was lower in the experimental group compared to those who received a placebo. Because their study utilized controls, the researchers can be confident that the groups differed only because of the effect of *Echinacea*. By reducing the likelihood that alternative hypotheses could explain their results, the researchers could strongly infer that they were measuring a real, positive effect of echinacea tea on colds (**FIGURE 1.7**).

The study described here supports the hypothesis that echinacea tea reduces the severity of colds. However, it is extremely rare that a single experiment will cause the scientific community to accept a hypothesis beyond a reasonable doubt. Dozens of studies, each using different experimental designs and many using extracts from different parts of the plant, have investigated the effect of *Echinacea* on common colds and other illnesses. Some of these studies have shown a positive effect, but many others have shown none. In the medical community as a whole, there is significant skepticism and disagreement regarding the effectiveness of this popular herb as a cold treatment. Only through continued controlled tests of the hypothesis will we discern an accurate answer to the question: Is *Echinacea* an effective cold treatment?

## Minimizing Bias in Experimental Design

Scientists and human research subjects may have strong opinions about the truth of a particular hypothesis even before it is tested. These opinions may cause participants to influence, or **bias,** the results of an experiment.

One potential source of bias is subject expectation. Individual experimental subjects may consciously or unconsciously model the behavior they feel the researcher expects from them. For example, an individual who knew she was receiving echinacea tea may have felt confident that she would recover more quickly. This might cause her to underreport her cold symptoms. This potential problem is avoided by designing a **blind experiment,** in which individual subjects are not aware of exactly what they are predicted to experience. In experiments on drug treatments, this means not telling participants whether they are receiving the drug or a placebo.

# Working with Data ▼

**Cover the bottom part of the lower graph to imagine that it was drawn so that the range of the vertical axis was from 2 to 5 rather than from 0 to 5. Does this affect how you might interpret the results?**

**(a)**

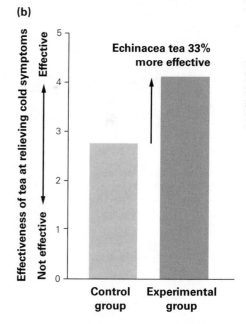

**(b)**

FIGURE 1.7 **A controlled experiment.** (a) In a controlled experiment testing echinacea tea as a treatment for colds, all 95 subjects were treated identically except for the type of tea they were given. (b) The results of the experiment indicated that echinacea tea was 33% more effective than the placebo in treating cold symptoms.

**What professional scientists do is a more refined version of everyday science.**

Another source of bias arises when a researcher makes consistent errors in the measurement and evaluation of results. This phenomenon is called *observer bias*. In the echinacea tea experiment, observer bias could take various forms. Expecting a particular outcome might lead a scientist to give slightly different instructions about which symptoms constituted a cold to subjects who received echinacea tea. Or, if the researcher expected people who drank echinacea tea to experience fewer colds, she might make small errors in the measurement of cold severity that influenced the final result.

To avoid the problem of experimenter bias, the data collectors themselves should be "blind." Ideally, the scientist, doctor, or technician applying the treatment does not know which group (experimental or control) any given subject is part of until after all data have been collected and analyzed (**FIGURE 1.8**). Blinding the data collector ensures that the data are **objective:** in other words, without bias.

We call experiments **double-blind** when both the research subjects and the technicians performing the measurements are unaware of either the hypothesis or whether a subject is in the control or experimental group. Double-blind experiments nearly eliminate the effects of human bias on results. When both researcher and subject have few expectations about the outcome, the results obtained from an experiment are more credible.

## Using Correlation to Test Hypotheses

Double-blind, placebo-controlled, randomized experiments represent the gold standard for medical research. However, well-controlled experiments can be difficult to perform when humans are the subjects. The requirement that both experimental and control groups be treated nearly identically means that some people unknowingly receive no treatment. In the case of healthy volunteers with head colds, the placebo treatment of sham tea did not hurt those who received it.

However, placebo treatments are impractical or unethical in many cases. For instance, imagine testing the effectiveness of a birth control drug using a

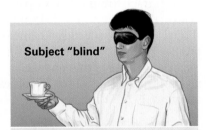

|  | Technician "blind" | Subject "blind" |
|---|---|---|
| What they know | • **Limited knowledge** of experimental hypothesis<br><br>• **No knowledge** of which group participants belong to | • **Limited knowledge** of experimental hypothesis<br><br>• **No knowledge** of which group he or she belongs to |
| How they behave | • **No difference** in instructions to participants<br><br>• **No difference** in treatment of participants<br><br>• **No difference** in data collection | • **Unbiased** reporting of symptoms or effects of treatment |

**FIGURE 1.8 Double-blind experiments.**   Double-blind experiments result in data that are more objective.

controlled experiment. This would require asking women to take a pill that may or may not prevent pregnancy while not using any other form of birth control!

## Experiments on Model Systems.
Scientists can use **model systems** when testing hypotheses that would raise ethical or practical problems when tested on people. For basic research that helps us understand how cells and genes function, the model systems are easily grown and manipulated organisms, such as certain species of bacteria, nematodes, and fruit flies, or even isolated cells from larger organisms that reproduce in dishes in the laboratory. In the case of research on human health and disease, model systems are typically other mammals and human cells. Nonhuman mammals are especially useful as model organisms in medical research because they are closely related to us. Like us, they have hair and produce milk for their young, and thus they also share with us similarities in anatomy and physiology (**FIGURE 1.9**).

The vast majority of animals used in biomedical research are rodents such as rats, mice, and guinea pigs, although some areas of research require animals that are more similar to humans in size, such as dogs or pigs, or share a closer evolutionary relationship, such as chimpanzees.

The use of model systems, especially animals, allows experimental testing on potential drugs and other therapies before these methods are employed on people. Research on model organisms such as lab rats has contributed to a better understanding of nearly every serious human health threat, including cancer, heart disease, Alzheimer's disease, and AIDS (acquired immunodeficiency syndrome). However, ethical concerns about the use of animals in research persist and can complicate such studies. In addition, the results of animal studies are not always directly applicable to humans; despite a shared evolutionary history, animals can have important differences from humans. Testing hypotheses about human health in human beings still provides the clearest answer to these questions.

## Looking for Relationships between Factors.
Scientists can also test hypotheses using correlations when controlled experiments on humans are difficult or impossible to perform. A **correlation** is a relationship between two variables.

Suggestions about using meditation to reduce susceptibility to colds are based on a correlation between psychological stress and susceptibility to cold virus infections (**FIGURE 1.10**, on the next page). This correlation was generated by researchers who collected data on subjects' psychological stress levels before giving them nasal drops that contained a cold virus. Doctors later reported on the incidence and severity of colds among participants in the study. Note that while the cold virus was applied to each participant in the study, the researchers had no influence on the stress level of the study participants—in other words, this was not a controlled experiment because people were not randomly assigned to different "treatments" of low or high stress.

> **STOP & STRETCH**  Even though the researchers did not manipulate the independent variable in this study, there are still an independent variable and a dependent variable. What are they?

Let's examine the results presented in Figure 1.10. The horizontal axis of the graph, or **x-axis**, illustrates the independent variable. This variable is stress, and the graph ranks subjects along a scale of stress level—from low stress on the

(a)

(b)

(c)

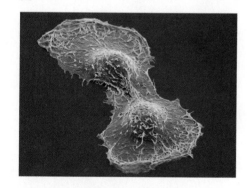

**FIGURE 1.9 Model systems in science.**  (a) Research on the nematode *Caenorhabditis elegans* (also known as *C. elegans*) has led to major advances in our understanding of animal development and genetics. (b) The classic "lab rat" is easy to raise and care for and as a mammal is an important model system for testing drugs and treatments designed for humans. (c) HeLa cells are derived from a cancerous growth biopsied from Henrietta Lacks in 1951. These cells have the unique property of being able to divide indefinitely, making them valuable for research on a wide variety of subjects from HIV to human growth hormone.

## Working with Data ▶

This graph groups people with similar, but not identical, stress index measures. Why might this have been necessary? If people with stress indices 3 and 4 have the same susceptibility to colds, does this call into question the correlation?

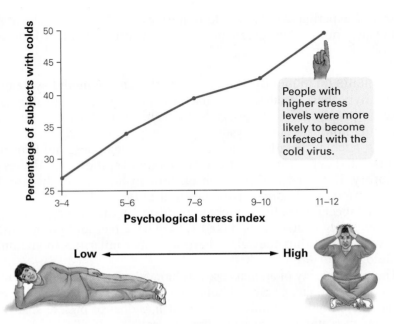

**FIGURE 1.10 Correlation between stress level and illness.** The graph indicates that people reporting higher levels of stress became infected after exposure to a cold virus more often than did people who reported low levels of stress.

left edge of the scale to high stress on the right. The vertical axis of the graph, the **y-axis**, is the dependent variable—the percentage of study participants who developed colds as reported by their doctors. Each point on the graph represents a group of individuals and tells us what percentage of people in each stress category had clinical colds.

The line connecting the five points on the graph illustrates a correlation—the relationship between stress level and susceptibility to cold virus infections. Because the line rises to the right, it illustrates a positive correlation. These data tell us that people who had higher stress levels were more likely to come down with colds. But does this relationship mean that high stress causes increased cold susceptibility?

To conclude that stress causes illness, we need the same assurances that are given by a controlled experiment. In other words, we must assume that the individuals measured for the correlation are similar in every way except for their stress levels. Is this a good assumption? Not necessarily. Most correlations cannot control for all alternative hypotheses. People who feel more stressed may have poorer diets because they feel time-limited and rely on fast food more often. In addition, people who feel highly stressed may be in situations where they are exposed to more cold viruses. These differences among people who differ in stress level may also influence their cold susceptibility (**FIGURE 1.11**). Therefore, even with a strong correlational relationship between the two factors, we cannot strongly infer that stress causes decreased resistance to colds.

Researchers who use correlational studies can eliminate a number of alternative hypotheses by closely examining their subjects. For example, this study on stress and cold susceptibility collected data from subjects on age, weight, sex, education, and their exposure to infected individuals. None of these factors differed consistently among low-stress and high-stress groups. While this analysis increases our confidence in the hypothesis that high

**(a) Does high stress cause high cold frequency?**

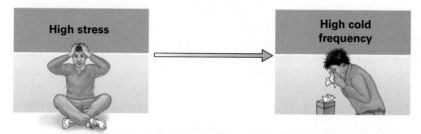

**(b) Or does one of the causes of high stress also cause high cold frequency?**

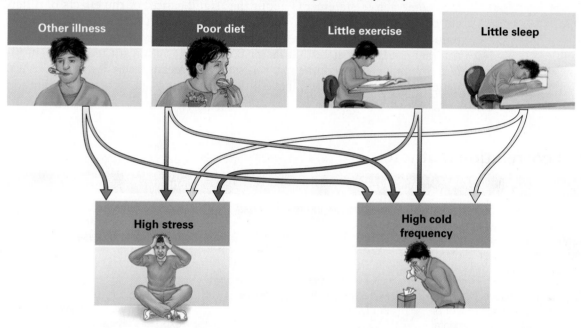

FIGURE 1.11 **Correlation does not signify causation.**    A correlation typically cannot eliminate all alternative hypotheses.

stress levels increase susceptibility to colds, people with high-stress lifestyles still may have important differences from those with low-stress lifestyles. It is possible that one of those differences is the real cause of disparities in cold frequency.

As you can see, it is difficult to demonstrate a cause-and-effect relationship between two factors simply by showing a correlation between them. In other words, correlation does not equal causation. For example, a commonly understood correlation exists between exposure to cold air and epidemics of the common cold. It is true that as outdoor temperatures drop, the incidence of colds increases. But numerous controlled experiments have indicated that chilling an individual does not increase his susceptibility to colds. Instead, cold outdoor temperatures mean increased close contact with other people (and their viruses), and this contact increases the number of colds we get during cold weather. Despite the correlation, cold air does not cause colds—exposure to viruses does.

Correlational studies are the main tool of epidemiology, the study of the distribution and causes of diseases. One commonly used epidemiological technique is a cross-sectional survey. In this type of survey, many individuals are both tested for the presence of a particular condition and asked about their exposure to various factors. The limitations of cross-sectional surveys include the effect of subject bias and poor recall by survey participants, in addition to all of the problems associated with interpreting correlations. **TABLE 1.1** provides an overview of the variety of correlational strategies employed to study the links between our environment and our health.

# 1.3 Understanding Statistics

During a review of scientific literature on cold prevention and treatment, you may come across statements about the "significance" of the effects of different cold-reducing measures. For instance, one report may state that factor A reduced cold severity, but that the results of the study were "not significant." Another study may state that factor B caused a "significant reduction" in illness. We might then assume that this statement means factor B will help us feel better, whereas factor A will have little effect. This is an incorrect assumption

**TABLE 1.1 Types of correlational studies.**

| Name | Description | Pros | Cons |
|------|-------------|------|------|
| Ecological studies | Examine specific human populations for unusually high levels of various diseases (e.g., documenting a "cancer cluster" around an industrial plant) | Inexpensive and relatively easy to do | Unsure whether exposure to environmental factor is actually related to onset of the disease |
| Cross-sectional surveys | Question individuals in a population to determine amount of exposure to an environmental factor and whether disease is present | More specific than ecological study | • Expensive<br>• Subjects may not know exposure levels<br>• Cannot control for other factors that may differ among subjects<br>• Cannot be used for rare diseases |
| Case-control studies | Compare exposures to specific environmental factors between individuals who have a disease and individuals matched in age and other factors who do not have the disease | • Relatively fast and inexpensive<br>• Best method for rare diseases | • Does not measure absolute risk of disease as a result of exposure<br>• Difficult to select appropriate controls<br>• Examines just one disease possibly associated with an environmental factor |
| Cohort studies | Follow a group of individuals, measuring exposure to environmental factors and disease prevalence | Can determine risk of various diseases associated with exposure to a particular environmental factor | • Expensive and time-consuming<br>• Difficult to control for alternative hypotheses<br>• Not feasible for rare diseases |
| Correlational experiment | Expose individuals who experience varying levels of an independent variable to an experimental treatment | Can control the amount of exposure to at least one environmental factor of interest | • Cannot eliminate alternative hypotheses<br>• Only feasible for hypotheses for which an experimental treatment can be applied |

because in scientific studies, *significance* is defined a bit differently from its usual sense. To evaluate the scientific use of this term, we need a basic understanding of statistics.

## What Statistical Tests Can Tell Us

We often use the term *statistics* to refer to a summary of accumulated information. For instance, a baseball player's success at hitting is summarized by a statistic: his batting average, the total number of hits he made divided by the number of opportunities he had to bat. The science of **statistics** is a bit different; it is a specialized branch of mathematics used to evaluate and compare data.

An experimental test utilizes a small subgroup, or **sample,** of a population. Statistical methods can summarize data from the sample—for instance, we can describe the average length of colds experienced by experimental and control groups. In statistics, this average is known as the **mean. Statistical tests** can then be used to extend the results from a sample to the entire population.

When scientists conduct an experiment, they hypothesize that there is a true, underlying effect of their experimental treatment on the entire population. An experiment on a sample of a population can only estimate this true effect because a sample is always an imperfect "snapshot" of an entire population.

Consider a hypothesis that the average hair length of women in a college class in 1965 was shorter than the average hair length at the same college today. To test this hypothesis, we could compare a sample of snapshots from college yearbooks. If hairstyles were very similar among the women in a snapshot, you could reasonably assume that the average hair length in the college class is close to the average length in the snapshot. However, what if you see that women in a snapshot have a variety of hairstyles, from short bobs to long braids? In this case, it is difficult to determine whether the average hair length in the snapshot is at all close to the average for the class. With so much variation, the snapshot could, by chance, contain a surprisingly high number of women with very long hair, causing the average length in the sample to be much longer than the average length for the entire class (**FIGURE 1.12**).

A statistical test calculates the likelihood, given the number of individuals sampled and the variation within samples, that the difference between two samples reflects a real, underlying difference between the populations from which these samples were drawn. A **statistically significant** result is one that is very unlikely to be due to chance differences between the experimental and control samples, so it likely represents a true difference between the groups.

> **STOP & STRETCH** Imagine that average hair length was 60 cm (23.6 in) in the class of 1965 and 65 cm (25.6 in) in the class of 2015. Given the variation in styles in 2015, however, this information is not statistically significant. Can we say that the average hairstyle in 2015 is longer than the average in 1965?

In the experiment with the echinacea tea, statistical tests indicated that the 33% reduction in cold severity observed by the researchers was statistically significant. In other words, there is a low probability that the difference in cold susceptibility between the two samples in the experiment is due to chance. If the experiment is properly designed, a statistically significant result allows researchers to infer that the treatment had an effect.

## Visualize This ▼

The snapshot from the 1960s is of a group of women who are considerably less racially and ethnically diverse than the more recent snapshot. How might this affect the comparison? How might a lack of racial and ethnic diversity in a sample of a cold treatment affect our interpretation of results?

**(a) Average hair length in this snapshot is shorter...**

Class of 1965

- Little variability
- High probability of reflecting average of all women in the class

**(b) ...than average hair length in this snapshot...**

Class of 2015

- High variability
- Low probability of reflecting average of all women in the class

**...so, is hair longer today than in 1965?**

**FIGURE 1.12 The role of statistics.** Statistical tests calculate the variability within groups to determine the probability that two groups differ only by chance.

# Working with Data ▼

Sometimes graphs have lines and sometimes they have bars. Why do you think these two data points (average of 7.5 and average of 4.5) are not connected with a line?

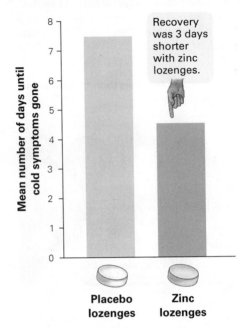

Recovery was 3 days shorter with zinc lozenges.

**FIGURE 1.13 Zinc lozenges reduce the duration of colds.** Fifty individuals taking zinc lozenges had colds lasting about 4½ days as opposed to approximately 7½ days for 50 individuals taking placebo.

# What Statistical Tests Can Tell Us: **A Closer Look** ▼

We can explore the role that statistical tests play in more detail by evaluating another study on cold treatments. This study examined the efficacy of zinc lozenges on reducing cold severity.

Some forms of zinc can block common cold viruses from invading cells in the nasal cavity. This observation led to the hypothesis that consuming zinc at the start of a cold could decrease cold severity. Researchers at the Cleveland Clinic tested this hypothesis using a sample of 100 of their employees who volunteered for a double-blind study within 24 hours of developing cold symptoms. The researchers randomly assigned subjects to the experimental group, who received zinc lozenges, or the control group, who received placebo lozenges. Members of both groups received the same instructions for using the lozenges and were asked to rate their symptoms until they had recovered.

When the data from the experiment were summarized, the statistics indicated that the mean recovery time was over 3 days shorter in the zinc group than in the placebo group (**FIGURE 1.13**). On the surface, this result appears to support the hypothesis. However, recall the example of the snapshot of women's hair length. A statistical test is necessary because of the effect of chance.

## The Problem of Sampling Error

The effect of chance on experimental results is known as **sampling error**—more specifically, sampling error is the difference between a sample and the population from which it was drawn. In any experiment, individuals in the experimental group will differ from individuals in the control group in random ways. Even if there is *no* true effect of an experimental treatment, the data from the experimental group will never be identical to the data from the control group.

For example, we know that people differ in their ability to recover from a cold infection. If we give zinc lozenges to one volunteer and placebo lozenges to another, it is likely that the two volunteers will have colds of different lengths. Even if the zinc-taker had a shorter cold than the placebo-taker, you would probably say that the test did not tell us much about our hypothesis—the zinc-taker might just have had a less severe cold for other reasons.

Now imagine that we had five volunteers in each group and saw a difference, or that the difference was only 1 day instead of 3 days. How would we determine if the lozenges had an effect? Statistical tests allow researchers to look at their data and determine how likely it is—that is, the **probability**—that the result is due to sampling error. In this case, the statistical test distinguished between two possibilities for why the experimental group had shorter colds: Either the difference was due to the effectiveness of zinc as a cold treatment or it was due to a chance difference from the control group. The results indicated that there was a low probability, less than 1 in 10,000 (0.01%), that the experimental and control groups were so different simply by chance. In other words, the result is statistically significant.

A statistical measure of the amount of variability in a sample is a number called the **standard error** of the mean; greater variability generates a larger standard error (**FIGURE 1.14**). Put simply, the true population mean is very likely to lie somewhere in the range between the mean of the sample minus the standard error and the mean of the sample plus the standard error. You have likely heard the term *standard error* used before—when public opinion polls are released, this range is often referred to *margin of error*. Results with a small standard error are more likely to be statistically significant if the hypothesis is true.

## Factors That Influence Statistical Significance

One characteristic of experiments that influences sampling error is **sample size**—the number of individuals in the experimental and control groups. If a

*A Closer Look, continued*

treatment has no effect, a small sample size could return results that appear different simply because of a large sampling error. This was the case with the vitamin C hypothesis described at the beginning of the chapter. Subsequent tests with larger sample sizes allowed scientists to reject the hypothesis that vitamin C prevents colds.

Conversely, if the effect of a treatment is real but the sample size of the experiment is small, a single experiment may not allow researchers to determine convincingly that their hypothesis has support. For example, one experiment performed at a Wisconsin clinic with 48 participants indicated that echinacea tea drinkers were 30% less likely to experience any cold symptoms after virus exposure as compared to individuals who received a placebo tea. However, the small sample size of the study meant that this result was not statistically significant—based on this experiment, we still cannot say whether echinacea tea has a protective effect.

The more participants there are in a study, the more likely it is that researchers will see a true effect of an experimental treatment, even if the effect is very small. For example, a study of over 21,000 men over 6 years in Finland demonstrated that those who took vitamin E supplements had 5% fewer colds than those who did not take these supplements. In this case, the large sample size allowed researchers to see that vitamin E has a real, but relatively tiny, effect on cold incidence. In other words, this statistically significant result has little real-world *practical* significance. The relationship among hypotheses, experimental tests, sample size, and statistical and practical significance is summarized in **FIGURE 1.15** on the next page.

There is one final caveat to this discussion. A statistically significant result is typically defined as one that has a *probability* of 5% or less of being due to chance alone. If all scientific research uses this same standard, as many as 1 in every 20 statistically significant results (that is, 5% of the total) is actually reporting an effect that is *not real*. In other words, some statistically significant results are "false positives," representing a surprisingly large sampling error. This potential explains why one supportive experiment is not enough to convince all scientists that a particular hypothesis is accurate.

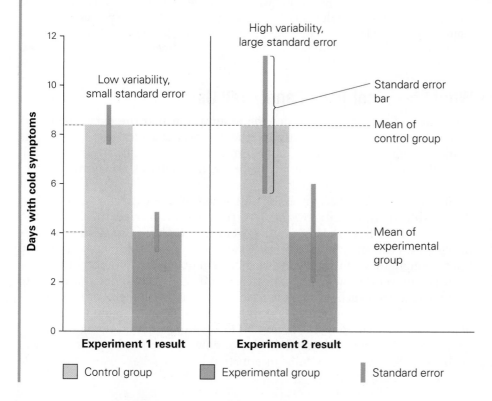

**FIGURE 1.14 Standard error.** The red lines on these bar graphs represent the standard error for each sample mean. A more variable sample has a larger standard error than a less variable sample. Even though the means are the same for both experimental and control groups in these two sets of bars, only the data summarized at left illustrate a significant difference because of the greater variability in the samples illustrated on the right side.

◀ **Visualize This**

**Increasing the size of the sample measured usually decreases the standard error. Why do you think this is the case?**

*A Closer Look, continued*

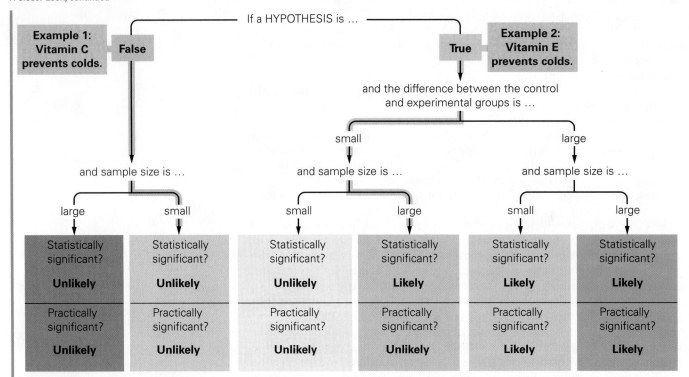

## Visualize This ▲

The Nurse's Health Survey, which has followed nearly 80,000 women for over 20 years, has found that a diet low in refined carbohydrates such as white flour and sugar but relatively high in vegetable fat and protein cuts heart disease risk by 30% relative to women with a high refined carbohy-drate diet. How does this result map out on the flowchart here?

**FIGURE 1.15 Factors that influence statistical significance.** This flowchart sum-marizes the relationship between the true effect of a treatment and the sample size of an experiment on the likelihood of obtaining statistical significance.

Even with a statistical test indicating that the result had a probability of less than 0.01% of occurring by chance, we should begin to feel assured that taking zinc lozenges will reduce the duration of colds only after additional hypothesis tests give similar results. In fact, scientists continue to test this hypothesis, and there is still no consensus about the effectiveness of zinc as a cold treatment. ▬

## What Statistical Tests Cannot Tell Us

All statistical tests operate under the assumption that the experiment was designed and carried out correctly. In other words, a statistical test evaluates the chance of sampling error, not observer error, and a statistically significant result is not the last word on an experimentally tested hypothesis. An examination of the experiment itself is required.

In the test of the effectiveness of zinc lozenges, the experimental design—double-blind, randomized, and controlled—minimized the likelihood that alternative hypotheses could explain the results. Given such a well-designed experiment, this statistically significant result allows researchers to infer strongly that consuming zinc lozenges reduces the duration of colds.

It is important to note that statistical significance is also not equivalent to *significance* as we usually define the term, that is, as "meaningful or important." Unfortunately, experimental results reported in the news often use the term *significant* without clarifying this important distinction. Understanding that prob-lem, as well as other misleading aspects of how science can be presented, will enable you to better use scientific information.

# 1.4 Evaluating Scientific Information

The previous sections should help you see why definitive scientific answers to our questions are slow in coming. However, a well-designed experiment can certainly allow us to approach the truth.

## Primary Sources

Looking critically at reports of experiments can help us make well-informed decisions about actions to take. Most of the research on cold prevention and treatment is first published as **primary sources**, written by the researchers themselves and reviewed within the scientific community (**FIGURE 1.16**). The process

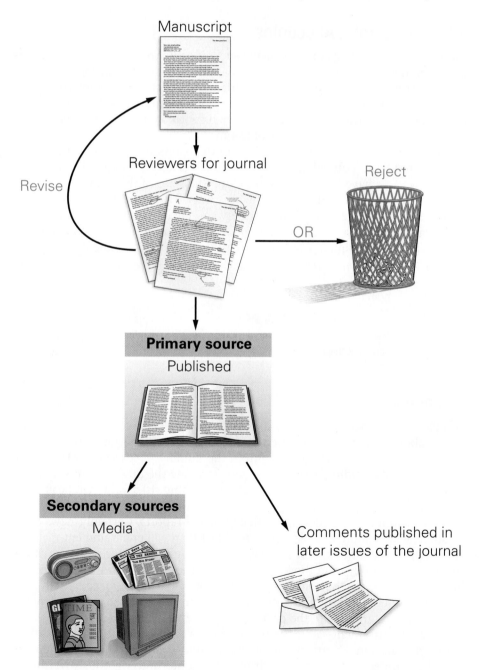

Manuscript

Revise

Reviewers for journal

Reject

OR

**Primary source**
Published

**Secondary sources**
Media

Comments published in later issues of the journal

**FIGURE 1.16 Primary sources: publishing scientific results.** Most scientific journals require papers to go through stringent review before publication.

◀ **Visualize This**

**How does having other anonymous experts review a draft of a scientist's paper make the final paper more reliable?**

of **peer review**, in which other scientists critique the results and conclusions of an experiment before it is published, helps increase confidence in scientific information. Peer-reviewed research articles in journals such as *Science, Nature*, the *Journal of the American Medical Association*, and hundreds of others represent the first and most reliable sources of current scientific knowledge.

However, evaluating the hundreds of scientific papers that are published weekly is a task not one of us can perform alone. Even if we focused only on a particular field of interest, the technical jargon used in many scientific papers may be a significant barrier to our understanding.

Instead of reading the primary literature, most of us receive our scientific information from **secondary sources** such as books, news reports, and advertisements. How can we evaluate information in this context? The following sections provide strategies for doing so, and the recurring Savvy Reader feature in this textbook will help you practice these evaluation skills.

## Information from Anecdotes

Information about dietary supplements such as echinacea tea and zinc lozenges is often in the form of **anecdotal evidence**—meaning that the advice is based on one individual's personal experience. A friend's enthusiastic plug for vitamin C, because she felt it helped her, is an example of a testimonial—a common form of anecdote. Advertisements that use a celebrity to pitch a product "because it worked for them" are classic forms of testimonials.

You should be cautious about basing decisions on anecdotal evidence, which is not at all equivalent to well-designed scientific research. For example, many of us have heard anecdotes along the lines of the grandpa who was a pack-a-day smoker and lived to the age of 94. However, hundreds of studies have demonstrated the clear link between cigarette smoking and premature death. Although anecdotes may indicate that a product or treatment has merit, only well-designed tests of the hypothesis can determine its safety and efficacy.

**STOP & STRETCH** One of the challenges of presenting science is that anecdotes make a much bigger impression on people than statistical summaries of data. Why do you think this is the case?

## Science in the News

Popular news sources provide a steady stream of science information. However, stories about research results often do not contain information about the adequacy of experimental design or controls, the number of subjects, or the source of the scientist's funding. How can anyone evaluate the quality of research that supports dramatic headlines such as those describing risks to human health and the environment?

First, you must consider the source of media reports. Certainly, news organizations will be more reliable reporters of fact than will entertainment tabloids, and news organizations with science writers should be better reporters of the substance of a study than those without. Television talk shows, which need to fill airtime, regularly have guests who promote a particular health claim. Too often these guests may be presenting information that is based on anecdotes or an incomplete summary of the primary literature.

Paid advertisements are a legitimate means of disseminating information. However, claims in advertising should be carefully evaluated. Advertisements of over-the-counter and prescription drugs must conform to rigorous government standards regarding the truth of their claims. Lower standards apply to advertisements for herbal supplements, many health-food products, and diet plans. Be sure to examine the fine print because advertisers often are required to clarify the statements made in their ads.

Another commonly used source for health information is the Internet. As you know, anyone can post information on this great resource. Typing in "common cold prevention" on a standard web search engine will return thousands of web pages—from highly respected academic and government sources to small companies trying to sell their products to individuals who have strong, sometimes completely unsupported, ideas about cures. Often it can be difficult to determine the reliability of a well-designed website, and even sites such as Wikipedia may contain erroneous or misleading information. Here are some things to consider when using the web as a resource for health information:

1. Choose sites maintained by reputable medical establishments, such as the National Institutes of Health (NIH) or the Mayo Clinic.
2. It costs money to maintain a website. Consider whether the site seems to be promoting a product or agenda. Advertisements for a specific product should alert you to possible bias.
3. Check the date when the website was last updated and see whether the page has been updated since its original posting. Science and medicine are disciplines that must frequently evaluate new data. A reliable website will be updated often.
4. Determine whether unsubstantiated claims are being made. Look for references and be suspicious of any studies that are not from peer-reviewed journals.

## Understanding Science from Secondary Sources

Once you are satisfied that a media source is relatively reliable, you can examine the scientific claim that it presents. Use your understanding of the process of science and of experimental design to evaluate the story and the science. Does the story about the claim present the results of a scientific study, or is it built around an untested hypothesis? Were the results obtained using the scientific method, with tests designed to reject false hypotheses? Is the story confusing correlation with causation? Does it seem that the information is applicable to nonlaboratory situations, or is it based on results from preliminary or animal studies?

Look for clues about how well the reporters did their homework. Scientists usually discuss the limitations of their research in their papers. Are these cautions noted in an article or television piece? If not, the reporter may be overemphasizing the applicability of the results.

Then, note if the scientific discovery itself is controversial. That is, does it reject a hypothesis that has long been supported? Does it concern a subject that is controversial? Might it lead to a change in social policy? In these cases, be extremely cautious. New and unexpected research results must be evaluated in light of other scientific evidence and understanding. Reports that lack comments from other experts may miss problems with a study or fail to place it in the context of other research. The Savvy Reader feature in this chapter provides a checklist of questions to ask and answer as you evaluate news reports.

Finally, the news media generally highlight only those science stories that editors find newsworthy. As we have seen, scientific understanding accumulates relatively slowly, with many tests of the same hypothesis finally leading to an accurate understanding of a natural phenomenon. News organizations are also more likely to report a study that supports a hypothesis rather than one that gives less-supportive results, even if both types of studies exist.

In addition, even the most respected media sources may not be as thorough as readers would like. For example, a recent review published in the *New England Journal of Medicine* evaluated the news media's coverage of new medications. Of 207 randomly selected news stories, only 40% that cited experts with financial ties to a drug told readers about this relationship. This potential conflict of interest calls into question the expert's objectivity. Another 40% of the news stories did not provide basic statistics about the drug's benefits. Most of the news reports also failed to distinguish between absolute benefits (how many people were helped by the drug) and relative benefits (how many people were helped by the drug relative to other therapies for the condition). The journal's review is a vivid reminder that we need to be cautious when reading or viewing news reports on scientific topics.

Even after following all of these guidelines, you will still find well-researched news reports on several scientific studies that seem to give conflicting and confusing results. As you now know, such confusion is the nature of the scientific process—early in our search for understanding, many hypotheses are proposed and discussed; some are tested and rejected immediately, and some are supported by one experiment but later rejected by more thorough experiments. It is only by clearly understanding the process and pitfalls of scientific research that you can distinguish "what we know" from "what we don't know."

## 1.5 Is There a Cure for the Common Cold?

So where does our discussion leave us? Will we ever find the best way to prevent a cold or reduce its effects? In the United States alone, over 1 billion cases of the common cold are reported per year, costing many more billions of dollars in medical visits, treatment, and lost workdays. Consequently, there is significant cause to search for effective protection from the different viruses that cause colds.

A search of medical publication databases indicates that every year nearly 100 scholarly articles regarding the biology, treatment, and consequences of common cold infection are published. This research has led to several important discoveries about the structure and biochemistry of common cold viruses, how they enter cells, and how the body reacts to these infections.

**STOP & STRETCH** Where would you look to find the latest research on common cold treatment and prevention?

Despite all of the research and the emergence of some promising possibilities, the best prevention method known for common colds is still the old standby—keep your hands clean. Numerous studies have indicated that rates of common cold infection are 20 to 30% lower in those who employ effective hand-washing procedures. Cold viruses can survive on surfaces for many hours; if you pick them up on your hands from a surface and transfer them to your mouth, eyes, or nose, you may inoculate yourself with a 7-day sniffle (**FIGURE 1.17**).

Of course, not everyone gets sick when exposed to a cold virus. The reason one person has more colds than another might not be due to a difference in personal hygiene. The correlation that showed a relationship between stress and cold susceptibility appears to have some merit. Research indicates that

**FIGURE 1.17 Preventing colds.**
According to the Centers for Disease Control and Prevention, the best defense against infection is effective hand washing. Scrub vigorously with a liquid soap for 20 seconds, then rinse under running water for 10 seconds. Alcohol-based hand gels (without water) containing at least 60% ethanol are effective if soap and water are unavailable but will not remove visible dirt.

among people exposed to viruses, the likelihood of ending up with an infection increases with high levels of psychological stress—something that many college students clearly experience.

Research also indicates that vitamin C intake, diet quality, exposure to cold temperatures, and exercise frequency appear to have no effect on cold susceptibility, although along with echinacea tea and zinc lozenges, there is some evidence that vitamin C may reduce cold symptoms a bit. Surprisingly, even though medical research has led to the elimination of killer viruses such as smallpox and polio, scientists are still a long way from "curing" the common cold.

# *savvy reader*

## A Toolkit for Evaluating Science in the News

The following checklist can be used as a guide to evaluating science information in the news. Although any issue that raises a red flag should cause you to be cautious about the conclusions of the story, no one issue should cause you to reject the story (and its associated research) out of hand. However, the fewer red flags raised by a story, the more reliable is the report of the significance of the scientific study. Use **TABLE 1.2** to evaluate this extract from an article by Terry Gupta in the *Nashua* (New Hampshire) *Telegraph,* describing one proposed cure for colds.

[Paula] Fortier, a licensed massage therapist, shared her special brew. To boiled water and tea, she adds cayenne pepper, honey and fresh squeezed lemon. Fortier's enthusiasm may be more contagious than the flu when she describes this heated, "unbeatable" combination. "Your throat opens up," she said. "You're less irritated and you sweat out some unwanted toxins." She varies the exact measures of the ingredients according to tolerance and taste. While on the topic of hot water, one of Fortier's favorite remedies for "an achy body that feels sore all over is a hot bath in a steeping tub with 2 cups of cider vinegar for 15 minutes. This draws off toxins, reduces inflammation and leaves you feeling better."

**TABLE 1.2 A guide for evaluating science in the news.**
**For each question, check the appropriate box.**

| Question | Preferred answer | | Raises a red flag | |
|---|---|---|---|---|
| **1.** What is the basis for the story? | Hypothesis test | ☐ | Untested assertion<br>No data to support claims in the article. | ☐ |
| **2.** What is the affiliation of the scientist? | Independent (university or government agency) | ☐ | Employed by an industry or advocacy group<br>Data and conclusions could be biased. | ☐ |
| **3.** What is the funding source for the study? | Government or non-partisan foundation (without bias) | ☐ | Industry group or other partisan source (with bias)<br>Data and conclusions could be biased. | ☐ |
| **4. If the hypothesis test is a correlation:** Did the researchers attempt to eliminate reasonable alternative hypotheses? | Yes | ☐ | No<br>Correlation does not equal causation. One hypothesis test provides poor support if alternatives are not examined. | ☐ |

*continued*

**TABLE 1.2 A guide for evaluating science in the news.** *continued*

| Question | Preferred answer | | Raises a red flag | |
|---|---|---|---|---|
| **If the hypothesis test is an experiment:** Is the experimental treatment the only difference between the control group and the experimental group? | Yes | ☐ | No<br>An experiment provides poor support if alternatives are not examined. | ☐ |
| 5. Was the sample of individuals in the experiment a good cross-section of the population? | Yes | ☐ | No<br>Results may not be applicable to the entire population. | ☐ |
| 6. Was the data collected from a relatively large number of people? | Yes | ☐ | No<br>Study is prone to sampling error. | ☐ |
| 7. Were participants blind to the group they belonged to or to the "expected outcome" of the study? | Yes | ☐ | No<br>Subject expectation can influence results. | ☐ |
| 8. Were data collectors or analysts blinded to the group membership of participants in the study? | Yes | ☐ | No<br>Observer bias can influence results. | ☐ |
| 9. Did the news reporter put the study in the context of other research on the same subject? | Yes | ☐ | No<br>Cannot determine if these results are unusual or fit into a broader pattern of results. | ☐ |
| 10. Did the news story contain commentary from other independent scientists? | Yes | ☐ | No<br>Cannot determine if these results are unusual or if the study is considered questionable by others in the field. | ☐ |
| 11. Did the reporter list the limitations of the study or studies on which he or she is reporting? | Yes | ☐ | No<br>Reporter may not be reading study critically and could be overstating the applicability of the results. | ☐ |

# SOUNDS RIGHT BUT IS IT?

Advertisements for skin care products often contain a phrase such as this one: "this product is clinically proven to…." That statement may convince you to purchase a topical cream that claims to reduce acne, eliminate rough patches of skin, or lighten dark spots. You would be making this decision based on your belief in the following statement:

**If a product is clinically proven to do what it advertises, that means it will work for you.**

Sounds right, but it isn't.

1. "Clinically" generally implies that the product was subject to an experiment of some type. Describe the ideal type of experiment that would test whether a cream reduces acne.

2. Imagine that the results of this experiment showed that that there was a statistically significant difference in the extent of acne between those patients using the "real" cream and those using the placebo. Is this difference in acne necessarily large?

3. Actually, a "clinically proven" claim often relies on consumer evaluation of the product—in other words, surveys that ask whether acne was reduced as a result of using the product. What is the problem with this type of experimental approach?

4. Even if the experiment is well controlled and double-blind, does the result of a single experiment prove a hypothesis?

5. Does this statistically significant result mean that everyone in the experiment who used the real cream had less acne than those who used the placebo?

6. Consider your answers to questions 1–5 and explain why the statement bolded above sounds right, but isn't.

# Chapter Review MasteringBiology®

Go to the Study Area in MasteringBiology® for practice quizzes, myeBook, BioFlix™ 3-D animations, MP3Tutor sessions, videos, current events, and more.

## Summary

## Section 1.1

Describe the characteristics of a scientific hypothesis.

- Science is a process of testing hypotheses—statements about how the natural world works. Scientific hypotheses must be testable and falsifiable (pp. 4–5).

Compare and contrast the terms *scientific hypothesis* and *scientific theory*.

- A scientific theory is an explanation of a set of related observations based on well-supported hypotheses from several different, independent lines of research (pp. 5–6).

Distinguish between inductive and deductive reasoning.

- Hypotheses are often developed via inductive reasoning, which consists of making a number of observations and then inferring a general principle to explain them (p. 6).
- Hypotheses are tested via the process of deductive reasoning, which allows researchers to make specific predictions about expected observations (pp. 6–7).

Explain why the truth of a hypothesis cannot be proven conclusively via deductive reasoning.

- Absolutely proving hypotheses is impossible because there may be other reasons besides the one hypothesized that could lead to the predicted result (p. 7).

## Section 1.2

Describe the features of a controlled experiment, and explain how these experiments eliminate alternative hypotheses for the results.

- Controlled experiments test hypotheses about the effect of experimental treatments by comparing a randomly assigned experimental group with a control group (p. 10).
- Controls are individuals who are treated identically to the experimental group except for application of the treatment (p. 11).

List strategies for minimizing bias when designing experiments.

- Bias in scientific results can be minimized with double-blind experiments that keep subjects and data collectors unaware of which individuals belong in the control or experimental group (pp. 11–12).

Define correlation, and explain the benefits and limitations of using this technique to test hypotheses.

- If performing controlled experiments on humans is considered unethical, scientists sometimes use correlations. The correlation of structure and function between humans and model organisms, such as other mammals, allows experimental tests of these hypotheses (pp. 12–13).
- Correlations cannot exclude alternative hypotheses. Thus, a correlation study can describe a relationship between two factors, but it does not strongly imply that one factor causes the other (pp. 13–15).

## Section 1.3

Describe the information that statistical tests provide.

- Statistics help scientists evaluate the results of their experiments by determining if results appear to reflect the true effect of an experimental treatment on a sample of a population (p. 17).
- A statistically significant result is one that is very unlikely to be due to chance differences between the experimental and control groups (pp. 17–19).
- Even when an experimental result is highly significant, hypotheses are tested multiple times before scientists come to a consensus on the true effect of a treatment (p. 20).

## Section 1.4

Compare and contrast primary and secondary sources.

- Primary sources of information are experimental results published in professional journals and peer-reviewed by other scientists before publication (pp. 21–22).
- Secondary sources are typically news, books, or websites that summarize the scientific research presented in primary sources. Secondary sources are not peer-reviewed and sometimes contain poorly supported information (p. 22).

Summarize the techniques you can use to evaluate scientific information from secondary sources.

- Anecdotal evidence is an unreliable means of evaluating information, and media sources are of variable quality; distinguishing between news stories and advertisements is important when evaluating the reliability of information. The Internet is a rich source of information, but users should look for clues to a particular website's credibility (pp. 22–23).

- Stories about science should be carefully evaluated for information on the actual study performed, the universality of the claims made by the researchers, and other studies on the same subject. Sometimes confusing stories about scientific information are a reflection of controversy within the scientific field itself (pp. 23–24).

## Roots to Remember

**The following roots of words come mainly from Latin and Greek and will help you to decipher terms:**

| | |
|---|---|
| **bio-** | means life. Chapter term: *biology* |
| **deduc-** | means to reason out, working from facts. Chapter term: *deductive reasoning* |
| **hypo-** | means under, below, or basis. Chapter term: *hypothesis* |
| **induc-** | means to rely on reason to derive principles (also, to cause to happen). Chapter term: *inductive reasoning* |
| **-ology** | means the study of or branch of knowledge about. Chapter term: *biology* |

## Learning the Basics

1. What is the value of a placebo in an experimental design?

2. Which of the following is an example of inductive reasoning?
   **A.** All cows eat grass; **B.** My cow eats grass and my neighbor's cow eats grass; therefore, all cows probably eat grass; **C.** If all cows eat grass, when I examine a random sample of all the cows in Minnesota, I will find that all of them eat grass; **D.** Cows may or may not eat grass, depending on the type of farm where they live.

3. A scientific hypothesis is _____.
   **A.** an opinion; **B.** a proposed explanation for an observation; **C.** a fact; **D.** easily proved true; **E.** an idea proposed by a scientist.

4. How is a scientific theory different from a scientific hypothesis?
   **A.** It is based on weaker evidence; **B.** It has not been proved true; **C.** It is not falsifiable; **D.** It can explain a large number of observations; **E.** It must be proposed by a professional scientist.

5. One hypothesis states that eating chicken noodle soup is an effective treatment for colds. Which of the following results does this hypothesis predict?
   **A.** People who eat chicken noodle soup have shorter colds than do people who do not eat chicken noodle soup; **B.** People who do not eat chicken noodle soup experience unusually long and severe colds; **C.** Cold viruses cannot live in chicken noodle soup; **D.** People who eat chicken noodle soup feel healthier than do people who do not eat chicken noodle soup; **E.** Consuming chicken noodle soup causes people to sneeze.

6. If I perform a hypothesis test in which I demonstrate that the prediction I made in question 5 is true, I have _____.
   **A.** proved the hypothesis; **B.** supported the hypothesis; **C.** not falsified the hypothesis; **D.** B and C are correct; **E.** A, B, and C are correct.

7. Control subjects in an experiment _____.
   **A.** should be similar in most ways to the experimental subjects; **B.** should not know whether they are in the control or experimental group; **C.** should have essentially the same interactions with the researchers as the experimental subjects; **D.** help eliminate alternative hypotheses that could explain experimental results; **E.** all of the above.

8. An experiment in which neither the participants in the experiment nor the technicians collecting the data know which individuals are in the experimental group and which ones are in the control group is known as _____.
   **A.** controlled; **B.** biased; **C.** double-blind; **D.** falsifiable; **E.** unpredictable.

9. A relationship between two factors, for instance, between outside temperature and the number of people with active colds in a population, is known as a(n) _____.
   **A.** significant result; **B.** correlation; **C.** hypothesis; **D.** alternative hypothesis; **E.** experimental test.

10. A primary source of scientific results is _____.
    **A.** the news media; **B.** anecdotes from others; **C.** articles in peer-reviewed journals; **D.** the Internet; **E.** all of the above.

11. A story on your local news station reports that eating a 1-ounce square of milk chocolate each day reduces the risk of heart disease in rats, and that this result is statistically significant. This means that _____.
    **A.** People who eat milk chocolate are healthier than those who do not; **B.** The difference between chocolate- and non-chocolate-eating rats in heart disease rates was greater than expected by chance; **C.** Rats like milk chocolate; **D.** Milk chocolate reduces the risk of heart disease; **E.** Two ounces of milk chocolate per day is likely to be even better for heart health than 1 ounce.

12. What features of the story on milk chocolate and heart health described in question 11 should cause you to consider the results less convincing?
    **A.** The study was sponsored by a large milk chocolate manufacturer; **B.** A total of 10 rats were used in the study;

**C.** The only difference between the rats was that subjects of the experimental group received chocolate along with their regular diets, while subjects of the control group received no additional food; **D.** The reporter notes that other studies indicate milk chocolate does not have a beneficial effect on heart health; **E.** All of the above.

## Analyzing and Applying the Basics

**1.** There is a strong correlation between obesity and the occurrence of a disease known as type 2 diabetes—that is, obese individuals have a higher instance of diabetes than nonobese individuals do. Does this mean that obesity causes diabetes? Explain.

**2.** In an experiment examining vitamin C as a cold treatment, students with cold symptoms who visited the campus medical center either received vitamin C or were treated with over-the-counter drugs. Students then reported on the length and severity of their colds. Both the students and the clinic health providers knew which treatment students were receiving. This study indicated that vitamin C significantly reduced the length and severity of colds. Which factors make this result somewhat unreliable?

**3.** Brain-derived neurotrophic factor (BDNF) is a substance produced in the brain that helps nerve cells (neurons) to grow and survive. BDNF also increases the connectivity of neurons and improves learning and mental function. A 2002 study that examined the effects of intense wheel-running on rats and mice found a positive correlation between BDNF levels and running distance. What could you conclude from this result?

## Connecting the Science

**1.** Much of the research on common cold prevention and treatment is performed by scientists employed or funded by drug companies. Often these companies do not allow scientists to publish the results of their research for fear that competitors at other drug companies will use this research to develop a new drug before they do. Should our society allow scientific research to be owned and controlled by private companies?

**2.** Should society restrict the kinds of research performed by government-funded scientists? For example, many people believe that research performed on tissues from human fetuses should be restricted; these people believe that such research would justify abortion. If most Americans feel this way, should the government avoid funding this research? Are there any risks associated with *not* funding research with public money?

Answers to **Stop & Stretch, Visualize This, Working with Data, Savvy Reader, Sounds Right, But Is It?**, and **Chapter Review** questions can be found in the **Answers** section at the back of the book.

CHAPTER **2**

# Science Fiction, Bad Science, and Pseudoscience

**Can dead humans come back as zombies?**

# Water, Biochemistry, and Cells

We are bombarded by information that is supposedly based in science. While some of it is harmless entertainment loosely based in science, like the science fiction zombies seen on television or in movies, and some bad science is believable only to the very young or uninformed, like planes and ships disappearing into the Bermuda Triangle, some of it is plausible enough that we have a hard time knowing whether to believe it.

If your roommate tells you that you will do better on exams if you bring a bottle of water along, you might believe her. Many parents believe that giving children candy or sugary drinks makes them hyperactive. Is the relative who

**Do sugary drinks cause hyperactivity?**

says he has to nap after eating turkey for Thanksgiving dinner just lazy, or does something in turkey make people tired?

These examples are pretty harmless ones. It won't hurt you to take water to an exam or to limit your intake of sweets or nap after Thanksgiving dinner, but what about spending money on crystals and magnets to treat disease or trying to cure cancer with alternative medicines instead of conventional ones?

Sometimes people will claim that there is science backing a product or an idea when this is not actually the case. This can be unintentional. For example, drawing conclusions from one study instead of waiting until a scientific

consensus emerges can cause one to come to erroneous conclusions, as can relying on bad science, experiments that are poorly designed or not properly controlled. In other cases, unscrupulous people will attempt to use science to lend legitimacy to a belief or ideology they want to spread. Dubious claims that are strengthened through the use of false or weakly supported assertions of scientific evidence are called pseudoscience.

In an effort to help develop the skill of assessing whether the science supporting an idea or product is sound, we will analyze examples of science fiction, weakly supported or bad science, and pseudoscience so that the qualities of each are more easily recognizable.

**Do planes disappear over the Bermuda Triangle?**

**Does eating turkey cause drowsiness?**

**FIGURE 2.1 Homeostasis.** The ability to maintain homeostasis requires complex feedback mechanisms between multiple sensory and physiological systems. Black-capped chickadees can maintain a core body temperature of 108°F (~42°C) during the day, even when the air temperature is well below zero.

---

**homeo-** means like or similar.

---

# 2.1 A Definition of Life

Zombies are supposedly human corpses that have been brought back to life. Often called the undead or walking dead, zombies are thought to survive by eating the flesh or drinking the blood of living humans. While we know that zombies are not actually alive, they do share many characteristics seen in other living organisms.

Living organisms contain a common set of biological molecules and are composed of cells. Other attributes found in most living organisms include growth, movement, reproduction, response to external environmental stimuli, and **metabolism.** Metabolism includes all of the chemical processes that occur in cells, including the breakdown of substances to produce energy, the synthesis of substances necessary for life, and the excretion of wastes generated by these processes.

At first glance, zombies seem to fit parts of this definition of life. They certainly can move, even though their movements tend to be hindered by various injuries they have sustained. It might even appear that zombies are reproducing, because they cause the production of more zombies when they attack living humans. Likewise, they even respond to a limited number of stimuli, lurching toward humans for their next meal.

However, closer inspection leads us to question just how much zombies resemble living organisms. They do not grow; a child zombie does not become an adult zombie over time. And, while the biting of humans by zombies does cause the production of more zombies, this is not the same as the kind of reproduction living organisms are capable of. Living organisms pass genetic information to their offspring when they reproduce. It is also unlikely that zombies actually metabolize the flesh they eat. It seems that they are driven to eat human flesh more to spread zombiism than to provide nourishment to themselves.

An additional quality of living organisms is the ability to achieve **homeostasis,** the maintenance of a constant internal environment despite a constantly changing external environment (**FIGURE 2.1**). Zombies are surprisingly good at maintaining their body temperature and blood pressure even with multiple flesh wounds and severed limbs. Unfortunately, their limited homeostatic abilities do not prevent them from being in a progressive state of decline, with each assault bringing new injuries, piled on top of previously obtained unhealed injuries.

One last feature of populations of living organisms is that they can evolve, that is, change in average physical characteristics over time. Because zombies don't mate and cannot pass on their genes, evolution is not possible.

**STOP & STRETCH** A popular theme in zombie culture is the zombie apocalypse. Such an event would occur if zombies were able to wipe out the human population. The agent that is transmitted during a zombie attack must be very virulent, because no one who is bitten by a zombie ever manages to escape becoming a zombie. How might this level of virulence end up limiting the duration of a zombie apocalypse?

Part of the reason we find zombies entertaining is because they do show some evidence of scientific plausibility. They exist in a state somewhere between the living and the dead and the transmission of zombiism through a bite from an infected zombie is a mode of transmission we know to be at work in other infectious diseases. We will see that this confusion of scientifically plausible notions with our willingness to accept things that seem like they could be true, can lead us to suspend our disbelief in cases where our entertainment is not the only thing at stake.

# 2.2   The Properties of Water

Does drinking water during an exam help students perform better? When this hypothesis was tested by researchers at the University of East London, they found that this might just be the case (**FIGURE 2.2**). Before we analyze this study, let's first learn about some of the properties of water and whether any of these properties can be used to help explain this result.

Water is made up of two elements: hydrogen and oxygen. **Elements** are the fundamental forms of matter and are composed of atoms that cannot be broken down by normal physical means such as boiling.

**Atoms** are the smallest units that have the properties of any given element. Ninety-two natural elemental atoms have been described by chemists, and several more have been created in laboratories. Hydrogen, oxygen, and calcium are examples of elements commonly found in biological systems. Each element has a one- or two-letter symbol: H for hydrogen, O for oxygen, and Ca for calcium, for example.

Atoms are composed of subatomic particles called **protons**, **neutrons**, and **electrons**. Protons have a positive electric charge; these particles and the uncharged neutrons make up the **nucleus** of an atom. All atoms of a particular element have the same number of protons, giving the element its **atomic number**. The negatively charged electrons are found outside the nucleus in an "electron cloud." Electrons are attracted to the positively charged nucleus. A *neutral atom* has equal numbers of protons and electrons (**FIGURE 2.3a–c**). Electrically charged **ions** do not have an equal number of protons and electrons (**FIGURE 2.3d**). In this case, the atom is not neutral but is instead charged.

**FIGURE 2.3 Atomic structure.**  An oxygen atom (a) contains a nucleus made up of 8 protons, 8 neutrons, and is orbited by 8 electrons. Carbon (b), also a neutral atom, is composed of 6 protons, 6 neutrons, and 6 electrons. Hydrogen can exist as a neutral atom (c) with 1 neutron and 1 electron or it can exist as a positively charged ion (d) composed of 1 proton only.

## Working with Data ▼

Which variable (bringing water to exam or exam score) is the independent variable (one whose value can be experimentally manipulated) and on which axis (x or y) do we find it? If there is a statistically different difference between the two treatment groups, how should that be shown on the graph?

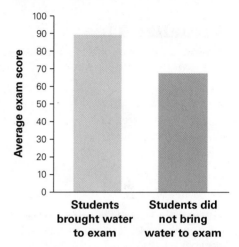

**FIGURE 2.2 Bringing water to exams.** This graph shows data similar to that obtained in a study that compared exam scores of students who did and did not bring water with them to an exam.

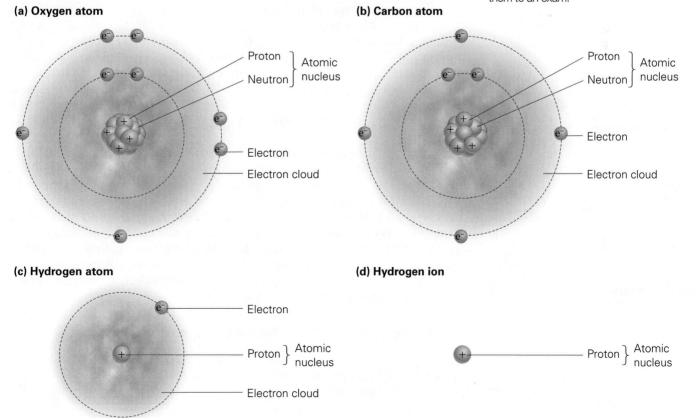

**(a) Oxygen atom**

**(b) Carbon atom**

**(c) Hydrogen atom**

**(d) Hydrogen ion**

## The Structure of Water

The chemical formula for water ($H_2O$) shows that it contains 2 hydrogen atoms for every 1 oxygen. Water, like other molecules, consists of 2 or more atoms joined by chemical bonds. A molecule can be composed of the same or different atoms. For example, a molecule of oxygen consists of 2 oxygen atoms joined to each other, while a molecule of carbon dioxide consists of 1 carbon and 2 oxygen atoms.

## Water Is a Good Solvent

Water has the ability to dissolve a wide variety of substances. A substance that dissolves when mixed with another substance is called a **solute.** When a solute is dissolved in a liquid, such as water, the liquid is called a **solvent.** Once dissolved, components of a particular solute can pass freely throughout the water, making a chemical mixture or solution.

Water is a good solvent because it is **polar,** meaning that different regions, or poles, of the molecule have different charges. The polarity arises because oxygen is more attractive to electrons—that is, it is more **electronegative**—than most other atoms, including hydrogen. As a result of oxygen's electronegativity, electrons in a water molecule spend more time near the nucleus of the oxygen atom than near the nuclei of the hydrogen atoms. With more negatively charged electrons near it, the oxygen in water carries a partial negative charge, symbolized by the Greek letter delta, $\delta^-$. The hydrogen atoms thus have a partial positive charge, symbolized by $\delta^+$ (**FIGURE 2.4**). When atoms of a molecule carry no partial charge, they are said to be **nonpolar.** The carbon-hydrogen bonds, for example, share electrons equally and are nonpolar.

Water molecules tend to orient themselves so that the hydrogen atom (with its partial positive charge) of one molecule is near the oxygen atom (with its partial negative charge) of another molecule (**FIGURE 2.5a**). The weak attraction between hydrogen atoms and oxygen atoms in adjacent molecules forms a **hydrogen bond.** Hydrogen bonding is a type of weak chemical bond that forms when a partially positive hydrogen atom is attracted to a partially negative atom. Hydrogen bonds can be intramolecular, involving different regions within the same molecule, or they can be intermolecular, between different molecules, as is the case in hydrogen bonding between different water molecules. **FIGURE 2.5b** shows the hydrogen bonding that occurs between water molecules in liquid form.

The ability of water to dissolve substances such as sodium chloride is a direct result of its polarity. Each molecule of sodium chloride is composed of 1 sodium ion ($Na^+$) and 1 chloride ion ($Cl^-$). In the case of sodium chloride, the negative pole of water molecules will be attracted to a positively charged sodium ion and separate it from a negatively charged chloride ion (**FIGURE 2.6**). Water can also dissolve other polar molecules, such as alcohol, in a similar manner. Polar molecules are called **hydrophilic** ("water-loving") because of their ability to dissolve in water. Nonpolar molecules, such as oil, do not contain charged atoms and are referred to as **hydrophobic** ("water-fearing") because they do not easily mix with water.

## Water Facilitates Chemical Reactions

Because it is such a powerful solvent, water can facilitate **chemical reactions,** which are changes in the chemical composition of substances. Solutes in a mixture, called **reactants,** can come in contact with each other, permitting the

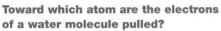

# Visualize This ▼

Toward which atom are the electrons of a water molecule pulled?

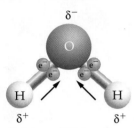

**FIGURE 2.4 Polarity in water.** Water is a polar molecule. Its atoms do not share electrons equally.

**hydro-** means water.

**-philic** means to love.

**-phobic** means to fear.

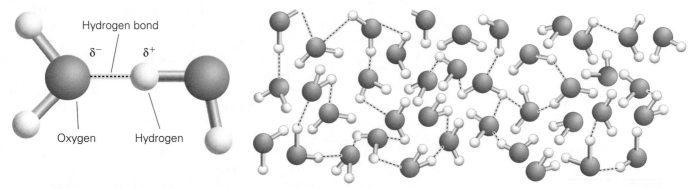

**(a) Bonds between two water molecules**

Hydrogen bond

$\delta^-$    $\delta^+$

Oxygen    Hydrogen

**(b) Bonds between many water molecules**

**FIGURE 2.5 Hydrogen bonding.**   Hydrogen bonding can occur when there is a weak attraction between the hydrogen and oxygen atoms between (a) two or (b) many different water molecules.

modification of chemical bonds that occur during a reaction. The molecules formed as a result of a chemical reaction are known as **products.**

## Water Is Cohesive

The tendency of like molecules to stick together is called **cohesion.** Cohesion is much stronger in water than in most liquids because of the sheer number of hydrogen bonds between water molecules. Cohesion is an important property of many biological systems. For instance, many plants depend on cohesion to help transport a continuous column of water from the roots to the leaves.

## Water Moderates Temperature

When heat energy is added to water, its initial effect is to disrupt the hydrogen bonding among water molecules. Therefore, this heat energy can be absorbed without changing the temperature of water. Only after the hydrogen bonds have been broken can added heat increase the temperature. In other words, the initial input of energy is absorbed.

**The Water Hypothesis Requires More Substantiation.**   We can now come back to our original question about whether drinking water can help students perform better on an exam. It is not too hard to imagine how some of the properties of water outlined earlier might lend it the ability to help regulate blood pressure, deliver dissolved nutrients to the body (including the brain), and maintain body temperature. It is also not hard to imagine that keeping blood pressure and temperature stable during a stressful exam and delivering nutrients to a hardworking brain might mean increased performance on an exam. Tempting as it is to stop there, a good scientist must not simply accept results because they are believable. A good scientist must also evaluate alternative hypotheses.

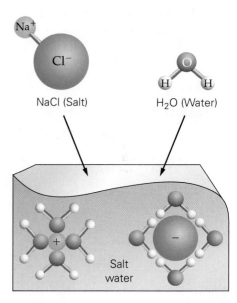

$Na^+$

$Cl^-$

NaCl (Salt)

$H_2O$ (Water)

Salt water

**FIGURE 2.6 Water as a solvent.** When salt is placed in water, the negatively charged regions of the water molecules surround the positively charged sodium ion, and the positively charged regions of the water molecules surround the negatively charged chlorine ion, breaking the bond holding sodium and chloride together and dissolving the salt.

**STOP & STRETCH**   Can you think of other explanations for the increased scores obtained by students who brought water to drink during an exam?

It might be the case that the more prepared students bring water to exams. Being more prepared for the exam could certainly increase exam scores whether or not someone drank water. The fact that alternate hypotheses may explain the data does not mean that the original hypothesis is incorrect, only that it requires more testing. In fact, the researchers that presented this study pointed out that the students bringing water to the exam tended to be upperclassmen, a variable that would need to be controlled for in subsequent studies. It is possible that upperclassmen, having gone through the experience of taking more exams, are more relaxed and will perform better than other students, regardless of how much knowledge of the subject they may have.

Rigorous testing of alternate hypotheses is one major difference between bad science, **pseudo**science, and good science. Good scientists don't perform only experiments that will help confirm their hypothesis. They actively seek out ways to falsify their hypotheses, by developing and testing alternate hypotheses.

**pseudo-** means false or fraudulent.

## Some bad science is believable only to the very young or uninformed, like planes and ships disappearing into the Bermuda Triangle.

# 2.3 Chemistry for Biology Students

Many people find chemistry intimidating and shy away from trying to understand chemical principles. Even the most basic understanding of chemistry can help one sort out fact from fiction. A basic understanding of chemistry will be used to help consider whether the claims made about the rate of disappearance of ships and planes over the so-called Bermuda Triangle are based in science or pseudoscience.

The chemistry of biological systems is based on the element carbon. The branch of chemistry that is concerned with carbon-containing molecules is called **organic chemistry.** Carbon interacts with other elements to produce the more complex molecules that comprise all living things. These interactions, or chemical bonds, follow a few simple patterns.

### Chemical Bonds

Chemical bonds between atoms and molecules involve attractions that help stabilize structures. In general, this involves the sharing or transfer of electrons. Carbon is often involved in chemical bonding because of its ability to make bonds with up to four other elements. Like a Tinkertoy™ connector, carbon has multiple sites for connections that allow carbon-containing molecules to take an almost infinite variety of shapes (**FIGURE 2.7**).

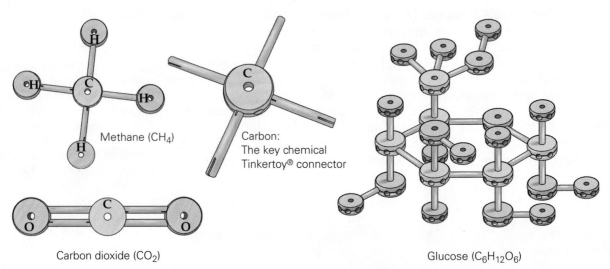

Methane ($CH_4$)

Carbon: The key chemical Tinkertoy® connector

Carbon dioxide ($CO_2$)

Glucose ($C_6H_{12}O_6$)

**FIGURE 2.7 Carbon, the chemical Tinkertoy™ connector.** Because carbon forms four covalent bonds at a time, carbon-containing compounds can have diverse shapes.

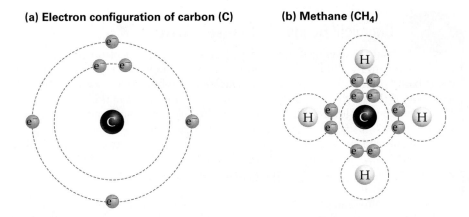

**(a) Electron configuration of carbon (C)**

**(b) Methane (CH$_4$)**

**FIGURE 2.8 Carbon bonding.** (a) Carbon has 4 unpaired electrons in its valence shell and is thus able to make up to 4 chemical bonds. (b) Methane consists of carbon covalently bonded to 4 hydrogens.

Carbon, with its 4 unpaired outer electrons (**FIGURE 2.8a**) can form up to 4 bonds. Carbon can form 4 single bonds, 2 double bonds, 1 double bond and 2 single bonds, and so on, depending on the number of electrons needed by the atom that is its partner. Methane (**FIGURE 2.8b**) is formed when carbon makes 4 single bonds to hydrogen.

**STOP & STRETCH** What does not make sense (chemically) about the structure below?

H–C=C–H

**The Bermuda Triangle.** The Bermuda Triangle is a roughly triangular area off the coast of Florida (**FIGURE 2.9**). The fact that there is actually no evidence that ships and planes go missing from this area at a higher rate than other areas has done little to dampen fears of some prospective travelers.

During the early days of sea exploration, speculation about the cause of these "disappearances" focused on sea monsters, and was later replaced by notions about disruption in gravitational or magnetic fields. As understanding of gravity and magnetic fields increased, the allure of these explanations decreased. One explanation circulating now involves methane. This idea posits that methane deposits under the ocean produce giant gas bubbles that rise up from the ocean floor and sink ships. If these gas bubbles escape the ocean, rise into the atmosphere, and are struck by lightning, they could cause fires that bring down airplanes. While there is no scientific evidence to support this hypothesis, its plausibility is enhanced, like that of all pseudoscience, by the inclusion of scientific principles, in this case chemistry.

**FIGURE 2.9 Underwater methane deposits.** Current pseudoscience explanations for the "disappearance" of ships and planes over the Bermuda Triangle include an untested hypothesis about methane gas bubbles rising from the ocean floor to knock ships out of the water or being ignited by lightning to cause planes to crash.

### ◀ Visualize This

**While methane deposits do exist under the Bermuda Triangle, they also exist in the same or higher concentration at many other undersea locations. For the methane hypothesis to be supported, what must be true of other oceanic locations with similarly sized methane deposits?**

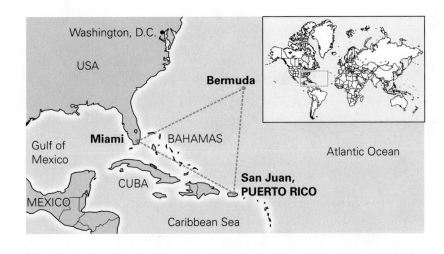

**(a) Methane**    **(b) Ethylene**

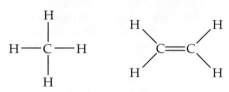

**FIGURE 2.10 Single and double bonds.**   (a) Covalent bonds are symbolized by a short line indicating a shared pair of electrons. (b) Double covalent bonds involve two pairs of shared electrons, symbolized by two horizontal lines.

# Chemical Bonds: A Closer Look ▼

Making chemical bonds depends on an atom's electron configuration. The electrons in the cloud that surround the atom's nucleus have different energy levels based on their distance from the nucleus. The first level, or **electron shell,** is closest to the nucleus, and the electrons located there have the lowest energy. The second energy level is farther away, and the electrons have more energy, and so on.

Each energy level can hold only a specific number of electrons. The first shell holds 2 electrons; the second shell holds up to 8. Electrons fill the lowest shell before filling a higher shell. For example, hydrogen with 1 electron needs only 1 more to fill its first shell.

Atoms with the same number of electrons in their outer shell, called the **valence shell,** exhibit similar chemical behaviors. When the valence shell is full, the atom won't normally form bonds with other atoms. Atoms whose valence shells are not full often combine via chemical bonds.

Atoms with 4 or 5 electrons in the outer shell tend to share electrons to complete their valence shells. When atoms share electrons, a **covalent bond** is formed.

Covalent bonds are shown by a short line indicating a shared pair of electrons (**FIGURE 2.10a**). Double bonds involve two pairs of shared electrons. A double bond is symbolized by two horizontal lines (**FIGURE 2.10b**).

Atoms with 1, 2, or 3 electrons in their valence shell tend to lose electrons and become positively charged ions, while atoms with 6 or 7 tend to gain electrons and become negatively charged ions. Positively and negatively charged ions associate into a type of bond called the ionic bond.

**Ionic bonds** form between charged atoms attracted to each other by similar, opposite charges. For example, a sodium atom forms an ionic bond with a chlorine atom to produce table salt (sodium chloride) when the sodium atom gives up an electron and the chlorine atom gains one (**FIGURE 2.11**). More than 2 atoms can be involved in an ionic bond: Calcium will react with 2 chlorine atoms to produce calcium chloride ($CaCl_2$), because calcium has 2 electrons in its valence shell.

Ionic bonds are about as strong as covalent bonds. They can be more easily disrupted, however, when mixed with liquids, such as water, that contain electric charges.

At any time, a small percentage of water molecules in a pure solution will be dissociated into $H^+$ and $OH^-$ ions. The **pH** scale measures the relative amounts of these ions in a solution (**FIGURE 2.12**). An **acid** is a substance that donates $H^+$ ions to a solution while a **base** is a substance that accepts $H^+$ ions. The more acidic a solution

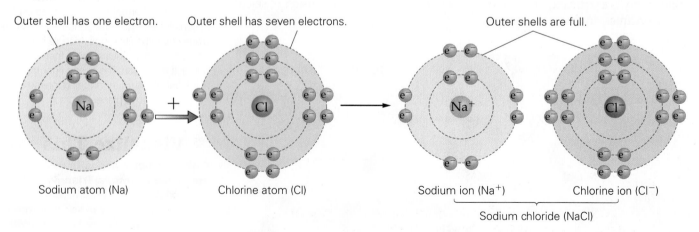

**FIGURE 2.11 Ionic bonding.**   Ionic bonds form when electrons are transferred between charged atoms.

*A Closer Look, continued*

is, the higher the $H^+$ concentration is relative to the $OH^-$ ions. The higher the $H^+$ concentration, the lower the pH. Basic solutions have fewer $H^+$ ions relative to $OH^-$ ions and thus a higher pH. There are equal numbers of both in pure water, so it is neutral, or pH 7. The pH of most cells is very close to 7.

# 2.4 Biological Macromolecules

**Macromolecules** are large organic molecules made of subunits. Those present in living organisms include carbohydrates, proteins, lipids, and nucleic acids. Some are synthesized by the body and some are obtained in the diet. Dietary macromolecules tend to be the focus of many myths related to how what we eat affects us.

## Carbohydrates

Sugars, or **carbohydrates**, are the major source of energy for cellular processes and play important structural roles in cells. The simplest carbohydrates are composed of carbon, hydrogen, and oxygen in the ratio ($CH_2O$). For example, glucose is symbolized as $6(CH_2O)$ or $C_6H_{12}O_6$. Glucose is a simple sugar, or **monosaccharide**, that consists of a single ring-shaped structure. Disaccharides are two rings joined together. Table sugar (sucrose) is a disaccharide composed of glucose and fructose, a sugar found in fruits.

Joining many individual **monomers** together produces **polymers**. Polymers of sugar monomers are called **polysaccharides** (**FIGURE 2.13**). Plants use tough polysaccharides in their cell walls as a sort of structural skeleton. The

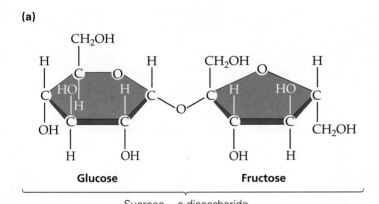

**(a)**

Glucose     Fructose

Sucrose—a disaccharide

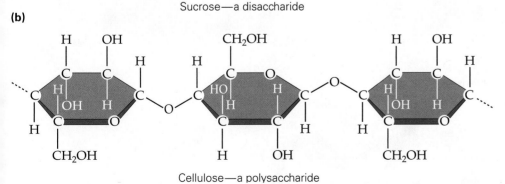

**(b)**

Cellulose—a polysaccharide

**FIGURE 2.13 Carbohydrates.**   Disaccharides are composed of two individual sugar monomers. (a) Polysaccharides are long chains of sugars joined by covalent bonds. (b) Cellulose is a polysaccharide composed of glucose monomers that plays a structural role in plant cell walls.

## Visualize This ▼

A substance with a pH of 5 would have how many times more $H^+$ ions than a substance with a pH of 7?

**FIGURE 2.12 The pH scale.**   The pH scale is a measure of hydrogen ion concentration ranging from 0 (most acidic) to 14 (most basic). Each pH unit actually represents a 10-fold (10x) difference in the concentration of $H^+$ ions.

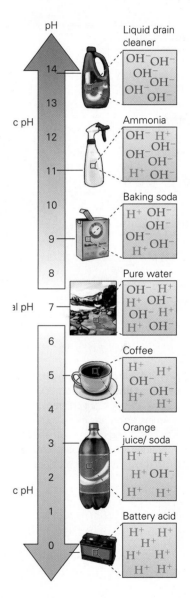

external skeletons of insects, spiders, and lobsters are composed of the polysaccharide chitin, and the cell walls that surround bacterial cells are rich in structural polysaccharides.

## Proteins

Living organisms require **proteins** for a wide variety of processes. Proteins are important structural components of cells; they are integral to the make-up of cell membranes and make up half the dry weight of most cells. Some cells, such as animal muscle cells, are primarily composed of proteins. Proteins called **enzymes** accelerate and help regulate all the chemical reactions that build up and break down molecules inside cells. The catalytic power of enzymes (their ability to help start chemical reactions) allows metabolism to occur under normal cellular conditions. Proteins can also serve as channels through which substances are brought into cells, and they can function as hormones that send chemical messages throughout an organism's body.

Proteins are large molecules made of monomer subunits called **amino acids.** There are 20 commonly occurring amino acids. Like carbohydrates, amino acids are made of carbons, hydrogens, and oxygens; these form the amino acid's carboxyl group ($-COO^-$). In addition, amino acids have nitrogen as part of an amino ($NH_2^+$) group along with various side groups. Side groups are chemical groups that give amino acids different chemical properties (**FIGURE 2.14a**).

Polymers of amino acids can be joined together in various sequences called polypeptides. A covalent bond joins adjacent amino acids to each other. **FIGURE 2.14b** shows three amino acids—valine, alanine, and phenylalanine—joined by peptide bonds. Precisely folded polypeptides produce specific proteins in much the same manner that children can use differently shaped beads to produce a wide variety of structures (**FIGURE 2.14c**). Each amino acid side group has unique chemical properties, including being polar or nonpolar. Because each protein is composed of a particular sequence of amino acids, each protein has a unique shape and therefore specialized chemical properties.

**macro-** means large.

**mono-** means one.

**-mer** means subunit.

**poly-** means many.

**Many parents believe that giving children candy or sugary drinks makes them hyperactive.**

**FIGURE 2.14 Amino acids, peptide bonds, and proteins.** (a) All amino acids have the same backbone but different side groups. (b) Amino acids are joined together by peptide bonds. Long chains of these are called polypeptides. (c) Polypeptide chains fold upon themselves to produce proteins.

**(a) General formula for amino acid**

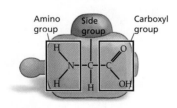

**(b) Peptide bond formation**

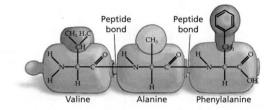

**(c) Protein**

# Lipids

There are three different types of lipids, which are grouped together because they are partially or entirely hydrophobic organic molecules made primarily of hydrocarbons. The three types of lipids include fats, steroids, and phospholipids.

**Fat.**  The structure of a **fat** is that of a 3-carbon glycerol molecule with 3 long hydrocarbon chains attached to it (**FIGURE 2.15a**). **Hydrocarbons** are carbon- and hydrogen-rich. Like the hydrocarbons present in gasoline, these can be burned to produce energy. The long hydrocarbon chains are called **fatty acid** tails. Fats are hydrophobic and function in storing energy within living organisms. Fatty acid tails are covalently bonded to glycerol through a type of chemical bond called an ester bond.

**Steroids.**  **Steroids** are composed of 4 fused carbon-containing rings. Cholesterol (**FIGURE 2.15b**) is one steroid that is probably familiar; its primary function in animal cells (plant cells do not contain cholesterol) is to help maintain the fluidity of membranes. Other steroids include the sex hormones testosterone, estrogen, and progesterone, which are produced by the sex organs and have effects throughout the body.

**Phospholipids.**  **Phospholipids** are similar to fats except that each glycerol molecule is attached to 2 fatty acid tails (not 3, as you would find in a dietary fat). The third bond in a phospholipid is to a phosphate head group. The phosphate head group consists of a phosphorous atom attached to 4 oxygen atoms and is hydrophilic. Thus, a phospholipid has a hydrophilic head and 2 hydrophobic tails (**FIGURE 2.15c**). Phospholipids often have an additional head

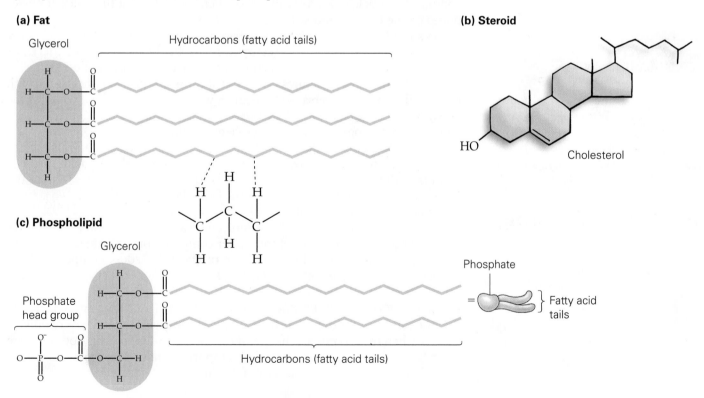

**FIGURE 2.15 Three types of lipids.**   (a) Fats are composed of a glycerol molecule with 3 hydrocarbon-rich fatty acid tails attached. (b) Cholesterol is a steroid common in animal cell membranes. (c) Phospholipids are composed of a glycerol backbone with 2 fatty acids attached and 1 phosphate head group. The cartoon drawing to the right shows how phospholipids are often depicted.

group, attached to the phosphate, which also confers unique chemical properties on the individual phospholipid. Phospholipids are important constituents of the membranes that surround cells and that designate compartments within cells.

## Nucleic Acids

Nucleic acids are composed of long strings of monomers called **nucleotides.** A nucleotide is made up of a sugar, a phosphate, and a nitrogen-containing base. There are two classes of nucleic acids in living organisms. **Ribonucleic acid (RNA)** plays a key role in helping cells synthesize proteins (and is discussed in detail in later chapters). The nucleic acid that serves as the primary storage of genetic information in nearly all living organisms is **deoxyribonucleic acid (DNA).** FIGURE 2.16 shows the three-dimensional structure of a DNA molecule and zooms inward to the chemical structure. You can see that DNA is composed of two curving strands that wind around each other to form a double helix. The sugar in DNA is the 5-carbon sugar deoxyribose. The nitrogen-containing bases, or **nitrogenous bases,** of DNA have one of four different chemical structures, each with a different name: **adenine (A), guanine (G), thymine (T),** and **cytosine (C).** Nucleotides are joined to each other along the length of the helix by covalent bonds.

Nitrogenous bases form hydrogen bonds with each other across the width of the helix. On a DNA molecule, an adenine (A) on one strand always pairs with a thymine (T) on the opposite strand. Likewise, guanine (G) always pairs with cytosine (C). The term **complementary** is used to describe these pairings. For example, A is complementary to T, and C is complementary to G. Therefore, the order of nucleotides on one strand of the DNA helix predicts the order of nucleotides on the other strand. Thus, if one strand of the DNA molecule is composed of nucleotides AACGATCCG, then we know that the order of nucleotides on the other strand is TTGCTAGGC.

As a result of this complementary base-pairing, the width of the DNA helix is uniform. There are no bulges or dimples in the structure of the DNA helix because A and G, called **purines,** are structures composed of 2 rings; C and T are single-ring structures called **pyrimidines.** A purine always pairs with a pyrimidine and vice versa, so there are always 3 rings across the width of the helix. A-to-T base pairs have 2 hydrogen bonds holding them together. G-to-C pairs have 3 hydrogen bonds holding them together.

Each strand of the helix thus consists of a series of sugars and phosphates alternating along the length of the helix, the **sugar-phosphate backbone.** The strands of the helix align so that the nucleotides face "up" on one side of the helix and "down" on the other side of the helix. For this reason, the two strands of the helix are said to be antiparallel.

The overall structure of a DNA molecule can be likened to a rope ladder that is twisted, with the sides of the ladder composed of sugars and phosphates (the sugar-phosphate backbone) and the rungs of the ladder composed of the nitrogenous-base sequences A, C, G, and T.

**Dietary Macromolecules and Behavior.** Most people accept as fact that hyperactivity in children can be induced by ingestion of sugary drinks or snacks, when there is not much scientific evidence to support this notion. There is, in fact, more scientific evidence to support the notion that parents expect their children to become hyperactive from ingesting sugar.

**Is the relative who says he has to nap after eating turkey for Thanksgiving dinner just lazy, or does something in turkey make people tired?**

**STOP & STRETCH** Consider the kinds of events where children are often given sugary drinks and snacks. Are there other factors that need to be taken into account when looking at the effects of sugar on behavior?

**(a) DNA double helix is made of two strands.**

**(b) Each nucleotide is composed of a phosphate, a sugar, and a nitrogenous base.**

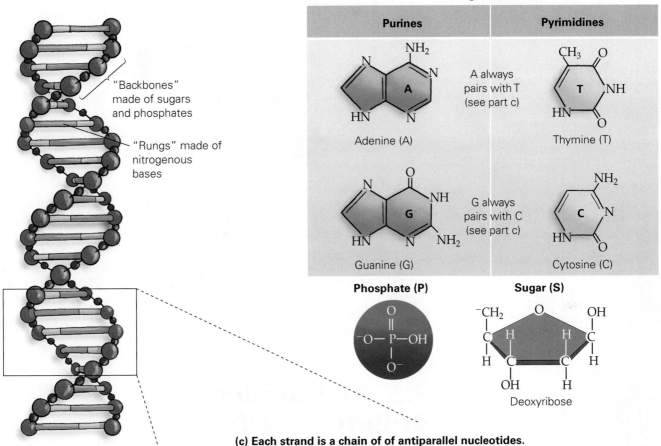

"Backbones" made of sugars and phosphates

"Rungs" made of nitrogenous bases

**Nitrogenous bases**

| Purines | Pyrimidines |
|---|---|
| $NH_2$ ... **A** ... Adenine (A) | $CH_3$ ... **T** ... Thymine (T) |
| A always pairs with T (see part c) | |
| **G** ... Guanine (G) | $NH_2$ ... **C** ... Cytosine (C) |
| G always pairs with C (see part c) | |

**Phosphate (P)**

**Sugar (S)**

Deoxyribose

# Visualize This ▲

**Look at part (a) of this figure. What do the red and purple spheres represent? Why are there two colors and sizes of nitrogenous bases?**

**(c) Each strand is a chain of of antiparallel nucleotides.**

**"Rung"**
The two strands are connected by hydrogen bonds between the nucleotides.

Sugar-phosphate "backbone"

Sugar-phosphate "backbone"

Nucleotides within the strand are connected by covalent bonds.

**FIGURE 2.16 DNA structure.** (a) DNA is a double-helical structure composed of nucleotides. (b) Each strand of the helix is composed of repeating units of sugars and phosphates, making the sugar-phosphate backbone, and of nitrogenous bases across the width of the helix. (c) Each nucleotide is composed of a phosphate, a sugar, and a nitrogenous base. Adenine and guanine are purines, which have a double-ring structure; cytosine and thymine are pyrimidines, which have a single-ring structure.

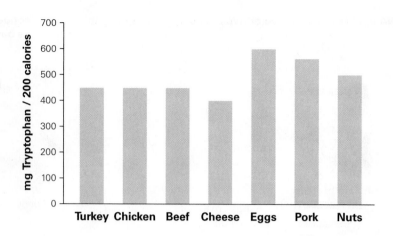

**FIGURE 2.17 Tryptophan concentration in food.** This graph shows that turkey is no more rich in the amino acid tryptophan than many other common foods.

Have you ever heard a relative claim that there is something in turkey that puts them to sleep, as they make their way to the couch after Thanksgiving dinner? It is commonly believed that the presence of the amino acid tryptophan in the protein-rich meat of turkey causes drowsiness. This myth may have gotten some traction when scientists showed that sleep and mood can be affected by tryptophan. However, one look at the data (**FIGURE 2.17**) showing levels of tryptophan in other protein-rich foods should put this myth to bed! An alternative hypothesis that seems to be better supported is that eating a large amount of food, which certainly happens on Thanksgiving, leads to drowsiness.

## 2.5 An Introduction to Evolutionary Theory

Are humans evolving into higher beings? The **theory of evolution** explains how a single common ancestor present on earth 4 billion years ago could give rise to the over 10 million different kinds of organisms found on Earth today. Evolutionary theory will be covered in much greater detail in later chapters. For now, we will try to understand whether the organisms on earth today represent a series of steps in the ladder toward humanness? Does this ladder have a few more rungs that humans can climb, maybe gaining an extra gene or two that will provide superpowers similar to those found in the X-Men?

To investigate this, let's see what happened to some of the earliest organisms to inhabit Earth—the tiny, single-celled bacteria (**FIGURE 2.18a**). Bacteria are classified as **prokaryotic** because they do not have a nucleus, a separate membrane-bound compartment to hold the genetic material. They also do not contain any membrane-bound internal compartments. They do, however, have a **cell wall** that helps them maintain their shape (**FIGURE 2.18b**).

The ancestor of bacteria also gave rise to a structurally more complex group of organisms called eukaryotes. **Eukaryotic** cells have a nucleus surrounding the genetic material (**FIGURE 2.18c**). In addition, eukaryotic cells contain membrane-bounded subcellular structures called organelles. Organelles will be covered in more detail later. For now, it is enough to understand that organelles work together to perform specific jobs inside cells that allow the cell to function properly. Eukaryotic organisms include single-celled amoebas and yeast as well as multicellular plants, fungi, and animals.

While there are differences between prokaryotic and eukaryotic cells, there are also many similarities. In fact, all living organisms are made of cells that

**(a) Different sizes:  prokaryotic (red) vs. eukaryotic (white) cells**

**(c) Eukaryotic cell features**

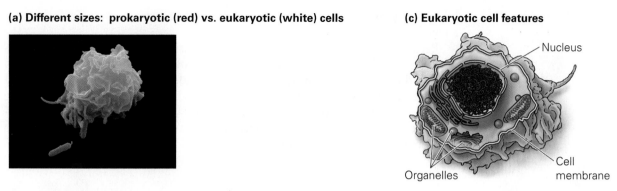

**(b) Prokaryotic cell features**

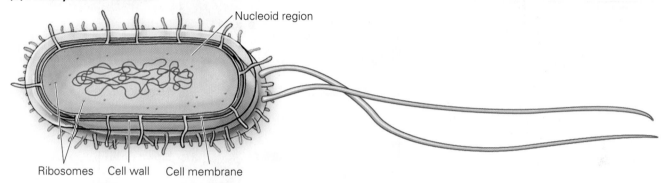

**FIGURE 2.18 Prokaryotic and eukaryotic cells.**   (a) Prokaryotic cells are typically about 1/10 the diameter of a eukaryotic cell, as evidenced by the size of the two bacterial cells and a white blood cell. (b) Prokaryotic cells are structurally simple cells. (c) Eukaryotic cells are more structurally complex cells.

contain the same kinds of macromolecules and other structural features, most likely because they once shared a common ancestor.

**STOP & STRETCH**   Organisms that share a common ancestor in their more recent evolutionary history would share more genes than organisms that are more distantly related. To bring this out of evolutionary time and into the present time, think about your own family. Why would you share more biological similarities with your first cousin than you would with a more distant relative?

The divergence and differences among groups of organisms arose as a result of **natural selection,** a process of gradual changes in the characteristics of populations over time. Natural selection was a major process in the diversification of life from a single ancestor because individual organisms vary from one another and because some of these variations increase the chances of survival and reproduction. A genetic trait that increases survival and reproduction should become more prevalent over time. In contrast, less successful variants should eventually be lost from the population.

So, was there selective pressure on "lower" organisms that led to the evolution of "higher" organisms like humans? Are humans currently evolving into super humans? This is not how scientists see the process of evolution because the real success, or failure, of an organism is measured by how well it is adapted to its environment. In many respects, bacteria are more successful than humans. Bacteria far outnumber humans and are found in many more environments than humans. They have also been on earth evolving and diversifying for billions of years, while humans and their ancestors have been on Earth for only a few million years.

Because evolution is not progressive, or a series of discrete steps toward increasing complexity, representations of the tree of life are usually horizontal and branched (**FIGURE 2.19**) instead of a vertical ladder. In this sense, bacteria that are well adapted to their environment are as highly evolved as humans that are equally well adapted to their environment. In fact, if a zombie apocalypse occurs and humans are wiped out, bacteria might prove to be better adapted to their environment than humans.

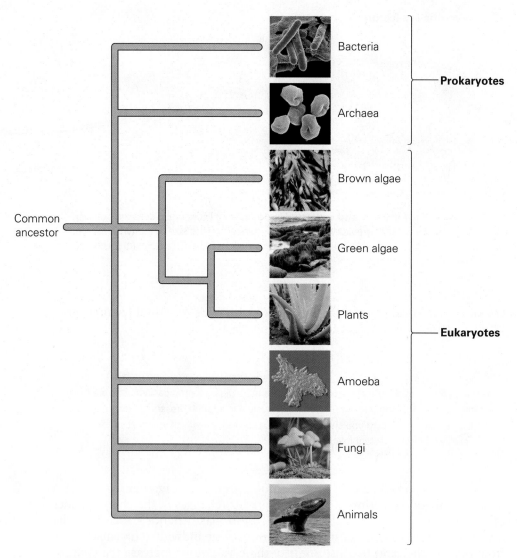

**FIGURE 2.19 A common ancestor on the tree of life.** All living organisms share basic characteristics and can be arranged into a branching tree of life based on more specific similarities. In this illustration, many groups are omitted for simplicity.

# savvy reader

Several U.S. companies sell products called water ionizers. These are water-filtering systems about the size and shape of a small suitcase, which can be installed on or under your sink at a cost of close to $1000. What follows is an example of the kinds of claims made by these manufacturers' websites:

Ionized water contains a negative electric charge that produces charged particles called hydroxyl ions. If you are looking for a natural solution that most conventional doctors don't know about, you need to try this amazing healing water. Drinking ionized water will:

- Help you to feel younger, and to have more energy.
- Relieve chronic pain without medication.
- Help you to sleep better through the night, every night.
- Rapidly skyrocket your energy and stamina so that you can do all the things you love, hugely improving your quality of life.
- Get rid of chronic illness, headaches, high cholesterol, diabetes, arthritis, and depression.

Some of these manufacturers even claim that their product can cure many types of cancer. Research purporting to support the anticancer properties of ionized water involves the purported ability of ionized water to clean up damaging oxygen-free radicals. To test this claim, manufacturers measured the concentration of oxygen-free radicals in various mouse tissues after exposure to ionized water or tap water (**FIGURE 2.20**).

**1.** Based on what you know about water, does the manufacturer's claim that ionized water contains a negative electric charge that produces hydroxyl ions make sense?

**2.** Evaluate these claims by listing the first two red flags that are raised. (Use the questions in the checklist provided in Chapter 1, Table 1.2.)

**3.** Give an example of language used on these websites that differ from the way scientists usually describe results.

**4.** Does the graph shown in Figure 2.20 show evidence that a decrease in oxygen-free radicals is associated with decreased cancer?

**5.** What key statistical element is missing from the graph shown in Figure 2.20?

## Ionized Water

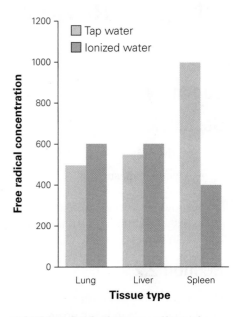

**FIGURE 2.20 Anticancer effect of ionized water.** The concentration of free radicals in various mouse tissues are displayed.

# SOUNDS RIGHT **BUT IS IT?**

Recently, a supercomputer named "Watson" beat two champion *Jeopardy* game contestants. The computer answered questions on a wide variety of topics and was able to make wagers or sit out when it did not know the answer. For many viewers, it seemed that this computer was not only able to retrieve information, but to reason in a manner very similar to the way humans do, leading some to wonder whether intelligence and reasoning are programmable skills.

**Will computers one day come to life and take over, controlling humans with their superior artificial intelligence?**

Sounds right, but it isn't.

**1.** Computers can retrieve data and calculate probabilities. Using the characteristics of life provided in Section 2.1, list features of living organisms that computers lack.

**2.** While chess-playing computers can respond to moves made by a human partner, to be autonomous, computers will need to be able to respond to many more stimuli. Do you think it is possible for programmers to predict all the different stimuli computers will need to be able to respond to in the future?

3. A key feature of human intelligence is the ability to learn. Retrieving stored data and learning differ in the same way that

memorizing some biological definitions or steps in a biological process differs from really understanding the definitions and processes. What happens when you memorize a definition for an exam and the exam asks a question that uses a definition in a slightly different way?

4. Whom do you think is better prepared to apply their knowledge to a new situation, a student that memorizes or a student that understands?

5. Consider your answers to questions 1–4 and explain why the original statement bolded above sounds right, but isn't.

# Chapter Review MasteringBiology®

Go to the Study Area in MasteringBiology® for practice quizzes, myeBook, BioFlix™ 3-D animations, MP3Tutor sessions, videos, current events, and more.

## Summary

### Section 2.1

Describe the properties associated with living organisms.

- Living organisms contain a common set of biological molecules and are composed of cells. They are able to grow, metabolize substances, reproduce, and respond to external stimuli. In addition, living organisms can maintain homeostasis and evolve (p. 32).

### Section 2.2

List the components of water and some of the properties that make it important in living organisms.

- Water consists of 1 oxygen and 2 hydrogen atoms (p. 34).
- Water is a good solvent in part because of its polarity and ability to form hydrogen bonds. Hydrogen bonding in water facilitates chemical reactions, promotes cohesion, and allows for heat absorption (pp. 34–35).

Describe how atomic structure affects chemical bonding.

- Chemical bonding depends on an element's electron configuration. Electrons closer to the nucleus have less energy than those that are farther away from the nucleus (p. 33).
- Atoms that have space in their valence shell form chemical bonds (p. 33).

### Section 2.3

What makes carbon an important component of living organisms?

- Life on Earth is based on the chemistry of the element carbon, which can make bonds with up to four other elements (pp. 36–37).

Compare and contrast hydrogen, covalent, and ionic bonds.

- Hydrogen bonds are weak attractions between hydrogen atoms and oxygen atoms in adjacent molecules (p. 34).
- Covalent bonds form when atoms share electrons. These tend to be strong bonds (p. 38).

- Ionic bonds form between positively and negatively charged ions. These tend to be weaker bonds (p. 38).

### Section 2.4

Describe the structure of carbohydrates, proteins, lipids, and nucleic acids and the roles these macromolecules play in cells.

- Carbohydrates function in energy storage and play structural roles. They can be single-unit monosaccharides or multiple-unit polysaccharides with sugar monomers joined by covalent bonds (pp. 39–40).
- Proteins play structural, enzymatic, and transport roles in cells. They are composed of amino acid monomers joined by covalent bonds (p. 40).
- Lipids are partially or entirely hydrophobic and come in three different forms. Fats are composed of glycerol covalently bonded to 3 fatty acids. Fats store energy. Phospholipids are composed of glycerol, 2 fatty acids, and a phosphate group. They are important structural components of cell membranes. Steroids are composed of 4 fused rings. Cholesterol is a steroid found in some animal cell membranes and helps maintain fluidity. Other steroids function as hormones (pp. 41–42).
- Nucleic acids are polymers of covalently bonded nucleotides, each of which is composed of a sugar, a phosphate, and a nitrogen-containing base (pp. 42–43).

### Section 2.5

Compare and contrast prokaryotic and eukaryotic cells.

- There are two main categories of cells: Those with nuclei and membrane-bound subcellular organelles are eukaryotes; those lacking a nucleus and membrane-bound organelles are prokaryotes (pp. 44–46).

Provide a general summary of the theory of evolution.

- All living organisms are composed of cells that share the same organic chemistry and basic cellular features. These similarities provide support for the theory of evolution, which states that all life on Earth derives from a single common ancestor (pp. 44–46).

# Roots to Remember

**The following roots of words come mainly from Latin and Greek and will help you to decipher terms:**

| | |
|---|---|
| **homeo-** | means like or similar. Chapter term: *homeostasis* |
| **hydro-** | means water. Chapter terms: *hydrophilic, hydrophobic* |
| **macro-** | means large. Chapter term: *macromolecule* |
| **-mer** | means subunit. Chapter terms: *polymer, monomer* |
| **mono-** | means one. Chapter terms: *monosaccharide, monomer* |
| **-philic** | means to love. Chapter term: *hydrophilic* |
| **-phobic** | means to fear. Chapter term: *hydrophobic* |
| **poly-** | means many. Chapter term: *polymer* |
| **pseudo-** | means false or fraudulent. Chapter term: *pseudoscience* |

# Learning the Basics

1. List the four biological molecules commonly found in living organisms.

2. List the structural features in a prokaryotic cell.

3. Define the basic tenets of evolutionary theory.

4. Water _____.

   A. is a good solute; B. facilitates chemical reactions; C. serves as an enzyme; D. makes strong covalent bonds with other molecules; E. consists of two oxygen and one hydrogen atoms.

5. Electrons _____.

   A. are negatively charged; B. along with neutrons comprise the nucleus; C. are attracted to the negatively charged nucleus; D. located closest to the nucleus have the most energy; E. all of the above are true.

6. Which of the following terms is least like the others?

   A. monosaccharide; B. phospholipid; C. fat; D. steroid; E. lipid

7. Different proteins are composed of different sequences of _____.

   A. sugars; B. glycerols; C. fats; D. amino acids; E. carbohydrates.

8. Proteins may function as _____.

   A. genetic material; B. cholesterol molecules; C. fat reserves; D. enzymes; E. all of the above.

9. A fat molecule consists of _____.

   A. carbohydrates and proteins; B. complex carbohydrates only; C. saturated oxygen atoms; D. glycerol and fatty acids.

10. Eukaryotic cells differ from prokaryotic cells in that _____.

    A. only eukaryotic cells contain DNA; B. only eukaryotic cells have a plasma membrane; C. only eukaryotic cells are considered to be alive; D. only eukaryotic cells have a nucleus; E. only eukaryotic cells are able to evolve.

11. Which of the following lists the chemical bonds from strongest to weakest?

    A. hydrogen, covalent, ionic; B. covalent, ionic, hydrogen; C. ionic, covalent, hydrogen; D. covalent, hydrogen, ionic; E. hydrogen, ionic, covalent

12. Which of the following is not consistent with evolutionary theory?

    A. All living organisms share a common ancestor; B. The environment affects which organism survives to reproduce; C. Natural selection always favors the same traits, regardless of environment; D. Humans are not necessarily the best adapted organisms.

# Analyzing and Applying the Basics

1. Consider a virus composed of a protein coat surrounding a small segment of genetic material (either DNA or RNA). Viruses cannot reproduce without taking over the genetic "machinery" of their host cell. Based on this description and biologists' definition of life, should a virus be considered a living organism?

2. It was a commonly held notion that humans should consume eight glasses of water a day until recent scientific evidence called this practice into question. What chemical properties of water do you think contributed to the acceptance of this hypothesis?

3. Carbon, oxygen, hydrogen, and nitrogen are common elements found in living organisms. Which two of the four types of macromolecules contain all four of these elements?

# Connecting the Science

1. Colon cleansing and juice fasting are practices that many celebrities endorse. How would you go about determining whether either of these practices was beneficial?

2. Why do you think people are often so willing to believe pseudoscientific explanations?

Answers to **Stop & Stretch, Visualize This, Working with Data, Savvy Reader, Sounds Right, But Is It?,** and **Chapter Review** questions can be found in the **Answers** section at the back of the book.

CHAPTER **3**

# Is It Possible to Supplement Your Way to Better Performance and Health?

**Do sports drinks enhance athletic performance?**

# Nutrients and Membrane Transport

Gingko to improve your memory, kava to reduce stress, ginseng to boost energy, and melatonin to help you sleep. Sounds like a recipe for success for a busy student. For good measure, chase those supplements down with some coconut water to slow aging and prevent cancer. Instead of going to the grocery store every weekend, should

**Are nutrition bars a healthy meal substitute?**

you stock your pantry with energy drinks, vitamin-enriched waters, protein powders, and nutrition bars? Dietary supplements can be bought in bulk and don't rot like fruits and vegetables. But are they as good for you? If so, you might even find time to join an intramural team or get back to playing the instrument you used to love to play, without having to sacrifice any study time.

Is it possible to supplement your way to enhanced academic

performance or better health? It seems that most Americans think so—we spend around 6 billion dollars a year on these items and over two-thirds of us are taking at least one such supplement. But are these products really doing what we think they are?

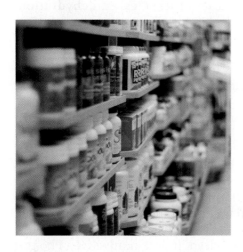

**Do nutritional supplements enhance academic performance or health?**

**Is it more healthful to eat whole foods?**

# 3.1  **Nutrients**

The food and drink that we ingest provide building-block molecules that can be broken down and used as raw materials for growth, maintenance, repair, and as a source of energy. Another name for the substances in food that provide structural materials or energy is **nutrients.**

## Macronutrients

Nutrients that are required in large amounts are called **macronutrients.** These include water, carbohydrates, proteins, and fats.

**Water and Nutrition.**  Most animals can survive for several weeks with no nutrition other than water. However, survival without water is limited to just a few days. Besides helping the body disperse other nutrients, water helps dissolve and eliminate the waste products of digestion.

A decrease below the body's required water level, called **dehydration,** can lead to muscle cramps, fatigue, headaches, dizziness, nausea, confusion, and increased heart rate. Severe dehydration can result in hallucinations, heat stroke, and death.

Sweating is evaporation of water from the skin. This helps maintain body temperature. When water is low and sweating decreases, the body temperature can rise to a harmful level.

Every day, humans lose about 3 liters of water as sweat, in urine, and in feces. A typical adult obtains about 1.5 liters of water per day from food consumption, leaving a deficit of about 1.5 liters that must be replaced.

While it works quite well to replace this water with tap water, many people drink bottled water, in part because of concerns about the quality of tap water. However, the U.S. Food and Drug Administration (FDA), the government agency that sets the standards for bottled water, uses the same standards applied by the Environmental Protection Agency (EPA) to tap water. In other words, water from both sources should be equally clean. In fact, nearly 40% of bottled waters are actually derived directly from municipal tap water.

Many people also choose to hydrate with sports drinks, many of which are high in simple sugars and calories. In fact, some of these drinks have so many additives that their solute concentration is higher than that of blood. When this happens, as you will learn later in this chapter, water will leave the tissues, actually increasing the risk of dehydration.

Bottled waters and sports drinks are also chosen for convenience and portability, but this comes at a cost to the environment. The billions of bottles used each year require 1.5 million barrels of oil to produce. Although the bottles are recyclable, 86% go to landfills each year. In addition, the energy required to transport water far exceeds the energy required to purify an equivalent amount of tap water.

Besides obtaining a healthy dose of water every day, people must consume foods that contain carbohydrates, proteins, and fats. (In Chapter 2, we explored the structure of these macromolecules, and now we focus on how they function in the body.)

**Carbohydrates as Nutrients.**  Foods such as bread, cereal, rice, and pasta, as well as fruits and vegetables, are rich in macronutrients called carbohydrates. Carbohydrates are the major source of energy for cells and can be composed of single-unit monosaccharides or long chains of polysaccharides.

The single-unit simple sugars are digested and enter the bloodstream quickly after ingestion. Sugars found in milk, juice, honey, and most refined foods are simple sugars. Sugars that are composed of many subunits and arranged in branching chains are called **complex carbohydrates.** Complex carbohydrates are found in fruits, vegetables, breads, legumes, and pasta.

The body digests complex carbohydrates more slowly than it does simpler sugars because complex carbohydrates have more chemical bonds to break. Endurance athletes will load up on complex carbohydrates for several days before a race to increase the amount of stored energy that will be available during competition. Many athletes will also drink sugary sports drinks to provide fuel during a competition (**FIGURE 3.1**).

Nutritionists agree that most of the carbohydrates in a healthful diet should be in the form of complex carbohydrates, and that it is best to consume only minimal amounts of processed sugars. A **processed food** is one that has undergone extensive refinement and, in doing so, has been stripped of much of its nutritive value. For example, unrefined raw brown sugar is made from the juice of the sugarcane plant and contains the minerals and nutrients found in the plant. Processing brown sugar to produce refined brown sugar results in the loss of these vitamins and minerals.

Foods that have not been stripped of their nutrition by processing are called **whole foods.** Whole grains, beans, and many fruits and vegetables are whole foods. In addition to being nutritious, whole foods also tend to be good sources of dietary fiber.

Fiber is an important part of a healthy diet. Also called *roughage,* fiber is composed mainly of those complex carbohydrates that humans cannot digest. For this reason, dietary fiber is passed into the large intestine, where some of it is digested by bacteria, and the remainder gives bulk to feces. Although fiber is not a nutrient because it is not absorbed by the body, it is still an important part of a healthful diet. Fiber helps maintain healthy cholesterol levels and may decrease your risk of certain cancers.

Fiber bars are a common sight on the shelves of grocery and convenience stores. While fiber bars can be rich in fiber, they are also often rich in processed sugars and non-nutritive additives. Because of this, it is better to get your fiber from whole food sources.

## Proteins as Nutrients.

Protein-rich foods include beef, poultry, fish, beans, eggs, nuts, and dairy products such as milk, yogurt, and cheese.

Your body is able to synthesize many of the commonly occurring amino acids that proteins are made of. Those your body cannot synthesize are called **essential amino acids** and must be supplied by the foods you eat. **Complete proteins** contain all the essential amino acids your body needs. Proteins obtained by eating meat are more likely to be complete than are those obtained by eating plants; plant proteins can often be missing one or more essential amino acids.

In the past, there was some concern that vegetarians might be at risk for deficiencies in certain amino acids. However, scientific studies have shown that there is little cause for concern. If a vegetarian's diet is rich in a wide variety of plant-based foods, the body will have little trouble obtaining all the amino acids it needs to build proteins.

**STOP & STRETCH**  Many cultures have characteristic meals that include food pairings. Common pairings include beans and rice, eggs and toast, and cereal and milk. From a nutrition standpoint, why might this practice of pairing different foods have evolved?

People looking to gain muscle mass will sometimes supplement their diet with protein powders that can be mixed with milk or water to help rebuild protein-rich muscle after a strenuous workout. However, extra protein does not always lead to extra muscle. Like any nutrient, excess protein is stored as fat.

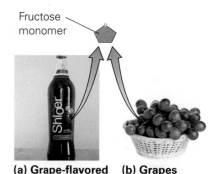

Fructose monomer

**(a) Grape-flavored drink**    **(b) Grapes**

**FIGURE 3.1 Carbohydrates in processed and whole foods.** Sugary sports drinks can be rich in rapidly digested high fructose corn syrup. Fructose is also found in fruit, where it is packaged along with complex carbohydrates and fiber, in addition to other nutrients.

**Should you stock your pantry with energy drinks, vitamin-enriched waters, protein powders, and nutrition bars?**

So, if you eat too much protein, you will end up increasing body fat. In addition, diets that are too rich in protein can lead to health problems such as bone loss and kidney damage. If you are looking to build muscle, the most sensible thing to do is to combine your workout regimen with a healthful diet. Even the most active people, for example endurance athletes, can easily obtain enough protein by eating a healthy diet (**FIGURE 3.2**).

**Fats as Nutrients.** Fat is a source of stored energy. In fact, gram for gram, fat contains around twice as much energy as carbohydrates or protein. Foods that are rich in fat include meat, milk, cheese, vegetable oils, and nuts. Most mammals, including humans, store fat just below the skin to help cushion and protect vital organs, to insulate the body from cold weather, and to store energy in case of famine.

Recall that fats have long hydrocarbon rich fatty acid tails. Your body can synthesize most of the fatty acids it requires. Those that cannot be synthesized are called **essential fatty acids.** Like essential amino acids, essential fatty acids must be obtained from the diet. Omega-3 and omega-6 fatty acids are essential fatty acids that can be obtained by eating fish. These fatty acids are thought to help protect against heart disease. Nutritionists recommend that people eat about 12 ounces of fish every week. Some people who don't eat much fish supplement their diets with fish oil capsules. However, fish contain other vitamins and minerals that fish oil supplements do not.

The fatty acid tails of a fat molecule can differ in the number and placement of double bonds (**FIGURE 3.3a**). When the carbons of a fatty acid are bound to as many hydrogen atoms as possible, the fat is said to be a **saturated fat** (saturated in hydrogen). When there are carbon-to-carbon double bonds, the fat is not saturated in hydrogen, and it is therefore an **unsaturated fat** (**FIGURE 3.3b**). The more double bonds there are, the higher the degree of unsaturation will be. When it contains many unsaturated carbons, the fat is referred to as **polyunsaturated.** The double bonds in unsaturated fats make the structures kink instead of lying flat. This form prevents the adjacent fat molecules from packing tightly together, so unsaturated fat tends to be liquid at room temperature. Cooking oil is an example of an unsaturated fat. Unsaturated fats are more likely to come from plant sources, while the fats found in animals are typically saturated. Saturated fats, with their absence of carbon-to-carbon double bonds, pack tightly together to make a solid structure. This is why saturated fats, such as butter, are solid at room temperature.

Commercial food manufacturers sometimes add hydrogen atoms to unsaturated fats by combining hydrogen gas with vegetable oils under pressure. This process, called **hydrogenation,** increases the level of saturation of a fat. This

**FIGURE 3.2 Protein in processed and whole foods.** Nutritionists recommend that most people consume around 50 grams of protein a day. This can be obtained by drinking two 8-ounce protein shakes or eating one small chicken breast and one cup of broccoli.

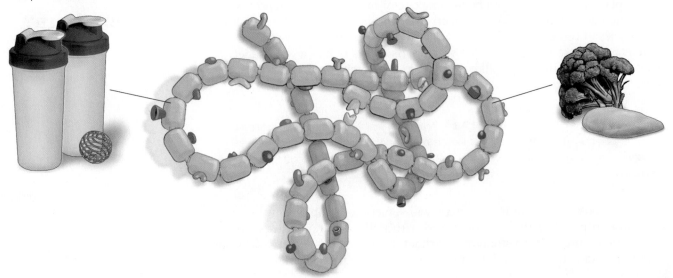

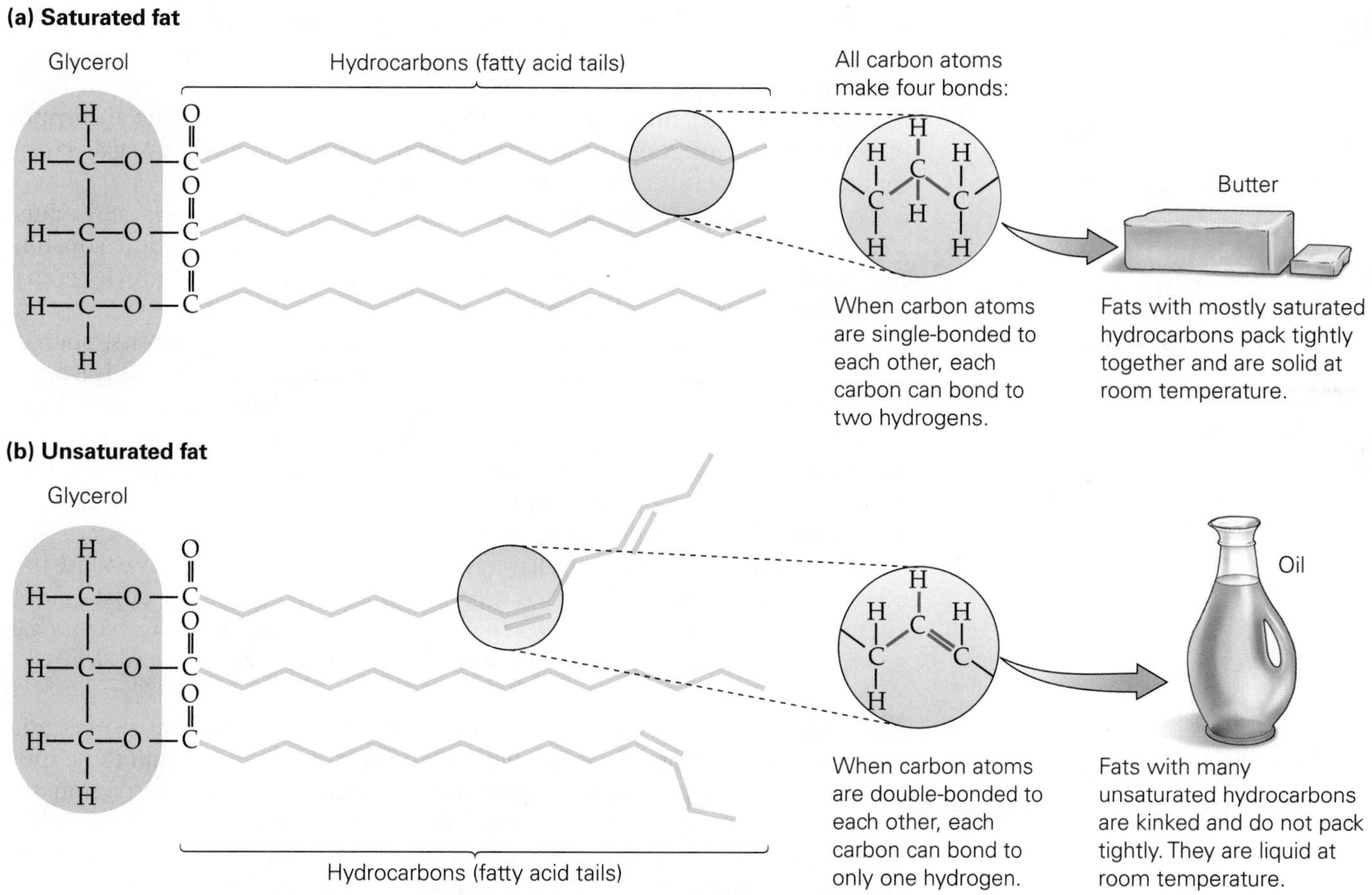

**(a) Saturated fat**

Glycerol

Hydrocarbons (fatty acid tails)

All carbon atoms make four bonds:

When carbon atoms are single-bonded to each other, each carbon can bond to two hydrogens.

Butter

Fats with mostly saturated hydrocarbons pack tightly together and are solid at room temperature.

**(b) Unsaturated fat**

Glycerol

Hydrocarbons (fatty acid tails)

When carbon atoms are double-bonded to each other, each carbon can bond to only one hydrogen.

Oil

Fats with many unsaturated hydrocarbons are kinked and do not pack tightly. They are liquid at room temperature.

**FIGURE 3.3 Saturated and unsaturated fats.** The types of bonds formed determine whether a fat will be (a) solid or (b) liquid at room temperature.

## Visualize This ▲

**Which of the fatty acid tails in this figure is polyunsaturated?**

process solidifies liquid oils, thereby making food seem less greasy and extending shelf life. Margarine is vegetable oil that has undergone hydrogenation.

**Trans fats** are produced by incomplete hydrogenation, which also changes the structure of the fatty acid tails in the fat so that, even though there are carbon-carbon double bonds, the fatty acids are flat and not kinked. In contrast to most fats, trans fats are not required or beneficial. If you choose to eat energy bars, protein bars, fiber bars, and other nutrition bars, it is best to look for those that do not contain trans fats.

The health risks of consuming foods rich in trans-fatty acids include increased risk of clogged arteries, heart disease, and diabetes. Because fat contains more stored energy per gram than carbohydrate and protein do, and because excess fat intake is associated with several diseases, nutritionists recommend that you limit the amount of fat in your diet.

## Micronutrients

Nutrients that are essential in minute amounts, such as vitamins and minerals, are called **micronutrients.** They are neither broken down by the body nor burned for energy.

**Vitamins.** Vitamins are organic substances, most of which the body cannot synthesize. Most vitamins function as **coenzymes,** or molecules that help enzymes,

**Should you take ginko to improve your memory, kava to reduce stress, ginseng to boost energy, and melatonin to help you sleep?**

and thus speed up the body's chemical reactions. Vitamin deficiencies can affect every cell in the body because many different enzymes, all requiring the same vitamin, are involved in numerous bodily functions. Vitamins also help with the absorption of other nutrients; for example, vitamin C increases the absorption of iron from the intestine. Vitamins may even help protect the body against cancer and heart disease and slow the aging process.

Vitamin D (calcitriol) is the only vitamin that cells can synthesize. Because sunlight is required for synthesis, people living in climates with little sunshine can develop deficiencies in vitamin D. In these areas, health-care providers may recommend vitamin D supplementation.

All other vitamins must be supplied by foods. Many vitamins, such as B vitamins and vitamin C, are water soluble, so boiling causes them to leach into water—this is why fresh vegetables are more nutritious than cooked ones. Water-soluble vitamins are not stored by the body so a lack of water-soluble vitamins is more likely the cause of dietary deficiencies than a lack of fat-soluble ones. Vitamins A, D, E, and K are fat-soluble and build up in stored fat; allowing an excess of these vitamins to accumulate in the body can be toxic.

The practice of taking multivitamin supplements has been called into question recently, especially when it comes to taking mega-doses of fat-soluble vitamins. Recent research shows that this practice may do more harm than good, potentially increasing risk of certain cancers, heart disease, and death.

**Minerals.** Minerals are substances that do not contain carbon but are essential for many cell functions. Because they lack carbon, minerals are said to be inorganic. They are important for proper fluid balance, in muscle contraction and conduction of nerve impulses, and for building bones and teeth. Calcium, chloride, magnesium, phosphorus, potassium, sodium, and sulfur are all minerals. Like some vitamins, minerals are water soluble and can leach out into the water during boiling. Also like most vitamins, minerals are not synthesized in the body and must be supplied through diet.

Calcium is one of the more commonly supplemented minerals. Bodies need calcium to help blood clot, muscles contract, nerves fire, and maintain healthy bone structure. When dietary calcium is low, calcium is removed from the bones, weakening them. If you are not getting enough calcium in your diet, around 1000 mg/day, many health-care providers will recommend supplements. Eight ounces of yogurt, 1.5 ounces of cheese, and one glass of milk together contain around 1000 mg of calcium.

## Antioxidants

In addition to vitamins and minerals, many whole foods contain molecules called **antioxidants** that prevent cells from damage caused by molecules that are generated by normal cell processes. These highly reactive molecules, called **free radicals,** have an incomplete electron shell, which makes them more chemically reactive than molecules with complete electron shells. Free radicals can damage cell membranes, the lining of arteries, and DNA. Antioxidants can inhibit the chemical reactions that involve free radicals and decrease the damage they do in cells. Antioxidants are abundant in fruits, vegetables, nuts, grains, and some meats.

After initial excitement about antioxidant supplements preventing heart disease, cancer, or slowing the aging process, we now know that the benefits of antioxidants seem to be limited to those consumed in whole foods. This may be because taking supplements shifts the balance so heavily against the formation of free radicals that there are not enough of them to perform necessary functions like killing new cancer cells, bacteria or other invaders.

**TABLE 3.1** lists some commonly supplemented vitamins, minerals, and antioxidants and describes current understanding of whether taking these supplements is advisable.

**TABLE 3.1** Common vitamin, mineral, and antioxidant supplements.

| Vitamin | Food Sources | Notes |
|---|---|---|
| B<sub>12</sub> | Chicken, fish, red meat, dairy | Because this vitamin is found in meat and dairy, some vegetarians and vegans are advised to supplement. |
| C | Most fruits, vegetables, and meats | Some people supplement vitamin C during cold season, although not much evidence supports this practice. |
| D | Milk, egg yolk, and soy | Because sunlight is required for the synthesis of this vitamin, people living in northern climates may be advised to supplement. |
| E | Almonds, many cooking oils, mangoes, broccoli, and nuts | Supplementing with high doses of vitamin E may increase risk of prostate cancer or stroke. |
| Folic Acid | Dark green vegetables, nuts, legumes (dried beans, peas, and lentils), and whole grains | Because deficiencies during pregnancy can lead to impaired spinal cord development in the fetus some women who are pregnant or planning on becoming pregnant are advised to supplement. |

| Mineral | Food Sources | Notes |
|---|---|---|
| Calcium | Milk, cheese, dark green vegetables, and legumes | Calcium is required for bone health. Those at risk for bone degeneration (osteoporosis) may be advised to supplement. |

*continued*

**TABLE 3.1 Common vitamin, mineral, and antioxidant supplements.** *continued*

| Magnesium |  | Spinach, fish, nuts, seeds, beans, and brown rice | Magnesium is required for skeletal and dental health. Supplementation is not typically recommended. |
|---|---|---|---|
| Potassium |  | Potatoes, tomatoes, bananas, avocados, and other fresh and dried fruits, dairy, whole grains, and meat | Potassium is required for muscle movement, nerve action, and proper kidney function. Potassium can help lower cholesterol, but a healthy diet does a better job of lowering cholesterol than supplements. |
| **Antioxidants** |  | **Food Sources** | **Notes** |
| Beta-carotene | | Orange fruits and vegetables, including carrots, cantaloupe, squash, mangoes, pumpkin, and apricots, collard greens, kale, and spinach | According to the National Center for Complementary and Alternative Medicine, supplementing with high doses of beta-carotene may increase the risk of lung cancer in smokers. |
| Flavanol |  | Cocoa and dark chocolate | Studies have not shown equivocally that supplementing provides any health benefit. |
| Lycopene |  | Red fruits and vegetables, including tomatoes and watermelon | To date, supplementing with lycopene has not been found to have any clear health benefit. |

# 3.2 **Cell Structure**

Some nutritional supplements are composed of extracts produced by grinding and breaking open plant cells. Others claim to affect particular subcellular structures in human cells. Because animal and plant cells evolved from a common ancestor, they share many of the same cellular structures and organelles. **Organelles** are to cells as organs are to the body. Each performs a specific job required by the cell and works in conjunction with other organelles to keep the cell functioning properly. Also inside cells is the **cytosol**, a watery matrix containing salts, and many of the enzymes required for cellular reactions. The cytosol houses the organelles. The **cytoplasm** includes both the cytosol and organelles.

Let's work our way in from the outside to the inside of these two types of cells (**FIGURE 3.4**) and then examine the structure and function of various internal cell components before investigating some supplements made from plant cells and some that inhibit organelle function.

**cyto-** and **-cyte** mean cell or a kind of cell.

**plasm** means fluid.

## Plasma Membrane

All cells are enclosed by a structure called a **plasma membrane** (**FIGURE 3.5**, on the next page). The plasma membrane defines the outer boundary of each cell, isolates the cell's contents from the environment, and serves as a barrier that determines which nutrients are allowed into and out of the cell. Membranes that enclose structures inside the cell are usually referred to as cell membranes, while the outer boundary is the plasma membrane.

Membranes are **semipermeable**, meaning that they allow some substances to cross while preventing others from doing so. This characteristic allows cells to maintain a different internal composition from the surrounding solution.

## **Visualize This ▼**

List three things plant cells contain that animal cells do not.

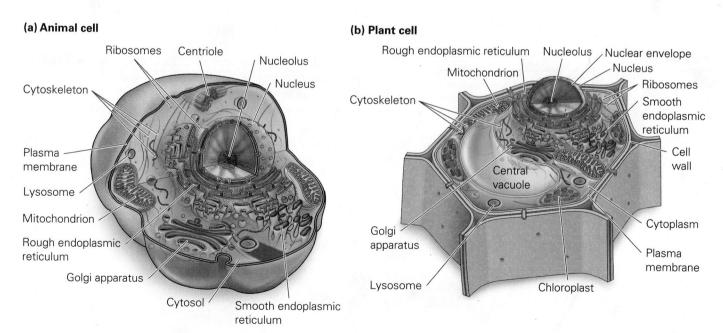

**(a) Animal cell**

Ribosomes · Centriole · Nucleolus · Nucleus · Cytoskeleton · Plasma membrane · Lysosome · Mitochondrion · Rough endoplasmic reticulum · Golgi apparatus · Cytosol · Smooth endoplasmic reticulum

**(b) Plant cell**

Rough endoplasmic reticulum · Nucleolus · Nuclear envelope · Nucleus · Mitochondrion · Ribosomes · Cytoskeleton · Smooth endoplasmic reticulum · Cell wall · Central vacuole · Golgi apparatus · Cytoplasm · Lysosome · Plasma membrane · Chloroplast

**FIGURE 3.4 Plant and animal cells.**   These drawings of a generalized (a) animal cell and (b) plant cell show the locations and sizes of organelles and other structures.

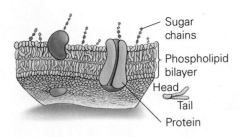

**FIGURE 3.5 The plasma membrane.** All cells are surrounded by a plasma membrane.

Internal and external membranes are composed primarily of phospholipids. The chemical properties of these lipids make membranes flexible and self-sealing. When phospholipid molecules are placed in a watery solution, such as in a cell, they orient themselves so that their hydrophilic heads are exposed to the water and their hydrophobic tails are away from the water. They cluster into a form called a **phospholipid bilayer** in which the tails of the phospholipids interact with themselves and exclude water, while the heads maximize their exposure to the surrounding water both inside and outside the membrane. The bilayer of phospholipids is stuffed with proteins that carry out enzymatic functions, serve as receptors for outside substances, and help transport substances throughout the cell.

All of the lipids and most of the proteins in the plasma membrane are free to bob about, sliding laterally. This fluidity allows the composition of any one location on the membrane to change. In the same manner that a patchwork quilt is a mosaic (different fabrics making up the whole quilt), so, too, is the membrane a mosaic with different regions of membrane composed of different types of phospholipids and proteins. Because of this, cell membranes are a **fluid mosaic** of lipids and proteins.

## Subcellular Structures

Inside the plasma membrane are the structures that allow a cell to maintain its structure and perform its designated functions.

**Cell Wall.** Some organisms, such as plants, fungi, and bacterial cells, have a **cell wall** (**FIGURE 3.6**) outside the plasma membrane that helps protect these cells and maintain their shape. The cell wall is rich in the polysaccharide cellulose, which is assembled into strong fibers that provide structural support.

Cellulose fibrils          Plant

**FIGURE 3.6 Cell wall.** Plants, fungi, and bacteria have a cell wall outside the plasma membrane.

**Nucleus.** All eukaryotic cells contain a **nucleus** (**FIGURE 3.7**)—a spherical structure surrounded by two membranes, which together are called the **nuclear envelope.** The nuclear envelope is studded with nuclear pores that regulate traffic into and out of the nucleus. Inside the nucleus is chromatin, composed of DNA and proteins. The **nucleolus** inside the nucleus is where ribosomes are synthesized.

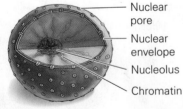

Nuclear pore
Nuclear envelope
Nucleolus
Chromatin

**FIGURE 3.7 Nucleus.** The nucleus houses the cell's genetic material.

**Mitochondrion.** Plant and animal cells contain **mitochondria,** energy-producing organelles surrounded by a double membrane (**FIGURE 3.8**). The inner and outer mitochondrial membranes are separated by the intermembrane space. The highly convoluted inner membrane carries many of the proteins involved in producing ATP. The matrix of the mitochondrion is the location of many of the reactions of cellular respiration.

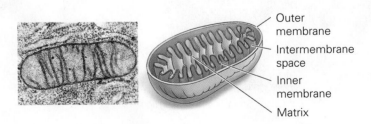

Outer membrane
Intermembrane space
Inner membrane
Matrix

**FIGURE 3.8 Mitochondrion.** Plant and animal cells both contain mitochondria.

**Chloroplast.** An important organelle present in plants and other photosynthetic eukaryotes, the **chloroplast** (**FIGURE 3.9**) uses the sun's energy to convert carbon dioxide and water into sugars. Each chloroplast has an outer membrane, an inner membrane, a liquid interior called the stroma, and a network of membranous sacs called **thylakoids** that stack on one another to form structures called grana. Chloroplasts also contain pigment molecules. The pigment **chlorophyll**, like the pigments in your clothes, reflects some wavelengths of light and absorbs others. Chlorophyll reflects green light.

**Lysosome.** A lysosome (**FIGURE 3.10**) is a membrane-enclosed sac of digestive enzymes that degrade proteins, carbohydrates, and fats. The low pH of the lysosome sequesters these protease, carbohydrase, and lipase enzymes from the rest of the cell. Lysosomes roam around the cell and engulf targeted molecules and organelles for recycling.

**Ribosomes.** Ribosomes (**FIGURE 3.11**) are workbenches where proteins are assembled. Ribosomes are built in the nucleolus and shipped out of the nucleus through nuclear pores to the cytoplasm. Composed of two subunits, they can be found floating in the cytoplasm or tethered to the **endoplasmic** reticulum.

**Endoplasmic Reticulum.** The endoplasmic reticulum (**FIGURE 3.12**), abbreviated ER, is a large network of membranes that begins at the nuclear envelope and extends into the cytoplasm of a eukaryotic cell. ER with ribosomes attached is called rough ER. Proteins synthesized on rough ER will be secreted from the cell or will become part of the plasma membrane. ER without ribosomes attached is called smooth ER. The function of the smooth ER depends on cell type but includes tasks such as detoxifying harmful substances and synthesizing lipids. Vesicles are pinched-off pieces of membrane that transport substances to the Golgi apparatus or plasma membrane.

**Golgi Apparatus.** The Golgi apparatus is a stack of membranous sacs (**FIGURE 3.13**). Vesicles from the ER fuse with the Golgi apparatus and empty their protein

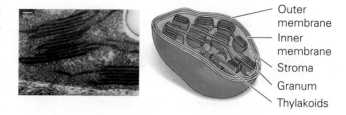

**FIGURE 3.9 Chloroplasts.** Cells and organisms that perform photosynthesis have chloroplasts.

Outer membrane
Inner membrane
Stroma
Granum
Thylakoids

**FIGURE 3.10 Lysosomes.** Lysosomes are digestive organelles.

Membrane
Digestive enzymes and digested material

**endo-** means inside.

Ribosomes

**FIGURE 3.11 Ribosomes.** Ribosomes are composed of two subunits.

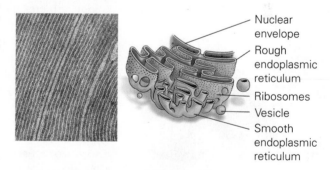

Nuclear envelope
Rough endoplasmic reticulum
Ribosomes
Vesicle
Smooth endoplasmic reticulum

**FIGURE 3.12 Endoplasmic reticulum.** The ER is composed of membranous sacs and tubules.

Vesicle from ER arriving at Golgi apparatus
Vesicle departing Golgi apparatus

**FIGURE 3.13 Golgi apparatus.** The Golgi is composed of membranous sacs.

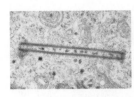

**FIGURE 3.14 Centrioles.** Centrioles help animal cells perform cell division.

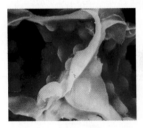
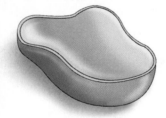

- Microfilaments
- Intermediate filaments
- Microtubules

**FIGURE 3.15 Cytoskeleton.** This framework gives cells structural support.

**FIGURE 3.16 Central vacuole.** This membranous organelle stores water and ions.

contents. The proteins are then modified, sorted, and sent to the correct destination in new transport vesicles that bud off from the sacs.

**Centrioles.** Centrioles are barrel-shaped rings composed of microtubules that anchor structures that help move chromosomes around when an animal cell divides (**FIGURE 3.14**). Centrioles are also involved in the formation of cilia and flagella. Plant cells do not have centrioles.

**Cytoskeletal Elements.** Cytoskeletal elements (**FIGURE 3.15**) are protein fibers that make up the structure of the cytoskeleton, a framework that gives shape and structural support to cells. Subcellular structures like the nucleus or mitochondria are anchored by these elements and some structures, like lysosomes, use the cytoskeleton like railroad tracks to travel from location to location inside the cell.

**Central Vacuole.** Plant cells have large fluid-filled central vacuoles (**FIGURE 3.16**) that contain a variety of dissolved molecules, including sugars and pigments that give color to flowers and leaves. Vacuoles also function to maintain pressure inside individual cells, which helps support the upright plant.

**Supplements and Cell Structures.** Chlorophyll supplements (**FIGURE 3.17**) are marketed as being able to energize, detoxify, and help heal wounds. There is no evidence to suggest that any of these claims are true. In fact, since human cells do not contain chlorophyll, there is nothing within them to supplement.

Some athletes use creatine to increase muscle mass or allow for more powerful bursts of energy. Creatine is used during protein synthesis on ribosomes and by mitochondria in the making of ATP. Although creatine is normally found in the body, the effects of supplementing have not been well studied. With side effects like kidney failure, it is safest not to supplement creatine above that which you normally obtain from your diet.

**STOP & STRETCH** Enzyme supplements can be found in most nutrition stores. Recall what enzymes are composed of and what would happen to them in the stomach. Is it likely that enzyme supplements have an effect on tissues outside the stomach? Why or why not?

◀ **Working with Data**

**A friend comes across a study stating that wounded rats treated with a topical solution of chlorophyll healed more rapidly than a control group left untreated. She plans to treat her leg wound by taking chlorophyll tablets. Discuss the judiciousness of her choice.**

**FIGURE 3.17 Chlorophyll.** Chlorophyll tablets are marketed as natural supplements.

# 3.3 Transport across Membranes

Whether ingested in food or supplemented in pills, once substances reach cells they must traverse the plasma membrane that surrounds the cell. Molecules must cross the plasma membrane to gain access to the inside of the cell, where they can be used to synthesize cell components or be metabolized to provide energy for the cell. The chemistry of the membrane facilitates the transport of some substances and prevents the transport of others.

**Dietary supplements can be bought in bulk and don't rot like fruits and vegetables. But are they as good for you?**

## Membrane Transport

The plasma membrane that surrounds cells is composed of a phospholipid bilayer. The interior of the bilayer is hydrophobic. Hydrophobic substances can dissolve in the membrane and pass through it more easily than hydrophilic ones. In this sense, the membrane of the cell is differentially permeable to the transport of molecules, allowing some to pass and blocking others from passing.

Substances that can cross the membrane will do so until the concentration is equal on both sides of the membrane, a condition called equilibrium. Carbon dioxide, water, and oxygen move freely across the membrane. Larger molecules, charged molecules, and ions cannot cross the lipid bilayer on their own. If these substances need to be moved across the membrane, they must move through proteins embedded in the membrane. Proteins in the membrane can serve as channels to allow molecules to cross until their concentration is equal on both sides of the membrane, or they can serves as pumps, driving molecules away from equilibrium (**FIGURE 3.18**).

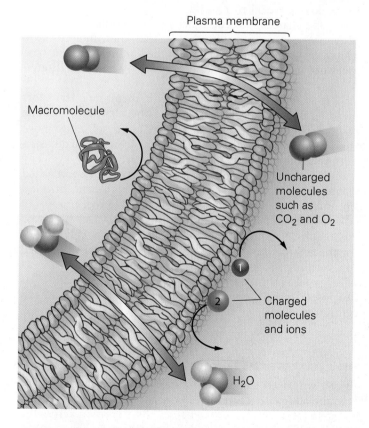

Plasma membrane

Macromolecule

Uncharged molecules such as $CO_2$ and $O_2$

① ②

Charged molecules and ions

$H_2O$

## ◀ Visualize This

**Cholesterol is largely hydrophobic but has a small region that is hydrophilic. This allows cholesterol to be a structural component of some animal cell membranes. Explain how the properties of the cholesterol molecule would allow it to stay lodged in cell membranes.**

**FIGURE 3.18 Transport of substances across membranes.** The ability of a substance to cross a membrane is, in part, a function of its size and charge.

**STOP & STRETCH** After a workout, you drink an energy drink composed of water, the large hydrophilic molecule glucose, and ionic electrolytes. Which of these three substances would require a transport protein to cross cell membranes and gain access to the inside of your cells?

## Membrane Transport A Closer Look ▼

The manner in which a substance is moved across a membrane depends on its particular chemistry and concentration in the cell. Different mechanisms for transporting substances have characteristic energy requirements.

### Passive Transport: Diffusion, Facilitated Diffusion, and Osmosis

All molecules contain energy that makes them vibrate and bounce against each other, scattering around like billiard balls during a game of pool. In fact, molecules will bounce against each other until they are spread out over all the available area. In other words, molecules will move from their own high concentration to their own low concentration. This movement of molecules from where they are in high concentration to where they are in low concentration is called **diffusion**. During diffusion, the net movement of molecules is from their own high to their own low concentration or *down* a concentration gradient. This movement does not require an input of outside energy; it is spontaneous. Diffusion will continue until there is equilibrium, at which time no concentration gradient exists, and there is no net movement of molecules.

Diffusion also occurs in living organisms. When substances diffuse across the plasma membrane, we call the movement **passive transport.** Passive transport is so named because it does not require an input of energy. The structure of the phospholipid bilayer that comprises the plasma membrane prevents many substances from diffusing across it. Only very small, hydrophobic molecules are able to cross the membrane by diffusion. In effect, these molecules dissolve in the membrane to slip from one side of the membrane to the other (**FIGURE 3.19**).

Hydrophilic molecules and charged molecules such as ions are unable to simply diffuse across the hydrophobic core of the membrane. Instead, these molecules are transported across membranes by proteins embedded in the lipid bilayer. This type of passive transport does not require an input of energy and is called **facilitated diffusion.**

Facilitated diffusion is so named because the specific membrane transport proteins make it easier for substances to diffuse across the plasma membrane (**FIGURE 3.20**). Although transport proteins are used to help the substance move across the plasma membrane, this form of transport is still considered to be diffusion because substances are traveling from high to low concentration.

The movement of water across a membrane is a type of passive transport called **osmosis.** Like other substances, water moves from its own high concentration to its own low concentration. Water can move through special protein pores in the membrane, called aquaporins, but even without these, water can still cross the membrane. When an animal cell is placed in a solution of salt water, water leaves the cell, causing the cell to shrivel (**FIGURE 3.21a**). When an animal cell is placed in a solution with less dissolved solute than the cell, water will enter the cell, and it will expand and may even break open. Likewise, plants that are overfertilized or exposed to road salt wilt because water leaves the cells to equilibrate the concentration of water on either side of the plasma membrane (**FIGURE 3.21b**).

### Active Transport: Pumping Substances across the Membrane

In some situations, a cell will need to maintain a concentration gradient. For example, nerve cells require a high concentration of certain ions inside the cell to transmit nerve impulses. To maintain this difference in concentration across the membrane requires the input of energy. Think of a hill with a steep incline or grade. Riding your bike down the hill requires no energy, but riding your bike up the grade requires energy. At the cellular level, that energy is in the form of ATP (adenosine triphosphate, the main source of energy in cell reactions). **Active transport** is transport that uses proteins, powered by ATP, to move substances up a concentration gradient (**FIGURE 3.22**).

---

**osmo-** means water.

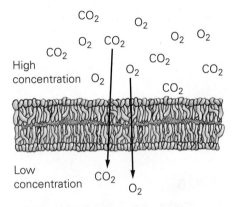

**FIGURE 3.19 Simple diffusion.** Simple diffusion of molecules across the plasma membrane occurs with the concentration gradient and does not require energy. Small hydrophobic molecules, carbon dioxide, and oxygen can diffuse across the membrane.

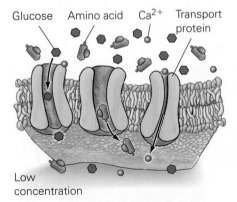

**FIGURE 3.20 Facilitated diffusion.** Facilitated diffusion is the diffusion of molecules assisted by substrate-specific proteins. Molecules move with their concentration gradient, which does not require energy.

*A Closer Look, continued*

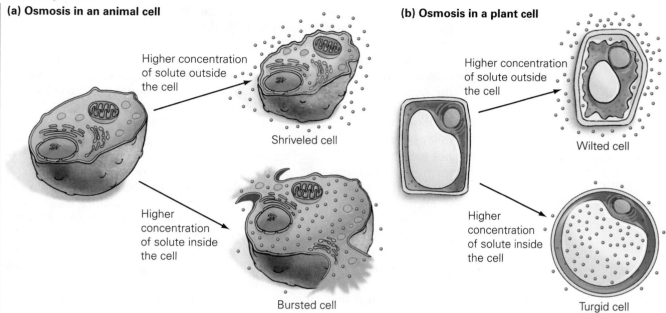

**(a) Osmosis in an animal cell**

Higher concentration of solute outside the cell

Shriveled cell

Higher concentration of solute inside the cell

Bursted cell

**(b) Osmosis in a plant cell**

Higher concentration of solute outside the cell

Wilted cell

Higher concentration of solute inside the cell

Turgid cell

**FIGURE 3.21 Osmosis.**   Osmosis is a special type of diffusion that involves the movement of water in response to a concentration gradient. Water moves toward a region that has more dissolved solute. (a) When water leaves an animal cell, it shrinks. (b) When water leaves a plant cell, the plant wilts instead of shrinks because of the support provided by the cell wall.

## Exocytosis and Endocytosis: Movement of Large Molecules across the Membrane

Larger molecules are often too big to diffuse across the membrane or to be transported through a protein, regardless of whether they are hydrophobic or hydrophilic. Instead, they must be moved around inside membrane-bound vesicles (small sacs) that can fuse with membranes. **Exocytosis** (**FIGURE 3.23a**) occurs when a membrane-bound vesicle, carrying some substance, fuses with the plasma membrane and releases its contents into the exterior of the cell. **Endocytosis** (**FIGURE 3.23b**) occurs when a substance is brought into the cell by a vesicle pinching the plasma membrane inward.

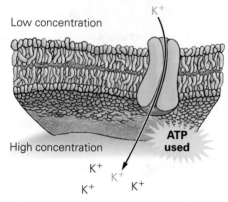

**Active transport**

Low concentration

$K^+$

High concentration

**ATP used**

$K^+$   $K^+$   $K^+$   $K^+$

**FIGURE 3.22 Active transport.**   Active transport moves substances against their concentration gradient and requires ATP energy to do so.

**exo-** means outside.

**(a) Exocytosis**

**(b) Endocytosis**

**FIGURE 3.23 Movement of large substances**   (a) Exocytosis is the movement of substances out of the cell. (b) Endocytosis is the movement of substances into the cell.

Once inside the cell, the supplement can finally do its job. But how do you know whether a supplement is actually doing something worthwhile for your health?

Before a prescription or nonprescription drug is released to the public, the FDA requires scientific testing to prove its safety and effectiveness. However, the same is not true of dietary supplements, even though people use supplements as medicines. In addition, claims made on the bottles and packages of such products have not necessarily been proven. This is why you may find the following asterisked footnote on supplement packaging: "These statements have not been evaluated by the Food and Drug Administration. This product is not intended to diagnose, treat, cure, or prevent any disease."

As with any product that makes claims about its benefits, you should carefully evaluate the claims before deciding to use the product. Use what you know about the process of science and the importance of well-controlled studies in your evaluation. If you don't have the time or desire to do the research yourself, consult your physician or a trustworthy website. (Use the guidelines presented in Chapter 1.) The FDA also keeps a record on their website of dietary supplements that are under review for causing adverse effects.

In the final analysis, it makes more sense to eat a healthy diet than it does to pop high doses of expensive pills, powders, and potions.

# *savvy reader*

## Probiotics

*This text was excerpted from a March 2013 U.C. Berkeley Wellness site, http://www.berkeleywellness.com /article/probiotics-pros-and-cons*

The large intestine is home to hundreds of trillions of bacteria. Fortunately, most are neutral or even beneficial, performing many vital body functions. For example, they help keep "bad" bacteria at bay, play a role in immunity, help us digest food, and absorb nutrients. But will consuming them as probiotics in foods or capsules make a notable difference to your health—especially if you are already healthy? That's what manufacturers want you to believe. Here's a look at the evidence.

**Digestive problems.** The best support for probiotics is for reducing diarrhea, especially infectious diarrhea and diarrhea associated with antibiotic use. Two reviews from 2012, one in the *Journal of the American Medical Association* (JAMA) and one in the *Annals of Internal Medicine*—which together included more than 80 studies and 14,000 people—found that probiotic therapy reduced the risk of antibiotic-associated diarrhea by 42% and 66%, respectively.

**Other uses.** Many other claims are made for probiotics—that they lower blood pressure and cholesterol, alleviate skin conditions like eczema, treat ulcers and urinary tract infections, improve vaginal health, prevent colon cancer, and ease anxiety and depression. But the evidence so far is preliminary, at best. Don't count on eating yogurt or douching with it to ward off or cure vaginal yeast infections.

Even if probiotics are beneficial for certain medical conditions, you'd have to take the right strain and right dose, and not everyone will even respond the same way to a given probiotic—much depends on the intestinal bacteria you have to begin with.

In addition, there's no guarantee that products contain the numbers of organisms claimed on labels—or that the organisms are even alive and survive digestion. And if they do survive, it's not certain they will colonize the intestines in sufficient numbers to have most of the proposed benefits. Some manufacturers claim to use processes that ensure that the bacteria stay alive and have therapeutic effects, but very few products undergo independent verification.

1. What are some indicators that this website might be a good source of information?

2. This article, and hundreds of other articles summarizing the research on dietary supplements, is put out by the University of California, Berkeley. Should you be more skeptical of claims made by a university putting together a database of resources for the public or a private company that manufactures an herbal supplement? Justify your answer.

3. Many product websites will include some data that seem to support their claims. Private companies can hire their own scientists to perform studies that often have results that differ from those of government and university-sponsored scientists. Would you be more skeptical of results produced by scientists hired by the company whose product they are testing or scientists who work for the government or a university? Justify your answer.

# SOUNDS RIGHT **BUT IS IT?**

Your parents gave you an African Violet plant as a "dorm warming" gift. The plant is not doing well. It is wilting and dropping leaves, even though you are watering it regularly. Your roommate believes that the plant must be short on nutrients and has been dousing the soil and leaves with a commercial plant fertilizer.

**Wilting plants should be fertilized.**

Sounds right, but it isn't.

1. Do you think a liquid fertilizer would have a high or low concentration of dissolved substances?

2. What effect would adding fertilizer to a plant have on the concentration of solute outside the plant cells?

3. Which direction would this cause water to move?

4. What happens to the cells of the plant when water moves in the direction described above?

5. Consider your answers to questions 1–4 and explain why the original statement bolded above sounds right, but isn't.

# Chapter Review    MasteringBiology®

Go to the Study Area in MasteringBiology® for practice quizzes, myeBook, BioFlix™ 3-D animations, MP3Tutor sessions, videos, current events, and more.

## Summary

### Section 3.1

Describe the role of nutrients in the body.

• Nutrients provide structural units and energy for cells (p. 52).

Describe the function of water in the body.

• Water is an important dietary constituent that helps dissolve and eliminate wastes and maintain blood pressure and body temperature (p. 52).

Describe the major dietary macronutrients and discuss the functions of each.

• Macronutrients are required in large amounts for proper growth and development. Macronutrients include carbohydrates, proteins, and fats. All of these molecules are composed of subunits that can be broken down for use by the cell (pp. 52–55).

List the major dietary micronutrients and describe their functions.

• Micronutrients are dietary substances required in minute amounts for proper growth and development; they include vitamins and minerals (pp. 55–56).

• Vitamins are organic substances, most of which the body cannot synthesize. Many vitamins serve as coenzymes to help enzymes function properly (pp. 55–56).

• Minerals are inorganic substances essential for many cell functions (p. 56).

## Section 3.2

Describe the structure and function of the plasma membrane.

- To gain access to cells, nutrients move across the plasma membrane, which functions as a semipermeable barrier that allows some substances to pass and prevents others from crossing (p. 59).
- The plasma membrane is composed of two layers of phospholipids, in which are embedded proteins and cholesterol (pp. 59–60).
- Some organisms, such as plants, fungi, and bacterial cells, have a cell wall outside the plasma membrane that helps protect these cells and maintain their shape (p. 60).

Describe the structure and function of the subcellular organelles.

- Subcellular organelles and structures perform many different functions within the cell. Mitochondria and chloroplasts are involved in energy conversions. Lysosomes are involved in the breakdown of macro- molecules. Ribosomes serve as sites for protein synthe- sis. Proteins can be synthesized on ribosomes attached to rough endoplasmic reticulum. Smooth endoplasmic reticulum synthesizes lipids. The Golgi apparatus sorts proteins and sends them to their cellular destination. Centrioles help cells divide. The plant cell central vacu- ole stores water and other substances (pp. 60–62).

## Section 3.3

Distinguish between passive transport and active transport.

- Passive transport mechanisms include unaided simple diffusion and diffusion facilitated by proteins. Passive transport always moves substances with their concen- tration gradient and does not require energy (p. 64).
- Osmosis, the diffusion of water across a membrane, can involve the movement of water through protein pores in the membrane (pp. 64–65).
- Active transport is an energy-requiring process that requires proteins in cell membranes to move substances against their concentration gradients (pp. 64–65).

Describe the processes of endocytosis and exocytosis.

- Larger molecules move into (endocytosis) and out (exo- cytosis) of cells enclosed in membrane-bound vesicles (p. 65).

## Roots to Remember

**The following roots of words come mainly from Latin and Greek and will help you to decipher terms:**

| | |
|---|---|
| **cyto-** | and **-cyte** mean cell or a kind of cell. Chapter terms: *cytosol, cytoplasm, cytoskeleton* |
| **endo-** | means inside. Chapter terms: *endocytosis, endoplasmic reticulum* |
| **exo-** | means outside. Chapter term: *exocytosis* |
| **osmo-** | means water. Chapter term: *osmosis* |
| **plasm** | means fluid. Chapter term: *cytoplasm, plasma membrane* |

## Learning the Basics

1. What are the two main functions of nutrients?

2. List three common cellular substances that can pass through cell membranes unaided.

3. Macronutrients _____.
   A. include carbohydrates and vitamins; B. should comprise a small percentage of a healthful diet; C. are essential in minute amounts to help enzymes function; D. include carbohydrates, fats, and proteins; E. are synthesized by cells and not necessary to obtain from the diet.

4. Which of the following is not a function of water?
   A. dispersing nutrients throughout the body; B. helping prevent cancer; C. helping to regulate body temperature; D. helping to regulate blood pressure.

5. Micronutrients _____.
   A. include vitamins and carbohydrates; B. are not metabolized to produce energy; C. contain more energy than fatty acids; D. can be synthesized by most cells.

6. The main constituents of the plasma membrane are

   _____.

   A. carbohydrates and lipids; B. proteins and phospho- lipids; C. fats and carbohydrates; D. fatty acids and nucleic acids.

7. A substance moving across a membrane against a concen- tration gradient is moving by _____.
   A. passive transport; B. osmosis; C. facilitated diffusion; D. active transport; E. diffusion.

**8.** A cell that is placed in salty seawater will _____.
A. take sodium and chloride ions in by diffusion; B. move water out of the cell by active transport; C. use facilitated diffusion to break apart the sodium and chloride ions; D. lose water to the outside of the cell via osmosis.

**9.** Which of the following forms of membrane transport require specific membrane proteins?
A. diffusion; B. exocytosis; C. facilitated diffusion; D. active transport; E. facilitated diffusion and active transport.

**10.** Water crosses cell membranes _____.
A. by active transport; B. through protein pores called aquaporins; C. against its concentration gradient; D. in plant cells but not in animal cells.

## Analyzing and Applying the Basics

**1.** A friend of yours does not want to eat meat, so instead she consumes protein shakes that she buys at a nutrition store. Can you think of a dietary strategy that would allow her to be a vegetarian while not consuming protein shakes?

**2.** Some studies have shown that as trans fat consumption increases, so does heart disease. If you were to use a line graph to show this relationship, which variable would be on the x-axis and what would the line showing this positive correlation look like?

**3.** A vitamin-enriched water manufacturer claims that their product has all the vitamin C of an orange. What substances would you not be getting from the vitamin water that you would by eating an orange?

## Connecting the Science

**1.** Select one supplement you have wondered about and spend a few minutes doing some web-based research on whether the claims made on its label are backed up by scientific evidence.

**2.** The New York City Board of Health adopted a city-wide ban on the use of trans fats in restaurant cooking. Do you think the government should be involved in making decisions about what private businesses can sell to consumers?

Answers to **Stop & Stretch, Visualize This, Working with Data, Savvy Reader, Sounds Right, But Is It?**, and **Chapter Review** questions can be found in the **Answers** section at the back of the book.

CHAPTER **4**

# Body Weight and Health

**Are overweight people less healthy than thin people?**

# Enzymes, Metabolism, and Cellular Respiration

Overweight people—are they unhealthy and lacking in self-control? If they could learn to control their appetites and exercise more, would their weight and health problems be cured? Why is it so hard for so many overweight and obese people to lose weight and keep it off with diet and exercise, even for those who demonstrate self-discipline and drive in many other areas of their lives?

The popular television series *America's Biggest Loser* chronicles the struggles of obese Americans who diet desperately and exercise excessively to lose weight rapidly.

**Can obesity be cured through willpower?**

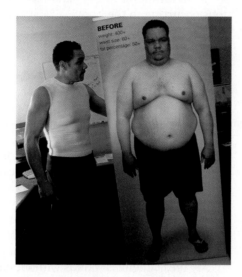

**Most Americans think so.**

While this makes for compelling television, a few years after appearing on the show, many of the contestants have gained back most or all of the weight they lost, and some have gained even more. This is also true of many people who lose weight on their own. Most diets fail and most dieters end up feeling like actual losers when they can't keep the weight off. Yet, we continue to encourage not only contestants on this show but also ourselves and those we care about to lose weight because we so strongly believe that being thin will

make us not only happier but also healthier.

No good can come of denying the negative health consequences of obesity, including increased risk of heart disease, stroke, and type 2 diabetes. But how overweight does one need to be to be at risk of developing these diseases? In addition, many people assume that it is better to be underweight than overweight. Is this true? To better understand what science tells us about weight and health, we will first look at the biological factors affecting how much body fat one stores.

**But is being underweight as bad as being overweight?**

# 4.1  Enzymes and Metabolism

The amount of fat that a given individual will store depends partly on how quickly or slowly he or she breaks down food molecules into their component parts. **Metabolism** is a general term used to describe all of the chemical reactions occurring in the body.

## Enzymes

All metabolic reactions are regulated by proteins called **enzymes** that speed up, or **catalyze,** the rate of biological reactions. Enzymes can help break down or build up substances, or build more complex substances from simpler ones. The enzymes that help your body break down the foods you ingest liberate the energy stored in the food's chemical bonds. Enzymes are usually named for the reaction they catalyze and end in the suffix **ase.** For example, sucrase is the enzyme that breaks down the table sugar sucrose.

_____
**-ase** is a common suffix in names of enzymes.
_____

To break chemical bonds, molecules must absorb energy from their surroundings, often by absorbing heat. This is why heating chemical reactants will speed up a reaction. However, heating cells to an excessively high temperature can damage or kill them, in part because proteins begin to break down at high temperatures. Enzymes do not require heat to catalyze the body's chemical reactions, so they break chemical bonds without damaging or killing cells. Furthermore, by eliminating the heat energy required to start a chemical reaction, enzymes allow the breakdown of chemical bonds to occur more quickly.

**Activation Energy.** The energy required to start the metabolic reaction serves as a barrier to catalysis and is called the **activation energy** (**FIGURE 4.1**). If not for the activation energy barrier, all of the chemical reactions in cells would occur relentlessly, whether the products of the reactions were needed or not. Because most metabolic reactions need to surpass the activation energy barrier before proceeding, they can be regulated by enzymes. In other words, a given chemical reaction will occur when the correct enzyme is present. How do enzymes decrease the activation energy barrier?

**Induced Fit.** The chemicals that are metabolized by an enzyme-catalyzed reaction are called the enzyme's **substrate.** Enzymes decrease activation energy by binding to their substrate and placing stress on its chemical bonds, decreasing the amount of initial energy required to break the bonds. The region of the enzyme where the substrate binds is called the enzyme's **active site.** Each active site has its own shape and chemical climate. When the substrate binds to the active site, the enzyme changes shape slightly to envelop the substrate. This shape change by the enzyme in response to substrate binding results in stress being placed on the bonds of the substrate. This is called the **induced fit** model of

**(a) No enzyme present**

Reactants                                        Products

**(b) Enzyme present**

Reactants                                        Products

**FIGURE 4.1 Activation energy.** (a) The activation energy barrier present in cells can be likened to an uphill bike ride. Once you are at the top of the hill, it takes much less energy to continue moving forward. (b) If you smooth out the grade of the hill, more people will make it. In cells, there is an energy barrier that prevents chemical reactions from occurring. Adding an enzyme helps lower this barrier.

enzyme catalysis. When the enzyme changes shape, it binds to the substrate more tightly, making it easier to break the substrate's chemical bonds. In this manner, the enzyme helps convert the substrate to a reaction product and then resumes its original shape so that it can perform the reaction again (**FIGURE 4.2**).

Each enzyme catalyzes a particular reaction—a property called **specificity.** The specificity of an enzyme is the result of its shape and the shape of its active site. Different enzymes have unique shapes because they are composed of amino acids in varying sequences. The 20 amino acids, each with its own unique side group, are arranged in distinct numbers and orders for each enzyme, producing enzymes of all shapes and sizes, each with an active site that can bind with its particular substrate. Although an infinite variety of enzymes could be produced, it is often the case that different organisms will utilize similar enzymes, likely due to their evolution from a common ancestor.

Enzymes mediate all of the metabolic reactions occurring in an organism's cells. Because enzymes, like all proteins, are coded for by genes, the amount of body fat a person stores is affected by many factors, some that can be controlled, and some that can't.

## Metabolism

The speed and efficiency of many different enzymes will lead to an overall increase or decrease in the rate at which a person can break down food. Thus, when you say that your metabolism is slow or fast, you are actually referring to the speed at which enzymes catalyze chemical reactions in your body.

A person's **metabolic rate** is a measure of his or her energy use. This rate changes according to the person's activity level. For example, we require less energy when asleep than we do when exercising. The **basal metabolic rate** represents the energy use of a wakeful person. The average basal metabolic rate is 70 calories per hour, or 1680 calories per day. However, this rate varies widely among individuals because many factors influence each person's basal metabolic rate, including exercise habits, biological sex, and genetics.

**Overweight people—are they unhealthy and lacking in self-control?**

## Visualize This ▼

**Is the enzyme itself permanently altered by the process of catalysis?**

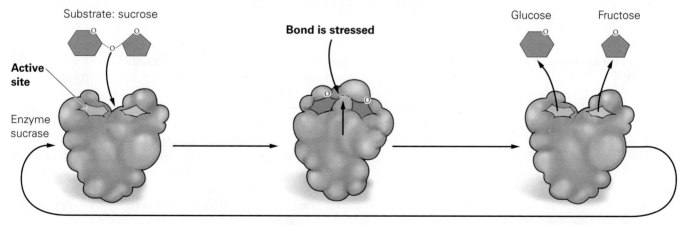

① The shape of the substrate matches the shape of the enzyme's active site.

② Initial substrate binding to the active site changes the shape of the active site, inducing the substrate to fit even more snugly in the active site, stressing the bonds of the substrate.

③ The shape change splits the substrate and releases the two subunits. The enzyme is able to perform the reaction again.

**FIGURE 4.2 Enzymes.** The enzyme sucrase is cleaving (splitting) the disaccharide sucrose into its monosaccharide subunits.

**If the obese could learn to control their appetites and exercise more, would their health problems be cured?**

During exercise, metabolic rate increases. In addition to burning calories during exercise, metabolic rate remains elevated for a period of time after exercise. The length of time the metabolic rate is elevated is a function of the intensity of the exercise.

Your biological sex also influences your metabolic rate. Males require more calories per day than females do because testosterone, a hormone produced in larger quantities in males, increases the rate at which fat breaks down. Men also have a higher percentage of muscle than women, which requires more energy to maintain than fat does.

Other genetic factors play a substantial role in determining body weight. Two people of the same size and sex, who consume the same number of calories and exercise the same amount, will not necessarily store the same amount of fat. Some people are simply born with lower basal metabolic rates. Genes that influence a person's rate of fat storage and utilization, like all genes, are passed from parents to children.

**STOP & STRETCH** Is the relationship between body weight and calories burned during exercise a negative or positive correlation?

To be completely metabolized, food must be broken down by the digestive system and then transported to individual cells via the bloodstream. Once inside cells, food energy can be converted into chemical energy by the process of cellular respiration.

## 4.2 Cellular Respiration

**Cellular respiration** is a series of metabolic reactions that converts the energy stored in chemical bonds of food into energy that cells can use, while releasing waste products. Energy is stored in the electrons of chemical bonds, and when bonds are broken in a multistage process, **adenosine triphosphate**, or **ATP**, is produced. ATP can supply energy to cells because it stores energy obtained from the movement of electrons that originated in food into its own bonds. Before trying to understand cellular respiration, it is important to have a better understanding of this chemical.

### Structure and Function of ATP

Structurally, ATP is a nucleotide triphosphate. It contains the nitrogenous base adenine, a sugar, and three phosphates (**FIGURE 4.3**). Each phosphate in the series of three is negatively charged. This series of negative charges repel each other. This energy is similar to the energy you can feel when you try to hold the negative poles of two magnets together.

Removal of the terminal phosphate group of ATP releases energy that can be used to perform cellular work. In this manner, ATP behaves much like a coiled spring. To think about this, imagine loading a dart gun. Pushing the dart into the gun requires energy from your arm muscles, and the energy you exert will be stored in the coiled spring inside the dart gun (**FIGURE 4.4**). When you shoot the dart gun, the energy is released from the gun and used to perform some work—in this case, sending a dart through the air.

The phosphate group that is removed from ATP can be transferred to another molecule. Thus, ATP can energize other compounds through **phosphorylation**, which means that it transfers a phosphate to another molecule. When a molecule, say an enzyme, needs energy, the phosphate group is transferred from ATP to the enzyme, and the enzyme undergoes a change in shape that allows the enzyme to perform its job. After removal of a phosphate group, ATP becomes **adenosine diphosphate (ADP)**

### Visualize This ▼

To get a feel for how negative charges repel each other, consider what happens when you place the negative poles of two magnets near each other. What do you think would happen if you placed the negative ends of three magnets near one another?

Nitrogenous base (adenine)

Sugar (ribose)    3 negatively charged phosphates

**FIGURE 4.3 The structure of ATP.** ATP is a nucleotide triphosphate.

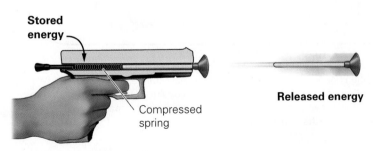

**FIGURE 4.4 Stored energy.** A dart gun uses energy stored in the coiled spring and supplied by the arm muscle to perform the work of propelling a dart.

(**FIGURE 4.5**). The energy released by the removal of the outermost phosphate of ATP can be used to help cells perform many different kinds of work. ATP helps power mechanical work such as the movement of cells, transport work such as the movement of substances across membranes during active transport, and chemical work such as the making of complex molecules from simpler ones (**FIGURE 4.6**).

Cells are continuously using ATP. Exhausting the supply of ATP means that more ATP must be regenerated. ATP is synthesized by adding back a phosphate group to ADP during the process of cellular respiration (**FIGURE 4.7**).

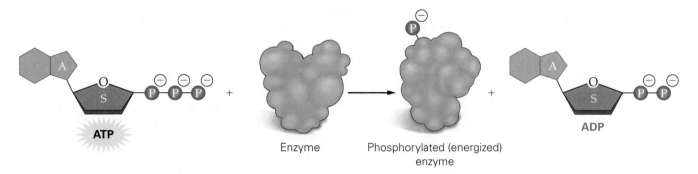

**FIGURE 4.5 Phosphorylation.** The terminal phosphate group of an ATP molecule can be transferred to another molecule, in this case an enzyme, to energize it. When ATP loses a phosphate, it becomes ADP.

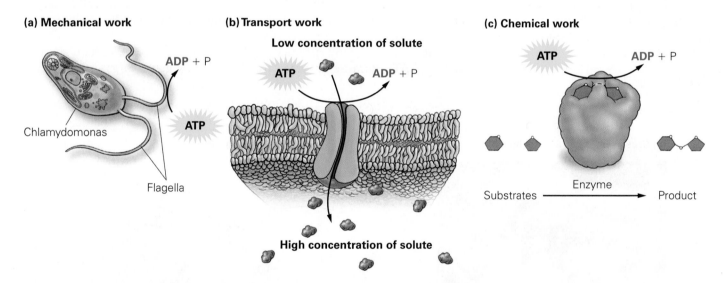

**FIGURE 4.6 ATP and cellular work.** ATP powers (a) mechanical work, such as the moving of the flagella of this single-celled green algae; (b) transport work, such as the active transport of a substance across a membrane from its own low to high concentration; and (c) chemical work, such as the enzymatic conversion of substrates to a product.

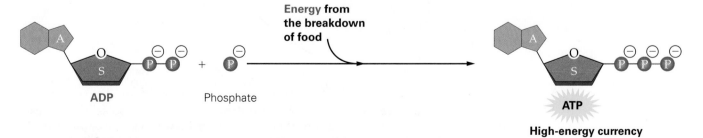

**FIGURE 4.7 Regenerating ATP.** ATP is regenerated from ADP and phosphate during the process of cellular respiration.

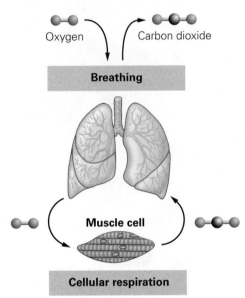

**FIGURE 4.8 Breathing and cellular respiration.** Inhalation brings oxygen into the lungs, from which it is delivered through the bloodstream to the tissues where it is used to drive cellular respiration. Carbon dioxide by-product is released by the tissues, diffuses into the blood, delivered to the lungs, and released during exhalation.

During this process, oxygen is consumed, and water and $CO_2$ are produced. Because some of the steps in cellular respiration require oxygen, they are said to be **aerobic** reactions, and this type of cellular respiration is called **aerobic respiration.**

## Cellular Respiration

The word *respiration* can also be used to describe breathing. When we breathe, we take oxygen in through our lungs and expel carbon dioxide. The oxygen we breathe in is delivered to cells, which undergo cellular respiration and release carbon dioxide (**FIGURE 4.8**).

Most foods can be broken down to produce ATP as they are routed through this process. Carbohydrate metabolism begins earliest in the pathway, while proteins and fats enter at later points.

For this reason, we usually study respiration by following the path of the sugar glucose. Glucose is an energy-rich sugar, but the products of its digestion—carbon dioxide and water—are energy poor. So where does the energy go? The energy released during the conversion of glucose to carbon dioxide and water is used to synthesize ATP.

Many of the chemical reactions in this process occur in the mitochondria, a subcellular organelle found in both plant and animal cells (**FIGURE 4.9a**). Through a series of complex reactions in the mitochondrion, a glucose molecule breaks apart, and carbon and oxygen are released from the cell as carbon dioxide. Hydrogen atoms from glucose combine with oxygen to produce water (**FIGURE 4.9b**).

There are many, many enzymes involved in the metabolism of food and production of ATP. Because enzymes are coded for by genes, organisms differ in

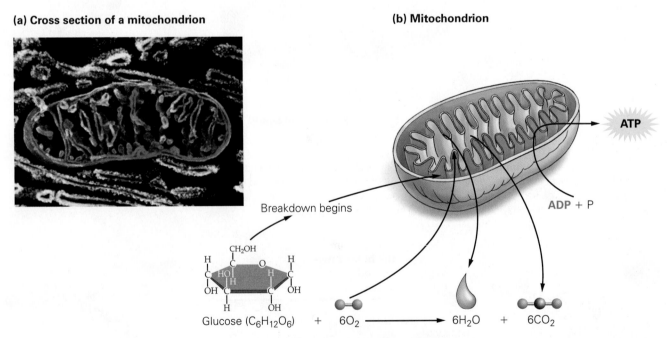

**FIGURE 4.9 Overview of cellular respiration.** (a) Much of this process of cellular respiration takes place within the mitochondrion. (b) The breakdown of glucose by cellular respiration requires oxygen and ADP plus phosphate. The energy stored in the bonds of glucose is harvested to produce ATP (from ADP and P), releasing carbon dioxide and water.

the rate at which they perform each step of cellular respiration and the rate at which they store unused energy as fat.

To gain a more thorough understanding of the complex reactions that comprise cellular respiration requires an even closer look at the inner workings of cells.

### Cellular Respiration: **A Closer Look** ▼

Cellular respiration occurs in four stages. Electrons are removed from glucose during the early stages and are used to make ATP in the final stage. These electrons do not simply float around because this would damage cell structures. Instead, they are carried by molecules called *electron carriers*. One of the electron carriers utilized by cellular respiration is a chemical called **nicotinamide adenine dinucleotide (NAD⁺).** You can think of this molecule as a sort of taxicab for electrons. The empty taxicab (NAD⁺) picks up electrons. A hydrogen atom contains one H⁺ ion and one electron. When NAD⁺ picks up a hydrogen atom it becomes NADH. The full taxicab (NADH) carries electrons to their destination, where they are dropped off, and the empty taxicab returns for more electrons (**FIGURE 4.10**). NADH will deposit its electrons for use in the final step of cellular respiration.

**Stage 1: Glycolysis.** To harvest energy, the 6-carbon glucose molecule is first broken down into two 3-carbon **pyruvic acid** molecules (**FIGURE 4.11**). This part of the process of cellular respiration actually occurs outside the

**glyco-** means sugar.

## Visualize This ▼

**NAD⁺ (the empty cab) picks up one positively charged hydrogen atom and how many electrons to give the neutral NADH molecule?**

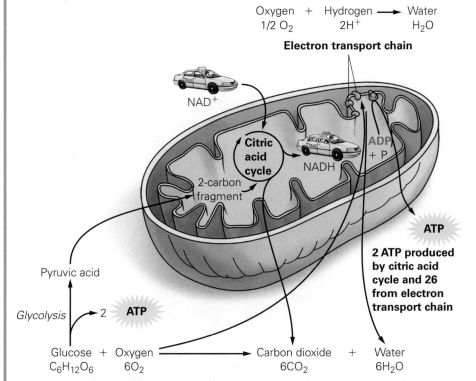

**FIGURE 4.10 Electron carriers.** Electron carriers in the cell are like taxicabs, shuttling electrons from the original glucose molecule to the final stage of respiration.

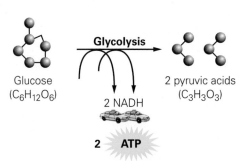

**FIGURE 4.11 Glycolysis.** Glycolysis, the enzymatic conversion of glucose to 2 pyruvic acids, also produces a small amount of NADH and ATP.

*A Closer Look, continued*

mitochondria, in the fluid cytosol. Glycolysis does not require oxygen and, after an initial input of energy, produces a net total of 2 molecules of ATP.

**Stage 2: The Citric Acid Cycle.** After glycolysis, the pyruvic acid is decarboxylated (loses a carbon dioxide molecule), and the 2-carbon fragment that is left is further metabolized inside the mitochondria. This fragment enters the **citric acid cycle**, a series of enzyme catalyzed reactions that take place in the matrix of the mitochondrion. Here, the original glucose molecule is further broken down. More of its electrons are harvested and the remaining carbons are released as carbon dioxide (**FIGURE 4.12**).

**Stage 3: Electron Transport and ATP Synthesis.** Electrons harvested during the citric acid cycle are now carried by NADH to the final stage in cellular respiration. The **electron transport chain** is a series of proteins embedded in the inner mitochondrial membrane that functions as a sort of conveyer belt for electrons, moving them from one protein to another. The electrons, dropped off by NADH molecules generated during glycolysis and the citric acid cycle, move toward the bottom of the electron transport chain toward the matrix of the mitochondrion, where they combine with oxygen to produce water.

Each time an electron is picked up by a protein or handed off to another protein, the protein moving it changes shape. This shape change allows the movement of hydrogen ions (H⁺) from the matrix of the mitochondrion to the intermembrane space. Therefore, while the proteins in the electron transport chain are moving electrons down the electron transport chain toward oxygen, they are also moving H⁺ ions across the inner mitochondrial membrane and into the intermembrane space. This decreases the concentration of H⁺ ions in the matrix and increases their concentration within the intermembrane space. Whenever a concentration gradient exists, molecules will diffuse from an area of their own high concentration to an area of their own low concentration. Because charged ions cannot diffuse across the hydrophobic core of the membrane, they escape through a protein channel in

# Visualize This ▼

**How many turns of the citric acid cycle would it take to release 4 of the original 6 carbons of glucose as carbon dioxide?**

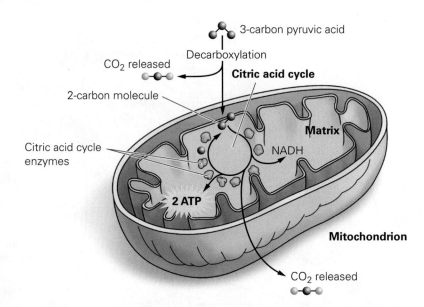

**FIGURE 4.12 The citric acid cycle.** The 3-carbon pyruvic acid molecules generated by glycolysis are decarboxylated, leaving a 2-carbon molecule that proceeds through a stepwise series of reactions that results in the production of more carbon dioxide, NADH, and ATP.

*A Closer Look, continued*

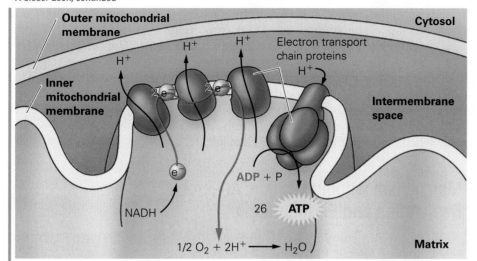

◀ **Visualize This**

What is meant by 1/2 $O_2$ in this figure?

**FIGURE 4.13 The electron transport chain of the inner mitochondrial membrane.** NADH brings electrons to the electron transport chain. As electrons move through the proteins of the electron transport chain, hydrogen ions are pumped into the intermembrane space. Hydrogen ions flow back through an ATP synthase protein, which converts ADP to ATP. In this manner, energy from electrons added to the electron transport chain is used to produce ATP.

the membrane called **ATP synthase.** This enzyme uses the energy generated by the rushing H⁺ ions to synthesize 26 ATP from ADP and phosphate in the same manner that water rushing through a mechanical turbine can be used to generate electricity (**FIGURE 4.13**).

## Metabolism of Other Nutrients

Most cells can break down not only carbohydrates but also proteins and fats. **FIGURE 4.14** shows the points of entry during cellular respiration for proteins and fats. Protein is broken down into component amino acids, which are then used to synthesize new proteins. Most organisms can also break down proteins to supply energy. However, this process takes place only when fats or carbohydrates are unavailable. In humans and other animals, the first step in producing

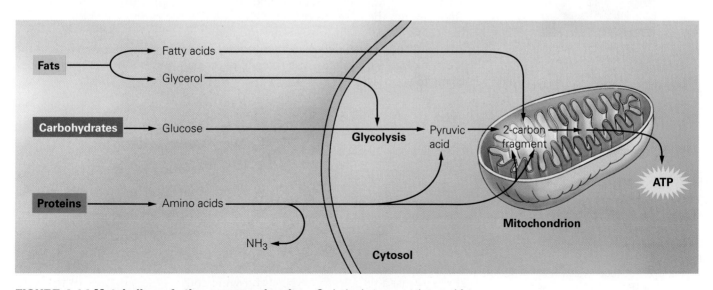

**FIGURE 4.14 Metabolism of other macromolecules.** Carbohydrates, proteins, and fats can all undergo cellular respiration; they just feed into different parts of the metabolic pathway.

energy from the amino acids of a protein is to remove the nitrogen-containing amino group of the amino acid. Amino groups are then converted to a compound called urea, which is excreted in the urine. The carbon, oxygen, and hydrogen remaining after the amino group is removed undergo further breakdown and eventually enter the mitochondria, where they are fed through the citric acid cycle and produce carbon dioxide, water, and ATP. The subunits of fats (glycerol and fatty acids) also go through the citric acid cycle and produce carbon dioxide, water, and ATP. Some cells will break down fat only when carbohydrate supplies are depleted.

## Metabolism without Oxygen: Anaerobic Respiration and Fermentation

Aerobic cellular respiration is one way for organisms to generate energy. It is also possible for some cells to generate energy in the absence of oxygen, by a metabolic process called **anaerobic respiration.**

Muscle cells normally produce ATP by aerobic respiration. However, oxygen supplies diminish with intense exercise. When muscle cells run low on oxygen, they must get most of their ATP from glycolysis, the only stage in the cellular respiration process that does not require oxygen. When glycolysis happens without aerobic respiration, cells can run low on the electron carrier NAD$^+$ because it is converted to NADH during glycolysis. When this happens, cells use a process called **fermentation** to regenerate NAD$^+$.

Fermentation cannot, however, be used for very long because one of the by-products of this reaction leads to the buildup of a compound called lactic acid. Lactic acid is produced by the actions of the electron acceptor NADH, which has no place to dump its electrons during fermentation because there is no electron transport chain and no oxygen to accept the electrons. Instead, NADH deposits its electrons by giving them to the pyruvic acid produced by glycolysis (**FIGURE 4.15a**). Lactic acid is transported to the liver, where liver cells use oxygen to convert it back to pyruvic acid.

This requirement for oxygen, to convert lactic acid to pyruvic acid, explains why you continue to breathe heavily even after you have stopped working out. Your body needs to supply oxygen to your liver for this conversion, sometimes referred to as paying back your oxygen debt. The accumulation of lactic acid also explains the phenomenon called "hitting the wall." Anyone who has ever felt as though their legs were turning to wood while running or biking knows this feeling. When your muscles are producing lactic acid by fermentation for a long time, the oxygen debt becomes too large, and muscles shut down until the rate of oxygen supply outpaces the rate of oxygen utilization, restoring proper feeling to your legs.

**STOP & STRETCH** Aerobic exercise (such as running, swimming, and biking) strengthens the heart, allowing it to pump more oxygen containing blood per beat. How does aerobic conditioning help prevent the buildup of lactic acid during exercise?

Some fungi and bacteria also produce lactic acid during fermentation. Certain microbes placed in an anaerobic environment transform the sugars in milk into yogurt, sour cream, and cheese. It is the lactic acid present in these dairy products that gives them their sharp or sour flavor. Yeast in an anaerobic environment produces ethyl alcohol instead of lactic acid. Ethyl alcohol is formed when carbon dioxide is removed from pyruvic acid (**FIGURE 4.15b**). The yeast used to help make beer and wine converts sugars present in grains (beer) or grapes (wine) into ethyl alcohol and carbon dioxide. Carbon dioxide, produced by baker's yeast, helps bread to rise.

---

**an-** means absence of.

**(a) Human muscle**

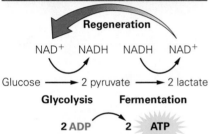

**(b) Yeast**

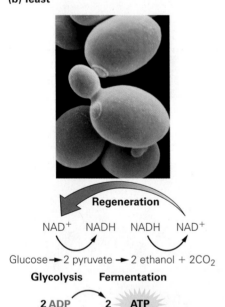

**FIGURE 4.15 Metabolism without oxygen.** Glycolysis can be followed by (a) lactate fermentation to regenerate NAD$^+$. This pathway also produces 2 ATP during glycolysis. Glycolysis followed by (b) alcohol fermentation also regenerates NAD$^+$ and produces 2 ATP.

# 4.3 Body Weight and Health

Now that we have gained an appreciation for the large number of different enzymes involved in metabolism, and the genes that code for them, we can turn our attention back to the question of health and weight. How does one determine whether their body weight is a healthy one?

## Body Mass Index

One method is to use a tool called the Body Mass Index, or BMI (**TABLE 4.1**). Your BMI is a value calculated by using your height and weight as an estimate of weight-related risk of illness and death.

However, this method is not as accurate as we would like it to be. Studies show that as many as 1 in 4 people may be misclassified by BMI tables because this measurement provides no means to distinguish between lean muscle mass and body fat. For example, an athlete with a lot of muscle will weigh more than a similar-sized person with a lot of fat because muscle is more dense than fat. Therefore, a person can be very fit, but be classified as overweight using this table.

## Overweight and Underweight Are Both Unhealthy

If your BMI is below 18.5, you are at risk for **anorexia,** or self-starvation, a disease rampant on college campuses. Estimates are that 1 in 5 college women and 1 in 20 college men restrict their food intake so severely that they are essentially starving themselves to death. Self-starvation can also occur when people allow themselves to eat—sometimes very large amounts of food (called binge eating)—but prevent the nutrients from being turned into fat by purging

**Many people assume that it is better to be underweight than overweight. Is this true?**

## Working with Data ▼

What is the BMI of a person who is 5 feet 11 inches and 200 pounds?

**TABLE 4.1** **Body mass index (BMI). BMI values are calculated based on height and weight.**

| | | | | | | | | | | | | | | |
|------|-----|-----|-----|-----|-----|-----|-----|-----|-----|-----|-----|-----|-----|-----|
| **4'10"** | 91 | 96 | 100 | 105 | 110 | 115 | 119 | 124 | 129 | 134 | 138 | 143 | 167 | 191 |
| **4'11"** | 94 | 99 | 104 | 109 | 114 | 119 | 124 | 128 | 133 | 138 | 143 | 148 | 173 | 198 |
| **5'0"** | 97 | 102 | 107 | 112 | 118 | 123 | 128 | 133 | 138 | 143 | 148 | 153 | 179 | 204 |
| **5'1"** | 100 | 106 | 111 | 116 | 122 | 127 | 132 | 137 | 143 | 148 | 153 | 158 | 185 | 211 |
| **5'2"** | 104 | 109 | 115 | 120 | 126 | 131 | 136 | 142 | 147 | 153 | 158 | 164 | 191 | 218 |
| **5'3"** | 107 | 113 | 118 | 124 | 130 | 135 | 141 | 146 | 152 | 158 | 163 | 169 | 197 | 225 |
| **5'4"** | 110 | 116 | 122 | 128 | 134 | 140 | 145 | 151 | 157 | 163 | 169 | 174 | 204 | 232 |
| **5'5"** | 114 | 120 | 126 | 132 | 138 | 144 | 150 | 156 | 162 | 168 | 174 | 180 | 210 | 240 |
| **5'6"** | 118 | 124 | 130 | 136 | 142 | 148 | 155 | 161 | 167 | 173 | 179 | 186 | 216 | 247 |
| **5'7"** | 121 | 127 | 134 | 140 | 146 | 153 | 159 | 166 | 172 | 178 | 185 | 191 | 223 | 255 |
| **5'8"** | 125 | 131 | 138 | 144 | 151 | 158 | 164 | 171 | 177 | 184 | 190 | 197 | 230 | 262 |
| **5'9"** | 128 | 135 | 142 | 149 | 155 | 162 | 169 | 176 | 182 | 189 | 196 | 203 | 236 | 270 |
| **5'10"** | 132 | 139 | 146 | 153 | 160 | 167 | 174 | 181 | 188 | 195 | 202 | 209 | 243 | 278 |
| **5'11"** | 136 | 143 | 150 | 157 | 165 | 172 | 179 | 186 | 193 | 200 | 208 | 215 | 250 | 286 |
| **6'0"** | 140 | 147 | 154 | 162 | 169 | 177 | 184 | 191 | 199 | 206 | 213 | 221 | 258 | 294 |
| **6'1"** | 144 | 151 | 159 | 166 | 174 | 182 | 189 | 197 | 204 | 212 | 219 | 227 | 265 | 302 |
| **6'2"** | 148 | 155 | 163 | 171 | 179 | 186 | 194 | 202 | 210 | 218 | 225 | 233 | 272 | 311 |
| **6'3"** | 152 | 160 | 168 | 176 | 184 | 192 | 200 | 208 | 216 | 224 | 232 | 240 | 279 | 319 |
| **6'4"** | 156 | 164 | 172 | 180 | 189 | 197 | 205 | 213 | 221 | 230 | 238 | 246 | 287 | 328 |
| **BMI** | **19** | **20** | **21** | **22** | **23** | **24** | **25** | **26** | **27** | **28** | **29** | **30** | **35** | **40** |

| 19 | | 25 | | 30 | 35 | 40 |
|----|----|----|----|----|----|----|
| <19 underweight | 19 to 24 normal weight | | 25 to 29 overweight | | 30 to 39 moderate obesity | >40 severe obesity |

# Working with Data ▼

**List the BMI categories in order of greatest to least risk of death.**

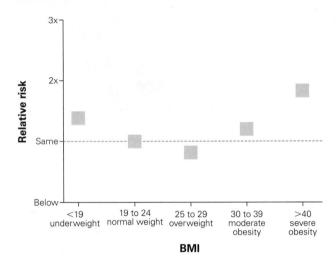

**FIGURE 4.16 Risk of death and BMI.** This graph shows the risk of death by any cause among adults in different BMI categories compared to the risk of death to normal weight adults. The blue boxes indicate the average relative risk for each BMI classification.

themselves, by vomiting or using laxatives. Binge eating followed by purging is called **bulimia.**

Anorexia has long-term health consequences. It can starve heart muscles to the point that altered rhythms develop. The lack of body fat in anorexia can also lead to the cessation of menstruation, a condition known as amenorrhea, which occurs when a protein called leptin, secreted by fat cells, signals the brain that there is not enough body fat to support pregnancy. Hormones (such as estrogen) that regulate menstruation are blocked, and menstruation ceases. Amenorrhea can be permanent and causes sterility in a substantial percentage of people with anorexia.

The damage done by the lack of estrogen is not limited to the reproductive system; bones are affected as well. Estrogen secreted by the ovaries during the menstrual cycle acts on bone cells to help them maintain their strength and size. Anorexics have reduced development of bone and put themselves at higher risk of broken bones, in a condition called **osteoporosis.**

If your BMI falls within the healthy range you have no reason to worry about health risks from body weight. In fact, studies show that even those in the obese category have less to worry about than we once thought. Many studies, including a recent meta-analysis (study of hundreds of studies), show that being overweight is actually associated with less mortality than being normal weight (**FIGURE 4.16**). In addition, moderate obesity is not associated with increased mortality.

**STOP & STRETCH** Recent research subdivides the moderate obesity category into two categories, with a BMI in the 30–34.9 range categorized as Grade 1 obesity and a BMI in the 35–39.9 range categorized as Grade 2 obesity. Under this scheme, a BMI of 40 or higher is Grade 3 obesity. Based on what you have learned about the health effects of being obese, why might these new categories make more sense than the old ones?

So what do we do with this information? Body weight is very difficult for individuals to control and the health consequences of overweight, while very real, don't seem to start until a person becomes obese. One way to prevent obesity in the first place may be to not participate in the kind of extreme calorie restriction and exercise regimens that contestants in *America's Biggest Loser* use to shed pounds. Study after study has shown that cycling through weight loss and gain (yo-yo dieting) actually causes a body to become more efficient at using calories. Over the long haul, this kind of dieting makes it more difficult to maintain a healthy body weight. A better approach seems to be simply working toward making good choices about food and exercise. This approach will help prevent most people from becoming obese and will help those who are make the lifestyle changes that can help them attain a healthy weight.

# *savvy reader*

Several websites claim that celebrities have used a substance derived from the Hoodia Gordonii (**FIGURE 4.17**) plant to help them lose weight. Summarized below are some of the claims made by these sites:

- Hoodia plants require over five years to grow to maturity. This cactus plant is also a fairly uncommon plant. These two factors limit its availability. Because the supply of Hoodia cannot meet the demand, Hoodia pills are more expensive than one would expect.
- Hoodia diet pills are natural. They contain no additives or preservatives, just parts of the plant. This natural substance acts on your brain to make you eat less.
- Hoodia has been used by indigenous people to help them traverse the Kalahari Desert.
- Drug companies have spent millions of dollars performing tests of Hoodia diet pills. No side effects have been found. Hoodia must be safe for everyone.
- Products do not contain fillers or toxic ingredients, and they are made in the United States

1. Many products are marketed as natural and many consumers let their guard down when they hear that a product is natural. Are all natural substances good for humans? Give an example of something that is natural that is not good for humans.

2. Dietary supplements are exempt from testing by the Food and Drug Administration (FDA). However, the FDA will step in to study a product that has proven harmful. In many cases, products claiming to be organic or natural contain toxins, prescription drugs, or drugs not approved for human use in the United States. In light of this, does the claim that the product is made in the United States offer any real proof of purity?

3. One site claims that a drug company performed tests on Hoodia and found that there were no side effects and that Hoodia appears to be safe for everyone. Is this the way scientists report data? What kinds of information should be here to substantiate this claim?

## Hoodia for Weight Loss

**FIGURE 4.17 Can a desert cactus plant help people lose weight?**

# SOUNDS RIGHT BUT IS IT?

One ill-conceived strategy employed for rapid weight loss is to consume laxatives, which are designed to soften fecal matter, making it easier to defecate. Because defecation can make one feel lighter or less full, some people think that using laxatives can help them to lose weight. This abuse of laxatives can permanently damage the large intestine, leading to dependence on laxatives for bowel movements, dehydration, and increased risk of colon cancer.

**Laxatives cause permanent weight loss.**

Sounds right, but it isn't.

1. Fecal matter is composed of fiber and other indigestible food remnants. It is present in the large intestine, until defecation. Laxatives act on the large intestine to hasten the expulsion of feces, which occurs after nutrients have been absorbed by the small intestine. Would the loss of food that cannot be digested lead to permanent weight loss?

2. If you step on a scale before defecating, you may weigh a bit more than after defecation. However, if the digestive tract is really just a tube with openings at the mouth and anus, is food that does not get absorbed by the small intestine ever really part of your body?

3. Losing weight involves the loss of fat. What gas do you breathe out when fat is metabolized?

4. In light of your answer to the previous question, how might exercise help with weight loss?

5. Consider your answers to questions 1–4 and explain why the original statement bolded above sounds right, but isn't.

# Chapter Review    MasteringBiology®

## Summary

## Section 4.1

Describe the structure and function of enzymes.

- Enzymes are proteins that catalyze specific cellular reactions. The substrate binds to the enzyme's active site and is converted to product (pp. 72–73).

Explain how enzymes decrease a reaction's activation energy barrier.

- The binding of a substrate to the enzyme's active site causes the enzyme to change shape (induced fit), placing more stress on the bonds of the substrate and thereby lowering the amount of energy required to start the reaction (pp. 72–73).

Define metabolism.

- Metabolism includes all the chemical reactions that occur in cells to build up or break down macromolecules (pp. 73–74).

## Section 4.2

Describe the structure and function of ATP.

- ATP is a nucleotide triphosphate. The nucleotide found in ATP contains a sugar and the nitrogenous base adenine (p. 74).
- The energy stored in the chemical bonds of food can be released and stored in the bonds of ATP. Cells use ATP to power energy-requiring processes (pp. 74–75).
- Breaking the terminal phosphate bond of ATP releases energy that can be used to perform cellular work and produces ADP plus a phosphate (pp. 75–76).

Describe the process of cellular respiration from the breakdown of glucose through the production of ATP.

- Cellular respiration begins in the cytosol, where a 6-carbon sugar is broken down into two 3-carbon pyruvic acid molecules during the anaerobic process of glycolysis (pp. 76–77).
- The pyruvic acid molecules then move across the two mitochondrial membranes, where they are decarboxylated. The remaining 2-carbon fragment then moves into the matrix of the mitochondrion, where the citric acid cycle strips them of carbon dioxide and electrons (p. 78).
- Electrons removed from chemicals that are part of glycolysis and the citric acid cycle are carried by electron carriers, such as NADH, to the inner mitochondrial membrane; there they are added to a series of proteins called the electron transport chain. At the bottom of the electron transport chain, oxygen pulls the electrons toward itself. As the electrons move down the electron transport chain, the energy that they release is used to drive protons ($H^+$) into the intermembrane space. Once there, the protons rush through the enzyme ATP synthase and produce ATP from ADP and phosphate (pp. 78–79).
- When electrons reach the oxygen at the bottom of the electron transport chain, they combine with the oxygen and hydrogen ions to produce water (pp. 78–79).

Explain how proteins and fats are broken down during cellular respiration.

- Proteins and fats are also broken down by cellular respiration, but they enter the pathway at later points than glucose (pp. 79–80).

Compare and contrast aerobic and anaerobic respiration.

- Anaerobic respiration is cellular respiration that does not use oxygen as the final electron acceptor (p. 80).

## Section 4.3

Discuss the relevance of body weight as a predictor of healthfulness.

- Obesity and severe underweight are associated with many negative health risks. Moderate overweight is not (pp. 81–82).

## Roots to Remember

**The following roots of words come mainly from Latin and Greek and will help you decipher terms:**

| | |
|---|---|
| **an-** | means absence of. Chapter term: *anaerobic* |
| **-ase** | is a common suffix in names of enzymes. Chapter term: *sucrase* |
| **glyco-** | means sugar. Chapter term: *glycolysis* |

## Learning the Basics

1. Define the term *metabolism*.
2. What is meant by the term *induced fit*?
3. What are the reactants and products of cellular respiration?

**4.** Which of the following is a *false* statement regarding enzymes?

A. Enzymes are proteins that speed up metabolic reactions; **B.** Enzymes have specific substrates; **C.** Enzymes supply ATP to their substrates; **D.** An enzyme may be used many times.

**5.** Enzymes speed up chemical reactions by _____.

A. heating cells; **B.** binding to substrates and placing stress on their bonds; **C.** changing the shape of the cell; **D.** supplying energy to the substrate.

**6.** What would happen if activation energy barriers didn't exist?

A. Substrates would not bind properly to enzymes; **B.** Chemical reactions in the body would never occur; **C.** Enzyme function would not be affected; **D.** Metabolic reactions would proceed even if their products were not needed.

**7.** Cellular respiration involves _____.

A. the aerobic metabolism of sugars in the mitochondria by a process called glycolysis; **B.** an electron transport chain that releases carbon dioxide; **C.** the synthesis of ATP, which is driven by the rushing of protons through an ATP synthase; **D.** electron carriers that bring electrons to the citric acid cycle; **E.** the production of water during the citric acid cycle.

**8.** The electron transport chain _____.

A. is located in the matrix of the mitochondrion; **B.** has the electronegative carbon dioxide at its base; **C.** is a series of nucleotides located in the inner mitochondrial membrane; **D.** is a series of enzymes located in the intermembrane space; **E.** moves electrons from protein to protein and moves protons from the matrix into the intermembrane space.

**9.** Most of the energy in an ATP molecule is released _____.

A. during cellular respiration; **B.** when the terminal phosphate group is hydrolyzed; **C.** in the form of new nucleotides; **D.** when it is transferred to NADH.

**10.** Anaerobic respiration _____.

A. generates proteins for muscles to use; **B.** occurs in yeast cells only; **C.** does not use oxygen as the final electron acceptor; **D.** utilizes glycolysis, the citric acid cycle, and the electron transport chain.

## Analyzing and Applying the Basics

**1.** A friend decides he will eat only carbohydrates because carbohydrates are burned to make energy during cellular respiration. He reasons that he will generate more energy by eating carbohydrates than by eating fats and proteins. Is this true?

**2.** If you could follow one carbon atom present in a carbohydrate you ingested and burned for energy, where would that carbon atom end up?

**3.** What would you say to a friend who qualifies as overweight on a BMI chart but who exercises regularly and eats a healthy, well-balanced diet?

## Connecting the Science

**1.** Early Earth was devoid of oxygen. What process could bacteria on early Earth use to generate energy?

**2.** Appetite suppressants are readily available but have not been shown to result in weight loss. Why might this be?

Answers to **Stop & Stretch, Visualize This, Working with Data, Savvy Reader, Sounds Right, But Is It?,** and **Chapter Review** questions can be found in the **Answers** section at the back of the book.

CHAPTER **5** | # Life in the Greenhouse

**Superstorm Sandy flooded large sections of New York City.**

# Photosynthesis and Climate Change

New Orleans. New York. Amsterdam. Alexandria. Mumbai. Bangkok. Hong Kong. Shanghai. According to predictions by climate scientists, some of the great cities of the world may share a future—portions of each may experience catastrophic flooding and become uninhabitable. New Orleans and New York City recently provided stark examples. In October 2012, the surge of water pushed ashore by Superstorm Sandy inundated sections of New York, killing 48 residents and leaving much of the city without power for a week. Floodwaters from Hurricane Katrina in 2005 covered 80% of New Orleans and killed 1,500 people.

The government spent over $13 billion rebuilding New Orleans and other parts of Louisiana after Katrina; the governors of New York and New Jersey asked for over $60 billion in aid after Sandy. Without this support, these areas would have likely remained destroyed and abandoned for decades. Imagine the impact of similar disasters on cities and

**Hurricane Katrina devastated 80% of New Orleans.**

**Global warming is causing Earth's ice to melt, raising sea levels by a meter over this century.**

communities that lack the resources of the United States.

Increasing catastrophic flooding in coastal cities is likely because Earth's sea levels are rising. Ice all over the planet's surface is melting and running off into the oceans at a dramatic rate. In Antarctica, Rhode Island–sized shelves of ice have fallen into the ocean. The Greenland ice cap is becoming thinner at its margins every year, and meltwater collecting underneath the ice threatens to cause large portions to slide off the island. Meanwhile, the dramatic glaciers in Glacier National Park, located in the northwest corner of Montana, are disappearing. Several of the park's glaciers have already shrunk to half their original size, and the number of glaciers has decreased from approximately 150 in 1850 to around 35 today. If this trend continues, scientists predict that by the year 2030 not a single glacier will be left in the park.

All of the water from the melting ice must go somewhere—it has

contributed to a sea-level rise of 10 to 20 cm (4 to 8 inches) in the last century. Like an overflowing bathtub, this excess water spills over the edges of ocean basins, inundating low-lying cities and coasts.

But why is the ice melting? Because Earth's average temperature is increasing. According to the U.S. National Climate Data Center, Earth has warmed by 0.25°C (0.5°F) each decade during the twentieth century. If the rate of warming continues, by the end of the twenty-first century the sea level will be between 50 and 100 centimeters (20 to 36 inches) higher than today—a much faster and more devastating change than already witnessed. Compounding this rapid raise in level is the fact that as ocean temperatures increase, the amount of energy available to be released in storms will rise, resulting in more powerful hurricanes and even more catastrophic flooding. This was exactly the scenario of Superstorm Sandy.

Why is Earth warming? And can anything be done to stop the coming flood?

**Rising sea levels threaten millions of people along ocean coasts.**

# Why is the ice melting? Because Earth's average temperature is increasing.

## ▼ Visualize This

How should this figure change when more carbon dioxide is added to the atmosphere?

# 5.1  The Greenhouse Effect

**Global warming** is the progressive increase of Earth's average temperature that has been occurring over the past century. Although the general public seems to believe that there is debate among scientists and government-appointed panels about global warming, that is not really the case. Scientists who publish in peer-reviewed journals and respected scientific panels and societies, such as the Intergovernmental Panel on Climate Control (IPCC), National Academy of Sciences, and the American Association for the Advancement of Science (AAAS), all agree that the average temperature on Earth is increasing and that most of the warming observed in the last century is attributable to human activities.

Global warming is contributing to **global climate change**, the local changes in average temperature, precipitation, and sea level relative to historical conditions that are occurring in locations all over the planet. While Earth's climate is constantly changing, human-caused global warning has dramatically increased the rate of change—so much so that it may be difficult for humans to adjust.

Global warming is caused by recent increases in the concentrations of particular gases in the atmosphere, including water vapor, carbon dioxide ($CO_2$), methane ($CH_4$), and ozone ($O_3$). The accumulation of many of these *greenhouse gases* is a direct result of human activity, namely, coal, oil, and natural gas combustion. The most abundant gas emitted by combustion of these fuels is carbon dioxide; for this reason, carbon dioxide is considered the most important greenhouse gas to control.

The presence of carbon dioxide and the other greenhouse gases in the atmosphere leads to a phenomenon called the **greenhouse effect.** Despite this name, the phenomenon caused by these gases is not exactly like that of a greenhouse, where panes of glass allow radiation from the sun to penetrate inside and then trap the heat that radiates from warmed-up surfaces. On Earth, the greenhouse effect works like this: Warmth from the sun heats Earth's surface, which then radiates the heat energy outward. Most of this heat is radiated back into space, but some of the heat warms up the greenhouse gases in the atmosphere and then is reradiated to Earth's surface.

In effect, greenhouse gases act like a blanket (**FIGURE 5.1**). When you sleep under a blanket at night, your body heat warms the blanket, which in turn keeps you warm. When the levels of greenhouse gases in the atmosphere increase, the effect is similar to sleeping under a thicker blanket—more heat is retained and reradiated, and therefore the temperature underneath the atmosphere is warmer.

The greenhouse effect is not in itself a dangerous phenomenon. If Earth's atmosphere did not have some greenhouse gases, too much heat would be lost to space, and Earth would be too

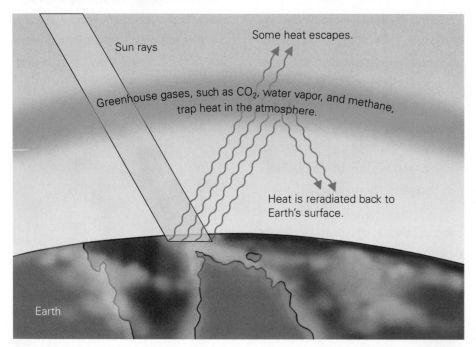

Sun rays

Some heat escapes.

Greenhouse gases, such as $CO_2$, water vapor, and methane, trap heat in the atmosphere.

Heat is reradiated back to Earth's surface.

Earth

**FIGURE 5.1 The greenhouse effect.**
Heat from the sun is absorbed in the atmosphere by water vapor, carbon dioxide, and other greenhouse gases and reradiated back to Earth.

cold to support life. In the past, greenhouse gases kept temperatures on Earth stable and hospitable to life. The danger posed by the greenhouse effect is in the excess warming caused by more and more carbon dioxide accumulation in the atmosphere as a result of coal, oil, and natural gas burning.

**STOP & STRETCH** The moon's average daytime temperature is 107°C (225°F), and its average nighttime temperature is –153°C (–243°F). However, the moon and Earth are the same distance from the sun. Think about the greenhouse gases on Earth to come up with a hypothesis for why temperatures on the moon might fluctuate so dramatically.

## Water, Heat, and Temperature

Bodies of water absorb energy and help maintain stable temperatures on Earth. You have perhaps noticed that when you heat water on a stove, the metal pot becomes hot before the water. This is because water heats more slowly than metal and has a stronger resistance to temperature change than most substances.

Heat and temperature are measures of energy. **Heat** is the total amount of energy associated with the movement of atoms and molecules in a substance. **Temperature** is a measure of the intensity of heat—for example, how fast the molecules in the substance are moving. When you are swimming in a cool lake, your body has a higher temperature than the water; however, the lake contains more heat than your body because even though its molecules are moving more slowly, the sum total of molecular movement in its large volume is much greater than the sum total of molecular movements in your much smaller body.

The formation of hydrogen bonds between neighboring molecules of water makes it more cohesive than other liquids (Chapter 2). The hydrogen bonds also make water resistant to temperature change, even when a large amount of heat is added. This phenomenon occurs because when water is first heated, the heat energy disrupts the hydrogen bonds. Only after enough of the hydrogen bonds have been broken can heat cause individual water molecules to move faster, thus increasing the temperature. As the temperature continues to rise, individual water molecules can move fast enough to break free of all hydrogen bonds and rise into the air as water vapor. This is the basis for the water cycle that moves water from land, oceans, and lakes to clouds and then back again to Earth's surfaces (**FIGURE 5.2**). When water cools, hydrogen bonds re-form between adjacent molecules, releasing heat into the atmosphere. A body of

## Visualize This ▼

What would happen to the amount of water in the ocean if all of the stored ice and snow melted? How would this melting affect the shoreline?

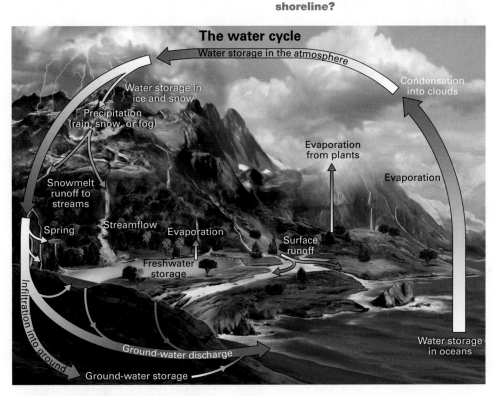

**FIGURE 5.2 The water cycle.** Water moves from the oceans and other surface water to the atmosphere and back, with stops in living organisms, underground pools and soil, and ice caps and glaciers on land.

## Visualize This ▶

Evaporation occurs when molecules at the surface of a liquid "escape" into a gaseous phase. Where would most of these escaped molecules appear on this figure and why?

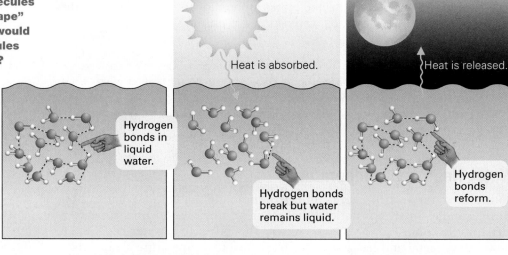

FIGURE 5.3 **Hydrogen bonding in water.** Hydrogen bonds break as they absorb heat and reform as water releases heat.

**Increasing catastrophic flooding in coastal cities is likely because Earth's sea levels are rising.**

water can release a large amount of heat into its surroundings while experiencing only a small drop in temperature (**FIGURE 5.3**).

Water's high heat-absorbing capacity has important effects on Earth's climate. The vast amount of water contained in Earth's oceans and lakes keeps temperatures moderate by absorbing huge amounts of heat radiated by the sun and releasing that heat during less-sunny times, warming the air and preventing large temperature swings.

The balance between the release and storage of heat energy is vital to climate conditions on Earth. This balance can be disrupted when increasing levels of carbon dioxide trap more heat in the atmosphere. Carbon dioxide in the atmosphere comes from many different sources, some of which are increasing.

## 5.2 The Flow of Carbon

The atoms that make up the complex molecules of living organisms move through the environment via biogeochemical cycles. **FIGURE 5.4** illustrates how carbon, like water, cycles back and forth between living organisms, the atmosphere, bodies of water, and rock. The carbon dioxide you exhale enters the atmosphere, where it can absorb heat; these molecules can return to Earth's surface, where they can dissolve in water or be absorbed by plants or certain other organisms.

Carbon dioxide taken up by plants, algae, and some types of bacteria is converted into carbohydrates using the energy from sunlight. Most living organisms depend on these carbohydrates as a source of cellular energy and rerelease the carbon dioxide into the atmosphere in the process of consuming them. Any unconsumed carbohydrates can become buried in the very rock of Earth for millennia; the carbon contained there can later be released through volcanic activity or by extraction and combustion by humans. It is the latter activity that is contributing to a buildup of carbon dioxide in the atmosphere.

The stored carbohydrates discussed in the previous paragraph are known as **fossil fuels** (**FIGURE 5.5**). These fuels—petroleum, coal, and natural gas—are "fossils" because they formed from the buried remains of ancient plants and microorganisms. Over a period of millions of years, the carbohydrates in

# Visualize This ▼

**Based on the observations that carbon dioxide levels have increased in the last 100 years and that volcanic activity has remained constant, which would you predict releases more carbon dioxide into the air: volcanic or human activity?**

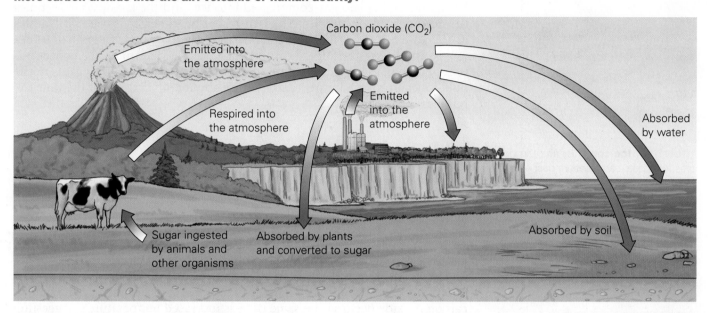

**FIGURE 5.4 The flow of carbon.** Living organisms, volcanoes, and fossil fuel emissions produce $CO_2$. Plants, oceans, and soil absorb $CO_2$ from the air.

these organisms were transformed by heat and pressure deep in Earth's crust into highly concentrated energy sources. Humans now tap these energy sources to power our homes, vehicles, and businesses, but as a result of our burning of these fuels, we have released millions of years of stored carbon as carbon dioxide.

Human use of fossil fuels is having a measurable effect; increases in carbon dioxide in the atmosphere are well documented by direct measurements over the past 50 years (**FIGURE 5.6**). Using *fossil air*, scientists have

**FIGURE 5.5 Burning fossil fuels.** The burning of fossil fuels by industrial plants and automobiles adds more carbon dioxide to the environment.

◀ **Working with Data**

**What evidence in the graph demonstrates the increased rate of carbon dioxide accumulation from 2000 to 2010 compared to the 1960s?**

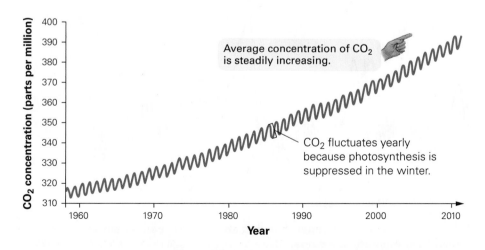

**FIGURE 5.6 Increases in atmospheric carbon dioxide.** Carbon dioxide levels from 1960 to present as measured by instruments at Mauna Loa observatory in Hawaii.

**FIGURE 5.7 Ice core.** By analyzing ice cores, scientists can measure past atmospheric concentrations of carbon dioxide.

also documented that the current level of carbon dioxide in the atmosphere greatly exceeds the levels present on Earth at any time in human history. Carbon dioxide concentrations in the past are measured by examining ice sheets that have existed for thousands of years. This measurement is possible because snow falling on an ice sheet surface traps air. As more accumulates, underlying snow is compressed into ice, and the trapped air becomes tiny ice-encased air bubbles. Thus, these bubbles are fossils—actual samples of the gases in the atmosphere at the time they formed. Cores can be removed from long-lived ice sheets and analyzed to determine the concentration of carbon dioxide trapped in fossil air (**FIGURE 5.7**). Other gases in the bubbles can provide indirect information about temperatures at the time the bubbles formed.

**STOP & STRETCH** Carbon stored by plants today or in the recent past can be burned for fuel in the form of logs or sawdust pellets. However, these types of "renewable" fuels are not considered contributors to increased global warming. What makes the carbon in these sources different from the carbon in fossil fuels?

Ice core data from Antarctica (**FIGURE 5.8**) indicate that although Earth has gone through cycles of high carbon dioxide, the concentration of carbon dioxide in the atmosphere today is higher than at any other time in the last 400,000 years. The ice core data also demonstrate that increased levels of carbon dioxide occur at the same time as increased temperatures, suggesting that carbon dioxide measurably warms Earth. Taken together, these data are quite worrisome; they tell us that Earth may soon be facing temperatures well above those we experience in the current climate, and warmer than Earth has seen in millennia.

## Working with Data ▼

**This graph has axes on both the left and right sides. Which line is indicated by the left axis and which by the right axis? Why are both lines included on the same graph?**

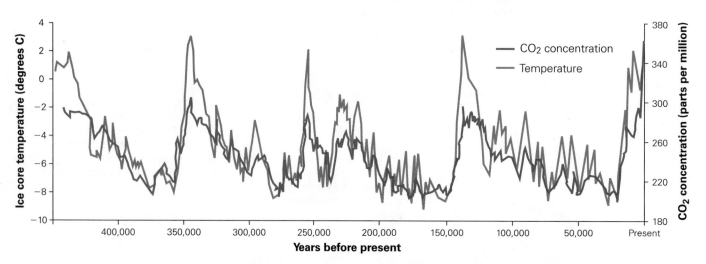

**FIGURE 5.8 Records of temperature and atmospheric carbon dioxide concentration from Antarctic ice cores.** These data indicate that increases in carbon dioxide levels are correlated with higher temperatures.

# 5.3  Can Photosynthesis Slow Down Global Climate Change?

Modern plants and certain microorganisms do as their ancient predecessors did—absorb carbon dioxide from the atmosphere. Is it possible that we could depend on these modern organisms to reduce the amount of greenhouse gases released by fossil fuel burning and thus the threat posed by global climate change?

**Photosynthesis** is the process that traps light energy from the sun and uses it to convert carbon dioxide and water into sugar. In other words, photosynthesis transforms solar energy into the chemical energy required by all living things.

**photo-** means light.

## Chloroplasts: The Site of Photosynthesis

**Chloroplasts** are the specialized organelles in plant cells where photosynthesis takes place. Chloroplasts are surrounded by two membranes (**FIGURE 5.9**). The inner and outer membranes together are called the chloroplast envelope (Chapter 2). The chloroplast envelope encloses a compartment filled with **stroma,** the thick fluid that houses some of the

**chloro-** means green.

**(a)**

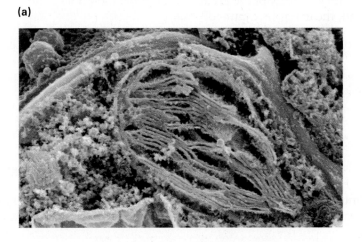

**(b)**

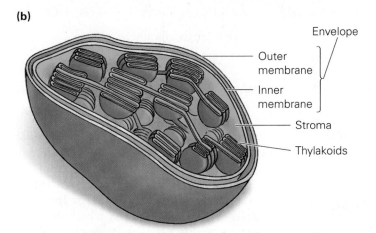

**FIGURE 5.9 Chloroplasts.**  The cross section (a) and drawing (b) of a chloroplast show the structures involved in photosynthesis.

enzymes of photosynthesis. Suspended in the stroma are disklike membranous structures called **thylakoids,** which are typically stacked in piles like pancakes. The large amount of membrane inside the chloroplast provides abundant surface area on which some of the reactions of photosynthesis can occur.

On the surface of the thylakoid membrane are millions of molecules of **chlorophyll,** a pigment that absorbs energy from the sun. It is chlorophyll that gives leaves and other plant structures their green color. Like all pigments, chlorophyll absorbs light. Light is made up of rays with different colors, and each color corresponds to a different wavelength—to the human eye, shorter and middle wavelengths appear violet to green, and longer wavelengths appear yellow to red (**FIGURE 5.10**). Chlorophyll looks green to human eyes because it absorbs the shorter (blue) and longer (red) wavelengths of visible light and reflects the middle (green) range of wavelengths.

**phyll-** means leaf.

When a pigment such as chlorophyll absorbs sunlight, electrons associated with the pigment become excited (that is, the electrons increase in energy level). In effect, light energy is transferred to the chlorophyll and becomes chemical energy. For most pigments, the molecule remains excited for a very brief amount of time before this chemical energy is lost as heat. This is why a surface that looks black (that is, one composed of pigments that absorb all visible light wavelengths) heats quickly in the sun. In contrast, a white surface, which does not absorb any light energy, remains relatively cool. Inside a chloroplast, however, the chemical energy of the excited chlorophyll molecules is not released as heat; instead, the energy is captured.

**STOP & STRETCH**   In temperate areas, leaves change color in the fall. This occurs because chlorophyll is reabsorbed by the plant, making other pigments in the leaf visible. Why do you suppose the chlorophyll is reabsorbed?

## Working with Data ▼

How would this graph appear for the pigments in the skin of a ripe red apple?

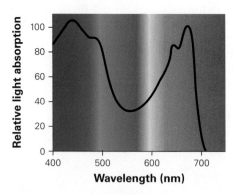

**FIGURE 5.10 The absorption spectra of leaf pigments.** Visible "white" light is actually made up of a series of wavelengths that appear as different colors to our eyes. The wavelengths absorbed by chlorophyll and other pigments in green plants are mainly red and blue and are thus removed from the spectrum of light that is reflected back to us from a leaf surface.

## The Process of Photosynthesis

In plants and other photosynthetic organisms, solar energy is used to rearrange the atoms of carbon dioxide and water absorbed from the environment into carbohydrates (initially, the sugar glucose). Photosynthesis produces oxygen as a waste product. The equation summarizing photosynthesis is as follows:

Carbon dioxide + Water + Light energy → Glucose + Oxygen;

or more technically

$$6\ CO_2 + 6\ H_2O + \text{Light energy} \rightarrow C_6H_{12}O_6 + 6\ O_2$$

Photosynthetic organisms use the carbohydrates that they produce to grow and supply energy to their cells. They, along with the organisms that eat them, liberate the energy stored in the chemical bonds of sugars via the process of cellular respiration. Any excess carbohydrates are stored in the body of the plant or other photosynthetic organism and could become the raw material for fossil fuel production.

The process of photosynthesis can be broken down into two steps (**FIGURE 5.11**). The first, or *photo*, step harvests energy from the sun during a series of reactions called the **light reactions,** which occur when there is sunlight. These reactions occur within the thylakoids of the chloroplast and generate the energy-carrying molecule ATP (the abbreviation for adenosine triphosphate, described in Chapter 4) as well as energized electrons, which are transported around the cell via the molecule NADPH (the abbreviation for nicotinamide adenine dinucleotide phosphate). The second, or *synthesis*, step, called

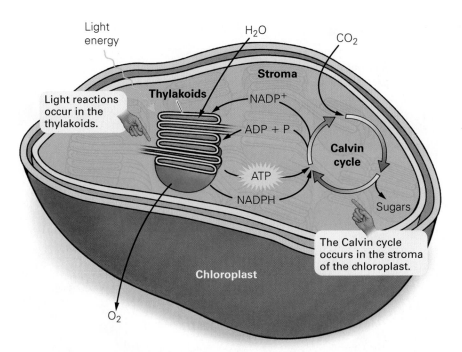

**FIGURE 5.11 Overview of photosynthesis.** Sunlight drives the synthesis of glucose and oxygen from carbon dioxide and water inside the chloroplasts of plants and other photosynthetic organisms. The reaction requires two steps: the first that generates cellular energy (ATP) and harvests electrons from water (carried by NADPH) and the second that uses those products to convert carbon dioxide into sugar.

the **Calvin cycle,** uses the ATP and NADPH generated by the light reactions to synthesize sugars from carbon dioxide. This step can occur in either the presence or the absence of sunlight, and for this reason, the Calvin cycle is also sometimes referred to as the light-independent reactions.

## The Process of Photosynthesis: A Closer Look ▼

### The Light Reactions

When sunlight strikes chlorophyll molecules in a thylakoid membrane, the excited electrons are trapped by another molecule (**FIGURE 5.12**, step 1, on the next page). The energized electrons are then transferred along a path to other molecules in an electron transport chain located in the thylakoid membrane. Some of the proteins within the electron transport chain use the energy of the electrons to pump protons from the exterior to the interior of the thylakoid, setting up a proton gradient between the inside and outside of this structure (**FIGURE 5.12**, step 2). The stockpiled protons diffuse back out of the thylakoid through an enzyme in its membrane, producing ATP. The newly synthesized ATP is released into the stroma of the chloroplast, where it can be used by the enzymes of the Calvin cycle to produce sugars and other organic molecules.

Oxygen is produced during the light reactions when water ($H_2O$) is split into two $H^+$ ions, two electrons, and a single oxygen atom (O) (**FIGURE 5.12**, step 3). Two oxygen atoms combine to produce $O_2$, which is released as a waste product from the chloroplast. The two electrons are transferred back to chlorophyll molecules in the thylakoid membrane to replace those passed along the electron transport chain. One of the $H^+$ ions will participate in step 4, described below, while the other becomes part of the proton gradient that drives ATP synthesis in step 2.

At the end of the electron transport chain, the electrons from chlorophyll are transferred to the electron carrier for plants, $NADP^+$. Just like the $NAD^+$ involved in cellular respiration, $NADP^+$ functions as an electron taxicab. The $NADP^+$ used

**◀ Visualize This**

**Which of the components in these reactions must be obtained by the chloroplast from the environment? Which are released from the chloroplast? Which components are contained within this organelle?**

*A Closer Look, continued*

## Visualize This ▼

**How is this set of reactions similar to the electron transport chain in cellular respiration? How is it different? (To review cellular respiration, see Chapter 4.)**

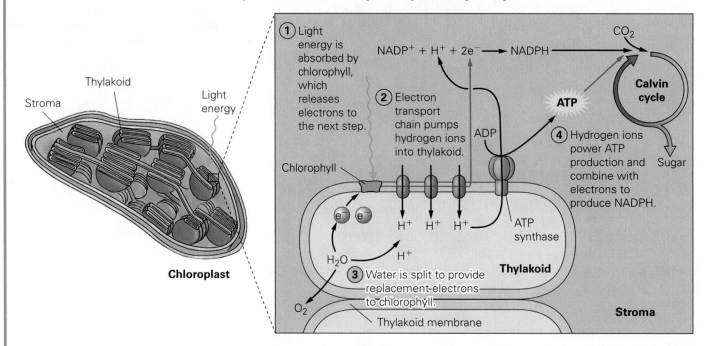

**FIGURE 5.12 The light reactions of photosynthesis.** (1) Sunlight strikes chlorophyll molecules located in the thylakoid membrane, exciting electrons, which then move to a higher energy level. (2) The electrons are captured by an electron transport chain, and their energy is used to pump hydrogen ions across the thylakoid membrane. (3) Water is split. Electrons removed from water are used to replace those lost from chlorophyll. Oxygen gas is released. (4) The movement of hydrogen ions out of the thylakoid power ATP production and generate NADPH. These molecules are produced in the stroma, where they will be available to the enzymes of the Calvin cycle.

## Visualize This ▼

**How would an absence of sunlight affect the Calvin cycle?**

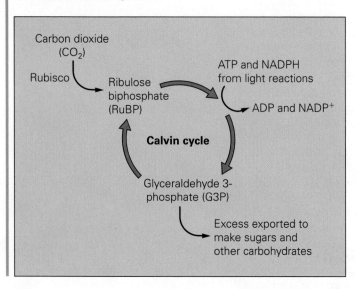

during photosynthesis picks up the two electrons and one $H^+$ ion released from water (**FIGURE 5.12**, step 4). The resulting NADPH ferries electrons to the stroma, where the enzymes of the Calvin cycle will use the electrons to assemble sugars.

Thus, the light reactions produce ATP and a source of electrons (NADPH), both of which are used in the synthesis step and release oxygen as a by-product.

### The Calvin Cycle

During the Calvin cycle, carbon dioxide from the environment reacts with 5-carbon molecules called ribulose bisphosphate, or RuBP (**FIGURE 5.13**). The enzyme that catalyzes this reaction is ribulose bisphosphate carboxylase

**FIGURE 5.13 Reactions of the Calvin cycle.** Carbon dioxide is incorporated into sugar in plants through a series of reactions that regenerate the initial carbon-containing starting product, the sugar ribulose bisphosphate (RuBP). The enzyme that attaches carbon dioxide to RuBP in the first step of the Calvin cycle is called rubisco. Glyceraldehyde 3-phosphate (G3P) is the simple sugar produced by these reactions. Excess G3P is exported to other pathways to produce organic molecules the plant needs.

*A Closer Look, continued*

oxygenase, or **rubisco,** the most abundant protein on the planet. The resulting 6-carbon molecules immediately break down into pairs of 3-carbon molecules. These 3-carbon molecules go through several rearrangements with the help of ATP and NADPH from the light reactions to produce the sugar glyceraldehyde 3-phosphate (or G3P). In the last step of the cycle, five G3P molecules are rearranged into three RuBP molecules. Excess G3P produced by this process is used by the cell to make glucose and other carbohydrate compounds. Because the first stable product of the Calvin cycle is a 3-carbon compound, this pathway is often called $C_3$ photosynthesis.

**STOP & STRETCH** Plants can make all of the macromolecules they need to survive from simple, inorganic components. They accumulate carbon from the atmosphere and hydrogen and oxygen from water to make the overwhelming majority of their mass. The remaining elements they require are from the soil. Consider the atoms found in proteins and nucleic acids (you may want to review Chapter 3) and list two or three soil nutrients that you think are most important to plant growth.

While photosynthesis does take carbon dioxide out of the atmosphere, it is important to realize that the fossil fuels that we have been using only for the last century or so took over 100 million years to form. In other words, right now, carbon dioxide is being released into the atmosphere many times faster than it can be absorbed via natural photosynthesis. We cannot simply rely on photosynthesis to remove excess greenhouse gases as a way to prevent global warming. In fact, rising temperatures may actually slow photosynthesis and reduce its effectiveness.

# 5.4 How High Temperatures Might Reduce Photosynthesis

Global warming can reduce photosynthesis in a number of ways, but primarily because it leads to higher temperatures, which in turn can lead to higher rates of evaporation. Carbon dioxide enters the leaves of land plants through tiny openings called **stomata** (**FIGURE 5.14**). Stomatal openings are surrounded by two kidney-bean-shaped cells called **guard cells.** When the guard cells are compressed against each other, the stomata are closed, thus restricting the flow of gases into or out of the plant. When the guard cells change shape to create a gap between them, the stomata are open, and carbon dioxide and oxygen gases can be exchanged.

*trans-* means across or to the other side.

In addition to the exchange of gases, water moves out of the plant through stomatal openings via a process called **transpiration** (**FIGURE 5.15,** on the next page). The transpired water is replaced in the plant by water from the soil. On hot, dry days, plants cannot absorb enough water from the soil and will close their stomata to reduce the rate of water lost to transpiration. However, closing stomatal openings also prevents carbon dioxide from entering the plant. Thus, with rising temperatures, the rate of photosynthesis declines.

Closing the stomatal openings may not just reduce the rate of photosynthesis; it may even counteract it. This occurs as the result of another series of reactions known as **photorespiration.** During photorespiration, the first enzyme in the Calvin cycle, rubisco, uses oxygen instead of carbon dioxide as a substrate for the reaction it catalyzes with the five-carbon sugar RuBP. Oxygen is used when carbon dioxide levels are low inside the leaf, such as when the stomata are closed. The compound produced by the reaction between RuBP and oxygen, called glycolate, cannot be used in the Calvin cycle. In fact, glycolate must be destroyed by the plant because high levels of this acid will further

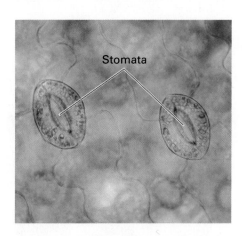

**FIGURE 5.14 Stomata.** Stomata are adjustable microscopic pores found on the surface of leaves that allow for gas exchange.

## Visualize This ▶

The stomata on a plant typically have a regular pattern of opening and closing in a 24-hour period. Using this figure and your understanding of the inputs plants need to perform photosynthesis, at what periods during a 24-hour period do you think most plant stomata are open, and why?

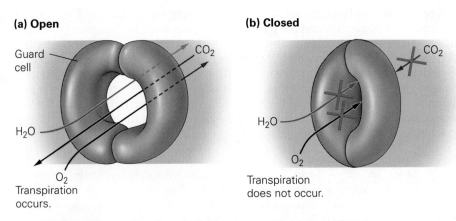

**(a) Open**

Guard cell

$CO_2$

$H_2O$

$O_2$
Transpiration occurs.

**(b) Closed**

$CO_2$

$H_2O$

$O_2$

Transpiration does not occur.

**FIGURE 5.15 Gas exchange and water loss.** (a) When stomata are open, oxygen gases and carbon dioxide can be exchanged, but water can be lost from a plant through a process called transpiration. (b) When the guard cells change shape to block the opening, gas exchange and transpiration do not occur.

inhibit photosynthesis. The breakdown of glycolate requires energy provided by cellular respiration—thus, instead of taking up carbon dioxide, plants that are photorespiring release it.

> **STOP & STRETCH** Photorespiration is a wasteful process, and thus it seems that evolution should have favored mutations in rubisco that prevent it from binding with oxygen. This hasn't happened, however. Can you think of a reason a mutation changing the binding site for carbon dioxide on this enzyme might actually be bad?

In warmer and drier environments, natural selection should favor plants that can minimize photorespiration despite having closed stomata much of the time. Two mechanisms for reducing photorespiration are known as $C_4$ and CAM photosynthesis, in which additional pathways that concentrate carbon dioxide in the plant ($C_4$) or capture it during the night when temperatures are cooler (CAM) occur before the $C_3$ photosynthetic pathway. $C_4$ and CAM, along with the $C_3$ process that occurs during the Calvin cycle, are briefly summarized in **TABLE 5.1**.

As Earth warms, it is possible that the abundance of $C_4$ and CAM plants will increase while the number of $C_3$ plants declines as a result of the increased burden of photorespiration. However, replacing $C_3$ plants, many of which are large trees, with $C_4$ plants, mostly grasses, will reduce the total amount of photosynthesis and the amount of carbon that can be taken out of the atmosphere. Because net rates of photosynthesis (as measured by grams of carbon dioxide removed from the atmosphere per acre per year) in grasslands are 30% to 60% less than rates in forests, the loss of trees significantly decreases the removal of carbon dioxide from the atmosphere.

The replacement of trees with grasses, **deforestation**, may happen naturally on a warming Earth as $C_4$ plants outperform $C_3$ types, but it is already happening at higher than natural rates thanks to human activities. It occurs when forests are cleared for logging, farming, and ever-expanding human settlements. Deforestation also contributes directly to the increase in carbon dioxide within the atmosphere; current estimates are that up to 25% of the carbon dioxide introduced into the atmosphere originates from the cutting and burning of forests in the tropics alone. Clearly, one effective way to reduce global warming is to reduce deforestation and promote photosynthesis by planting trees through reforestation projects.

In the worst case, rising temperatures due to global warming will cause vegetated land to become desert, completely eliminating photosynthesis in certain

**TABLE 5.1 C₃, C₄, and CAM plant photosynthesis.**

Plants have evolved adaptations that prevent water loss.

| Type of plant and example | Stomata status | Description |
|---|---|---|
| C₃ Soybean | | The Calvin cycle converts two 3-carbon sugars into glucose. |
| C₄ Corn | | Enzymes scavenge $CO_2$ to produce 4-carbon sugars, even when stomata are closed. The 4-carbon sugars "pump" carbon dioxide molecules to the Calvin cycle when they break down. |
| CAM Jade | | Water loss is slowed by opening stomata only at night. Carbon dioxide is stored as an organic acid in the vacuole. The acid's breakdown releases this carbon dioxide to the Calvin cycle during the day. |

areas. Warming temperatures will expose more snow- and ice-covered regions, allowing additional photosynthesis there, but this additional carbon dioxide uptake is likely to be offset by carbon dioxide released from the more rapid decay of carbohydrates in the soil in these formerly frozen regions. As we have seen, photosynthesis is not likely to soak up much of the excess carbon dioxide released by fossil fuel combustion and, in fact, the overall rate of photosynthesis may even decline on a warmer planet.

# 5.5  Decreasing the Effects of Climate Change

The devastation Superstorm Sandy caused in New York and that Hurricane Katrina caused in New Orleans is only a small part of the environmental damage seen worldwide as a result of climate change. This process has resulted in unprecedented droughts in some areas and torrential rains in others. It will cause the extinction of animals and plants that have narrow temperature requirements. It may contribute to the spread of infectious disease and dangerous pests that are now limited by temperature to the tropics. And outside of the other effects of global warming, the carbon dioxide absorbed by the ocean

**If the rate of warming continues as predicted, by the end of the twenty-first century the sea level will be between 50 and 100 centimeters (20 to 36 inches) higher than today.**

is causing it to acidify, threatening coral reefs, algae in the open ocean, and the organisms that depend on them and which we, in turn, depend on. Clearly, excess carbon dioxide in the atmosphere is already imposing a cost on people.

The United States has a disproportionate impact on the rate of carbon dioxide emissions into the atmosphere. Home to only 4% of the world's population, the United States produces close to 25% of the carbon dioxide emitted by fossil fuel burning. The emissions rate of carbon dioxide for an average American is twice that of a Japanese or German individual, three times that of the global average, four times that of a Swede, and 20 times that of the average Indian.

**STOP & STRETCH** Based on the statistic described above, is there a clear correlation between carbon dioxide emissions and standard of living?

Most of the emissions for an individual country come from industry, followed by transportation and then by commercial, residential, and agricultural emissions. All of us can work to reduce our personal contribution to global warming by decreasing residential and transportation emissions. Most residential emissions are from energy used to heat and cool homes and to power electrical appliances. Transportation emissions are affected by our choice of the vehicles we use for transport, the fuel economy of cars, and the number of miles traveled. **TABLE 5.2** describes many ways that you can decrease your greenhouse gas emissions and indicates the number of pounds of carbon dioxide that each action would save annually. These reductions may seem trivial in comparison to the scope of the problem, but when they are multiplied by the more than 310 million people in the United States, the savings become significant.

Having an effect on industrial, commercial, and agricultural sectors is difficult for any one individual. Changes in these areas will instead take leadership from the policy makers who are committed to reducing emissions. Making these changes requires that our leaders, and all of us, understand that even though the implications of and solutions to global warming may be open to debate, the fact that it is occurring at an unprecedented rate is not.

**TABLE 5.2 Decreasing your greenhouse gas emissions.**
Here are some ideas that you can use to help slow the rate of global warming.

| Action | | Annual decrease in carbon dioxide production |
|---|---|---|
| Drive an energy-efficient vehicle. SUVs average 16 miles per gallon, while smaller cars average 25 miles per gallon. | | 13,000 pounds |
| Carpool 2 days per week. | | 1,590 pounds |
| Recycle glass bottles, aluminum cans, plastic, newspapers, and cardboard. | | 850 pounds |
| Walk 10 miles per week instead of driving. | | 590 pounds |
| Buy high-efficiency appliances. | | 400 pounds per appliance |
| Buy food and other products with reusable or recyclable packaging, or reduced packaging, to save the energy required to manufacture new containers. | | 230 pounds |
| Use a push mower instead of a power mower. | | 80 pounds |
| Plant shade trees around your home to decrease energy consumption and to remove carbon dioxide by photosynthesis. | | 50 pounds |

# savvy reader

## Is Climate Change More Good Than Bad?

There are many likely effects of climate change: positive and negative, economic and ecological, humanitarian and financial. And if you aggregate them all, the overall effect is positive today—and likely to stay positive until around 2080. That was the conclusion of Professor Richard Tol of Sussex University after he reviewed 14 different studies of the effects of future climate trends.

To be precise, Prof. Tol calculated that climate change would be beneficial up to 2.2°C of warming from 2009 (when he wrote his paper).... Now Prof. Tol has a new paper, published as a chapter in a new book, called *How Much Have Global Problems Cost the World?*, which is edited by Bjørn Lomborg, director of the Copenhagen Consensus Centre, and was reviewed by a group of leading economists. In this paper he casts his gaze backwards to the last century. He concludes that climate change did indeed raise human and planetary welfare during the 20th century.

...If you wish to accept the consensus on temperature models, then you should accept the consensus on economic benefit.

Overall, Prof. Tol finds that climate change in the past century improved human welfare. By how much? He calculates by 1.4 per cent of global economic output, rising to 1.5 per cent by 2025. For some people, this means the difference between survival and starvation....

1. The first paragraph of this excerpt lists the systems that could be affected by climate change and then states that the research described indicates that its overall effect is positive. Which of these systems appear to have been addressed by the research described in the article? Does the conclusion that the overall effect is positive follow from the research that was cited?

2. What is the definition of consensus? Does the author of this excerpt make a convincing case for a consensus among economists that climate change will have a net economic benefit for the next several decades?

3. Review the checklist for analyzing science in the news found in Chapter 1 (Table 1.1). What aspects of well reported scientific research are missing from this article?

*Source:* http://www.spectator.co.uk/features/9057151/carry-on-warming/

This text was excerpted from Matt Ridley, "Why Climate Change is Good for the World," *The Spectator*, October 19, 2013. ▬

# SOUNDS RIGHT BUT IS IT?

After comparing the chemical equations for aerobic respiration and photosynthesis, students are sometimes tempted to say:

**Animals breathe by taking in oxygen and releasing carbon dioxide, while plants breathe by taking in carbon dioxide and releasing oxygen.**

Sounds right, but it isn't.

1. Why do plants need energy?
2. What is the molecule that provides cellular energy for these activities?
3. What is the equation for photosynthesis?
4. Does photosynthesis produce ATP that is used in the cellular activities described in your answer to question 1?
5. Where do plants get the ATP for cellular activities?
6. What is the equation for aerobic respiration?
7. Reflect on your answers to questions 1–6 and explain why the above statement in bold sounds right, but isn't.

# Chapter Review  MasteringBiology®

Go to the Study Area in MasteringBiology® for practice quizzes, myeBook, BioFlix™ 3-D animations, MP3Tutor sessions, videos, current events, and more.

## Summary

### Section 5.1

Describe the greenhouse effect.

- Greenhouse gases, particularly carbon dioxide, increase the amount of heat retained in Earth's atmosphere, which then leads to increased surface temperatures (pp. 88–89).

Explain why water is slow to change temperature.

- Water can absorb large amounts of heat without undergoing rapid or drastic changes in temperature because heat must first be used to break hydrogen bonds between adjacent water molecules. A high heat-absorbing capacity is a characteristic of water (pp. 89–90).

### Section 5.2

Compare and contrast the water cycle and the carbon cycle.

- Both water and carbon cycle among animals, plants, soil, oceans, and the atmosphere. Most of the movement of water occurs outside living organisms, while carbon is mostly cycled due to biological activity (pp. 89–90).

Discuss the origin of fossil fuels and their relationship to the carbon cycle.

- Fossil fuels are the buried remains of ancient plants, which took carbon dioxide out of the atmosphere and transformed it into carbohydrates. The burning of fossil fuels is returning carbon to the atmosphere, leading to rising concentrations of carbon dioxide and global warming (pp. 89–92).

### Section 5.3

State the basic equation of photosynthesis.

- Photosynthesis utilizes carbon dioxide from the atmosphere to make sugars and other substances. During photosynthesis, energy from sunlight is used to rearrange the atoms of carbon dioxide and water to produce sugars and oxygen. Photosynthesis is described by the equation: $6\ CO_2 + 6\ H_2O + \text{Light energy} \rightarrow C_6H_{12}O_6 + 6\ O_2$ (p. 94).

Describe the light reactions of photosynthesis.

- Photosynthesis occurs in chloroplasts. Sunlight strikes the chlorophyll molecule within chloroplasts, boosting electrons to a higher energy level. These excited electrons are dropped down an electron transport chain located in the thylakoid membrane, and ATP is made (pp. 95–96).
- Electrons are also passed to electron carriers (NADPH). The electrons that are lost from chlorophyll become replaced by electrons acquired during the splitting of water, and oxygen is released (pp. 95–96).

Explain the events that occur during the Calvin cycle, and describe the relationship between the light reactions and the Calvin cycle.

- The Calvin cycle utilizes the products of the light reactions (ATP and the electron carrier NADPH) to incorporate carbon dioxide into sugars, regenerating the starting products of the cycle and exporting excess sugars to be used as chemical building blocks for plant compounds (pp. 96–97).

### Section 5.4

Explain the role of stomata in balancing photosynthesis and water loss.

- Stomata on a plant's surface not only allow in carbon dioxide for photosynthesis but also allow water to escape from the plant. Guard cells surrounding the stomata can change shape to close the stomata and restrict water loss (pp. 97–98).

Define photorespiration, and explain why this process is detrimental to plants that experience it.

- Photorespiration occurs when stomata are closed, carbon dioxide declines in the plant, and oxygen is incorporated into the first step of the Calvin cycle. The resulting product is poisonous to the plant, and energy must be expended to reduce it (p. 97).
- $C_4$ and CAM plants have evolved to perform photosynthesis while reducing the risk of photorespiration in dry conditions (pp. 98–99).

### Section 5.5

Discuss how our own activities contribute to or reduce the risk of global climate change.

- Humans are deforesting Earth's land surface, reducing the rate of photosynthesis and thus the uptake of atmospheric carbon dioxide, so reforestation is one strategy to reduce global warming (pp. 98–99).
- Humans can reduce carbon dioxide emissions by increasing energy efficiency and reducing energy use (pp. 100–101).

## Roots to Remember

**The following roots of words come mainly from Latin and Greek and will help you decipher terms:**

**chloro-** means green. Chapter terms: *chloroplast, chlorophyll*

**photo-** means light. Chapter terms: *photosynthesis, photorespiration*

**phyll-** means leaf. Chapter term: *chlorophyll*

**trans-** means across or to the other side. Chapter term: *transpiration*

## Learning the Basics

1. What are the reactants and products of photosynthesis?

2. Carbon dioxide functions as a greenhouse gas by _____.
   A. interfering with water's ability to absorb heat; B. increasing the random molecular motions of oxygen; C. allowing radiation from the sun to reach Earth and absorbing the reradiated heat; D. splitting into carbon and oxygen and increasing the rate of cellular respiration.

3. Water has a high heat-absorbing capacity because _____.
   A. the sun's rays penetrate to the bottom of bodies of water, mainly heating the bottom surface; B. the strong covalent bonds that hold individual water molecules together require large inputs of heat to break; C. it has the ability to dissolve many heat-resistant solutes; D. initial energy inputs are first used to break hydrogen bonds between water molecules and only after these are broken, to raise the temperature; E. all of the above are true.

4. The burning of fossil fuels _____.
   A. releases carbon dioxide to the atmosphere; B. primarily occurs as a result of human activity. C. is contributing to global warming; D. is possible thanks to photosynthesis that occurred millions of years ago; E. all of the above are correct.

5. Stomata on a plant's surface _____.
   A. prevent oxygen from escaping; B. produce water as a result of photosynthesis; C. cannot be regulated by the plant; D. allow carbon dioxide uptake into leaves; E. are found in stacks called thylakoids.

6. Which of the following **does not** occur during the light reactions of photosynthesis?
   A. Oxygen is split, releasing water; B. Electrons from chlorophyll are added to an electron transport chain; C. An electron transport chain drives the synthesis of ATP for use by the Calvin cycle; D. NADPH is produced and will carry electrons to the Calvin cycle; E. Oxygen is produced when water is split.

7. Which of the following is a **false** statement about photosynthesis?
   A. During the Calvin cycle, electrons and ATP from the light reactions are combined with atmospheric carbon dioxide to produce sugars; B. The enzymes of the Calvin cycle are located in the chloroplast stroma; C. Oxygen produced during the Calvin cycle is released into the atmosphere; D. Sunlight drives photosynthesis by boosting electrons found in chlorophyll to a higher energy level; E. Electrons released when sunlight strikes chlorophyll are replaced by electrons from water.

8. Which human activity generates the most carbon dioxide?
   A. driving; B. cooking; C. bathing; D. using aerosol sprays.

9. Photorespiration occurs _____.
   A. under hot and dry conditions; B. when oxygen is incorporated in the first step of the Calvin cycle; C. when carbon dioxide levels are high inside the plant; D. A and B are correct; E. A, B, and C are correct.

10. Select the **true** statement regarding metabolism in plant and animal cells.
    A. Plant and animal cells both perform photosynthesis and aerobic respiration; B. Animal cells perform aerobic respiration only, and plant cells perform photosynthesis only; C. Plant cells perform aerobic respiration only, and animal cells perform photosynthesis only; D. Plant cells perform cellular respiration and photosynthesis, and animal cells perform aerobic respiration only.

## Analyzing and Applying the Basics

1. During the last Ice Age, global temperatures were 4°C to 5°C lower than they are today. How would this temperature difference affect the water cycle?

2. Imagine an Earth without living organisms on it. How does this difference change the water cycle and the carbon cycle?

**3.** Before life evolved on Earth, there was very little oxygen in the air, mostly because oxygen is a very reactive molecule and tends to combine with other compounds. Explain why oxygen can be maintained at 16% in our atmosphere today.

## Connecting the Science

**1.** A "carbon footprint" is the amount of carbon dioxide you produce as a result of your lifestyle. List five realistic actions you can take to reduce your carbon footprint. Which of these actions will have the greatest impact on your footprint? How difficult for you would it be to take this action?

**2.** The impact of global warming is not uniformly negative—some regions will benefit from a change in the climate. In addition, technological fixes exist that can help humans deal with some of the consequences of sea-level rise, and people can move from threatened coastlines and areas experiencing the negative effects of global warming. Given these facts, should humans commit to trying to stop global warming, or should we instead focus on mitigating its consequences?

Answers to **Stop & Stretch, Visualize This, Working with Data, Savvy Reader, Sounds Right, But Is It?,** and **Chapter Review** questions can be found in the Answers section at the back of the book.

CHAPTER **6** | # Cancer

**Actress Angelina Jolie carries a cancer-causing gene which puts her, and some of her children, at high risk of getting cancer.**

# DNA Synthesis, Mitosis, and Meiosis

Cancer is a disease that will affect most people at some point in their lives. Of those who live an average life span, more than one in three will be directly affected with a diagnosis, and those who escape the disease themselves are likely to be affected by the diagnosis of a loved one.

Some types of cancer are genetically inherited, and others are acquired during an individual's lifetime. Actress and humanitarian

**The origin of some cancers, like the one that originated prior to Lance Armstrong's ban from competitive cycling for doping, can be harder to explain.**

**Lifestyle changes, like quitting smoking, can dramatically affect the risk of some cancers.**

Angelina Jolie's battle against a genetically inherited form of breast cancer led to her decision to have both of her breasts surgically removed. Prior to his fall from grace, Lance Armstrong acquired testicular cancer, requiring the surgical removal of one testicle. The difficult medical choices faced by these individuals, while frightening to consider, can serve as an impetus

for each of us to attempt to better understand the biological basis of cancer.

A better understanding of cancer can help us to both protect ourselves and support those we love. Lifestyle changes can lead to the prevention, delayed onset, or slowed progression of many types of cancer. When a cancer diagnosis does occur, understanding the biological mechanisms of various treatments can help us support those we love as they make medical decisions and undergo treatments.

**Understanding cancer can help us take better care of ourselves and support a loved one with the diagnosis.**

# 6.1 What Is Cancer?

**Cancer** is a disease that occurs when a cell makes copies of itself, or replicates, when it should not. Cell replication occurs during **cell division,** a process by which the original parent cell divides to form two daughter cells. This process is regulated so that a cell divides only when more cells are required.

## Tumors Can Be Cancerous

Unregulated cell division leads to a pileup of cells that form a lump or **tumor.** A tumor is a mass of cells that has no apparent function in the body. Tumors that stay in one place and do not affect surrounding structures are said to be **benign.** Some benign tumors remain harmless; others become cancerous. Invasive tumors, or those that infiltrate surrounding tissues, are **malignant** cancers. **Metastasis** occurs when the cells of a malignant tumor break away and start new cancers at distant locations (**FIGURE 6.1**).

Cancer cells can travel virtually anywhere in the body via the lymphatic and circulatory systems. The lymphatic system collects fluid, called lymph, lost from blood vessels. The lymph is then returned to the blood vessels, a process that also allows cancer cells access to the bloodstream. Lymph nodes are structures that filter fluids released from blood vessels. When a cancer patient is undergoing surgery, the surgeon will often remove a few lymph nodes that will be analyzed for the presence of cancer cells. The presence of cancer cells in the nodes is an indication that some cells have broken away from the original tumor and might present elsewhere in the body. One of the reasons it was so incredible that Lance Armstrong survived cancer was that the original cancer in his testicle spread to his brain and lungs. Metastatic cancers are much more difficult to treat than cancers that are detected before they spread.

**mal-** means bad or evil.

**meta-** means change or between.

## Risk Factors for Cancer

A **risk factor** is a condition or behavior that increases the likelihood of developing a disease. These risk factors can be impacted by genetics or environmental exposures.

**Inherited Cancer Risk.** Angelina Jolie carries a version of the *BRCA1* gene (for *breast cancer* susceptibility) that makes her five times more likely to get breast cancer than other women. After being told that she had an 87% risk for breast cancer, she decided to undergo a double mastectomy—a procedure to remove both of her

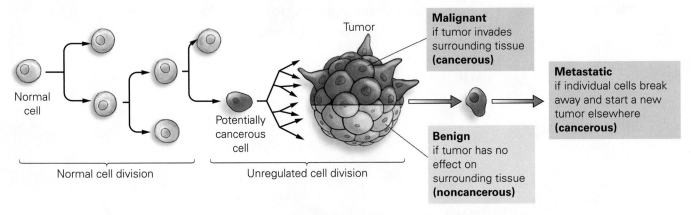

**FIGURE 6.1 What is cancer?** A tumor is a clump of cells with no function. Tumors may remain benign, or they can invade surrounding tissues and become malignant. Tumor cells may move, or metastasize, to other locations in the body. Malignant and metastatic tumors are cancerous.

breasts. The version of the gene that she carries also increases her risk for ovarian cancer—the disease that killed her mother. To eliminate her chances of developing ovarian cancer, Jolie is planning to have her ovaries removed as well. How did Jolie know her risks? Because several members of Jolie's family had been diagnosed with breast and ovarian cancers, she was tested for the presence of this gene. Since only about 1% of people carry this particular gene, unless several immediate family members have been diagnosed with cancer, this testing is not usually performed.

**Lifestyle changes can lead to prevention, delayed onset, or slowed progression of cancer.**

**STOP & STRETCH**   Around 12% of women will be diagnosed with breast cancer at some point in their lifetime. If you took a random sample of 1000 women, how many would you predict would get breast cancer? If you took a random sample of 1000 women with the *BRCA1* gene, how many of them would you expect to get breast cancer?

**Environmental Exposures.**   Exposure to particular substances, called **carcinogens**, is correlated with development of particular cancers. For example, exposure to the human papilloma virus (HPV) can cause anal and oral cancer in females and males, cervical cancer in females, and penile cancer in males. HPV-caused cancer risk can be all but eliminated by vaccination; vaccines are available to both males and females. You can also limit your risk for HPV-caused cancers by limiting the number of people you have sex with and by following safer sex guidelines.

Everyone knows that smoking increases risk of cancer. What is less well known is that smoking combined with excessive alcohol consumption increases cancer risk at a greater level than one would expect. This is because some carcinogens enhance the activity of other carcinogens. When this occurs, the substances involved are said to be acting in a **synergistic** manner. Cigarette smoking and alcohol consumption, two practices commonly combined by college students, have a far greater effect on cancer risk than the sum of each separate risk factor combined (**FIGURE 6.2**).

## Working with Data ▼

**Is the cancer risk associated with smoking and drinking additive or multiplicative? Explain your answer.**

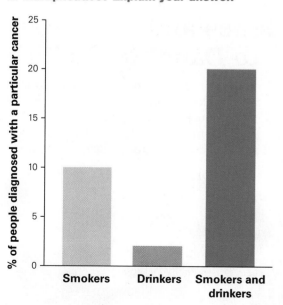

**FIGURE 6.2 Alcohol and tobacco are synergists.** Smoking cigarettes while drinking is an unhealthy practice.

While we have seen that exposures to particular agents can increase risks of particular cancers, there are general risk factors that increase risk of virtually every type of cancer. These include tobacco use, excessive alcohol consumption, a high-fat and low-fiber diet, lack of exercise, and obesity (**TABLE 6.1**).

Limiting these exposures can help prevent cancers from developing in our cells and from being passed on to daughter cells produced when cells divide.

**TABLE 6.1** Decreasing Your Cancer Risk

| Risk-Reducing Behavior | Specific Risk Reduction Information | Biological Mechanism of Risk Reduction |
|---|---|---|
| **Don't Use Tobacco.** | The use of tobacco of any type, whether delivered via cigarettes, cigars, pipes, or chewing tobacco, increases your risk of many cancers. Electronic cigarettes, which contain nicotine only, are new enough that risks from their use have not yet been determined. | Tobacco and tobacco cigarette smoke contain more than 20 known cancer-causing substances. Chemicals present in tobacco and cigarette smoke have been shown to increase cell division, damage DNA, inhibit a cell's ability to repair damaged DNA, and prevent cells from dying when they should. |
| **Limit Alcohol Consumption.** | Men who want to decrease their cancer risk should have no more than two alcoholic drinks a day, and women one or none. | When harmful chemicals dissolve in alcohol, they are able to traverse cell membranes and can damage DNA. |
| **Eat a Low-Fat, High-Fiber Diet.** | Eat at least 5 servings of fruits and vegetables every day as well as 6 servings of food from other plant sources, such as breads, cereals, grains, rice, pasta, or beans. | Fruits and vegetables are rich in antioxidants which help prevent DNA damage. |
| **Exercise Regularly.** | Engage in physical activity for at least 30 minutes 5 days a week. | Exercise keeps the immune system functioning effectively, allowing it to recognize and destroy cancer cells. |
| **Maintain a Healthy Weight.** | Avoid becoming obese. If you are obese, consult a physician for a weight loss program. | Because fatty tissues can store hormones, the abundance of fatty tissue has been hypothesized to increase the risk of hormone-sensitive cancers such as breast, uterine, ovarian, and prostate cancer. |

# 6.2 Passing Genes and Chromosomes to Daughter Cells

Cells have evolved to divide for a variety of reasons having nothing to do with cancer. Cell division produces new cells to allow an organism to grow, to replace damaged cells, and, in some cases, to reproduce (**FIGURE 6.3**).

Some organisms reproduce by making exact copies of themselves. Reproduction of this type, called **asexual reproduction**, results in offspring

**(a)**

**(b)**

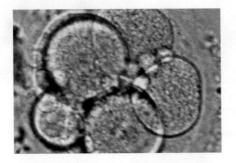

**FIGURE 6.3 Why do cells divide?** Cells divide in order to make more cells. This can allow an organism to grow (a). Each of us begins life as a single fertilized egg cell that underwent millions of rounds of cell division to produce all the cells that comprise the tissues and organs of our bodies. Cells also divide in order to heal wounds (b). As this cut heals, new cells will replaced those damaged by the injury.

# Visualize This ▶

**(a) If one amoeba divides one time, how many offspring are produced? (b) If you removed one stem and leaf of the ivy plant on this tree and replanted it, would you expect it to grow into a larger stem and leaf or into two stems and leaves?**

**(a) Amoeba**

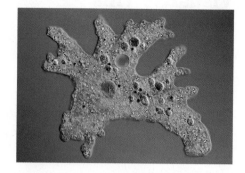

**(b) English ivy**

**FIGURE 6.4 Asexual reproduction.**
(a) This single-celled amoeba divides by copying its DNA and producing offspring that are genetically identical to the original, parent amoeba. (b) Some multicellular organisms, such as this English ivy plant, can reproduce asexually from cuttings.

that are genetically identical to the original parent cell. Single-celled organisms, such as bacteria and amoeba, reproduce in this manner (**FIGURE 6.4a**). Some multicellular organisms can reproduce asexually also. For example, some plants can grow from clippings of stems, leaves, or roots. Reproduction from such cuttings is also a form of asexual reproduction (**FIGURE 6.4b**). Organisms whose reproduction requires genetic information from two parents undergo **sexual reproduction**. Humans reproduce sexually when sperm and egg cells each contribute genetic information at fertilization.

## Genes and Chromosomes

Whether reproducing sexually or asexually, all dividing cells must first make a copy of their genetic material, the **DNA (deoxyribonucleic acid)**. DNA carries the instructions, called **genes**, for building all of the proteins that a cell requires. The DNA in the nucleus is wrapped around proteins to produce structures called **chromosomes**.

Chromosomes are in an uncondensed, string-like form when a cell is not preparing to divide (**FIGURE 6.5a**). In order for cell division to occur, the DNA in each chromosome is compressed into a more compact linear structure that is easier to maneuver during cell division. Condensed chromosomes are less likely to become tangled or broken than are the uncondensed and string-like structures.

Each chromosome carries hundreds of genes. When a chromosome is replicated, a copy is produced that carries those same genes. The copied chromosomes, now called **sister chromatids**, are attached to each other at a region toward the middle of the replicated chromosome, called the **centromere** (**FIGURE 6.5b**). Because the centromere is not always located precisely in the center of the chromosome, it can subdivide the chromosome into one long and one short arm. Scientists have mapped the location of the *BRCA1* gene to the long arm of chromosome number 17.

> **STOP & STRETCH**   If humans have 23 pairs of chromosomes, each carrying hundreds of genes, roughly how many genes are there in the human genome?

## DNA Replication

During the process of **DNA replication** that precedes cell division, the double-stranded DNA molecule is copied, first by splitting the molecule in half up the middle of the helix. New nucleotides are added to each side of the original parent molecule, maintaining the A-to-T and G-to-C base pairings. This process results in two daughter DNA molecules, each composed of one strand of

**(a) Uncondensed DNA**

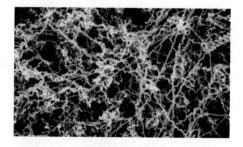

**(b) DNA condensed into chromosomes**

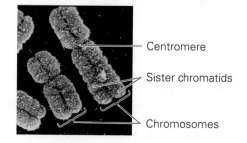

— Centromere

— Sister chromatids

— Chromosomes

**FIGURE 6.5 DNA condenses during cell division.**  (a) DNA in its replicated but uncondensed form prior to cell division. (b) During cell division, each copy of DNA is wrapped neatly around many small proteins, forming the condensed structure of a chromosome. After DNA replication, two identical sister chromatids are produced and joined to each other at the centromere.

## Visualize This ▶

Assume another round of replication were to occur to one of the half purple, half red DNA molecules shown in part (a). How many total DNA molecules would be produced? If the nucleotides being added to the newly synthesized strand are purple, what proportion of each DNA molecule would be purple?

**FIGURE 6.6 DNA replication.**
(a) DNA replication results in the production of two identical daughter DNA molecules from one parent molecule. Each daughter DNA molecule contains half of the parental DNA and half of the newly synthesized DNA. (b) The DNA polymerase enzyme moves along the unwound helix, tying together adjacent nucleotides on the newly forming daughter DNA strand. Free nucleotides have three phosphate groups, two of which are cleaved to provide energy for this reaction before the nucleotide is added to the growing chain.

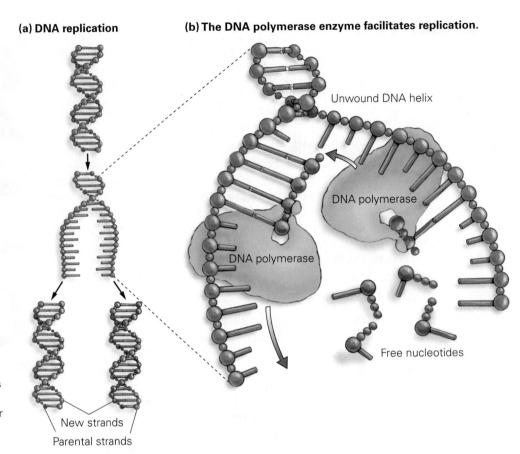

**(a) DNA replication**

**(b) The DNA polymerase enzyme facilitates replication.**

Unwound DNA helix

DNA polymerase

DNA polymerase

Free nucleotides

New strands

Parental strands

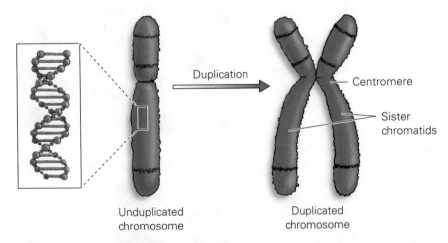

Duplication

Centromere

Sister chromatids

Unduplicated chromosome

Duplicated chromosome

**FIGURE 6.7 Unduplicated and duplicated chromosomes.** An unreplicated chromosome is composed of one double-stranded DNA molecule. A replicated chromosome is X-shaped and composed of two identical double-stranded DNA molecules. Each DNA molecule of the duplicated chromosome is a copy of the original chromosome and is called a sister chromatid.

parental nucleotides and one newly synthesized strand (**FIGURE 6.6a**). Because each newly formed DNA molecule consists of one-half conserved parental DNA and one-half new daughter DNA, this method of DNA replication is referred to as **semiconservative replication.**

Replicating the DNA requires the assistance of enzymes, in particular, a **DNA polymerase.** The DNA polymerase moves along the length of the unwound parental DNA strand to facilitate synthesis of the newly formed strand (**FIGURE 6.6b**). When free nucleotides floating in the nucleus have an affinity for each other (A for T and G for C), these complementary nucleotides bind to each other across the width of the helix and then the DNA polymerase catalyzes the formation of the covalent bond between adjacent nucleotides along the length of the helix. When an entire chromosome has been replicated, the newly synthesized sister chromatids are identical to each other and attached to each other at the centromere (**FIGURE 6.7**).

The DNA polymerase can make mistakes when facilitating base pairing. While uncommon, pairing an A with a G for instance, such mistakes can alter the sequence of the original gene. Changes in the DNA of a gene are called **mutations.** Normally, replication of the *BRCA1* gene produces an exact copy of the gene. If mistakes in the copying occur, a mutant version of the gene can be produced, which appears to be the case in the version of the gene that Angelina Jolie carries.

# 6.3  The Cell Cycle and Mitosis

After a cell's chromosomes and DNA have been replicated, the cell is able to undergo cell division and produce daughter cells. One type of cell division, **mitosis**, is an asexual division that produces two daughter cells that are identical to their original parent cell and to each other. Mitosis occurs in the type of body cells called somatic cells. **Somatic cells** include any cell type that does not produce sex cells. In plants, for example, the leaves and stem are composed of somatic cells and undergo mitosis. The reproductive organs of the plant produce pollen and egg cells, which are non-somatic cells called sex cells.

    For cells that divide by mitosis, the cell cycle includes three steps: (1) **interphase**, when the DNA replicates; (2) *mitosis*, when the copied chromosomes split and move into the daughter nuclei; and (3) **cytokinesis**, when the **cytoplasm** of the parent cell splits (**FIGURE 6.8a**). As you will see, interphase and mitosis are further subdivided.

**mito-** means a thread.

**soma-** and **-some** mean body.

**cyto-** and **-cyte** relate to cells.

**-kinesis** means motion.

## Interphase

A normal cell spends most of its time in interphase (**FIGURE 6.8b**). During this phase of the cell cycle, the cell performs its typical functions and produces the proteins required for the cell to do its particular job. For example, during interphase, a muscle cell produces proteins required for muscle contraction. Different cell types spend varying amounts of time in interphase. Cells that frequently divide, like skin cells, spend less time in interphase than do those that seldom divide, such as some nerve cells. A cell that will divide also begins preparations for division during interphase. Interphase can be separated into three phases: $G_1$, S, and $G_2$.

**(a) Copying and partitioning DNA**

**(b) Steps in the cell cycle**

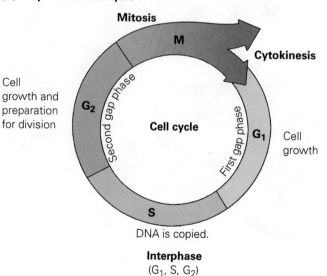

| Interphase | Mitosis | Cytokinesis |
|---|---|---|
| DNA is copied. | DNA is split equally into two daughter cells. | Parent cell is cleaved in half. |

Cell growth and preparation for division

$G_2$

Mitosis

M

Cytokinesis

Cell cycle

Second gap phase

First gap phase

$G_1$

Cell growth

S

DNA is copied.

Interphase
($G_1$, S, $G_2$)

**FIGURE 6.8 The cell cycle.** (a) During interphase, the DNA is copied. Separation of the DNA into two daughter nuclei occurs during mitosis. Cytokinesis is the division of the cytoplasm, creating two daughter cells. (b) During interphase, there are two stages when the cell grows in preparation for cell division, $G_1$ and $G_2$ stages, and one stage where the DNA replicates, the S stage. The chromosomes are separated and two daughter cells are formed during the M phase.

During the $G_1$ (first gap or growth) phase, most of the cell's organelles duplicate. Consequently, the cell grows larger during this phase. During the S (synthesis) phase, the DNA composing the chromosomes replicates. During the $G_2$ (second gap) phase of the cell cycle, the cell continues to grow and prepares for the division of chromosomes that will take place during mitosis.

**STOP & STRETCH** If a cell at $G_1$ contains 4 picograms of DNA, how many picograms of DNA will it contain at the end of the S phase of the cell cycle?

## Mitosis

The movement of chromosomes from the original parent cell into two daughter cells occurs during mitosis. Whether these phases occur in an animal or a plant, the outcome of mitosis and the next phase, **cytokinesis**, is the same:

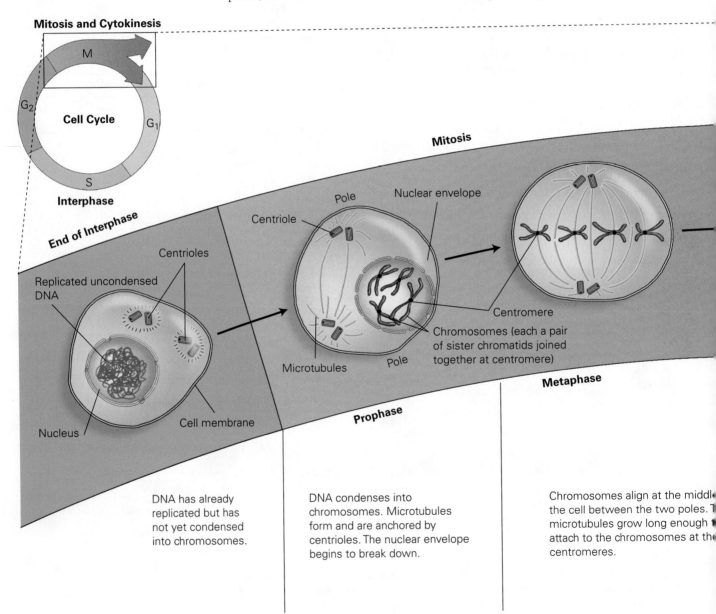

**Mitosis and Cytokinesis**

M

$G_2$

**Cell Cycle**

$G_1$

S

**Interphase**

**End of Interphase**

Replicated uncondensed DNA

Centrioles

Nucleus

Cell membrane

DNA has already replicated but has not yet condensed into chromosomes.

**Prophase**

Pole

Centriole

Microtubules

Pole

**Mitosis**

Nuclear envelope

Centromere

Chromosomes (each a pair of sister chromatids joined together at centromere)

DNA condenses into chromosomes. Microtubules form and are anchored by centrioles. The nuclear envelope begins to break down.

**Metaphase**

Chromosomes align at the middl· the cell between the two poles. T microtubules grow long enough · attach to the chromosomes at th· centromeres.

the production of genetically identical daughter cells. To achieve this outcome, the sister chromatids of a replicated chromosome are pulled apart, and one copy of each is placed into each newly forming nucleus. Mitosis is accomplished during four stages: prophase, metaphase, anaphase, and **telophase**. **FIGURE 6.9** summarizes the cell cycle in animal cells. The four stages of mitosis are nearly identical in plant cells.

During **prophase**, the replicated chromosomes condense, allowing them to move around in the cell without becoming entangled. Protein structures called **microtubules** also form and grow, ultimately radiating out from opposite ends, or **poles**, of the dividing cell. The growth of microtubules helps the cell to expand. Motor proteins attached to microtubules also help pull the chromosomes around during cell division. The membrane that surrounds the nucleus, called the **nuclear envelope**, breaks down so that the microtubules can gain access to the replicated chromosomes. At the poles of each dividing animal cell,

**telo-** means end or completion.

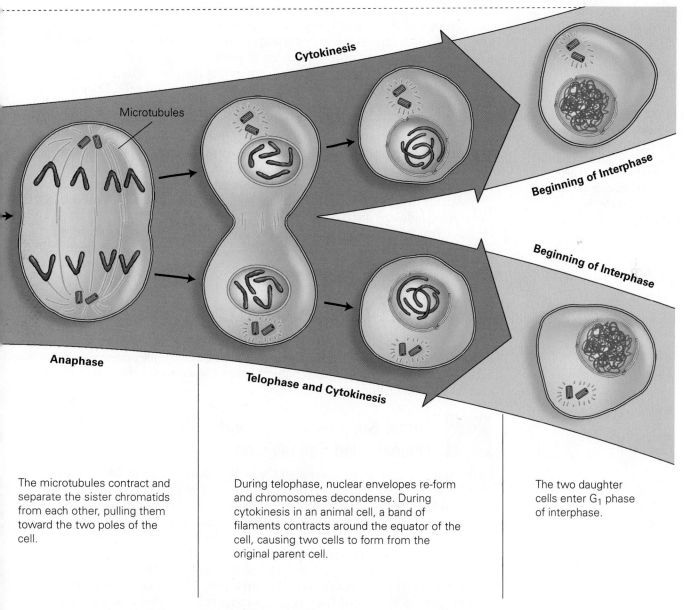

The microtubules contract and separate the sister chromatids from each other, pulling them toward the two poles of the cell.

During telophase, nuclear envelopes re-form and chromosomes decondense. During cytokinesis in an animal cell, a band of filaments contracts around the equator of the cell, causing two cells to form from the original parent cell.

The two daughter cells enter $G_1$ phase of interphase.

**FIGURE 6.9 Cell division in animal cells.** This diagram illustrates how cell division proceeds from interphase through mitosis and cytokinesis.

**(a) Cytokinesis in an animal cell**

Band of microfilaments

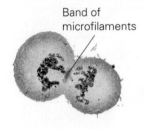

**(b) Cytokinesis in a plant cell**

Cell plate

**FIGURE 6.10 A comparison of cytokinesis in animal and plant cells.** (a) Animal cells produce a band of filaments that tightens like a belt to pinch the cell in half. (b) Plant cells form a cell plate down the middle of the parent cell that gives rise to the cell wall.

structures called **centrioles** physically anchor one end of each forming microtubule. Plant cells do not contain centrioles, but microtubules in these cells do remain anchored at a pole.

During **metaphase,** the replicated chromosomes are aligned across the middle, or equator, of each cell. To do this, the microtubules, which are attached to each chromosome at the centromere, line up the chromosomes in single file across the middle of the cell.

During **anaphase,** the centromere splits, and the microtubules shorten to pull each sister chromatid of a chromosome to opposite poles of the cell.

In the last stage of mitosis, **telophase,** the nuclear envelopes re-form around the newly produced daughter nuclei, and the chromosomes revert to their uncondensed form.

## Cytokinesis

Cytokinesis is the division of the cytoplasm that takes place directly after telophase. During cytokinesis in animal cells, a band of proteins encircles the cell at the equator and divides the cytoplasm. This band of proteins contracts to pinch apart the two nuclei and the surrounding cytoplasm, creating two daughter cells from the original parent cell. Cytokinesis in plant cells requires that cells build a new **cell wall,** an inflexible structure surrounding the plant cells. **FIGURE 6.10** shows the difference between cytokinesis in animal and plant cells. During telophase of mitosis in a plant cell, membrane-bound vesicles deliver the materials required for building the cell wall to the center of the cell. These materials include a tough, fibrous carbohydrate called **cellulose** as well as some proteins. The membranes surrounding the vesicles gather in the center of the cell to form a structure called a **cell plate.** The cell plate and forming cell wall grow across the width of the cell and form a barrier that eventually separates the products of mitosis into two daughter cells. After cytokinesis, the cell reenters interphase, and if the conditions are favorable, the cell can divide again.

Any tissue that undergoes mitotic cell division can give rise to a tumor if a cell begins to divide when it should not. That tumor can increase in size if cell division remains uncontrolled.

# 6.4 Cell Cycle Control

When cell division is working properly, it is a tightly controlled process and tumor formation is prevented. Even if one renegade cell escapes these regulatory controls and forms a tumor, mechanisms are in place to get rid of the tumor. How is such regulation accomplished?

 ## Tumor Suppressors Prevent Uncontrolled Cell Division

Cells do not simply proceed through the cell cycle from start to finish. Instead, proteins are continuously surveying cells to make sure that, among other things, the DNA has been replicated properly. Proteins called **tumor suppressors** inspect newly replicated DNA. If it is damaged in any way, for example, if a G:T base pair exists, the cell does not continue the process of cell division.

The normal *BRCA1* gene encodes a protein that functions as a tumor suppressor. The mutant version of the *BRCA1* gene that Angelina Jolie inherited is not able to carry out this function. Therefore, cells where this mutant gene is expressed (i.e., cells composing breast and ovarian tissues), could divide more than they are supposed to. If this happens, tumors can form (**FIGURE 6.11**).

**Mutations to tumor-suppressor genes**

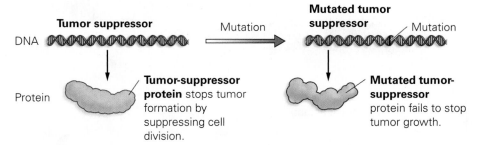

FIGURE 6.11 **Mutations to tumor-suppressor genes.** Mutations to tumor-suppressor genes can increase the likelihood of cancer developing.

## Regulation of the Cell Cycle: A Closer Look ▼

Tumor suppressors are not the only proteins guarding checkpoints. After every major event in the cell cycle, additional proteins determine whether the cell should be allowed to continue through the cell cycle (**FIGURE 6.12**). At the $G_1$ checkpoint, proteins check to determine if the cell has grown enough to be able to subdivide into two daughter cells. After the DNA has undergone replication during the S phase, the success of that replication is assessed at the $G_2$ checkpoint. Late in the M phase, proteins at the third checkpoint double check that each chromosome is present in the proper duplicated configuration and attached to microtubule.

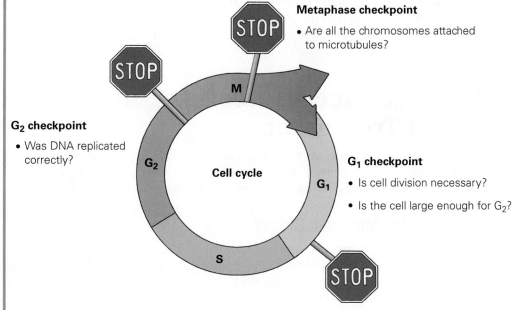

◀ **Visualize This**

At which checkpoint will a cell that has been treated with a chemical that prevents microtubule formation be stopped?

FIGURE 6.12 **Controls of the cell cycle.** Checkpoints at $G_1$, $G_2$, and metaphase determine whether a cell will be allowed to divide.

The proteins that participate in this regulation are coded for cell cycle control genes that we all carry. These cell cycle genes, called **proto-oncogenes** (proto meaning before, and onco meaning cancer) can give rise to cancer-causing **oncogenes** if they are mutated. Proto-oncogenes encode proteins that stimulate cell division when conditions are right. Oncogenes stimulate cell division when they should not (**FIGURE 6.13**, on the next page). Proto-oncogenes and tumor suppressors work together to facilitate the passage of healthy cells, and to impede the passage of damaged cells, through cellular checkpoints.

**proto-** means before.

**onco-** means cancer.

*A Closer Look, continued*

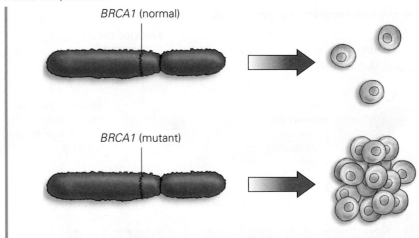

**FIGURE 6.13** The normal version of the *BRCA1* gene leads to controlled cell division. The mutant version of *BRCA1* is unable to perform this job.

**STOP & STRETCH** Our current understanding of how cancers arise is sometimes called the "multiple hit model" of carcinogenesis. Based on your understanding of how cancers arise, describe what the phrase "multiple hit" means.

Most of us will inherit few, if any, mutant cell-cycle control genes. Instead, our level of exposure to environmental risk factors will determine whether enough mutations will accumulate during our lifetime to cause cancer. This is why cancer is more common in the elderly. Older people have tissues that have undergone more cell division and cells that have been exposed to more carcinogens, leading to increased likelihood of mutation to genes controlling the cell cycle.

# 6.5 Cancer Detection and Treatment

When cancers are detected and treated early, their progression can be halted and the odds of survival increase. Being on the lookout for warning signs (**FIGURE 6.14**) can help alert individuals that a cancer is developing.

## Detection Methods: Biopsy

Many cancers are first detected when a lump is discovered either by self-examination or during a medical exam. Once a lump has been discovered, a physician might perform a **biopsy** to surgically remove and analyze some of the cells. If the lump is composed of fluid, not cells, it is most likely a benign cyst. Cysts can arise in response to hormones or inflammation caused by injury. They do not progress to cancers and often go away without any form of treatment.

When a solid mass of cells is found, the cells can be viewed under a microscope. Benign tumors consist of orderly growths of cells that resemble the cells of the tissue from which they were taken. Malignant or cancerous cells do not resemble other cells found in the same tissue; they divide so rapidly that they do not have time to produce all the proteins necessary to build normal cells, which leads to an abnormal appearance that a trained technician can identify.

Since Angelina Jolie chose to have her breasts removed before cancer developed, she did not have to be concerned about whether the cancer had spread and there was no reason to undergo further treatment.

**C** hange in bowel or bladder habits

**A** sore that does not heal

**U** nusual bleeding or discharge

**T** hickening or lump

**I** ndigestion or difficulty swallowing

**O** bvious change in wart or mole

**N** agging cough or hoarseness

**FIGURE 6.14 Warning signs of cancer.** Self-screening for cancer can save your life. If you experience one or more of these warning signs, see your doctor.

## Treatment Methods: Chemotherapy and Radiation

Unfortunately, a biopsy of Lance Armstrong's tumor showed that it was cancerous, and subsequent tests showed that the cancer had spread. The day after his cancer diagnosis, Lance Armstrong had one testicle surgically removed. Because the cancer had spread, he also underwent chemotherapy and radiation.

**Chemotherapy.**   During **chemotherapy**, chemicals that selectively kill dividing cells are injected into the bloodstream. Because cancer is a disease caused by mutations, some cancer cells carry mutations that will allow them to be resistant to various chemotherapeutic agents. Therefore, treating a cancer patient with a combination of chemotherapeutic agents aimed at different cell cycle events increases the chances of destroying all the cancerous cells in a tumor. For example, some chemotherapeutic agents prevent the chromosomes from being pulled to the equator during cell division and others prevent DNA synthesis.

Unfortunately, normal cells that divide rapidly can also be affected by chemotherapy. Hair follicles, cells that produce red and white blood cells, and cells that line the intestines and stomach are often damaged or destroyed. The effects of chemotherapy therefore include temporary hair loss (**FIGURE 6.15**), anemia (dizziness and fatigue due to decreased numbers of red blood cells), and lowered protection from infection due to decreases in the number of white blood cells. In addition, damage to the cells of the stomach and intestines can lead to nausea, vomiting, and diarrhea.

**Radiation Therapy.**   After a tumor is surgically removed, there is always a chance that some cancer cells from the tumor were not excised and remain in the body. Radiation therapy is the use of high-energy particles aimed at the location of the tumor in an effort to kill any cancer cells that might remain. This therapy is typically used only when cancers are located close to the surface of the body because it is difficult to focus a beam of radiation on internal organs and tissue damage can be quite severe.

Because a testicle can be removed in its entirety, radiation therapy is usually not necessary for men with testicular cancer, like Lance Armstrong. However, the cancer that had spread to his brain would normally have been subject to radiation treatment. It is very difficult for a surgeon to remove a brain tumor without affecting surrounding tissues or leaving some cancer cells behind. Because he was still allowed to participate in professional cycling events, and because he feared that radiation to his brain would affect his balance and ability to ride a bike, Lance Armstrong chose not to undergo radiation therapy on the tumors that metastasized to his brain. He hoped that surgical removal and chemotherapy would kill those cancer cells.

**FIGURE 6.15 Chemotherapy.**
Chemotherapy caused Lance Armstrong to lose his hair.

**Of those who live an average lifespan, more than one in three will be diagnosed with cancer.**

**STOP & STRETCH**   One risk of radiation therapy is an increased likelihood of new tumors emerging 5 to 15 years later. Why might this treatment increase cancer risk?

Angelina Jolie carries a gene that increased her risk of cancer. Could she pass that risk on to her children? What about Lance Armstrong? At the time of his diagnosis, he did not have children but now he has five. Are they at increased risk of cancer because he underwent cell-damaging chemotherapy prior to their birth?

An understanding of the type of cell division that produces sperm and eggs will help you answer these questions.

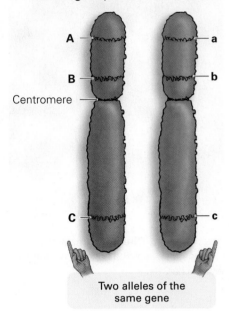

**Homologous** pair of chromosomes

**FIGURE 6.16 A homologous pair of chromosomes.** Homologous pairs of chromosomes have the same genes (shown here as A, B, and C) but may have different alleles. The dominant allele is represented by an uppercase letter, while the recessive allele is shown with the same letter in lowercase. Note that the chromosomes of a homologous pair each have the same size, shape, and positioning of the centromere.

**meio-** means to make smaller.

# 6.6 Meiosis

**Meiosis** is a form of cell division that produces specialized cells called **gametes** that contain half the number of chromosomes of the parent cell. Gametes are produced only within the **gonads**, or sex organs.

In humans, the male gonads are testes, and the female gonads are ovaries. The male gametes are called sperm cells, female are called egg cells. Because human somatic cells have 46 chromosomes and meiosis reduces that number by half, the gametes produced during meiosis contain 23 chromosomes each.

Chromosomes in somatic cells occur in pairs. The 46 chromosomes in human somatic cells are actually 23 different pairs of chromosomes. In somatic cells, one member of each pair was inherited from the mother and one from the father. The members of a **homologous pair** of chromosomes are the same size and shape and carry the same genes, although not necessarily the same versions (**FIGURE 6.16**). Different versions of the same gene are called **alleles** of a gene. We have seen that there are different versions of the *BRCA1* gene. The normal version of the gene helps control the cell cycle, while a mutant version does not.

A **karyotype** is a highly magnified photograph of the chromosomes, arranged in pairs. The 46 human chromosomes can be arranged into 22 pairs of nonsex chromosomes, or **autosomes**, and one pair of **sex chromosomes** (the X and Y chromosomes) to make a total of 23 pairs. Human males have an X and a Y chromosome, while females have two X chromosomes (**FIGURE 6.17**).

Once meiosis is completed, there is one copy of each chromosome (1–23) in every gamete. When only one member of each homologous pair

## Visualize This ▼

**Which sex chromosome is larger and therefore carries more genetic information?**

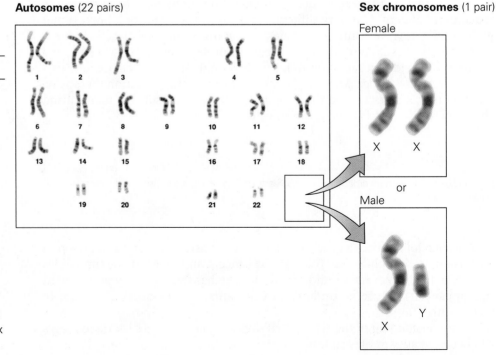

**FIGURE 6.17 Karyotype.** The pairs of chromosomes in this karyotype are arranged in order of decreasing size and numbered from 1 to 22. The X and Y sex chromosomes are the 23rd pair.

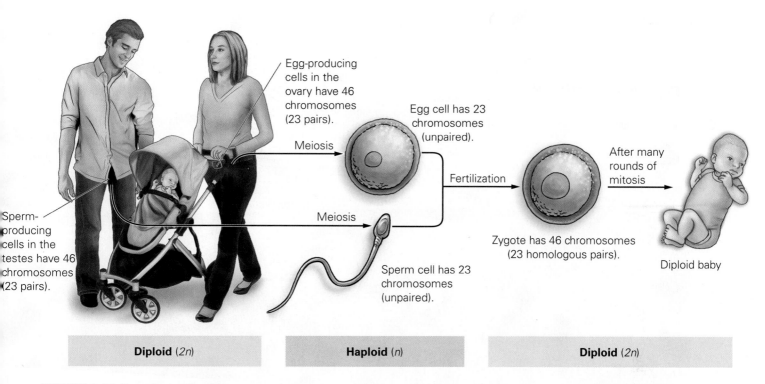

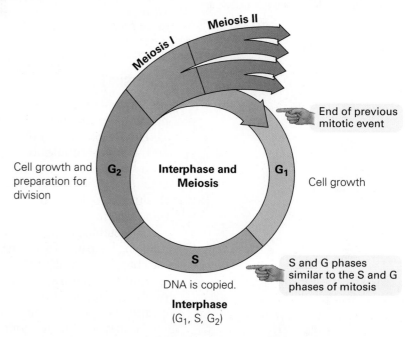

**FIGURE 6.18 Gamete production.**
The diploid cells of the ovaries and testes undergo meiosis and produce haploid gametes. When fertilization occurs, the diploid condition is restored.

is present in a cell, we say that the cell is **haploid (n)**—both egg cells and sperm cells are haploid. After the sperm and egg fuse, the fertilized cell, or **zygote**, will contain two sets of chromosomes and is said to be **diploid (2n)** (**FIGURE 6.18**). Like mitosis, meiosis is preceded by an interphase stage that includes $G_1$, S, and $G_2$. Interphase is followed by two phases of meiosis, meiosis I and meiosis II, in which divisions of the nucleus take place (**FIGURE 6.19**). Meiosis I separates the members of a homologous pair from each other. Meiosis II separates the chromatids from each other. Both meiotic divisions are followed by cytokinesis, during which the cytoplasm is divided between the resulting daughter cells.

## Interphase

The interphase that precedes meiosis consists of $G_1$, S, and $G_2$. This interphase of meiosis is similar in most respects to the interphase that precedes mitosis. The centrioles from which the microtubules will originate are present. The G phases are times of cell growth and preparation for division. The S phase is when DNA replication occurs. Once the cell's DNA has been replicated, it can enter meiosis I.

**FIGURE 6.19 Interphase and meiosis.** Interphase consists of $G_1$, S, and $G_2$ and is followed by two rounds of nuclear division, meiosis I and meiosis II.

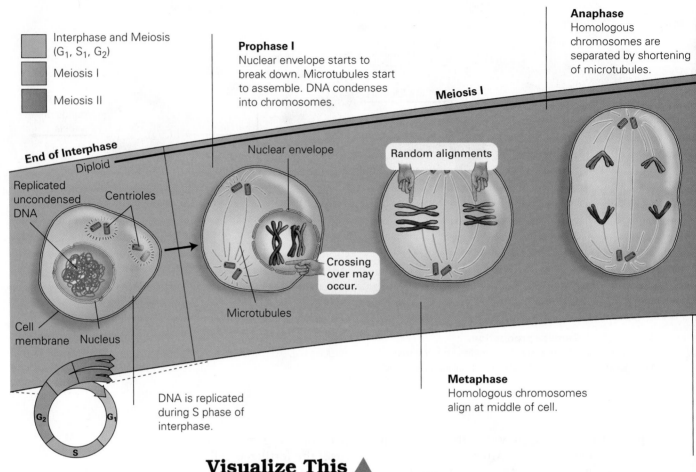

Interphase and Meiosis
($G_1$, $S_1$, $G_2$)

Meiosis I

Meiosis II

**Prophase I**
Nuclear envelope starts to
break down. Microtubules start
to assemble. DNA condenses
into chromosomes.

**Anaphase**
Homologous
chromosomes are
separated by shortening
of microtubules.

Meiosis I

End of Interphase
Diploid

Nuclear envelope

Random alignments

Replicated
uncondensed
DNA

Centrioles

Crossing
over may
occur.

Cell
membrane    Nucleus

Microtubules

$G_2$    $G_1$

S

DNA is replicated
during S phase of
interphase.

**Metaphase**
Homologous chromosomes
align at middle of cell.

FIGURE 6.20 **Meiosis.** This diagram
illustrates interphase, meiosis I, meiosis II,
and cytokinesis in an animal cell.

# Visualize This ▲

**There is a short interphase between meiosis I and meiosis II. Look at the
structure of chromosomes as they leave meiosis I and again at prophase II.
Does DNA duplication occur during the interphase preceding meiosis II?**

## Meiosis I

The first meiotic division, meiosis I, consists of prophase I, metaphase I, anaphase I, and telophase I (**FIGURE 6.20**).

During prophase I of meiosis, the nuclear envelope starts to break down,
and the microtubules begin to assemble. The previously replicated chromosomes condense so that they can be moved around the cell without becoming
entangled. The condensed chromosomes can be seen under a microscope. At
this time, the homologous pairs of chromosomes exchange genetic information in a process called *crossing over,* which will be explained in a moment.

At metaphase I, the chromosomes line up at the equator, but they do so in
homologous pairs. This is the key difference between meiosis and mitosis, where
the chromosomes align single file at the equator. Homologous pairs are arranged
arbitrarily regarding which member faces which pole. This process is called
*random alignment.* At the end of this section, you will find detailed descriptions of
crossing over and random alignment along with their impact on genetic diversity.

At anaphase I, the homologous pairs are separated from each other by
the shortening of the microtubules, and at telophase I, nuclear envelopes re-
form around the chromosomes. DNA is then partitioned into each of the two

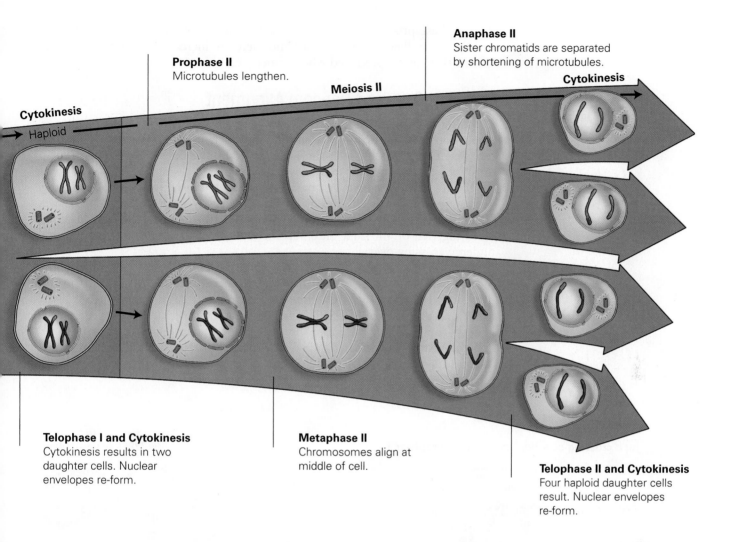

**Cytokinesis**

Haploid

**Prophase II**
Microtubules lengthen.

**Meiosis II**

**Anaphase II**
Sister chromatids are separated
by shortening of microtubules.

**Cytokinesis**

**Telophase I and Cytokinesis**
Cytokinesis results in two
daughter cells. Nuclear
envelopes re-form.

**Metaphase II**
Chromosomes align at
middle of cell.

**Telophase II and Cytokinesis**
Four haploid daughter cells
result. Nuclear envelopes
re-form.

daughter cells by cytokinesis. Because each daughter cell contains only one copy of each member of a homologous pair, at this point the cells are haploid. Now both of these daughter cells are ready to undergo meiosis II.

**STOP & STRETCH**   How is meiosis I similar to mitosis? How is it different?

## Meiosis II

Meiosis II consists of prophase II, metaphase II, anaphase II, and telophase II. This second meiotic division is virtually identical to mitosis and serves to separate the sister chromatids of the replicated chromosome from each other.

At prophase II of meiosis, the cell is readying for another round of division, and the microtubules are lengthening again. At metaphase II, the chromosomes align in single file across the equator in much the same way that they do during mitosis. At anaphase II, the sister chromatids separate from each other and move to opposite poles of the cell. At telophase II, the separated chromosomes each become enclosed in their own nucleus.

Each individual can produce millions of different types of gametes due to two events that occur during meiosis I—crossing over and random alignment.

Both of these processes greatly increase the number of different kinds of gametes that an individual can produce and therefore increase the variation in individuals that can be produced when gametes combine.

## Crossing Over and Random Alignment

**Crossing over** occurs during prophase I of meiosis I. It involves the exchange of portions of chromosomes from one member of a homologous pair to the other member. Crossing over can occur several times on each homologous pair during each occurrence of meiosis.

To illustrate crossing over, consider an example using genes involved in the production of flower color and pollen shape in sweet pea plants. These two genes are on the same chromosome and are called **linked genes.** Linked genes move together on the same chromosome to a gamete, and they may or may not undergo crossing over.

If a pea plant has red flowers and long pollen grains, the chromosomes may appear as shown in **FIGURE 6.21**. It is possible for this plant to produce four different types of gametes with respect to these two genes. Two types of gametes would result if no crossing over occurred between these genes—the gamete containing the red flower and long pollen chromosome and the gamete containing the white flower and short pollen chromosome. Two additional types of gametes could be produced if crossing over did

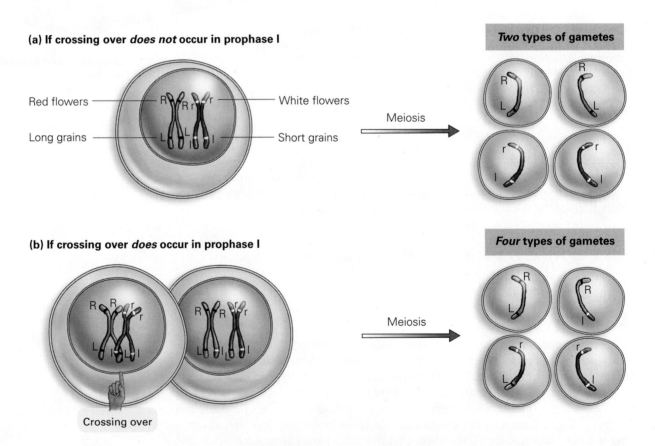

**(a) If crossing over *does not* occur in prophase I**

Red flowers — Long grains

White flowers — Short grains

Meiosis

*Two* types of gametes

**(b) If crossing over *does* occur in prophase I**

Crossing over

Meiosis

*Four* types of gametes

**FIGURE 6.21 Crossing over.** If a flower with the above arrangement of alleles undergoes meiosis, it can produce (a) two different types of gametes for these two genes if crossing over does not occur or (b) four different types of gametes for these two genes if crossing over occurs at L.

occur—one type containing the red flower and short pollen grain chromosome and the other type containing the reciprocal white flower and long pollen grain chromosome. Therefore, crossing over increases genetic diversity by increasing the number of distinct combinations of genes that may be present in a gamete.

**Random alignment** of homologous pairs also increases the number of genetically distinct types of gametes that can be produced. The arrangement of homologous pairs of chromosomes at metaphase I determines which chromosomes will end up together in a gamete. If we consider only two homologous pairs of chromosomes, then two different alignments are possible, and four different gametes can be produced. As the number of chromosomes in an organism's genome increases, so does the number of possible alignments and the number of genetically distinct gametes that organism can produce (**FIGURE 6.22**).

Random alignment does not occur during mitosis, because the chromosomes align single file across the equator in order to produce identical daughter

# Visualize This ▼

**The cell below has two homologous pairs. How many different alignments are possible with three homologous pairs of chromosomes?**

**(a) One possible metaphase I alignment**

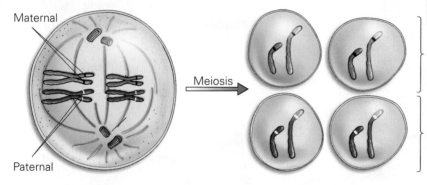

Maternal

Paternal

Meiosis

**Two combinations of chromosomes in gametes**

These gametes contain only chromosomes this organism inherited from its mother.

These gametes contain only chromosomes this organism inherited from its father.

**(b) Another possible metaphase I alignment**

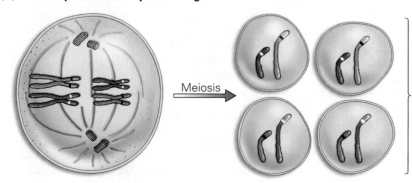

Meiosis

**Two additional combinations of chromosomes in gametes**

These gametes contain one maternal and paternal chromosome.

**FIGURE 6.22 Random alignment.** In this example, the organism undergoing meiosis has only four chromosomes. The organism inherited the blue chromosomes from its father and the red chromosomes from its mother. When there are two homologous pairs of chromosomes, two possible alignments, (a) and (b), can occur. These different alignments can lead to novel combinations of genes in the gametes.

cells. For a summary of the differences between mitosis and meiosis, see **FIGURE 6.23**.

Now that you have a clear understanding of cell division, we can turn our attention to the question of whether people can pass cancer on to their children. In the cases of Angelina Jolie and Lance Armstrong, the answer to that question will differ.

Genetic testing showed that Angelina Jolie inherited the *BRCA1* gene from her mother. Typically, it takes more than one mutation to cause a cancer, but this particular mutation of the *BRCA1* will cause breast or ovarian cancer in the vast majority of women that carry the gene. Since Jolie inherited one mutant version of the *BRCA1* gene, she can produce egg cells carrying the mutant version or the normal version. Therefore, each of her biological children—daughters Shiloh and Vivienne and son Knox—has a 50% chance of carrying the mutant version of this gene. If they did inherit the mutant

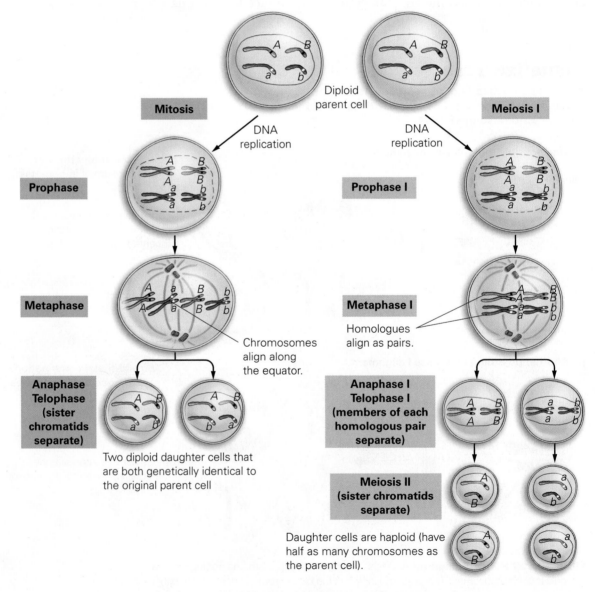

**FIGURE 6.23 Comparing mitosis and meiosis.** Mitosis is a type of cell division that occurs in somatic cells and gives rise to daughter cells that are exact genetic copies of the parent cell. Meiosis occurs in cells that will give rise to gametes and decreases the chromosome number by one-half. Each gamete receives one member of each homologous pair.

version of the gene, Jolie's daughters are at increased risk of breast and ovarian cancer and her son is at increased risk of breast and prostate cancer. Any of her biological children who did not receive the mutant version of the *BRCA1* gene have a cancer risk that is equal to that of their adopted siblings Maddox, Zahara, and Pax, assuming that they didn't inherit the mutant version from their biological parents.

The causes of Lance Armstrong's cancer are not as clear-cut as Jolie's genetic predisposition. His family history of cancer is not nearly as pronounced as hers, so his cancer could have developed in response to some combination of inherited mutations and mutations caused by carcinogens he was exposed to during his lifetime. When it comes to cancer risk in children of those diagnosed with cancer, there is also concern about mutations caused by chemotherapy. When chemotherapeutic agents injure somatic cells, that damage is not transmissible to offspring. However, if the chemotherapy damages cells that undergo meiosis to produce sperm, mutations could be passed to children. Because of this risk, Armstrong stored sperm in a sperm bank before undergoing chemotherapy. Of his five children, three with his first wife were conceived with sperm stored before his surgery and chemotherapy, and two with his current wife were conceived naturally. Because the specific cause of Lance Armstrong's cancer is not known, it is not possible to determine the risk of cancer for any of his five children. Likewise, it is not possible to determine whether the two children born after his chemotherapy are at increased risk since we don't know whether the sperm involved in their conception carried chemotherapy-damaged, cell-cycle, control genes. Therefore, for any children that Armstrong (or any of us) might have, the combined effects of inherited mutant alleles and any mutations induced by environmental exposures in conjunction with lifestyle factors will determine whether and when cancers may develop.

**Prior to his fall from grace Lance Armstrong acquired testicular cancer.**

# *savvy reader*

## Alternative Cancer Treatments

One website claims that 80% of cancer therapies are blocked by the patient's emotions and that these blocks can be removed by a technique called tapping. Just download the file, follow the tapping instructions, and you will feel better in no time. Another website claims that sharks do not get cancer and that taking shark-cartilage supplements can help treat cancer in humans. In a study of the effectiveness of shark cartilage, the study authors report that 15 of 29 patients diagnosed with terminal cancer were still alive one year after beginning to take the supplements, which is "a remarkable result by any measure," according to the study authors.

Websites for alternative cancer treatment centers offer unproven treatments to patients that cost tens of thousands of dollars and are not covered by insurance. According to advertisements, these treatments are often supervised by medical doctors. The sites are filled with testimonials, but no data, about the effectiveness of such treatments.

1. The website that suggested that emotions can block cancer therapies presented no evidence to back up this claim. What would you do if you wanted to know whether that claim had any merit?

2. The study on shark cartilage was published in a non-peer reviewed journal. Does this fact add or subtract from the credibility of this article? Why?

3. What other information do you need to determine whether the one-year survival rate of the shark cartilage study has any real meaning?

4. Would the fact that the author of the shark cartilage study owns a company that sells the product make you more or less skeptical of his findings? What about the fact that sharks actually do get cancer?

5. While most medical doctors have dedicated their lives to helping people in need, a few will promote products and services simply to make money. Should you always believe the word of someone with an advanced degree?

6. Why do you think that so many people fall prey to dubious cancer cures?

# SOUNDS RIGHT BUT IS IT?

A tanning salon located near campus gives a 20% discount to students. Two girls on your dorm floor have been using the tanning beds. When you question the safety of this practice, your friends claim that being exposed to 20 minutes of ultraviolet light from tanning beds is actually safer than being in the sun for a few hours. When you point out that tanning beds can cause skin cancer, they say that most salons have switched to bulbs that have only one kind of ultraviolet light, making them much safer than before. They also claim that using tanning beds makes them look and feel more healthful, in part because using tanning beds helps them to get the recommended amount of vitamin D.

**The use of tanning beds is not only safe, it improves health.**

Sounds right, but it isn't.

1. Even if there were evidence that tanning for 20 minutes was safer than being in the sun for a few hours, would this provide evidence that tanning beds are safe? Why or why not?

2. Ultraviolet light (UV) from the sun occurs in two different DNA wavelengths, longer UVA rays and shorter UVB rays. UVB rays cause the surface of the skin to burn while the longer UVA rays penetrate further into the skin. Early tanning beds used UVB bulbs, but most have now switched to UVA bulbs. From an economic standpoint, why might it benefit tanning bed manufacturers and salon owners to switch to UVA?

3. When light-skinned people are exposed to damaging levels of UV light, their skin temporarily darkens in an attempt to prevent further damage from occurring. (Dark-skinned people have naturally high levels of pigments in the skin, an evolutionary adaptation that occurred when their ancestors were living in a geographic range with high UV light exposure.) A tan in light-skinned people is evidence that the skin has been injured. Should a tan in light-skinned people be considered evidence of good health?

4. While sunlight is required for the body to synthesize vitamin D, the American Cancer Society recommends that no one be exposed to more than 15 minutes of sunlight before applying sunscreen. If you wanted to ensure that you were getting enough vitamin D without exposing yourself to ultraviolet light from any source, what could you do?

5. Consider your answers to questions 1–4 and explain why the original statement bolded above sounds right, but isn't.

# Chapter Review MasteringBiology®

## Summary

### Section 6.1

Describe the cellular basis of cancer.

- Unregulated cell division can lead to the formation of a tumor (p. 108).

Compare and contrast benign and malignant tumors.

- Benign or noncancerous tumors stay in one place and do not prevent surrounding tissues and organs from functioning. Malignant tumors are those that are invasive or those that metastasize to surrounding tissues, starting new cancers (p. 108).

List several risk factors for cancer development that are under your control.

- Apart from genetics and aging, most cancer risk factors are things you can control. These include not smoking, eating a healthy diet, exercising, maintaining a healthy weight, and minimizing alcohol consumption (pp. 108–110).

### Section 6.2

List the normal functions of cell division.

- Cell division is a process cells undergo to produce new cells for growth, repair, and asexual reproduction (pp. 110–111).

Describe the structure and function of chromosomes.

- Chromosomes are composed of DNA wrapped around proteins. They can be uncondensed and string-like or condensed, depending on whether the cell is actively dividing. Chromosomes carry genes (pp. 111–113).

Outline the process of DNA replication.

- During DNA replication, one strand of the double-stranded DNA molecule is used as a template for the synthesis of a new daughter strand of DNA. The newly synthesized DNA strand is complementary to the parent strand. The enzyme DNA polymerase ties together the nucleotides on the forming daughter strand (pp. 111–113).

### Section 6.3

Describe the events that occur during interphase of the cell cycle.

- Interphase consists of two gap phases of the cell cycle ($G_1$ and $G_2$), during which the cell grows and prepares to enter mitosis or meiosis, and the S (synthesis) phase, during which time the DNA replicates. The S phase of interphase occurs between $G_1$ and $G_2$ (pp. 113–114).

Diagram two chromosomes as they proceed through mitosis of the cell cycle.

- During prophase, the replicated chromosomes condense into linear X-shaped chromosomes. At metaphase, these replicated chromosomes align across the middle of the cell. At anaphase, the sister chromatids separate from each other and align at opposite poles of the cells. At telophase, separate nuclear envelopes re-form around the linear chromosomes present at both poles of the cell (pp. 114–116).

Describe the process of cytokinesis in animal and plant cells.

- Cytokinesis is the last phase of the cell cycle. During cytokinesis, the cytoplasm is divided into two portions, one for each daughter cell. In animal cells, this involves the pinching of one cell into two cells by a band of filaments. In plant cells, this involves the construction of a cell wall in the middle of the subdivided plant cell (p. 116).

### Section 6.4

Describe how the cell cycle is regulated and how disregulation can lead to tumor formation.

- When cell division is working properly, it is a tightly controlled process. Normal cells divide only when conditions are favorable. Proteins survey the cell and its environment at checkpoints as the cell moves through the cell cycle, and can halt cell division if conditions are not favorable. Mistakes in regulating the cell cycle arise when genes that control the cell cycle are mutated. Tumor suppressors are normal genes that can encode proteins that stop cell division if conditions are not favorable and can repair damage to the DNA. Proto-oncogenes encode genes to stimulate cell division when conditions are favorable. Oncogenes are mutated versions of these genes that stimulate cell division when they should not (pp. 116–118).

## Section 6.5

Describe why chemotherapy and radiation are used to treat cancer.

- Chemotherapy selectively targets rapidly dividing cells. Radiation kills cells by exposing them to high-energy particles (p. 119).

## Section 6.6

Explain what types of cells undergo meiosis, the end result of this process, and how meiosis increases genetic diversity.

- Meiosis is a type of cell division that occurs in cells that give rise to gametes. Gametes contain half as many chromosomes as somatic cells do. The reduction of chromosome number that occurs during meiosis begins with diploid cells and ends with haploid cells (p. 120).
- Meiosis is preceded by an interphase stage in which the DNA is replicated. During meiosis I, the members of a homologous pair of chromosomes are separated from each other. During meiosis II, the sister chromatids are separated from each other (pp. 120–121).

Diagram four chromosomes from a diploid organism undergoing meiosis.

- Homologues align in pairs during meiosis I and as individual chromosomes at metaphase II (pp. 122–123).

Explain the significance of crossing over and random alignment in terms of genetic diversity.

- Homologous pairs of chromosomes exchange genetic information during crossing over at prophase I of meiosis, thereby increasing the number of genetically distinct gametes that an individual can produce. The alignment of members of a homologous pair at metaphase I is random with regard to which member of a pair faces which pole. This random alignment of homologous chromosomes increases the number of different kinds of gametes an individual can produce (pp. 124–125).

## Roots to Remember

**The following roots of words come mainly from Latin and Greek and will help you decipher terms:**

| | |
|---|---|
| **cyto-** and **-cyte** | relate to cells. Chapter terms: *cytoplasm, cytokinesis* |
| **-kinesis** | means motion. Chapter term: *cytokinesis* |
| **mal-** | means bad or evil. Chapter term: *malignant* |
| **meio-** | means to make smaller. Chapter term: *meiosis* |
| **meta-** | means change or between. Chapter terms: *metastasis, metaphase* |
| **mito-** | means a thread. Chapter term: *mitosis* |
| **onco-** | means cancer. Chapter term: *oncogene* |
| **proto-** | means before. Chapter term: *proto-oncogene* |
| **soma-** | and **-some** mean body. Chapter terms: *somatic, chromosome* |
| **telo-** | means end or completion. Chapter term: *telophase* |

## Learning the Basics

1. List the ways in which mitosis and meiosis differ.

2. What property of cancer cells do chemotherapeutic agents attempt to exploit?

3. A cell that begins mitosis with 46 chromosomes produces daughter cells with _____.
   A. 13 chromosomes; B. 23 chromosomes; C. 26 chromosomes; D. 46 chromosomes.

4. The centromere is a region at which _____.
   A. sister chromatids are attached to each other; B. metaphase chromosomes align; C. the tips of chromosomes are found; D. the nucleus is located.

5. Mitosis _____.
   A. occurs in cells that give rise to gametes; B. produces haploid cells from diploid cells; C. produces daughter cells that are exact genetic copies of the parent cell; D. consists of two separate divisions, mitosis I and mitosis II.

6. At metaphase of mitosis, _____.
   A. the chromosomes are condensed and found at the poles; B. the chromosomes are composed of one sister chromatid; C. cytokinesis begins; D. the chromosomes are composed of two sister chromatids and are lined up along the equator of the cell.

7. Sister chromatids _____.
   A. are two different chromosomes attached to each other; B. are exact copies of one chromosome that are attached to each other; C. arise from the centrioles; D. are broken down by mitosis; E. are chromosomes that carry different genes.

8. DNA polymerase _____.
   A. attaches sister chromatids at the centromere; B. synthesizes daughter DNA molecules from fats and phospholipids; C. is the enzyme that facilitates DNA synthesis; D. causes cancer cells to stop dividing.

**9.** After telophase I of meiosis, each daughter cell is _____.
**A.** diploid, and the chromosomes are composed of one double-stranded DNA molecule; **B.** diploid, and the chromosomes are composed of two sister chromatids; **C.** haploid, and the chromosomes are composed of one double-stranded DNA molecule; **D.** haploid, and the chromosomes are composed of two sister chromatids.

**10.** List two things that happen during meiosis that cause gametes to differ from one another.

**11.** Define the terms *proto-oncogene* and *oncogene*.

**12.** In what ways is the cell cycle similar in plant and animal cells, and in what ways does it differ?

**13.** Describe three ways that cancer cells differ from normal cells.

## Analyzing and Applying the Basics

**1.** If your father obtained a mutation to his skin cell from ultraviolet light exposure during his youth, could he have passed that mutation on to you?

**2.** Will all tumors progress to cancers?

**3.** Why are some cancers treated with radiation therapy while others are treated with chemotherapy?

## Connecting the Science

**1.** Should members of society be forced to pay the medical bills of smokers when the cancer risk from smoking is so evident and publicized? Explain your reasoning.

**2.** Would you want to be tested for the presence of cell-cycle control mutations? How would knowing whether you had some mutated proto-oncogenes be beneficial or harmful?

Answers to **Stop & Stretch**, **Visualize This**, **Working with Data**, **Savvy Reader**, **Sounds Right, But Is It?**, and **Chapter Review Questions** can be found in the **Answers** section at the back of the book

# CHAPTER 7 | Are You Only as Smart as Your Genes?

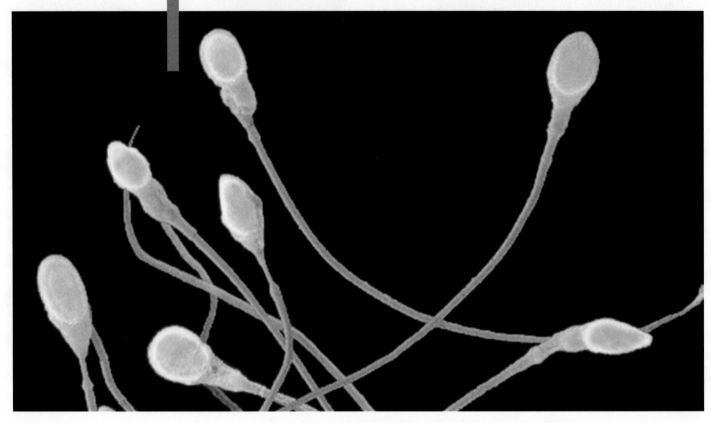

**Can a woman create the perfect child if she chooses the right sperm?**

# Mendelian and Quantitative Genetics

The Fairfax Cryobank is a nondescript brick building located in a quiet, tree-lined suburb of Washington, DC. Stored inside this unremarkable edifice are the hopes and dreams of thousands of women and their partners. The Fairfax Cryobank is a sperm bank; inside its many freezers are vials containing sperm collected from hundreds of men. Women can order these sperm for a procedure called artificial insemination, which may allow them to conceive a child despite the lack of a fertile male partner.

Women who purchase sperm from the Fairfax Cryobank can choose from hundreds of potential donors. The donors are categorized into three classes, and their sperm is priced accordingly. Most women who choose artificial insemination want detailed information about the donor before they purchase a

**If she chooses a donor with the right genes, will her child be a genius?**

sample; while all Fairfax Cryobank donors submit to comprehensive physical exams, disease testing, and provide a detailed family health history, not all provide childhood pictures, audio CDs of their voices, or personal essays. Sperm samples from men who did not provide this additional information are sold at a discount because most women seek a donor who seems compatible in interests and aptitudes.

However, in addition to the information-rich donors, there is also a set of premium sperm donors referred to by Fairfax Cryobank as its "Graduate" category. These men either are in the process of earning or have completed a post-graduate degree in medicine, law, or academia. A sperm sample from this donor category can be 30% more expensive than sperm from the standard donor. And because several samples are typically

needed, ensuring pregnancy from one of these donors is likely to cost hundreds of additional dollars.

Why would some women be willing to pay significantly more for sperm from a donor who has an advanced degree? Because academic achievement is associated with intelligence. These women want intelligent children, and they are willing to pay more to provide their offspring with "extra-smart" genes. But are these women putting their money in the right place? Is intelligence about genes, or is it a function of the environmental conditions in which a baby is raised? In other words, is who we are a result of our "nature" or our "nurture"? As you read this chapter, you will see that the answer to this question is not a simple one—our characteristics come from both our biological inheritance and the environment in which we developed.

**If a woman chooses a donor with the right genes, her child may look like her partner.**

**Or is a child's intelligence more influenced by her environment?**

**Why would some women be willing to pay significantly more for sperm from a donor who has an advanced degree?**

# 7.1 The Inheritance of Traits

Most of us recognize similarities between our birth parents and ourselves. Family members also display resemblances—for instance, all the children of a single set of parents may have dimples. However, it is usually quite easy to tell siblings apart. Each child of a set of parents is unique, and none of us is simply the "average" of our parents' traits. We are each more of a combination—one child may be similar to her mother in eye color and face shape, another similar to mom in height and hair color.

To understand how your parents' traits were passed to you and your siblings, you need to understand the human life cycle. A **life cycle** is a description of the growth and reproduction of an individual (**FIGURE 7.1**).

A human baby is typically produced from the fusion of a single sperm cell produced by the male parent and a single egg cell produced by the female parent. Egg and sperm (called *gametes*) fuse at **fertilization,** and the resulting cell, called a zygote, duplicates all of the genetic information it contains and undergoes mitosis to produce two identical daughter cells. Each of these daughter cells divides dozens of times in the same way. The cells in this resulting mass then differentiate into specialized cell types, which continue to divide and organize to produce the various structures of a developing human, called an embryo. Continued division of this single cell and its progeny leads to the production of a full-term infant and eventually an adult.

We are made up of trillions of individual cells, all of them the descendants of that first product of fertilization and nearly all containing exactly the same information originally found in the zygote. All of our traits are influenced by the information contained in that tiny cell.

## Genes and Chromosomes

Each normal sperm and egg contains information about "how to build an organism." A large portion of that information is in the form of genes, segments of DNA that generally code for proteins.

**FIGURE 7.1 Instructions inside.** Both parents contribute genetic information to their offspring via their sperm and egg. The single cell that results from the fusion of one sperm with one egg contains all of the instructions necessary to produce an adult human.

## Visualize This ▼

**At what stage would exposure to a mutagen that damages DNA have the greatest effect on an individual? At what stage would exposure have the least effect?**

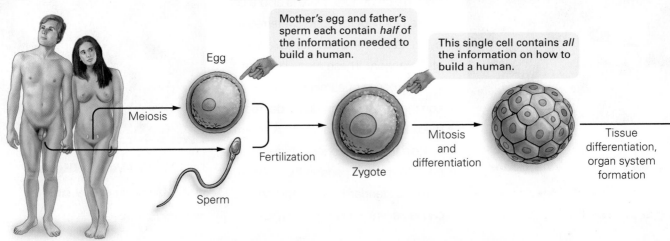

Mother's egg and father's sperm each contain *half* of the information needed to build a human.

This single cell contains *all* the information on how to build a human.

Egg

Meiosis

Sperm

Fertilization

Zygote

Mitosis and differentiation

Tissue differentiation, organ system formation

Adult  Gametes  Single-celled embryo  Multicellular embryo

Imagine genes as roughly equivalent to the words used in an instruction manual. These words are contained on chromosomes, which are roughly analogous to pages in the manual.

Prokaryotes such as bacteria typically contain a single, circular chromosome that floats freely inside the cell and is passed in its entirety to each offspring. In contrast, eukaryotes carry their genes on more than one linear chromosome. The number of chromosomes in eukaryotes can vary greatly, from 2 in the jumper ant (*Myrmecia pilosula*) to an incredible 1260 in the stalked adder's tongue, a species of fern (*Ophioglossum reticulatum*, **FIGURE 7.2**).

Human cells contain 46 chromosomes, most of which carry thousands of genes. Thus, each cell has 46 pages of instructions, with each page containing thousands of words.

The instruction manual inside a cell is different from the instruction manual that comes with, for instance, a kit for building a model car. You would read the manual for building the car beginning at page 1 and follow an orderly set of steps to produce the final product. The human instruction manual is much more complicated—the pages and words to be read are different for different types of cell and may even change according to the situation. The final product of any given cell depends on the words used and the order in which the words are read from this common instruction manual.

For instance, eye cells and pancreas cells in mammals both carry instructions for the protein rhodopsin, which helps detect light, but rhodopsin is produced only in eye cells, not in pancreas cells. Rhodopsin requires assistance from another protein, called transducin, to translate the light that strikes it into the actions of the eye cell. Transducin is also produced in cells in the tongue, but there it functions in translating the binding of certain molecules from food into the sensation of bitter flavor.

Thus, a protein may serve two or more different functions depending on its context. Because genes, like words, can be used in many combinations, the instruction manual for building a living organism is very flexible (**FIGURE 7.3**, on the next page).

## Producing Diversity in Offspring

During reproduction, genes from both parents are copied and transmitted to the next generation. The copying and transmittal of genes from one generation to the next introduces variation among genes. It is this **genetic variation** that most interests individuals seeking sperm donors; the genes in the sperm they select will provide information about the traits of their offspring.

**(a) Stalked adder's tongue (a fern)**

**(b) Single fern cell**

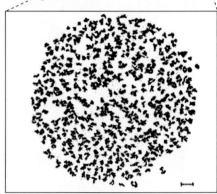

**FIGURE 7.2 Variation in chromosome number.** The amount of genetic information in an organism does not correlate to its complexity, as can be seen by examination of the 1260 chromosomes contained in a single cell of the stalked adder's tongue fern.

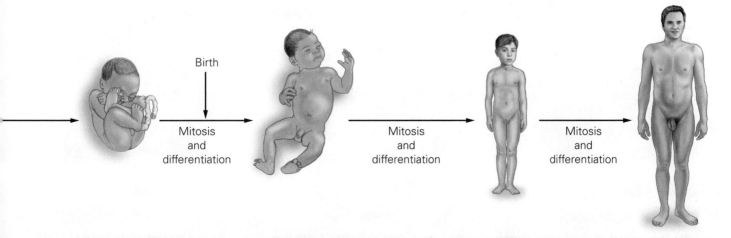

Birth

Mitosis and differentiation

Mitosis and differentiation

Mitosis and differentiation

Fetus            Baby            Child            Adult

## Visualize This ▶

Write at least two more instructions that can be extracted from this group of 14 words.

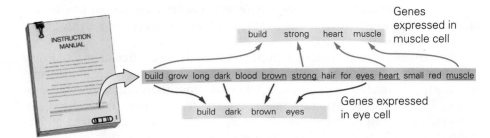

**FIGURE 7.3 Genes as words in an instruction manual.**   Different words from the manual are used in different parts of the body, and identical words may be used in distinctive combinations in different cells.

**Gene Mutation Creates Genetic Diversity.**   Recall the process of DNA replication. Like a retyped instruction manual page, copies of chromosomes are rewritten rather than "photocopied." As a result, there is a chance of a typographical error, or mutation, every time a cell divides. Mutations in genes lead to different versions, or **alleles**, of the gene. The various types of mutation are described in **FIGURE 7.4**.

As you can see from this figure, many mutations result in nonsensical instructions, that is, dysfunctional alleles, and thus are often harmful to the individual who possesses them. Dysfunctional alleles tend to be lost over time because individuals with them do not function as well as those without the mutation. In short, individuals with dysfunctional mutations may not survive or may reproduce at very low rates.

However, some mutations are neutral in effect, or even beneficial in certain situations, while some harmful mutations can be hidden if the individuals who carry them also carry a functional allele. These mutations tend to persist over generations. Because mutations occur at random and are not expected to occur in the same genes in different individuals, each of us should have a unique set of alleles reflecting the mutations passed along to us by our unique set of ancestors.

By contributing to differences among families over many generations, mutation creates genetic variation in a population in the form of new alleles. When

**FIGURE 7.4 The formation of different alleles.**   Different alleles for a gene may form as a result of copying errors. In this analogy, misspellings (mutations) do not change the meaning of the word (allele), but some may result in altered meanings (different allele function) or have no meaning at all (no allele function).

## Visualize This ▼

Which type of mutation do you think is most likely, and why?

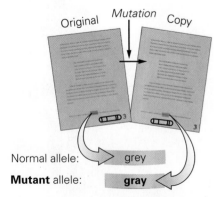

**(a) The mutant allele has the same meaning** (mutant allele function the same as the original allele).

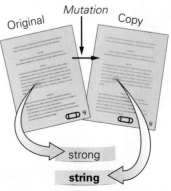

**(b) The mutant allele has a different meaning** (mutant allele functions differently than the original allele).

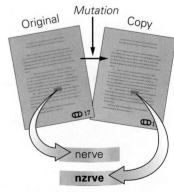

**(c) The mutant allele has no meaning** (mutant allele is no longer functional).

a novel characteristic increases an individual's chance of survival and reproduction, the mutation contributes to a population's adaptation to its environment (Chapter 11). Genetic misspellings are thus the engine that drives evolution itself.

> **STOP** & **STRETCH**   Not all of the DNA in a cell codes for proteins—some DNA function as binding sites for gene promoters (proteins that "turn on" genes), other segments may provide structural support for chromosomes, and other parts may be meaningless. What do you think the effect of "misspellings" is on each of these nongene segments of DNA?

**Segregation and Independent Assortment Create Gamete Diversity.** Both parents contribute genetic instructions to each child, but they do not contribute their entire manual. If they did, the genetic instructions carried in human cells would double every generation, making for a pretty crowded cell. Instead, the process of meiosis reduces the number of chromosomes carried in gametes by one-half (Chapter 6).

Although they are only transmitting half of their genetic information in a gamete, each parent actually gives a complete copy of the instruction manual to each child (**FIGURE 7.5**).

This can occur because, in effect, our body cells each contain two copies of the manual—that is, each has two versions of each page, with each version containing essentially the same words. More technically speaking, the 46 chromosomes each cell contains are actually 23 pairs of chromosomes, with each member of a pair containing essentially the same genes. Each set of two equivalent chromosomes is referred to as a **homologous pair.** The members of a homologous pair are equivalent, but not identical, because even though both have the same genes, each contains a unique set of alleles inherited from one or the other parent.

The process of meiosis separates homologous pairs of chromosomes and also places chromosomes independently into each gamete. These two events explain why siblings are not identical (with the exception of identical twins). It is because parents do not give all of their offspring exactly the same set of alleles.

When homologous pairs are separated during meiosis, the alleles carried on the members of the pair are separated as well. The separation of pairs of alleles during the production of gametes is called **segregation.** Thus, a parent with two different alleles of a gene will produce gametes with a 50% probability of containing one version of the allele and a 50% probability of containing the other version.

**These women want intelligent children, and they are willing to pay more to provide their offspring with "extra-smart" genes.**

---

**homo-** means the same.

---

The 23 pages of each instruction manual are roughly equivalent to the 23 chromosomes in each egg and sperm.

Egg | Sperm | Zygote

The zygote has 46 pages, equivalent to 46 chromosomes.

**FIGURE 7.5 Equivalent information from parents.**  Each parent provides a complete set of instructions to each offspring.

## Visualize This ▼

This man could produce four distinctly different kinds of sperm cell when considering these two genes and four alleles. List the other two possibilities not pictured here.

Parent cells have 2 copies of each chromosome — that is, 2 full sets of instruction manual pages, 1 from each parent.

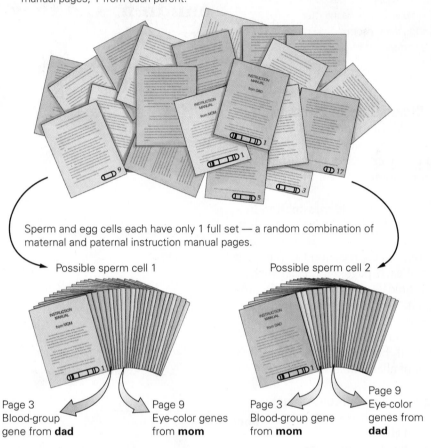

Sperm and egg cells each have only 1 full set — a random combination of maternal and paternal instruction manual pages.

Possible sperm cell 1

Possible sperm cell 2

Page 3
Blood-group
gene from **dad**

Page 9
Eye-color genes
from **mom**

Page 3
Blood-group gene
from **mom**

Page 9
Eye-color
genes from
**dad**

**FIGURE 7.6 Each egg and sperm is unique.** Because each sperm is produced independently, the set of chromosomes in each sperm nucleus will be a unique combination of the chromosomes that the man inherited from his mom and dad.

The segregation of chromosomes during meiosis leads to **independent assortment,** in which the alleles for each gene are inherited (mostly) independently of each other. Independent assortment arises from the *random alignment* of chromosomes during meiosis, which is the uncoordinated "lining up" of chromosome pairs before the first division of meiosis. As a result of random alignment, each homologous chromosome pair is segregated into daughter cells independently of all the other pairs during the production of gametes. Therefore, genes that are on different chromosomes are inherited independently of each other.

Due to independent assortment, the instruction manual contained in a single sperm cell is made up of a unique combination of pages from the manuals a man received from each of his parents. In fact, almost every sperm he makes will contain a unique subset of chromosomes—and thus a unique subset of his alleles. **FIGURE 7.6** illustrates this. In the figure, you can see that independent assortment causes an allele for an eye-color gene to end up in a sperm cell independently from an allele for the blood-group gene.

The independent assortment of segregated chromosomes into daughter cells is repeated every time a sperm is produced, and thus the set of alleles that a child receives from a father is different for all of his offspring. The sperm that contributed half of your genetic information might have carried an eye-color allele from your father's mom and a blood-group allele from his dad, while the sperm that produced your sister might have contained both the allele for eye color and the allele for blood group from your paternal grandmother. As a result of independent assortment, only about 50% of an individual's alleles are identical to those found in another offspring of the same parents—that is, for each gene, you have a 50% chance of being like your sister or brother.

**STOP & STRETCH** In Figure 7.6, you can see that a man can produce four different combinations of alleles in his sperm when considering two genes on two different chromosomes. How many different allele combinations within gametes can be produced when considering three genes on three different chromosomes?

**Random Fertilization Results in a Large Variety of Potential Offspring.** As a result of the independent assortment of 23 pairs of chromosomes, each individual human can make at least 8 million different types of either egg or sperm. Consider now that each of your parents was able to produce such an enormous diversity of gametes. Further, any sperm produced by your father had an equal chance (in theory) of fertilizing any egg produced by your mother.

In other words, gametes combine without regard to the alleles they carry, a process known as **random fertilization.** Hence, the odds of your receiving your particular combination of chromosomes are 1 in 8 million times 1 in 8 million— or 1 in 64 trillion. Remarkably, your parents together could have made more than 64 trillion genetically different children, and you are only one of the possibilities.

Mutation creates new alleles, and independent assortment and random fertilization result in unique combinations of alleles in every generation. These processes help to produce the diversity of human beings.

### A Special Case—Identical Twins.
Although the process of sexual reproduction can produce two siblings who are very different from each other, an event that may occur after fertilization can result in the birth of children who share 100% of their genes.

Identical twins are referred to as **monozygotic twins** because they develop from one zygote—the product of a single egg and sperm. Recall that after fertilization the zygote grows and divides, producing an embryo made up of many daughter cells containing the same genetic information. Monozygotic twinning occurs when cells in an embryo separate from each other. If this happens early in development, each cell or clump of cells can develop into a complete individual, yielding twins, or in very rare cases, triplets or quadruplets who carry identical genetic information (**FIGURE 7.7a**).

**mono-** means one.

In contrast to identical twins, nonidentical twins (also called fraternal twins) occur when two separate eggs fuse with different sperm. These twins are called **dizygotic,** and although they develop simultaneously, they are genetically no more similar than siblings born at different times (**FIGURE 7.7b**).

**di-** means two.

◀ **Visualize This**

Conjoined twins form when the two individual twins share some body parts. Which type of twin formation is more likely to result in conjoined twins?

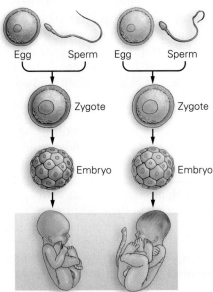

**(a) Monozygotic (identical) twins**

Egg    Sperm

Zygote

Embryo

Embryo splits

Two embryos

100% genetically identical

**(b) Dizygotic (fraternal) twins**

Egg    Sperm    Egg    Sperm

Zygote    Zygote

Embryo    Embryo

50% identical (no more similar than siblings born at different times)

**FIGURE 7.7 The formation of twins.**   (a) Monozygotic twins form from one fertilization event and thus are genetically identical. (b) Dizygotic twins form from two independent fertilizations, resulting in two embryos who are only as genetically similar as any other siblings.

In humans, about 1 in every 80 pregnancies produces dizygotic twins, while only approximately 1 of every 285 pregnancies results in identical twins.

Identical twins provide a unique opportunity to study the relative effects of our genes and environment in determining who we are. Because identical twins carry the same genetic information, researchers are able to study how important genes are in determining health, tastes, intelligence, and personality.

We will examine the results of some twin studies later in the chapter, as we continue to explore the question of predicting the heredity of the complex genetic traits possessed by a particular sperm donor. Our review of the relationship among parents, offspring, and genetic material has now prepared us to examine the inheritance of traits controlled by a single gene.

## 7.2 Mendelian Genetics: When the Role of Genes Is Clear

A few human genetic traits have easily identifiable patterns of inheritance. These traits are said to be "Mendelian" because Gregor Johann Mendel (**FIGURE 7.8**) was the first person to accurately describe their inheritance.

Mendel was born in Austria in 1822. Because his family was poor and could not afford private schooling, he entered a monastery to obtain an education. After completing his monastic studies, Mendel attended the University of Vienna. There he studied math and botany in addition to other sciences. Mendel attempted to become an accredited teacher but was unable to pass the examinations. After leaving the university, he returned to the monastery and began his experimental studies of inheritance in garden peas.

Mendel studied close to 30,000 pea plants over a 10-year period. His careful experiments consisted of controlled matings between plants with different traits. Mendel was able to control the types of mating that occurred by hand-pollinating the peas' flowers—that is, by taking pollen, which produces sperm, from the anthers of one pea plant and applying it to the stigma of the carpel (the egg-containing structure) of another pea plant. By growing the seeds that resulted from these controlled matings, he could evaluate the role of each parent in producing the traits of the offspring (**FIGURE 7.9**).

Although Mendel did not understand the chemical nature of genes, he was able to determine how traits were inherited by carefully analyzing the appearance of parent pea plants and their offspring. His patient, scientifically sound experiments demonstrated that both parents contribute equal amounts of genetic information to their offspring.

Mendel published the results of his studies in 1865, but his scientific contemporaries did not fully appreciate the significance of his work. Mendel eventually gave up his genetic studies and focused his attention on running the monastery until his death in 1884. His work was independently rediscovered by three scientists in 1900; only then did its significance to the new science of genetics become apparent.

The pattern of inheritance Mendel described occurs primarily in traits that are the result of a single gene with a few distinct alleles. **TABLE 7.1** lists some of the traits Mendel examined in peas; we will examine the principles he discovered, such as dominance and recessiveness, by looking at human disease genes that are of interest to prospective parents.

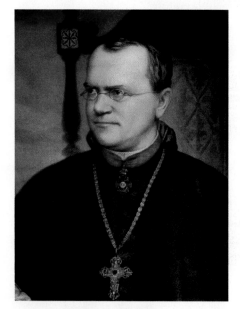

**FIGURE 7.8 Gregor Mendel.** The father of the science of genetics.

### Genotype and Phenotype

pheno- comes from a verb meaning "to show."

We call the genetic composition of an individual his **genotype** and his physical traits his **phenotype**. The genotype is a description of the alleles

**TABLE 7.1 Pea traits studied by Mendel.**

| Character Studied | Dominant Trait | Recessive Trait |
|---|---|---|
| Seed shape | Round | Wrinkled |
| Seed color | Yellow | Green |
| Flower color | Purple | White |
| Stem length | Tall | Dwarf |

## Visualize This

Each pea develops as a result of the fertilization of a single egg. Are the peas in a single pod genetically identical?

**1** A pea flower normally self-pollinates.

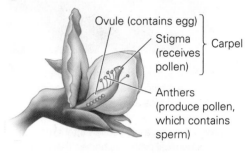

Ovule (contains egg)
Stigma (receives pollen)
Carpel
Anthers (produce pollen, which contains sperm)

**2** Pollen-containing structures can be removed to prevent self-fertilization.

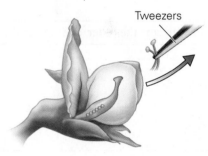

Tweezers

**3** Pollen from another flower is dabbed on to stigma.

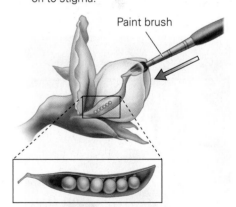

Paint brush

The resulting seeds will contain information on flower color, seed shape and color, and plant height from both parents.

**FIGURE 7.9 Peas and genes.** Pea plants were an ideal study organism for Mendel because their reproduction is easy to control, they complete their life cycle in a matter of weeks, and a single plant can produce thousands of offspring.

for a particular gene carried on each member of a homologous pair of chromosomes (see Chapter 6, Figure 6.17). An individual who carries two different alleles for a gene has a **heterozygous** genotype. An individual who carries two copies of the same allele has a **homozygous** genotype.

The effect of an individual's genotype on her phenotype depends on the nature of the alleles she carries. Some alleles are **recessive,** meaning that their effects can be seen only if a copy of a dominant allele (described below) is not also present. For example, in pea plants the allele that codes for wrinkled seeds is recessive to the allele for round seeds. Wrinkled seeds will only appear when seeds carry only the wrinkled allele and no copies of the round allele.

Often a recessive allele is one that codes for a nonfunctional protein. Homozygotes having two copies of such an allele produce no functional protein. In contrast, heterozygotes carrying one copy of the functional allele have normal phenotypes because the normal protein is still produced. The functional allele in the case of the pea plant with round seeds prevents water from accumulating in the seed. When seeds containing one or two copies of this allele dry, they look much the same as when they first matured. However, wrinkled seeds result from two recessive, nonfunctional alleles (that is, the absence of a functional allele). In this case, water flows into the seed, inflating it and causing its coat to increase in size. When this seed dries, it deflates and wrinkles, much like the surface of a balloon becomes wrinkly after it is blown up once and then deflated.

**Dominant** alleles are so named because their effects are seen even when a recessive allele is present. In the wrinkled seed example, the dominant allele is the one that produces a functional protein. The dominant allele does not always code for the "normal" condition of an organism, however. Sometimes mutations can create abnormal dominant alleles that essentially mask the effects of the recessive, normal allele. For example, "American Albino" horses—known for

**FIGURE 7.10 A dominant allele.**
American Albino horses occur when an individual carries an allele that prevents normal coat color development. Because the product of this allele actively interferes with a biochemical pathway, horses that have only one copy of the allele are albino. (Photo by Linda Gordon.)

**hetero-** means the other, another, or different.

**-zygous** derives from **zygote**, the "yoked" cell resulting from the union of an egg and sperm.

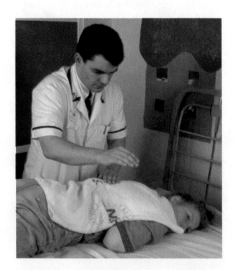

**FIGURE 7.11 Treatment for cystic fibrosis.** Percussive therapy consists of pounding the back of a patient with CF to loosen mucus inside the lungs. The mucus is then coughed up, reducing opportunities for bacterial infections to take hold.

their snow-white coats, pink skin, and dark eyes—result from an allele that stops a horse's hair-color genes from being expressed during the horse's development. Because the allele prevents normal coat color development, it has its effect even if the animal carries only one copy—in other words, albinism in horses is dominant to the normal coat color (**FIGURE 7.10**). Dominant conditions in humans include cheek dimples and, surprisingly, the production of six fingers and toes.

Mendel worked with traits in peas that expressed only simple dominance and recessive relationships. However, for some genes, more than one dominant allele may be produced, and for others, a dominant allele may have different effects in a **hetero**zygote than a homozygote. These situations are referred to as *codominance* and *incomplete dominance*, respectively (and are discussed in Chapter 8). For now, like Mendel, we will focus only on simple dominant and recessive traits.

## Genetic Diseases in Humans

Most alleles in humans do not cause disease or dysfunction; they are simply alternative versions of genes. The diversity of alleles in the human population contributes to diversity among us in our appearance, physiology, and behaviors.

While women who use sperm banks are likely interested in a wide variety of traits in donors, sperm banks are primarily concerned with providing sperm that do not carry a risk of common genetic diseases. Some genetic diseases are produced by recessive alleles, while others are the result of dominant alleles.

**Cystic Fibrosis Is a Recessive Condition.** Individuals with cystic fibrosis (CF) cannot transport chloride ions into and out of cells lining the lungs, intestines, and other organs. As a result of this dysfunction, the balance between sodium and chloride in the cell is disrupted, and the cell produces a thick, sticky mucus layer instead of the thin, slick mucus produced by cells with the normal allele. Affected individuals suffer from progressive deterioration of their lungs and have difficulties absorbing nutrients across the lining of their intestines. Most children born with CF suffer from recurrent lung infections and have dramatically shortened life spans (**FIGURE 7.11**).

Cystic fibrosis results from the production of a mutant chloride ion transporter protein that cannot embed in a cell membrane. The CF allele is recessive because individuals who carry only one copy of the normal allele can still produce the functional chloride transporter protein. The disease affects only homo**zygous** individuals with two mutant alleles and thus no functional proteins.

Cystic fibrosis is among the most common genetic diseases in European populations; nearly 1 in 2500 individuals in these populations is affected with the disease, and 1 in 25 is heterozygous for the allele. Heterozygotes for a recessive disease are called **carriers** because even though they are unaffected, these individuals can pass the trait to the next generation. Sperm banks can test donor sperm for several recessive disorders, including CF; any men who are carriers of CF are excluded from most donor programs, including Fairfax Cryobank. Other relatively well-known recessive conditions in humans include Tay-Sachs disease, a fatal condition that causes the relentless deterioration of mental and physical abilities of affected infants as a result of the accumulation of wastes in the brain, and spinal muscular atrophy (SMA), a condition that causes death of the nerves that control muscles, rendering affected individuals unable to walk and eventually, breathe. The Fairfax Cryobank tests for both Tay-Sachs and SMA in donors who are likely carriers.

**Huntington's Disease Is Caused by a Dominant Allele.** Early symptoms of Huntington's disease include restlessness, irritability, and difficulty in walking, thinking, and remembering. These symptoms typically begin to manifest in middle age. Huntington's disease is progressive and incurable—the nervous, mental, and muscular symptoms gradually become worse and eventually result in the death of the affected individual. Huntington's disease is an example of a fatal genetic condition caused by a dominant allele.

The Huntington's allele causes production of a protein that forms clumps inside the nuclei of cells. Nerve cells in areas of the brain that control movement are especially likely to contain these protein clumps, and these cells gradually die off over the course of the disease (**FIGURE 7.12**). Because this dysfunctional allele produces a mutant protein that damages cells, the presence of the normal allele cannot compensate or correct for this mutant version. An individual needs only one copy of the Huntington's allele to be affected by the disease; that is, even heterozygotes exhibit the symptoms of Huntington's.

**STOP & STRETCH**   In nearly all cases, dominant mutations that result in a fatal condition cannot be passed on from one generation to the next. What characteristic of Huntington's disease allows this allele to persist in the human population?

Only since the mid-1980s has genetic testing allowed people with a family history of Huntington's disease to learn whether they are affected before they show signs of the disease. Although most sperm banks do not test for the presence of the Huntington's allele because it is a rare condition, the detailed family medical histories required of sperm bank donors enable Fairfax Cryobank to exclude men with a family history of Huntington's disease from their donor list.

## Using Punnett Squares to Predict Offspring Genotypes

Traits such as cystic fibrosis and Huntington's disease are the result of a mutation in a single gene, and the inheritance of these conditions and of other single-gene traits is relatively easy to understand. We can predict the likelihood of inheritance of small numbers of these single-gene traits by using a tool developed by Reginald Punnett, a British geneticist. A **Punnett square** is a table that lists the different kinds of sperm or eggs parents can produce relative to the gene or genes in question and then predicts the possible outcomes of a **cross**, or mating, between these parents (**FIGURE 7.13**).

# Visualize This ▼

**In some mutations, a sperm carrying the mutant allele is less mobile than one carrying the normal allele. How would this affect the likelihood of having a child who carried two mutant alleles?**

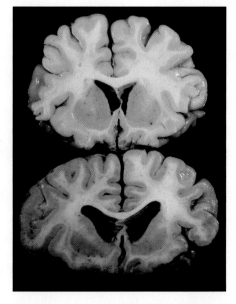

**FIGURE 7.12 Huntington's and the brain.**   Cells containing mutant proteins die, shrinking and deforming the brain. This image compares the brain of a Huntington's sufferer (top) with an unaffected individual (bottom).

**FIGURE 7.13 A cross between heterozygote individuals.**   What are the risks associated with sperm donated by a man who carries one copy of the cystic fibrosis allele? This Punnett square illustrates the likelihood that a woman who carries the cystic fibrosis allele would have a child with cystic fibrosis if the sperm donor were also a carrier.

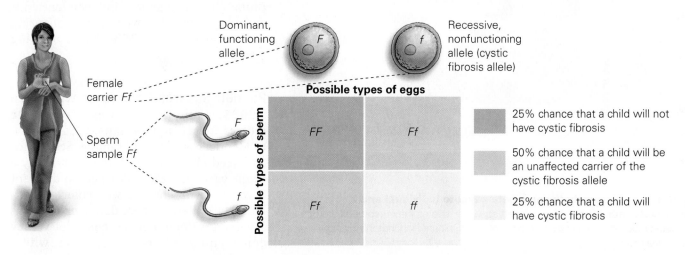

**Using a Punnett Square with a Single Gene.** Imagine a woman and a sperm donor who are both carriers of the CF allele. Different alleles for a gene are symbolized with letters or number codes that refer to a trait that the gene affects. For instance, the CF gene is symbolized *CFTR* for *cystic fibrosis transmembrane regulator*. The dysfunctional *CFTR* allele is called *CFTR-DF508*, so both carriers would have the genotype *CFTR/CFTR-DF508*. However, to make this easier to follow, we will use a simpler key: the letters *F* and *f*, representing the dominant functional allele and recessive nonfunctional allele, respectively. Thus, a carrier for CF has the genotype *Ff*. A genetic cross between two carriers could then be symbolized as follows:

$$Ff \times Ff$$

We know that the female in this cross can produce eggs that carry either the *F* or *f* allele since the process of meiosis will segregate the two alleles from each other. We place these two egg types across the horizontal axis of the Punnett square (Figure 7.13). The male can also make two types of sperm, containing either the *F* or the *f* allele. We place these along the vertical axis. Thus, the letters on the horizontal and vertical axes represent all the possible types of eggs and sperm that the mother and father can produce by meiosis if we consider only the gene that codes for the chloride transport protein.

Inside the Punnett square are all the genotypes that can be produced from a cross between these two heterozygous individuals. The content of each box is determined by combining the alleles from the egg column and the sperm row.

Note that for a cross involving a single gene with two different alleles, there are three possible offspring types. The chance of this cross producing a child affected with CF is one in four, or 25%, because the *ff* combination of alleles occurs once out of the four possible outcomes. You can see why sperm banks would exclude carriers of CF from their donor rolls: The risk of an affected child born to a woman who might not know she is a carrier is too high. The *FF* genotype is also represented once out of four times, meaning that the probability of a homozygous unaffected child is also 25%. The probability of producing a child who is a **carrier** of cystic fibrosis is one in two, or 50%, since two of the possible outcomes inside the Punnett square are unaffected heterozygotes—one produced by an *F* sperm and an *f* egg and the other produced by an *f* sperm and an *F* egg.

When parents know which alleles they carry for a single-gene trait, they can easily determine the probability that a child they produce will have the disease phenotype (**FIGURE 7.14**). You should note that this probability is generated independently for each child. In other words, each offspring of two carriers has a 25% chance of being affected.

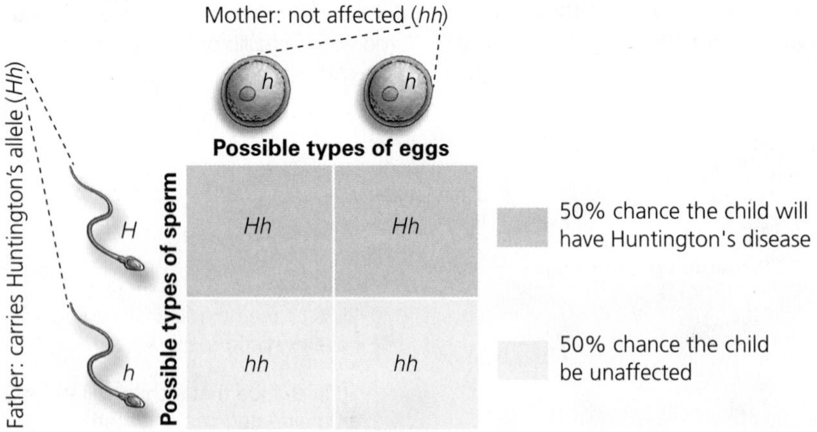

## Punnett Squares for Crosses with More Genes.

Dihybrid crosses are genetic crosses involving two traits. Let's go back to Mendel's peas as an example. Seed color and seed shape are each determined by a single gene, and each is carried on different chromosomes. The two seed-color gene alleles Mendel studied are designated here as *Y*, which is dominant and codes for yellow color, and *y*, the recessive allele, which

# Working with Data ▼

If one child produced by these parents carries the Huntington's allele, what is the likelihood that a second child produced by this couple will have Huntington's disease?

**FIGURE 7.14 A cross between a heterozygote individual and a homozygote individual.** This Punnett square illustrates the outcome of a cross between a man who carries a single copy of the dominant Huntington's disease allele and an unaffected woman.

results in green seeds when homozygous. The two seed-shape alleles Mendel studied are designated as *R*, the dominant allele, which codes for a smooth, round shape, and *r*, which is recessive and codes for a wrinkled shape.

Because the genes for seed color and seed shape are on different chromosomes, they are placed in eggs and sperm independently of each other. In other words, a pea plant that is heterozygous for both genes (genotype *YyRr*) can make four different types of eggs: one carrying dominant alleles for both genes (*YR*), one carrying recessive alleles for both genes (*yr*), one carrying the dominant allele for seed color and the recessive allele for seed shape (*Yr*), and one carrying the recessive allele for color and the dominant allele for shape (*yR*).

As with the Punnett square discussed above, the analysis of a dihybrid cross places all possible sperm genotypes on one axis of the square and all possible egg genotypes on the other axis. Thus, a Punnett square for a cross between two individuals who are heterozygous for both seed-color and seed-shape genes would have four columns representing the four possible egg genotypes and four rows representing the four possible sperm genotypes, resulting in 16 boxes within the square describing four different possible phenotypes (**FIGURE 7.15**).

The phenotypes produced by a dihybrid cross result in a 9:3:3:1 **phenotypic ratio**. In this case, 9/16 include those genotypes produced by both dominant alleles (Y_R_ where the dashes indicate that the second allele for each gene could be either dominant or recessive); 3/16 include those produced by dominant alleles of one gene only (Y_rr); 3/16 include those produced by dominant alleles of the other gene (yyR_); and 1/16 have the phenotype produced by possessing recessive alleles only (yyrr).

**STOP & STRETCH**    Punnett squares can be produced for any number of genes, each of which has two alleles. Imagine a cross between two heterozygous tall, yellow, and round-seeded pea plants (TtYyRr). How many different types of gametes can each plant make? How many different cells would the resulting Punnett square table contain?

# Working with Data ▼

**Curly hair (*CC*) is dominant over wavy (*Cc*) or straight hair (*cc*), and darkly pigmented eyes (*DD* or *Dd*) are dominant over blue eyes (*dd*). (Eye color is actually determined by several different genes, but we'll imagine it is controlled by only one in this example.) What fraction of the offspring of a mother and father who are heterozygous for both of these traits will have curly hair and dark eyes?**

**FIGURE 7.15 A dihybrid cross.** Punnett squares can be used to predict the outcome of a cross involving two different genes. This cross involves two pea plants that are both heterozygous for the seed-color and seed-shape genes.

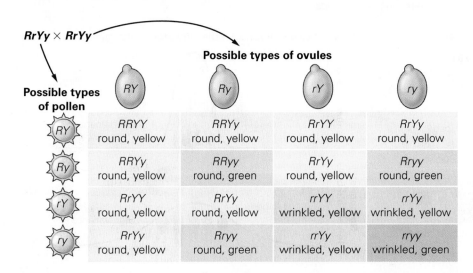

**Proportion of each phenotype in offspring and their associated genotypes**

| | Phenotype | Genotypes |
|---|---|---|
| 9 | Round, yellow | *RRYY, RrYY, RRYy, RrYy* |
| 3 | Round, green | *RRyy, Rryy* |
| 3 | Wrinkled, yellow | *rrYY, rrYy* |
| 1 | Wrinkled, green | *rryy* |

**Is intelligence about genes, or is it a function of the environmental conditions in which a baby is raised?**

As you might imagine, as the number of genes in a Punnett square analysis increases, the number of boxes in the square increases, as does the number of possible genotypes. With two genes, each with two alleles, the number of unique gametes produced by a heterozygote is four, the number of boxes in the Punnett square is 16, and the number of unique genotypes of offspring that can be produced is nine. With three genes, each with two alleles, the Punnett square has 64 boxes and 22 different possible genotypes. With four genes, the square has 256 boxes, and with five genes, there are over 1000 boxes! Predicting the outcome of a cross becomes significantly more difficult as the number of genes we are following increases.

As scientists identify more genes and alleles, the amount of information about the genes of sperm donors—or any potential parent—will also increase. Identifying and testing for particular genes in potential parents will allow us to predict the *likelihood* of numerous genotypes in their offspring. Unfortunately, this increase in genetic testing is not necessarily equaled by an increase in our understanding of how more complex traits develop, as we shall see in the next section.

## 7.3 Quantitative Genetics: When Genes and Environment Interact

The single-gene traits discussed in the previous section have a distinct off-or-on character; individuals have either one phenotype (for example, the pea seed is round) or the other (the pea seed is wrinkled). Traits like this are known as *qualitative* traits. However, many of the traits that interest women who are choosing a sperm donor, such as height, weight, eye color, musical ability, susceptibility to cancer, and intelligence, do not have this off-or-on character. These traits are called **quantitative traits** and show **continuous variation**; that is, we can see a large range of phenotypes in a population—for instance, from very short people to very tall people. Wide variation in quantitative traits leads to the great diversity we see in the human population.

The distribution of phenotypes of a quantitative trait in a population can be displayed on a graph. These data often take the form of a curve called a *normal distribution*. **FIGURE 7.16a** illustrates the normal distribution of heights

**Working with Data** ▼

Examine Figure 7.16a closely: Does an average height of 5 feet, 10 inches in this particular population imply that most men were this height? Were most men in this population close to the mean, or was there a wide range of heights?

**FIGURE 7.16 A quantitative trait.**    (a) This photo of people arranged by height illustrates a normal distribution. The highest point of the bell curve is also the mean height of 5 feet, 10 inches. (b) Fourteen-year-old boys and professional jockeys have the same average weight—approximately 114 pounds. However, to be a jockey, you must be within about 4 pounds of this average. Thus, the variance among jockeys in weight is much smaller than the variance among 14-year-olds.

**(a) Normal distribution of student height in one college class**

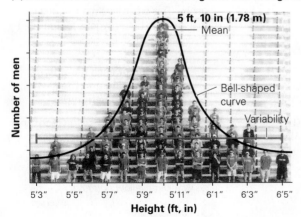

**(b) Variance describes the variability around the mean.**

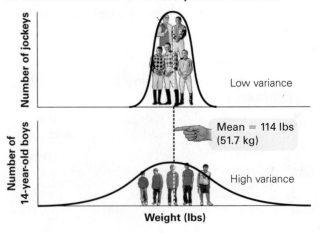

in a college class. Each individual is standing behind the label (at the bottom) that indicates their height. The curved line drawn across the photo summarizes these data—note the similarity of this curve to the outline of a bell, leading to its common name, a bell curve.

A bell curve contains two important pieces of information. The first is the highest point on the curve, which generally corresponds to the average, or **mean,** value for data. The mean is calculated by adding all of the values for a trait in a population and dividing by the number of individuals in that population. The second is in the width of the bell itself, which illustrates the variability of a population. The variability is described with a mathematical measure called **variance,** which is essentially the average distance any one individual in the population is from the mean. If a low variance for a trait indicates a small amount of variability in the population, a high variance indicates a large amount of variability (**FIGURE 7.16b**).

## Why Traits Are Quantitative

A range of phenotypes may exist for a trait because numerous genotypes are found among the individuals in the population. This happens when a trait is influenced by more than one gene; traits influenced by many genes are called **polygenic traits.**

As we saw above, when a single gene with two alleles determines a trait, only three possible genotypes are present: *FF, Ff,* and *ff,* for example. But when more than one gene, each with more than one allele, influences a trait, many genotypes are possible. Some of the genes involved in polygenic traits may produce **transcription regulators,** proteins that determine if, and when, another gene is expressed (in other words, is "turned on"). In addition, some of the alleles for certain polygenic traits may have little or no response to typical transcription factors for the gene. For example, eye color in humans is a polygenic trait influenced by several genes that help produce and distribute the protein melanin, a brown pigment, to the iris. When different alleles for these genes interact, a range of eye colors, from dark brown (lots of melanin) to pale blue (very little melanin), is found in humans. This continuous variation in eye color is a result, in part, of the interaction of different alleles for all of the following: genes for melanin production, genes for transcription factors that determine whether melanin is produced, and genes for melanin distribution.

Continuous variation also may occur in a quantitative trait due to the influence of environmental factors. In this case, each genotype is capable of producing a range of phenotypes depending on outside influences. For a clear example of the effect of the environment on phenotype, see **FIGURE 7.17**. These identical twins share 100% of their genes but are quite different in appearance. Their difference is entirely due to variations in their environment—one twin smoked cigarettes and had high levels of sun exposure, whereas the other did not smoke and spent less time in the sun.

Most traits that show continuous variation are influenced by both genes and the environment. Skin color in humans is an example of this type of trait. The shade of an individual's skin is dependent on the amount of melanin present near the skin's surface. As with eye color, a number of genes have an effect on skin-color phenotype—those that influence melanin production and those that affect melanin distribution. However, the environment, particularly the amount of exposure to the sun during a season or lifetime, also influences the skin color of individuals (**FIGURE 7.18**). Melanin production increases, and any melanin that is present darkens in sun-exposed skin. In fact, after many years of intensive sun exposure, skin may become permanently darker.

**FIGURE 7.17 The effect of the environment on phenotype.** These identical twins have exactly the same genotype, but they are quite different in appearance due to environmental factors. The twin on the right was a life-long smoker and sun tanner, while the one on the left never smoked and spent less time in the sun.

**poly-** means many.

**(a) Genes**

**(b) Environment**

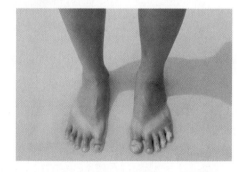

**FIGURE 7.18 Skin color is influenced by genes and environment.** (a) The difference in skin color between these two young women is due primarily to variations in several alleles that control skin pigment production. (b) The difference in color between the sun-protected and sun-exposed portions of the individual in this picture is entirely due to environmental effects.

## Working with Data ▼

How would a graph that showed a *low correlation* between parents and offspring differ from this graph?

Points represent parent-offspring pairs with matching immunity levels.
• Weak • Average • Strong

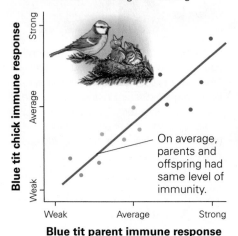

On average, parents and offspring had same level of immunity.

**FIGURE 7.19 Using correlation to calculate heritability.** The close correlation of immune system response between parents and offspring in the blue tit, a European bird, indicates that immune response is highly heritable.

Women choosing Graduate category sperm donors from Fairfax Cryobank are presumably interested in having smart, successful children, but intelligence has both a genetic and an environmental component. With an important role for both influences, how can we predict if the child of a father with a graduate degree will also be capable of earning a similar degree?

### Calculating Heritability

To determine the role of genes in determining quantitative traits, scientists must calculate the **heritability** of the trait. To estimate heritability for quantitative traits in most populations, researchers use correlations between individuals with varying degrees of genetic similarity. A correlation determines how accurately one can predict the measure of a trait in an individual when its measure in a related individual is known. For example, **FIGURE 7.19** shows a correlation between parent birds and their offspring in the strength of their response to tetanus vaccine. An individual that responded strongly produced a large number of antitetanus proteins, called antibodies, while ones that responded weakly produced a lower number of antibodies.

As you can see from the graph, parent birds with weak responses tended to have offspring with weak responses, and parents with strong responses had offspring with strong responses. This strong correlation indicates that the ability to respond to tetanus is highly heritable—most of the difference between birds in their immune system response results from genetic differences.

Heritability in humans is typically measured by examining correlations between groups. These studies calculate how similar or different parents are to their children, or siblings are to each other, in the value of a particular trait. When examined across an entire population, the strength of a correlation provides a measure of heritability (**FIGURE 7.20**). The uses and limitations of heritability are addressed more specifically in Section 7.4.

### Calculating Heritability of Human Intelligence: A Closer Look ▼

Intelligence is often measured by performance on an IQ test. The French psychologist Alfred Binet developed the intelligence quotient (or IQ) in the early 1900s to identify schoolchildren who were in need of remedial help. Binet's IQ test was not based on any theory of intelligence and was not meant to comprehensively measure mental ability, but the tests remain a commonly used way to measure innate or "natural" intelligence. There is still significant controversy over the use of IQ tests in this way.

Even if IQ tests do not really measure general intelligence, IQ scores have been correlated with academic success—meaning that individuals at higher academic levels usually have higher IQs. So, even without knowing their IQ scores, we can reasonably expect that donors in the Graduate category have higher IQs than do other available sperm donors. However, the question of whether the high IQ of a prospective sperm donor has a genetic basis still remains.

*A Closer Look, continued*

For most traits, such as body size, people come in a wide variety of types.

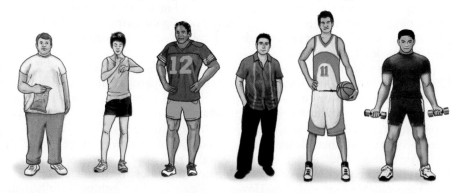

FIGURE 7.20 **Determining heritability in humans.**    Comparisons between parents and children can help us estimate the genetic component of a quantitative trait.

How many of our differences are due to the environment, and how many are a result of different genes?

Correlations between relatives (for instance, here between fathers and sons) can provide information about the importance of genes in determining variation among individuals.

The average correlation between IQs of parents and their children—that is, the estimated heritability of IQ by this method—is 0.42 (42%). In other terms, by this estimate, 42% of the variability in IQ among humans is due to differences in genotypes. However, parents and the children who live with them are typically raised in a similar social and economic environment. As a result, correlations of IQ between the two groups cannot distinguish the relative importance of genes from the importance of the environment. This is the problem found in most arguments about "nature versus nurture"—do children resemble their parents because they are "born that way" or because they are "raised that way"?

To avoid the problem of overlap in environment and genes between parents and children, researchers seek situations that remove one or the other overlap. These situations are called **natural experiments** because one factor is "naturally" controlled, even without researcher intervention. Human twins are one source of a natural experiment to test hypotheses about the heritability of quantitative traits in humans.

By comparing monozygotic (identical) twins, who share all of their alleles, to dizygotic (fraternal) twins, who share, on average, 50% of their alleles, researchers can begin to separate the effects of shared genes from the effects of shared environments. Because twins raised in the same family have similar childhood experiences, one would expect that the only real difference between monozygotic and dizygotic twins is their genetic similarity. The average heritability of IQ calculated from a number of twin studies of this type is about 52%, meaning that these results indicate that 52% of the variability in IQ is due to genes. Surprisingly, this value is even higher than the 42% calculated from the correlation between parents and children.

However, monozygotic twins and dizygotic twins likely *do* differ in more than just genotype. In particular, identical twins are treated more similarly than

*A Closer Look, continued*

**FIGURE 7.21 Twins separated at birth.** Tamara Rabi (left) and Adriana Scott were reunited at age 20 when a mutual friend noticed their remarkable resemblance. Tamara was raised on the Upper West Side of Manhattan and Adriana in suburban Long Island, but they both have similar outgoing personalities, love to dance, and even prefer the same brand of shampoo.

nonidentical twins. This occurs both because they look very similar and because other people may presume that they are identical in all other respects. If monozygotic twins are *expected* to be more alike than dizygotic twins, their IQ scores may be similar because they are encouraged to have the same experiences and to achieve at the same level.

There is one natural experiment that can address this problem, however—comparing identical and nonidentical twins raised apart. In this way, the problem of differential treatment of the two *types* of twins is minimized because no one in their social environment knows that the individual members of a pair have a twin (**FIGURE 7.21**). If variation in genes does not explain much of the variation among peoples' IQ scores, then identical twins raised apart should be no more similar than any two unrelated people of the same age and sex.

Unfortunately, for scientists (but fortunately for children), the frequency of early twin separation is extremely low. However, researchers have estimated the heritability of IQ at a remarkable 72% in a small sample of twins raised apart. This study and the other correlations appear to support the hypothesis that differences in our genes explain the majority of the variation in IQ among people. **TABLE 7.2** summarizes the estimates of IQ heritability but previews the cautions discussed in the next section of this chapter.

# 7.4 Genes, Environment, and the Individual

Perhaps we can now determine the importance of a sperm donor father who has earned a doctorate to his child's intellectual development. We know that a sperm donor will definitely influence some of his child's traits—eye and skin

**TABLE 7.2 To what extent is IQ heritable?** A summary of various estimates of IQ heritability, their shortcomings, and the problems with using them to understand the role of genes in determining an individual's potential intelligence.

| Method of Measurement | Estimated Percentage of Phenotype Determined by Genes | Warnings When Interpreting This Result | Warnings That Apply to All Measurements of Heritability |
|---|---|---|---|
| Correlation between parents' IQ and children's IQ in a population | 42% | When parents and children live together, a correlation can't rule out environmental influence. | • Heritability values are specific to the populations for which they were measured. • High heritability for a trait does not mean that the trait will not respond to a change in the environment. • Heritability is a measure of a population, not an individual. |
| Natural experiment comparing IQ in pairs of identical twins versus nonidentical twins | 52% | Because identical twins are treated as more alike than nonidentical twins the heritability value could be an overestimate. | |
| Natural experiment comparing IQ of identical twins raised apart versus nonidentical twins raised apart | 72% | Small sample size may skew results. | |

color and susceptibility to genetic diseases. In addition, according to the studies discussed above, the donor will probably pass on some intellectual traits to the child. In fact, with a heritability of IQ at above 50%, it appears that genes are primary in determining an individual's intelligence. Perhaps it is a good idea to pay a premium price for Graduate category sperm after all.

However, we need to be very careful when applying the results of twin studies to questions about the individual sperm donors. To understand why, we will take a closer look at the practical significance of heritability.

## The Use and Misuse of Heritability

A calculated heritability value is unique to the population in which it was measured and to the environment of that population. We should be very cautious when using heritability to measure the *general* importance of genes to the development of a trait. The following sections illustrate why.

### Differences Between Groups May Be Entirely Environmental.

A "thought experiment" can help illustrate this point. Body weight in laboratory mice has a strong genetic component, with a calculated heritability of about 90%. In a population of mice in which weight is variable, bigger mice have bigger offspring, and smaller mice have smaller offspring.

Imagine that we randomly divide a population of variable mice into two groups—one group is fed a rich diet, and the other group is fed a poor diet. Otherwise, the mice are treated identically. As you might predict, regardless of their genetic predispositions, the well-fed mice become fat, while the poorly fed mice become thin. Consider the outcome if we were to keep the mice in these same conditions and allowed the two groups to reproduce. Not surprisingly, the second generation of well-fed mice is likely to be much heavier than the second generation of poorly fed mice. Now imagine that another researcher came along and examined these two populations of mice without knowing their diets. Knowing that body weight is highly heritable, the researcher might logically conclude that the groups are genetically different. However, we know this is not the case—both are grandchildren of the same original source population. It is the environment of the two populations that differs (**FIGURE 7.22**, on the next page).

Now extend the same thought experiment to human groups. Imagine that we have two groups of humans, and we have determined that IQ had high heritability. In this case, people in one group were affluent, and their average IQ was higher. The other group was impoverished, and their average IQ was lower. What conclusions could you draw about the genetic differences between these two populations? The answer to the question above is none—as with the laboratory mice, these differences could be entirely due to environment. The high heritability of IQ cannot tell us if two human groups in differing social environments vary in IQ because of variations in genes or because of differences in environment.

### A Highly Heritable Trait Can Still Respond to Environmental Change.

A high heritability for IQ might seem to imply that IQ is not strongly influenced by environmental conditions. However, intelligence in other animals can be demonstrated to be both highly heritable and strongly influenced by the environment.

Rats can be bred for maze-running ability, and researchers have produced rats that are "maze bright" and rats that are "maze dull." Maze-running ability is highly heritable in the laboratory environment; that is, bright rats have

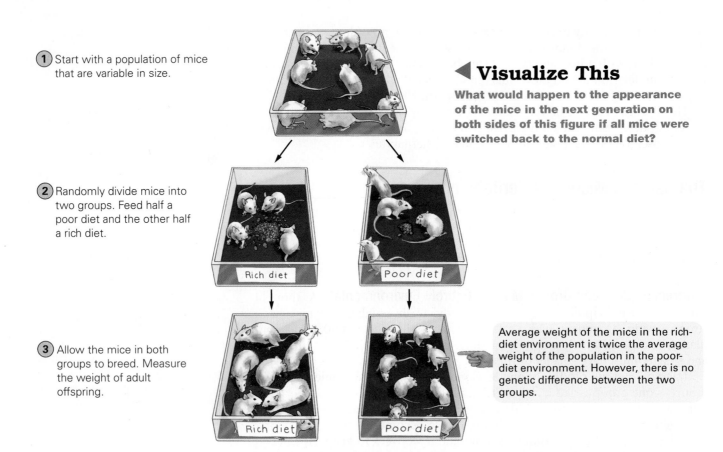

① Start with a population of mice that are variable in size.

**◀ Visualize This**

What would happen to the appearance of the mice in the next generation on both sides of this figure if all mice were switched back to the normal diet?

② Randomly divide mice into two groups. Feed half a poor diet and the other half a rich diet.

Rich diet

Poor diet

③ Allow the mice in both groups to breed. Measure the weight of adult offspring.

Average weight of the mice in the rich-diet environment is twice the average weight of the population in the poor-diet environment. However, there is no genetic difference between the two groups.

Rich diet

Poor diet

**FIGURE 7.22 The environment can have powerful effects on highly heritable traits.** If genetically similar populations of mice are raised in radically diverse environments, then differences between the populations are entirely due to environment.

bright offspring, and dull rats have dull offspring. The results of an experiment that measured the number of mistakes made by maze-bright and maze-dull rats raised in different environments are presented in **TABLE 7.3**.

In the typical lab environment, bright rats were much better at maze running than dull rats. But in both a very boring or restricted environment and a very enriched environment, the two groups of rats did about the same. In fact, no rats excelled in a restricted environment, and all rats did better at maze running in enriched environments, with the duller rats improving most dramatically.

**TABLE 7.3 A highly heritable trait is not identical in all environments.**

| | | Number of Mistakes in … | | |
|---|---|---|---|---|
| | Phenotype | Normal Environment | Restricted Environment | Enriched Environment |
| | Maze-bright rats | 115 | 170 | 112 |
| | Maze-dull rats | 165 | 170 | 122 |
| **Explanation of Results** | | Maze-dull rats made more mistakes than maze-bright rats when running a maze. | Both groups made the same number of mistakes when running a maze. | Both groups made fewer mistakes when running the maze. The maze-dull rats improved the most. |

What this example demonstrates is that we cannot predict the response of a trait to a change in the environment, even when that trait is highly heritable. Thus, even if IQ has a strong genetic component, environmental factors affecting IQ can have big effects on an individual's intelligence.

> **STOP & STRETCH**   Some commentators have argued that given IQ's high heritability, policies that increase financial resources to failing schools will ultimately fail to increase achievement because such a predominantly genetic trait will not respond well to environmental change. Use your understanding of the proper application of heritability to refute this argument.

### Heritability Does Not Tell Us Why Two Individuals Differ.
High heritability of a trait is often presumed to mean that the difference between two individuals is mostly due to differences in their genes. However, even if genes explain 90% of the population variability in a particular environment, the reason one individual differs from another may be entirely a function of environment (as an example of this, look back at the identical twins in Figure 7.18).

Currently, there is no way to determine if a particular child is a poor student because of genes, a poor environment, or a combination of both factors. There is also no way to predict whether a child produced from the sperm of a man with a doctorate will be an accomplished scholar. All we can say is that given our current understanding of the heritability of IQ and the current social environment, the alleles in Graduate category sperm *may* increase the probability of having a child with a high IQ.

It is also important to recognize that very few genes are limited in their effect to a single trait. Even Mendel noticed that pea plants he studied with white flowers also always had pale seeds. A gene that influences several unrelated traits is said to exhibit **pleiotropy.** A variant of a pleiotropic gene may have both desirable and undesirable effects. For example, a mutation in a single gene involved in the production of connective tissue in humans leads to individuals that are unusually tall with long arms and fingers—traits that could be quite beneficial to some athletes (**FIGURE 7.23**). However, individuals with this condition, known as Marfan syndrome, also experience vision problems and sometimes serious heart defects. Some scientists have found a link between high IQ and severe nearsightedness, and it is reasonable to expect that other less-desirable traits may be associated with higher IQ.

### How Do Genes Matter?

We know that genes can have a strong influence on eye color, risk of genetic diseases such as cystic fibrosis, and even the structure of the brain. But what really determines who we are—nature or nurture?

Even with single-gene traits, the outcome of a cross between a woman and a sperm donor is not a certainty; it is only a probability. Couple this with traits being influenced by more than one gene, and independent assortment greatly increases the offspring types possible from a single mating (recall how the number of cells in a Punnett square increased exponentially when more genes were added). Knowing the phenotype of potential parents gives you relatively little information about the phenotype of their children. So, even if genes have a strong effect on traits, we cannot "program" the traits of children by selecting the traits of their parents (**FIGURE 7.24**).

In truth, we are really asking ourselves the wrong question when we wonder if nature or nurture has a more powerful influence on who we are. Both nature and nurture play an important role. Our cells carry instructions for all the essential characteristics of humanity, but the process of developing from embryo to

**FIGURE 7.23 Genes can have both positive and negative effects.**   Flo Hyman was a star volleyball player for the US national team, leading them to a silver medal at the 1984 Olympics. Her unusual height of 1.96 m (6'5") and long arm span were a result of Marfan syndrome, which also led to weakness of the vessels carrying blood from her heart. She died during a match in 1986 as a result of the failure of these vessels.

**FIGURE 7.24 Genes are not destiny.** Even when the traits of both parents are known, the children they produce may be very different in interests and aptitudes.

adult takes place in a physical and social environment that influences how these genes are expressed. Scientists are still a long way from answering questions about how all of these complex, interacting circumstances result in who we are.

What is the lesson for women and couples who are searching for a sperm donor from Fairfax Cryobank? Donors in the Graduate category may indeed have higher IQs than donors in the cryobank's other categories, but there is no real way to predict if a particular child of one of these donors will be smarter than average. According to the current data on the heritability of IQ, sperm from high-IQ donors may increase the odds of having an offspring with a high IQ, but only if parents provide them with a stimulating, healthy, and challenging environment in which to mature. This, of course, would be good for children with any alleles.

# *savvy reader*

## Can Your Genes Be Bullied?

[Researcher Isabelle] Ouellet-Morin and her team looked at 28 pairs of identical twins born between 1994 and 1995.... In each of these 28 pairs, one twin had been a victim of bullying while the other had not.

Part of the survey included an analysis of the kids' DNA methylation of SERT, a gene that's responsible for transporting serotonin, a neurotransmitter involved in mood regulation and depression. (DNA methylation is a chemical process that affects whether or not a gene gets expressed in response to social and physical cues.)

Bullied twins had higher SERT DNA methylation at age 10 compared with their non-bullied twins, the study found. What's more, the children with higher SERT methylation levels had blunted ... responses to stress.

"Many people think that our genes are immutable; however this study suggests that environment, even the social environment, can affect their functioning," Ouellet-Morin said. "This is particularly the case for victimization experiences in childhood, which change not only our stress response but also the functioning of genes involved in mood regulation."

1. Did the twins in this study vary in the genes they were born with? Explain.
2. What is the effect of gene methylation? How would differences in gene methylation cause a difference in phenotype between two people with identical genes?
3. In the last paragraph, the researcher appears to be arguing that a correlation between being bullied and SERT methylation implies that bullying causes personality changes in children. Explain why she hasn't proved that bullying causes a change in behavior.

Source: http://www.livescience.com/25840-bullying-may-alter-genes.html

# SOUNDS RIGHT **BUT IS IT?**

In the Harry Potter books and movies, many of the characters who knew Harry's parents tell him that he resembles his mother or note his similarity to his father in his willingness to bend the rules. To many fans, these comments make sense, since a child receives half of his genetic information from his mother and half from his father. Thus, it seems fair to say that:

**Harry Potter has his mother's eyes.**

Sounds right, but it isn't.

1. Do you think it is more likely that the color and shape of a person's eyes are determined by one gene or many genes?
2. Did Harry receive copies of genes that determine eye color and shape from his mother?
3. Did he receive copies of genes that determine eye color and shape from his father?
4. Think back to the Punnett squares you've viewed and drawn. Do genes for only one or both parents likely influence eye color and shape?
5. Reflect on your answers to questions 1–4. Explain why the statement bolded above sounds right, but isn't.

# Chapter Review MasteringBiology®

Go to the Study Area in MasteringBiology® for practice quizzes, myeBook, BioFlix™ 3-D animations, MP3Tutor sessions, videos, current events, and more.

## Summary

### Section 7.1

Describe the relationship between genes, chromosomes, and alleles.

- Children resemble their parents in part because they inherit their parents' genes, segments of DNA that contain information about how to make proteins (pp. 134–135).
- Chromosomes contain genes. Different versions of a gene are called alleles (p. 136).
- Mutations in genes generate a variety of alleles. Each allele typically results in a slightly different protein product (p. 136).

Explain why cells containing identical genetic information can look different from each other.

- Although nearly all cells in an individual contain the exact same genetic information, the genes that are "read" in those cells differ. Even when the same gene is expressed in two different cells, their products may have very different effects (pp. 135–137).

Define *segregation* and *independent assortment* and explain how these processes contribute to genetic diversity.

- Segregation separates alleles during the production of eggs and sperm, meaning that a parent contributes only half of its genetic information to an offspring (p. 137).
- In independent assortment, individual chromosomes are sorted into eggs and sperm independent of each other, meaning that the subset of information passed on by parents is unique for each gamete (pp. 137–138).

### Section 7.2

Distinguish between homozygous and heterozygous genotypes and describe how recessive and dominant alleles produce particular phenotypes when expressed in these genotypes.

- An individual heterozygous for a particular gene carries two different alleles for the gene, while one who is homozygous carries two identical alleles (p. 141).
- A dominant allele is expressed even when the genotype is heterozygous, while a recessive allele is only expressed when the individual carries no copies of the dominant allele—that is, when it is homozygous recessive (pp. 141–142).

Demonstrate how to use a Punnett square to predict the likelihood of a particular offspring genotype and phenotype from a cross of two individuals with known genotype.

- A Punnett square helps us determine the probability that two parents of known genotype will produce a child with a particular genotype (pp. 143–146).

### Section 7.3

Define *quantitative trait* and describe the genetic and environmental factors that cause this pattern of inheritance.

- Many traits—such as height, IQ, and musical ability—show quantitative variation, which results in a range of values for the trait within a given population (p. 146).
- Quantitative variation in a trait may be generated because the trait is influenced by several genes, because the trait can be influenced by environmental factors, or due to a combination of both factors (p. 147).

Describe how heritability is calculated and what it tells us about the genetic component of quantitative traits.

- The role of genes in determining the phenotype for a quantitative trait is estimated by calculating the heritability of the trait (p. 148).
- Heritability is calculated by examining the correlation between parents and offspring or by comparing pairs of monozygotic twins to pairs of dizygotic twins (pp. 148–150).

### Section 7.4

Explain why a high heritability still does not always mean that a given trait is determined mostly by the genes an individual carries.

- Calculated heritability values are unique to a particular population in a particular environment. The environment may cause large differences among individuals, even if a trait has high heritability (pp. 150–152).
- Our current understanding of the relationship between genes and complex traits does not allow us to predict the phenotype of a particular offspring from the phenotype of its parents (p. 153).

# Roots to Remember

**The following roots of words come mainly from Latin and Greek and will help you decipher terms:**

| | |
|---|---|
| **di-** | means two. Chapter term: *dizygotic* |
| **hetero-** | means the other, another, or different. Chapter terms: *heterozygous, heterozygote* |
| **homo-** | means the same. Chapter terms: *homozygous, homologous* |
| **mono-** | means one. Chapter term: *monozygotic* |
| **pheno-** | comes from a verb meaning "to show." Chapter term: *phenotype* |
| **poly-** | means many. Chapter term: *polygenic* |
| **-zygous** | derives from *zygote*, the "yoked" cell resulting from the union of an egg and sperm. Chapter terms: *monozygotic, heterozygote* |

# Learning the Basics

**1.** What is the relationship between genotype and phenotype?

**2.** What factors cause quantitative variation in a trait within a population?

**3.** Which of the following statements correctly describe the relationship between genes and chromosomes?
A. Genes are chromosomes; B. Chromosomes contain many genes; C. Genes are made up of hundreds or thousands of chromosomes; D. Genes are assorted independently during meiosis, but chromosomes are not; E. More than one of the above is correct.

**4.** An allele is a _____.
A. version of a gene; B. dysfunctional gene; C. protein; D. spare copy of a gene; E. phenotype.

**5.** Sperm and eggs in humans always _____.
A. each have 2 copies of every gene; B. each have 1 copy of every gene; C. each contain either all recessive alleles or all dominant alleles; D. are genetically identical to all other sperm or eggs produced by that person; E. each contain all of the genetic information from their producer.

**6.** Scientists have recently developed a process by which a skin cell from a human can be triggered to develop into a human heart muscle cell. This is possible because _____.
A. most cells in the human body contain the genetic instructions for making all types of human cells; B. a skin cell is produced when all genes in the cell are expressed; turning off some genes in the cell results in a heart cell; C. scientists can add new genes to old cells to make them take different forms; D. a skin cell expresses only recessive alleles, so it can be triggered to produce

dominant heart cell alleles; E. it is easy to mutate the genes in skin cells to produce the alleles required for other cell types.

**7.** What is the physical basis for the independent assortment of alleles into offspring?
A. There are chromosome divisions during gamete production; B. Homologous chromosome pairs are separated during gamete production; C. Sperm and eggs are produced by different sexes; D. Each gene codes for more than one protein; E. The instruction manual for producing a human is incomplete.

**8.** Among heritable diseases, which genotype can be present in an individual without causing a disease phenotype in that individual?
A. heterozygous for a dominant disease; B. homozygous for a dominant disease; C. heterozygous for recessive disease; D. homozygous for a recessive disease; E. all of the above.

**9.** A quantitative trait _____.
A. may be one that is strongly influenced by the environment; B. varies continuously in a population; C. may be influenced by many genes; D. has more than a few values in a population; E. all of the above.

**10.** When a trait is highly heritable, _____.
A. it is influenced by genes; B. it is not influenced by the environment; C. the variance of the trait in a population can be explained primarily by variance in genotypes; D. A and C are correct; E. A, B, and C are correct.

# Genetics Problems

**1.** A single gene in pea plants has a strong influence on plant height. The gene has two alleles: tall (*T*), which is dominant, and short (*t*), which is recessive. What are the genotypes and phenotypes of the offspring of a cross between a *TT* and a *tt* plant?

**2.** What are the genotypes and phenotypes of the offspring of *Tt* × *Tt*?

**3.** The "*P*" gene controls flower color in pea plants. A plant with either the *PP* or *Pp* genotype has purple flowers, while a plant with the *pp* genotype has white flowers. What is the relationship between *P* and *p*?

**4.** Albinism occurs when individuals carry two recessive alleles (*aa*) that interfere with the production of melanin, the pigment that colors hair, skin, and eyes. If an albino child is born to two individuals with normal pigment, what is the genotype of each parent?

**5.** Pfeiffer syndrome is a dominant genetic disease that occurs when certain bones in the skull fuse too early in the development of a child, leading to distorted head and face shape. If a man heterozygous for the allele that

causes Pfeiffer syndrome marries a woman who is homo-zygous for the nonmutant allele, what is the chance that their first child will have this syndrome?

**6.** A cross between a pea plant that produces yellow peas and a pea plant that produces green peas results in 100% yellow pea offspring. **a.** Which allele is dominant in this situation? **b.** What are the likely genotypes of the yellow pea and green pea plants in the initial cross?

**7.** A cross between a pea plant that produces round (*R*), yellow (*Y*) peas and a pea plant that produces wrinkled (*r*), green (*y*) peas results in 50% yellow, round pea offspring and 50% green, wrinkled pea offspring. What are the genotypes of the plants in the initial cross?

**8.** A woman who is a carrier for the cystic fibrosis allele marries a man who is also a carrier.

**A.** What percentage of the woman's eggs will carry the cystic fibrosis allele? **B.** What percentage of the man's sperm will carry the cystic fibrosis allele? **C.** The probability that this couple will have a child who carries two copies of the cystic fibrosis allele is equal to the percentage of eggs that carry the allele times the percentage of sperm that carry the allele. What is this probability? **D.** Is this the same result you would generate when doing a Punnett square of this cross?

**9.** The allele *BRCA2* was identified in families with unusually high rates of breast and ovarian cancer. Up to 80% of women with one copy of the *BRCA2* allele develop one of these cancers in their lifetime. **a.** Is *BRCA2* a dominant or a recessive allele? **b.** How is *BRCA2* different from the typical pattern of Mendelian inheritance?

## Analyzing and Applying the Basics

**1.** Two parents both have brown eyes, but they have two children with brown eyes and two with blue eyes. How is it possible that two people with the same eye color can have children with different eye color? If eye color in this family is determined by differences in genotype for a single gene with two alleles, what percentage of the children are expected to have blue eyes? If the ratio of brown to blue eyes in this family does not conform to expectations, why does this result not refute Mendelian genetics?

**2.** Does a high value of heritability for a trait indicate that the average value of the trait in a population will not change if the environment changes? Explain your answer.

**3.** The heritability of IQ has been estimated at about 72%. If John's IQ is 120 and Jerry's IQ is 90, does John have stronger "intelligence" genes than Jerry does? Explain your answer.

## Connecting the Science

**1.** If scientists find a gene that is associated with a particular "undesirable" personality trait (for instance, a tendency toward aggressive outbursts), will it mean greater or lesser tolerance toward people with that trait? Will it lead to proposals that those affected by the "disorder" should undergo treatment to be "cured" and that measures should be taken to prevent the birth of other individuals who are also afflicted?

**2.** The higher price for Graduate sperm at the Fairfax Cryobank seems to imply that these donors are rare and highly desirable. If you were a woman who was looking for a sperm donor, would you focus your selection process on the Graduate donors? What might you miss by focusing only on these donors?

**3.** Down syndrome is caused by a mistake during meiosis and results in physical characteristics such as a short stature and distinct facial features as well as cognitive impairment (also known as mental retardation). Does the fact that Down syndrome is a genetic condition that results in low IQ mean that we should put fewer resources into education for people with Down syndrome? How does your answer to this question relate to questions about how we should treat individuals with other genetic conditions?

Answers to **Stop & Stretch, Visualize This, Working with Data, Savvy Reader, Sounds Right, But Is It?,** and **Chapter Review** questions can be found in the **Answers** section at the back of the book.

CHAPTER **8** | # DNA Detective

The Romanov family ruled Russia until their overthrow, exile, and 1918 execution.

# Complex Patterns of Inheritance and DNA Profiling

On the night of July 16, 1918, Nicholas II, the tsar of Russia, and his wife, Alexandra Romanov, along with their five children and four family servants, were executed in a small room in the basement of the house to which they had been exiled. These murders ended three centuries of rule by the Romanov family over the Russian Empire.

In February 1917, in the wake of protests throughout Russia, Nicholas II had relinquished his power by abdicating for both himself and on behalf of his only son and heir to the throne, Alexis, then 13 years old.

The tsar hoped that by abdicating he might quell the escalating unrest among Russian citizens. Russian soldiers were dying in service of their country during World War I and citizens at home were suffering from

**The fall of the communist Soviet Union prompted the desire for a proper burial of the Romanov family.**

**Many believed that bones found in a grave in Ekaterinburg were those of the slain Romanovs.**

severe food shortages and poverty, while the tsar and his family lived in opulent luxury. In November 1917, members of the Communist Party took power in Russia and immediately sought to rid the country of every last vestige of the Romanovs and all they represented.

Shortly after midnight on July 16, the family was awakened and asked to dress. Nicholas, Alexandra, and their children—Olga, Tatiana, Maria, Anastasia, and Alexis—along with the family physician, cook, maid, and valet, were escorted to a room in the basement of the house in which they had been exiled. Armed men entered the room and killed the royal family and their entourage.

After the murders, the men loaded the bodies of the Romanovs

and their servants into a truck and drove to a remote wooded area in Ekaterinburg. Historical accounts differ regarding whether the bodies were dumped down a mineshaft, later to be removed, or were immediately buried. There is also some disagreement regarding the burial of two of the people who were executed. Some reports indicated that all 11 people were buried together, and two of them were either badly decomposed by acid placed on the ground of the burial site or burned to ash. Other reports indicated that two members of the family were buried separately. Some people even believe that two victims escaped the execution. In any case, the bodies of at least nine people were buried in a

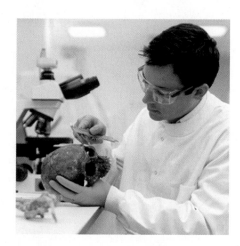

**A variety of scientific techniques were used to confirm that the bones buried in the shallow grave belonged to the Romanovs.**

shallow grave, where they lay undisturbed until 1991.

After the dissolution of the Soviet Union, postcommunist leaders allowed the bones to be exhumed so that they could be given a proper burial. This exhumation took on powerful political meaning because the people of Russia hoped to do more than just give the family a proper burial; they hoped to also lay to rest the brutality of the communist regime that took power after the murders of the Romanov family.

Because all that remained of the bodies when they were exhumed was a pile of bones, it was difficult to know if they were actually those of the Romanovs. The bodies were badly decayed, leaving investigators only teeth and bones to work with. Dental evidence showed that five of the bodies had gold, porcelain, and platinum dental work, which had been available only to aristocrats in the early part of the century.

The bones seemed to indicate that they belonged to six adults and three children. Investigators electronically superimposed the photographs of the skulls on archived photographs of the family. They compared the skeletons' measurements with clothing known to have belonged to the family. These and other data were consistent with the hypothesis that the bodies could be those of the tsar, his wife the tsarina, three of their five children, and the four servants.

Scientists also tried to determine the sex of the buried individuals based on pelvic bone structure. Females have evolved to have wider pelvic openings to accommodate the passage of a child through the birth canal. Russian scientists thought that all three of the children's skeletons and two of the adults' skeletons were probably female (and four of the adult skeletons were male). However, the pelvises had decayed, so it was impossible to be certain.

By using the evidence available to them, Russian scientists had shown only that these skeletons might be the Romanovs. They had not yet shown with any degree of certainty that these bodies *did* belong to the slain royals. The new Russian leaders did not want to make a mistake when symbolically burying a former regime. Unassailable proof was necessary.

Before more sophisticated techniques became available to scientists, solving the puzzle of the Ekaterinburg bones required that scientists be able to show relatedness between the skeletons.

# 8.1 Extensions of Mendelian Genetics

Patterns of inheritance are fairly simple to predict when genes are inherited in a straightforward manner, such as a trait controlled by a single gene with a dominant and a recessive allele. Patterns of inheritance that are a little more complex are said to be *extensions* of Mendelian genetics.

## Incomplete Dominance, Codominance, Multiple Alleles, and Pleiotropy

For some genes, the phenotype of the heterozygote is intermediate between both homozygotes—a situation called **incomplete dominance**. For example, the alleles that determine flower color in snapdragons are incompletely dominant: One homozygote produces red flowers, the other, presumably carrying two nonfunctional copies of a color gene, produces white flowers; the heterozygote, carrying one allele encoding red color and one allele encoding white color, produces pink flowers (**FIGURE 8.1**).

In some cases, the phenotype of a heterozygote is actually a combination of both fully expressed traits, instead of a mixture. This situation, by which two different alleles of a gene are both expressed in an individual, is known as **codominance** (**FIGURE 8.2**). In cattle, for example, the allele that codes for red hair color and the allele that codes for white hair color are both expressed in a

**Flower color in snapdragons**

 ×  =

Red = *RR*
Homozygote

White = *rr*
Homozygote

Pink = *Rr*
Heterozygote

**FIGURE 8.1 Incomplete dominance.** Snapdragons show incomplete dominance in the inheritance of flower color. The heterozygous flower has a phenotype that is in between that of the two homozygotes.

◀ **Visualize This**

If flower color were inherited such that *R* was completely dominant over *r*, what color would heterozygous flowers be?

**Coat color in cattle**

 ×  =

Red = $R^1R^1$

White = $R^2R^2$

Roan = $R^1R^2$
(patchy red and white coat)

**FIGURE 8.2 Codominance.** Roan coat color in cattle is an example of codominance. Both alleles are equally expressed in the heterozygote, so the conventional uppercase and lowercase nomenclature for alleles no longer applies.

heterozygote. These individuals have patchy coats that consist of an approximately equal mixture of white hairs and red hairs.

Because all that was left of the Romanovs was a pile of bones, scientists could study only a few genetic traits to show the relatedness of the adult skeletons to two of the four children's skeletons. Genetic traits that were obvious, such as bone size and structure, are controlled by many genes. Traits that are affected by the interactions among many genes are called **polygenic traits.** Bone structure is also affected by environmental factors like nutrition and physical activity level. These complexities made using bone size and structure to predict which of the adults' skeletons were related to the children's skeletons a matter of guesswork. The scientists had to use more sophisticated analyses.

A technique that scientists use to help determine relatedness of people is blood typing, which involves determining if certain carbohydrates are located on the surface of red blood cells. These surface markers are part of the **ABO blood system.** The ABO blood system displays two extensions of Mendelism—codominance and **multiple allelism,** which occurs when there are more than two alleles of a gene in the population. In fact, three distinct alleles of one blood-group gene code for the enzymes that synthesize the sugars found on the surface of red blood cells. Two of the three alleles display codominance to each other, and one allele is recessive to the other two.

**poly-** comes from the Greek word for many.

## Visualize This ▼

How many different ABO blood system alleles are there in the human population? How many different ABO blood system alleles could one person carry?

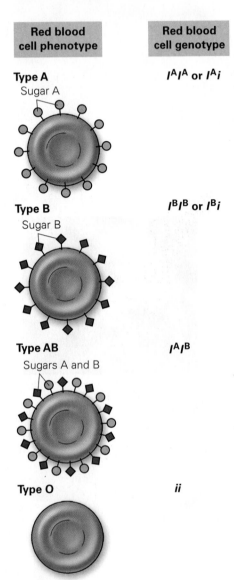

| Red blood cell phenotype | Red blood cell genotype |
|---|---|
| **Type A**<br>Sugar A | $I^A I^A$ or $I^A i$ |
| **Type B**<br>Sugar B | $I^B I^B$ or $I^B i$ |
| **Type AB**<br>Sugars A and B | $I^A I^B$ |
| **Type O** | $ii$ |

**FIGURE 8.3 ABO blood system.** Red blood cell phenotypes and corresponding genotypes. Alleles $I^A$ and $I^B$ are codominant, and both are dominant to $i$.

**FIGURE 8.3** summarizes the possible genotypes and phenotypes for the ABO blood system. The three alleles of this blood-type gene are $I^A$, $I^B$, and $i$. A given individual will carry only two alleles, even though three alleles are present in the entire population. In other words, one person may carry the $I^A$ and $I^B$ alleles, and another might carry the $I^A$ and $i$ alleles. There are three different alleles, but each individual can carry only two alleles.

The symbols used to represent these alternate forms of the blood-type gene tell us something about their effects. The lowercase $i$ allele is recessive to both the $I^A$ and $I^B$ alleles. Therefore, a person with the genotype $I^A i$ has type A blood, and a person with the genotype $I^B i$ has type B blood. A person with both recessive alleles, genotype $ii$, has type O blood. The uppercase $I^A$ and $I^B$ alleles display codominance in that neither masks the expression of the other. Both of these alleles are expressed. Thus, a person with the genotype $I^A I^B$ has type AB blood.

**STOP & STRETCH** Imagine a new allele $I^C$ that is codominant with $I^A$ and $I^B$ and dominant to $i$. List all of the possible blood-group genotypes and phenotypes in a population containing all four alleles.

Another molecule on the surface of red blood cells is called the **Rh factor.** Someone who is positive (+) for this trait has the Rh factor on his or her red blood cells, while someone who is negative (−) does not. This trait, unlike the ABO blood system, is inherited in a straightforward two-allele, completely dominant manner with $Rh^+$ dominant to $Rh^-$. Persons who are Rh positive can have the genotype $Rh^+ Rh^+$ or $Rh^+ Rh^-$. An Rh-negative individual has the genotype $Rh^- Rh^-$.

Blood typing is often used to help establish whether a given set of parents could have produced a particular child. For example, a child with type AB blood and parents who are type A and type B could be related, but a child with type O blood could not have a parent with type AB blood. Likewise, if a child has blood type B and the known mother has blood type AB, then the father of that child could have type AB, A, B, or O blood, which does not help to establish parentage.

If a child has a blood type consistent with alleles that he or she may have inherited from a man who *might* be his or her father, this finding does not mean that the man *is* the father. Instead, it is only an indication that the man could be. In fact, many other men would also have that blood type (**TABLE 8.1**). Therefore, blood-type analysis can be used only to eliminate people from consideration. Blood typing cannot be used to positively identify someone as the parent of a particular child.

Clinicians must take ABO blood groups into account when performing blood transfusions. Persons receiving transfusions from incompatible blood groups will mount an immune response against those sugars that they do not

**TABLE 8.1 Frequency of blood types in U.S. population.** Percentages of the population with a given blood type are listed from most to least common. The negative and positive superscripts refer to the absence or presence of the Rh factor.

| Blood Type | $O^+$ | $A^+$ | $B^+$ | $O^-$ | $A^-$ | $AB^+$ | $B^-$ | $AB^-$ |
|---|---|---|---|---|---|---|---|---|
| **Frequency in U.S. Population (%)** | 40 | 32 | 11 | 7 | 5 | 3 | 1.5 | 0.5 |

carry on their own red blood cells. The presence of these foreign red blood cell sugars causes a severe reaction in which the donated, incompatible red blood cells form clumps. This clumping can block blood vessels and kill the recipient. **TABLE 8.2** shows the types of blood transfusions individuals of various blood types can receive.

Blood-typing analysis can sometimes provide scientists with some information about potential relatedness of victims. However, it was not an option in the case of the Romanovs. The remains contained no blood.

The Romanov family was known to carry a trait that demonstrates another extension of Mendelism: pleiotropy. **Pleiotropy** is the ability of a single gene to cause multiple effects on an individual's phenotype. Alexis was the last of several Romanovs to have **hemophilia,** a pleiotropic blood-clotting disorder. A person with the most common form of hemophilia cannot produce a protein called clotting factor VIII. When this protein is absent, blood does not form clots to stop bleeding from a cut or internal blood vessel damage. Affected individuals bleed excessively, even from small cuts.

Additional effects of hemophilia include excessive bruising, pain and swelling in the joints, anemia, and extreme fatigue. Historical records indicate that Alexis, heir to the throne, was so ill with hemophilia that his father had to carry him to the basement room where he was executed.

**pleio-** comes from the Greek word for many.

**hemo-** means blood.

After the dissolution of the Soviet Union, postcommunist leaders allowed the bones to be exhumed so the Romanovs could be given a proper burial.

# 8.2 **Sex Determination and Sex Linkage**

It appears that Alexis inherited the hemophilia allele from his mother. We can deduce this pattern of inheritance because we now know that the clotting factor gene is inherited in a sex-specific manner. The clotting factor VIII gene (the gene that, when mutated, causes hemophilia) is located in the X chromosome.

**TABLE 8.2** **Blood transfusion compatibilities.**

| Recipient | Recipient Can Receive | Recipient Cannot Receive |
|---|---|---|
| Type O | Type O | Type A |
| | | Type B |
| | | Type AB |
| Type A | Type O | Type B |
| | Type A | Type AB |
| Type B | Type O | Type A |
| | Type B | Type AB |
| Type AB | Type O | None |
| | Type A | |
| | Type B | |
| | Type AB | |

# Visualize This ▼

How and why would the X chromosomes in two egg cells shown below differ from each other?

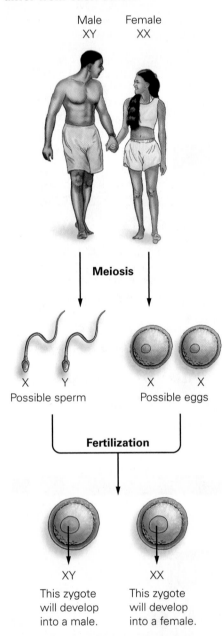

**FIGURE 8.4 Sex determination in humans.** Sperm and eggs each carry one sex chromosome and 22 autosomes.

Of the 23 pairs of chromosomes present in the cells of human males, 22 pairs are **autosomes**, or nonsex chromosomes, and one pair, X and Y, are the **sex chromosomes**. Males have 22 pairs of autosomes and one X and one Y sex chromosome. Females also have 22 pairs of autosomes, but their sex chromosomes are comprised of two X chromosomes.

## Chromosomal Sex Determination

The X and Y chromosomes are involved in establishing the sex of an individual through a process called **sex determination**. Men produce sperm cells containing one of each autosome and either an X or a Y chromosome. Females also produce gametes with 22 unpaired autosomes, but the sex chromosome in an egg cell is always one of two X chromosomes. Therefore, the sperm cell determines the sex of the offspring. If an X-bearing sperm unites with an egg cell, the resulting offspring will be female (XX). If a sperm bearing a Y chromosome unites with an egg cell, the resulting offspring will be male (XY) (**FIGURE 8.4**). The manner in which sex is determined in several other organisms is outlined in **TABLE 8.3**.

## Sex Linkage

Genes located on the X or Y chromosome are called **sex-linked genes** because biological sex is inherited along with, or linked to, the X or Y chromosome. Sex-linked genes found on the X chromosomes are said to be X linked, while those on the Y chromosome are Y linked. The X chromosome is much larger than the Y chromosome, which carries very little genetic information.

**X-Linked Genes.** The fact that males have only one X chromosome leads to some peculiarities in inheritance of sex-linked genes. Males are more likely to suffer from diseases caused by recessive alleles on the X chromosome because they have only one copy of any X-linked gene. Females are less likely to suffer from these diseases because they carry two copies of the X chromosome and thus have a greater likelihood of carrying at least one functional version of each X-linked gene.

A **carrier** of a recessively inherited trait has one copy of the recessive allele and one copy of the normal allele and will not exhibit symptoms of the disease. Only females can be carriers of X-linked recessive traits because males with a copy of the recessive allele will exhibit the trait. Both males and females can be carriers of nonsex-linked, autosomal traits.

Even though female carriers of an X-linked recessive trait will not display the recessive trait, they can pass the trait on to their offspring. For this reason, most women carrying the hemophilia allele will not even realize that they are a carrier until their son becomes ill (**FIGURE 8.5a**). Hemophilic males will always and only pass the hemophilia allele to their daughters, who can then pass it to their sons (**FIGURE 8.5b**). In the United States, there are over 20,000 hemophiliacs, the vast majority of whom are male.

**STOP** & **STRETCH**   Color blindness is an X-linked recessive trait. The normal allele codes for the production of proteins called opsins that help absorb different wavelengths of light. A lack of opsins causes insensitivity to light of red or green wavelengths. From a genetic standpoint, what must be true of the parents of a female who is color blind?

**TABLE 8.3 Sex determination strategies in some nonhuman organisms.**

| Type of Organism | | Mechanism of Sex Determination |
|---|---|---|
| **Vertebrates (fish, amphibians, reptiles, birds, and mammals)** |  | In some vertebrates, the male has two of the same chromosomes and the female has two different chromosomes. In these cases, the female determines the sex of the offspring. |
| **Egg-laying reptiles** |  | In many egg-laying species, two organisms with the same suite of sex chromosomes could become different sexes. Sex depends on which genes are activated during embryonic development. For example, the sex of some reptiles is determined by the incubation temperature of the egg. |
| **Wasps, ants, bees** |  | In bees, sex is determined by the presence or absence of fertilization. Males (drones) develop from unfertilized eggs. Females (workers and queens) develop from fertilized eggs. |
| **Bony fishes** |  | Some species of bony fishes change their sex after maturation. All individuals will become females unless they are deflected from that pathway by social signals such as displays of dominance. |
| **Worms** |  | Worms have both male and female reproductive organs, a condition referred to as intersex. |

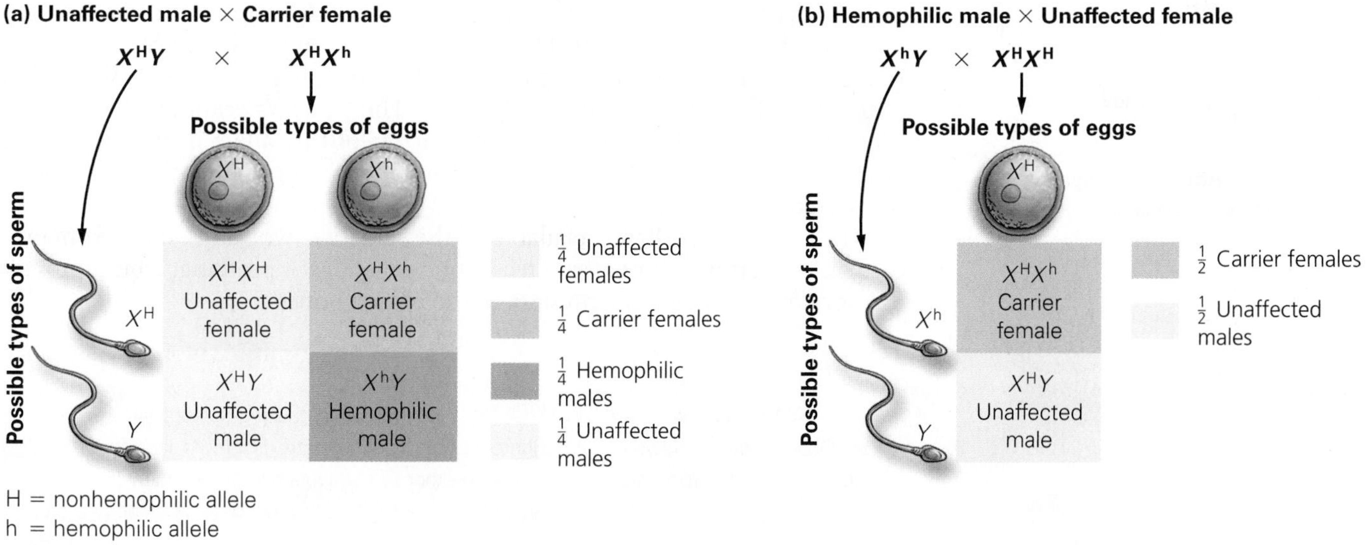

**(a) Unaffected male × Carrier female**

$X^H Y$  ×  $X^H X^h$

H = nonhemophilic allele
h = hemophilic allele

**(b) Hemophilic male × Unaffected female**

$X^h Y$  ×  $X^H X^H$

**FIGURE 8.5 Genetic crosses involving the X-linked hemophilia trait.** Cross (a) shows possible outcomes and probabilities of a mating between a nonhemophilic male and a female carrier. (b) shows possible outcomes and probabilities of a cross between a hemophilic male and an unaffected female.

**Visualize This ▲**

**What genetic cross would result in the highest frequency of affected males?**

## Scientists tried to determine the sex of the buried individuals based on pelvic bone structure.

### Visualize This ▼

Use the symbols below to draw a pedigree of a woman who had a son with her first husband, followed by two boys with her second husband.

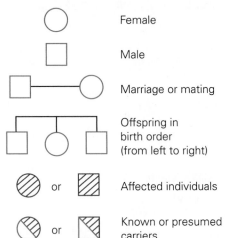

**Pedigree analysis symbols**

○   Female

□   Male

□—○   Marriage or mating

Offspring in birth order (from left to right)

⊘ or ▨   Affected individuals

⊘ or ▧   Known or presumed carriers

**FIGURE 8.6 Pedigree analysis.**
Symbols used in pedigrees.

**Y-Linked Genes.** Genes on the Y chromosome are passed from fathers to sons. The Y chromosome is very short and carries few genes. One gene known to be located exclusively on the Y chromosome is called the *SRY* gene (for sex-determining region of the Y chromosome). The expression of this gene triggers a series of events leading to development of the testes around 8 weeks after conception.

Scientists examined DNA from the bones thought to be the Romanovs for sequences known to be present only on the Y chromosome. When DNA that was isolated from the children's remains was analyzed, it became clear that the children's bones all belonged to girls. If these bones did belong to the Romanovs, one of the two missing children was Alexis, the Romanovs' only son.

## 8.3 Pedigrees

Another line of evidence was provided by the extensive family trees of the Romanovs and their relatives. We can trace the lineage of hemophilia through the Romanov family by using a chart called a pedigree. A **pedigree** is a family tree that follows the inheritance of a genetic trait for many generations of relatives. Pedigrees are often used in studying human genetics because it is impossible and unethical to set up controlled matings between humans the way one can with fruit flies or plants. Pedigrees allow scientists to study inheritance by analyzing matings that have already occurred. **FIGURE 8.6** identifies some of the symbols used in pedigrees, and **FIGURE 8.7** shows how scientists can use pedigrees to determine whether a trait is inherited as autosomal dominant or recessive, or as a sex-linked allele.

Information is available about the Romanovs' ancestors from genealogies kept of royal European families. In addition, scientists interested in hemophilia had kept very good records of the inheritance of this trait. Hemophilia was common among European royal families but rare among the rest of the population. This was because members of the royal families intermarried to preserve the royal bloodlines.

The tsarina must have been a carrier of the hemophilia allele because her son Alexis had the trait. Her mother, Alice, must also have been a carrier because the tsarina's brother Fred had the disease, as did two of her sister Irene's sons, Waldemar and Henry. The tsarina's grandmother, Queen Victoria, seems to have been the first carrier of this allele in the family, because there is no evidence of this disease before her eighth child, Leopold, was diagnosed with it (**FIGURE 8.8**).

The extensive pedigree available to the scientists working on the Romanov case, in concert with a powerful technique called DNA profiling, would provide the key data in solving the mystery of the buried bones.

**STOP & STRETCH**   The mutation causing hemophilia was passed from Queen Victoria to two of her daughters and one of her sons. Is it more likely that this mutation originally arose in Queen Victoria's mother or that Queen Victoria underwent three separate mutations while she was making the egg cells that produced her affected and carrier children? Justify your answer.

**(a) Dominant trait: Polydactyly**

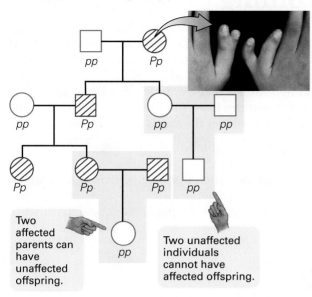

Two affected parents can have unaffected offspring.

Two unaffected individuals cannot have affected offspring.

**(b) Recessive trait: Attached earlobes**

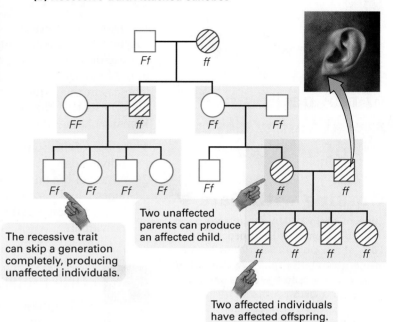

The recessive trait can skip a generation completely, producing unaffected individuals.

Two unaffected parents can produce an affected child.

Two affected individuals have affected offspring.

**(c) Sex-linked trait: Muscular dystrophy**

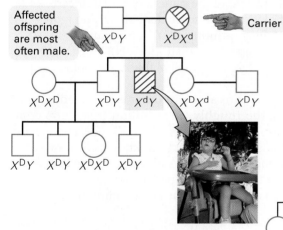

Affected offspring are most often male.

Carrier

**FIGURE 8.7 Pedigrees showing different modes of inheritance.** (a) Polydactyly is a dominantly inherited trait. People with this condition have extra fingers or toes. (b) Having attached earlobes is a recessively inherited trait. (c) Muscular dystrophy, a progressive X-linked disease that causes muscle wasting, is inherited as an X-linked recessive trait.

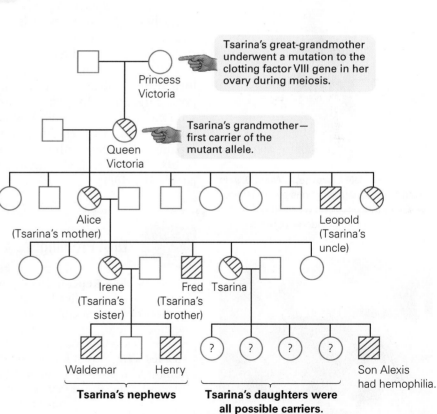

Princess Victoria

Tsarina's great-grandmother underwent a mutation to the clotting factor VIII gene in her ovary during meiosis.

Queen Victoria

Tsarina's grandmother— first carrier of the mutant allele.

Alice (Tsarina's mother)

Leopold (Tsarina's uncle)

Irene (Tsarina's sister)

Fred (Tsarina's brother)

Tsarina

Waldemar     Henry

Tsarina's nephews

Tsarina's daughters were all possible carriers.

Son Alexis had hemophilia.

**FIGURE 8.8 Origin and inheritance of the hemophilia allele.** This abbreviated pedigree shows the origin of the hemophilia allele and its inheritance among the tsarina's family. It appears that the tsarina's great-grandmother underwent a mutation that she passed on to her daughter, who then passed it on to three of her nine children—one of whom was the tsarina's mother, Alice.

**Dental evidence showed that five of the bodies had metal fillings typical of aristocrats.**

## Working with Data ▼

Each of the 13 DNA sequences used in profiling occurs in approximately 10% of the population. What is the rough probability that two unrelated individuals would have the same 13 bands on a DNA profile?

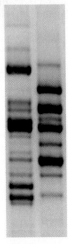

FIGURE 8.9 **DNA profile.** When DNA is isolated and separate regions are analyzed, the size of those regions will differ between individuals who are not identical twins.

FIGURE 8.10 **A thermal cycler.** This machine undergoes temperature changes to allow DNA to unwind and serve as a template for the synthesis of new DNA.

# 8.4 DNA Profiling

Limits on the effectiveness of conventional techniques such as blood typing and karyotyping to identify the bones found in the Ekaterinburg grave necessitated the use of more sophisticated techniques. To do so, scientists took advantage of the fact that any two individuals who are not identical twins have small differences in the sequences of nucleotides that comprise their DNA. To test the hypothesis that the bones buried in the Ekaterinburg grave belonged to the Romanov family, the scientists had to answer the following questions:

1. Which of the bones from the pile are actually different bones from the same individuals?
2. Which of the adult bones could have been from the Romanovs, and which bones could have belonged to their servants?
3. Are these bones actually from the Romanovs, not some other related set of individuals?

## DNA Profiling

All of these questions were answered using **DNA profiling,** a technique that allows unambiguous identification of individuals, based on DNA sequences. When it first came into use, the technique was called DNA fingerprinting, because it helps identify people in the same way actual fingerprinting can. However, because this technology can do more than identify an individual, scientists started referring to DNA fingerprinting as DNA profiling.

To begin this process, it is necessary to isolate the DNA to be analyzed. Scientists can isolate DNA from blood, semen, vaginal fluids, hair follicles, skin, and degraded skeletal remains.

The term **genome** is used to refer to all of an organism's hereditary information. Instead of analyzing the roughly three billion base pairs composing the entire human genome, scientists hone in on 13 different specific sequences of DNA that all humans carry and that are known to show variation in length between different individuals. DNA of different lengths can be separated by size, producing a banding pattern specific to a given individual (**FIGURE 8.9**). Related individuals have more bands in common than unrelated individuals.

### DNA Profiling: **A Closer Look** ▼

The 13 different regions that scientists use for DNA profiling are called **Short Tandem Repeats (STRs)** because they are adjacent repeats of short DNA sequences (four or five bases). STRs are sprinkled between the gene coding sequences of different chromosomes. For example, in a specific location of a particular chromosome, one person may have two repeats of the sequence GATC. At the same chromosomal location, another individual might have five such repeats. These 13 sites were chosen for use in DNA profiling because they are known to vary in number of repeats in the human population. For a particular repeat, maybe GATC, scientists know the percentage of the population that carries one repeat (GATC), two repeats (GATCGATC), three repeats (GATCGATCGATC), and so on.

To work with an STR, scientists first make many copies of the region using a technique called **polymerase chain reaction (PCR).** PCR is performed in test tubes inserted in an automated machine that continuously cycles through periods of heating and cooling (**FIGURE 8.10**). Heat will **denature** DNA by breaking the hydrogen bonds between the double-stranded DNA molecule, resulting in two individual single strands of DNA.

Primer

**①** The required components are placed in a test tube and inserted into the thermal cycler.

Double-stranded DNA
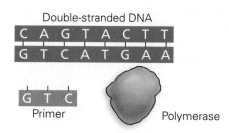

Primer     Polymerase

**②** Heating the DNA separates the double-stranded DNA into two separate single strands, allowing primers to bind to complementary sequences.

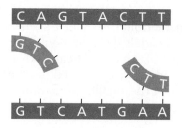

**③** *Taq* polymerase uses the free nucleotides to initiate DNA synthesis, using the primer as a start site.

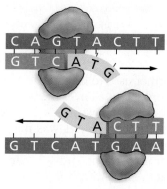

**④** Two double-stranded molecules of DNA are produced from the one original DNA molecule.

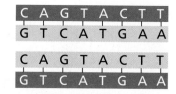

**⑤** This process is repeated many times, with each cycle doubling the amount of DNA.

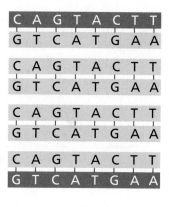

**FIGURE 8.11 The polymerase chain reaction (PCR).** PCR is used to make copies of DNA. Each round of PCR doubles the number of DNA molecules, yielding millions of copies of DNA in a short period of time.

*A Closer Look, continued*

Once the STR DNA is denatured, a special heat-tolerant enzyme called *Taq* polymerase uses nucleotides included in the test tube to build new DNA using each of the separated strands as a template. This enzyme was given the first part of its name (*Taq*) because it was first isolated from ***Thermus aquaticus,*** a bacterium that lives in hydrothermal vents and can withstand very high temperatures. The second part of the enzyme's name (polymerase) describes its synthesizing activity—it acts as a DNA polymerase. The polymerase requires a short sequence of DNA known as a primer to be present at the beginning of the region to be copied. Scientists know the sequence of the STR to be amplified, so finding the correct primer is not difficult. This cycle of heating and cooling the tube is repeated many times, with each round of PCR doubling the amount of double-stranded DNA present in the tube, and producing millions of copies of each STR region for scientists to analyze in just a few hours (**FIGURE 8.11**). The mixture of an individual's copied STRs can then be separated by allowing the mixture to migrate through a solid support called **gel,** which is similar in consistency to gelatin. When an electric current is applied, the gel impedes the progress of the larger DNA fragments more than it does the smaller ones (**FIGURE 8.12**). This type of size-based separation of DNA fragments using electric current is called **gel electrophoresis.**

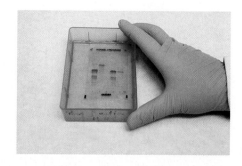

**FIGURE 8.12 Agarose gel.** Fragments of DNA, such as those produced by PCR, can be separated by placing them into an agarose gel and applying an electric current. Smaller DNA samples will migrate through the gel more quickly than larger ones. The DNA in this gel has been stained to show its location in the gel.

## Mystery Solved

In 1992, a team of Russian and English scientists used DNA profiling to confirm that the pile of decomposed bones in the Ekaterinburg grave belonged to nine different individuals.

Once scientists had established that the bones from nine different people were buried in the grave, they tried to determine which bones might belong to the adult Romanovs and which belonged to the servants. For the answer to these questions, scientists took advantage of the fact that Romanov family members would have more DNA sequences in common with each other than they would with the servants.

Each DNA region that a child carries is inherited from one of his or her parents. Therefore, each band produced in a DNA profile of a child must be present in the DNA profile of one of that child's parents (**FIGURE 8.13**). By comparing DNA profiles made from the smaller skeletons, scientists were able to determine which of the six adult skeletons could have been the tsar and tsarina. **FIGURE 8.14** shows part of a hypothetical DNA profile that illustrates how the banding patterns produced can be used to determine which of the bones belonged to the parents of the smaller skeletons and which bones may have belonged to the unrelated servants.

> **STOP & STRETCH** Consider DNA profiles involving the analysis of eight DNA sequences. If the profiles of two siblings are compared, how many bands would you expect them to share in common? How many bands would the profiles of identical twins have in common?

DNA evidence was further used to help put to rest claims made by many pretenders to the throne. People from all over the world had alleged that they were either a Romanov who had escaped execution or a descendant of an escapee.

The most compelling of these claims was made by a young woman who was rescued from a canal in Berlin, Germany, two years after the murders. This young woman suffered from amnesia and was cared for in a mental hospital, where the staff named her Anna Anderson (**FIGURE 8.15**). She later came to

## Visualize This ▼

Compare the bands found in the child's profile. Is every band in the child's profile present in each parent or in one of the two parents?

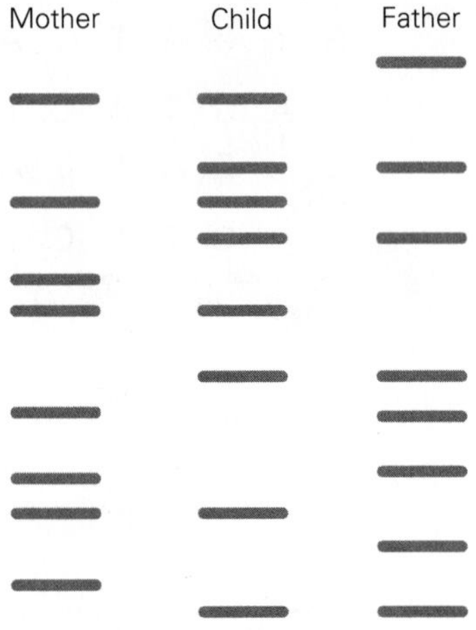

Mother        Child        Father

**FIGURE 8.13 DNA profile.** This photograph shows the DNA profiles of a mother, a father, and their child.

**FIGURE 8.14 Hypothetical DNA profiles of adult- and child-sized skeletons.** A hypothetical DNA profile of the bone cells of individuals found in the Ekaterinburg grave is shown. From the results of this profile, it is evident that children 1, 2, and 3 are the offspring of adults 1 and 3. The remaining DNA from adults does not match any of the children, so these adults are not the parents of any of these children.

**(a) Anna Anderson**

**(b) Anastasia Romanov**

**FIGURE 8.15 Anna Anderson and Anastasia Romanov.**   DNA evidence was more useful than photographs in helping to determine whether Anna Anderson was, as she claimed, Anastasia Romanov.

believe that she was Anastasia Romanov, a claim she made until her death in 1984. The 1997 animated film *Anastasia* served to bolster her story in the minds of many who saw the film.

DNA profiling had been done in the early 1990s on intestinal tissue removed during a surgery performed before Anna Anderson's death. The analysis showed that Anna was not related to anyone buried in the Ekaterinburg grave and could not be Anastasia.

Thus far in our narrative, scientists have answered two of the questions posed. They have determined that (1) nine different individuals were buried in the Ekaterinburg grave and (2) two of the adult skeletons were the parents of the three children. We still haven't answered the last question: How did the scientists show that these were bones from the Romanov family, not just some other set of related individuals?

To answer this question, scientists turned to living relatives of the Romanovs. DNA testing was performed on England's Prince Philip, who is a grandnephew of Tsarina Alexandra. In addition, Nicholas II's dead brother George was exhumed, and his DNA was tested. **FIGURE 8.16** (on the next page) shows the Romanov family pedigree; you can see how these individuals are related to one another.

These DNA tests showed that George was genetically related to the adult male skeleton and that Prince Philip was genetically related to the adult female skeleton, thus tying the buried skeletons to family members of the tsar and tsarina. This evidence strongly supported the hypothesis that the adult skeletons were indeed those of the tsar and tsarina. In 1998, Russian authorities provided a funeral and proper burial for the Romanov family, some 80 years after their execution. Ten years after that, bone fragments unearthed in a forest near where the family was killed were subjected to DNA profiling and found to belong to Alexis and his sister.

## Visualize This ▼

Would the baby prince George, new heir of the royal family in England, be likely to have hemophilia? Why or why not?

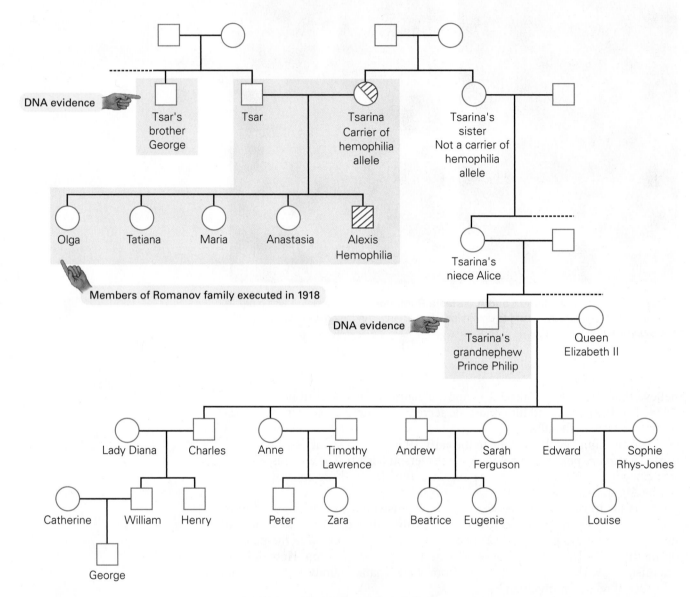

**FIGURE 8.16 Romanov family pedigree.** This pedigree shows only the pertinent family members. DNA from the tsar's brother George showed that he was related to the tsar. Prince Philip, the tsarina's grandnephew, married Queen Elizabeth II. Together they had four children, Charles, Anne, Andrew, and Edward, the current British royal family.

# *savvy reader*

## Choosing the Sex of Your Child

While browsing the products on a shelf in a pharmacy waiting for a prescription to be filled, you come across a product that claims to help you conceive a child of a particular sex. This product costs thirty dollars and purports to have a success rate of 60%. Companies that sell these kits claim to increase the odds of having a child of a particular gender by altering the acidity of the female reproductive tract.

1. Compare the odds of having a child of a particular sex by chance with the success rate reported by the product manufacturer.

2. Is it possible to know whether the success rate of the sex selection product is actually better than chance if we don't have any information about the statistical significance of the reported success rate?

3. Aside from carrying different sex chromosomes, would an X-bearing sperm differ in any known way from a Y-bearing sperm?

4. Did you learn any biology in this chapter that would suggest that the pH of the reproductive tract would determine whether an X-bearing or a Y-bearing sperm reached the egg cell?

5. In China, South Korea, and India, selective abortion of female fetuses has led to altered sex ratios. There are now about 11 boys born in these countries for every 10 girls. What are some possible biological consequences of a sex ratio that is not 1:1?

# SOUNDS RIGHT **BUT IS IT?**

A couple with two boys is considering having another child. While they are grateful to be fertile and to have had two healthy boys, they do think it would be fun to have a girl. In investigating their odds, they came across some data on sibships, or groups of siblings with the same parents. The study they saw showed that three-fourths of sibships of three contain members of both genders versus containing all boys or all girls. The couple now believes that their next child will very likely be a girl.

**If a couple has two boys, the odds are higher than normal that their next child will be a girl.**

Sounds right, but it isn't.

1. The probabilities of independent events, that is, events not affected by previous events, are multiplied to determine their combined probability. For example, when you flip a coin twice, the outcome of the second flip is independent of the outcome of the first flip. Therefore, the likelihood of flipping heads twice is one-half times one-half or one-fourth. What is the likelihood of flipping heads three times in a row?

2. Each fertilization of an egg by a sperm is an independent event. What is the probability that a couple will have three boys in a row?

3. The total probabilities of related independent events equal one. This means that the probability that a couple with three kids will have some outcome aside from three boys is seven-eighths. Describe, in terms of gender, what those sibships could contain.

4. While it is true that in sibships with three children seven-eighths would have at least one girl, how is looking at all the possible sibships that could be produced when there are three children different than the scenario outlined above?

5. Consider your answers to questions 1–4 and explain why the original statement bolded above sounds right, but isn't.

# Chapter Review MasteringBiology®

Go to the Study Area in MasteringBiology® for practice quizzes, myeBook, BioFlix™ 3-D animations, MP3Tutor sessions, videos, current events, and more.

## Summary

### Section 8.1

Differentiate incomplete dominance from codominance.

- Incomplete dominance is an extension of Mendelian genetics in which the phenotype of the progeny is intermediate to that of both parents (pp. 160–161).
- Codominance occurs when both alleles of a given gene are expressed (pp. 160–161).

Explain how polygenic and pleiotropic traits differ.

- Polygenic inheritance occurs when many genes control one trait (p. 161).
- Pleiotropy occurs when a single gene leads to multiple effects (p. 163).

Outline the pattern of inheritance seen in the multiple allelic ABO blood system.

- The ABO blood system displays both multiple allelism (alleles $I^A$, $I^B$, and $i$) and codominance because both $I^A$ and $I^B$ are expressed in the heterozygote (pp. 161–163).

### Section 8.2

Describe the mechanism of sex determination in humans.

- In humans, males have an X and a Y chromosome and can produce gametes containing either sex chromosome, and females have two X chromosomes and always produce gametes containing an X chromosome. When an X-bearing sperm fertilizes an egg cell, a female baby will result. When a Y-bearing sperm fertilizes an egg cell, a male baby will result (p. 164).

Explain the pattern of inheritance exhibited by sex-linked genes.

- Genes linked to the X and Y chromosomes show characteristic patterns of inheritance. Males need only one recessive X-linked allele to display the associated phenotype. Females can be carriers of an X-linked recessive allele and may pass an X-linked disease on to their sons. Y-linked genes are passed from fathers to sons (pp. 164–166).

### Section 8.3

Explain the utility of a genetic pedigree.

- Pedigrees are charts that scientists use to study the transmission of genetic traits among related individuals (pp. 166–167).

### Section 8.4

Explain the significance of DNA profiling and how the process works.

- DNA profiling can be used to unambiguously identify a person and can show the relatedness of individuals based on similarities in their DNA sequences (p. 168).
- Short tandem repeats of DNA are found at specific chromosomal locations in all humans. The number of times a short fragment of DNA is repeated, however, is highly variable (pp. 168–169).
- Many copies of the repeat sequences can be produced using the polymerase chain reaction (pp. 168–169).
- When the amplified repeated sequences are subject to gel electrophoresis, they will be separated by size and a banding pattern specific to the individual being profiled will emerge (p. 169).
- Related individuals will share similar banding patterns (p. 170).

## Roots to Remember

**The following roots of words come mainly from Latin and Greek and will help you decipher terms:**

| | |
|---|---|
| **hemo-** | means blood. Chapter term: *hemophilia* |
| **pleio-** and **poly-** | both come from the Greek word for many. Chapter terms: *pleiotropy and polygenic* |

## Learning the Basics

1. Compare and contrast the terms *codominance* and *incomplete dominance*.

2. What does it mean for a trait to have multiple alleles? Can an individual carry more than two alleles of one gene?

3. How is sex determined in humans?

4. Describe the technique of DNA profiling.

5. If a man with blood type A and a woman with blood type B have a child with type O blood, what are the genotypes of each parent?

6. A man with type A⁺ blood whose father had type O⁻ blood and a woman with type AB⁻ blood could produce children with which phenotypes relative to these blood-type genes?

**7.** Which of the following is *not* part of the procedure used to make a DNA profile?

**A.** Short tandem repeat sequences are amplified by PCR; **B.** DNA is placed in a gel and subjected to an electric current; **C.** The genes that encode DNA patterns are cloned into bacteria; **D.** DNA from blood, semen, vaginal fluids, or hair root cells can be used for analysis.

**8.** Which of the following statements is consistent with the DNA profile shown in **FIGURE 8.17**?

**A.** B is the child of A and C; **B.** C is the child of A and B; **C.** D is the child of B and C; **D.** A is the child of B and C; **E.** A is the child of C and D.

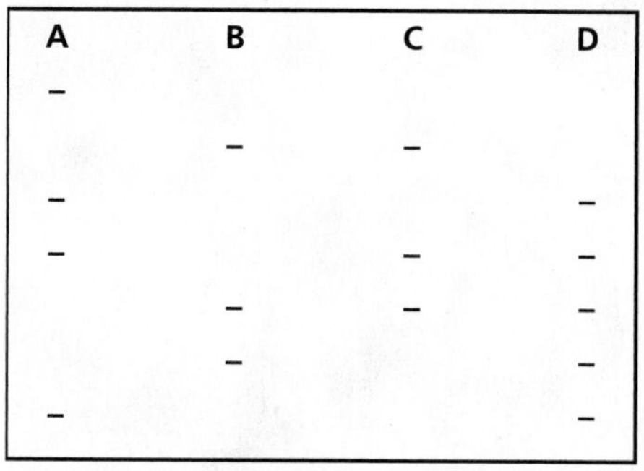

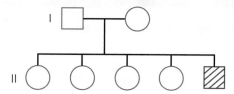

**FIGURE 8.17 DNA profile.**

**9.** The pedigree in **FIGURE 8.18** illustrates the inheritance of a sex-linked recessive trait. What is the genotype of individual II-5?

**A.** $X^H X^H$; **B.** $X^H X^h$; **C.** $X^h X^h$; **D.** $X^H Y$; **E.** $X^h Y$

**FIGURE 8.18 Pedigree.**

**10.** A woman is a carrier of the X-linked recessive color blindness gene. She has children with a man with normal color vision. Which of the following is true of their offspring?

**A.** All the males will be color blind; **B.** All the females will be carriers; **C.** Half the females will be color blind; **D.** Half the males will be color blind.

**11.** Which of the following traits is most likely to be a polygenic trait?

**A.** A blood disease caused by one mutation; **B.** A complex trait like intelligence; **C.** A disease you get when exposed to radiation; **D.** Having type AB blood.

## Analyzing and Applying the Basics

**1.** Why is DNA profiling analysis a much more powerful way to identify an individual than ABO blood type analysis?

**2.** Draw a DNA profile showing a pattern that could be produced if samples from two sisters and their parents were analyzed.

**3.** Draw a pedigree of a mating between first cousins and use the pedigree to explain why matings between relatives can lead to an increased likelihood of offspring with rare recessive diseases.

## Connecting the Science

**1.** Sometimes chromosomal sex does not match biological sex. For instance, males can sometimes have two X chromosomes and no Y chromosome. Most of these males carry a copy of the SRY gene on one X chromosome. What event that occurs during meiosis can explain this phenomenon?

**2.** The Innocence Project is a nonprofit legal clinic at the Benjamin Cardozo School of Law in New York City that attempts to help prisoners whose claims of innocence can be verified by DNA testing because biological evidence from their cases still exists. Over 300 inmates have been exonerated since the project started in 1992. Thousands of inmates await the opportunity for testing. Does the fact that the Innocence Project has shown that the criminal justice system sometimes convicts innocent people change your opinion about the death penalty? Why or why not?

Answers to **Stop & Stretch, Visualize This, Working with Data, Savvy Reader, Sounds Right, But Is It?,** and **Chapter Review** questions can be found in the **Answers** section at the back of the book.

CHAPTER **9**

# Genetically Modified Organisms

**Many people are concerned about consuming genetically modified foods.**

# Gene Expression, Mutation, Stem Cells, and Cloning

Would you want your children to drink milk or consume fruits, vegetables, or meats that have been produced using gene-altering technologies? Do you worry about how genetically modified foods will affect the environment now and in the future? What if your beloved dog died; would you want to be able to clone him? If your grandmother had a debilitating disease, would you want her to be able to

**Some have embraced genetic technologies that allow them to clone a favorite pet.**

**Is it acceptable to modify animals to make them more interesting as pets, such as these fluorescent GloFish$^R$?**

help treat the disease using cells procured from human embryos? Do you think it is ethical to use animals as biological factories for the production of substances helpful in treating human diseases, or to modify pets to make them glow? How about human cloning; do you think the practice should be embraced or outlawed?

Answering these questions requires an understanding of

genetic engineering, the use of laboratory techniques to alter genetic traits in an organism. Once modified, these organisms are referred to as genetically modified organisms or GMOs.

During your lifetime, it is likely that you will need to make many personal decisions involving GMOs. To make sound decisions, it will help to understand how genes typically produce their products and then how and why they are modified to suit human desires.

**Others draw the line at cloning humans.**

**Pet fish have been genetically modified so that they glow.**

# 9.1 Protein Synthesis and Gene Expression

To genetically modify foods, scientists can move a gene known to produce a certain protein from one organism to another. Alternatively, scientists can change the amount of protein a gene produces. Regulating the amount of protein produced by a cell is also referred to as *regulating gene expression.*

One of the first examples of scientists controlling gene expression occurred in the early 1980s when genetic engineers began to produce **recombinant bovine growth hormone (rBGH)** in their laboratories. Recombinant (r) bovine growth hormone is a protein that is made by genetically engineered bacteria. These bacterial cells have had their DNA manipulated so that it carries the instructions for, or encodes, a cow growth hormone that can be produced in the laboratory. Growth hormones act on many different organs to increase the overall size of the body. Bovine growth hormone that is produced in a laboratory can be injected into dairy cows to increase their milk production.

Production of growth hormone protein, or any protein, in the lab or in a cell, requires the use of the genetic information coded in the DNA.

## From Gene to Protein

**Protein synthesis** involves using the instructions carried by a gene to build a particular protein. Genes do not build proteins directly; instead, they carry the instructions that dictate how a protein should be built. Understanding protein synthesis requires that we review a few basics about DNA, genes, and RNA. First, DNA is a polymer of **nucleotides** that make chemical bonds with each other based on their complementarity: adenine (A) to thymine (T) and cytosine (C) to guanine (G). Second, a gene is a sequence of DNA that encodes a protein.

**nucleo-** or **nucl-** refers to a nucleus.

Proteins are large molecules composed of amino acids. Each protein has a unique function that is dictated by its particular structure. The structure of a protein is the result of the order of amino acids that constitute it because the chemical properties of amino acids cause a protein to fold in a particular manner. Before a protein can be built, the instructions carried by a gene are first copied. When the gene is copied, the copy is made up not of DNA (deoxyribonucleic acid) but of **RNA (ribonucleic acid).**

**-ic** is a common ending of acids.

RNA, like DNA, is a polymer of nucleotides. A nucleotide is composed of a sugar, a phosphate group, and a nitrogen-containing base. Whereas the sugar in DNA is deoxyribose, the sugar in RNA is **ribose.** RNA has the nitrogenous base uracil (U) in place of thymine. RNA is usually single stranded, not double stranded like DNA (**FIGURE 9.1**).

**-ose** is a common ending for sugars.

When a cell requires a particular protein, a strand of RNA is produced using DNA as a guide or template. RNA nucleotides are able to make base pairs with DNA nucleotides. C pairs with G, and A pairs with U. The RNA copy then serves as a blueprint that tells the cell which amino acids to join together to produce a protein. Thus, the flow of genetic information in a eukaryotic cell is from DNA to RNA to protein (**FIGURE 9.2**).

How does this flow of information actually take place in a cell? Going from gene to protein involves two steps. The first step, called **transcription,** involves producing the copy of the required gene. In the same way that a transcript of a speech is a written version of the oral presentation, transcription inside a cell produces a transcript of the original gene, with the RNA nucleotides substituted for DNA nucleotides. The second step, called **translation,** involves decoding the copied RNA sequence and producing the protein for which it codes. In the same way that a translator deciphers one language into another, translation in a cell involves moving from the language of nucleotides (DNA and RNA) to the language of amino acids and proteins.

# Visualize This ▼

**Point out the chemical difference between the sugar in DNA and the one in RNA and the difference between the nitrogenous bases thymine and uracil.**

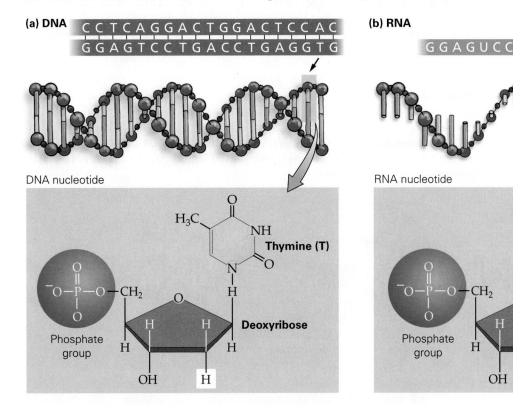

**(a) DNA**

C C T C A G G A C T G G A C T C C A C
G G A G T C C T G A C C T G A G G T G

DNA nucleotide

Thymine (T)

Phosphate group

Deoxyribose

**(b) RNA**

G G A G U C C U G A C C U G A G G U G

RNA nucleotide

Uracil (U)

Phosphate group

Ribose

**FIGURE 9.1 DNA and RNA.** (a) DNA is double stranded. Each DNA nucleotide is composed of the sugar deoxyribose, a phosphate group, and a nitrogen-containing base (A, G, C, or T). (b) RNA is single stranded. RNA nucleotides are composed of the sugar ribose, a phosphate group, and a nitrogen-containing base (A, G, C, or U).

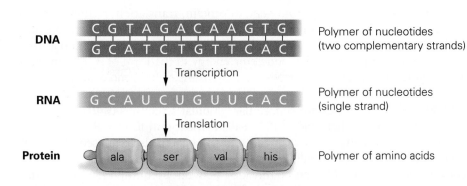

**DNA**

C G T A G A C A A G T G
G C A T C T G T T C A C

Polymer of nucleotides (two complementary strands)

↓ Transcription

**RNA**

G C A U C U G U U C A C

Polymer of nucleotides (single strand)

↓ Translation

**Protein**

ala — ser — val — his

Polymer of amino acids

**FIGURE 9.2 The flow of genetic information.** Genetic information flows from DNA to an RNA copy of the DNA gene, to the amino acids that are joined together to produce the protein coded for by the gene.

◀ **Visualize This**

**Divide the number of nucleotides by the number of amino acids to determine how many nucleotides are required to encode each amino acid.**

## Transcription

Transcription is the copying of a DNA gene into RNA. The copy is synthesized by an enzyme called **RNA polymerase.** To begin transcription, the RNA polymerase binds to a nucleotide sequence at the beginning of every gene, called the **promoter.** Once the RNA polymerase has located the beginning of the gene by binding to the promoter, it then rides along the strand of the DNA helix that comprises the gene (**FIGURE 9.3**). As it is traveling along the gene, the RNA polymerase unzips the DNA double helix and ties together RNA nucleotides that are complementary to the DNA strand it is using as a template. This results in the production of a single-stranded RNA molecule that is complementary to the DNA sequence of the gene. This complementary RNA copy of the DNA gene is called **messenger RNA (mRNA)** because it carries the message of the gene that is to be expressed.

## Visualize This ▼

**Propose a sequence for the DNA strand that is being used to produce the mRNA. Keep in mind that purines (A and G) are composed of two rings and thus are represented by longer pegs in this illustration. Once you have proposed a DNA sequence, determine the mRNA sequence that would be produced by transcription.**

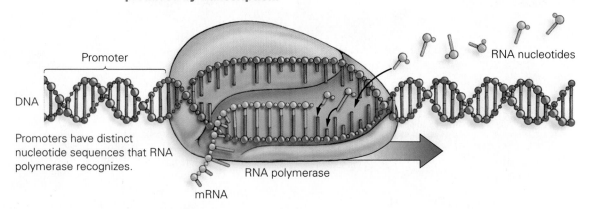

Promoter

DNA

Promoters have distinct nucleotide sequences that RNA polymerase recognizes.

RNA nucleotides

RNA polymerase

mRNA

**FIGURE 9.3 Transcription.** RNA polymerase ties together nucleotides within the growing RNA strand as they form hydrogen bonds with their complementary base on the DNA. When the RNA polymerase reaches the end of the gene, the mRNA transcript is released.

## Translation

The second step in moving from gene to protein, translation, requires that the mRNA be used to produce the actual protein for which the gene encodes. For this process to occur, a cell needs mRNA, a supply of amino acids to join in the proper order, and some energy in the form of ATP. Translation also requires structures called ribosomes and transfer RNA molecules.

**-some** and **somal** relate to a body.

**Ribosomes.** **Ribosomes** are subcellular, globular structures (**FIGURE 9.4**) that are composed of another kind of RNA called **ribosomal RNA (rRNA),** which is wrapped around many different proteins. Each ribosome is composed of two sub-units—one large and one small. When the large and small subunits of the ribosome come together, the mRNA can be threaded between them. In addition, the ribosome can bind to structures called **transfer RNA (tRNA)** that carry amino acids.

**Transfer RNA (tRNA).** Transfer RNA (**FIGURE 9.5**) is yet another type of RNA found in cells. An individual transfer RNA molecule carries one specific amino acid and interacts with mRNA to place the amino acid in the correct location of the growing polypeptide.

As mRNA moves through the ribosome, small sequences of nucleotides are sequentially exposed. These sequences of mRNA, called **codons,** are three nucleotides long and encode a particular amino acid. Transfer RNAs also have a set of three nucleotides, which will bind to the codon if the right sequence is present. These three nucleotides at the base of the tRNA are called the **anticodon**

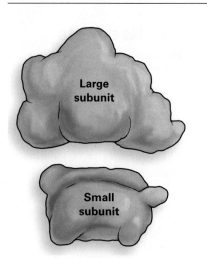

Large subunit

Small subunit

**FIGURE 9.4 Ribosome.** Ribosomes are composed of two subunits. Each subunit in turn is composed of rRNA and protein.

because they complement a codon on mRNA. The anticodon on a particular tRNA binds to the complementary mRNA codon. In this way, the codon calls for the incorporation of a specific amino acid. The ribosome moves along the mRNA sequentially exposing codons for tRNA binding.

When a tRNA anticodon binds to the mRNA codon, a peptide bond is formed. The ribosome adds the amino acid that the tRNA is carrying to the growing chain of amino acids that will eventually constitute the finished protein. The transfer RNA functions as a sort of cellular translator, fluent in both the language of nucleotides (its own language) and the language of amino acids (the target language).

To help you understand protein synthesis, let us consider its similarity to an everyday activity such as baking a cake (**FIGURE 9.6**). To bake a cake, you would consult a recipe book (genome) for the specific recipe (gene) to make your cake (protein). You may copy the recipe (mRNA) out of the book so that the original recipe (gene) does not become stained or damaged. The original recipe (gene) is left in the book (genome) on a shelf (nucleus) so that you can make another copy when you need it. The original recipe (gene) can be copied again and again. The copy of the recipe (mRNA) is placed on the kitchen counter (ribosome) while you assemble the ingredients (amino acids). The ingredients (amino acids) for your cake (protein) include flour, sugar, butter, milk, and eggs. The ingredients are measured in measuring spoons and cups (tRNAs). (While in baking you might use the same cups and spoons for several ingredients, in protein synthesis we use tRNAs that are dedicated to one specific ingredient.) The measuring spoons and cups bring the ingredients to the kitchen counter. Like the ingredients in a cake that can be used in many ways to produce a variety of foods, amino acids can be combined in different orders to produce different proteins. The ingredients (amino acids) are always added according to the instructions specified by the original recipe (gene). Within cells, the sequence of bases in the DNA dictates the sequence of bases

# Visualize This ▼

The structure of a tRNA molecule involves regions where the RNA strand forms complementary bonds with itself, causing the RNA to fold up on itself. What nitrogenous bases might be involved in bonding in such regions of internal complementarity?

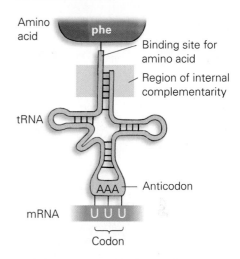

**FIGURE 9.5 Transfer RNA (tRNA).** Transfer RNAs translate the language of nucleotides into the language of amino acids. The tRNA that binds to UUU carries only one amino acid (phe, the three-letter abbreviation for phenylalanine).

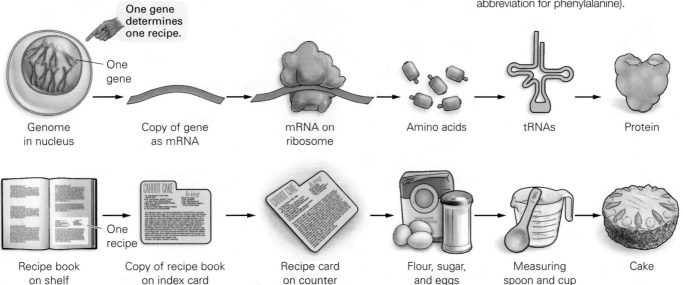

**FIGURE 9.6 Protein synthesis and cake baking.** A recipe book on a shelf resembles the genome in the nucleus of a cell. Copying one recipe onto an index card is similar to copying the gene to produce mRNA. The recipe card is placed on the kitchen counter, and the mRNA is placed on the ribosome. The ingredients for a cake are flour, sugar, butter, milk, and eggs. The ingredients to make a protein are various amino acids. The ingredients are measured in measuring spoons and cups (tRNAs) that are dedicated to one specific ingredient. The measuring spoons and cups bring the ingredients to the kitchen counter. The ingredients (amino acids) are always added according to the instructions specified by the original recipe (gene).

in the RNA, which in turn dictates the order of amino acids that will be joined together to produce a protein. Protein synthesis ends when a codon that does not code for an amino acid, called a **stop codon**, moves through the ribosome. When a stop codon is present in the ribosome, no new amino acid can be added, and the growing protein is released. Once released, the protein folds up on itself and moves to where it is required in the cell. A summary of the process of translation is shown in **FIGURE 9.7**.

The process of translation allows cells to join amino acids in the sequence coded by the gene. Scientists can determine the sequence of amino acids that a gene calls for by looking at the **genetic code**.

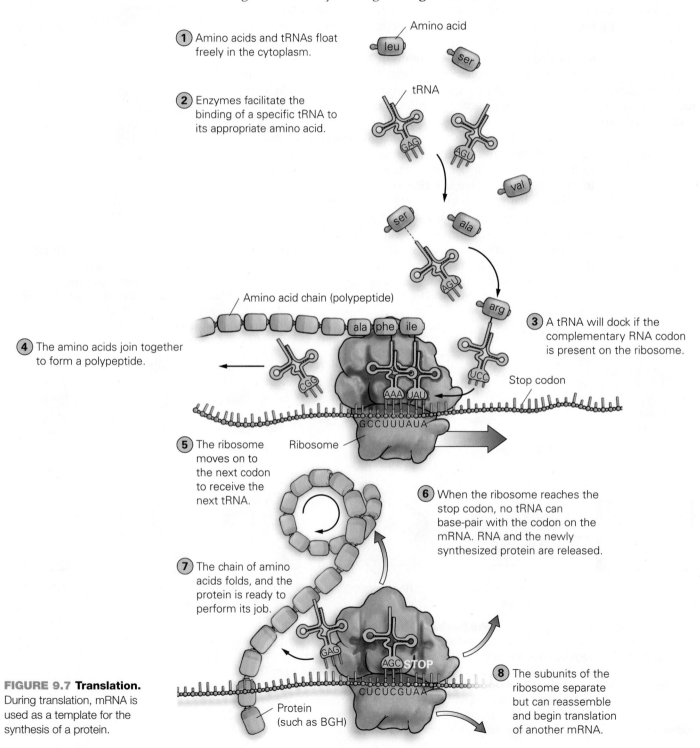

1) Amino acids and tRNAs float freely in the cytoplasm.

2) Enzymes facilitate the binding of a specific tRNA to its appropriate amino acid.

Amino acid

tRNA

Amino acid chain (polypeptide)

4) The amino acids join together to form a polypeptide.

3) A tRNA will dock if the complementary RNA codon is present on the ribosome.

Stop codon

5) The ribosome moves on to the next codon to receive the next tRNA.

Ribosome

6) When the ribosome reaches the stop codon, no tRNA can base-pair with the codon on the mRNA. RNA and the newly synthesized protein are released.

7) The chain of amino acids folds, and the protein is ready to perform its job.

8) The subunits of the ribosome separate but can reassemble and begin translation of another mRNA.

Protein (such as BGH)

**FIGURE 9.7 Translation.**
During translation, mRNA is used as a template for the synthesis of a protein.

**Genetic Code.**   The genetic code shows which mRNA codons code for which amino acids. As **TABLE 9.1** shows, there are 64 codons, 61 of which code for amino acids. Three of the codons are stop codons that occur near the end of an mRNA. Because stop codons do not code for an amino acid, protein synthesis ends when a stop codon enters the ribosome. In the table, you can see that the codon AUG functions both as a start codon (and thus is found near the beginning of each mRNA) and as a codon dictating that the amino acid methionine (met) be incorporated into the protein being synthesized. This initial methionine is often removed later.

The genetic code has some additional properties to take note of. The code is redundant without being ambiguous, and it is also universal. The redundancy of the code can be seen in examples where the same amino acid is coded for by more than one codon. For example, the amino acid threonine (thr) is incorporated into a protein in response to the codons ACU, ACC, ACA, and ACG. There is, however, no situation where a given codon can call for more than one amino acid. For example, AGU codes for serine (ser) and nothing else. Therefore, there is no ambiguity in the genetic code regarding which amino acid any codon will call for. The genetic code is also universal in the sense that organisms typically decode the same gene to produce the same protein. This is why genes can be moved from one organism to another.

**STOP & STRETCH**   What amino acids would be coded for by the mRNA CCU-AAU? Are there other codons that code for these amino acids?

**TABLE 9.1  The genetic code.**

To determine which amino acid is coded for by each mRNA codon, first look at the left-hand side of the chart for the first-base nucleotide in the codon; there are four rows, one for each possible RNA nucleotide—A, C, G, or U. Then look at the intersection of the second-base columns at the top of the chart and the first-base rows to narrow your search. Finally, the third-base nucleotide in the codon on the right-hand side of the chart determines the amino acid that a given mRNA codon codes for. Note the three codons UAA, UAG, and UGA that do not code for an amino acid; these are stop codons. The codon AUG is a start codon, found at the beginning of most protein-coding sequences.

| First base | Second base U | Second base C | Second base A | Second base G | Third base |
|---|---|---|---|---|---|
| **U** | UUU UUC Phenylalanine (phe) / UUA UUG Leucine (leu) | UCU UCC UCA UCG Serine (ser) | UAU UAC Tyrosine (tyr) / UAA **Stop codon** UAG **Stop codon** | UGU UGC Cysteine (cys) / UGA **Stop codon** UGG Tryptophan (trp) | U C A G |
| **C** | CUU CUC CUA CUG Leucine (leu) | CCU CCC CCA CCG Proline (pro) | CAU CAC Histidine (his) / CAA CAG Glutamine (gln) | CGU CGC CGA CGG Arginine (arg) | U C A G |
| **A** | AUU AUC AUA Isoleucine (ile) / AUG Methionine (met) **Start codon** | ACU ACC ACA ACG Threonine (thr) | AAU AAC Asparagine (asn) / AAA AAG Lysine (lys) | AGU AGC Serine (ser) / AGA AGG Arginine (arg) | U C A G |
| **G** | GUU GUC GUA GUG Valine (val) | GCU GCC GCA GCG Alanine (ala) | GAU GAC Aspartic acid (asp) / GAA GAG Glutamic acid (glu) | GGU GGC GGA GGG Glycine (gly) | U C A G |

**FIGURE 9.8 Substitution mutation.**
A single nucleotide change from the normal DNA sequence (a) to the mutated sequence (b) can result in the incorporation of a different amino acid. If the substituted amino acid has chemical properties different from those of the original amino acid, then the protein may assume a different shape and thus lose its ability to perform its job.

**(a) Normal DNA sequence**

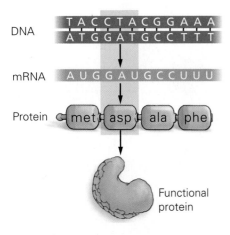

**(b) Mutated DNA sequence**

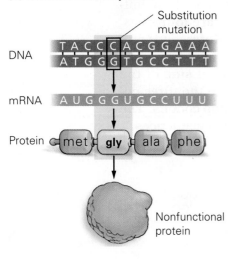

# Mutations

Changes to the DNA sequence, called **mutations,** can affect the order or types of amino acids incorporated into a protein during translation. Mutations to a gene can result in the production of different forms, or alleles, of a gene. Different alleles result from changes in the DNA that alter the amino acid order of the encoded protein. Mutations can result in the production of either a nonfunctional protein or a protein different from the one previously required. If this protein does not have the same amino acid composition, it may not be able to perform the same job (**FIGURE 9.8**). For instance, a substitution of a single nucleotide results in the incorporation of a new amino acid in the hemoglobin protein and compromises the ability of cells to carry oxygen, producing sickle-cell disease.

There are also cases when a mutation has no effect on a protein. These cases may occur when changes to the DNA result in the production of an mRNA codon that codes for the same amino acid as was originally required. Due to the redundancy of the genetic code, a mutation that changes the mRNA codon from, say, ACU to ACC will have no impact because both of these codons code for the amino acid threonine. This is called a **neutral mutation** (**FIGURE 9.9a**). In addition, mutations can result in the substitution of one amino acid for

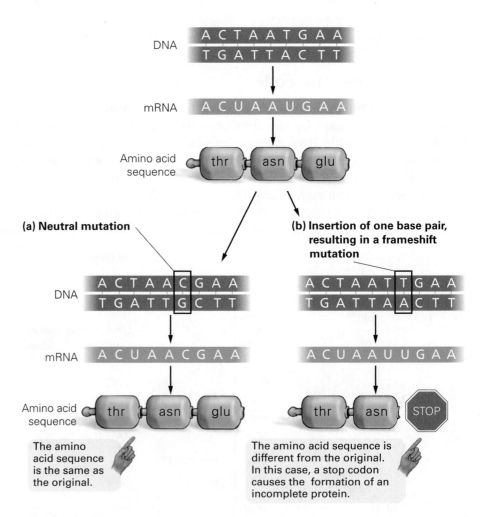

**FIGURE 9.9 Neutral and frameshift mutations.** (a) Neutral mutations result in the incorporation of the same amino acid as was originally called for. (b) The insertion (or deletion) of a nucleotide can result in a frameshift mutation.

another with similar chemical properties, which may have little or no effect on the protein.

Inserting or deleting a single nucleotide can have a severe impact because the addition (or deletion) of a nucleotide can change the groupings of nucleotides in every codon that follows (**FIGURE 9.9b**). Changing the triplet groupings is called altering the **reading frame.** All nucleotides located after an insertion or deletion will be regrouped into different codons, producing a **frameshift mutation.** For example, inserting an extra letter *H* after the fourth letter of the sentence, "The dog ate the cat," could change the reading frame to the nonsensical statement, "The dHo gat eth eca t." Inside cells, this often results in the incorporation of a stop codon and the production of a shortened, nonfunctional protein.

Cells in all organisms undergo this process of protein synthesis, with different cell types selecting different genes from which to produce proteins (**FIGURE 9.10**). In eukaryotic cells, transcription and translation are spatially separate, with transcription occurring in the nucleus and translation occurring in the cytoplasm. Cells lacking a membrane-bound nucleus and organelles are called prokaryotic cells. Prokaryotic cells (such as bacterial cells) also undergo protein synthesis, but transcription and translation occur at the same time and in the same location instead of occurring in separate places. As an mRNA is being transcribed, ribosomes attach and begin translating.

## Gene Expression

Each cell in your body, except sperm or egg cells, has the same complement of genes you inherited from your parents but expresses only a small percentage of those genes. For example, because your muscles and nerves each perform a specialized suite of jobs, muscle cells turn on or express one suite of genes and nerve cells another (**FIGURE 9.11**). Turning a gene on or off, or modulating it more subtly, is called regulating gene expression. The expression of a given gene is regulated so that it is turned on and turned off in response to the cell's needs.

**(a) Eukaryotic protein synthesis**

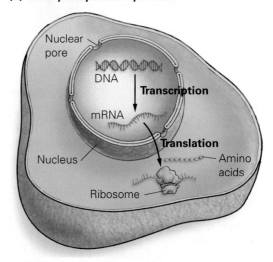

**(b) Prokaryotic protein synthesis**

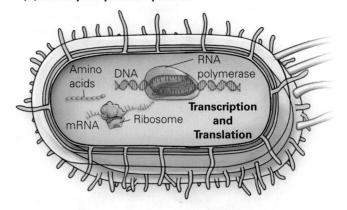

**FIGURE 9.10 Protein synthesis in eukaryotic and prokaryotic cells.**
(a) In eukaryotes, transcription occurs in the nucleus and translation in the cytoplasm.
(b) In prokaryotic cells, which lack nuclei, transcription and translation occur simultaneously.

**(a) Muscle cells**

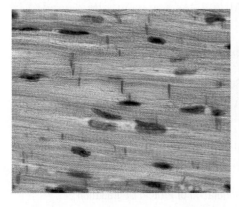

**(b) Nerve cells**

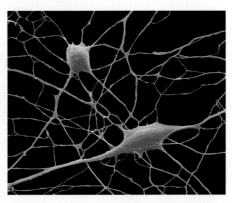

**FIGURE 9.11 Gene expression differs from cell to cell.** (a) A muscle cell performs different functions and expresses different genes than (b) a nerve cell. Both of these cells have the same suite of genes in their nucleus but express different subsets of genes.

## Regulating Gene Expression: **A Closer Look** ▼

Gene expression can be regulated through the rate of transcription or translation. It is also possible to regulate how long an mRNA or protein remains functional in a cell.

### Regulation of Transcription

Gene expression is most commonly regulated by controlling the rate of transcription. This can occur at the promoter. Prokaryotic cells typically regulate gene expression by blocking transcription via proteins called **repressors** that bind to the promoter and prevent the RNA polymerase from binding. When the gene needs to be expressed, the repressor will be released from the promoter so that the RNA polymerase can bind (**FIGURE 9.12a**). This is the main mechanism by which simple single-celled prokaryotes regulate gene expression.

The more complex eukaryotic cells have evolved more complex mechanisms to control gene expression. To control transcription, eukaryotic cells more commonly enhance gene expression using proteins called **activators** that help the RNA polymerase bind to the promoter, thus facilitating gene expression (**FIGURE 9.12b**). The rate at which the polymerase binds to the promoter is also affected by substances that are present in the cell. For example, the presence of alcohol in a liver cell might result in increased transcription of a gene involved in the breakdown of alcohol.

**FIGURE 9.12 Regulation of gene expression.** Gene expression can be regulated by (a) repression or (b) activation. Prokaryotes typically use repression of transcription, and eukaryotes use activation of transcription, to regulate gene expression.

**(a) Repression of transcription**

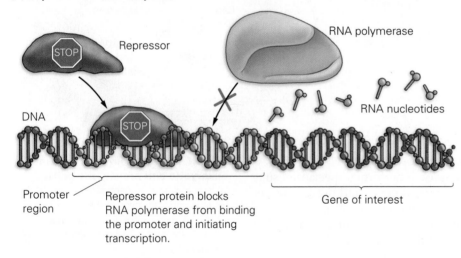

Repressor protein blocks RNA polymerase from binding the promoter and initiating transcription.

**(b) Activation of transcription**

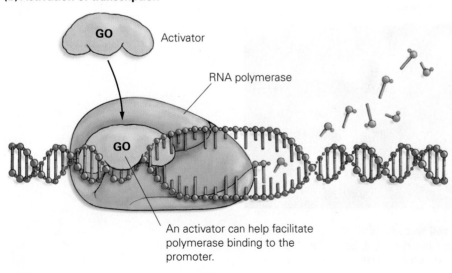

An activator can help facilitate polymerase binding to the promoter.

### Regulation by Chromosome Condensation

It is also possible to regulate gene expression by condensing all or part of a **chromosome.** This prevents RNA polymerase from being able to access genes. Entire chromosomes are also inactivated when they condense during mitosis.

### Regulation by mRNA Degradation

Eukaryotic cells can also regulate the expression of a gene by regulating how long a messenger RNA is present in the cytoplasm. Enzymes called *nucleases* roam the cytoplasm, cutting RNA molecules by binding to one end and breaking the bonds between nucleotides. If a particular mRNA has a long adenosine nucleotide "tail," it will survive longer in the cytoplasm and be translated more times. All mRNAs are eventually degraded in this manner; otherwise, once a gene had been transcribed one time, it would be expressed forever.

### Regulation of Translation

It is also possible to regulate many of the steps of translation. For example, the binding of the mRNA to the ribosome can be slowed or hastened, as can the movement of the mRNA through the ribosome.

### Regulation of Protein Degradation

Once a protein is synthesized, it will persist in the cell for a characteristic amount of time. Like the mRNA that provided the

*A Closer Look, continued*

instructions for its synthesis, the life of a protein can be affected by cellular enzymes called *proteases* that degrade the protein. Speeding up or slowing down the activities of these enzymes can change the amount of time that a protein is able to be active inside a cell.

**chromo-** means color.

Genetic engineering permits precise control of gene expression in many different circumstances. In the case of rBGH, farmers can simply decide how much protein to inject into the bloodstream of a cow. However, they must first synthesize the protein.

# 9.2 Producing Recombinant Proteins

The first step in the production of the rBGH protein is to transfer the BGH gene from the nucleus of a cow cell into a bacterial cell. Bacteria are single-celled prokaryotes that copy themselves rapidly. They can thrive in the laboratory if they are allowed to grow in a liquid broth containing the nutrients necessary for survival. Bacteria with the BGH gene can serve as factories to produce millions of copies of this gene and its protein product. Making many copies of a gene is called **cloning** the gene.

## Cloning a Gene Using Bacteria

The following three steps are involved in moving a BGH gene into a bacterial cell (**FIGURE 9.13**, on the next page).

**Step 1. Remove the Gene from the Cow Chromosome.** The gene is sliced out of the cow chromosome on which it resides by exposing the cow DNA to enzymes that cut DNA. These enzymes, called **restriction enzymes**, act like highly specific molecular scissors. Most restriction enzymes cut DNA only at specific sequences, called *palindromes*, such as

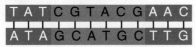

Note that the bottom middle sequence is the reverse of the top middle sequence. Many restriction enzymes cut the DNA in a staggered pattern, leaving "sticky ends," such as

 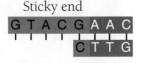

The unpaired bases form bonds with any complementary bases with which they come in contact. The enzyme selected by the scientist cuts on both ends of the BGH gene but not inside the gene.

A particular restriction enzyme cuts DNA at a specific sequence. Therefore, scientists need some information about the entire suite of genes present in a particular organism, called the **genome,** to determine which restriction enzyme cutting sites surround the gene of interest. Cutting the DNA generates many different fragments, only one of which will carry the gene of interest.

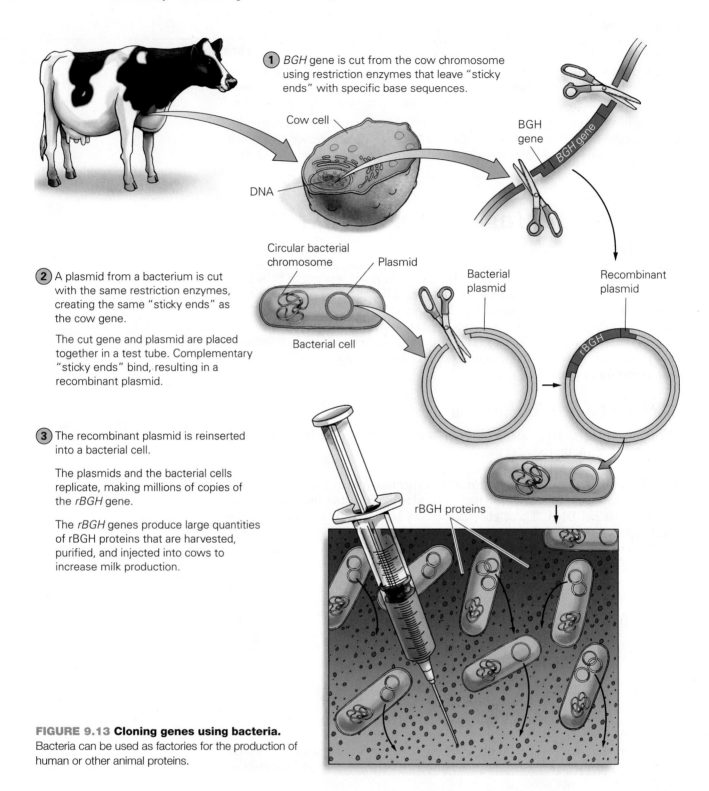

① *BGH* gene is cut from the cow chromosome using restriction enzymes that leave "sticky ends" with specific base sequences.

Cow cell

DNA

BGH gene

BGH gene

Circular bacterial chromosome

Plasmid

Bacterial cell

Bacterial plasmid

Recombinant plasmid

rBGH

② A plasmid from a bacterium is cut with the same restriction enzymes, creating the same "sticky ends" as the cow gene.

The cut gene and plasmid are placed together in a test tube. Complementary "sticky ends" bind, resulting in a recombinant plasmid.

③ The recombinant plasmid is reinserted into a bacterial cell.

The plasmids and the bacterial cells replicate, making millions of copies of the *rBGH* gene.

The *rBGH* genes produce large quantities of rBGH proteins that are harvested, purified, and injected into cows to increase milk production.

rBGH proteins

**FIGURE 9.13 Cloning genes using bacteria.**
Bacteria can be used as factories for the production of human or other animal proteins.

**Step 2. Insert the BGH Gene into the Bacterial Plasmid.** Once the gene is removed from the cow genome, it is inserted into a bacterial structure called a plasmid. A plasmid is a circular piece of DNA that normally exists separate from the bacterial chromosome and can replicate independently of the bacterial chromosome. Think of the plasmid as a molecular ferryboat, able to carry a gene into a bacterial cell, where it can be replicated. To incorporate the BGH gene into the plasmid, the plasmid is also cut with the same restriction enzyme used to cut the gene. Cutting both the plasmid and gene with the same enzyme allows the sticky ends that are generated to base-pair with each other (A to T and G to C). When the cut plasmid and the cut gene are placed together in a test tube, they re-form into a circular plasmid with the extra gene incorporated.

The bacterial plasmid has now been genetically engineered to carry a cow gene. At this juncture, the BGH gene is referred to as the rBGH gene, with the *r* indicating that this product is genetically engineered, or recombinant, because it has been removed from its original location in the cow genome and recombined with the plasmid DNA.

**Step 3. Insert the Recombinant Plasmid into a Bacterial Cell.** The recombinant plasmid is now inserted into a bacterial cell. Bacteria can be treated so that their cell membranes become porous. When they are placed into a suspension of plasmids, the bacterial cells allow the plasmids back into the cytoplasm of the cell. Once inside the cell, the plasmids replicate themselves, as does the bacterial cell, making thousands of copies of the rBGH gene. Using this procedure, scientists can grow large amounts of bacteria capable of producing BGH.

Once scientists successfully clone the BGH gene into bacterial cells, the bacteria produce the protein encoded by the gene. Then the scientists are able to break open the bacterial cells, isolate the BGH protein, and inject it into cows. Bacteria can be genetically engineered to produce many proteins of importance to humans. For example, bacteria are now used to produce the clotting protein missing from people with hemophilia, as well as human insulin for people with diabetes.

Close to one-third of all dairy cows in the United States now undergo daily injections with rBGH. These injections increase the volume of milk that each cow produces by around 20%. Prior to marketing the recombinant protein to dairy farmers, scientists had to demonstrate that its product would not be harmful to cows or to humans who consume the cows' milk. This involved obtaining approval from the U.S. Food and Drug Administration (FDA), the governmental organization charged with ensuring the safety of all domestic and imported foods and food ingredients (except for meat and poultry, which are regulated by the U.S. Department of Agriculture). According to the FDA, there is no detectable difference between milk from treated and untreated cows and no way to distinguish between the two.

FDA approval has not stopped the debate. Many involved in the debate argue that milk from treated cows does differ in composition, and may be harmful. Whether these differences will be substantiated and whether they have an effect on human health remains to be seen. As we wait for definitive answers, many consumers want their milk and milk products to indicate whether it came from treated cows.

**STOP & STRETCH**   If you saw two containers of milk in a grocery cooler, one labeled "hormone and rBGH free" and one with no label, would you infer that one product is safer than the other? Why or why not?

But what about the welfare of the cows? There is some evidence that cows treated with rBGH are more susceptible to certain infections than untreated cows. Due in part to concerns about the health of these animals, many developed countries have banned rBGH use.

The rBGH story is a little different from that of most GMOs because rBGH protein is produced by bacteria and then administered to cows. When organisms are genetically modified, the genome itself is altered.

**FIGURE 9.14 Artificial selection in corn.** Selective-breeding techniques resulted in the production of modern corn (right) from ancient teosinite corn.

# 9.3 Genetically Modified Plants and Animals

Unless you are making a concerted effort not to, you are probably eating genetically modified foods every day. Close to 85% of the corn and soy grown in the United States has been genetically modified and the rate of modification of other crops is increasing as well. Most processed foods contain corn syrup, soy, or vegetable oils from genetically modified plants. Some consumers find this troubling, and others find that their concerns are tempered by the fact that plants and animals have been subject to human modification by artificial selection for thousands of years (**FIGURE 9.14**). Every time a farmer chooses which plants or animals to use in producing the next generation, he or she is modifying the frequency of alleles in that population of organisms. While breeding those cattle that produce the most milk or crossing crop plants that are easiest to harvest does not involve moving a gene from one organism to another, it does select for the propagation of particular genes.

If artificial selection works to propagate desired genes, why do we bother with laboratory techniques? As you will see, the selection process is sped up considerably when genetic technologies are used and the suite of genes available for use is also increased. Two corn plants that are bred together can only draw from traits coded by genes contained in their genomes. With genetic engineering, genes from other organisms can be introduced to the genome. It is this "splicing" of genes from unrelated organisms that is the source of most of the concerns about modifying foods. To understand whether these concerns are justified, we must first consider how crops are modified.

## Modifying Crop Plants

To move a novel gene into a plant, the gene must be able to gain access to the plant cell, which means it must be able to move through the plant's rigid outer cell wall. Moving genes into many agricultural crops such as corn, barley, and rice can be accomplished by using a device called a **gene gun.** A gene gun shoots metal-coated pellets covered with foreign DNA into plant cells (**FIGURE 9.15**). A small percentage of these genes may be incorporated into the plant's genome. When a gene from one species is incorporated into the genome of another species, a **transgenic organism** is produced. When the embryonic plant so treated grows into an adult plant, all of its cells contain the inserted gene and the gene will be passed to its offspring.

Crop plants are genetically modified to increase their shelf life. Tomatoes, for example, have been engineered to soften and ripen more slowly. The longer ripening time means that tomatoes stay on the vine longer, thus making them taste better. The slower ripening also increases the amount of time that tomatoes will be appealing to consumers who do not want to purchase fruit that is overripe or bruised.

Genetic engineering techniques increase crop yield when plants are manipulated to be resistant to pesticides and herbicides. For centuries, farmers have tried to increase yields by killing the pests that damage crops and by controlling the growth of weeds that compete for nutrients, rain, and sunlight. In the United States, farmers typically spray high volumes of chemical pesticides and herbicides directly onto their fields. This practice concerns people worried about the health effects of eating foods that have been treated by these often toxic or cancer-causing chemicals. In addition, both pesticides and herbicides may leach through the soil and contaminate drinking water.

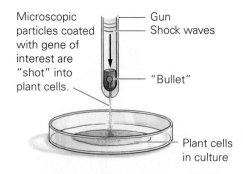

**FIGURE 9.15 Genetically modifying plants using a gene gun.** A gene gun shoots a plastic bullet loaded with tiny DNA-coated pellets into a plant cell. The bullet shells are prevented from leaving the gun, but the DNA-covered pellets enter the nucleus of some cells, producing a genetically modified cell, which can give rise to a modified plant that can pass the modification to its offspring.

There is hope that plants modified to be resistant to herbicides and pesticides would allow farmers to use less of these chemicals. However, weeds and pests continue to evolve resistances to genetic modifications designed to thwart them, in some cases causing the evolution of highly resistant "superweeds" and "superbugs" that may result in the use of even more herbicides and pesticides (**FIGURE 9.16**).

**(a) Rogue weed in a soybean field**

**(b)**

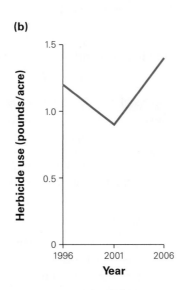

◀ **Working with Data**

**By roughly what percent did herbicide use increase in this 10-year period? State a conclusion consistent with these data that explains why this pattern of an initial drop, followed by an increase in herbicide use, might occur.**

**FIGURE 9.16 Superweeds.** In 1996 farmers began planting seeds genetically modified to resist herbicides. (a) A herbicide-resistant weed growing in a field of soybeans. (b) The change in the use of herbicide on soybeans in the United States between 1996 and 2006, the last year soybeans were surveyed by the USDA.

The nutritive value of crops can also be increased through genetic engineering. A gene that regulates the synthesis of beta-carotene has been inserted into rice, a staple food for many of the world's people. Scientists hope that the engineered rice will help decrease the number of people in underdeveloped nations who become blind due to a deficiency of beta-carotene, which is necessary for the synthesis of vitamin A. Eating this genetically modified rice, called *golden rice*, increases a person's ability to synthesize the vitamin (**FIGURE 9.17a**). Concerns about the health effects of eating golden rice have slowed its distribution and only a few countries are currently growing it and testing it for human consumption. Scientists have even modified potatoes that may some day serve as edible vaccines. Potatoes that carry genes to help prevent the waterborne bacterial disease cholera could save hundreds of thousands of lives each year (**FIGURE 9.17b**).

There is also concern that GM crop plants may transfer engineered genes from modified crop plants to their wild or weedy relatives. Wind, rain, birds, and bees carry genetically modified pollen to related plants near fields containing GM crops (or even to farms where no GM crops are being grown). Many cultivated crops have retained the ability to interbreed with their wild relatives; in these cases, genes from farm crops can mix with genes from the wild crops. This transfer of genes from GM to wild organisms is also a concern when animals are genetically modified.

## Modifying Fish for Human Consumption

Animals as well as plants undergo genetic modifications. Wild Alaskan salmon have been engineered to carry genes from other fish that make them grow larger

**(a)**

**(b)**

**FIGURE 9.17 The promise of GM foods.** Undernourished people at increased risk for blindness may be helped by (a) golden rice, which has been genetically engineered to produce beta-carotene. Those living where access to clean water is limited may be helped by (b) potatoes engineered to help prevent cholera.

**FIGURE 9.18 GM salmon.** The larger salmon carries an inserted DNA sequence that causes it to produce more growth hormone than the unmodified salmon below it.

and faster than normal (**FIGURE 9.18**). Eggs from wild Alaskan salmon are injected with the gene that codes for a growth hormone from another faster growing fish that is regulated via a promoter sequence that is very efficient at binding RNA polymerase. Because this gene is injected into a fertilized egg cell, all the cells of the grown salmon would express this enhanced version of the growth factor.

While breeders of the transgenic salmon indicate that the fish are sterile and will not be released into the wild, there is concern that introduction of engineered salmon into streams could result in reproduction of transgenic fish with wild fish. If this occurs, the offspring of such matings might outcompete non-modified fish and disrupt aquatic food webs.

**STOP & STRETCH** How might the introduction of very large salmon into streams containing wild salmon disrupt the food web in that stream?

In addition to being genetically modified for human dietary consumption, plants and animals are also modified to help humans in other ways.

## Pharming

**Pharming**, a blend of the words *pharmaceutical* and *farming*, is the production of GMOs that produce drugs used to treat or prevent disease. Tobacco plants have been engineered to produce proteins normally found in human blood. The FDA recently approved the first human product produced by an animal for human use. Goats have been engineered to produce a human blood-clotting protein in their milk (**FIGURE 9.19**). Concerns about pharming include the ethics of using animals in this manner and the unregulated spread of modified plant or animal genes via reproduction.

If your beloved dog died, would you want to be able to clone him?

# 9.4 Genetically Modified Humans

Genetic modification of humans has been even more controversial than modification of other living things. From the use of stem cells to the cloning of humans, the potential risks and benefits are still emerging.

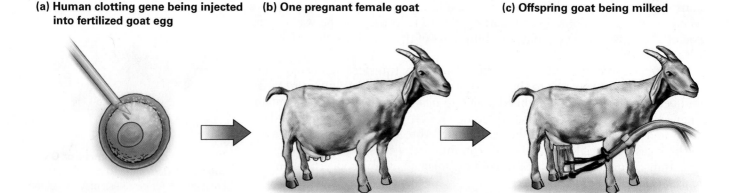

**(a) Human clotting gene being injected into fertilized goat egg**

**(b) One pregnant female goat**

**(c) Offspring goat being milked**

**FIGURE 9.19 Goat's milk.** Herds of goats that produce human blood-clotting proteins are engineered by (a) microinjecting a fertilized goat egg with a human clotting gene; (b) implanting eggs that incorporate the gene into a pregnant female goat; and (c) milking the goat to obtain the human protein. Each goat can produce more clotting protein than can be isolated from tens of thousands of human blood donations.

## Stem Cells

**Stem cells** are unspecialized or **undifferentiated** precursor cells that have not yet been programmed to perform a specific function. They are **totipotent,** meaning they can become any other cell type (**FIGURE 9.20**). Imagine that you are remodeling an old home, and you have a type of material that you can mold into anything you might need—brick, tile, pipe, plaster, and so forth. Scientists believe that stem cells, when nudged in a particular direction, can become any type of cell and therefore may serve as this type of all-purpose repair material in the body.

As the embryo develops, its cells become less and less able to produce other cell types. As a human embryo grows, the early cells start dividing and forming different specialized cells such as heart cells, bone cells, and muscle cells. Once formed, specialized non-stem cells can divide only to produce replicas of themselves. They cannot backtrack and become a different type of cell.

Stem cells can be isolated from early embryos that are left over after fertility treatments. *In vitro* (Latin, meaning "in glass") fertilization procedures often result in the production of excess embryos because many egg cells are harvested from a woman who wishes to become pregnant. These egg cells are then mixed with her partner's sperm in a petri dish, resulting in the production of many fertilized eggs that grow into embryos. A few of the embryos are then implanted into the woman's uterus. The remaining embryos are stored so that more attempts can be made if pregnancy does not result or if the couple desires more children. When the couple achieves the desired number of pregnancies, they are faced with a choice of what to do with the remaining embryos. They may choose to discard them, donate them to other couples experiencing fertility problems, or allow them to be used in stem cell research.

Stem cells can also be found in non-embryonic tissues, including the umbilical cord of a newborn and in the primary teeth of children. Adult stem cells have been found in a variety of organs and tissues, including bone marrow, some blood vessels and muscles, the brain, and the liver. The role of stem cells in the adult body is to help maintain tissues and replace damaged or diseased cells. Adult stem cell research has, in some ways, progressed more rapidly than expected due to the wide variety of tissues in which these cells can be found but has been hampered because these cells also occur in small numbers in a particular tissue and will divide only a limited number of times. These properties make establishing useful cultures of stem cells somewhat more difficult when using stem cells derived from adult tissues versus those derived from embryos.

Whether derived from embryos or the tissues of children and adults, stem cells may someday be used to replace organs damaged in accidents or organs that are gradually failing due to **degenerative diseases.** Degenerative diseases, like liver and lung diseases, heart disease, multiple sclerosis, Alzheimer's, and Parkinson's start with the slow breakdown of an organ and progress to organ failure.

Stem cells could provide healthy tissue to replace those tissues damaged by spinal cord injury or burns. Using stem cells to produce healthy tissues as replacements for damaged tissues is a type of **therapeutic cloning.** New heart muscle could be produced to replace muscle damaged during a heart attack. A diabetic could have a new pancreas, and people suffering from some types of arthritis could have replacement cartilage to cushion their joints. Thousands of people waiting for organ transplants might be saved if new organs were grown in the lab.

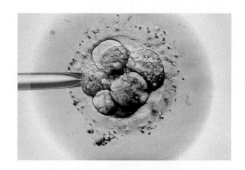

**FIGURE 9.20 Human embryo.** Stem cells can be obtained from an early embryo such as the one cell being removed from this early embryo.

**STOP & STRETCH**   The debate surrounding stem cell research hinges on whether embryonic or non-embryonic stem cells are used. Most individuals find the use of non-embryonic stem cells acceptable. Focusing on when human life begins, summarize the arguments made for and against the use of embryonic stem cells in scientific research.

## Gene Therapy

Scientists who try to replace defective human genes with functional genes are performing **gene therapy.** One type of gene therapy, called **somatic cell gene therapy,** can be performed on body cells to fix or replace the defective protein in only the affected cells. Using this method, scientists introduce a functional version of a defective gene into an affected individual cell in the laboratory, allow the cell to reproduce, and then place the copies of the cell bearing the corrected gene into the diseased person.

Gene therapy to date has focused on diseases caused by single genes for which defective cells can be removed from the body, treated, and reintroduced to the body. For example, studies are underway to determine whether individuals with the genetically inherited lung disease cystic fibrosis may be able to breathe in harmless viruses carrying a functioning copy of the cystic fibrosis gene. If that gene gains access to lung cells and replaces the nonfunctional version with the functional one, lung function could improve (**FIGURE 9.21**).

One additional type of genetic engineering involves making an exact copy of an entire organism by a process called **reproductive cloning.** This process has proven very controversial, especially when the cloning involves humans.

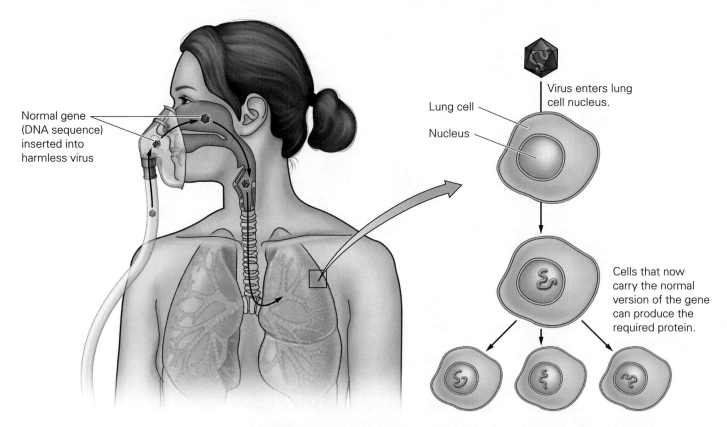

Normal gene (DNA sequence) inserted into harmless virus

Lung cell

Nucleus

Virus enters lung cell nucleus.

Cells that now carry the normal version of the gene can produce the required protein.

**FIGURE 9.21 Gene therapy.** A person with cystic fibrosis has defective genes that prevent normal lung function. Inhaling copies of the normal gene, inserted into a harmless virus, can help restore lung function.

## Cloning Humans

Although we are not used to thinking of identical twins as human clones, they are actually clones of each other. Identical twins arise when an embryo subdivides itself into two separate embryos early in development. When animals are cloned in the laboratory a different process is used, but the result is the same in the sense that the cloned organisms are twins; they are just born many years apart.

Many different animals have been cloned in the laboratory, including sheep, cattle, goats, pigs, horses, mice, cats, rabbits, and monkeys. Initial attempts at cloning focused on livestock that had genetic traits that made them beneficial to farmers. Sheep that produced high-quality wool were the very first cloned animals. Because the animal's genome is duplicated, cloning prize livestock leaves less to chance than simply allowing two animals to breed.

The laboratory process used to clone animals, **nuclear transfer**, involves taking the nucleus from an adult cell and fusing it with an egg cell that has had its nucleus removed (**FIGURE 9.22**). It is important to note that only a small percentage of attempted nuclear transfers result in births and that those animals that are born tend to have health problems and age more rapidly than their naturally conceived counterparts.

It appears likely that the same nuclear transfer technique can be used to clone humans. In 2013, scientists in the United States transferred the nucleus of a human connective tissue cell into an egg cell with its nucleus removed. The resulting cell was allowed to undergo cell division until it produced an embryo composed of around 200 cells, at which time the scientists harvested the embryo for use in therapeutic cloning. It is not clear whether that embryo would have survived gestation to be born, nor is it known whether humans produced by cloning might suffer from the same health-related issues as their cloned livestock predecessors.

## Should human cloning be embraced or outlawed?

### ◀ Visualize This

**Would the cloned baby in this figure be genetically more similar to the man's natural child or to his identical twin brother?**

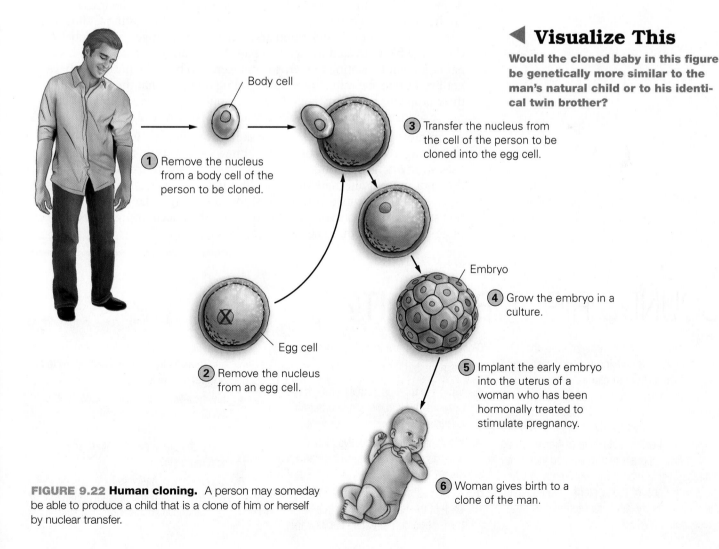

1. Remove the nucleus from a body cell of the person to be cloned.

Body cell

2. Remove the nucleus from an egg cell.

Egg cell

3. Transfer the nucleus from the cell of the person to be cloned into the egg cell.

Embryo

4. Grow the embryo in a culture.

5. Implant the early embryo into the uterus of a woman who has been hormonally treated to stimulate pregnancy.

6. Woman gives birth to a clone of the man.

**FIGURE 9.22 Human cloning.** A person may someday be able to produce a child that is a clone of him or herself by nuclear transfer.

# *savvy reader*

## Labeling GMOs

The following was excerpted from the GMO FREE NY (http://gmofreeny.net/thecaseforgmolabeling.html) website dated 2014 that makes the case for labeling genetically modified foods. No author of this essay is named.

"Genetic modification (GM; also called genetic engineering or GE) is biotechnology used to create new varieties of plants and animals that exhibit traits found in unrelated species, such as bacteria and viruses."

The author then points out that many countries do have laws requiring the labeling of GMO foods, and goes on to say, "Americans have been eating GMOs without their knowledge or consent since 1996. We are the ultimate guinea pigs."

1. The author chose to focus on modifications involving the transfer of genes from bacteria and viruses to crop foods instead of the transfer of a gene from one crop food to another. Why do you think the author focused on bacteria and viruses instead of other kinds of organisms?

2. Assume a food was modified with bacterial or viral DNA. What happens to the DNA of that, or any, food after it is ingested: Is it used to produce bacterial and viral proteins or is it broken down?

3. If you read the entire essay, you would find that the author focuses only on the potential negative outcomes of producing and ingesting GMOs. For example, consider the suggestion above that humans ingesting unlabeled GMOs are being treated like guinea pigs. If the author had softened his or her argument by stating that crops engineered to be more nutritious have not lived up to the initial hype surrounding them, or that the benefits of these crops may not outweigh the risks, would his or her argument for labeling foods seem more credible to you? Why or why not?

4. An opinion piece does not have the same requirements for presenting substantiating evidence as a conventional news story. In this piece, the author uses strong language, often punctuates sentences with exclamation points, does not present alternate opinions, and possibly overstates conclusions in an attempt to convince you that GM foods should be labeled as such. Did this strategy work on you, or would a more balanced approach have been more likely to get you to agree with his or her opinion?

Source: http://gmofreeny.net/thecaseforgmolabeling.html

# SOUNDS RIGHT **BUT IS IT?**

Human cloning is often presented as though clones grow and develop inside a test tube until the adult human clones produced are indistinguishable from the original human they were cloned from, and from each other. If you want to produce five clones of WNBA player Lisa Leslie, place five of her treated cells in five separate test tubes and incubate until the fully formed clones are ready to take the court. Each of these clones would have the athletic ability, drive, and temperament to succeed in the WNBA.

**Cloned humans are identical in every way.**
Sounds right, but it isn't.

1. What has to happen to an embryo produced by human cloning in order for continued development to occur?

2. Would human reproductive cloning using cells from an adult produce an adult clone?

3. If Lisa Leslie was cloned five times, and each of the babies grew up in a household with different eating habits, ranging from very unhealthy to very healthy, do you think all five clones would grow to be six feet five inches tall?

4. Identical twins are clones of each other. What nongenetic factors account for the fact that clones like identical twins do not share the same personalities, ambitions, and interests?

5. Consider your answers to questions 1–4 and explain why the original statement bolded above sounds right, but isn't.

# Chapter Review MasteringBiology®

## Summary

### Section 9.1

Define the term *gene expression*.

- Gene expression is the regulation of protein-producing genes. This can include lowering or increasing the amount of transcription and translation of the gene (p. 178).

Describe how messenger RNA is synthesized during the process of transcription.

- Transcription occurs in the nucleus of eukaryotic cells when an RNA polymerase enzyme binds to the promoter, located at the start site of a gene, and makes an mRNA that is complementary to the DNA gene (pp. 178–180).

Describe how proteins are synthesized during the process of translation.

- Translation occurs in the cytoplasm of eukaryotic cells and involves mRNA, ribosomes, and transfer RNA. Messenger RNA carries the code from the DNA, and ribosomes are the site where amino acids are assembled to synthesize proteins. Transfer RNA carries amino acids, which bind to triplet nucleotide sequences on the mRNA called codons. A particular tRNA carries a specific amino acid. Each tRNA has its unique anticodon that binds to the codon and carries instructions for its particular amino acid (pp. 180–182).

Use the genetic code to identify the amino acid for which a DNA segment or a codon codes.

- The amino acid coded for by a particular codon can be determined using the genetic code (p. 183).

Define the term *mutation*, and explain how mutations can affect protein structure and function.

- Mutations are changes to DNA sequences that can affect protein structure and function (p. 184).
- Neutral mutations are changes to the DNA that do not result in a different amino acid being incorporated. Insertions or deletions of nucleotides can result in frameshift mutations that change the protein more drastically (pp. 184–185).

Explain why and how some genes are expressed in some cell types but not others.

- A given cell type expresses only a small percentage of the genes that an organism possesses (p. 185).
- Turning the expression of a gene up or down is accomplished in different ways in prokaryotes and eukaryotes.

Prokaryotes typically block the promoter with a repressor protein to keep gene expression turned off. Eukaryotes regulate gene expression in any of five ways: (1) increasing transcription through the use of proteins that stimulate RNA polymerase binding; (2) varying the time that DNA spends in the uncondensed, active form; (3) altering the mRNA life span; (4) slowing down or speeding up translation; and (5) affecting the protein life span (pp. 186–187).

### Section 9.2

Outline the process of cloning a gene using bacteria.

- Bacteria can be used to clone genes by placing the gene of interest into a plasmid, which then makes millions of copies of the gene as the plasmid replicates itself inside its bacterial host. Bacteria can then express the gene by transcribing an mRNA copy and translating the mRNA into a protein (pp. 187–189).

### Section 9.3

Describe how plants and animals are genetically modified.

- A gene gun can be used to insert a particular gene into embryonic plant cells (p. 190).
- Animal cells can be modified by injection of genes into an early embryo (pp. 192–193).

List some of the health and environmental concerns surrounding genetic engineering.

- Although there have been no documented incidents of negative health effects from GM food consumption, there is concern that GM foods may be unhealthy. Concerns about GM crops include their impacts on surrounding organisms, the evolution of resistances, and transfer of modified genes to wild or weedy relatives (pp. 190–191).

### Section 9.4

Describe the science behind, and significance of, stem cells and gene therapy.

- Stem cells are undifferentiated cells that can be reprogrammed to act as a variety of cell types. Stem cells may someday allow scientists to treat degenerative diseases (p. 193).
- Gene therapy involves replacing defective genes with normal genes (p. 194).

How and why does cloning occur?

- Cloning occurs via nuclear transfer. Typically, cloning occurs to propagate animals with desirable agricultural traits. It may someday be possible to clone humans, but it is unclear how cloned humans would fare health wise (pp. 194–195).

## Roots to Remember

**The following roots of words come mainly from Latin and Greek and will help you decipher terms:**

| | |
|---|---|
| **chromo-** | means color. Chapter term: *chromosome* |
| **-ic** | is a common ending of acids. Chapter terms: *deoxyribonucleic acid, ribonucleic acid* |
| **nucleo-** | or **nucl-** refers to a nucleus. Chapter term: *nucleotide* |
| **-ose** | is a common ending for sugars. Chapter terms: *ribose, dexyribose* |
| **-some** | or **somal** relates to a body. Chapter terms: *chromosome, ribosome* |

## Learning the Basics

1. List the order of nucleotides on the mRNA that would be transcribed from the following DNA sequence: CGATTACTTA.

2. Using the genetic code (Table 9.1 on p. 183), list the order of amino acids encoded by the following mRNA nucleotides: CAACGCAUUUUG.

3. Describe the process of nuclear transfer used in cloning.

4. Transcription _____.
   **A.** synthesizes new daughter DNA molecules from an existing DNA molecule; **B.** results in the synthesis of an RNA copy of a gene; **C.** pairs thymines (T) with adenines (A); **D.** occurs on ribosomes.

5. Transfer RNA (tRNA) _____.
   **A.** carries monosaccharides to the ribosome for synthesis; **B.** is made of messenger RNA; **C.** has an anticodon region that is complementary to the mRNA codon; **D.** is the site of protein synthesis.

6. During the process of transcription, _____.
   **A.** DNA serves as a template for the synthesis of more DNA; **B.** DNA serves as a template for the synthesis of RNA; **C.** DNA serves as a template for the synthesis of proteins; **D.** RNA serves as a template for the synthesis of proteins.

7. Translation results in the production of _____.
   **A.** RNA; **B.** DNA; **C.** protein; **D.** individual amino acids; **E.** transfer RNA molecules.

8. The RNA polymerase enzyme binds to _____, initiating transcription.
   **A.** amino acids; **B.** tRNA; **C.** the promoter sequence; **D.** the ribosome.

9. A particular triplet of bases in the coding sequence of DNA is TGA. The anticodon on the tRNA that binds to the mRNA codon is _____.
   **A.** TGA; **B.** UGA; **C.** UCU; **D.** ACU.

10. RNA and DNA are similar because _____.
    **A.** both are double-stranded helices; **B.** uracil is found in both of them; **C.** both contain the sugar deoxyribose; **D.** both are made up of nucleotides consisting of a sugar, a phosphate, and a base.

11. **True/False:** A kidney cell and a heart cell from the same cat would have the same genes.

12. **True/False:** Scientists know that genetically modified crops do not hurt the environment.

## Analyzing and Applying the Basics

1. Take another look at the genetic code (Table 9.1, p. 183). Do you see any similarities between codons that code for the same amino acid? Based on this difference, why might a mutation that affects the nucleotide in the third position of the codon be less likely to affect the structure of the protein than a mutation that affects the codon in the first position?

2. Genes encode RNA polymerase molecules. What would happen to a cell that has undergone a mutation to its RNA polymerase gene?

3. Draw a box around the six-base-pair site at which a restriction enzyme would most likely be cut:

   ATGAATTCCGTCCG

   TACTTAAGGCAGGC

# Connecting the Science

**1.** The first "test-tube baby," Louise Brown, was born over 30 years ago. Sperm from her father was combined with an egg cell from her mother. The fertilized egg cell was then placed into the mother's uterus for the period of gestation. At the time of Louise's conception, many people were very concerned about the ethics of scientists performing these in vitro fertilizations. Do you think human cloning will eventually be as commonplace as in vitro fertilizations are now? Why or why not?

**2.** Do you think it is acceptable to grow genetically modified foods if health risks turn out to be low but environmental effects are high?

Answers to **Stop & Stretch, Visualize This, Working with Data, Savvy Reader, Sounds Right, But Is It?**, and **Chapter Review** questions can be found in the **Answers** section at the back of the book.

# CHAPTER 10

# Where Did We Come From?

**Should public school students be taught about alternative hypotheses to evolution?**

# The Evidence for Evolution

In 2006, the highly respected journal *Science* published an article in which the authors compared the results of a simple survey among populations around the world. The survey asked whether the statement, "Human beings, as we know them, developed from earlier species of animals," is true or false, or whether the respondent is not sure or does not know. In the United States, 39% of respondents thought the statement was false, and another 21% were unsure—leaving only 40% who thought the statement was true. In similarly developed countries, such as Iceland, Denmark, Sweden, Great Britain, and France, up to 80% of the population agree with the statement. In fact, compared with the 32 countries discussed in the article, acceptance of the idea that humans are the product of evolution was lower only in Turkey.

Public opinion in the United States has not changed much since

**Another idea: evolution.**

the *Science* article was published. In 2009, a similar poll asked respondents to choose a statement that best expressed their point of view regarding the origin and development of human beings, offering the choices: "Human beings evolved from less advanced life forms over millions of years" and "God created human beings in their present form in the last 10,000 years." In the United States, 35% chose the evolution statement, while 47% chose the creation statement. In Canada, acceptance of the evolution statement was 61%, and in Great Britain, 68%.

To biologists in the United States, these poll results are puzzling. The theory of evolution, including the concept that humans evolved from nonhuman ancestors, forms the bedrock of biological science. Eighty-seven percent of scientists of all types surveyed agreed with the statement that humans have evolved via natural processes. The United States has a world-class higher education system in the sciences, one that draws students from all over the world. So why is the American public seemingly so

skeptical of an idea that is a fundamental principle of biological science?

Could it be that many Americans, expressing the national tradition of questioning authority, reflexively reject ideas that seem counter to their traditions and experience? The majority of American citizens are practicing Christians, and many believe that the creation story in the Bible is literally true. It also seems clear that humans are set apart from all other animals, in our domination of nearly all terrestrial environments and their native organisms. Because of this conflict between how many people experience the world and what science tells us, some argue that the idea of evolution is as much a matter of faith as their own religious doctrine. In this chapter, we examine this debate by exploring the theory of evolution and the origin of humans as a matter of science.

**One idea about the origin of humans: special creation.**

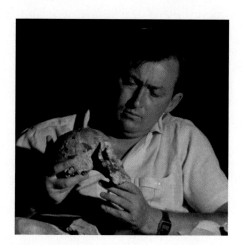

**Why do biologists insist that evolution is the correct explanation?**

# 10.1  What Is Evolution?

**Evolution** really has two different meanings to biologists. The term can refer to either a process or an organizing principle, that is, a theory.

## The Process of Evolution

Generally, the word *evolution* means "change," and the process of evolution reflects this definition as it applies to populations of organisms. A **biological population** is a group of individuals of the same species that is somewhat independent of other groups, often isolated from them by geography. **Biological evolution,** then, is a change in the characteristics of a biological population that occurs over the course of generations. The changes in populations that are considered evolutionary are those that are passed from parent to offspring via genes.

**evol-** means to unroll.

Changes that may take place in populations due only to short-term changes in their environment are not evolutionary. For example, the average dress size for women in the United States has increased from 8 to 14 over the past 50 years. This change is not genetic, but occurred because of an increase in average calorie intake in the population. Thus, it is not an evolutionary change.

> **STOP & STRETCH**  From 1982 to 2008, the number of teenagers in the United States with orthodontic braces increased 99%. Was this likely the result of an evolutionary change in children's teeth? Explain.

As an example of the process of biological evolution, consider the species of organism commonly known as head lice. Some populations of head lice in the United States have become resistant to the pesticide permethrin, found in over-the-counter delousing shampoos. Initially, lice infections were readily controlled through treatment with these products; however, over time, populations of lice evolved to become less susceptible to the effects of these chemicals. Unlike the example of the change in women's dress sizes resulting from increased food availability, this change in the susceptibility of the lice population to permethrin is a result of a change in their genes. The evolution of pesticide resistance can occur rapidly—a study in Israel demonstrated that populations of head lice were four times less susceptible to permethrin only 30 months (or 40 lice generations) after the pesticide was introduced in that country. Note that in this example, individual head lice did not "evolve" or change; instead, the population as a whole changed from one in which most lice were susceptible to the pesticide to one in which most lice were resistant to it.

The differences in resistance to permethrin among individuals in the population resulted from genetic variation. In particular, some lice carried gene variants (that is, alleles) that conferred resistance, and others carried nonresistant alleles. Because individuals with the resistant alleles survived permethrin treatment, they passed these resistant alleles on to their offspring. As a result, a population made up primarily of individual lice that carried the susceptible alleles changed into one in which most of the lice carried the resistant alleles. This change in the characteristics of the population took multiple generations (**FIGURE 10.1**).

According to the definition of evolution presented in this chapter, the population of lice has evolved. In this case, the process of **natural selection,** the differential survival and reproduction of individuals in a population, brought about the evolutionary change. Natural selection is the process by which populations adapt to their changing environment (Chapter 11). Other forces,

# Visualize This ▼

**How would this figure be different if none of the lice in this child's hair were resistant to pesticide? Could evolution occur?**

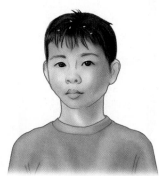

Initial lice infestation consists of both susceptible and resistant lice.

After permethrin treatment, most lice are dead, but a few that are resistant to the pesticide survive.

Reinfestation with the offspring of the resistant lice. The population of lice is now more resistant to permethrin.

**FIGURE 10.1 The process of evolution.** The evolution of pesticide resistance in lice occurs as a result of natural selection, one of the mechanisms by which traits in a population can change over time.

including chance, can cause evolutionary changes in the genetic makeup of populations as well (Chapter 12).

Most people accept that traits in populations can evolve. Evolutionary change in biological populations, such as the development of pesticide resistance in insects and antibiotic resistance in bacteria (Chapter 11), has been observed multiple times. Changes that occur within a biological population are referred to as **microevolution.** The controversy about teaching evolution instead comes from discussions of **macroevolution,** the changes that result in the origin of new species.

## The Theory of Evolution

Although some Americans do not believe that microevolutionary changes occur within a species, theirs is a minority view. However, as the surveys described at the beginning of the chapter illustrate, many dispute whether the theory of evolution, which includes both the process of microevolution and its macroevolutionary results, is "scientific truth."

Some of the controversy is generated by the use of the word *theory*. When people use *theory* in everyday conversation, they often are referring to a tentative explanation with little systematic support; for example, a sports fan might have a theory about why her team is losing, or a gardener might have a theory about why his roses fail to bloom. Usually, these explanations amount to a "best guess" regarding the cause of some phenomenon. A **scientific theory** is much more substantial—it is a statement that provides the current best explanation of how the universe works (Chapter 1). Scientific theories are supported by numerous lines of evidence and have withstood repeated experimental tests.

**Many believe that the creation story in the Bible is literally true.**

**micro-** means extremely small.

**macro-** means large scale.

## Visualize This ▶

Why do some branches on the tree stop short of the right side of the tree?

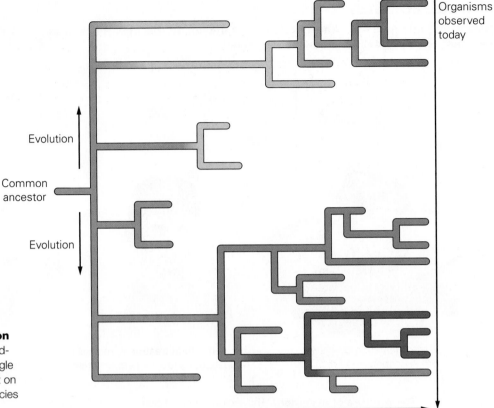

Organisms observed today

Evolution

Common ancestor

Evolution

Time (thousands of generations)

**FIGURE 10.2 The theory of common descent.** This theory states that all modern organisms are descended from a single common ancestor. Each branching point on the tree represents the origin of new species from an ancestral form.

**FIGURE 10.3 Charles Darwin.**
Beginning when he was a young naturalist, Charles Darwin conceived of and developed a theory of evolution that remains one of the most well-supported ideas in biology.

For instance, the theory of gravity explains the motion of the planets, and germ theory explains the relationship between microorganisms and human disease. The theory of evolution is a principle for understanding how species originate and why they have the characteristics that they exhibit. The **theory of evolution** can thus be stated:

> All species present on Earth today are descendants of a single common ancestor, and all species represent the product of millions of years of accumulated microevolutionary changes.

In other words, modern animals, plants, fungi, bacteria, and other living things are related to each other and have been diverging from their common ancestor, by various processes, since the origin of life on this planet. While the processes that cause evolution—such as natural selection—are not the subject of controversy among the general public, the part of the theory of evolution that states that all life shares a common ancestor, called the theory of **common descent**, is (**FIGURE 10.2**). It is this controversy that underlies the opposition to what is often termed *Darwinism*.

## 10.2 Charles Darwin and the Theory of Evolution

The theory of evolution is sometimes called "Darwinism" because Charles Darwin is credited with bringing it into the mainstream of modern science (**FIGURE 10.3**).

The youngest son of a wealthy physician, Darwin spent much of his early life as a lackluster student. After dropping out of medical school, Darwin entered Cambridge University to study for the ministry at the urging of his father. Darwin hated most of his classes but did strike up friendships with several scientists at the college. One of his closest companions was Professor John Henslow, an influential botanist who nurtured Darwin's deep curiosity about the natural world. It was Henslow who secured Darwin his first "job" after graduation. In 1831, at age 22, Darwin set out on what would become his life-defining journey—the voyage of the *HMS Beagle*.

The *Beagle's* mission was to chart the coasts and harbors of South America. The ship's personnel were to include a naturalist for "collecting, observing, and noting anything worthy to be noted in natural history." Henslow recommended Darwin to be the unpaid assistant naturalist (and socially appropriate dinner companion) to the *Beagle's* aristocratic captain after two other candidates had turned it down.

## Early Views of Evolution

The hypothesis that organisms change over time was not new when Darwin embarked on his voyage. The Greek poet Anaximander (611–546 B.C.) seems to have been the first Western philosopher to postulate that humans evolved from fish that had moved onto land.

The first modern evolutionist, Jean Baptiste Lamarck, had published his ideas about evolution in 1809, the year of Darwin's birth. Lamarck was the first scientist to state clearly that organisms adapted to their environments. He proposed that all individuals of every species had an innate, inner drive for perfection, and that the traits they acquired over their lifetimes could be passed on to their offspring. Lamarck used this principle in an attempt to explain the long legs of wading birds, which he argued arose when the ancestors of those animals attempted to catch fish. As they waded in deeper water, they would stretch their legs to their full extent, resulting in gradual lengthening of the limbs. These longer legs would be passed on to the next generation, which would in turn stretch their legs while fishing and pass on even longer legs to the next generation.

Lamarck's contemporaries were unconvinced by his proposed mechanism for species change—for instance, it was easily seen that the children of muscular blacksmiths had similar-sized biceps to those of bankers' children and had not inherited their fathers' highly developed muscles (**FIGURE 10.4**). Lamarck's critics were also unwilling to question the more socially acceptable alternative hypothesis that Earth and its organisms were created in their current forms by God and did not change over time. It was this more acceptable hypothesis that Darwin found himself questioning as a result of his around-the-world voyage.

**FIGURE 10.4 Offspring do not inherit their parents' acquired traits.** Arnold Schwarzenegger was a champion body builder before he became an actor and politician. Lamarck's theory of the inheritance of acquired characteristics suggests that Arnold's son Patrick would also have very developed muscles. Patrick's more average physique demonstrates that Lamarck's ideas were incorrect.

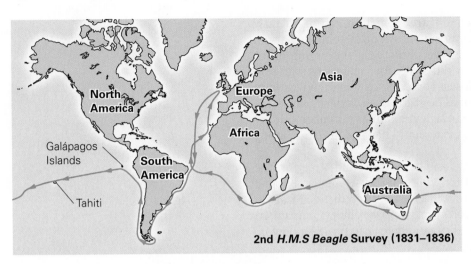

**FIGURE 10.5 The voyage of *HMS Beagle.*** Darwin's expedition took him to tropical locales from South America to Tahiti.

**(a) Dome-shelled tortoise from Santa Cruz Island, an island with abundant ground-level vegetation**

**(b) Flat-shelled tortoise from Española Island, an island with tall vegetation**

**FIGURE 10.6 Giant tortoises of the Galápagos.** The subspecies of giant tortoises on the Galápagos Islands from different environments look distinct from one another.

## The Voyage of the *Beagle*

In the 5 years that he spent on the expedition of the *HMS Beagle,* Darwin spent most of his time on land—luckily for him as it turns out, because he was nearly constantly seasick on board the ship. The trip was a dramatic awakening for the young man, who was awed at the sight of the Brazilian rain forest, amazed by the scantily clothed natives in the chilly climate of Tierra del Fuego, and intrigued by the diversity of animals and plants he collected (**FIGURE 10.5**).

On the ship, Darwin had ample time to read, including Charles Lyell's book *Principles of Geology,* which put forth the hypothesis that the geological processes working today are no different from those working in the past, and that large geological features result from the accumulated effect of these geological forces. In supporting this idea, Lyell argued that deep canyons resulted from the gradual erosion of rock by rivers and streams over enormous timescales. Lyell's hypothesis called into question the belief that Earth was less than 10,000 years old, an age that was deduced from a literal reading of the Bible.

Darwin was also strongly influenced by a stop in the Galápagos, a small archipelago of volcanic islands off the coast of Ecuador. While at first look the islands seemed nearly lifeless, during the month that the *Beagle* spent sailing them, Darwin collected an astonishing variety of organisms. Many of the birds and reptiles he observed appeared to be unique to each island. For instance, while all islands had populations of tortoises, the type of tortoise found on one island was different from the types found on other islands (**FIGURE 10.6**).

Darwin wondered why God would place different, unique subtypes of tortoise on islands in the same small archipelago. On his return to England, Darwin reflected on his observations and concluded that the populations of tortoises on the different islands must have descended from a single ancestral tortoise population. He noted a similar pattern of divergence between other groups of species on the closest mainland and the Galápagos Islands—for instance, prickly pear cacti in mainland Ecuador have the ground-hugging shape familiar to many of us, whereas on the Galápagos, these plants are tree size, but both the mainland and island cacti are clearly related (**FIGURE 10.7**).

## Developing the Hypothesis of Common Descent

Darwin's observations and portions of his fossil collection, sent periodically back to Henslow via other ships, made Darwin a scientific celebrity even before the *Beagle* returned to England. A journal of his travels was a bestseller, and on returning home to England, he settled into a comfortable life with his wife Emma, heiress to a large fortune. After the voyage and the publication of his journal, Darwin reflected on his journal entries and recognized that the observations he had made on living and fossilized plants and animals supported the hypothesis of common descent. However, knowing that this hypothesis was still considered radical, Darwin shared his ideas with only a few close friends.

Instead of publishing his ideas about evolution, Darwin spent the next two decades carefully collecting evidence and further developing his theory. He was finally spurred into publishing his ideas in 1858, after receiving a letter from fellow scientist Alfred Russel Wallace. With Wallace's letter was a manuscript detailing a mechanism for evolutionary change—nearly identical to Darwin's theory of natural selection.

Concerned that his years of scholarship would be forgotten if Wallace published his ideas first, Darwin had excerpts of both his and Wallace's work presented in July 1858 at a scientific meeting in London, and the next year he published the book, *On the Origin of Species by Means of Natural Selection, or the Preservation of Favoured Races in the Struggle for Life.* The main point of this text was to put forward the hypothesis of natural selection. But Darwin devoted the last several chapters of *The Origin of Species* to describing the evidence he had accumulated supporting common descent.

The evidence put forth in *The Origin of Species* was so complete, and from so many different areas of biology, that the hypothesis of common descent no longer appeared to be a tentative explanation. In response, scientists began to refer to this idea and its supporting evidence as the *theory* of common descent. Indeed, most biologists today would agree that common descent is a scientific fact. Darwin's careful catalogue of evidence in his book had revolutionized the science of biology.

## Alternative Ideas on the Origins and Relationships among Organisms

Let us explore the statement "common descent (or evolution) is a fact" more closely. When *The Origin of Species* was published, most Europeans believed that special creation explained how organisms came into being. According to this belief, God created organisms during the 6 days of creation described in the first book of the Bible, Genesis. This belief also states that organisms, including humans, have not changed significantly since creation. According to some biblical scholars, the Genesis story indicates that creation also occurred fairly recently, within the last 10,000 years.

For a hypothesis to be testable by science, we must be able to evaluate it through observations or measurements made within the material universe (Chapter 1). Because a supernatural creator is not observable or measurable, there is no way to determine the existence or predict the actions of such an entity through the scientific method. Therefore, as it is stated, special creation is not a scientific hypothesis. In fact, any statement that supposes a supernatural cause—including intelligent design, which argues that while evolution is possible, some specific features of organisms must have been designed by a creator—cannot be considered science.

**STOP & STRETCH** *Faith* can be defined as the acceptance of ideals or beliefs that are not necessarily demonstrable through experimentation or direct observation of nature. How does faith differ from science? Can statements of faith be tested scientifically? Explain.

The idea of special creation does provide *some* scientifically testable hypotheses, however. For instance, the assertion that organisms came into being within the last 10,000 years and that they have not changed substantially since their creation is testable through observations of the natural world. We can call this hypothesis about the origin and relationships among living organisms the *static model hypothesis,* indicating that organisms are unchanging, in addition

## Visualize This

Consider these photos along with Figure 10.6. What environmental factors might have caused the difference in evolution between prickly pears on the mainland and prickly pears on the islands?

**(a) Prickly pear cactus in mainland South America**

Cactus pad

**(b) Prickly pear cactus in Galápagos**

FIGURE 10.7 **Divergence from a common ancestor.** Prickly pear cacti have very different forms in South America compared to the Galápagos, but they clearly share ancestry, as evidenced by similar pad and flower structures.

to being recently derived (**FIGURE 10.8a**). There are also several intermediate hypotheses between the static model and common descent. One intermediate hypothesis is that all living organisms were created, perhaps even millions of years ago, and that changes have occurred by microevolution in these species, but that brand-new species have not arisen. We will call this the *transformation hypothesis* (**FIGURE 10.8b**). Another intermediate hypothesis is that different *types* of organisms (for example, plants, animals with backbones, or insects) arose separately and since their origin have diversified into numerous species. We will call this the *separate types hypothesis* (**FIGURE 10.8c**). Are these three alternatives to the theory of common descent (**FIGURE 10.8d**) equally likely and reasonable explanations for the origin of biological diversity?

**(a) Static model**
Species arise separately and do not change over time.

**(b) Transformation**
Species arise separately and change over time in order to adapt to the changing environment.

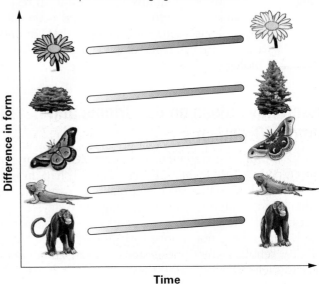

**(c) Separate types**
Species change over time, and new species can arise, but not from a common ancestor. Each group of species derives from a separate ancestor that arose independently.

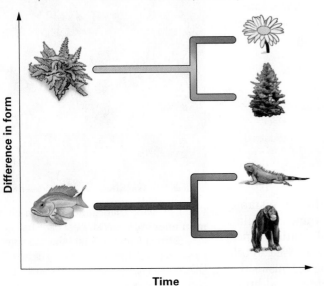

**(d) Common descent**
Species do change over time, and new species can arise. All species derive from a common ancestor.

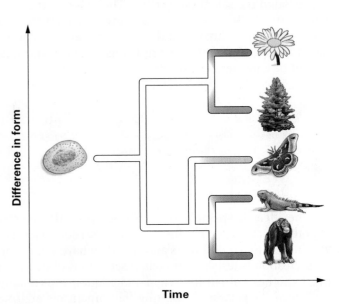

**FIGURE 10.8 Four hypotheses about the origin of modern organisms.** A graphical representation of the four hypotheses.

It appears from the polls that many Americans think so. But what about all those scientists who insist that the theory of common descent is fact? Why would so many maintain this position? As you will soon see, the three alternative hypotheses are not equivalent to the theory of common descent. To understand why, we must examine the observations that help us test these hypotheses.

# 10.3 Examining the Evidence for Common Descent

The evidence that all organisms share a common ancestor comes from several areas of biology and geology. According to many critics of evolution, it seems highly unlikely that humans, with our highly developed brains, could be "just another branch" in the animal family tree. Because this criticism is so common, we will address it head on—by examining the hypothesis that humans share a common ancestor with apes, including the chimpanzee and gorilla.

Any zookeeper will tell you that the primate house is their most popular exhibit. People love apes and monkeys. It is easy to see why—primates are curious, playful, and agile. But something else drives our fascination with primates: We see ourselves reflected in them. The forward placement of their eyes and their reduced noses appear human-like. They have hands with fingernails instead of paws with claws. Some can stand and walk on two legs for short periods. They can finely manipulate objects with their fingers and opposable thumbs. They exhibit extensive parental care, and even their social relations are similar to ours—they tickle, caress, kiss, pout, and grin (**FIGURE 10.9**).

Why are primates, particularly the great apes (gorillas, orangutans, chimpanzees, and bonobos) so similar to humans? Scientists contend that it is because humans and apes are recent descendants of a common biological ancestor.

## Linnaean Classification

As the modern scientific community was developing in the sixteenth and seventeenth centuries, various methods for organizing biological diversity were proposed. Many of these classification systems grouped organisms by similarities in habitat, diet, or behavior; some of these classifications placed humans with the great apes, and others did not.

Into the classification debate stepped Carl von Linné, a Swedish physician and botanist. Von Linné gave all species of organisms a two-part, or binomial, name in Latin, which was the common language of science at the time. In fact, he adopted a Latinized name for himself—Carolus Linnaeus. The Latin names that Linnaeus assigned to other organisms typically contained information about the species' traits—for instance, *Acer saccarhum* is Latin for "maple tree that produces sugar," the sugar maple, whereas *Acer rubrum* is Latin for "red maple." The scientific name of a species contains information about its classification as well. For example, all species with the generic name *Acer* are all maples, and all species with the generic name *Ursus* are bears.

Linnaeus also created a logical and useful system for organizing diversity—grouping organisms in a hierarchy. From broadest to narrowest groupings, the classification levels he designated are

KINGDOM
    PHYLUM
        CLASS
            ORDER
                FAMILY
                    GENUS
                        SPECIES

**Some argue that the idea of evolution is as much a matter of faith as their own religious doctrine.**

FIGURE 10.9 **Are humans related to apes?** Biologists contend that apes and humans are similar in appearance and behavior because we share a relatively recent common ancestor.

Since Linnaeus's time, biologists have added a broader level—the Domain—as the largest grouping of organisms.

Although Linnaeus himself preceded the theory of evolution and believed in special creation, Darwin noted that Linnaeus's hierarchal classification system implied evolutionary relationships among organisms (**FIGURE 10.10**). Linnaeus's classification system was so effective at capturing these relationships that, even after the widespread acceptance of evolutionary theory, it has remained the standard for biological classification. The system has only been modified slightly, with the addition of the domain level and with "sub" and "super" levels between his categories—such as superfamily between family and order—to better represent the relationships among groups of organisms (Chapter 13).

While his hierarchical categories are still in use, many of the actual groupings Linnaeus proposed in the eighteenth century have been overturned or radically altered by the accumulation of additional data. However, when it comes to humans, his original classification remains largely supported by data. Linnaeus placed humans, monkeys, and apes in the same order, which he called Primates. Among the primates, humans are most similar to apes. Humans and apes share a number of characteristics, including relatively large brains, erect posture, lack of a tail, and

## Visualize This ▶

List one or more traits that all members of the order Primates share.

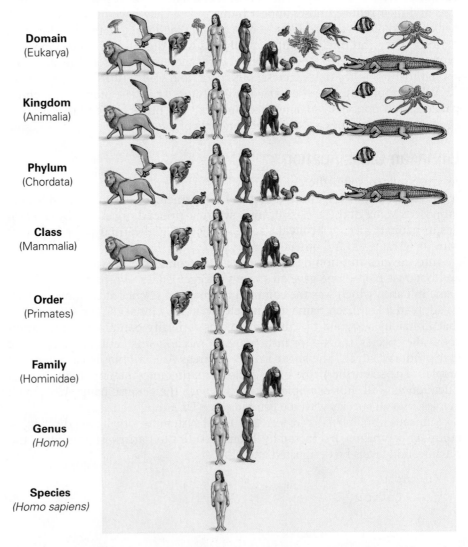

Domain (Eukarya)

Kingdom (Animalia)

Phylum (Chordata)

Class (Mammalia)

Order (Primates)

Family (Hominidae)

Genus (*Homo*)

Species (*Homo sapiens*)

**FIGURE 10.10 The Linnaean classification of humans.** All organisms within a category share basic characteristics, and as the groups become smaller subdivisions toward the bottom of the figure, the organisms have more similarities.

increased flexibility of the thumb. Scientists now place humans and apes in the same family, Hominidae.

Humans and the African great apes share even more characteristics, including elongated skulls, short canine teeth, and reduced hairiness; they are placed together in the same *sub*family, Homininae. When the classification of humans, great apes, and other primates is shown in the form of a tree diagram (**FIGURE 10.11**), it is easy to see why Darwin concluded that humans and modern apes probably evolved from the same ancestor.

## Anatomical Homology

The tree of relationships implied by Linnaeus's classification forms a hypothesis that can be tested. If modern species represent the descendants of ancestors that also gave rise to other species, we should be able to observe other, less obvious similarities between humans and apes in anatomy, behavior, and genes. These similarities are referred to as **homology.** Homology in skeletal anatomy among a variety of organisms can be seen in comparing

**homolog-** indicates similar or shared origin; from a word meaning "in agreement."

## Visualize This ▼

Add "squirrel" to this tree. How can you indicate the shared ancestor of squirrels and other members of this group?

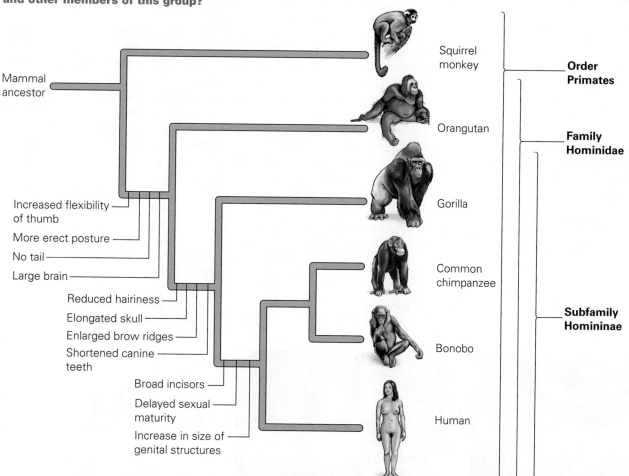

**FIGURE 10.11 Shared characteristics among humans and apes imply shared ancestry.** This tree diagram represents the current classification of humans and apes. Characteristics noted on the side of the evolutionary tree are shared by all of the species on that branch and those to the right.

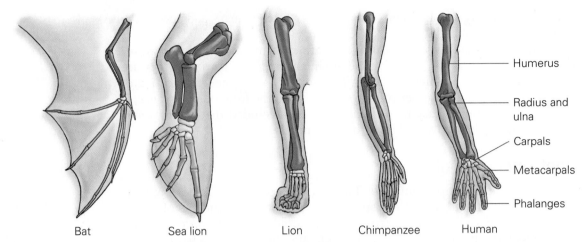

Bat   Sea lion   Lion   Chimpanzee   Human

Humerus

Radius and ulna

Carpals

Metacarpals

Phalanges

## Visualize This ▲

How is the structure of the green bones different among these species? How is the function of these bones different?

**FIGURE 10.12 Homology of mammal forelimbs.** The bones in the forelimbs of these mammals are very similar; equivalent bones in each organism are shaded the same color. The similarity in underlying structure despite great differences in function is evidence of shared ancestry.

mammalian forelimbs, including the presence of a single bone in the upper part of the limb, paired bones in the lower portion, multiple small bones in the wrist, and a number of elongated "finger" bones (**FIGURE 10.12**). The similarities between human and chimpanzee forelimbs are striking, especially the presence of a distinct, opposable thumb.

Note, however, that not all superficial similarities are evidence of evolutionary relationship. Both birds and bats have wings and can fly but do not share a recent common ancestor. Their similarities are due to **convergent evolution,** the development of similar structures in unrelated organisms with similar lifestyles. Another example of convergence is the football-like shape found among animals who spend a significant amount of time hunting food in the water—from tuna fish and penguins, to otters and dolphins. One challenge associated with the study of evolution is describing which similarities result from shared ancestry and distinguishing them from similarities that evolved in parallel in different groups.

Even more compelling evidence of shared ancestry comes from similarities between functional traits in one organism and seemingly nonfunctional or greatly reduced features, or **vestigial traits,** in another. These traits represent a vestige, or remainder, of biological heritage. For example, flightless birds such as ostriches produce functionless wings, and flowering plants still produce a tiny "second generation" (called a gametophyte) within a developing flower ovule, a vestige of their relationship with ferns (**FIGURE 10.13**).

**FIGURE 10.13 Vestigial structures in plants.** The ancestors of flowering plants were similar to modern ferns. Ferns have two independent stages in their life cycle: (a) the familiar form, which is known as the sporophyte stage and (b) a tiny, independent, gametophyte stage. (c) Flowering plants no longer have two independent stages but do produce a tiny gametophyte within the flower itself.

**(a) Fern sporophyte**

**(b) Fern gametophyte (an independent plant)**

**(c) Flowering plant sporophyte containing microscopic gametophyte**

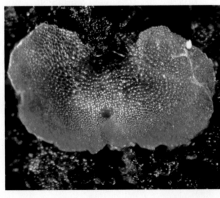

Gametophyte generation is found here.

At least two vestigial traits in humans link us to other primates and great apes (**FIGURE 10.14**). Great apes and humans have a tailbone like other primates, yet neither great apes nor humans have a tail. In addition, all mammals possess tiny muscles called arrector pili at the base of each hair. When the arrector pili contract under conditions of emotional stress or cold temperatures, the hair is elevated. In furry mammals, the arrector pili help to increase the perceived size of the animal, and they increase the insulating value of the hair coat. In humans, the same emotional or physical conditions produce only goose bumps, which provide neither benefit.

Darwin maintained that the hypothesis of evolution provided a better explanation for vestigial structures than did the hypothesis of special creation represented by the static model. A useless trait such as goose bumps is better explained as the result of inheritance from our biological ancestors than as a feature that appeared—or was created—independently in our species.

**STOP & STRETCH**    Some vestigial structures persist because they may still have a function or did in the very recent past. Wisdom teeth, which erupt in early adulthood and often have to be removed because they do not "fit" in a modern person's jaw, are sometimes considered vestigial. However, they may have had a function in the recent past. Why might wisdom teeth have persisted over the course of human evolution?

| "Useful" trait in primate relative | Vestigial trait in human |
| --- | --- |

**(a) Tail bone**

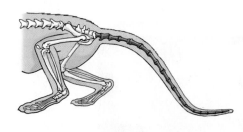
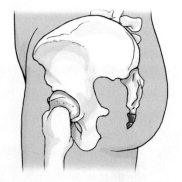

**(b) Goose bumps**

**FIGURE 10.14 Vestigial traits reflect our evolutionary heritage.** (a) Humans and other great apes do not have tails, but they do have a vestigial tailbone. (b) Goose bumps are reminders of our relatives' hairier bodies.

## Developmental Homologies

Multicellular organisms demonstrate numerous similarities in the process of development from fertilized egg to adult organism. Genes that control development are similar among animals that appear as different as humans and fruit flies. As a consequence of these shared developmental pathways, early embryos of very different species often look very similar. For example, all chordates—animals that have a backbone or closely related structure—produce structures called pharyngeal slits (pharyngeal means "pertaining to the throat," referring to their location) and most have tails as early embryos (**FIGURE 10.15**). These structures are even seen in human embryos early in development. The similarities suggest that all chordates derive from a single common ancestor with a particular developmental pathway that they all inherited.

## Molecular Homology

Scientists now understand that differences among individuals arise largely from differences in their genes. It stands to reason that differences among species must also derive from differences in their genes. If the hypothesis of common descent is correct, then species that appear to be closely related must have more similar genes than do species that are more distantly related. The most direct way to measure the overall similarity of two species' genes is to evaluate similarities in the DNA sequences of genes found in both organisms. Species that share a more common ancestor should have more similar DNA sequences than species that share a more distant ancestor (**FIGURE 10.16**).

Many genes are found in nearly all living organisms. For instance, genes that code for histones, the proteins that help store DNA neatly inside cells, are found in algae, fungi, fruit flies, humans, and all other organisms that contain linear chromosomes. Among organisms that share many aspects of structure and function, such as humans and chimpanzees, many genes are

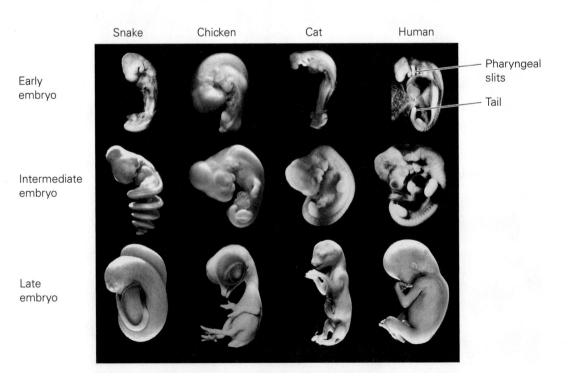

**FIGURE 10.15 Similarity among chordate embryos.** These diverse organisms appear very similar in the first stages of development (shown in the top row), evidence that they share a common ancestor that developed along the same pathway.

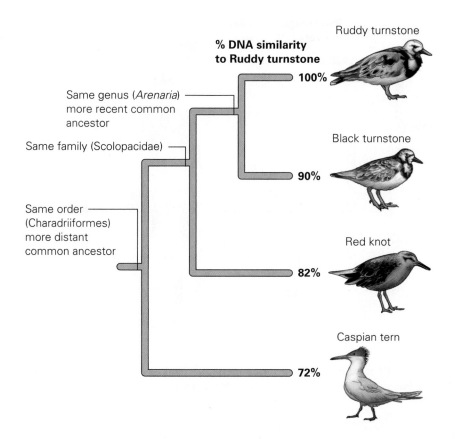

% DNA similarity
to Ruddy turnstone

Ruddy turnstone

100%

Same genus (*Arenaria*)
more recent common
ancestor

Same family (Scolopacidae)

Black turnstone

90%

Same order
(Charadriiformes)
more distant
common ancestor

Red knot

82%

Caspian tern

72%

◄ **Working with Data**
Which bird is more closely related to the
red knot: the Caspian tern or the black
turnstone? Explain your answer.

**FIGURE 10.16 DNA evidence of relation-
ship.** Scientists used the physical characteristics of these four bird species to put them
into different taxonomic categories, shown here
in tree form. DNA studies later supported this
hypothesis, because species that share a more
recent common ancestor have more similar
DNA sequences than species that share a more
distant ancestor.

shared. However, because amino acids can be coded for by more than one
DNA codon, the sequences of genes, even those that produce proteins with
the same amino acid sequence, are typically not identical from one species
to the next.

A comparison of the sequences of dozens of genes that are found in humans and other primates demonstrates the relationship between classification
and gene sequence similarity (**FIGURE 10.17**, on the next page). The DNA
sequences of these genes in humans and chimpanzees are 99.01% similar,
whereas the DNA sequences of humans and gorillas are identical over 98.9%
of their length. More distantly related primates are less similar to humans in
DNA sequence. This pattern of similarity in DNA sequence exactly matches the
biological relationships implied by physical similarity in this group. This result
supports the hypothesis of common descent among the primates.

Interestingly, DNA evidence has called into question groupings that once
seemed obvious based on morphological similarity. For instance, reptiles were
once considered a single group consisting of all the land animals that have scales
as skin covering and lay shelled eggs, while birds were considered a distinct
group based on their production of feathers. It is now clear from DNA evidence
that some living reptiles—the crocodiles and their relatives—are more closely
related to birds than they are to other reptiles. Despite this dramatic exception,
DNA evidence generally confirms the broad groupings of organisms based on
similar morphology and development.

At first, a finding of similarities in DNA sequence may not seem especially
surprising. If genes are instructions, then you would expect the instructions for
building a human and a chimpanzee to be more similar than the instructions
for building a human and a monkey. After all, humans and chimpanzees have
many more physical similarities than humans and monkeys do, including reduced hairiness and lack of a tail.

## Working with Data ▶

Looking at the percentages on this graph might lead a reader to suspect that chimpanzees and gorillas are more closely related to each other than chimpanzees and humans. Explain why this is misreading the data.

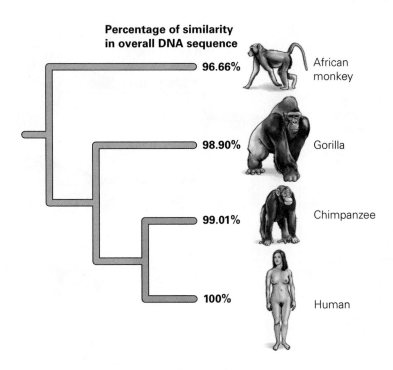

**Percentage of similarity in overall DNA sequence**

96.66% — African monkey

98.90% — Gorilla

99.01% — Chimpanzee

100% — Human

**FIGURE 10.17 DNA evidence for human ancestry.** Similarities in DNA sequences among the primates parallels the hypothesis of relationship implied by similar morphology and development.

However, remember that the genes being compared perform the same function in all of these species. For example, one of the genes in this analysis is *BRCA1*, which in all organisms has the general function of helping repair damage to DNA. (In humans, *BRCA1* is also associated with increased risk for breast cancer, which explains the source of the gene's name.) Given the identical function of *BRCA1* in all organisms, there is no reason to expect that differences in *BRCA1* sequences among different species should display a pattern—unless the organisms are related by descent. But there is a pattern; the *BRCA1* gene of humans is more similar to the *BRCA1* gene of chimpanzees than to the same gene in monkeys. The best explanation for this observation is that humans and chimpanzees share a more recent common ancestor than either species shares with monkeys.

Differences in DNA sequence between humans and chimpanzees can also allow us to estimate when these two species diverged from their common ancestor. The estimate is based on a **molecular clock.** The principle behind a molecular clock is that the rate of change in certain DNA sequences, due to the accumulation of mutations that affect the DNA sequence but not the protein sequence, seems to be relatively constant within a species. According to one application of a molecular clock, the amount of time it takes for a 1% difference in DNA sequence (about the difference between humans and chimpanzees) to accumulate in diverging species is 5 to 6 million years. Scientists have examined the fossil record in an effort to confirm this hypothetical date.

## Biogeography

The distribution of species on Earth is known as the **biogeography** of life. As discussed in this chapter, Darwin observed during his travels on the *Beagle* that each island in the Galápagos had a unique species of tortoise, and that the islands had species of prickly pear cactus that resembled, but were not identical to, the prickly pear found on the mainland. He also noted that the species on the Galápagos were very different from species found on other, similar tropical islands that he visited. If the alternative hypotheses that species appeared independently were true, we would predict either that all tropical islands have the same suite of species or that all tropical islands have completely unique sets of species. The observations of biogeography therefore suggest that species in a geographic location are generally descended from ancestors in nearby geographic locations, supporting the theory of branching descent from common ancestors.

By the time Darwin made his observations of tortoise and prickly pear biogeography, humans were distributed over the entire Earth. The modern biogeography of our species cannot help us determine our relationship to other organisms. However, Darwin reasoned that if humans and apes share a common ancestor, then we highly mobile humans must have first appeared where our less-mobile relatives can still be found. He predicted that evidence of early human ancestors would be found in Africa—the home of chimpanzees and other great apes.

## The Fossil Record

Evidence of human ancestors comes from **fossils,** the remains of living organisms left in soil or rock. These fossils and those of other species form a record of ancient life and provide direct evidence of change in organisms over time. There are many examples of fossil series that show a progression from more ancient forms to more modern forms, as in the transition between ancient and modern horses (**FIGURE 10.18**).

The chronological appearance of other groups of fossils also supports the theory of evolution. For example, evidence from anatomy and developmental biology indicates that modern mammals and other four-legged land animals evolved from fish ancestors. Correspondingly, we find the first fossil fish in very ancient rocks and the first fossil mammals in much younger layers of rock.

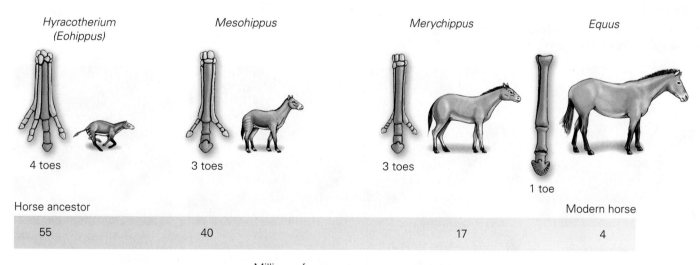

**FIGURE 10.18 The fossil record of horses.** Horse fossils provide a fairly complete sequence of evolutionary change from small, catlike animals with four toes to the modern horse with one massive toe.

Fossils typically form when the organic material decomposes and minerals fill the space left behind (**FIGURE 10.19**). However, fossilized impressions can also form—for instance, of shells, animal burrows, the soft tissues surrounding bones, or footprints. Fossilization is more likely to occur when organisms or their traces are quickly buried by sediment. Fortunately for scientists looking for fossils of **hominins**—humans and human ancestors—there is a relatively good record.

Hominin fossils can be distinguished from other primate fossils by some key characteristics. One essential difference between humans and other apes is the way we move. While chimpanzees and gorillas use all four limbs, humans are bipedal; that is, they walk upright on only two limbs. This difference in locomotion results in several anatomical differences between humans and other apes (**FIGURE 10.20**). In hominins, the face is on the same plane as the back instead of at a right angle to it; thus, the foramen magnum, the hole in the skull through which the spinal cord passes, is found on the back of the skull in other apes but at the base of the skull in humans. In addition, the structures of the pelvis and knee are modified for an upright stance; the foot is changed from being grasping to weight bearing; and the lower limbs are elongated relative to the front limbs.

The first hominin fossils were found not in Africa, but in Europe. Remains of *Homo neanderthalensis* (Neanderthal man) were discovered in 1856 in a small cave within the Neander Valley of Germany. In 1891, fossils of older, human-like creatures now called *Homo erectus* (standing man) were discovered in Java (Indonesia) in 1891. It was not until 1924 that the first *African* hominin fossil, the Taung child, was discovered in South Africa. This fossil was later determined to be a much older species than Neanderthals and *Homo erectus* and was placed in a new genus, *Australopithecus*. Paleontologists continue to discover new hominin fossils in southern and eastern Africa, including the famous "Lucy," a remarkably complete skeleton of the species *Australopithecus afarensis*, discovered in 1974 in Ethiopia. Lucy's fossil skeleton included a large section of her pelvis, which clearly indicated that she walked upright.

*homini-* means human-like.

# Visualize This ▼

**Bones that decompose quickly do not form high-quality fossils. Based on the information from this figure, why do you think that is the case?**

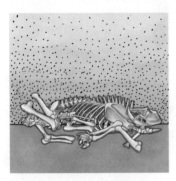 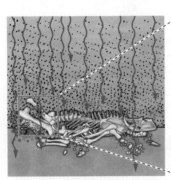 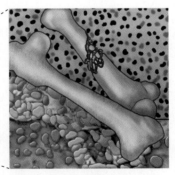 

**①** An organism is rapidly buried in water, mud, sand, or volcanic ash. The tissues begin to decompose very slowly.

**②** Water seeping through the sediment picks up minerals from the soil and deposits them in the spaces left by the decaying tissue.

**③** After thousands of years, most or all of the original tissue is replaced by very hard minerals, resulting in a rock model of the original bone.

**④** When erosion or human disturbance removes the overlying sediment, the fossil is exposed.

**FIGURE 10.19 Fossilization.** A fossil is commonly a rock "model" of an organic structure.

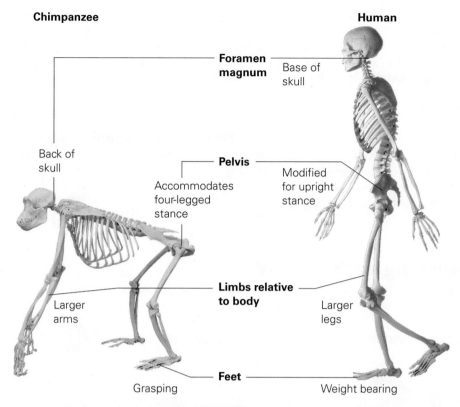

**FIGURE 10.20 Anatomical differences between humans and chimpanzees.**
Humans are bipedal animals, while chimpanzees typically travel on all fours. If any bipedal features are present in a fossil primate, the fossil is classified as a hominin.

By determining the age of these fossils and many other hominin species, scientists have confirmed Darwin's predictions—the earliest human ancestors arose in Africa.

## The Record of Our Ancestors

Scientists can determine the date when an ancient fossil organism lived by estimating the age of the rock that surrounds the fossil. **Radiometric dating** relies on radioactive decay, which occurs as radioactive elements in rock spontaneously break down into different, unique elements known as daughter products. Each radioactive element decays at its own unique rate. The rate of decay is measured by the element's **half-life**—the amount of time required for one-half of the amount of the element originally present to decay into the daughter product. Radiometric dating has led to an estimate of the age of Earth—4.6 billion years—based on the age of the oldest rocks on the planet.

**radio-** is the combining form of radiation.

**-metric** means to measure.

Using this technique, scientists have determined that the most ancient hominin fossil, the species *Ardepithecus ramidus*, is 5.2 to 5.8 million years old. (Two even older fossil species, *Orrorin tugenensis* and *Sahelanthropus tchadensis*, have been described as 6- and 7-million-year-old human ancestors, respectively. However, most scientists are reserving judgment about whether these animals were bipedal until more examples are found.)

As the number of described hominin fossils has increased, a tentative genealogy of humans has emerged. These fossil species can be arranged in a pedigree of relationships, indicating ancient and more modern species determined by radiometric dating and grouping similar species and dividing unique ones by comparing the organisms' anatomy (**FIGURE 10.21**). While some details of the pedigree in the figure are still subject to debate among scientists, the basic story is clear from the fossil record; modern humans are the last remaining branch of a once-diverse group of hominins.

**STOP & STRETCH** In this chapter, you learned how scientists have used a molecular clock to estimate when humans and chimpanzees diverged. How does the age estimate for the fossils of our oldest human ancestors compare to the age estimated by the molecular clock? Do these two pieces of evidence about the timing of human evolution agree or disagree?

## Working with Data ▼

**Approximately how many years ago did the common ancestor of modern humans and *Homo habilis* exist?**

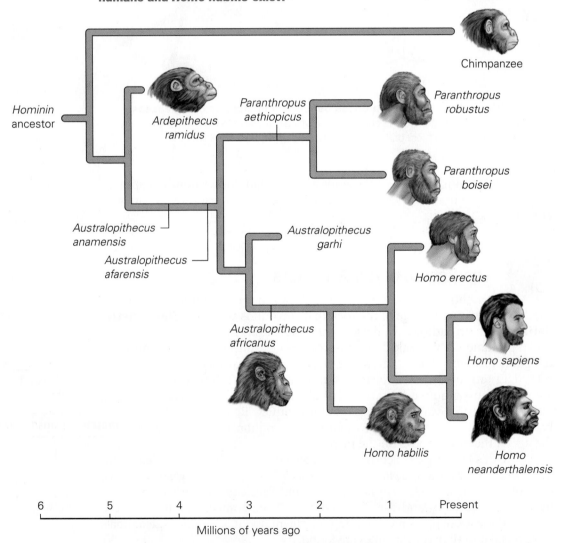

**FIGURE 10.21 The evolutionary relationships among hominin species.** This tree represents the current consensus among scientists who are attempting to uncover human evolutionary history.

## The Record of Our Ancestors: A Closer Look ▼

When rock is newly formed from the liquid underlying Earth's crust, it contains a fixed amount of any radioactive element. When the rock hardens, some of these radioactive elements become trapped. As a trapped element decays over time, the amount of radioactive material in the rock declines, and correspondingly, the amount of daughter product increases. By determining the ratio of radioactive element to daughter product in a rock sample and knowing the half-life of the radioactive element, scientists can estimate the number of years that have passed since the rock formed (**FIGURE 10.22**).

Like the fossil record of horses or the transition from fish to land animals, the hominin fossil record shows a clear progression—in this case, from more "apelike" to more "human-like" ancestors over time. Besides being bipedal, humans differ from other apes in having a relatively large brain, a flatter face, and a more extensive culture. The oldest hominins are bipedal but are otherwise similar to other apes in skull shape, brain size, and probable lifestyle. More recent hominin fossils show greater similarity to modern humans, with flattened faces and increased brain size (**FIGURE 10.23**, on the next page). Evidence collected with fossils of the most recent human ancestors indicates the existence of symbolic culture and extensive tool use, trademarks of modern humans. But do these observations provide convincing evidence that modern humans evolved from a common ancestor with other apes?

The common ancestor of humans and chimpanzees is often called the "missing link" because it has not been identified. However, finding the fossilized common ancestor between chimpanzees and humans, or between any two species for that

## Working with Data ▼

How old is a rock that contains 12.5% of the original parent element?

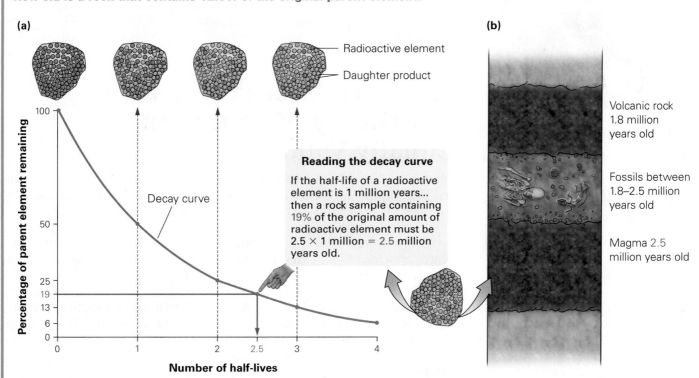

**FIGURE 10.22 Radiometric dating.** (a) The age of a fossil can be estimated when it is found between two layers of magma-formed rock. (b) The age of rocks can be estimated by measuring the amount of radioactive material (designated by dark purple circles) with a known half-life and the amount of daughter material (designated by light blue circles) in a sample of rock.

*A Closer Look, continued*

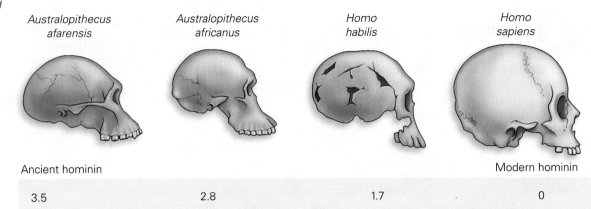

| | Australopithecus afarensis | Australopithecus africanus | Homo habilis | Homo sapiens |
|---|---|---|---|---|
| Age of fossil as determined by radiometric dating (million years ago) | Ancient hominin | | | Modern hominin |
| | 3.5 | 2.8 | 1.7 | 0 |

**FIGURE 10.23 From ancient to modern.** Fossils of ancient hominins display numerous ape-like characteristics, including a large jaw, small braincase, and receding forehead. More recent hominin fossils have a reduced jaw, larger braincase, and smaller brow ridge, much like modern humans.

**FIGURE 10.24 The common ancestor of humans and chimpanzees?** A "missing link" between humans and chimpanzees would not look half human and half ape, as this cartoon suggests.

matter, is extremely difficult, if not impossible. To identify a common ancestor, the evolutionary history of both species since their divergence must be clear. Like humans, modern chimpanzees have been evolving over the 5 million years since they diverged from humans. In other words, a missing link would not look like a modern chimpanzee with some human features or a cross between the two species—as suggested by nineteenth-century cartoons depicting Charles Darwin as an "ape man" (**FIGURE 10.24**). The fact that scientists cannot conclusively identify a fossil species that was the common ancestor of modern humans and modern chimpanzees is not evidence that these two species are not related. The vast majority of the evidence supports the hypothesis that chimpanzees are our closest living relatives.

# 10.4 Are Alternatives to the Theory of Evolution Equally Valid?

Observations of anatomical and genetic similarities among modern organisms and biogeographical patterns provide good evidence to support the theory of evolution. As with nearly all evidence in science, these observations allow us to infer the accuracy of the hypothesis of common descent but do not prove the hypothesis correct. This type of evidence is similar to the "circumstantial evidence" presented in a murder trial, such as finding the murder weapon in a car belonging to a suspect or the presence of the suspect's fingerprints on the door of the victim's home. The vast majority of the European scientific community was convinced by the indirect evidence that Darwin had catalogued, and they embraced the theory of common descent as the best explanation for the origin of species by the late nineteenth century. Since Darwin's time, scientists have accumulated additional indirect evidence, such as the DNA sequence similarities discussed earlier, to support this theory.

However, as in a murder trial, direct evidence is always preferred to establish the truth—for instance, the testimony of an eyewitness or a recording of the crime by a security camera. Of course, there are no human eyewitnesses to the evolution of humans, but we have a type of "recording" in the form of the fossil record. Because fossilization requires specific conditions, the fossil record

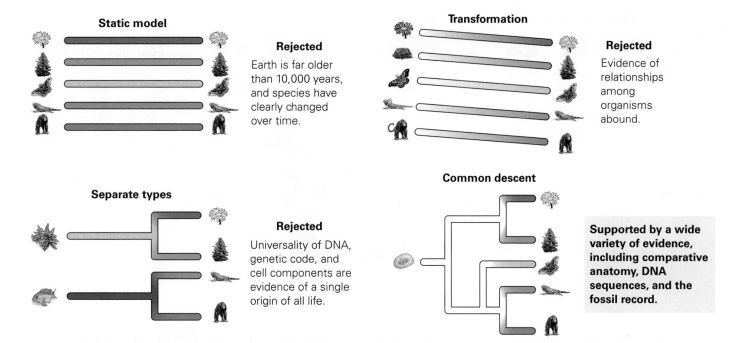

**FIGURE 10.25 Four hypotheses about the origin of modern organisms.** A scientific evaluation of these hypotheses leads to the rejection of all of them except for the theory of common descent.

is not a complete recording of the history of life—it is more like a security video that captures only a small portion of the action, with many blank segments. Just as the blank segments in a security video do not make the video an incorrect record of events, "gaps" in the fossil record do not diminish its value. The evidence present in the fossil record provides convincing support for the theory of common descent. All of the lines of evidence for common descent are summarized in **TABLE 10.1** on pages 224–225.

Now we return to the three competing hypotheses: static model, transformation, and separate types (**FIGURE 10.25**). Do the observations described in the previous section allow us to reject any of these hypotheses?

## Weighing the Alternatives

The physical evidence we have discussed thus far allows us to clearly reject only one of the hypotheses—the static model. The fossil record provides unambiguous evidence that the species that have inhabited this planet have changed over time, and radiometric dating indicates that Earth is far older than 10,000 years.

Of the remaining three hypotheses, transformation is the poorest explanation of the observations. If organisms arose separately and each changed on its own path, there is no reason to expect that different species would share structures—especially if these structures are vestigial in some of the organisms. There is also no reason to expect similarities among species in DNA sequence. The hypothesis of transformation predicts that we will find little evidence of biological relationships among living organisms. As our observations have indicated, evidence of relationships abounds.

Both the hypothesis of common descent and the hypothesis of separate types contain a process by which we can explain observations of relationships. That is, both hypothesize that modern species are descendants of common ancestors. The difference between the two theories is that common descent

Why is the American public seemingly so skeptical of an idea that is a fundamental principle of biological science?

**TABLE 10.1 The evidence for evolution.**

| Observation | Example | Why It Suggests Common Descent |
|---|---|---|
| **Biological classification.** The most logical and useful system groups organisms hierarchically. | The levels of classification can identify degrees of relationship. 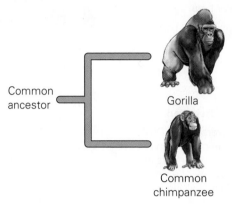 | All species in the same family share a relatively recent common ancestor, while all families in the same class share a more distant common ancestor. |
| **Anatomical homology.** Organisms that look quite different have surprisingly similar structures. | Mammalian forelimbs share a common set of bones organized in the same way, despite their very different functions. 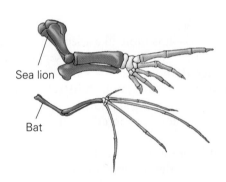 | The simplest explanation is that each species inherited the basic structure from the same common ancestor, and evolution led to their modification in each group. |
| **Vestigial traits.** Some species display traits that are nonfunctional but have a functional equivalent in other species. | Flightless birds such as ostriches produce functionless wings.  | The simplest explanation is that the trait was functional in an ancestral species but lost its function over time in one branch of the evolutionary line. |

*continued*

**TABLE 10.1** **The evidence for evolution.** *continued*

| Observation | Example | Why It Suggests Common Descent |
|---|---|---|
| **Biogeography.** The distribution of organisms on Earth corresponds in part to the relationship implied by biological classification. | The different species of tortoises on the Galápagos Islands are clearly related.  | Similarities among species in a geographic location imply divergence from ancestors in that geographic location. |
| **Homology in development.** Early embryos of different species often look similar. | All chordates—animals that have a backbone or closely related structure—produce structures called pharyngeal slits, and most have tails as early embryos. 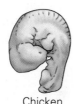  Chicken      Cat      Human | Similarities in early development suggest that these organisms derived from a single common ancestor that developed along a similar pathway. |
| **Homology of DNA.** The DNA sequences of species that are closely related in a taxonomic grouping are more similar than those from more distantly related groups. | The closer two organisms are in classification, the more similar is the DNA sequence for the same genes, and vice versa. 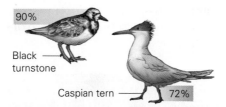  90% Black turnstone  Caspian tern 72%  % DNA sequence similarity to Ruddy turnstone | Similar DNA sequences in different species imply that the species evolved from a common ancestor with a particular sequence. |
| **The fossil record.** The remains of extinct organisms show progression from more ancient forms to more modern forms. | The transition between ancient and modern horses 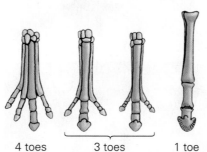  4 toes      3 toes      1 toe | Fossils provide direct evidence of change in organisms over time and suggest relationships among modern species. |

hypothesizes a single common ancestor for all living things, whereas separate types hypothesizes that ancestors of different groups arose separately and then gave rise to different types of organisms. Separate types seems more reasonable than common descent to many people. It seems impossible that organisms as different as pine trees, mold, ladybugs, and humans share a common ancestor. However, several observations indicate that all these disparate organisms are related.

The most compelling evidence for the single origin of all life is the universality of both DNA and of the relationship between DNA and proteins. For example, genes from bacteria can be transferred to plants, and the plants will make a functional bacterial protein (Chapter 9). This is possible only because both bacteria and plants translate genetic material into functional proteins in a nearly identical manner. If bacteria and plants arose separately, we could not expect them to translate genetic information similarly.

The fact that organisms as different as pine trees, mold, ladybugs, and humans contain cells with nearly all of the same components and biochemistry is also evidence of shared ancestry (**FIGURE 10.26**). A mitochondrion could have many different possible structural forms and still perform the same function; the fact that the mitochondria in a plant cell and an animal cell are essentially identical implies that both groups of organisms inherited these mitochondria from a common ancestor.

**STOP & STRETCH** Can you list other cellular structures or processes discussed in this text that are common to all eukaryotic organisms?

Pine trees, mold, ladybugs, and humans *are* very different. Proponents of the hypothesis of separate types argue that the differences among these organisms could not have evolved in the time since they shared a common ancestor. But the length of time during which these organisms have been diverging is immense—nearly 2 billion years. The remaining basic similarities among all living organisms serve as evidence of their ancient relationship.

**FIGURE 10.26 The unity and diversity of life.** The theory of evolution, including the theory of common descent, provides the best explanation for how organisms as distinct as pine trees, mold, ladybugs, and humans can look very different while sharing a genetic code and many aspects of cell structure and cell division.

## The Best Scientific Explanation for the Diversity of Life

Scientists favor the theory of common descent because it is the best explanation for how modern organisms came about. The theory of evolution—including the theory of common descent—is robust, meaning that it is a good explanation for a variety of observations and is well supported by a wide variety of evidence from anatomy, geology, molecular biology, and genetics. Evidence for the theory of common descent demonstrates **consilience,** meaning that there is agreement among observations derived from different sources. Consilience is a feature of all strongly supported scientific theories.

The theory of common descent is no more tentative than is atomic theory; few scientists disagree with the models that describe the basic structures of atoms, and few disagree that the evidence for the theory of common descent is overwhelming. Most scientists would say that both of these theories are so well supported that we can call them fact.

Evolutionary theory helps us understand the functions of human genes, comprehend the interactions among species, and predict the consequences of a changing global environment for modern species. Describing evolution as "*just a theory*" vastly understates the importance of evolutionary theory as a foundation of modern biology. People who do not have a grasp of this fundamental biological principle may lack an appreciation of the basic unity and diversity of life and fail to understand the effects of evolutionary history and change on the natural world and on ourselves.

## *savvy reader*

## Is There Still Scientific Debate about Evolution?

The following excerpt is from a publication of the Discovery Institute, a prominent antievolution organization.

Another 100 scientists have joined the ranks of scientists from around the world publicly stating their doubts about the adequacy of Darwin's theory of evolution. "Darwinism is a trivial idea that has been elevated to the status of the scientific theory that governs modern biology," says dissent list signer Dr. Michael Egnor. Egnor is a professor of neurosurgery and pediatrics at State University of New York, Stony Brook, and an award-winning brain surgeon named one of New York's best doctors by *New York* magazine. Discovery Institute's Center for Science and Culture today announced that over 700 scientists from around the world have now signed a statement expressing their skepticism about the contemporary theory of Darwinian evolution. The statement, located online at www.dissentfromdarwin.org, reads: "We are skeptical of claims for the ability of random mutation and natural selection to account for the complexity of life. Careful examination of the evidence for Darwinian theory should be encouraged."

1. The "dissent from Darwin" project can be thought of as an argument against the theory of evolution. Describe the essence of this argument.

2. Critique the argument. What counterpoints can be made?

3. What is the background of Dr. Egnor, the scientist quoted in the article? Given his background, does it appear that this is a debate among biologists who study evolution?

4. The Internet contains volumes of information on topics that promote controversy, such as evolution. How would you determine if the website this article was drawn from was credible? If a group that sponsors a website has a clear agenda on an issue, how might that have an impact on the material posted there?

*Discovery Institute, February 8, 2007* ▬▬

# SOUNDS RIGHT **BUT IS IT?**

State legislatures throughout the United States have passed laws that encourage public high school science teachers to teach the "controversy" associated with certain topics. For example, a law passed in Tennessee in 2012 encourages teachers to present the "scientific strengths and scientific weaknesses" of topics that arouse "debate and disputation" such as biological evolution (as well as the chemical origins of life, global warming, and human cloning). Supporters of these types of laws make the following argument:

**Teaching that evolution is scientifically controversial promotes student understanding of science.**

Sounds right, but it isn't.

1. According to the evidence we have, is there a reasonable scientific alternative to the theory of evolution? Explain.

2. According to the evidence we have, are there any observations that appear to falsify the theory of evolution?

3. Recall Chapter 1. Is the statement "a supernatural creator designed living organisms, or some biological processes, or some biological structures" a scientific hypothesis? Explain.

4. What do you think science teachers could present as the "scientific strengths and weaknesses" of biological evolution, as suggested by the Tennessee law?

5. Reflect on your answers to questions 1–4 above. Explain why the original statement bolded above sounds right, but isn't.

---

# Chapter Review MasteringBiology®

Go to the Study Area in MasteringBiology® for practice quizzes, myeBook, BioFlix™ 3-D animations, MP3Tutor sessions, videos, current events, and more.

## Summary

## Section 10.1

Define *biological evolution*, and distinguish it from other forms of non-evolutionary change in organisms.

- The process of evolution is the change that occurs in the characteristics of organisms in a population over time (p. 202).
- The theory of evolution, as described by Charles Darwin, is that all modern organisms are related to one another and arose from a single common ancestor (pp. 203–204).

Illustrate the theory of common descent using a tree diagram.

- Modern organisms can be arranged on a "tree" of relationships based on similarities in morphology, development, and genes (p. 204).

## Section 10.2

Summarize how Darwin's experiences led him to develop the outline of the theory of evolution.

- Scientists before Charles Darwin had hypothesized that species could change over time (p. 205).

- Darwin's voyage on the *HMS Beagle* led him to suspect that this hypothesis was correct. Over the course of 20 years, he was able to gather enough evidence to support this hypothesis and his hypothesis of common descent so that most scientists accepted these theories as the best explanation for the diversity of life on Earth (pp. 206–207).

# Section 10.3

Detail the modern biological classification system, and explain how it supports the theory of evolution.

- Linnaeus classified organisms based on physical similarities between them, for instance, placing humans and monkeys in the same order of animals. Darwin argued that the pattern of biological relationships illustrated by Linnaeus's classification provided strong support for the theory of common descent (pp. 209–210).

Explain how homologies in anatomy and genetics, even in useless traits, support the theory of evolution.

- Similarities in the underlying structures of a variety of organisms and the existence of vestigial structures, such as the appendix or goose bumps in humans, are difficult to explain except through the theory of common descent (pp. 211–213).
- Modern data on similarities of DNA sequences among organisms match the hypothesized evolutionary relationships suggested by anatomical similarities and provide an independent line of evidence supporting the hypothesis of common descent (pp. 214–216).

Describe how details of embryonic development support the theory of evolution.

- Similarities in embryonic development and embryonic structures among diverse organisms—for instance, the presence of pharyngeal slits in the embryos of all chordates—are best explained as a result of their common ancestry (p. 214).

Define *biogeography*, and explain how it supports the theory of evolution.

- Biogeography is the study of the geographical distribution of organisms (p. 216).
- Biogeographical patterns support the hypothesis of common descent because species that appear related physically are also often close to each other geographically: for example, the tortoises and mockingbirds on the Galápagos Islands (pp. 216–217).

Explain how the fossil record provides direct evidence of evolutionary change in species over time.

- The fossilized remains of extinct species in many groups of organisms demonstrate a progression of forms from more ancient to more modern types. For example, early hominins appear much more ape-like than more recent hominin fossils (pp. 217–222).

# Section 10.4

Articulate why the theory of evolution is considered the best explanation for the origin of humans and other organisms.

- Evidence strongly supports the hypothesis that organisms have changed over time and are related to each other (pp. 223–226).
- Shared characteristics of all life, especially the universality of DNA and the relationship between DNA and proteins, provide evidence that all organisms on Earth descended from a single common ancestor rather than from multiple ancestors (p. 226).

## Roots to Remember

**These roots come from Greek or Latin and will help you decode the meaning of words:**

| | |
|---|---|
| **evol-** | means to unroll. Chapter term: *evolution* |
| **homini-** | means human-like. Chapter terms: *hominid, hominin* |
| **homolog-** | indicates similar or shared origin; from a word meaning "in agreement." Chapter terms: *homologous, homology* |
| **macro-** | means large scale. Chapter term: *macroevolution* |
| **-metric** | means to measure. Chapter term: *radiometric* |
| **micro-** | means extremely small. Chapter term: *microevolution* |
| **radio-** | is the combining form of radiation. Chapter term: *radiometric* |

## Learning the Basics

1. Describe the theory of common descent.

2. What observations did Charles Darwin make on the Galápagos Islands that helped convince him that evolution occurs?

   **A.** the existence of animals that did not fit into Linnaeus's classification system; **B.** the similarities and differences among cacti and tortoises on the different islands; **C.** the presence of species he had seen on other tropical islands far from the Galápagos; **D.** the radioactive age of the rocks of the islands; **E.** fossils of human ancestors.

**3.** The process of biological evolution _____.
**A.** is not supported by scientific evidence; **B.** results in a change in the features of individuals in a population; **C.** takes place over the course of generations; **D.** B and C are correct; **E.** A, B, and C are correct.

**4.** In science, a theory is a(n) _____.
**A.** educated guess; **B.** inference based on a lack of scientific evidence; **C.** idea with little experimental support; **D.** a body of scientifically acceptable general principles; **E.** statement of fact.

**5.** The theory of common descent states that all modern organisms _____.
**A.** can change in response to environmental change; **B.** descended from a single common ancestor; **C.** descended from one of many ancestors that originally arose on Earth; **D.** have not evolved; **E.** can be arranged in a hierarchy from "least evolved" to "most evolved."

**6.** The DNA sequence for the same gene found in several species of mammals _____.
**A.** is identical among all species; **B.** is equally different between all pairs of mammal species; **C.** is more similar between closely related species than between distantly related species; **D.** provides evidence for the hypothesis of common descent; **E.** more than one of the above is correct.

**7.** Marsupial mammals give birth to young that complete their development in a pouch on the mother's abdomen. All the native mammals of Australia are marsupials, while these types of mammals are absent or uncommon on other continents. This observation is an example of _____.
**A.** developmental evidence for evolution; **B.** biogeographic evidence for evolution; **C.** genetic evidence for evolution; **D.** fossil evidence for evolution; **E.** not useful evidence for evolution.

**8.** Even though marsupial mammals give birth to live young, an eggshell forms briefly early in their development. This is evidence that _____.
**A.** marsupials share a common ancestor with some egg-laying species; **B.** marsupials are not really mammals; **C.** all animals arose from a common ancestor; **D.** marsupial mammals were separately created by God; **E.** the fossil record of marsupial mammals is incorrect.

**9.** A species of crayfish that lives in caves produces eyestalks like its above-ground relatives, but no eyes. Eyestalks in cave-dwelling crayfish are thus _____.
**A.** an evolutionary error; **B.** a dominant mutation; **C.** biogeographical evidence of evolution; **D.** a vestigial trait; **E.** evidence that evolutionary theory may be incorrect.

**10.** Which of the following taxonomic levels contains organisms that share the most recent common ancestor?
**A.** family; **B.** order; **C.** phylum; **D.** genus; **E.** class

# Analyzing and Applying the Basics

**1.** The classification system devised by Linnaeus can be "rewritten" in the form of an evolutionary tree. Draw a tree that illustrates the relationship among the flowering species listed, given their classification (note that *subclass* is a grouping between class and order):

Pasture rose (*Rosa carolina*, Family Rosaceae, Order Rosales, Subclass Rosidae)

Live forever (*Sedum purpureum*, Family Crassulaceae, Order Rosales, Subclass Rosidae)

Spring avens (*Geum vernum*, Family Rosaceae, Order Rosales, Subclass Rosidae)

Spring vetch (*Vicia lathyroides*, Family Fabaceae, Order Fabales, Subclass Rosidae)

Multiflora rose (*Rosa multiflora*, Family Rosaceae, Order Rosales, Subclass Rosidae)

**2.** DNA is not the only molecule that is used to test for evolutionary relationships among organisms. Proteins can also be used, and the sequences of their building blocks (called amino acids) can be compared in much the same way that DNA sequences are compared. *Cytochrome c* is a protein found in nearly all living organisms; it functions in the transformation of energy within cells. The percentage difference in amino acid sequence between humans and other organisms can be summarized as follows:

| Cytochrome c | Percentage difference from sequence of human sequence |
| --- | --- |
| Chimpanzee | 0.0 |
| Mouse | 8.7 |
| Donkey | 10.6 |
| Carp | 21.4 |
| Yeast | 32.7 |
| Corn | 33.3 |
| Green algae | 43.4 |

Draw the evolutionary tree implied by these data that illustrates the relationship between humans and the other organisms listed.

**3.** Look at the tree you generated for question 2. It does not imply that yeast is more closely related to corn than it is to green algae. Why not?

## Connecting the Science

**1.** Humans and chimpanzees are more similar to each other genetically than many very similar-looking species of fruit fly are to each other. What does this similarity imply regarding the usefulness of chimpanzees as stand-ins for humans during scientific research? What do you think it implies regarding our moral obligations to these animals?

**2.** Creationists have argued that if students learn that humans descended from animals and are, in fact, a type of animal, these impressionable youngsters will take this fact as permission to act on their "animal instincts." What do you think of this claim?

Answers to **Stop & Stretch, Visualize This, Working with Data, Savvy Reader, Sounds Right, But Is It?**, and **Chapter Review** questions can be found in the **Answers** section at the back of the book.

CHAPTER **11** | # An Evolving Enemy

A man captured crossing the Mexico/Texas border in 2012 was carrying
an almost untreatable strain of an infectious disease.

# Natural Selection

The Rio Grande Valley marks part of the southern border of the United States. This long and remote landform is a common crossing for immigrants who seek a better life in the U.S. but do not have permission to enter. Many are caught by officers from the Department of Immigration and Customs Enforcement (ICE). Of the 100,000 undocumented immigrants detained by ICE at this part of the border in 2012, one stood out. Not because he was from Nepal—which was unusual, but not unprecedented—but because he carried some very dangerous baggage with him.

The Nepalese migrant was infected with a strain of bacteria known as XDR-TB. According to the *Wall Street Journal*, the bacterium the man carried, which causes the disease tuberculosis (TB), was resistant to 8 of the 15 drugs commonly used to eliminate it. In medical terminology, the strain was extensively drug resistant (XDR). Doctors working with ICE learned of the man's infection within four days of his arrest—but by then

**On his journey across the globe, he exposed thousands of his fellow travelers.**

**The potential for spread of this disease has alarmed health officials.**

he had been in contact with dozens of their employees as well as fellow detainees. Of even more concern, the man had traveled to the U.S. through 13 countries, including Nepal, Brazil, Mexico, and areas of South Asia and the Persian Gulf. In his journey, he had traveled by airplane, boat, and bus. Fortunately, of the 12 ICE officers exposed, none tested positive for XDR-TB infection. But officials may never know exactly who else was exposed.

The story of the Nepalese migrant mirrors another, more widely publicized, event. In the summer of 2007, the U.S. Centers for Disease Control and Prevention (CDC), the federal agency responsible for monitoring and protecting public health, ordered Andrew Speaker, a 31-year-old lawyer from Atlanta, to be placed under involuntary quarantine in a hospital. That spring, Speaker was diagnosed with a tuberculosis infection. However, he had no apparent symptoms, and the subsequent drug treatment

didn't interfere with his plans to fly to Greece to get married and then to Rome, Italy, for his honeymoon. But while Speaker was in Rome, the CDC contacted him to inform him they had determined he had XDR-TB.

Depending on which side tells the tale, the CDC either ordered—or asked—Speaker and his wife to remain in Rome until they could arrange travel back to the U.S. Speaker instead flew to Prague, and from there to Montreal, before driving a rental car back to Atlanta. In his travels, everyone he had contact with incurred a risk of infection.

People around the world were outraged at Speaker's seemingly reckless behavior. How could he put so many at risk? As it turns out, few were really threatened by his actions. Instead, the recklessness of millions of ordinary people around the world puts us at risk of this, and other deadly microbes.

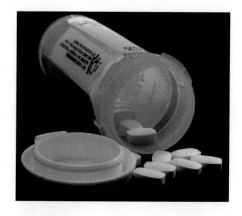

**But it is the unconsciously reckless behavior of many of us that is the ultimate cause of the problem.**

**233**

# He carried some very dangerous baggage with him.

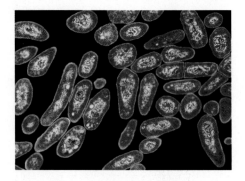

**FIGURE 11.1 The organism that causes tuberculosis.** The bacterial species *Mycobacterium tuberculosis.*

---

**tubercul-** means a small swelling.

---

**-osis** indicates a condition.

---

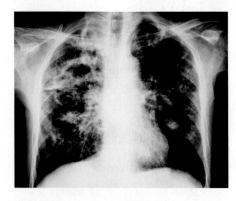

**FIGURE 11.2 Lung tubercles.** The white spots on this lung X-ray are nodules, or tubercles, within the lung tissue.

# 11.1 Return of a Killer

Tuberculosis has been plaguing humans for thousands of years. Tubercular decay has been found in the spines of Egyptian mummies from as long ago as 3000 B.C. In 460 B.C., the Greek physician Hippocrates identified a condition that appears to be tuberculosis as the most widespread disease of the times. As recently as 1906 in the United States, 2 of every 1000 deaths were due to TB infection.

Thanks to relatively recent advances in science and medicine, TB now accounts for only 1.5 of every 100,000 deaths in the United States. In the 1980s, the dramatic drops in infection and death due to TB around the world had many public health professionals convinced that this ancient scourge could be completely eliminated from the human population. Today, those hopes have largely faded as a result of both our own missteps and the powerful force of natural selection.

## What Is Tuberculosis?

The bacterium *Mycobacterium tuberculosis* causes the disease tuberculosis (**FIGURE 11.1**). Tuberculosis is a public health problem not because it is especially deadly, but instead because it affects so many. Two billion people, over one-third of the world's population, carry *M. tuberculosis,* and new infections are estimated to occur at a rate of one per second. Ninety percent of *M. tuberculosis* infections are symptomless, and most of these resolve as the bacteria are destroyed by the infected individual's immune system. However, the remaining 10% of infected individuals develop active disease, and more than one-half of these individuals will die without treatment. Tuberculosis now causes approximately 2 million deaths worldwide every year.

The symptoms of **tuberculosis** include a cough that produces blood, fever, fatigue, and a long, relentless wasting in which the patient gradually becomes weaker and thinner. These symptoms explain the antiquated name for this disease, *consumption,* because the infection seemed to consume people from within.

We now understand that the consumptive symptoms of TB arise from destruction of lung tissue caused by the body's reaction to active *M. tuberculosis* infection. Colonies of the bacteria in the lung are walled off by immune system cells, creating structures called tubercles (**FIGURE 11.2**); while this response does slow the spread of the disease within the lungs, it permanently damages the tissue. The reduction in the capacity of the lungs to provide oxygen to the body causes the physical wasting characteristic of tuberculosis (**FIGURE 11.3**). The infection is difficult to cure because *M. tuberculosis* can lie dormant for months inside these tubercles. When the tubercles later degrade, they release bacteria back into the lung tissue. Once released from the tubercle, these bacteria can cause additional infections within the lungs and be spread to other individuals.

Although you can be infected without symptoms, transmission of tuberculosis is almost entirely from people with active disease. When individuals with an active infection cough, sneeze, speak, or spit, they expel infectious droplets. A single sneeze can release about 40,000 of these droplets (**FIGURE 11.4**). People with prolonged, frequent, or intense contact with infected individuals are at highest risk of becoming infected themselves.

Most individuals can fight off infection with *M. tuberculosis,* but many cannot. Those at highest risk of active TB are young children; the elderly; individuals in poor overall health due to poor nutrition, other illnesses, or drug abuse; and people with AIDS. Until about 60 years ago, these individuals had little chance of surviving tuberculosis. Today, their prognosis—especially in Western countries where high-quality health care is readily available—is much better. But that may be changing.

## Treatment—and Treatment Failure

The treatments for tuberculosis in the nineteenth and early twentieth centuries, at least among the wealthy, consisted primarily of long stays in rural facilities where the air was fresh and unpolluted. These tuberculosis sanatoriums (**FIGURE 11.5**) were useful for two reasons. By moving patients from areas where the air was thick with lung-damaging particles and chemicals to sanatoriums, doctors were able to preserve patients' lung function for longer periods of time than would otherwise be possible. And because sanatoriums isolated patients, they reduced the spread of TB to the rest of the community. In poorer communities, individuals with active TB were often forcibly isolated in much more grim conditions.

The discovery of **antibiotics,** drugs that kill microbes, including bacteria, revolutionized tuberculosis treatment in the 1940s. Since then, infected patients with active TB are typically kept in isolation for only 2 weeks until antibiotics kill off most of the *M. tuberculosis* in the lungs. At this point, the patient is no longer contagious and can return to the community. However, because *M. tuberculosis* can hide inside immune system cells for long periods, antibiotic treatment must be maintained for 6 to 12 months to completely eliminate the organism.

Since the 1980s, however, scientists have chronicled a disturbing rise in the number of **antibiotic-resistant** tuberculosis infections—ones that cannot be cured by the standard drug treatment. According to the CDC, 20% of reported TB cases from 2000 to 2004 did not respond to standard treatments (such cases are called multidrug-resistant TB, or MDR-TB), and 2% were resistant to treatment with second-line drugs (such cases are called extensively drug-resistant TB, or XDR-TB). Even in the United States, with abundant access to resources and drugs, one-third of individuals diagnosed with active XDR-TB have died of the disease.

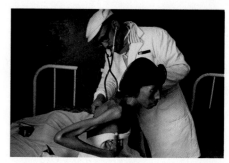

**FIGURE 11.3 The effects of tuberculosis infection.** As infected individuals lose more and more lung tissue to tuberculosis infection, they have difficulty obtaining enough oxygen. As a result, their tissues begin to waste away.

---

**anti-** means in opposition to.

---

**-biotic** indicates pertaining to life.

---

**FIGURE 11.4 Transmission through the air.** Like cold and flu viruses, the tuberculosis bacteria can be transmitted in droplets that are emitted in the sneeze or cough of an infected individual.

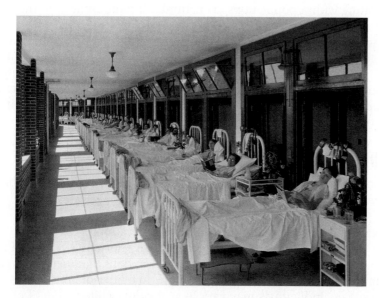

**FIGURE 11.5 A tuberculosis sanatorium.** Typically, patients at sanatoriums spent many hours every day exposed to fresh air, either outside or, like these patients, in large open porches.

## How could he put so many at risk?

As control has become less effective, the number of cases of TB in both developed and developing countries has begun to rise. In countries with fewer resources, the toll of XDR-TB could be much greater. In a recent XDR-TB outbreak in South Africa, 52 of 53 individuals diagnosed with the strain died within a month of showing signs of active disease. The resurgence of tuberculosis has now been declared a global health emergency by the World Health Organization.

> **STOP & STRETCH** Why might it be a problem if someone who has an infectious disease and seeks treatment receives an ineffective treatment?

Why are we losing the battle against tuberculosis? And what can be done to stop it? Answering these questions requires an understanding of an important force for evolutionary change: natural selection.

# 11.2 Natural Selection Causes Evolution

In *The Origin of Species,* Charles Darwin put forth two major ideas: the theory of common descent (Chapter 10) and the theory of **natural selection.** Darwin's presentation of the theory of common descent—that all species living today appear to have descended from a single ancestor—was thorough and convincing. Within 20 years of the publication of his book, the theory of common descent had been accepted by most scientists. However, it was another 60 years before the scientific community accepted Darwin's theory of natural selection, which explains in large part *how* organisms evolved from a common ancestor to become the great variety we see today. Darwin proposed that through the process of natural selection, the physical or behavioral traits of organisms that lead to increased survival or reproduction in a particular environment become common within their population, while less favorable traits are lost. The changes accumulating within populations via natural selection can lead to the development of new species.

Darwin reasoned that the process of natural selection is an inevitable consequence of the competition for survival among variable individuals in a population. Today, natural selection is considered one of the most important causes of evolution (although others, such as the processes of genetic drift and sexual selection as described in Chapter 12, also cause populations to change over time).

## Darwin's Observations

The theory of natural selection is elegantly simple. It is an inference based on four general observations:

**1. Individuals within Populations Vary.** Observations of groups of humans support this statement—people do come in an enormous variety of shapes, sizes, colors, and facial features. It may be less obvious that there is variation in nonhuman populations as well. For example, in a litter of gray wolves born to a single female, individuals may vary in coat color, while in a field of flowers, one plant may bloom earlier than others (**FIGURE 11.6**). We can add all kinds of less obvious differences to this visible variation; for example, the amount of

**(a) Variation in coat color**

**(b) Variation in blooming time**

**FIGURE 11.6 Observation 1: Individuals within populations vary.** (a) Gray wolves vary in coat color, even within a single litter of animals. (b) Flowers may vary in blooming time, with some individual plants blooming much earlier than others of the same species.

## Visualize This ▲

**Under what conditions might it be an advantage to bloom earlier than other nearby flowers?**

caffeine produced in the seeds of a coffee plant varies among individuals in a wild population. Each different type of individual in a population is called a **variant.**

### 2. Some of the Variation among Individuals Can Be Passed on to Their Offspring.
Although Darwin did not understand how it occurred, he observed many examples of the general resemblance between parents and offspring. He also noticed that people took advantage of the inheritance of variation in other species. Pigeon fanciers in Darwin's time clearly recognized the inheritance of variation; they could see that, for instance, pigeons with neck ruffs were more likely to produce offspring with neck ruffs than were pigeons without ruffs. Thus, when they wanted to produce a ruffed variety of pigeon, they encouraged breeding among the birds with this trait (**FIGURE 11.7**). Darwin hypothesized that offspring tend to have the same characteristics as their parents in natural populations as well.

For several decades after *The Origin of Species* was published, the observation that some variations were inherited was the most controversial part of the theory of natural selection. Because scientists could not adequately explain the origin and inheritance of variation, many were unwilling to accept that natural selection could be a mechanism for evolutionary change. When Gregor Mendel's work on inheritance in pea plants (Chapter 7) was rediscovered in the 1900s, the mechanism for this observation became clear—natural selection operates on genetic variation that is passed from one generation to the next.

### 3. Populations of Organisms Produce More Offspring than Will Survive.
This observation is clear to most of us—the trees in the local park make literally millions of seeds every summer, but only a small fraction of these survive to germinate, and only a few of the seedlings live for more than a year or two.

**FIGURE 11.7 Observation 2: Some of the variation among individuals can be passed on to their offspring.** Darwin noted that breeders could create flocks of pigeons with fantastic traits by using as parents of the next generation only those individuals that displayed these traits.

If a female elephant (colored pink) lives a full fertile lifetime, she will bear about six calves in about 90 years. On average, half of her calves will be female.

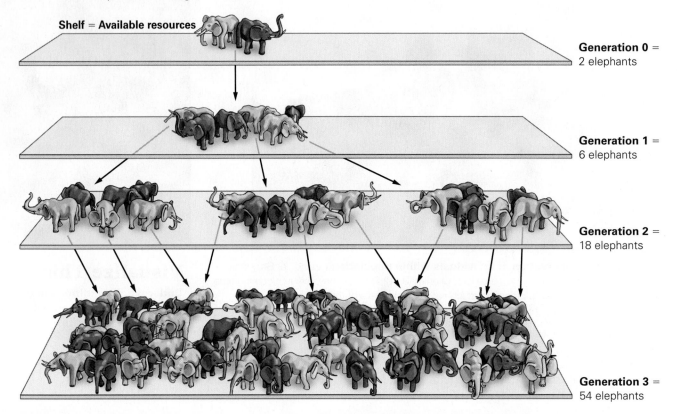

**Shelf = Available resources**

**Generation 0 =** 2 elephants

**Generation 1 =** 6 elephants

**Generation 2 =** 18 elephants

**Generation 3 =** 54 elephants

**FIGURE 11.8 Observation 3: Populations of organisms produce more offspring than will survive.** Even slow-breeding animals like elephants are capable of producing huge populations relatively quickly.

# Visualize This ▲

**Predict what will happen to the elephant population as a result of limited resource availability when Generation 4 is produced.**

In *The Origin of Species,* Darwin gave a graphic illustration of the difference between offspring production and survival. In his example, he used elephants, animals that live long lives and are very slow breeders. A female elephant does not begin breeding until age 30, and she produces about 1 calf every 10 years until around age 90. Darwin calculated that even at this very low rate of reproduction, if all the descendants of a single pair of African elephants survived and lived full, fertile lives, after about 500 years their family would have more than 15 million members (**FIGURE 11.8**)—many more than can be supported by all the available food resources on the African continent!

**4. Survival and Reproduction Are Not Random.**  In other words, the subset of individuals that survives long enough to reproduce is not an arbitrary group. Some variants in a population have a higher likelihood of survival and reproduction than other variants do; that is, there is differential survival and reproduction among individuals in the population. The survival and reproduction of one variant compared to others in the same population is referred to as its relative **fitness**. Traits that increase an individual's relative fitness in a particular environment are called **adaptations**. Individuals with adaptations to a particular environment are more likely to survive and reproduce than are individuals lacking such adaptations; in other words, these individuals have higher relative fitness.

Darwin referred to the results of differential survival and reproduction as natural selection. Adaptations are "naturally selected" in the sense that

**adapt-** means to fit in.

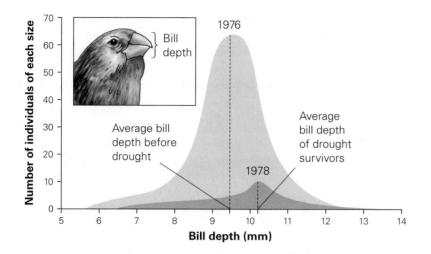

**FIGURE 11.9 Observation 4: Survival and reproduction are not random.** The pale purple curve summarizes bill depth in ground finches on Daphne Island in the Galápagos in 1976. The dark purple curve below it represents the population in 1978, after the drought of 1977. These data indicate that survivors of the drought had a larger average bill depth than the predrought population. The change in the population's average bill size occurred because finches with larger-than-average bills had higher fitness than did small-billed birds during the drought.

## Working with Data ▲

**Use the graph to determine how total population size of ground finches changed between 1976 and 1978.**

individuals possessing them survive and contribute offspring to the next generation. Although Darwin used the word *selection,* which implies some active choice, natural selection is a passive process that is simply determined by differences among individuals and their success in their particular environment. For example, among the birds called medium ground finches living on an island in the Galápagos archipelago, scientists have observed that when rainfall is scarce, a large bill is an adaptation. This is because birds with larger bills are able to crack open large, tough seeds—the only food available during severe droughts. As shown in **FIGURE 11.9**, the 90 survivors of a 1977 drought had an average bill depth that was 6% greater than the average bill depth of the original population of 751 birds. In these environmental conditions, a large bill increases survival.

Adaptations are not only traits that increase survival. Any trait possessed by an individual that increases the number of offspring it produces relative to other individuals in a population is also an adaptation. For example, flowers in a crowded mountain meadow may have a relatively limited number of potential insect pollinators. More pollinator visits generally result in more seeds being produced by a single flower, so any trait that increases a flower's attractiveness to pollinators, such as a brighter color or greater nectar production, should be favored by natural selection (**FIGURE 11.10**).

## Darwin's Inference: Natural Selection Causes Evolution

Based on his observations, Darwin reasoned that the result of natural selection is that favorable inherited variations tend to increase in frequency in a population over time, while unfavorable variations tend to be lost. In other words, adaptations become more common in a population as the individuals who possess them contribute larger numbers of their offspring to the succeeding generation. Natural selection results in a change in the traits of individuals in a population over the course of generations—that is, evolution. While there are other factors, such as genetic drift and migration of individuals, that can cause populations to evolve over time, natural selection is the only force that can lead to adaption of a population to its environment.

It is a testament to the power of the theory of natural selection that today it seems self-evident to us—we might even wonder why Darwin is considered a

**FIGURE 11.10 Adaptations are not about survival only.** Variations that increase a flower's attractiveness to a pollinator can increase its reproductive success by increasing the number of seeds it produces.

## Visualize This ▼

What would this sequence of changes look like if the selection was for dogs with a particular behavioral trait, for instance, pointing at a prey animal?

Artificial selection for dogs with short legs

Only these two dogs are allowed to breed.

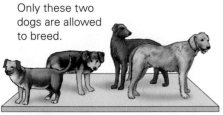

**Generation 1**

Artificial selection for dogs with short legs

Only these two dogs are allowed to breed.

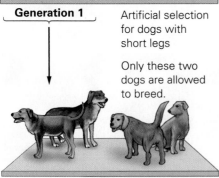

**Generation 2**

Artificial selection for dogs with short legs

Only these two dogs are allowed to breed.

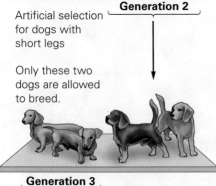

**Generation 3**

Dachshunds

brilliant scientist for explaining something that seems so obvious. But the theory of natural selection only became so powerful after it was tested and shown to work—in nature—in the manner Darwin described. Natural selection proved such a powerful idea that it has influenced how we think about many phenomena, from the success of particular brands of soft drinks to the relationships among nations.

> **STOP & STRETCH** Some biologists argue that human evolution as a result of natural selection on our physical traits has nearly stopped because of our ability to use technology to compensate for poor environmental conditions or to overcome physical handicaps. What examples can you give that support this view? In what ways might humans still be subject to natural selection?

## Testing Natural Selection

Darwin proposed a scientific explanation of how evolution occurs, and like all good hypotheses, it needed to be tested. All of the tests described next illustrate that natural selection is an effective mechanism for evolutionary change.

**Artificial Selection.** Selection imposed by human choice is called *artificial selection*. It is artificial in the sense that humans deliberately control the survival and reproduction of individual plants and animals to change the characteristics of the population. Individuals with preferred traits are permitted to breed, whereas those that lack preferred traits are not allowed to breed.

The fancy pigeons that Darwin studied arose by artificial selection, and the great variety of domestic dogs also resulted from this process. In each case, different breeds evolved through selection by breeders for various traits (**FIGURE 11.11**). These examples demonstrate that differential survival and reproduction change the characteristics of populations. However, due to the direct intervention of humans on the survival and reproduction of these organisms, artificial selection is not exactly equivalent to natural selection. Can change in populations occur without direct human intervention?

**Natural Selection in the Lab.** Another test of the effectiveness of natural selection is to examine whether populations living in artificially manipulated laboratory environments change over time. An example of this kind of experiment is one performed on fruit flies placed in environments containing different concentrations of alcohol.

High concentrations of alcohol cause cell death. Many organisms, including fruit flies and humans, produce enzymes that metabolize alcohol—that is, they break it down, extract energy from it, and modify it into less-toxic chemicals. There is variation among fruit flies in the rate at which they metabolize alcohol. In a typical laboratory environment, most flies process alcohol relatively slowly, but about 10% of the population possesses an enzyme variant that allows those flies to metabolize alcohol twice as rapidly as the more common variant.

In an experiment (**FIGURE 11.12**), scientists divided a population of fruit flies into two randomized groups. Initially, these two groups had the same percentage of fast and slow alcohol metabolizers. One group of flies was placed in

**FIGURE 11.11 Artificial selection can cause evolution.** When breeders select dogs with certain traits to produce the next generation of animals, they increase the frequency of that trait in the population. Over generations, the trait can become quite exaggerated. Dachshunds are descendants of dogs that were selected for the production of very short legs.

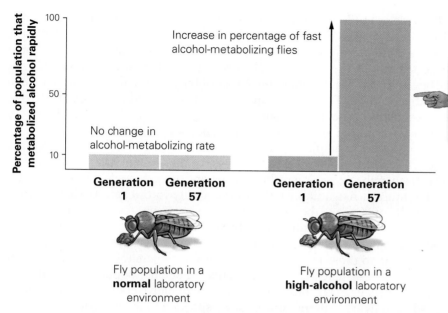

FIGURE 11.12 **Natural selection in laboratory conditions.** When fruit flies are placed in an alcohol-spiked environment, the percentage of flies that can rapidly metabolize alcohol increases over many generations because of natural selection. In the normal laboratory environment, there is no selection for faster alcohol processing, so the average rate of alcohol metabolism does not change.

◀ **Working with Data**

This data can be represented as a line graph as well as a bar graph. Sketch the line graph.

an environment containing typical food sources; the other group was placed in an environment containing the same food spiked with alcohol. After 57 generations, the percentage of fast alcohol-metabolizing flies in the environment with only typical food sources was the same as at the beginning of the experiment—10%. But after the same number of generations, the percentage of fast alcohol-metabolizing flies in the alcohol-spiked environment was 100%. Because all of the flies in this environment were now of the fast alcohol-metabolizing variety, the *average* rate of alcohol metabolism in the population in this environment was much higher in generation 57 than in generation 1. The population had evolved.

The evolution of the fruit flies in this experiment was a result of natural selection. In an environment where alcohol concentrations were high, individuals that were able to metabolize alcohol relatively rapidly had higher fitness. Because they lived longer and were less affected by alcohol, the fast alcohol-metabolizing flies left more offspring than the slow alcohol-metabolizing flies did. Thus, each generation had a higher frequency of fast alcohol-metabolizing individuals than the previous generation did. After many generations, flies that could rapidly metabolize alcohol predominated in the population.

Selection can change populations in highly regulated laboratory environments. But does it have an effect in natural wild populations?

**Natural Selection in Wild Populations.** The evolution of *M. tuberculosis* from being susceptible to antibiotics to being resistant is one example of natural selection in a wild population; clearly, a change in the environment (that is, the introduction of antibiotics) caused a change in the bacteria population. Dozens of other **pathogens**, organisms that cause disease, have become resistant to drugs and pesticides in the last 50 years as well. But even these changes may seem suspect because the adaptation is to a human-imposed environmental change. Although studying adaptation to natural environmental changes in the field is a significant challenge, the effects of natural selection have been observed in dozens of wild populations.

**patho-** means disease.

A classic example of natural selection in a natural setting is the evolution of bill size in Galápagos finches in response to drought (review Figure 11.9). The survivors of the drought tended to be those with the largest bills, which could more easily handle the tough seeds that were available in the dry environment.

The survival of this nonrandom subset of birds resulted in a true change in the next generation. The population of birds that hatched from eggs in 1978—the descendants of the drought survivors—had an average bill depth 4% to 5% larger than that of the predrought population.

A more recent example of natural selection causing evolution has occurred in the last few decades on the eastern coast of the United States, where an invasive Asian crab species is wreaking havoc on native mussels. But one species, the blue mussel, has quickly evolved the ability to thicken its shell when it grows in the presence of the Asian crab, thwarting their attacks. Scientists at the University of New Hampshire were able to demonstrate that this was an evolutionary change by comparing blue mussel populations in regions invaded by the Asian crab to those in more northerly waters, where the Asian crab cannot survive. These researchers demonstrated that while both populations of mussels thicken their shells in response to the presence of native crabs, only the mussels that had been living with the Asian crabs responded to the presence of this species. Clearly, natural selection for individual crabs that could recognize this new species as a predator had caused a change in the mussel population.

# 11.3 Natural Selection Since Darwin

While tests of natural selection seemed to support the theory, it took 60 years for the rest of biological science to catch up to Darwin's insight and adequately explain it. Scientists working in the new field of genetics in the 1920s began to recognize that genetic principles could explain how natural selection causes the evolution of populations.

## The Modern Synthesis

The union between genetics and evolution was termed "the modern synthesis" during its development in the 1930s and 1940s. The modern synthesis outlines the model of evolutionary change accepted by the vast majority of biologists.

The modern synthesis is predicated on a number of the genetic principles discussed in Chapters 6 through 9, namely:

- Genes are segments of genetic material (typically DNA) that contain information about the structure of molecules called proteins.
- The actions of proteins within an organism help determine its physical traits.
- Different versions of the same gene are called alleles, and variation in physical traits among individuals in a population is often due to variation in the alleles they carry.
- Different alleles for the same gene arise through mutation—changes in the DNA sequence.
- Half of the alleles carried by a parent are passed to their offspring via their eggs or sperm.

We can apply these genetic principles to the fruit flies exposed to a high-alcohol environment. In this population there are two variants, or alleles, of the gene that controls alcohol processing. One allele produces a fast alcohol-metabolizing enzyme, and the other produces a slow alcohol-metabolizing enzyme. In the high-alcohol environment, flies that made mostly fast alcohol-metabolizing enzyme had more offspring than did flies that made primarily the slow alcohol-metabolizing enzyme. Therefore, in the next generation, produced primarily by fast-alcohol metabolizers, a higher percentage of flies in the population inherited and thus carried at least one allele for the fast alcohol-metabolizing enzyme. This exemplifies why we can now describe the evolution of a population as an increase or decrease in the frequency of an allele for a particular gene.

The existence of two different alleles for alcohol metabolism in fruit flies suggests that one of these alleles is a mutated version of the other. In the normal laboratory environment, neither of these alleles appears to have a strong effect on fitness. Because the slow alcohol metabolizers are more numerous than the fast alcohol metabolizers, it appears that there might be a slight disadvantage to carrying the fast alcohol-metabolizing enzyme in a low-alcohol environment. However, in the high-alcohol environment, the mutation resulting in the fast alcohol-metabolizing allele gives a strong advantage, and its presence in the population allows for the population's evolution (**FIGURE 11.13**). Scientists now understand that the random process of gene mutation generates the raw

**FIGURE 11.13 Mutation and natural selection.** When a gene has mutated, its product may have a slightly different activity. If this new activity leads to increased fitness in individuals carrying the mutated gene, it will become more common in the population through the process of natural selection.

## Visualize This ▼

How would the ratio of fast to slow metabolizers differ on each "shelf" of this figure if the flies were not in a high-alcohol environment?

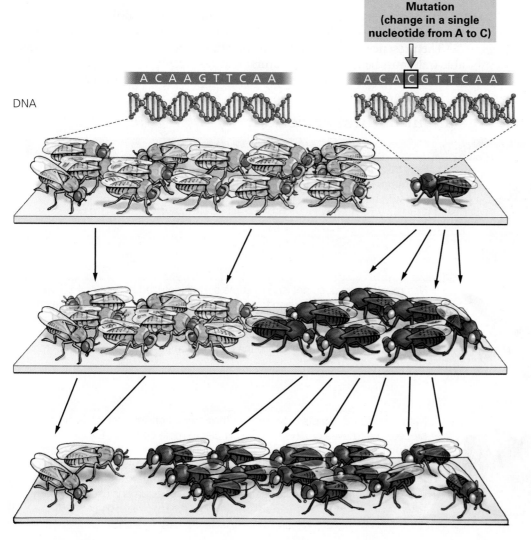

**Mutation (change in a single nucleotide from A to C)**

A C A A G T T C A A

A C A C G T T C A A

DNA

1. Mutation arises in the gene for alcohol-metabolizing enzyme, creating a new, mutant allele.

2. The mutation results in a protein that increases the fly's alcohol-metabolism rate.

3. In an alcohol-spiked environment, the flies with the fastest alcohol-metabolism rate produce the most offspring, many of which carry the mutant allele.

4. Over several generations, the frequency of the mutant allele increases within the population.

material—variations—for evolution, and that natural selection acts as a filter that selects for or against new alleles produced by mutation.

> **STOP & STRETCH** The allele responsible for cystic fibrosis (discussed in Chapter 7) appears to protect carriers (individuals with one copy of the allele) from *M. tuberculosis* infection. In addition, it appears that the cystic fibrosis allele is more common in humans with ancestors from crowded, urban environments in Northern Europe. Use the modern understanding of evolution via natural selection to develop a hypothesis about why the cystic fibrosis allele is more common in these populations.

## The Subtleties of Natural Selection

Because the idea of natural selection has been applied to the realm of human society—such as to the success or failure of a particular company or technology—misunderstandings of how it works in nature are common. Common misunderstandings of natural selection fall into three categories: the relationship between the individual and the population, the limitations on the traits that can be selected, and the ultimate result of selection. **TABLE 11.1** provides examples of statements that you might hear in reference to natural selection and a brief explanation of how each demonstrates a misunderstanding of this process.

In short, the subtleties of natural selection can be described in three statements: (1) Natural selection only acts on traits present in the population. (2) The presence of any adaptation sometimes means that a slightly less valuable trait must be forgone. (3) Natural selection results in the fit of an organism to the environment it is currently experiencing, not some future environment.

**TABLE 11.1 Misunderstandings about natural selection.**

| A Misunderstanding of Natural Selection | | How Natural Selection Really Works |
| --- | --- | --- |
| "The Dodo was too stupid to adapt to human hunters, so it had to go extinct." |  | Only traits that are present in the population can be selected for. The Dodo was not "stupid" or unworthy of survival; the population simply did not contain variants with hunter-avoiding traits. |
| "Some animals are not adapted to their environments very well. For instance, if swallows were well adapted to North America, they wouldn't have to migrate to the tropics every winter." |  | Natural selection does not lead to some sort of idealized perfect state. There are always trade-offs between traits. A swallow is very well adapted to catching flying insects. Any swallow variant with physical changes that allow it to eat the seeds that are available in winter would be less able to catch flying insects. These variants would thus lose in the competition for food during the summer. |
| "If natural selection improves populations, why aren't chimpanzees evolving into humans?" |  | Natural selection adapts organisms to their current environment; it is not a process that moves organisms from simpler to more complex. Evolution in chimpanzees fits these animals to their current situation. In their environment, bipedal humans already exist and fully exploit the resources in which our ancestors were able to first specialize. Traits that make chimpanzees more "human-like" would not be successful because the resources humans use are already taken. |

## The Subtleties of Natural Selection: **A Closer Look** ▼

Natural selection does not result in the evolution of "perfect" organisms. To make sense of this, you need to have a good understanding of not only natural selection's power but also its limitations.

### Natural Selection Cannot Cause New Traits to Arise

One mistaken idea about natural selection is that certain species are doomed to extinction because they "will not" adapt to a new environment. Polar bears are threatened with extinction because they rely on sea ice for hunting, and sea ice in the Arctic is rapidly diminishing as a result of global warming. Why don't they just switch to another food? After all, grizzly bears eat a wide variety of foods.

This idea that species "should" change to survive results from our understanding of the selection process in human culture. A car company will fail if it does not change its designs to meet the changing needs of its customers. But change in the company occurs because businesses can *invent* new "adaptations" that respond to rapid changes, say, in tastes or gas prices. In contrast, in living populations, evolution can occur only when traits that influence survival are already present in a population and have a genetic basis.

The example of the alcohol-metabolizing fruit flies illustrates this point. Selection did not cause change in individual flies; either a fly could rapidly metabolize alcohol or it could not. It was the differential survival and reproduction of these types of flies in an alcohol-laced environment that caused the population to change.

### Natural Selection Does Not Result in Perfection

Natural selection does cause populations to become better fit to their environment, but the result is organisms that are fit for certain conditions, but not all.

Changes in traits that increase survival and reproduction in one environment may be liabilities in another environment. In other words, most adaptations result in trade-offs. For example, Richard Lenski and his coworkers at Michigan State University found that certain bacteria would adapt to an environment where food levels were low by evolving chemicals that were deadly to other bacteria. Individual bacteria that produced this chemical, by killing off their nearby competitors, had more of the very limited food available to them and thus became prevalent in the population. However, when the poisonous bacteria were grown in a food-rich environment, they did not grow as well as nonpoisonous bacteria did. This occurs because the poisonous bacteria use energy to produce their toxin—a trade-off that reduces the amount of energy that can be used for growth and reproduction. In a food-rich environment, individual bacteria that did not produce toxin had more offspring and thus were better adapted.

Adaptations of organisms are also constrained by their underlying biology. The evolutionary biologist Stephen Jay Gould described one of the most famous examples of this—the thumb found on the front paws of a giant panda (**FIGURE 11.14**). These animals appear to have six digits: five fingers composed of the same bones as our fingers, and a thumb constructed from an enlarged wrist bone. The muscles that operate this opposable thumb are rerouted hand muscles. The thumb is an adaptation that increases pandas' ability to strip leaves from bamboo shoots, their primary food source. A more effective opposable thumb is our own, adapted from one of the basic five digits. However, in the panda population, the variation of a digit that was separate from the rest of the paw did not exist. Individuals with enlarged wrist bones did exist, so what evolved in giant pandas was a thumb that does its limited job but is not as flexible as our own thumb. While pandas differ from other bears in that they have a thumb, they will never be able to text with it.

Another, more familiar, example of the limitation of adaptation is our own upright posture—judging by the number of lower back and knee injuries reported every year, humans are not perfectly formed for bipedal walking. This is because our skeleton evolved from a four-legged ancestor, and the changes that make us upright are modifications of that ancestral design.

**FIGURE 11.14 The panda's thumb.**
In addition to the five digits on its paw, the giant panda has a partially opposable "thumb" made up of elongated wrist bones.

*A Closer Look, continued*

**(a)**                    **(b)**

**FIGURE 11.15 Natural selection does not imply progression.** (a) Natural selection favored the evolution of flowers after the appearance of flying insects made insect pollination possible. (b) Grasses lost the showy parts of their flowers when natural selection favored wind pollination in their environment. The spiderwort in (a) is a close relative of grasses.

### Natural Selection Does Not Cause Progression toward a Goal

You may have heard natural selection described as "survival of the fittest"; however, it is important to recognize that natural selection favors those variants with the most appropriate adaptations to the current environment. Natural selection thus does not result in the progress of a population toward a particular predetermined goal; instead, it depends on the particular environment in which a population lives.

The example of the alcohol-metabolizing flies again helps to illustrate this point. Only the population of flies in the high-alcohol environment evolved a faster rate of alcohol metabolism. Without a change in the environment, the alcohol-metabolizing rate of the population of flies in the normal environment did not evolve.

The situational nature of natural selection can lead to evolutionary patterns that defy our sense that species are evolving toward a "more perfect" condition. For example, flowering plants evolved from nonflowering plants, and the flower is, in part, an adaptation to attract bees and other pollinators to help increase seed production. However, some species of flowering plants, especially grasses, have adapted to environments where wind pollination is particularly effective and the need to attract insect pollinators is much reduced (**FIGURE 11.15**). In these plants, natural selection has favored individuals that have very reduced flower parts and generate primarily pollen-producing and egg-producing structures—much like the reproductive structures of their distant nonflowering ancestors. Grasses have not regressed but instead have simply adapted to the environment they experience.

## Patterns of Selection

As Darwin noted, natural selection is a force that causes the traits in a population to change over time. A more modern understanding of the process has helped scientists recognize that different environmental conditions may lead to no change in the population or even cause it to split into two species.

The type of natural selection experienced by the flies in the alcohol-laden environment and by bacteria responding to the use of antibiotics is called **directional selection** because it causes the population traits to move in a particular direction (**FIGURE 11.16a**). Directional selection is typically the type of selection that leads to change in a population over time—in the case of the flies, to a more alcohol-tolerant population, and in the case of bacteria, to a more antibiotic-resistant strain.

In certain environments, however, the average variant in the population may have the highest fitness. This results in **stabilizing selection,** in which the

## Working with Data ▼

**How would the graph in part (a) differ if the red flower was avoided by the pollinator?**

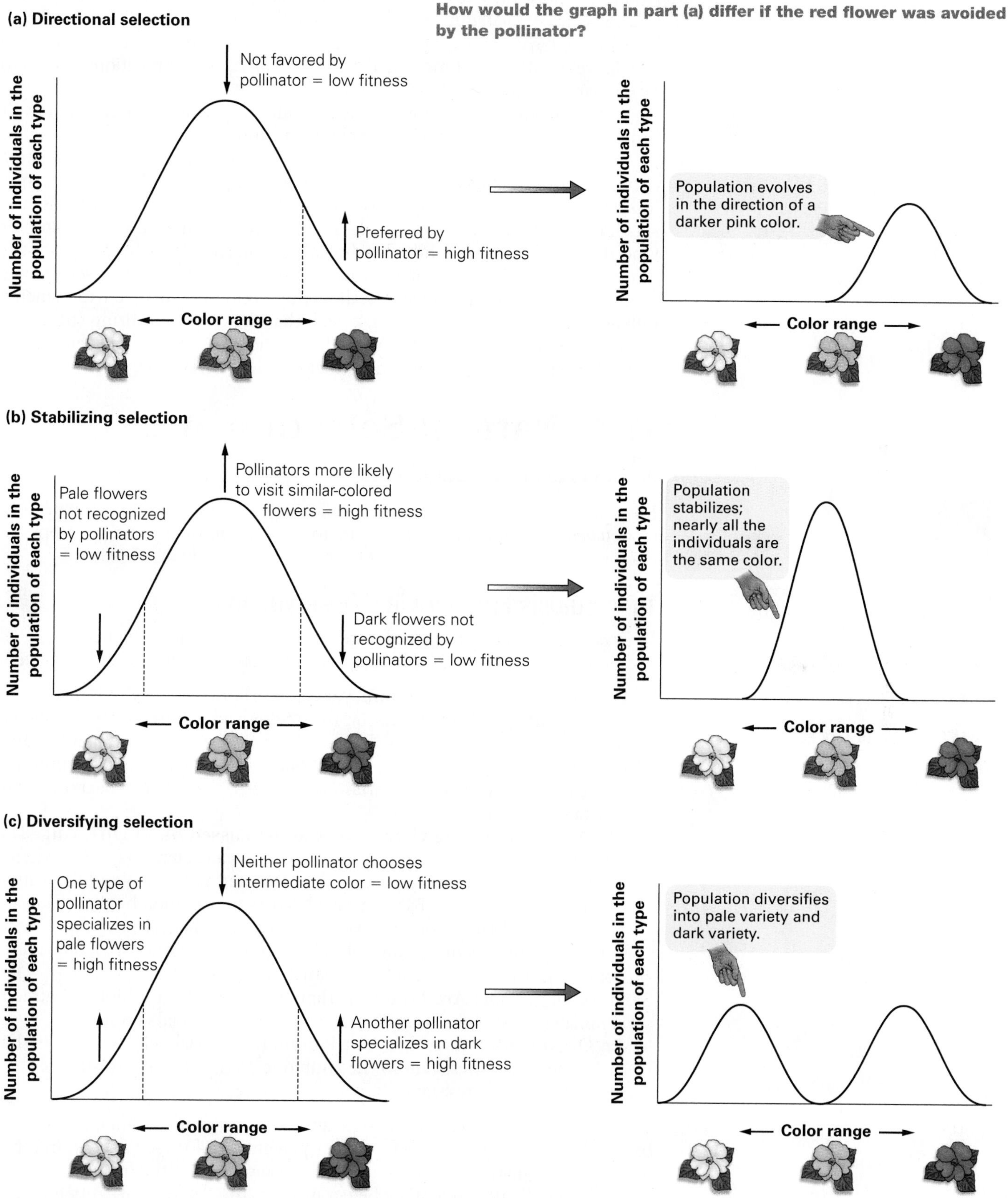
(a) Directional selection
Not favored by pollinator = low fitness
Preferred by pollinator = high fitness
Number of individuals in the population of each type
Color range
Population evolves in the direction of a darker pink color.
(b) Stabilizing selection
Pollinators more likely to visit similar-colored flowers = high fitness
Pale flowers not recognized by pollinators = low fitness
Dark flowers not recognized by pollinators = low fitness
Population stabilizes; nearly all the individuals are the same color.
(c) Diversifying selection
Neither pollinator chooses intermediate color = low fitness
One type of pollinator specializes in pale flowers = high fitness
Another pollinator specializes in dark flowers = high fitness
Population diversifies into pale variety and dark variety.

**FIGURE 11.16 Directional, stabilizing, and diversifying selection.** In a variable population, different environmental conditions cause different types of selection.

extreme variants in a population are selected against, and the traits of the population stay the same (**FIGURE 11.16b**). For example, in humans, the survival of newborns is correlated to birth weight—both extremely small and extremely large babies have lower survival, causing the average birth weight of babies to be relatively stable over time. Stabilizing selection causes populations to tend to resist change in unchanging environments.

Finally, in some situations, the most common variant may have the lowest fitness, resulting in **diversifying selection**, known sometimes as *disruptive selection*. Diversifying selection causes the evolution of a population consisting of two or more variants (**FIGURE 11.16c**). For example, spadefoot toad tadpoles in New Mexican ponds come in two forms—large meat-eating individuals, and smaller, more vegetarian individuals. Because there is intense competition for food in these ponds, individuals who can specialize in one food type or the other, but not both, are favored by natural selection. Diversifying selection is especially likely within a species if different subpopulations are experiencing different environmental conditions, such that the traits that bring success in one environment are not so successful in another. The primary mechanism by which new species arise (explored in Chapter 12) is diversifying selection.

# 11.4  Natural Selection and Human Health

Knowing how natural selection works allows us to understand how populations of *M. tuberculosis* have become resistant to all of our most effective antibiotics. It also provides insight into how best to combat these evolving enemies.

## Tuberculosis Fits Darwin's Observations

*Mycobacterium tuberculosis* evolved to become resistant to our antibiotics via natural selection because it fulfills all the necessary conditions Darwin observed and noted.

1. **Organisms in the Population Vary.** Bacteria hidden in the lungs reproduce as they feed on the tissue there; any time there is reproduction, mutation can occur. As a result, even during drug treatment, new variants of *M. tuberculosis* continually arise. Some of these variants have proteins that disable or counteract certain antibiotics, making the bacteria more resistant to these drugs.
2. **The Variation among Organisms Can Be Passed on to Offspring.** The traits that cause drug resistance are coded for in a bacterium's DNA. When a cell divides to reproduce, it copies this DNA and passes it—and the trait—on to its daughter cells. Bacteria can also pass on variation by direct transfer of fragments of DNA from one bacterial cell to another. This type of "inheritance" does not occur in multicellular organisms like ourselves, but it does not change the basic principle that variation is heritable.
3. **More Organisms Are Produced than Survive.** The antibiotic treatment eliminates most of the bacteria in the infected individual's body.
4. **An Organism's Survival Is Not Random.** Bacterial cells with traits that make them more resistant to the antibiotic are more likely to survive than those that are less resistant.

Because of the increased fitness of antibiotic-resistant variants, subsequent bacterial generations consist of a greater percentage of these variants. In other words, the population evolves to become resistant to the drug treatment.

Immediately after scientists began using antibiotics against tuberculosis, they noticed that some individuals would become ill again after seemingly successful treatment. Even more puzzling was that these recurrent infections were

**The recklessness of millions of ordinary people around the world puts us at risk.**

much more difficult to treat than the initial one. It was only until scientists incorporated their understanding of natural selection into tuberculosis treatments that effective, long-term therapies were developed.

Two characteristics of the early treatment strategies for tuberculosis actually sped up the development of drug resistance. The first was using the drugs for too short a time period. The second was using only one type of antibiotic at a time in patients with active tuberculosis disease.

## Selecting for Drug Resistance

Antibiotics are effective because they buy a patient time to control a bacterial infection with his or her own immune system. By keeping the bacterial population low, antibiotics allow the body to devote energy to developing an immune response instead of losing energy to the effects of the infection.

Because most of the *M. tuberculosis* cells in an infected individual are susceptible to antibiotic, a few days of treatment will eliminate the majority of these organisms. Once most of the bacteria are dead, the patient feels much better as the most debilitating aspects of the disease (for example, fever, severe cough) are reduced or eliminated. However, a small number of bacteria that are more resistant to the antibiotic take longer to kill. If drug treatment stops as soon as the patient feels better—the typical pattern in early treatment protocols—any of the more resistant bacteria that remain can multiply and restart the infection. Thus, a drug-resistant strain is born. And because this new population is more resistant, the resurgent infection is much more difficult to control (**FIGURE 11.17**). As the number of resistant bacteria increase in the patient, the chances increase that one will appear with a mutation that makes it strongly resistant to the antibiotic. If the patient returns to the community at this point, a disease that is highly resistant to the most commonly effective drugs could begin to spread.

**STOP & STRETCH** Why would some individual bacteria carry a mutation that makes them more resistant to an antibiotic even though they have never been exposed to this antibiotic?

## Stopping Drug Resistance

Combating the development of drug resistance in *M. tuberculosis* required removing the factors that promoted its development in the first place. One strategy is to maintain drug therapy for months, until all signs of the bacterial infection are cleared from the body. The other is to use multiple drugs on active infections to avoid selecting for strongly drug-resistant variants that already exist in the population.

**Combination drug therapy,** also called *drug cocktail therapy,* is commonly used on diseases for which resistance to a single drug can develop rapidly. HIV, the virus that causes AIDS, is one example. The effectiveness of combination therapy is based on the following fact: The greater the number of drugs used, the greater the number of changes that are required in the bacterial genome for resistance to develop.

The likelihood of a bacterial variant arising that is resistant to a single drug is relatively small but still very

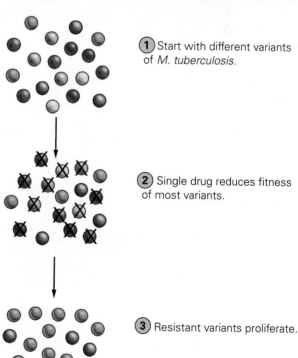

**FIGURE 11.17 Directional selection in tuberculosis.** In a variable population of *M. tuberculosis* bacteria, some cells may be more resistant to a particular antibiotic. When this antibiotic is used, these resistant variants survive and continue to reproduce. The result is a population that is resistant to the antibiotic.

**Single drug therapy**

① Start with different variants of *M. tuberculosis*.

② Single drug reduces fitness of most variants.

③ Resistant variants proliferate.

possible in a patient with 1 billion different bacterial variants. However, the likelihood of a bacterial variant arising with resistance to two or three drugs in a cocktail is extremely small. Put another way, the chance that any bacterium exists that is resistant to a single drug is analogous to the likelihood that in 1 billion lottery ticket holders, one person will hold the winning combination—in other words, relatively likely. The likelihood of a variant being resistant to several different drugs is analogous to that same ticket holder winning the lottery several times in a row—incredibly unlikely. Just as it is exceedingly rare to win the lottery twice in a row, it is very difficult for *M. tuberculosis* to adapt to an environment where it faces two "killer drugs" at once (**FIGURE 11.18**).

If scientists learned these lessons about drug resistance in the 1940s, why is drug-resistant *M. tuberculosis* making a comeback only now? Primarily, it is because public health researchers let down their guard and didn't follow TB patients to ensure that they continued taking their drugs for the 6 to 12 months required to clear out all of the invading bacteria. It wasn't just

**FIGURE 11.18 Combination drug therapy prevents antibiotic resistance.** Using multiple antibiotics makes the environment much harsher for *M. tuberculosis* and decreases the likelihood that a variant with multiple resistances will evolve.

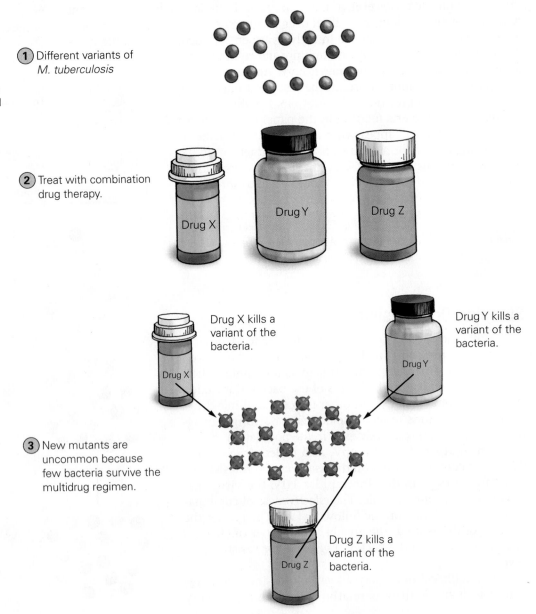

1 Different variants of *M. tuberculosis*

2 Treat with combination drug therapy.

Drug X

Drug Y

Drug Z

Drug X kills a variant of the bacteria.

Drug Y kills a variant of the bacteria.

Drug X

Drug Y

3 New mutants are uncommon because few bacteria survive the multidrug regimen.

Drug Z kills a variant of the bacteria.

Drug Z

Andrew Speaker's recklessness that exposed his fellow airline passengers to XDR-TB. It was thousands of ordinary TB patients who failed to complete their treatment regimen that led to the development of this dangerous variant. With this understanding, public health strategies that focus on maintaining the long-term treatment of individuals, and bringing that treatment to underserved communities, can go a long way in stemming the development of more cases of XDR-TB.

Not surprisingly, tuberculosis is not the only bacterial disease in which resistance has developed, largely as a result of inadequate treatment with antibiotics. MRSA, methicillin-resistant *Staphylococcus aureus,* is another formerly easily treated bacteria that has evolved into a more dangerous and deadly type (**FIGURE 11.19**). Dozens of other bacterial **pathogens** that were once easily treated by antibiotics now contain strains resistant to one or more antibiotics, including organisms responsible for ear infections, sexually transmitted diseases, and pneumonia.

The rise of antibiotic resistance stems not only from failure to follow drug treatment but also from the overuse of these drugs. Antibiotics are often erroneously prescribed to individuals with viral infections, such as the common cold, on which they have no effect and only serve to increase the likelihood that formerly easily controlled bacteria may develop resistance. Antibiotics are also commonly and liberally used in animal agriculture; they are given to poultry, cows, and pigs to prevent bacterial disease outbreaks in crowded feedlots. Drug resistance is also not confined to bacteria; the same evolutionary process has occurred in viruses, like HIV, and other pathogens, such as the protozoan that causes malaria.

To stop the development of these antibiotic-resistant "superbugs," patients and physicians must use antibiotics judiciously and wisely. But perhaps evolution has another trick up its sleeve that gives us some hope as well.

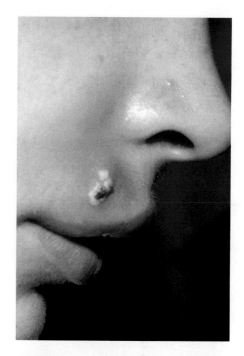

**FIGURE 11.19 MRSA.** Staph infections of the skin are relatively common, causing boils like these. However, if untreatable by antibiotics, staph can become a systemic disease that may be deadly.

## Can Natural Selection Save Us from Superbugs?

Bacteria can rapidly evolve resistance to our antibiotics. Why can't we evolve resistance to the bacteria? Can't natural selection save us from these pathogens?

There are clearly differences among healthy individuals in their susceptibility to long-term tuberculosis infection and even to active disease. Given the existence of this heritable variation and differences in survival among those exposed to this killer bacteria, can we expect natural selection to cause the human population to evolve resistance to *M. tuberculosis*?

Eventually, perhaps—but remember that natural selection occurs because of differential survival and reproduction of individuals over time. In short, for the human population as a whole to become resistant to this bacterium, nonresistant variants must die out of the population. This process has not occurred in the more than 6000 years since tuberculosis has been in human populations. Because a majority of nonresistant people are not even exposed to *M. tuberculosis* and thus will continue to survive and reproduce, nonresistant variants will probably never be lost from the population. Clearly, future *human* evolution is not a likely solution to the problem of antibiotic-resistant superbugs.

However, recall that natural selection does not result in perfect organisms, only those with traits that are effective in the current environmental conditions. In *M. tuberculosis,* variants that are resistant to multiple antibiotics are also much less likely to spread to other individuals. In other words, antibiotic resistance is a trade-off that reduces a bacterial cell's ability to survive and reproduce under normal conditions. This raises hopes that quick response to TB outbreaks in vulnerable populations can contain the spread of this dangerous pathogen.

Natural selection has assisted in our fight against tuberculosis in another way as well. By shaping the human brain in response to environmental challenges, it has provided us with a powerful tool for fighting this disease—our own ingenuity. Understanding how the disease is spread has helped prevent thousands of new infections, and the development of antibiotics has fought off millions more. A vaccine that can stimulate the immune system to prevent initial infection is in development that may eventually eliminate the disease entirely. Before that happens, we need to apply our ingenuity fully to address the alarming spread of this resistant disease.

While the story of the XDR-TB infected traveler at the Mexican border is still unfolding, the good news for Andrew Speaker was that later tests of the *M. tuberculosis* variant in his lungs confirmed that it was the more treatable MDR-TB and not XDR-TB. He received antibiotic treatment and lung surgery and was relatively quickly released from the hospital. This outcome was good news for the people on Andrew Speaker's airplane flights, as was the fact that in the absence of active disease, his ability to infect other individuals was very low. Tuberculosis does not need to return to the unbeatable enemies list. And with attention to the process of natural selection guiding our use and application of antibiotics, it shouldn't become one.

# *savvy reader*

## Blogging the TB Story
### *What gall!*

Who does Andrew Speaker think he is? Everyone knows that tuberculosis is a pretty bad disease and is spread through the air. Everyone knows that the air on an airplane is pretty polluted on the best days, and that one is more likely than not to come off an airplane with the virus du jour. So he knows he has TB, gets on a plane anyway, or make that several planes, long international flights. Runs from the CDC when they track him down—on another plane, no less. And his comment now is that he really didn't think he was putting anyone at risk?

Right. Maybe he should have stopped at "I didn't think" or maybe changed the sentence to, "I didn't care." What incredible disregard for other people. What arrogance! I'm afraid it rather sums up the way too many people have become these days. The thought is ME, ME, ME. Speaker is perhaps the quintessential ME guy. He is even willing to put his bride at risk, for goodness' sakes! Not to mention all the hundreds of people he came in contact with. Shame on him.

1. List the facts asserted within this opinion piece. Does the author provide any supporting evidence for them?

2. No opinion piece presents evidence for all the statements contained within it. What resources do you have to evaluate the credibility of this author?

3. This blog posting is a reasonable summary of the reactions many people had to initial reports about Andrew Speaker. How do these initial reactions appear now that more of the facts of the case have been revealed?

From the Wordpress blog site of "fwtandy" (no further identification) posted June 8, 2007

# SOUNDS RIGHT BUT IS IT?

It can be difficult to imagine how complicated organisms have come about without some designer being involved. Even Charles Darwin understood that it is difficult to see how an organ as complex as the human eye, with a pupil that regulates light, a lens that can flex to focus at different distances, and the sophisticated nervous tissue of the retina, could have evolved through the process of natural selection. But he did conceive of how it was possible. You might hear critics of evolutionary theory make a statement like this:

**The human eye is too complex to have evolved by chance from nothing.**

Sounds right, but it isn't.

1. Where does the variation that natural selection acts on arise from?

2. Scientists estimate that each individual has three "new" mutations that could affect their anatomy and physiology. In a population of only 10,000 people and using our low estimate of mutations that could be subject to selection, about how many "new" variants of genes appear every generation?

3. Question 2 gives you a sense of how much genetic variety appears by chance. Is natural selection "chance"?

4. Consider the evolution of the panda's thumb described in the chapter. What did it evolve from?

5. Some simple single-celled organisms can sense light and will avoid it (or move toward it). Do these organisms have eyes?

6. Simple animals, such as earthworms, can change the shape of parts of their bodies, like their mouths. Is it possible that the same mechanisms that change mouth shape could change lens or pupil shape?

7. Reflect on your answers to questions 1–6. Explain why the statement bolded above sounds right, but isn't.

# Chapter Review  MasteringBiology®

Go to the Study Area in MasteringBiology® for practice quizzes, myeBook, BioFlix™ 3-D animations, MP3Tutor sessions, videos, current events, and more.

## Summary

## Section 11.1

Describe the history of tuberculosis in human populations, and explain why current treatments are ineffective against some strains of the disease.

- Tuberculosis is an ancient and widespread disease caused by bacterial infection and can result in lung disease and death in about 10% of cases (pp. 234–235).

- The advent of antibiotics turned tuberculosis from a possible death sentence to a readily curable condition. However, beginning about 25 years ago, variants of the tuberculosis bacteria appeared that are resistant to most antibiotics (pp. 235–236).

## Section 11.2

List the four observations that led to the inference of natural selection.

- Individuals in a population vary, and some of this variation can be passed on to offspring (pp. 236–237).

- Not all individuals born in a population survive to adulthood, and not all adults produce the maximum number of offspring possible (pp. 237–238).

- Survival and reproduction are not random. Advantageous traits, called adaptations, increase an individual's fitness, which is his or her chance of survival or reproduction (pp. 238–239).

Explain how natural selection causes evolutionary change.

- The increased fitness of individuals with particular adaptations causes the adaptation to become more prevalent in a population over generations (pp. 239–240).

Provide examples of evidence that supports the hypothesis that natural selection leads to the evolution of populations.

- Artificial selection, when humans deliberately control an organism's fitness, causes the evolution of different breeds of animals and varieties of plants (p. 240).

- Populations exposed to environmental changes, both in the lab and in nature, have been shown to evolve traits that make them better fitted to the environment (pp. 240–242).

## Section 11.3

Describe how natural selection works on allele frequencies in a population.

- The modern definition of evolution is a genetic change in a population of organisms (pp. 242–244).
- Alleles that code for adaptations become more common in a population over generations as a result of natural selection (pp. 243–244).

Discuss why natural selection does not result in "perfectly adapted" organisms or drive organisms toward some ideal state.

- Natural selection can act only on the variants currently available in the population. Natural selection results in a population that is better adapted to its environment but usually not perfectly adapted as a result of trade-offs. Natural selection does not push a population in the direction of a predetermined "goal" (pp. 244–246).

List the three patterns of selection, provide examples of each, and explain how they lead to different outcomes.

- Selection can cause the traits in a population to change in a particular direction. However, in some environments it may cause certain traits to resist change and in other environments cause multiple variants to evolve (pp. 246–248).

## Section 11.4

Using your understanding of natural selection, explain why combination drug therapy is an effective tool to combat drug resistance.

- Disease-causing organisms can evolve resistance to antibiotics because they consist of multiple variants that have differential survival when exposed to various drugs; thus, a population of disease-causing organisms can evolve drug resistance via natural selection (pp. 248–249).
- A mutant organism that is resistant to several different antibiotics is relatively unlikely, so combination drug therapy can reduce the risk of antibiotic resistance evolving (pp. 249–251).
- As a result of trade-offs, varieties of disease-causing organisms that are multiple-drug-resistant are less likely to survive and reproduce in normal conditions than are non-drug-resistant varieties. Resistant organisms, therefore, are less transmissible than are nonresistant varieties (p. 251).

## Roots to Remember

**These roots come mainly from Latin and Greek and will help you understand terms:**

| | |
|---|---|
| **adapt-** | means to fit in. Chapter term: *adaptation* |
| **anti-** | means in opposition to. Chapter term: *antibiotic* |
| **-biotic** | indicates pertaining to life. Chapter term: *antibiotic* |
| **-osis** | indicates a condition. Chapter term: *tuberculosis* |
| **patho-** | means disease. Chapter term: *pathogen* |
| **tubercul-** | means a small swelling. Chapter term: *tuberculosis* |

## Learning the Basics

1. What types of drugs have helped reduce the death rate due to tuberculosis infection, and why have they become less effective more recently?

2. Define *artificial selection*, and compare and contrast it with natural selection.

3. Describe how *Mycobacterium tuberculosis* evolves when it is exposed to an antibiotic.

4. Which of the following observations is not part of the theory of natural selection?

   A. Populations of organisms have more offspring than will survive; B. There is variation among individuals in a population; C. Modern organisms are unrelated; D. Traits can be passed on from parent to offspring; E. Some variants in a population have a higher probability of survival and reproduction than other variants do.

5. The best definition of *evolutionary fitness* is _____.

   A. physical health; B. the ability to attract members of the opposite sex; C. the ability to adapt to the environment; D. survival and reproduction relative to other members of the population; E. overall strength.

6. An adaptation is a trait of an organism that increases _____.

   A. its fitness; B. its ability to survive and replicate; C. in frequency in a population over many generations; D. A and B are correct; E. A, B, and C are correct.

7. The heritable differences among organisms are a result of _____.

   A. differences in their DNA; B. mutation; C. differences in alleles; D. A and B are correct; E. A, B, and C are correct.

# Species and Races

January 20, 2009, represented a watershed moment in American history, the inauguration of the first African American president, Barack Obama. Regardless of how you feel about his presidency, nearly everyone can appreciate how Obama's election represented the culmination of a major transformation in American society, less than six generations removed from the legal enslavement of Africans. When the African American scholar and poet Elizabeth Alexander read a verse she penned for the inauguration entitled "Praise Song for the Day," she explicitly referenced how far the nation had come:

> Say it plain: that many have died
>    for this day.
> Sing the names of the dead who
>    brought us here,
> who laid the train tracks, raised the
>    bridges,
> picked the cotton and the lettuce,
>    built

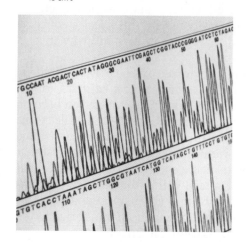

She later learned that she was genetically more European than African.

In fact, she shares a recent ancestor with the comedian Stephen Colbert.

> brick by brick the glittering edifices
> they would then keep clean and
>    work inside of.

African slavery was in part justified by a belief that black Africans were somehow biologically inferior to white Europeans. Even after slavery ended, racism did not. Dr. Alexander's professional career has been dedicated to improving race relations and reducing the effects of racism that she and other African Americans have experienced. Imagine Dr. Alexander's surprise when she learned that according to genetic analysis, her ancestry is actually only 27% African—and 66% European. The analysis was performed as part of a documentary series examining the diversity of America, and her bemused reaction to the results was caught on film.

Moments later, the host of the documentary, Harvard professor of African American studies Henry Louis Gates, Jr., revealed to Dr. Alexander that the analysis identified a distant cousin: the comedian Stephen Colbert, whose genetic ancestry is 100% European. Alexander was more closely related to Colbert than to Gates—and given these results, she likely shares little genetic history with other African Americans to whom her work seeks to give voice.

But what does a genetic analysis that places a noted African American poet closer to a white entertainer than to other African Americans tell us about race? Is she more white than black? Or is race more a social construction than a biological reality? This chapter examines the biological definition of race as a step on the road to the formation of new species and addresses the question of whether humans can be biologically grouped into distinct races.

How does this new genetic information fit into our experience of human races?

**(a) Lion**

**(b) Leopard**

**FIGURE 12.1 Same genus, different species.** (a) *Panthera leo*, the lion, and (b) *Panthera pardus*, the leopard.

# 12.1 What Is a Species?

All humans belong to the same species. Before we can understand the concept of race, we first need to understand both what is meant by this statement and what is known about how species originate.

In the mid-1700s, the Swedish scientist Carolus Linnaeus began the task of cataloging all of nature. Linnaeus developed a classification scheme that grouped organisms according to shared traits (Chapter 10). The primary category in his classification system was the **species**, a group whose members have the greatest resemblance. Linnaeus assigned a two-part name to each species—the first part of the name indicates the **genus**, or broader group; the second part is specific to a particular species within that genus. For example, lions, the species *Panthera leo*, are classified in the same genus with other species of roaring cats, such as the leopard, *Panthera pardus* (**FIGURE 12.1**).

Linnaeus coined the binomial name *Homo sapiens* (*Homo* meaning "man" and *sapiens* meaning "knowing or wise") to describe the human species. Although Linnaeus recognized variability among humans, by placing all of us in the same species he acknowledged our basic unity.

Modern biologists have kept the basic Linnaean classification, although they have added a **subspecies** name, *Homo sapiens sapiens*, to distinguish modern humans from earlier humans that appeared approximately 250,000 years ago. Other subspecies of human include the Neanderthals, known as *Homo sapiens neanderthalensis*.

## The Biological Species Concept

While most people intuitively grasp the differences between most species—both lions and leopards are definitely cats but not the same species—biologists have had difficulty finding a single definition that can be consistently applied. Several useful concepts have been proposed and used. The definition most commonly used is the biological species concept.

**Biological Species Are Reproductively Isolated.** According to the **biological species concept**, a species is defined as a group of individuals that, in nature, can interbreed and produce fertile offspring but cannot reproduce with members of other species. In practice, this definition can be difficult to apply. For example, species that reproduce asexually (such as most bacteria) and species known only via fossils do not easily fit into this species concept. However, the biological species concept does help us understand why species are distinct from one another.

Recall that differences in traits among individuals arise partly from differences in their genes, and that the different forms of a gene that exist are known as *alleles* of the gene. By the process of evolution, a particular allele can become more common in a species. If interbreeding does not occur, then this allele cannot spread from one species to the other. In this way, two species can evolve differences from each other. For example, various lines of evidence suggest that the common ancestor of lions and leopards had a spotted coat. The allele that eliminated the spots from lions arose and spread within this species, presumably because it provided an advantage to lions in their environment, but this allele is not found in leopards. The spotless allele has not been transferred to the leopard population because lions and leopards cannot interbreed.

**STOP & STRETCH** Which environmental conditions experienced by lions may have made the allele for spotlessness advantageous in their population?

Scientists refer to the sum total of the alleles found in all the individuals of a species as the species' **gene pool.** A change in the frequency of an allele in a gene pool can take place only within a biological species.

### The Nature of Reproductive Isolation.

The spread of an allele throughout a species' gene pool is called **gene flow.** Gene flow cannot occur between different biological species because pairing between them fails to produce fertile offspring. This reproductive isolation can take two general forms: prefertilization barriers or postfertilization barriers, summarized in **TABLE 12.1**, on page 261.

As you can see in the table, most mechanisms of reproductive isolation are invisible, occurring before mating or fertilization can occur or soon after fertilization. In the instances when a mating between two different species results in live offspring—a **hybrid** descendant—the offspring are often sterile. A well-known example of an interspecies hybrid is the mule, resulting from a cross between a horse and a donkey. Mules have a well-earned reputation as sturdy farm animals, but they cannot produce their own offspring.

Hybrid sterility often occurs because hybrid individuals cannot produce viable sperm or egg cells. During the production of eggs and sperm, homologous chromosomes pair up and separate during the first cell division of meiosis (Chapter 6). Because a hybrid forms from the chromosome sets of two different species, its chromosomes are not homologous and thus cannot pair up correctly during this process.

In the case of mules, the horse parent has 64 chromosomes and therefore produces eggs or sperm with 32; the donkey parent has 62 chromosomes, producing eggs or sperm with 31. The mule will therefore have 63 chromosomes and no way to effectively sort these into pairs during the first division of meiosis (**FIGURE 12.2**).

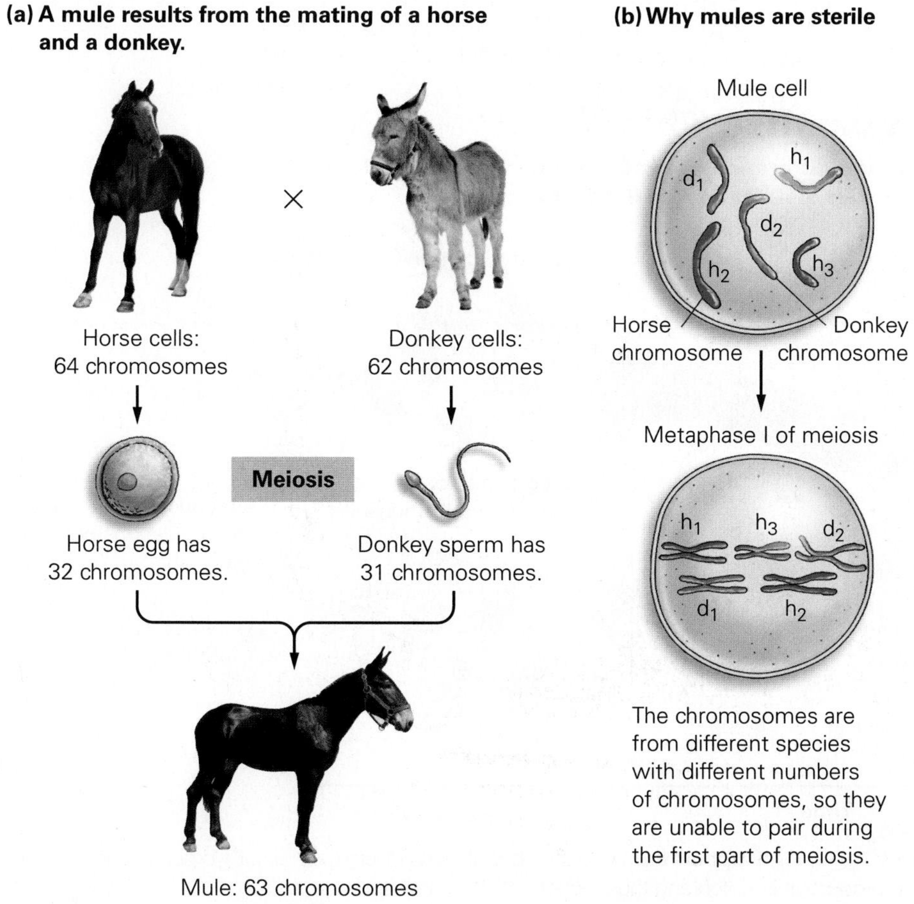

**(a) A mule results from the mating of a horse and a donkey.**

Horse cells:
64 chromosomes

Donkey cells:
62 chromosomes

Meiosis

Horse egg has 32 chromosomes.

Donkey sperm has 31 chromosomes.

Mule: 63 chromosomes

**(b) Why mules are sterile**

Mule cell

$d_1$  $h_1$  $d_2$  $h_2$  $h_3$

Horse chromosome     Donkey chromosome

Metaphase I of meiosis

$h_1$  $h_3$  $d_2$  $d_1$  $h_2$

The chromosomes are from different species with different numbers of chromosomes, so they are unable to pair during the first part of meiosis.

◄ **Visualize This**

If donkeys had 60 chromosomes so that their sperm contained 30, a mule would have 62 chromosomes. Would having an even number of chromosomes permit mules to be fertile?

**FIGURE 12.2 Reproductive isolation between horses and donkeys.** (a) A cross between a female horse and a male donkey produces a mule with 63 chromosomes. (b) Mules produce only very few eggs or sperm because their chromosomes cannot pair properly during meiosis. Only a small number of chromosomes are illustrated to simplify the drawing.

While a tiny number of female mules, surprisingly, have produced offspring, this event is so rare that the gene pools of donkeys and horses have remained separate.

---

**STOP & STRETCH** The corn varieties planted by most farmers are called "hybrids" because each is produced when two different varieties of the species *Zea mays* (for example, a short variety and a sweet variety) are crossed. Do you expect that the corn produced by these crosses is sterile? Why or why not?

---

Given the definition of a biological species, it is clear that all humans belong to the same biological species. To understand the concept of races within a species, however, we must first examine how species form.

## Speciation

According to the theory of common descent, all modern organisms descended from a common ancestral species. The evolution of one or more species from an ancestral form is called **speciation.**

For one species to give rise to a new species, most biologists agree that three steps are necessary (**FIGURE 12.3**).

1. Isolation of the gene pools of subgroups, or **populations,** of the species;
2. Evolutionary changes in the gene pools of one or both of the isolated populations; and
3. The evolution of reproductive isolation between these populations, preventing any future gene flow.

## Visualize This ▼

**Is the difference between two diverging populations always visible, as it is in this example?**

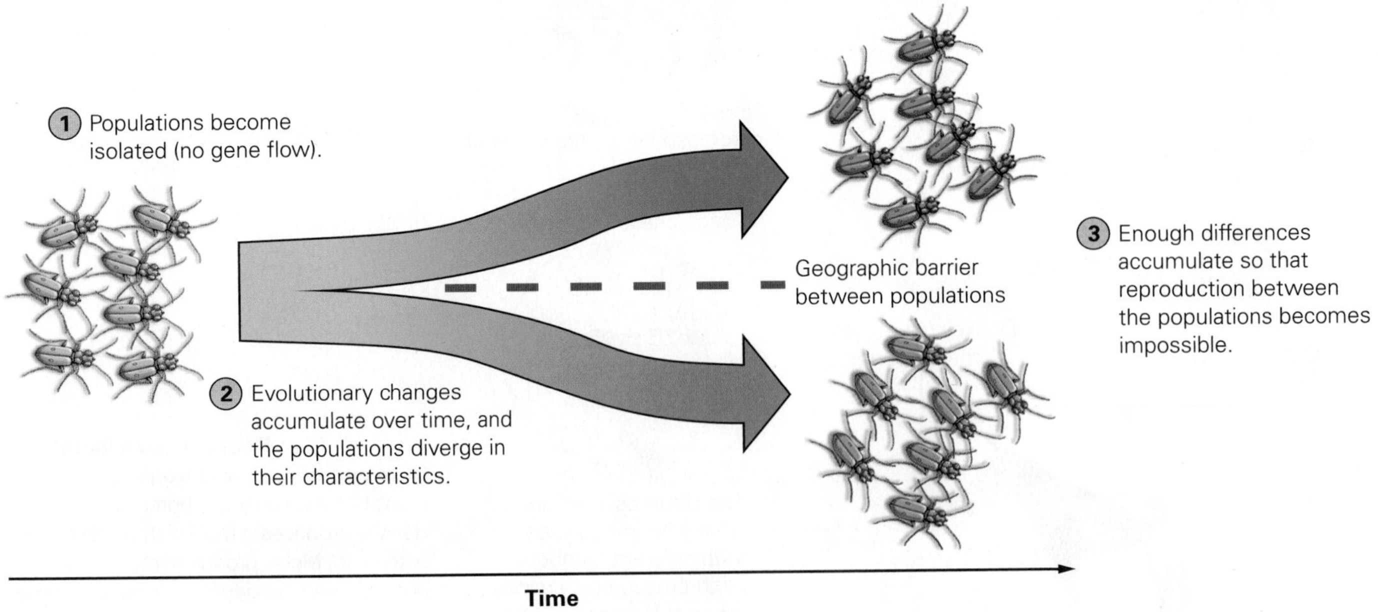

**1** Populations become isolated (no gene flow).

**2** Evolutionary changes accumulate over time, and the populations diverge in their characteristics.

Geographic barrier between populations

**3** Enough differences accumulate so that reproduction between the populations becomes impossible.

Time

**FIGURE 12.3 Speciation.** Isolated populations diverge in traits. Divergence can lead to reproductive isolation and thus the formation of new species.

Recall that gene flow occurs when reproduction is occurring within a species. Now imagine what would happen if two populations of a species became physically isolated from each other, so that the movement of individuals between these two populations was impossible. Even without genetic or behavioral barriers to mating between these two populations, gene flow between them would cease.

**TABLE 12.1 Mechanisms of reproductive isolation.**

| Type | Effect | Example |
|---|---|---|
| **Prefertilization barriers prevent fertilization from occurring.** | | |
| **Spatial isolation** | Individuals from different species do not come in contact with each other. | Polar bear (Arctic) and spectacled bear (South America) never encounter each other in natural settings. |
| **Behavioral** | Ritual behaviors that prepare partners for mating are different in different species. | Many birds with premating songs or "dances" will not mate with individuals who do not know the ritual.<br> |
| **Mechanical** | Sex organs are incompatible between different species, so sperm cannot reach egg. | Many insects with "lock-and-key" type genitals physically prevent sperm from contacting eggs of a different species. |
| **Temporal** | Timing of readiness to reproduce is different in different species. | Plants with different flowering periods cannot fertilize each other. |
| **Gamete incompatibility** | Proteins on egg that allow sperm binding do not bind with sperm from another species. | Animals with external reproduction, such as sponges, have specific proteins on their eggs that will bind only to sperm from the same species.<br> |
| **Postfertilization barriers: Fertilization occurs, but hybrid cannot reproduce.** | | |
| **Hybrid inviability** | Zygote cannot complete development because genetic instructions are incomplete. | A sheep crossed with a goat can produce an embryo, but the embryo dies in the early developmental stages. |
| **Hybrid sterility** | Hybrid organism cannot produce offspring because chromosome number is odd. | Mules (see Figure 12.2) |

## Isolation and Divergence of Gene Pools

The gene pools of populations may become isolated from each other for several reasons. Often, a small population becomes isolated when it migrates to a location far from the main population. This is the case on many oceanic islands. Bird, reptile, plant, and insect species on these islands appear to be the descendants of species from the nearest mainland. The original ancestral migrants arrived on the islands by chance.

Because they are separated by hundreds of miles of open ocean, populations on these islands are isolated from their ancestral population (**FIGURE 12.4**). In addition, because migrant populations are small, their gene pools can change rapidly via the process of genetic drift, as described in Section 12.3.

The establishment of a new population far from an original population may lead to the evolution of several new species—a process known as **adaptive radiation.** According to this hypothesis, the diversity of unique species on oceanic islands, as well as in isolated bogs, caves, and lakes, resulted from colonization of these once "empty" environments by one founding species that rapidly diversified into many species. Populations may also become isolated from each other by the intrusion of a geologic barrier. This could be an event as slow as the rise of a mountain range or as rapid as a sudden change in the course of a river. The emergence of the Isthmus of Panama separating the Pacific Ocean from the Caribbean Sea between 3 and 6 million years ago represents one such intrusion event. Scientists have described dozens of pairs of aquatic species—each made up of one member living on either side of the isthmus. Over the 6 million years since their ancestral species was divided by the isthmus, genetic changes accumulated independently on each side—so much so that the members of each pair are now separate species. Populations that are isolated from each other by distance or a barrier are known as **allopatric.**

However, separation between the gene pools of two populations may occur even if the populations are living near each other, that is, if they are **sympatric.** This appears to be the case in populations of the apple maggot fly. Apple maggot flies are notorious pests of apples grown in northeastern North America.

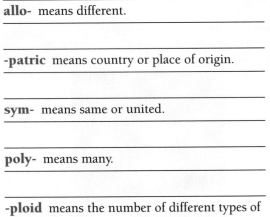

**allo-** means different.

**-patric** means country or place of origin.

**sym-** means same or united.

**poly-** means many.

**-ploid** means the number of different types of chromosomes.

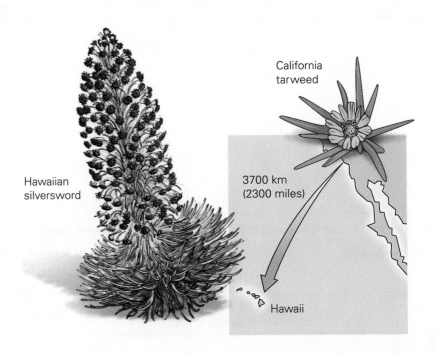

**FIGURE 12.4 Migration leads to speciation.** California tarweed seeds were blown or carried by birds to the Hawaiian Islands, creating an isolated population. With no gene flow between California and Hawaii, a very different group of species, Hawaiian silverswords, evolved from the tarweed colonists.

Hawaiian silversword

California tarweed

3700 km (2300 miles)

Hawaii

However, apple trees are not native; they were first introduced to this continent less than 300 years ago. Apple maggot flies also infest the fruit of hawthorn shrubs, a group of species that are native to North America.

Even though apples and hawthorns live in close proximity, it is clear that flies from these two plants actually have little opportunity to mate. Flies mate on the fruit where they lay their eggs, and hawthorns produce fruit approximately 1 month after apples do. Each population of fly has a strong preference regarding the fruit on which it will mate, and flies that lay eggs on hawthorns develop much faster than flies that lay eggs on apples. As a result, there appears to be little gene flow between the apple-preferring and hawthorn-preferring populations.

Because there is no gene flow, the two groups of apple maggot flies have begun to diverge—the gene pools of the two groups now differ strongly in the frequency of some alleles (**FIGURE 12.5**). While not yet considered separate biological species, these two populations may continue to diverge and eventually become reproductively incompatible.

In plants, isolation of gene pools can occur instantaneously and without any barriers between populations. A simple hybrid between two plant species is typically infertile because it cannot make gametes (the same problem that occurs in mules; review Figure 12.2). However, some hybrid plants can become fertile again—if a mistake during mitosis produces a cell containing duplicated chromosomes. The process of chromosome duplication is called **polyploidy**, and it results in a cell that contains two copies of each chromosome from each parent species. If polyploidy occurs inside a plant bud, all the cells of the branch that arises from that bud will be polyploid, containing two identical copies of every chromosome (**FIGURE 12.6**, on the next page).

Because polyploid cells now contain pairs of identical chromosomes, meiosis can proceed, and thus flowers produced on the branch can produce eggs and sperm. A polyploid flower can then self-fertilize and give rise to hundreds of offspring, representing a brand new species that is isolated from its parent plants. Recent research suggests that this process of "instantaneous speciation" may have been a key factor in the evolution of as many as 50% of flowering plant species. Polyploidy occurs in some animal groups, such as insects and frogs, as well.

## The Evolution of Reproductive Isolation

To become truly distinct biological species, diverging populations must become reproductively isolated either by their behavior or by genetic incompatibility. In the case of canola, genetic incompatibility occurs immediately—a cross between canola and kale does not result in offspring. In most animals, the process may be more gradual, occurring when the amount of divergence has caused numerous genetic differences between two populations.

There is no hard-and-fast rule about how much divergence is required; sometimes, a difference in a single gene can lead to incompatibility, while at other times, populations demonstrating great physical differences can produce healthy and fertile offspring (**FIGURE 12.7**, on the next page). Exactly how reproductive isolation evolves on a genetic level is still unknown and is an actively researched question in biology.

Once reproductively isolated, species that derived from a common ancestor can accumulate many differences, even completely new genes.

### Is Speciation Gradual or Sudden?  Darwin assumed that speciation occurred over millions of years as tiny changes gradually accumulated. This hypothesis

## ▼ Working with Data

The Honeycrisp variety of apple was first made commercially available in the 1990s—it flowers later than many other apple trees and its fruit does not complete ripening until late September. How do you think introduction of the Honeycrisp might affect the process of divergence of the apple maggot fly?

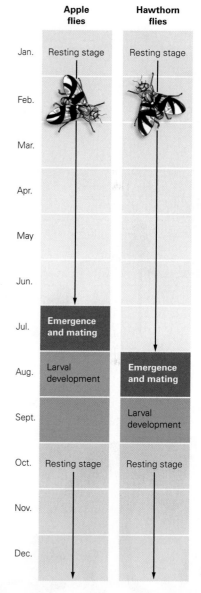

**FIGURE 12.5 Differences in the timing of reproduction can lead to speciation.** This graph illustrates the life cycle of two populations of the apple maggot fly: one that lives on apple trees and one that lives on hawthorn shrubs. The mating period for these two populations differs by a month, resulting in little gene flow between them.

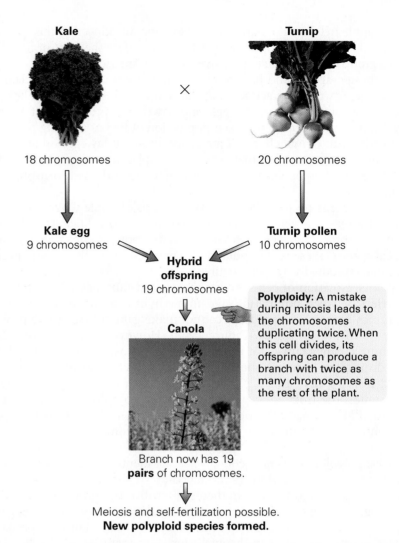

**Kale** × **Turnip**

18 chromosomes        20 chromosomes

**Kale egg**        **Turnip pollen**
9 chromosomes        10 chromosomes

**Hybrid offspring**
19 chromosomes

**Canola**

**Polyploidy:** A mistake during mitosis leads to the chromosomes duplicating twice. When this cell divides, its offspring can produce a branch with twice as many chromosomes as the rest of the plant.

Branch now has 19 **pairs** of chromosomes.

Meiosis and self-fertilization possible.
**New polyploid species formed.**

**FIGURE 12.6 Instantaneous speciation.** Canola evolved from a hybrid of kale and turnip. Although the hybrid initially was sterile because its chromosomes could not line up during meiosis, a mistake in mitosis led to chromosome duplication in one of the plant's cells. When branches produced by this cell produced flowers, new seeds with double the number of chromosomes could form, growing into whole plants with this new chromosome number. Because it has a different number of chromosomes from either of its parents, canola pollen cannot fertilize kale or turnip plant eggs and vice versa, so this plant was instantly reproductively isolated from its parents.

**(a)**

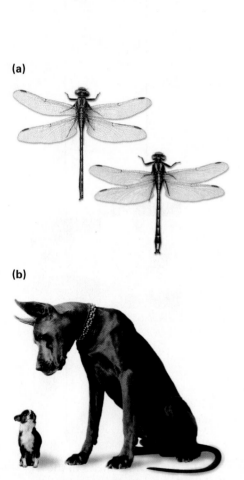

**(b)**

**FIGURE 12.7 How different are two species?** (a) These two species of dragonfly look alike but cannot interbreed. (b) Dog breeds provide a dramatic example of how the evolution of large physical differences does not always result in reproductive incompatibility.

is known as **gradualism.** Other biologists have argued that most speciation events are sudden, result in dramatic changes in form within the course of a few thousand years, and are followed by many thousands or millions of years of little change—a hypothesis known as **punctuated equilibrium.** The hypothesis of punctuated equilibrium is supported by observations of the fossil record, which seems to reflect just this pattern (**FIGURE 12.8**). Although the tempo of evolutionary change may not match Darwin's predictions, the process of natural selection he described can still explain many instances of divergence.

Regardless of the tempo of speciation, the period after the separation of the gene pools of two populations but before the evolution of reproductive isolation could be thought of as a period during which races of a species may form.

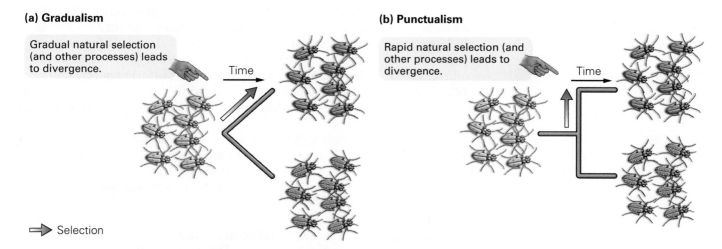

**(a) Gradualism**

Gradual natural selection (and other processes) leads to divergence.

Time

**(b) Punctualism**

Rapid natural selection (and other processes) leads to divergence.

Time

➡ Selection

**FIGURE 12.8 Gradualism versus punctuated equilibrium.** The pattern of evolutionary change in groups of species may be (a) gradual, representing a constant level of small changes, or (b) more punctuated, with hundreds of thousands of years of stasis followed unpredictably by rapid, large changes occurring within a few thousand years.

# 12.2   Are Human Races Biological?

Biologists do not agree on a standard definition of *biological race*. In fact, not all biologists feel that *race* is a useful term; many prefer to use the term *subspecies* to describe subgroups within a species, and others feel that race is not a useful biological concept at all. When the term is applied, it is often inconsistent. For example, populations of birds with slightly different colorations might be called different races by some bird biologists, while other biologists would argue that the same variations in color are meaningless.

However, our question about how race matters in practical terms leads us to a specific definition of race. What we want to know is: If an individual is identified as a member of a particular race, does that mean she is more closely related and thus biologically *more similar* to individuals of the *same* race than she is to individuals of *other* races? The definition of **biological race** that addresses this question is the following: Races are populations of a single species that have diverged from each other as a result of isolation of their gene pools. Biologists may alternatively refer to this as the **genealogical species concept** because it reflects closer shared ancestry—a genealogy—among certain individuals within a biological species. With little gene flow among races, evolutionary changes that occur in one race may not occur in a different race. However, to understand if the racial categories people in the United States identify with have a biological basis—that is, are true genealogical species— we must first understand how these racial categories came to be identified in American society.

## The History of Human Races

Until the height of the European colonial period in the seventeenth and eighteenth centuries, few cultures distinguished between groups of humans according to shared physical characteristics. People primarily identified themselves and others as belonging to particular cultural groups with different customs, diets, and languages.

**African slavery was justified by a belief that black Africans were biologically inferior to white Europeans.**

As northern Europeans began to contact people from other parts of the world, being able to set groups of people "apart" made colonization and slavery less morally troublesome. Thus, when Linnaeus classified all humans as a species, he was careful to distinguish definitive varieties (what we would now call races) of humans. Linnaeus named five races of *Homo sapiens*. Linnaeus not only described physical characteristics, he ascribed particular behaviors and aptitudes to each race; reflecting his Western bias, he set the European race as the superior form.

Linnaeus's classification of human races is one of dozens of examples of how scientists' work has been used to legitimize cultural practices. In this case, hundreds of years of injustice were supported by a scientific classification that made the lower status of nonwhite groups in European and American society appear "natural."

Scientists since Linnaeus have also proposed hypotheses about the number of races of the human species. A still common number is five, reflected in the race categories listed on the U.S. census form: white, black, Pacific Islander, Asian, and Native American. However, some scientists have described as many as 26 different races of the human species, and some personal genotyping services, companies that analyze the genes of individuals to determine their ancestry, claim to be able to distinguish more than 50 population groups. Does any of these groups represent a biological race?

To answer our question about the biology of race, we can try to determine if the physical characteristics used to delineate supposed human races—skin color, eye shape, and hair texture, for instance—developed because these groups evolved independently of each other. We can test this hypothesis by looking at the fossil record for evidence of isolation during human evolution and by looking at the gene pools of modern populations of these supposed races for consistent differences among groups.

## The Morphological Species Concept

The ancestors of humans are known only through the fossil record. We cannot delineate fossil species using either the biological or genealogical species concepts. Instead, **paleontologists,** scientists who study fossils, use a more practical definition: A species is defined as a group of individuals with some reliable physical characteristics distinguishing them from all other species. In other words, individuals in the same species have similar morphology—they look alike in some key feature. This is known as the **morphological species concept.** The morphological differences among species are assumed to correlate with isolation of gene pools. **TABLE 12.2** compares and contrasts the three species concepts.

**paleo-** means ancient.

**morpho-** means shape or form.

**STOP & STRETCH**   Which species concept, biological, genealogical, or morphological, do you think is more likely to be used by biologists who are trying to count the number of bird species in a local park? Why?

Using the morphological species concept, scientists have identified the fossils of our direct human ancestors. This has allowed them to reconstruct the movement of humans since our species' first appearance.

## Modern Humans: A History

The immediate predecessor of *Homo sapiens* was *Homo erectus*, a species that first appeared in east Africa about 1.8 million years ago and spread to Asia and Europe over the next 1.65 million years. Fossils identified as early *H. sapiens*

**TABLE 12.2 Comparison of three species concepts.**

| Species Concept | Definition | Benefits of Using This Concept | Disadvantages of Using This Concept |
|---|---|---|---|
| **Biological** | Species consist of organisms that can interbreed and produce fertile offspring and are reproductively isolated from other species. | Useful in identifying boundaries between populations of similar organisms. Relatively easy to evaluate for sexually reproducing species. | Cannot be applied to organisms that reproduce asexually or to fossil organisms. May not be meaningful when two populations of the same species are separated by large geographical distances. |
| **Genealogical** | Species consist of organisms that can interbreed, are all descendants of a common ancestor, and represent independent evolutionary lineages. | Most evolutionarily meaningful because each species has its own unique evolutionary history. Can be used with asexually reproducing species. | Difficult to apply in practice. Requires detailed knowledge of gene pools of populations within a biological species. Cannot be applied to fossil organisms. |
| **Morphological** | Species consist of organisms that share a set of unique physical characteristics that is not found in other groups of organisms. | Easy to use in practice on both living and fossil organisms. Only a few key features are needed for identification. | Does not necessarily reflect evolutionary independence from other groups. |

appear in Africa in rocks that are approximately 250,000 years old. The fossil record shows that these early humans rapidly replaced *H. erectus* populations in Africa, Europe, and Asia.

Most data support the hypothesis that all modern human populations descended from these African *H. sapiens* ancestors within the last few hundred thousand years. One line of evidence supporting this is that humans have much less genetic diversity (measured by the number of different alleles that have been identified for any gene) than any other great ape, indicating that there has been little time to accumulate many different gene variants. Using the same reasoning, we know that African populations must be the oldest human populations because they are more genetically diverse than others around the world. Thus, African populations are likely the source of all other human populations.

Given the evidence of recent African ancestry, the physical differences we see among human populations must have arisen in the last 150,000 to 200,000 years or in about 10,000 human generations. In evolutionary terms, this is not much time. All humans shared a common ancestor very recently; thus, the defined human races cannot be very different from each other.

## Genetic Evidence of Divergence

Even with little genetic difference among populations, our question about the meaning of race is still relevant. After all, even if two races differ from each other only slightly, if the difference is consistent, then people are biologically more similar to members of their own race than to people of a different race. The ancestry results produced by the comparison of Elizabeth Alexander and Stephen Colbert seem to indicate that a person can measure what percentage of their genome comes from any particular racial group. Do these results indicate

**What does a genetic analysis that places a noted African American poet closer to a white entertainer tell us about race?**

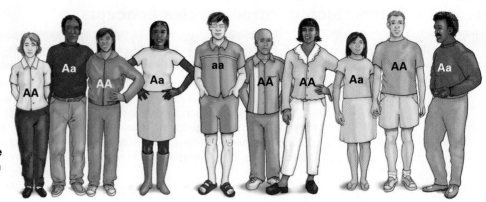

**FIGURE 12.9 How to calculate allele frequency.** The frequency of any allele in a population of adults can be calculated if individuals' genotypes are known.

that someone who has 74% European ancestry has more biological similarities to other Europeans than to Africans or Asians?

Researchers can examine the gene pool of populations described as a single race to determine if the race was once truly isolated from other races. Remember that when populations are isolated from each other, little gene flow occurs between them. If an allele appears in one population, it cannot spread to another. As a result, isolated populations should contain some unique alleles.

In addition to finding unique alleles, researchers should be able to observe differences among isolated populations in the frequency of particular alleles. When a trait becomes more common in a population due to evolution, it is because the allele for that trait has become more common (Chapter 10). In other words, evolution results in a change in **allele frequency** in a population—that is, in the percentage of copies of any given gene that are a particular allele.

## Calculating Allele Frequencies

Many genes have several different forms; for example, the blood group gene has at least three alleles—O, A, and B (discussed in Chapter 8), and most populations contain all three alleles. Populations may differ in the frequency of these alleles; one may have more individuals with type B blood, while another has more individuals with type A blood.

We can compare allele frequencies among populations (or over time) if we know individuals' genotypes. **FIGURE 12.9** illustrates the relationship between individual genotypes and a population's gene frequency in a simpler case than blood type—a single gene with two alleles.

If populations are isolated from each other, evolutionary changes—that is, an increase or decrease in the frequency of the *A* allele over time—which occur in one population will not necessarily occur in another. These changes would show up as differences between populations in allele frequency.

## Calculating Allele Frequencies: A Closer Look ▼

Allele frequencies in a population can be determined using the mathematical relationship independently discovered by Godfrey Hardy of Cambridge University and German physician Wilhelm Weinberg. The **Hardy–Weinberg theorem** states that allele frequencies will remain stable in populations that are large in size, randomly mating, and experiencing no migration or natural selection. This rule provides a baseline for predicting how allele frequencies will change if any of these conditions are violated. In other words, the Hardy–Weinberg theorem enables scientists to quantify the effect of evolutionary change on allele frequencies. Today, the Hardy–Weinberg theorem forms the basis of the modern science of **population genetics.**

*A Closer Look, continued*

In the simplest case, the Hardy–Weinberg theorem (which we abbreviate to HW) describes the relationship between allele frequency and genotype frequency for a gene with two alleles in a stable population. The frequencies of these two alleles are written as $p$ and $q$.

In Figure 12.9, 70% of the alleles in the population are dominant ($A$), and 30% are recessive ($a$). Thus, $p = 0.7$ and $q = 0.3$. Each gamete produced by members of the population carries one copy of the gene. Therefore, 70% of the gametes produced by this entire population will carry the dominant allele, and 30% will carry the recessive allele. The frequency of gametes produced of each type is equal to the frequency of alleles of each type.

HW assumes that every member of the population has an equal chance of mating with any member of the opposite sex. The fertilizations that occur in this situation are analogous to the result of a lottery drawing. In this analogy, we can imagine individuals in a population each contributing an equal number of gametes to a "bucket." Fertilizations result when a gamete drawn from the sperm bucket fuses with another drawn from the egg bucket. Because the frequency of gametes carrying the dominant allele in the bucket is equal to the frequency of the dominant allele in the population, the chance of drawing an egg that carries the dominant allele is 70%.

In **FIGURE 12.10**, a modified Punnett square illustrates the relationship between allele frequency and genotype frequency in a stable population. On the horizontal axis of the square, we place the two types of gametes that can be produced by females in the population ($A$ and $a$), while on the vertical axis we place the two types of gametes that can be produced by males. In addition, on each axis is an indication of the frequency of these types of egg and sperm in the population: 0.7 for $A$ eggs and $A$ sperm, 0.3 for $a$ eggs and $a$ sperm.

Used like the typical Punnett square (Chapter 7), the grid of the square also shows the frequency of each genotype in this population. The frequency of the

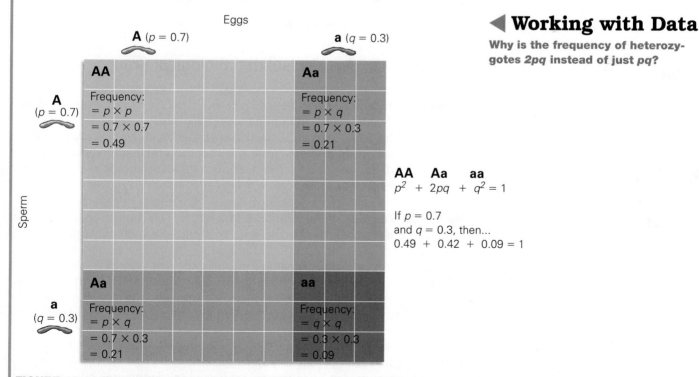

**◀ Working with Data**

Why is the frequency of heterozygotes *2pq* instead of just *pq*?

AA   Aa   aa
$$p^2 + 2pq + q^2 = 1$$

If $p = 0.7$
and $q = 0.3$, then...
$0.49 + 0.42 + 0.09 = 1$

**FIGURE 12.10 The relationship between allele frequency and gamete frequency.** The frequency of any allele in a population of adults is equal to the frequency of that allele in gametes produced by that population. Knowing the allele frequency of gametes allows us to predict the frequency of various genotypes in the next generation.

*A Closer Look, continued*

*AA* genotype in the next generation will be equal to the frequency of *A* sperm being drawn (0.7) times the frequency that *A* eggs will be drawn (0.7), or 0.49. The frequency of the *AA* genotype is thus $p \times p$ ($= p^2$). This calculation can be repeated for each genotype. The frequency of the *aa* genotype $q \times q$ ($= q^2$), and the *Aa* genotype $p \times q \times 2$ ($= 2pq$). By this theorem, Hardy and Weinberg mathematically proved that the frequency of genotypes in one generation of a population depends on the frequency of genotypes in the previous generation.

Once this relationship between allele and phenotype frequency became clear, scientists could use the known frequency of a phenotype produced by a recessive allele to calculate the allele frequency. For instance, if the frequency of individuals with blood type O is 1 in 100 births (0.01), then $q^2$—the frequency of homozygous recessive individuals in the population—is equal to 0.01. Therefore, *q* is simply the square root of this number, or 0.1. In this population, the frequency of the type O allele is 0.1, or 10%.

The Human Genome Project has made the HW theorem obsolete for calculating allele frequencies from genotype frequencies in a population—now scientists can use DNA technology to determine the frequency of an allele. However, the Hardy–Weinberg theorem is still useful in studies of human variation for identifying populations that are surprisingly different in allele frequency. Recall that one of the assumptions of the theorem is that the population is randomly mating. If groups of the population have been (or are) isolated, this assumption is not true—it is impossible to mate randomly if certain individuals can never come in contact with certain other individuals. A statistical test can identify when an allele does not conform to HW expectations and should be useful for tracing human ancestry. ▬

We can now make two predictions to test a hypothesis of whether biological races exist within a species. If a race has been isolated from other populations of the species for many generations, it should have these two traits:

1.  Some unique alleles
2.  Differences in allele frequency, relative to other races, for some genes

The next section describes the evidence thus far.

## Human Races Are Not Isolated Biological Groups

Recall the five major human races described in the census: white, black, Pacific Islander, Asian, and Native American. Do these groups show the predicted pattern of race-specific alleles and unique patterns of allele frequency? Yes, but not in a way that provides convincing evidence of consistent difference among these groups. Instead, most evidence indicates that the genetic differences between unrelated individuals within a race are much greater than the average genetic differences between races.

**No Alleles Are Found in All Members of a Race.** The Human Genome Project has allowed researchers to scan the genome of thousands of individuals looking for evidence of alleles that are found in a single human group and not in other groups. The types of alleles that are most commonly used in this analysis are called **single nucleotide polymorphisms,** abbreviated **SNPs** ("snips"). A SNP is a single base pair in the DNA sequence of humans that can differ from one individual to another. Humans are remarkably similar, having identical gene sequences for 99% of our genome. The 1% of the genome where there is variability is primarily made up of SNPs. We can think of the different DNA bases found at a particular SNP site as different alleles.

The primary reason scientists are interested in identifying SNPs in the human genome is to understand human diversity in disease susceptibility and other traits that affect health and well-being. However, because 99% of SNPs

**Is race more a social construction than a biological reality?**

appear to be in parts of the genome that do not produce proteins, they may have little or no effect on evolutionary fitness and thus can be readily passed on to future generations. You can imagine that when an SNP allele arises via mutation that has no negative effects on fitness, there may be no barrier to its spread throughout a human population. For this reason, it is the more neutral SNPs that are most useful for understanding ancestry.

Researchers have identified a number of SNP alleles that are unique to particular human populations. Most genetic ancestry testing companies identify three major groups: African, European, and Asian (including Native American). When Elizabeth Alexander learned that she is genetically 27% African, the results were based on a search of her DNA for African-specific SNP alleles. Within those larger groups, certain populations display unique SNP alleles that may help identify an individual's ancestry more specifically. However, it is important to note that no SNP allele is found in every individual in any population, and among groups classified in the same major race, some populations may have no individuals who have a particular SNP allele considered unique to a race.

What is true of SNP alleles may be more clearly illustrated with alleles associated with readily apparent phenotypes. For example, sickle-cell anemia has long been thought of as a "black" disease. This illness occurs in individuals who carry two copies of the sickle-cell allele, resulting in red blood cells that deform into a sickle shape under certain conditions. The consequences of these sickling attacks include severe pain, and heart, kidney, lung, and brain damage. Many individuals with sickle-cell anemia do not live past childhood.

Nearly 10% of African Americans and 20% of Africans carry one copy of the sickle-cell allele, whereas the allele is almost completely absent in European Americans. However, if we examine the distribution of the sickle-cell allele more closely, we see that this seemingly race-specific pattern is not so straightforward. Not all human populations classified as black have a high frequency of the sickle-cell allele. In fact, in populations from southern and north-central Africa, which are traditionally classified as black, this allele is very rare or absent. Among populations who are classified as white or Asian, there are some in which the sickle-cell allele is relatively common, such as white populations in the Middle East and Asian populations in northeast India (**FIGURE 12.11**, on the next page). Thus, the sickle-cell allele is not a characteristic of all black populations or unique to a supposed "black race."

Similarly, cystic fibrosis, a disease that results in respiratory and digestive problems and early death, was often thought of as a "white" disease. Cystic fibrosis occurs in individuals who carry two copies of the cystic fibrosis allele. As with sickle-cell anemia, it has become clear that the allele that causes cystic fibrosis is not found in all white populations and is found, in low frequency, in some black and Asian populations.

These examples of the sickle-cell allele and cystic fibrosis allele demonstrate the typical pattern of gene distribution. Scientists have not identified a single allele that is found in all (or even most) populations of a commonly described race but not found in other races. Only a tiny number of SNPs have been identified as unique to a particular human racial group, and these are never found in all populations of the race or within every individual in a population.

The hypothesis that human races represent independent evolutionary groups is not supported by these observations.

## Populations within a Race Are Often as Different as Populations Compared across Races.
Certain SNP alleles are useful in that they link individuals to particular population groups. This is not surprising since an ancestral population tends to be associated with a particular geography, and people living close

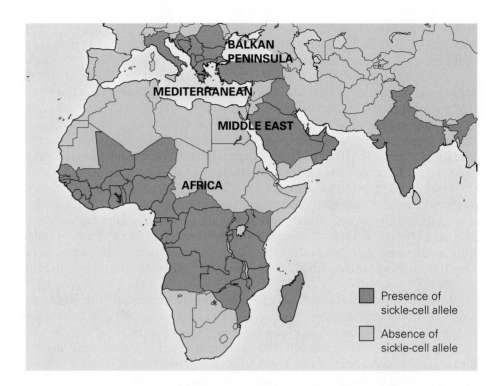

**FIGURE 12.11 The sickle-cell allele: Not a "black gene."** The map illustrates where the sickle-cell allele is found in human populations. Note that it is not found in all African populations but is found in some European and Asian populations.

to each other are likely to share more ancestors (and thus be more genetically similar) than people living far apart. However, if race is to be biologically meaningful, then the allele frequency for *many different* SNPs and genes should be more similar among populations within a race than between populations of different races.

Again, the pattern of SNP allele frequency among populations can be illustrated by more obvious alleles. The human populations in each part of **FIGURE 12.12** are listed by increasing frequency of a particular allele in the population. The color coding of each population group in each of the graphs corresponds to the racial category in which the population is typically placed. For example, at the top of Figure 12.12a, the Taiwanese population is categorized as Asian, while the Cree Indian population of eastern Canada is categorized as Native American. If the hypothesis that human racial groups have a biological basis is correct, then populations from the same racial group should be clustered together on each bar graph.

Figure 12.12a shows the frequency of the allele that interferes with an individual's ability to taste the chemical phenylthiocarbamide (PTC) in several populations. People who carry two copies of this recessive allele cannot detect PTC, which tastes bitter to people who carry one or no copies of the allele. Note that members of a single "race," such as Asian, vary widely in the frequency of this allele.

Figure 12.12b lists the frequency of one allele for the gene *haptoglobin 1* in a number of different human populations. Haptoglobin 1 is a protein that helps scavenge the blood protein hemoglobin from old, dying red blood cells. Again, we see a wide distribution in allele frequency within the race categories.

Figure 12.12c illustrates variation among human populations in the frequency of a repeating DNA sequence on chromosome 8. Repeating sequences are common in the human genome, and differences among individuals in the number of repeats create the unique signatures of DNA fingerprints. The frequency of one pattern of repeating sequence in a segment of

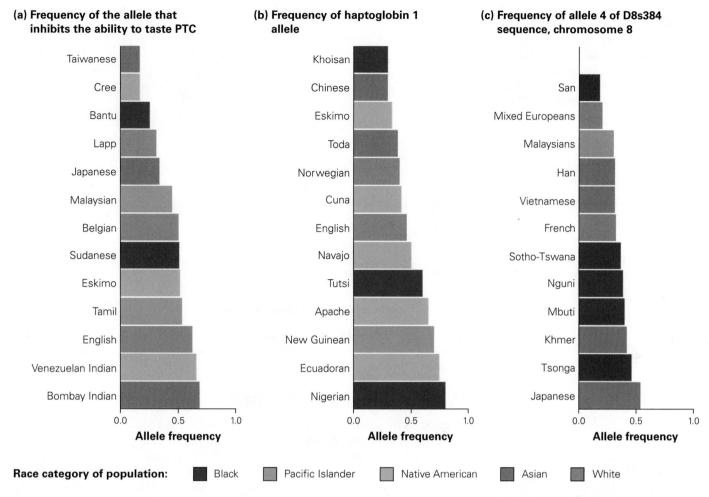

**(a) Frequency of the allele that inhibits the ability to taste PTC**

Taiwanese
Cree
Bantu
Lapp
Japanese
Malaysian
Belgian
Sudanese
Eskimo
Tamil
English
Venezuelan Indian
Bombay Indian

0.0   0.5   1.0
**Allele frequency**

**(b) Frequency of haptoglobin 1 allele**

Khoisan
Chinese
Eskimo
Toda
Norwegian
Cuna
English
Navajo
Tutsi
Apache
New Guinean
Ecuadoran
Nigerian

0.0   0.5   1.0
**Allele frequency**

**(c) Frequency of allele 4 of D8s384 sequence, chromosome 8**

San
Mixed Europeans
Malaysians
Han
Vietnamese
French
Sotho-Tswana
Nguni
Mbuti
Khmer
Tsonga
Japanese

0.0   0.5   1.0
**Allele frequency**

**Race category of population:**   ■ Black   ■ Pacific Islander   ■ Native American   ■ Asian   ■ White

**FIGURE 12.12 Do human races show genetic evidence of isolation?** The bars on each of these histograms illustrate the frequency of the described allele in many different human populations. These histograms illustrate that populations within these "races" are not necessarily more similar to each other than they are to populations in different races.

chromosome 8, called allele 4 of the D8s384 sequence, is illustrated for a number of populations.

What we see in the three graphs in Figure 12.12 is that allele frequencies for these genes are *not* more similar within racial groups than between racial groups. In fact, in two of the three graphs, Figure 12.12a and Figure 12.12b, the populations with the highest and lowest allele frequencies belong to the same race—for these genes, there is more variability *within* a race than there are average differences *among* races. Even though certain SNPs can help us track an individual to a particular ancestral population, the pattern of diversity in other genes tells us that these "ancestral" patterns are not evidence of deep biological similarities among populations within a racial group.

**STOP & STRETCH**   We are accustomed to looking at differences in skin color and eye shape and presuming they signify underlying biological relationships. Can you think of another physical trait that varies greatly in human populations but that is rarely used as evidence for biological relationship? How would including this trait affect the classification of races?

Human races fail to meet the criteria for identifying populations as consistently isolated from each other. Both the fossil evidence and genetic evidence indicate that the six commonly listed human racial groups do *not* represent biological races.

## Working with Data ▶

Type O is the most common blood type in Europe. Given the pattern of B frequency here, what do you predict the pattern of frequency of type O blood would appear on the same graph?

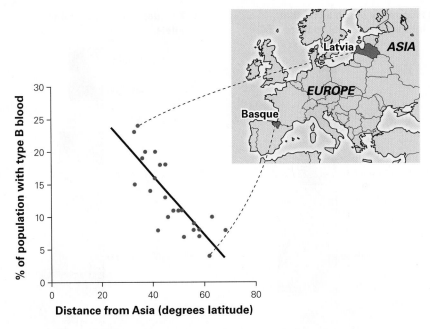

**FIGURE 12.13 Genetic mixing in humans.** The decline in frequency of type B blood from west to east in Europe reflects the movement of alleles from Asian populations into European populations over the past 2000 years.

## Human Races Have Never Been Truly Isolated

The results of genetic ancestry tests on individuals in the United States typically reflect the mixing that has occurred as a result of the past 300 years of European and Asian immigration and the history of the African slave trade. Recent research indicates that most African Americans have at least 20% European SNPs, and that 30% of white U.S. college students have less than 90% European SNPs. But the "melting pot" of the United States is not unique in human history. In fact, the evidence that human populations have been "mixing" since modern humans first evolved is contained within the gene pool of human populations. For instance, the frequency of the B blood group decreases from east to west across Europe (**FIGURE 12.13**). The allele that codes for this blood type apparently evolved in Asia, and the pattern of blood group distribution seen in Figure 12.13 corresponds to the movement of Asians into Europe beginning about 2000 years ago. As the Asian immigrants mixed with the European residents, their alleles became a part of the European gene pool. Populations closest to Asia experienced a large change in their gene pools, while populations who were more distant encountered a more "diluted" immigrant gene pool made up of the offspring between the Asian immigrants and their European neighbors.

Other genetic analyses have led to similar maps. For example, one indicates that populations that practiced agriculture arose in the Middle East and migrated throughout Europe and Asia about 10,000 years ago. These data indicate that there are no clear boundaries within the human gene pool. Interbreeding of human populations over hundreds of generations has prevented the isolation required for the formation of distinct biological races.

## 12.3 Why Human Groups Differ

Human races are not true biological races. However, as is clear to all of us, human populations do differ from each other in many traits. In this section, we explore what is known about why populations share certain superficial traits and differ in others.

**Malaria sickle-cell overlap**

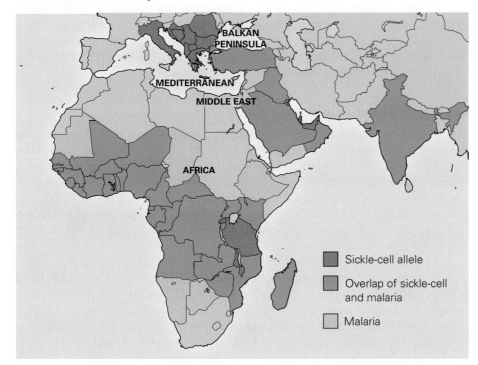

FIGURE 12.14 **The sickle-cell allele is common in malarial environments.** This map shows the distributions of the sickle-cell allele and malaria in human populations.

## Natural Selection

Recall the distribution of the sickle-cell allele in human populations shown in Figure 12.11. This allele is found in some populations of at least three of the typically described races. Why is it higher in certain populations?

The sickle-cell allele is higher in certain populations because in particular environments natural selection favors individuals who carry one copy of it. The sickle-cell anemia allele is an *adaptation,* a feature that increases fitness, within populations in malaria-prone areas. Malaria is a disease caused by a parasitic, single-celled organism that spends part of its life cycle feeding on red blood cells, eventually killing the cells. Because their red blood cells are depleted, people with severe malaria suffer from anemia, which may result in death. When individuals carry a single copy of the sickle-cell allele, their blood cells deform when infected by a malaria parasite. These deformed cells quickly die, reducing the ability of the parasite to reproduce and infect more red blood cells and therefore reducing a carrier's risk of anemia.

The sickle-cell allele reduces the likelihood of severe malaria, so natural selection has caused it to increase in frequency in susceptible populations. The protection that the sickle-cell allele provides to heterozygote carriers is demonstrated by the overlap between the distribution of malaria and the distribution of sickle-cell anemia (**FIGURE 12.14**).

Another physical trait that has been affected by natural selection is nose form. The pattern of nose shape in populations generally correlates to climate factors—populations in dry climates tend to have narrower noses than do populations in moist climates. A long, narrow nose has a greater internal surface area, exposing inhaled air to more moisture, reducing lung damage, and increasing the fitness of individuals in dry environments. Among tropical Africans, people living at drier high altitudes have much narrower noses than do those living in humid rain-forest areas (**FIGURE 12.15**).

◀ **Visualize This**

Given the pattern pictured here, in which population groups in the United States would you expect to find individuals who carry the sickle-cell allele?

**(a) Ethiopian with a narrow nose**

**(b) Bantu with a broad nose**

FIGURE 12.15 **Nose shape is affected by natural selection.** (a) Long, narrow noses are more common among populations in cold, dry environments because they conserve heat and water. (b) Broad, flattened noses are more common in warm, wet environments because they effectively radiate body heat.

Interestingly, our preconception puts two populations of Africans (represented by the images in Figure 12.15) in the same race and explains differences in their nose shape as a result of natural selection, but we place white and black populations into different races and explain their skin color differences as evidence of long isolation from each other. However, like nose shape, skin color is a trait that is strongly influenced by natural selection.

## Convergent Evolution

Traits that are shared by unrelated populations because they share similar environmental conditions are termed *convergent*. **Convergent evolution** occurs when natural selection for similar environmental factors causes unrelated organisms to resemble each other. For example, the similarity in shape between white-sided dolphins and reef sharks is a result of convergent evolution. We know by their anatomy and reproductive characteristics that sharks are most closely related to other fish and dolphins to other mammals (**FIGURE 12.16**).

The pattern of skin color in human populations around the globe also appears to be the result of convergent evolution, in which unrelated human populations appear similar as a result of evolution in similar environmental conditions. When scientists compare the average skin color in a native human population to the level of ultraviolet (UV) light to which that population is exposed, they see a close correlation—the lower the UV light level, the lighter the skin, regardless of the race in which the population is classified (**FIGURE 12.17**).

UV light is high-energy radiation in a range that is not visible to the human eye. Among its many effects, UV light interferes with the body's ability to store the vitamin folate. Folate is required for proper development in babies and for adequate sperm production in males. Men with low folate levels have low fertility, and women with low folate levels are more likely to have children with severe birth defects. Therefore, individuals with adequate folate have higher fitness than individuals without. Because darker-skinned individuals absorb less UV light, they have higher folate levels in high-UV environments than light-skinned individuals do. In other words, in environments where UV light levels are high, dark skin is favored by natural selection (**FIGURE 12.18**).

Human populations in low-UV environments face a different challenge. Absorption of UV light is essential for the synthesis of vitamin D. Vitamin D is

## Visualize This ▼

Penguins are also oceanic predators of fish. How do these animals demonstrate convergence with dolphins and sharks?

**FIGURE 12.16 Convergence.** The similarity in shape between dolphins and sharks results from similar adaptations to life as an oceanic predator of fish, not shared ancestry.

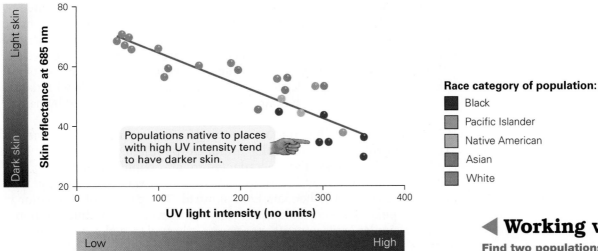

**FIGURE 12.17 There is a strong correlation between skin color and exposure to UV light.** Reflectance is an indication of color—higher reflectance indicates lighter skin. The color of the dots on the graph specifies the racial category of each population.

◀ **Working with Data**

Find two populations classified as the same race that have very different skin color. Find two populations classified in different races that have the same skin color.

## Visualize This ▼

**What could an individual with dark skin in a low-UV environment do to compensate? What about an individual with light skin in a high-UV environment?**

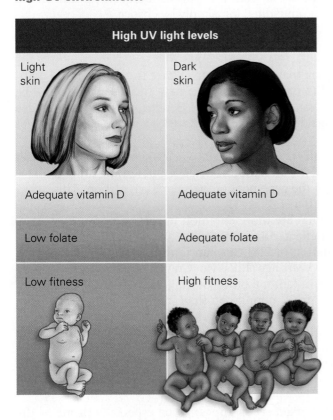

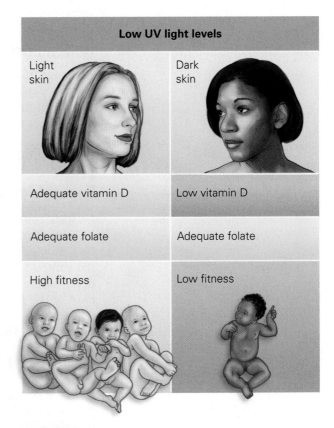

**FIGURE 12.18 The relationship between UV light levels, folate, vitamin D, and skin color.** Darker, UV-resistant skin is an advantage within populations in regions where UV light levels are high because individuals with darker skin have higher fitness. Lighter, UV-transparent skin is favored by natural selection where UV levels are low.

crucial for the proper development of bones. Women are especially harmed by low vitamin D levels—inadequate development of the pelvic bones can make giving birth deadly. There is no risk of not making enough vitamin D when UV light levels are high, regardless of skin color. However, in areas where levels of UV light are low, individuals with lighter skin absorb a larger fraction of UV light and thus have higher levels of vitamin D than do individuals with darker skin. Thus, in less sunny environments, light skin has been favored by natural selection.

Because UV light has important effects on human physiology, it has driven the evolution of skin color in human populations. Where UV light levels are high, dark skin was an adaptation, and populations became dark skinned. Where UV light levels are low, light skin was an adaptation, and populations evolved to become light skinned. The pattern of skin color in human populations is a result of the convergent evolution of different populations in similar environments, not necessarily evidence of separate races of humans.

**STOP & STRETCH** The word for *mother* in many different languages is a variation of the sound "ma." This appears to be an example of convergent evolution in language. Think about the first word-like sounds babies make. Why might many different cultures share this characteristic sound for *mother*?

Natural selection has caused differences among human populations, but it has also resulted in some populations superficially appearing more similar to some other human populations. In contrast, populations who appear on the surface to be similar may be quite different, simply by chance.

## Genetic Drift

A change in allele frequency that occurs due to chance is called **genetic drift.** Human populations tend to travel and colonize new areas, so we seem to be especially prone to evolution via genetic drift. Genetic drift occurs in three different types of situations, described below (**FIGURE 12.19**).

**Founder Effect.** Genetic differences can occur when a small sample of a larger population establishes a new population. The gene pool of the immigrants is rarely an exact reflection of the gene pool of the source population. This difference leads to the **founder effect.**

Rare genetic diseases that are more common in certain populations may be a result of the founder effect. For example, the Amish of Pennsylvania are descended from a population of 200 German founders who immigrated to the United States over 200 years ago. Ellis–van Creveld syndrome, a recessive disease that causes dwarfism (among other effects), is 5000 times more common in the Pennsylvania Amish population than in other German American populations. This difference is a result of a single founder in that original population who carried the allele. Since the Pennsylvania Amish usually marry others within their small religious community, the allele has stayed at a high level—1 in 8 Pennsylvania Amish are carriers of the Ellis–van Creveld allele, compared to fewer than 1 in 100 non–Amish Americans of German ancestry.

Plants with animal-dispersed seeds appear to be especially prone to the founder effect. For example, cocklebur, a widespread weed that produces hitchhiker fruit designed to grab onto the fur of a passing mammal (**FIGURE 12.20**, on page 280), consists of populations that are quite variable in size and shape.

**(a) Founder effect:** A small sample of a large population establishes a new population.

Frequency of red allele is low in original population.

Several travelers happen to carry the red allele.

Frequency of red allele is much higher in new population.

**(b) Population bottleneck:** A dramatic but short-lived reduction occurs in population size.

Frequency of red allele is low in original population.

Many survivors of tsunami happen to carry red allele.

Frequency of red allele is much higher in new population.

**(c) Chance events in small populations:** The carrier of a rare allele does not reproduce.

Frequency of red allele is low in original population.

The only lizard with red allele happens to fall victim to an accident and dies.

Red allele is lost.

**FIGURE 12.19 The effects of genetic drift.** A population may contain a different set of alleles because (a) its founders were not representative of the original population; (b) a short-lived drop in population size caused a change in allele frequency in one of the populations; or (c) one population is so small that low-frequency alleles are lost by chance.

The variation among populations appears to have been caused by differences in the hitchhikers that happened to be carried to new locations, where they founded new populations.

A variant of the founder effect is the bottleneck effect, in which a disaster wipes out most of the population, leaving only a small subset of survivors. A sixteenth-century bottleneck on the island of Puka Puka in the South Pacific resulted in a human population that is clearly different from other Pacific Island populations: The 17 survivors of a tsunami on Puka Puka were all relatively petite, and modern Puka Pukans are significantly shorter in stature compared to people native to nearby islands. Bottlenecks are experienced by nonhuman

**FIGURE 12.20 An example of the founder effect in plants.** Cocklebur is a hitchhiker plant, dispersing by hooking the spikes on its fruits to the fur (or sock) of a passing animal. When these burrs are removed at a distant location, the plant can found a new population that might be quite different in size and shape from the source population.

populations as well. For example, the genetic similarity among individuals in a large population of Galápagos tortoises on the island of Isabela seems to suggest a bottleneck. Geological evidence indicates that most of the tortoise population was wiped out during a volcanic eruption about 88,000 years ago. The current population apparently descended from a tiny group of survivors and thus has little genetic diversity.

**Genetic Drift in Small Populations.** Even without a population bottleneck, allele frequencies may change in a population due to chance events. When an allele is in low frequency within a small population, only a few individuals carry a copy of it. If one of these individuals fails to reproduce, or if it passes on only the more common allele to surviving offspring, the frequency of the rare allele may drop in the next generation. If the population is small enough, even relatively high-frequency alleles may be lost after a few generations by this process.

A human population that illustrates the effects of genetic drift in small populations is the Hutterites, a religious sect with communities in South Dakota and Canada. Modern Hutterite populations trace their ancestry back to 442 people who migrated from Russia to North America between 1874 and 1877. Hutterites tend to marry other members of their sect, so the gene pool of this population is small and isolated from other populations. Genetic drift in this population over the last century has resulted in a near absence of type B blood among the Hutterites, as compared to a frequency of 15 to 30% in other European immigrants in North America.

Genetic drift in populations that remain small for many generations can lead to a rapid loss of many different alleles. While this problem is uncommon in humans, the effects of genetic drift on small populations of endangered species can lead to extinction (Chapter 15).

Humans are a highly mobile species, and we have been founding new populations for millennia. Most early human populations were also probably quite small. These factors make human populations especially susceptible to the founder effect, genetic bottlenecks, and genetic drift and have contributed to the differences among modern human groups. However, in addition to natural selection and random genetic change, humans' highly social nature may have enhanced certain superficial differences in appearance among different populations.

## Sexual Selection

Men and women within a population may have preferences for particular physical features in their mates. These preferences can cause populations to differ in appearance. When a trait influences the likelihood of mating, that trait is under the influence of a form of natural selection called **sexual selection.**

Darwin proposed the hypothesis of sexual selection in 1871 as an explanation for differences between males and females within a species. For instance, the enormous tail on a male peacock results from female peahens that choose mates with showier tails. Because large tails require so much energy to display and are more conspicuous to their predators, peacocks with the largest tails must be both physically strong and smart to survive. Peahens can use the size of the tail, therefore, as a measure of the "quality" of the male. Tail length does appear to be a good measure of overall fitness in peacocks; the offspring of well-endowed males are more likely to survive to adulthood than are the offspring of males with scanty tails. When a peahen chooses a male with a large tail, she is ensuring that her offspring will receive high-quality genes. Sexual selection explains the differences between males and females in many species (**FIGURE 12.21**).

**(a) Peacock**

In humans, there is some evidence that the difference in overall body size between men and women is a result of sexual selection—namely, a widespread female preference for larger males—perhaps again because size may be an indication of overall fitness. However, some apparently sexually selected traits seem to have little or no relationship to fitness and reflect simply a "social preference." In our highly social species, this type of sexual selection may be common.

For example, some scientists have suggested that lack of thick facial and body hair in many Native American and Asian populations resulted from selection by both men and women for less hairy mates. Some scientists even hypothesize that many physical features that are unique to particular human populations evolved as a result of these socially derived preferences. While intriguing, there is as yet little evidence to support this idea and no simple way to test it.

## Assortative Mating

Differences between human populations may be reinforced by the ways in which people choose their mates. Individuals usually prefer to mate with someone who is like themselves, by a process called positive **assortative mating**. For example, there is a tendency for people to mate assortatively by height—that is, tall women tend to marry tall men—and by skin color (**FIGURE 12.22**).

When two populations differ in obvious physical characteristics, the number of matings between them may be small if the traits of one population are considered culturally unattractive to members of the other population. Assortative mating has been observed in other organisms as well; for instance, sea horses choose mates that are similar in size to themselves, and in some species of fruit flies, females will mate only with males who have the same body color. Positive assortative mating tends to maintain and even exaggerate physical differences between populations. In highly social humans, assortative

**(b) Lion**

**(c) Blue morpho butterfly**

**FIGURE 12.22 Assortative mating.** People tend to choose mates who look like them in socially important characteristics.

**FIGURE 12.21 The effects of sexual selection.** Sexual selection is responsible for many unique and fantastic characteristics of organisms from (a) the peacock's tail to (b) the male lion's mane to (c) the bright colors of butterflies.

mating may be an important amplifier of superficial physical differences between groups.

While human populations may display differences due to natural selection in certain environments, genetic drift, sexual selection, and assortative mating, the genetic evidence indicates that many of these differences are literally no more than skin deep. Beneath a veneer of physical differences, humans are basically the same.

# *savvy reader*

## Race and Health Guidelines

On Jan. 31, the departments of Agriculture and Health and Human Services released the newest version of the official Dietary Guidelines for Americans. ... [O]ne particular piece of advice has been making headlines: ... Americans [should] reduce their daily intake of sodium to 2,300 milligrams. ... This alone poses a remarkable challenge; less than 15 percent of the population currently meets this target. But the Dietary Guidelines don't stop there. They also recommend reducing salt intake to 1,500 mg for people who are 51 and older or have hypertension, diabetes, or chronic kidney disease. And they set the same, more stringent goal for anyone—anyone at all—who happens to be African-American.

There may be reasonable, if not entirely obvious, reasons why older individuals and those with certain pre-existing medical conditions should make stronger efforts to reduce their salt intake. ... But should the same rules apply to all black people? If these guidelines are evidence-based, what's the evidence that race—in and of itself, regardless of age, ailment, or other considerations—is a risk factor of the same consequence as, say, diabetes? ... The guidelines [note] that, for whatever reason, blacks' "blood pressure ... tends to be even more responsive to the blood pressure–raising effects of sodium than others."

Without a doubt, African-Americans have higher rates of hypertension than U.S. whites; research shows that the age-adjusted prevalence in blacks is 41.8 percent, versus 29.8 percent for whites. However, epidemiologist Richard Cooper has placed this and other racial disparities in an international context, showing that U.S. whites have a higher prevalence of hypertension than Nigerians, while U.S. blacks have a lower prevalence than Germans and Finns. ...

1. What is the evidence presented in this article excerpt that supports the suggestion that African Americans should reduce salt intake?

2. What evidence does the author of the article present that does not seem to support the suggestion?

3. What more would you need to know to determine if the recommendation that all African Americans need to reduce salt intake to levels lower than healthy white Americans is well justified?

http://www.slate.com/articles/health_and_science/medical_examiner/2011/04/black_salt.html

# SOUNDS RIGHT BUT IS IT?

The Hawaiian Islands are home to a group of birds found nowhere else—the Hawaiian honeycreepers. This group of about 50 species (many of which are now extinct) appears to have descended from a single species of Asian rosefinch that made it to the islands between 4 and 5 million years ago. The ancestral rosefinch likely had a short, fat bill and fed on seeds—however, many honeycreeper species have long, curved bills and feed on flower nectar; the bill shape of particular species fits perfectly with the flower shape of the plants they feed upon. This fact might lead you to say:

**Some species of honeycreepers evolved with long, curved bills in order to feed on the nectar of particular flowers.**

Sounds right, but it isn't.

**1.** Is it likely that variation in bill sizes and shapes existed within the population of rosefinches that inhabited the Hawaiian Islands 4–5 million years ago?

**2.** Is it likely that variation in foraging (that is, how an individual hunts for food) existed within this population of rosefinches?

**3.** Would all birds in the original population demonstrate these traits?

**4.** In an environment where few animals feed on a particular flower's nectar, would an individual with a slightly longer and thinner bill and the behavior of flower foraging have high evolutionary fitness?

**5.** How do differences among individuals in evolutionary fitness affect the traits of a population?

**6.** Reflect on your answers to questions 1–5, and explain why the statement bolded above sounds right, but isn't.

---

# Chapter Review MasteringBiology®

Go to the Study Area in MasteringBiology® for practice quizzes, myeBook, BioFlix™ 3-D animations, MP3Tutor sessions, videos, current events, and more.

## Summary

### Section 12.1

Define *biological species*, and list the mechanisms by which reproductive isolation is maintained by biological species.

- All humans belong to the same biological species, *Homo sapiens sapiens*. A biological species is defined as a group of individuals that can interbreed and produce fertile offspring. Biological species are reproductively isolated from each other, thus separating the gene pools of species (p. 258).

- Reproductive isolation is maintained by prefertilization factors, such as differences in mating behavior or timing, or postfertilization factors, such as hybrid inviability or sterility (pp. 259–261).

Describe the three steps in the process of speciation.

- Speciation occurs when populations of a species become isolated from each other. These populations diverge from each other, and reproductive isolation between the populations evolves (pp. 260–264).

### Section 12.2

Explain how a "race" within a biological species can be defined using the genealogical species concept.

- Biological races are populations of a single species that have diverged from each other but have not become reproductively isolated. This definition of race corresponds to the genealogical species concept (p. 265).

List the evidence that modern humans are a young species that arose in Africa.

- The morphology of human ancestors in the fossil record provides evidence that the modern human species is approximately 200,000 years old (p. 267).

- Genetic evidence indicates that modern humans have limited genetic diversity, a characteristic of young species, and that the oldest populations (containing the most genetic diversity) are found in Africa (p. 267).

List the genetic evidence expected when a species can be divided into unique races.

- The genetic evidence for biological race include (1) alleles that are unique to a particular race,

(2) similar allele frequencies for a number of genes among populations within races, and (3) differences in allele frequencies among populations in different races (pp. 267–268).

Describe how the Hardy–Weinberg theorem is used in studies of population genetics.

- The Hardy–Weinberg theorem provides a rule for calculating allele frequency from genotype frequency (pp. 268–270).

- If the assumptions of Hardy–Weinberg (no natural selection, no migration, and random mating of individuals within a population) are violated, the allele frequency will not match expectations. Scientists look for alleles that are not typical to identify those that might provide information about ancestry (pp. 268–270).

Summarize the evidence that indicates human races are not deep biological divisions within the human species.

- Although some single nucleotide polymorphism (SNP) alleles are more common in certain human population groups, modern human groups do not show evidence that they have been isolated from each other long enough to have formed distinct races (pp. 270–273).

- Other genetic evidence indicates that human groups have been mixing for thousands of years (p. 274).

# Section 12.3

Provide examples of traits that have become common in certain human populations due to the natural selection these populations have undergone.

- Similarities among human populations may evolve as a result of natural selection. The sickle-cell allele is selected for in populations in which malaria incidence is high, and light skin is selected for in areas where the UV light level is low (pp. 275–278).

Define *genetic drift* and provide examples of how it results in the evolution of a population.

- Human populations may show differences due to genetic drift, which is defined as changes in allele frequency due to chance events such as founder effects or population bottlenecks (pp. 278–280).

Describe how human and animal behavior can cause evolution via sexual selection and assortative mating.

- Sexual selection, by which individuals—typically females—choose mates that display some "attractive" quality, may also be responsible for creating differences among human populations (pp. 280–281).

- Positive assortative mating, in which individuals choose mates who are like themselves, can reinforce differences between human populations (pp. 281–282).

# Roots to Remember

**The following roots come mainly from Latin and Greek and will help you decipher terms:**

| | |
|---|---|
| **allo-** | means different. Chapter term: *allopatric* |
| **morpho-** | means shape or form. Chapter term: *morphological* |
| **paleo-** | means ancient. Chapter term: *paleontologists* |
| **-patric** | means country or place of origin. Chapter term: *allopatric, sympatric* |
| **-ploid** | means the number of different types of chromosomes. Chapter term: *polyploid* |
| **poly-** | means many. Chapter term: *polyploid* |
| **sym-** | means same or united. Chapter term: *sympatric* |

# Learning the Basics

**1.** Describe the three steps of speciation.

**2.** Why is a relative lack of genetic diversity in humans evidence that we are a young species?

**3.** Describe three ways that evolution can occur via genetic drift.

**4.** Which of the following is an example of a prefertilization barrier to reproduction?

**A.** A female mammal is unable to carry a hybrid offspring to term; **B.** Hybrid plants produce only sterile pollen; **C.** A hybrid between two bird species cannot perform a mating display; **D.** A male fly of one species performs a "wing-waving" display that does not convince a female of another species to mate with him; **E.** A hybrid embryo is not able to complete development.

**5.** According to the most accepted scientific hypothesis about the origin of two new species from a single common ancestor, most new species arise when _____.

**A.** many mutations occur; **B.** populations of the ancestral species are isolated from one another; **C.** there is no natural selection; **D.** a supernatural creator decides that two new species would be preferable to the old one; **E.** the ancestral species decides to evolve.

**6.** For two populations of organisms to be considered separate biological species, they must be _____.

**A.** reproductively isolated from each other; **B.** unable to produce living offspring; **C.** physically very different from each other; **D.** A and C are correct; **E.** A, B, and C are correct.

**7.** Cystic fibrosis, a recessive disease, affects 1 of every 2500 white babies born in the United States, a frequency of 0.0004. Use the Hardy–Weinberg theorem to calculate the frequency of the cystic fibrosis allele in this population.

**A.** 0.0004; **B.** 0.04; **C.** 0.02; **D.** 0.00000016; **E.** There is not enough information in the question to calculate the answer.

**8.** All of the following statements support the hypothesis that humans cannot be classified into biological races *except*:

**A.** There is more genetic diversity within a racial group than average differences between racial groups; **B.** Alleles that are common in one population in a racial group may be uncommon in other populations of the same race; **C.** Geneticists can use particular SNP alleles to identify the ancestral group(s) of any individual human; **D.** There are no alleles found in all members of a given racial group; **E.** There is genetic evidence of mixing among human populations occurring thousands of years ago until the present.

**9.** Similarity in skin color among different human populations appears to be primarily the result of _____.

**A.** natural selection; **B.** convergent evolution; **C.** which biological race they belong to; **D.** A and B are correct; **E.** A, B, and C are correct.

**10.** The tendency of individuals to choose mates who are like themselves is called _____.

**A.** natural selection; **B.** sexual selection; **C.** assortative mating; **D.** the founder effect; **E.** random mating.

## Analyzing and Applying the Basics

**1.** Wolf populations in Alaska are separated by thousands of miles from wolf populations in the northern Great Lakes of the lower 48 states. Wolves in both populations look similar and have similar behaviors. However, the U.S. government has treated these two populations quite differently, listing the Great Lakes populations as endangered until recently but allowing hunting of wolves in Alaska. Some opponents of wolf protection have argued that the "wolf" should not be considered endangered at all in the United States because of the large population in Alaska, while supporters of wolf protection state that the Great Lakes population represents a unique population that deserves special status. Should these two populations be considered different races or species? What information would you need to test your answer?

**2.** Phenylketonuria (PKU) is the inability to metabolize the amino acid phenylalanine. The frequency of PKU in Irish populations is 1 in every 7000 births, while the frequency in urban British populations is 1 in 18,000 and only 1 in 36,000 in Scandinavian populations. PKU is found only in individuals who are homozygous recessive for the disease allele. Use the Hardy–Weinberg theorem to calculate the frequency of the disease allele in each population. Give two reasons that this allele, which can result in severe mental retardation in homozygous individuals, may be found in different frequencies in these populations.

**3.** A species of aster normally blooms between mid-August and late September. The range of the aster is located in a climate that usually produces freezing temperatures in late October. The pollinators (bees, ants, and other insects) of the aster are active between late April and early October. Mutations can occur in the gene that controls flowering time for some individuals of this aster, causing them to flower earlier or longer than normal. The table shows possible flowering times.

| Aster Type | Flowering Time |
| --- | --- |
| Normal | August 15–September 30 |
| Early mutation | July 1–August 15 |
| Expanded mutation | July 15–September 30 |

Describe how this population of asters might split into two species.

## Connecting the Science

**1.** According to the most recent data available, the rate at which the white population commits suicide is almost twice as much as most minority groups. Is this difference likely to be biological? How could you test your hypothesis?

**2.** The only information that was collected from every household in the United States in the 2010 census was the names of residents of the household, their relationship to each other, sex, age, and race and ethnicity. Do you think it is important to collect race and ethnicity data from all citizens? Why or why not? Is there some other piece of information that you think is more useful to the government?

Answers to **Stop & Stretch, Visualize This, Working with Data, Savvy Reader, Sounds Right, But Is It?,** and **Chapter Review** questions can be found in the **Answers** section at the back of the book.

# CHAPTER 13

# The Greatest Species on Earth?

**Giant sequoia are the largest individual organisms that have ever existed on Earth.**

# Biodiversity and Classification

It is a tried-and-true conversation starter: "Who was the best of all time?" Ask it and hear sports fans debating the greatest athlete of all time, literature fans defending their favorite authors, students of history ranking distinguished politicians, and musicians arguing about the world's greatest rock bands. These are topics with plenty of room for friendly disagreement and with no obvious answer. But if you interrupted any of these conversations to ask "What is the greatest species of all time?" you'd likely get a nearly universal answer. Humans, of course. What other species built the pyramids, wrote *The Odyssey,* sent men to the moon, or wrote *Let It Be*?

It probably seems self-evident to many people that humans are the

**Horseshoe crabs have been around for 440 million years longer than us.**

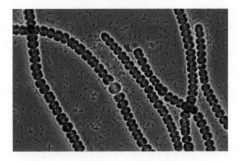

**Cyanobacteria are the most self-sufficient species, needing only water, sunlight, and air.**

greatest species. We humans are highly intelligent; we use our communication skills, cooperation, and technology to solve problems. We have superior sensory capabilities and use them to learn about how the world works and then employ that knowledge to take advantage of natural resources. We are ingenious in finding ways to survive in different environments. With these skills, we have adapted nearly every part of Earth's surface for human habitation and modified the environment to meet our needs. We dominate other species, even using the evolutionary process to shape domesticated animals and plants. What species can compete with this?

One of the few places you'd find disagreement on this question is at

a gathering of biologists. These scientists know that humans are puny compared to the largest individuals ever—giant sequoia trees—many of which are thousands of years old. And are modern humans, a species only 200,000 years old, that much more impressive than horseshoe crabs, survivors of 445 million years of change? Or more inventive than cyanobacteria, organisms that can make all the components they need to survive from just water, sunlight, and air? Or better adapted than Argentine ants, a single colony of which can spread across the globe and consist of trillions of individuals? Maybe not. In this chapter, we'll survey the range of living organisms on the planet to really test the assumption that humans are the best species on Earth.

**But are any of these species really greater than humans?**

# 13.1 Biological Classification

Before we begin evaluating "greatness," we must know something about the multiple forms of living organisms. Biologists are still discovering living species and describing the fossil remnants of extinct species. But even without a complete accounting of all the different forms of life on Earth, we do have a good understanding of the range of biological diversity, both present and past.

## How Many Species Exist?

A characteristic of life on Earth is that it is full of variety. Scientists refer to the variety within and among living species as **biodiversity.** Studies of biodiversity don't just provide interesting stories and fodder for arguments about the "best" species. Much more importantly to many biologists, these studies help us understand the evolutionary origins of different groups of organisms and their role in healthy biological systems.

Chapter 12 detailed the challenges inherent in identifying which groups of organisms should be considered discrete **species**—in general, a species is a group of individuals that regularly breed together and are generally distinct from other species in appearance or behavior. **Systematists** are scientists who specialize in describing and categorizing a particular group of organisms. Typically, for an organism to be considered a new species by the scientific community, a systematist must create a description of the species that clearly distinguishes it from similar species, and he or she must publish this description in a professional journal. The scientist also must collect individual specimens for storage in a specialized museum. Most animal collections are found in natural history museums, while plant repositories are called herbaria (**FIGURE 13.1**); microbes and fungi are kept in facilities called type-collection centers.

Thanks to advances in computing power and communication, systematists' descriptions of species are now being collected in centralized databases; the most recent count of these databases lists 1.3 million identified species. Using strategies that extrapolate from the well-described groups of organisms to those

**(a) Natural history museum**

**(b) Herbarium**

**FIGURE 13.1 Biological collections.** Most collections contain several examples of each species to show the range of variation within the species. (a) A collection of mollusks in a natural history museum. (b) Plant specimens are stored in herbaria.

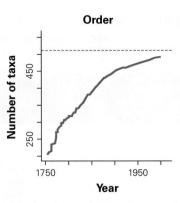

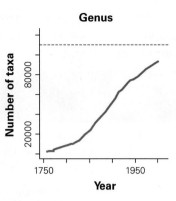

**FIGURE 13.2 Estimating biodiversity.** Scientists use extrapolation to determine how many species are yet to be discovered. These graphs indicate the number of new groups of animals within each taxonomic level over time. Leveling off of the graph indicates that there are probably few new groups left to discover. The dotted lines on each graph provide an estimate of where the graph will level off, based on the historical pattern.

less well-known (**FIGURE 13.2**) leads to an estimate that the total number of species on Earth is around 8.7 million, meaning that still less than 20% of the total are known to science. Scientists who specialize in particular groups of species provide even larger estimates—the simple fact is that we still don't know how many.

Scientists do know that the diversity of life on Earth today is very different from the diversity in the past. Paleontologists have been able to piece together the history of life by examining fossils and other ancient evidence. Early in this reconstruction, they recognized distinct "dynasties" of groups of organisms that appeared during different periods. The record of the rise and fall of these dynasties has allowed scientists to subdivide life's history into **geologic periods.** Each period is defined by a particular set of fossils. **TABLE 13.1** (on the next page) gives the names of major geologic periods, their age and length, and the major biological events that occurred during each period.

Despite the long history of diversity, it is very rare for scientists to describe either a modern or fossil organism that appears unrelated to all other species. In fact, species can be grouped into a few broad categories based on shared characteristics. The most general categories are domains and kingdoms.

## What species can compete with this?

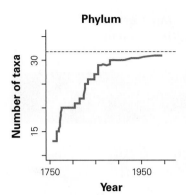

### Kingdoms and Domains

Modern species are divided into three large groups, called **domains,** which are unified by similarities in cell structure. The three domains are named **Bacteria, Archaea,** and **Eukarya.** Within each of these domains are smaller groups, called **kingdoms. TABLE 13.2** (on page 291) summarizes the kingdoms and domains discussed in Section 13.2.

**eu-** means true.

### Kingdoms and Domains: A Closer Look ▼

Systematists work in the field of **biological classification,** in which they attempt to organize biodiversity into discrete and logical categories. The task of classifying life is much like categorizing books in a library—books can be divided into fiction or nonfiction, and within each of these divisions, more precise categories can be made (for example, nonfiction can be divided into biography, history, science, and so forth). The book-cataloging system used in most public libraries, the Dewey

**TABLE 13.1 Geological periods.** The history of life is divided into four major eras, with all but the first era divided into several periods. Periods are marked by major changes in the dominant organisms present on Earth.

| Era | Period | Millions of Years Ago | Features of Life on Earth |
|---|---|---|---|
| Pre Cambrian | | 4500 | Formation of Earth. The first organisms, primitive bacteria, appear within 1000 million years. |
| | | 543 | Life is dominated by single-celled organisms in the ocean. Ediacaran fauna appear at the end of the era. |
| Paleozoic | Cambrian | 495 | All modern animal groups appear in the oceans. Algae are abundant. |
| | Ordovician | 439 | Life is diverse in the oceans. Cephalopods like squid appear, and trilobites are common. |
| | Silurian | 408 | Life begins to invade land. The first colonists are small seedless plants, primitive insects, and soft-bodied animals. |
| | Devonian | 354 | Known as the age of fishes. Sharks and bony fish appear. Large trilobites are abundant in the oceans. |
| | Carboniferous | 290 | Land is dominated by dense forests of seedless plants. Insects become abundant. Large amphibians appear. |
| | Permian | 251 | Early reptiles appear on land. Seedless plants abundant. Coral and trilobites abundant in oceans. Permian ends with extinction of 95% of living organisms. |
| Mesozoic | Triassic | 206 | Early dinosaurs, mammals, and cycads appear on land. Life "restarts" in the oceans. |
| | Jurassic | 144 | Huge plant-eating dinosaurs evolve. Forests are dominated by cycads and tree ferns. |
| | Cretaceous | 65 | Massive carnivorous and flying dinosaurs are abundant. Large cone-bearing plants dominate forests. Flowering plants appear. |
| Cenozoic | Tertiary | 1.8 | After the extinction of the dinosaurs, mammals, birds, and flowering plants diversify. |
| | Quaternary | 0 | Most modern organisms present. |

**TABLE 13.2** **The classification of life.** Until recently, most biologists used the five-kingdom system to organize life's diversity. Now many use a six-category system, which better reflects evolutionary relationships by acknowledging the existence of three major domains as well as four of the kingdoms.

| Kingdom Name | Kingdom Characteristics | Examples | Approximate Number of Known Species | Domain Name and Characteristics |
|---|---|---|---|---|
| **Plantae** | Multicellular, make own food, largely stationary | Pines, wheat, moss, ferns | 300,000 | **Eukarya** All organisms contain eukaryotic cells. |
| **Animalia** | Multicellular, rely on other organisms for food, mobile for at least part of life cycle | Mammals, birds, fish, insects, spiders, sponges | 1,000,000 | |
| **Fungi** | Multicellular, rely on other organisms for food, reproduce by spores, body made up of thin filaments called hyphae | Mildew, mushrooms, yeast, *Penicillium*, rusts | 100,000 | |
| **Protista** | Mostly single-celled forms, wide diversity of lifestyles, including plant-like, fungus-like, and animal-like types | Green algae, *Amoeba, Paramecium*, diatoms, chytrids | 15,000 | |
| **Bacteria** | Prokaryotic, mostly single-celled forms, although some form colonies or filaments | *Escherichia coli, Salmonella, Bacillus anthracis, Anabena*, sulfur bacteria | 4,000 | **Bacteria** Prokaryotes with cell wall containing peptidoglycan. Wide diversity of lifestyles, including many that can make their own food. |
| **Archaea** | Prokaryotic, mostly single-celled forms, although some form colonies or filaments | *Thermus aquaticus, Halobacteria halobium*, methanogens | 1,000 | **Archaea** Prokaryotes without peptidoglycan and with similarities to Eukarya in genome organization and control. Many known species live in extreme environments. |

*A Closer Look, continued*

decimal system, is only one way of shelving books. For instance, academic and research libraries use a different system, developed by the U.S. Library of Congress. Librarians use the cataloging system that is appropriate to the collection of books they work with and the needs of the library's users. Just as there are alternative methods of organizing books, there is more than one way to organize biodiversity to meet differing needs.

Biologists have traditionally subdivided living organisms into large groups that share some basic characteristic. Sixty years ago, most biologists divided life into two categories: plants, for organisms that were immobile and apparently made their own food; and animals, for organisms that could move about and relied on other organisms for food.

Many organisms did not fit easily into this neat division of life, so beginning in 1969 scientists began to use a system of five kingdoms, which categorized organisms according to cell type and method of obtaining energy: the Monera, a group of all single-celled organisms without a nucleus (consisting of what we now know as the domains Archea and Bacteria); Protista, a group containing organisms with nuclei that spend some of their life cycle as single, mobile cells; and three kingdoms of multicellular organisms: Plantae, which make their own food; Animalia, which rely on other organisms for food; and Fungi, which digest dead organisms. The five-kingdom system is not perfect either; for instance, the Protista kingdom contains a wide diversity of organisms, from amoebas to seaweeds, with only superficial similarities.

More recently, many biologists have argued that the most appropriate way to classify life is according to evolutionary relationships among organisms. Recall that the theory of evolution states that all modern organisms represent the descendants of a single common ancestor. The process of divergence from early ancestors into the diversity of modern species has resulted in the modern "tree of life" (**FIGURE 13.3**). When life is classified according to the relationships among organisms, major groupings correspond to divergences that occurred very early in life's history, and minor groupings correspond to more recent divergences.

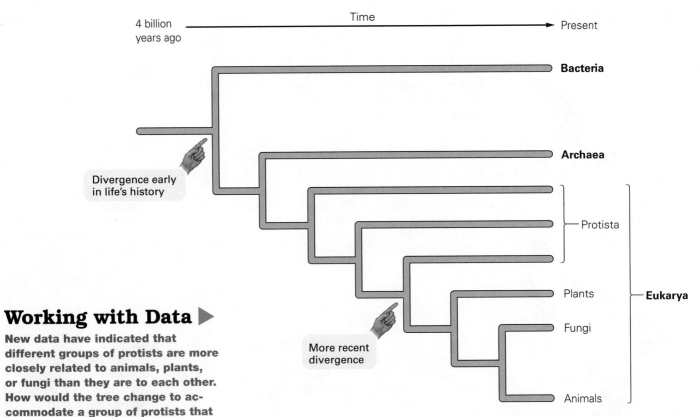

## Working with Data ▶

New data have indicated that different groups of protists are more closely related to animals, plants, or fungi than they are to each other. How would the tree change to accommodate a group of protists that share a recent common ancestor with, for example, animals?

**FIGURE 13.3 The tree of life.** This tree is a simplification of the data generated by rRNA analysis describing the possible evolutionary relationships among living organisms.

*A Closer Look, continued*

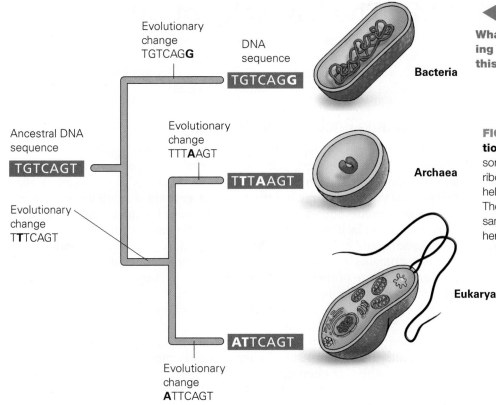

Evolutionary change TGTCAG**G**

DNA sequence

TGTCAG**G** — **Bacteria**

Ancestral DNA sequence

TGTCAGT

Evolutionary change TTT**A**AGT

TTT**A**AGT — **Archaea**

Evolutionary change T**T**TCAGT

ATTCAGT — **Eukarya**

Evolutionary change **A**TTCAGT

**◄ Working with Data**

**What information are scientists using to infer the ancestral sequence in this figure?**

**FIGURE 13.4 Determining the relationships among organisms.** Comparisons of the DNA sequences that code for ribosomal RNA in various organisms have helped scientists construct the tree of life. The actual comparison consisted of thousands of base pairs; only a few are shown here for simplicity.

Determining the evolutionary relationship among *all* living organisms requires comparisons of their DNA. Because each species is unique, the sequence of nucleotides within the DNA of each species is unique. However, because all species share a common ancestor, all organisms also have basic similarities in their DNA sequences. As evolutionary lineages diverged from one another, mutations in DNA sequences occurred independently in each lineage and are now a record of evolutionary relationship among organisms. In short, the DNA sequences of closely related organisms should be more similar than the DNA sequences of more distantly related organisms (**FIGURE 13.4**).

When comparing all modern species, scientists must examine genes that perform a similar function among organisms as diverse as humans, Argentine ants, and cyanobacteria. The DNA sequence that best fits this requirement contains instructions for making ribosomal RNA (rRNA), a structural part of ribosomes. Recall that ribosomes are the factories found in all cells, where genes are translated into proteins. Each ribosome contains several rRNA molecules in both its large and small subunits. A comparison of the DNA coding for small-subunit rRNAs from myriad organisms yielded a tree diagram similar to that shown in Figure 13.3.

Note from Figure 13.3 that three of the kingdoms (Fungi, Animalia, and Plantae) represent relatively recently diverged groups of organisms. Single-celled organisms that do not contain a nucleus for storing DNA were once all placed in the same kingdom but are actually found in two different, quite distinct groups, the Archaea and Bacteria. And the kingdom Protista is a hodgepodge of many very different organisms. To better reflect such biological relationships, starting in about 1990 biologists began to categorize life into three domains: Bacteria, Archaea, and Eukarya. These domains represent the most ancient divergence of living organisms.

**STOP & STRETCH** Biologists could classify organisms by other shared characteristics, such as similarity in foods consumed or habitat. Can you think of a situation for which one of these types of classifications may be more helpful than an evolutionary classification?

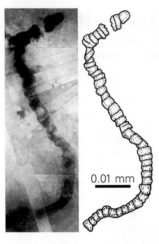

**FIGURE 13.5 The oldest form of life.**
This photograph of a fossil is accompanied by an interpretive drawing showing the fossil's likely living form. The fossil was found in rocks dated at 3.465 billion years old.

---

**-karyo-** means kernel.

---

**FIGURE 13.6 A diversity of prokaryotes.** (a) *Escherichia coli*, the "lab rat" of basic genetic studies, lives on the partially digested food in our intestines. (b) *Desulforudis audaxviator,* whose name means "bold traveler," was discovered 2 miles below Earth's surface living on the energy released by radioactive decay. (c) *Halobacterium*, a salt-loving archaean, is found in high populations in these salty ponds. The red pigment in this bacterium's cells is used in photosynthesis.

# 13.2 The Diversity of Life

Dividing life into six categories—those described in Table 13.2—simplifies our discussion of the "best of" biodiversity. In this section, we describe the six categories. In addition, we discuss entities that have many characteristics of living organisms but are not considered *alive*—the viruses.

## The Domains Bacteria and Archaea

Life on Earth arose at least 3.6 billion years ago, according to the fossil record. The most ancient fossilized cells are remarkably similar in external appearance to modern bacteria and archaea (**FIGURE 13.5**). Both bacteria and archaea are **prokaryotes**; this means they do not contain a nucleus, which provides a membrane-bound, separate compartment for the DNA in other cells. Prokaryotes also lack other internal structures bounded by membranes, such as mitochondria and chloroplasts, which are found in more complex **eukaryotes.** The two domains of prokaryotes differ in many fundamental ways, including the very structure of their cell membranes, but to simplify our discussion, we will consider both groups together.

Although some species may be found in chains or small colonies as in Figure 13.5, most prokaryotes are **unicellular,** meaning that each cell is an individual organism. Individual prokaryotic cells are hundreds of times smaller than the cells that make up our bodies. Because they are microscopic, they are often called microorganisms or **microbes,** and biologists who study these organisms (as well as unicellular eukaryotes) are known as **microbiologists.**

The relatively simple structure of prokaryotes belies their incredible complexity and diversity. We may think of humans as remarkably adaptable in our capacity to settle anywhere from the driest deserts to the coldest tundra, but as a group, prokaryotes put us to shame (**FIGURE 13.6**). Representatives of these organisms are found anywhere humans live, but some prokaryotes can survive in rocks thousands of meters beneath Earth's surface, others feed on hydrogen sulfide emitted by hydrothermal vents on the bottom of the ocean, and some live in lakes trapped underneath the Antarctic ice sheet. Many of the known Archaea specialize in surviving extreme environments, including high-salt and high-temperature habitats. Perhaps even more remarkably, many prokaryotes in both domains live on or in other organisms, evading their hosts' infection-fighting systems.

While there are some amazing stories of human survival in the direst conditions—the explorer Ernest Shackleton and his crew trapped for over two

**(a)** *Escherichia coli*

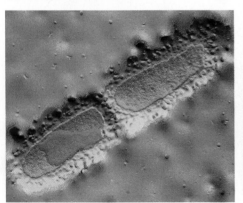

**(b)** *Desulforudis audaxviator*

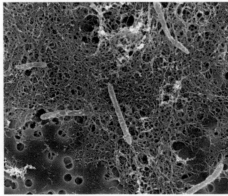

**(c)** *Halobacterium*

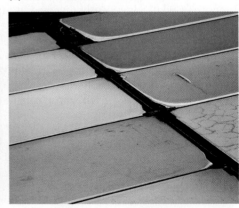

years in the Antarctic, an Uruguayan rugby team surviving a plane crash in the high Andes and hiking to safety, a young girl left on a remote island surviving alone for 18 years—prokaryotes also best humans in this regard. Some bacteria have the ability to form **endospores,** resistant structures containing DNA, ribosomes, and a little bit of cytoplasm. Endospores are resistant to extreme temperatures, drying, radiation, and even the vacuum of outer space; but these structures can generate new living cells eons after they form, as demonstrated by spores revived after being recovered from 250-million-year-old salt deposits.

Cyanobacteria (sometimes called blue-green algae) have the unique characteristic of needing no other living organism to survive—they can manufacture all of the components of life simply from water, sunlight, and air through the processes of photosynthesis and nitrogen fixation. In fact, it is thanks to oxygen-generating activities of cyanobacteria starting around 3.6 billion years ago that the Earth can now support the diverse living forms that are here today. Humans aren't the only species to have changed the very atmosphere of Earth!

Fine, so prokaryotes are more adaptable and may be more influential than humans. But humans still dominate all other species, so that must be our advantage over these microscopic competitors. Scientists who study *Yersinia pestis,* the bacterium that causes bubonic plague and nearly wiped out the human population of Europe in the fourteenth century might disagree, as would those who investigate the bacteria that cause tuberculosis, salmonella, typhoid, and syphilis. And sure, while humans have discovered **antibiotics** that can protect us from these often-fatal diseases by killing or disabling the dangerous bacteria that cause them, we didn't invent these drugs. In fact, more than half of antibiotics are derived from bacteria that produce them as weapons against other, competing bacteria.

Bacterial defense against the viruses that can attack them has resulted in another class of valuable molecules called **restriction enzymes,** proteins that can chop up DNA at specific sequence sites and thus interfere with, or restrict, the growth of these viruses. Restriction enzymes are a key factor that has made much of our advances in genetic technology possible—and evolved within these seemingly simple organisms.

Finally, while Earth now teems with over 7 billion human beings, prokaryotes are much more numerous. In fact, there are likely more prokaryotes living in your mouth right now than the total number of humans who have ever lived. And scientists still have relatively weak understanding of exactly how many species of prokaryotes even exist—some microbiologists estimate that the number of undescribed species could range up to 100 million.

## The Origin of the Domain Eukarya

The third domain of life contains all of the organisms that keep their genetic material within a nucleus inside their cells—that is, the eukaryotes. The most ancient fossils of eukaryotic cells are approximately 2 billion years old, nearly 1.5 billion years younger than the oldest prokaryotic fossils.

The earliest eukaryotic cells likely developed from prokaryotes that produced excess cell membrane that folded into the cell itself. In some cells, these internal membranes may have segregated the genetic material into a primitive nucleus and formed channels for translating, rearranging, and packaging proteins, much like modern eukaryote's endoplasmic reticulum and Golgi bodies. According to the **endosymbiotic theory,** the mitochondria and chloroplasts found in eukaryotic cells appear to have descended from bacteria that took up residence inside larger primitive eukaryotes. When organisms live together, the relationship is known as a **symbiosis.** In this case, the

**Are we more inventive than cyanobacteria, organisms that can make all the components they need to survive from just water, sunlight, and air?**

symbiosis was mutually beneficial, and over time the cells became inextricably tied together (**FIGURE 13.7**).

When biologist Lynn Margulis first popularized the endosymbiotic hypothesis in the United States in 1981, many of her colleagues were skeptical. But an examination of the membranes, reproduction, and ribosomes of mitochondria and chloroplasts shows clear similarities to the same features in certain bacteria. Even more convincingly, some of the DNA sequence within mitochondria (mtDNA) is similar to the DNA sequence found in certain bacterial species.

After the evolution of mitochondria, independent endosymbioses between eukaryotic cells and various species of photosynthetic bacteria appeared. These relationships led to the evolution of chloroplasts—one symbiosis led to green algae and land plants, while it appears that other relationships led to the unique chloroplasts in other modern groups, including red and brown algae.

## Visualize This ▼

The bacteria ancestors of chloroplasts and mitochondria have a single membrane, but chloroplasts and mitochondria have two membranes. Use the diagram here to explain how the second membrane came about.

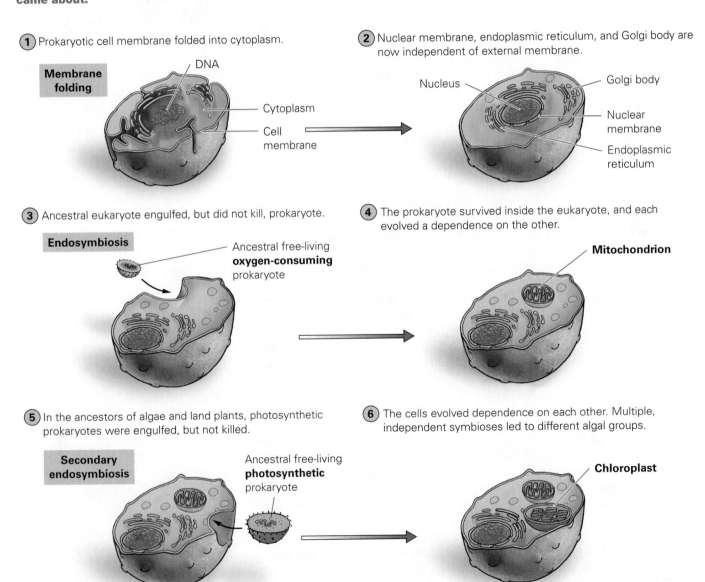

① Prokaryotic cell membrane folded into cytoplasm.

**Membrane folding**

DNA

Cytoplasm

Cell membrane

② Nuclear membrane, endoplasmic reticulum, and Golgi body are now independent of external membrane.

Nucleus

Golgi body

Nuclear membrane

Endoplasmic reticulum

③ Ancestral eukaryote engulfed, but did not kill, prokaryote.

**Endosymbiosis**

Ancestral free-living **oxygen-consuming** prokaryote

④ The prokaryote survived inside the eukaryote, and each evolved a dependence on the other.

**Mitochondrion**

⑤ In the ancestors of algae and land plants, photosynthetic prokaryotes were engulfed, but not killed.

**Secondary endosymbiosis**

Ancestral free-living **photosynthetic** prokaryote

⑥ The cells evolved dependence on each other. Multiple, independent symbioses led to different algal groups.

**Chloroplast**

**FIGURE 13.7 The evolution of eukaryotes.** Mitochondria and chloroplasts appear to be descendants of once free-living bacteria that took up residence within an ancient nucleated cell.

**STOP** & **STRETCH** Mitochondria are thought to have evolved before chloroplasts. What is the evidence that supports this hypothesis?

## Kingdom Protista

The kingdom **Protista** is made up of the simplest known eukaryotes. Most protists are single-celled creatures, although several have enormous **multicellular** (many-celled) forms.

The modern kingdom Protista contains organisms resembling animals, fungi, and plants. As with Bacteria and Archaea, most members of kingdom Protista remain unknown. There is currently no agreement among scientists regarding how many **phyla,** that is, groups just below the level of kingdom, are contained within Protista. Some argue as few as eight, and others propose as many as 80. There is also considerable evidence that the different groups of protists do not share a single common ancestor that separates them from plants, animals, and fungi. **TABLE 13.3** (on the next page) lists a few of the more common phyla within the kingdom.

The plant-like protists that make food via photosynthesis are called **algae,** and this group is actually made up of several distinct, quite divergent, categories of organisms. Each of these algal phyla has its own methods of producing and storing food. As the source of one-quarter of Earth's oxygen and the basis for most aquatic food chains, algae certainly should make the top ten list of greatest organisms on Earth.

Unlike algae, animal-like and fungus-like protists cannot make their own food. Like us, these phyla consume organic molecules in order to survive. The most abundant organic molecule on Earth is cellulose, the carbohydrate that makes up plant cell walls. While the range of food items eaten by humans is wide, we cannot digest cellulose—in fact, no animals can break this molecule down directly. However, several groups of protists (as well as a variety of bacteria) can. Maybe we should not be so sure that humans make superior use of the available resources.

Another group of protists call into question humans' superiority in communication and cooperative behavior (**FIGURE 13.8**). Slime molds, fungus-like protists that grow primarily as single cells on the surface of soil or dead plant material, can call for reinforcements and congregate when times are tough. This phenomenon occurs when individual cells begin to dehydrate or starve, causing them to release a chemical signal that acts as a homing signal for other slime mold cells. Once about 100,000 cells have gathered together, the slime mold transforms into something that appears and behaves like a single multicellular organism, and which is called a slug. The slug travels as a unit until it finds suitable conditions, then transforms again into a spore-producing structure. Individual cells become encapsulated in cellulose and the spore head shatters, scattering these cells far into the new environment. Humans did cooperate to travel to the moon, but these slime molds do something awfully similar as just single-celled organisms.

## Kingdom Animalia

From the origin of the first prokaryote until approximately 1.2 billion years ago, life on Earth consisted only of single-celled creatures. Then multicellular organisms first began to appear in the fossil record. The ancient, many-celled creatures of 600 million years ago, called the *Ediacaran fauna,* were organisms unlike any modern species and included giant fronds and ornamented disks (**FIGURE 13.9**).

Biologists are unsure which of these ancient species is the common ancestor of modern **animals**—defined as multicellular organisms that

**(a)**

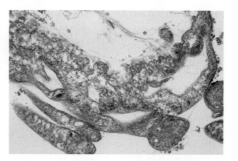

**(b)**

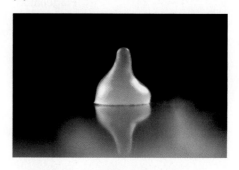

**FIGURE 13.8 Communication and cooperation.** Cellular slime molds (a) live as single cells in good conditions, but (b) cooperate to travel to new areas when conditions deteriorate.

**FIGURE 13.9 Ediacaran fauna.** This reconstruction of multicellular organisms that lived before the Cambrian explosion is based on 580-million-year-old fossil remains.

anima- means breath or soul.

**TABLE 13.3** **The diversity of Protista.** Protista contains animal-like, fungus-like, and plant-like organisms. A sampling of protistan phyla is described here.

| Kingdom Protista: Common Names and Characteristics of Select Phyla | | Example | |
|---|---|---|---|
| **Animal-like protists** | **Ciliates** Free-living, single-celled organisms that use hair-like structures to move. | *Paramecium* | |
| | **Flagellates** Use one or more long whip-like tail for locomotion. Most are free living, but some cause disease by infecting human organs. | *Giardia* | |
| | **Amoebas** Flexible cells that can take any shape and move by extending pseudopodia ("false feet"). | *Amoeba* | |
| **Fungus-like protists** | **Slime molds** Feed on dead and decaying material by growing net-like bodies over a surface or by moving about as single amoeba-like cells. | *Physarum* | |
| **Plant-like protists** | **Diatoms** Single cells encased in silica (glass). | Diatom | |
| | **Brown algae** Large multicellular seaweeds. | Kelp | |
| | **Green algae** Closest relatives to land plants. Single-celled to multicellular forms. | *Volvox* | |

make their living by ingesting other organisms and that are motile (have the ability to move) during at least one stage of their life cycle. What is clear from the fossil record is that by about 530 million years ago, *all* modern animal groups had emerged.

The remarkably sudden appearance of the modern forms of animals—a period comprising little more than 1% of the history of life on Earth—is referred to as the **Cambrian explosion,** named for the geologic period during which it occurred. Some scientists hypothesize that the evolution of the animal lifestyle itself—that is, as predators of other organisms—led to the Cambrian explosion.

It can be difficult to conceive of an animal as complex as a human evolving from a simple eukaryotic ancestor. However, humans are not very different from other eukaryotes. When the first cell containing a nucleus appeared, all of the complicated processes that take place in modern cells, such as cell division and cellular respiration, must have evolved. When the first multicellular animals appeared, many of the processes required to maintain these larger organisms, such as communication systems among cells and the formation of organs and organ systems, arose. Although a human and a sea star appear to be very different, the way we develop and the structures and functions of our cells and common organs are nearly identical.

In fact, there appears to be surprisingly little genetic difference between humans and starfish; most of that difference occurs in a group of genes that control **development,** the process of transforming from a fertilized egg into an adult creature. In addition, the amount of time since the divergence of the major evolutionary lineages of animals—530 million years—is quite a long time for dramatic differences among species in different phyla to evolve. Put in more familiar terms, if the time since the Cambrian explosion were one 24-hour day, all of human history would fit in only the last 2 seconds.

Zoologists, scientists who study the kingdom **Animalia,** have described more than 25 phyla in this kingdom. Some of the more well-known groups are illustrated in **TABLE 13.4,** on the next page.

Most people typically picture mammals, birds, and reptiles when they think of animals, but species with backbones (including these animals as well as fish and amphibians), known as **vertebrates,** represent only 4% of the total species in the kingdom. Other vertebrates provide competition for the greatest species. Many of the traits we think of as uniquely human are found in other vertebrates—and sometimes are even better developed. Beavers have the ability to dramatically modify the environment to fit their needs. Numerous vertebrates, including crows and chimpanzees, use technology in the form of tool making to better exploit the resources that surround them. Dolphins have highly sophisticated methods of communication, using not only auditory and visual signals like in humans, but also their senses of touch and smell. Certain fish have the ability to switch sex in order to maximize opportunities for reproduction. Some vertebrates even appear to have a creative impulse—as evidenced by certain whale songs and the nest of the bowerbird (**FIGURE 13.10**).

(a)

(b)

(c)

(d)

**FIGURE 13.10 Vertebrate skills.** (a) Beavers are mammals that can radically transform an environment by damming streams to create ponds and removing large numbers of trees. (b) This crow is demonstrating advanced tool use by shaping a stick in order to use it as a fishing pole for grubs. (c) Clownfish can switch from male to female depending on the number of individuals of each sex in a group. (d) This bowerbird nest is built to attract a female, but the décor shows an artistic flair that rivals human creativity.

**TABLE 13.4** **Phyla in the kingdom Animalia.** **A sampling of the diversity of animals. The rows are arranged generally in order of appearance in evolutionary time— from the more ancient sponges to the more recent chordates.**

| Kingdom Animalia: Major Phyla | Description | Example | |
|---|---|---|---|
| **Porifera** | Fixed to underwater surface and filter bacteria from water that is drawn into their loosely organized body cavity. | Sponge | |
| **Cnidaria** | Radially symmetric (like a wheel) with tentacles. Some are fixed to a surface as adults (e.g., corals), while others are free floating in marine environments (e.g., jellyfish). | Anenome | |
| **Platyhelminthes** | Flatworms with a ribbon-like form. Live in a variety of environments on land and sea or as parasites of other animals. | Tapeworm | |
| **Mollusca** | Soft-bodied animals often protected by a hard shell. Body plan consists of a single muscular foot and body cavity enclosed in a fleshy mantle. Phylum includes snails, clams, and squid. | Octopus | |
| **Annelida** | Segmented worms. Body divided into a set of repeated segments. | Earthworm | |

*continued*

**TABLE 13.4 Phyla in the kingdom Animalia.** *continued*

| Kingdom Animalia: Major Phyla | Description | Example |
|---|---|---|
| **Nematoda** | Roundworms with a cylindrical body shape. Very diverse and widespread in many environments. | Roundworm |
| **Arthropoda** | Segmented animals in which the segments have become specialized into different roles (such as legs, mouthparts, and antennae). Body completely enclosed in an external skeleton that molts as the animal grows. Phylum includes insects and spiders as well as crabs and lobsters. | Shrimp |
| **Echinodermata** | Slow-moving or immobile animals without segmentation and with radial symmetry. Internal skeleton with projections gives the animal a spiny or armored surface. | Sea urchin |
| **Chordata** | Animals with a spinal cord (or spinal cord-like structure). Includes all large land animals, as well as fish, aquatic mammals, and salamanders. | Duck-billed platypus |

The remaining 96% of known organisms in Animalia are the **inverte-brates** (animals without backbones, as shown in **FIGURE 13.11**). In fact, the vast majority of all multicellular organisms on Earth are invertebrates. As a result of this diversity, there are innumerable examples of invertebrate capabilities that can outshine those of humans. Argentine ants rival humans in terms of their distribution across Earth's surface, being found on six continents and numerous oceanic islands, but outnumber humans by orders of magnitude. The jellyfish *Turritopsis nutricula* is the only animal known to be able to revert to an earlier developmental stage after becoming sexually mature, making it theoretically immortal. And octopi have the ability to change their shape and skin to camouflage themselves in a variety of environments.

**in-** means without.

**zoo-** means pertaining to animals.

Some **zoologists** estimate that there may be as many as 30 million unknown invertebrate species, especially in the oceans. While our ignorance about the diversity of animals is great, another kingdom of multicellular eukaryotes is even less well known—the fungi.

**(a) Ant (*Linepithema humile*)**

**(b) Jellyfish (*Turritopsis nutricula*)**

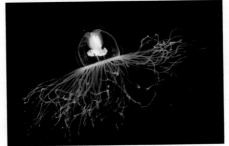

**(c) Cuttlefish (*Sepia officinalis*)**

**FIGURE 13.11 Invertebrates.** Most animals are invertebrates, including these examples.

## Kingdom Fungi

Like plants, fungi are immobile, and many produce fruit-like organs that disperse **spores**, cells that are analogous to plant seeds in that they germinate into new individuals. In fact, for generations biologists placed fungi in the plant kingdom. However, while plants make their own food via photosynthesis, fungi feed on other organisms by secreting digestive chemicals into their environment; these chemicals break down complex organic food sources into small molecules, which the fungi then absorb. Because they rely on other organisms for their food sources, fungi are more like animals than plants. In fact, DNA sequence analysis by **mycologists**, biologists who study fungi, indicates that Fungi and Animalia are more closely related to each other than either kingdom is to the plants.

**myco-** means fungal.

**Are we more impressive than horseshoe crabs, survivors of 445 million years of change?**

The mushroom you think of when you imagine a fungus is a misleading image of the kingdom. Most of the functional part of fungi is made up of very thin, stringy material called **hyphae**, which grows over and within its food source—the mushroom is simply the reproductive organ that appears on the surface of the food (**FIGURE 13.12**). The string-like form of hyphae maximizes the surface over which feeding takes place. Some fungi feed on living tissue, while others decompose dead organisms. This latter role of fungi is key to recycling nutrients in biological systems.

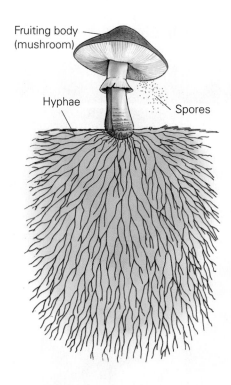

Fruiting body (mushroom)

Hyphae

Spores

◀ **Visualize This**

**Why is a mushroom produced outside the food resource for the hyphae?**

**FIGURE 13.12 Fungi.** Hyphae can extend over a large area. The familiar mushroom, as well as the fruiting structures of less-familiar fungi, primarily functions only as methods of dispersing spores.

The phyla of fungi are generally distinguished from each other by their method of spore formation (**TABLE 13.5**, on the next page). However, convergent evolution, in which unrelated species take similar forms because of similar environments, has led to the appearance of similar body shapes and lifestyles—which we call "fungal forms"—among these different phyla. One of the most commercially important fungal forms is **yeast,** a single-celled type of fungi, which lives in liquids and is found in at least two different fungal phyla. Unlike most eukaryotes, including ourselves, yeasts have a metabolic process, called alcoholic **fermentation,** that can extract significant energy from carbohydrates even in the absence of oxygen. The activity of yeasts in oxygen-poor but sugar-rich environments results in the formation of ethyl alcohol. The metabolism of yeasts within flour batter also leads to the production of carbon dioxide, which is trapped by wheat protein fibers and allows dough to "rise" during bread making.

Another form known as **mold** is found in all fungal phyla and is also commercially important. This quickly reproducing, fast-growing form can spoil fruits and other foods, although some produce certain types of flavorful cheese, including blue and Camembert, as they are "spoiling" milk. The antibiotic penicillin was derived from a species of mold. In fact, about one-third of the bacteria-killing antibiotics in widespread use today are derived from fungi, which produce them because they compete with bacteria for food sources.

While we may think humans have the monopoly in forcing other species to do our will, certain fungi have this ability as well (**FIGURE 13.13**).

**FIGURE 13.13 Slave and master.** The fungus growing out of the head of this ant controlled the ant's behavior before the ant died in order to maximize its own dispersal.

**TABLE 13.5 Fungal diversity.** Fungi are classified into phyla based on their mode of spore production. The most common phyla are listed here.

| Kingdom Fungi: Major Phyla | Description | Example | |
|---|---|---|---|
| **Zygomycota** | Sexual reproduction occurs in a small resistant structure called a zygospore. Most reproduction is asexual—directly via mitosis. | *Rhizopus stolonifera*—bread mold | |
| **Glomeromycota** | Probably do not reproduce sexually, spores formed in various manners. Unique in that all members of this phylum are mycorrhizal. | Mycorrhizal fungi | |
| **Ascomycota** | Spores are produced in sacs on the tips of hyphae in fruiting structures. | Morel | |
| **Basidiomycota** | Spores are produced in specialized club-shaped appendages on the tips of hyphae in fruiting structures. | *Amanita muscaria*—poisonous fly agaric | |

*Ophiocordyceps unilateralis* is a fungal parasite of ants that causes these animals to change their behavior completely. The ants the fungus infects normally live in trees—but when infected with the fungus, the ant turns into a virtual zombie. It drops from the tree, climbs the stem of a plant and firmly attaches itself to a leaf with its jaws. There, the ant is slowly consumed from within by the fungus. When the fungus eventually produces a mushroom, it is now in a prime spot for maximal distribution of its spores. That ant is a beast of burden in the most dramatic way.

Fungi may also be the original farmers. Mycorrhizae are a symbiotic relationship between fungal hyphae and plant roots that benefit both partners

(**FIGURE 13.14**). Perhaps 90% of all plants have mycorrhizae. A relationship between plants and fungi is evident even from the oldest fossils of land plants, and the symbiosis may have evolved when some species of fungi that were parasitizing plants did not kill these plants outright. Instead, by allowing their hosts to survive on less fertile soils, these less-deadly fungi did millions of years ago what human farmers learned to do only in the last 20,000 years—expand and maintain habitat for a species that is their food source.

## Kingdom Plantae

The kingdom **Plantae** consists of multicellular eukaryotic organisms that make their own food via photosynthesis. Plants have been present on land for over 400 million years, and their evolution is marked by increasingly effective adaptations to the terrestrial environment (**TABLE 13.6**).

The first plants to colonize land were small and close to the ground. Their small size was necessary, since they had no way to transport water much distance from the surface of the soil. The evolution of **vascular tissue**, made up of

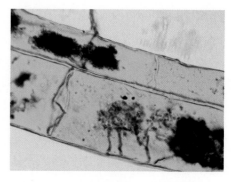

**FIGURE 13.14 Fungi as farmers.**
Mycorrhizal fungi (seen here as dark bodies within the plant cells) benefit by obtaining carbohydrates produced by the plant's photosynthesis. Plants are better able to absorb certain nutrients—and thus survive on poorer soils—thanks to these fungi.

**TABLE 13.6 Plant diversity.** The four major phyla of plants are listed here in order of their appearance in evolutionary history.

| Kingdom Plantae: Major Phyla | Description | Example | |
|---|---|---|---|
| **Bryophyta** | Mosses. Lacking vascular tissue, these plants are very short and typically confined to moist areas. Reproduce via spores. | Moss | |
| **Pteridophyta** | Ferns and similar plants. Contain vascular tissue and can reach tree size. Reproduce via spores. | Fern | |
| **Coniferophyta** | Cone-bearing plants resembling the first seed producers | Cycad | |
| **Anthophyta** | Flowering plants. Seeds produced within fruits, which develop from flowers. Advances in vascular tissues and chemical defenses contribute to their current dominance on Earth. | Orchid | |

specialized cells that can transport water and other substances, allowed plants to reach tree-sized proportions and to colonize much drier areas. The evolution of **seeds**, structures that protect and provide a food source for young plants, represented another adaptation to dry conditions on land.

Several species of nonflowering seed plants would definitely rival humans for the title of Earth's greatest species. Bristlecone pine trees are among the longest living individual organisms on Earth; the oldest known bristlecone sprouted over 5000 years ago, near the beginning of the Egyptian dynasty. The largest organisms on Earth are Giant Sequoias at as much as 94 m (300 ft) and 9 m (29 ft) in diameter—16 times bigger than the largest blue whale, the largest animal ever known to exist.

Most modern plants belong to a group that appeared about 140 million years ago, the **flowering plants.** Like their ancestors, flowering plants possess vascular tissue and produce seeds; in addition, these plants evolved a specialized reproductive organ, the flower. Over 90% of the known plant species are flowering plants (**FIGURE 13.15**).

From about 100 million to 80 million years ago, the number of distinct groups, or families, of flowering plants increased from around 20 to over 150. During this time, flowering plants became the most abundant plant type in nearly every habitat. Among the eukaryotic kingdoms, plants are also the most well-described; many botanists believe that the number of unknown plant species is relatively small—probably a few thousand.

The rapid expansion of flowering plants is called **adaptive radiation**— the diversification of one or a few species into a large and varied group of descendant species. Adaptive radiation typically occurs either after the appearance of an evolutionary breakthrough in a group of organisms or after the extinction of a competing group. The radiation of animals during the Cambrian explosion may be a result of the evolution of predation, and the radiation of mammals beginning about 65 million years ago occurred after the extinction of dinosaurs. The radiation of flowering plants must be due to an evolutionary breakthrough—some advantage they had allowed them to assume roles that were already occupied.

**(a) Hammer orchids**

**(b) Viburnum**

**(c) Curare vine**

**FIGURE 13.15 Diversity of flowering plants.** These plants may rival humans for greatest species. (a) Hammer orchids manipulate their pollinators and provide them no reward. (b) Like all flowering plants, this viburnum acts like a mammal in that it only commits major resources to an egg that is fertilized. (c) The curare vine produces a deadly toxin that rivals humans' most dangerous chemical weapons.

Plant biologists, or **botanists,** still debate which traits of flowering plants give them an advantage over nonflowering. Some believe that the reproductive characteristics of flowering plants led to their radiation—including the ability to employ other species, something we might think of as unique to humans. Flowering plants rely on the assistance of animals in transferring gametes for the process of **pollination.** For this strategy to work, plants have changed forms to attract reliable pollination partners, including producing vivid and fragrant **petals** that draw insects to the plant's sugar-rich nectar. As an insect or animal becomes a reliable visitor to a flower type, the plant evolves to become more effective at "loading" the visitor with the male gamete, pollen. In turn, the plant's modifications drive the evolution of the pollinator so that it becomes exclusive in its relationship to the flower—by causing selection in the pollinator's traits, the flowering plant ensures that its pollen is always delivered to other flowers of the same species. The pattern of both species in a relationship evolving adaptations to each other is known as **coevolution.** This has been occurring for at least 140 million years, and may have been a key factor in increasing the diversity of flowering plants.

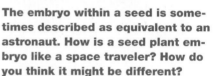

botan- means pertaining to plants.

Some flowering plant species have taken advantage of this coevolutionary process to create pollination "slaves" of male insects that will deliver pollen between flowers that look like female insects. The males receive nothing from this interaction, and in fact, waste a lot of time and energy in the process. Humans may have service animals, but we provide them with food, water, and shelter. It is unclear that we have ever been as effective as this in tricking another animal to do our will with no reward at all!

Another trait that may have led to success among flowering plants is **double fertilization,** wherein sperm from a single pollen grain fertilize both the egg and a specialized food-producing tissue (**FIGURE 13.16**). This is similar to how mammals like humans reproduce, in that energy is only committed to offspring in large amounts after successful fertilization. This strategy protects individuals from wasting energy in the case of failed fertilization. Flowering plants evolved this conservative reproductive strategy millions of years before mammals even appeared.

Flowering plants may also beat humans in their ability to defer and confound natural enemies. Because plants cannot physically escape from their

# Visualize This ▼

The embryo within a seed is sometimes described as equivalent to an astronaut. How is a seed plant embryo like a space traveler? How do you think it might be different?

**FIGURE 13.16 Sexual reproduction in flowering plants.** The reproductive differences between flowering plants and other plants, including the production of fruit for dispersal, may have led to their adaptive radiation.

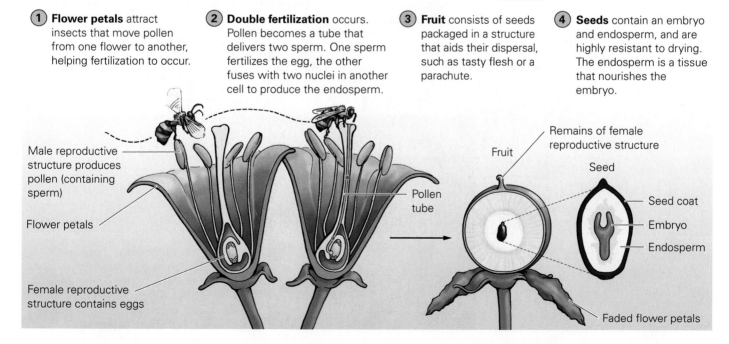

**1** **Flower petals** attract insects that move pollen from one flower to another, helping fertilization to occur.

**2** **Double fertilization** occurs. Pollen becomes a tube that delivers two sperm. One sperm fertilizes the egg, the other fuses with two nuclei in another cell to produce the endosperm.

**3** **Fruit** consists of seeds packaged in a structure that aids their dispersal, such as tasty flesh or a parachute.

**4** **Seeds** contain an embryo and endosperm, and are highly resistant to drying. The endosperm is a tissue that nourishes the embryo.

Male reproductive structure produces pollen (containing sperm)

Flower petals

Female reproductive structure contains eggs

Pollen tube

Fruit

Remains of female reproductive structure

Seed

Seed coat

Embryo

Endosperm

Faded flower petals

predators, natural selection has favored the production of predator-deterring toxins via secondary reactions to primary biochemical pathways—a technological innovation, in a way. The chemicals produced are known as **secondary compounds.** For instance, curare vines produce toxins that block the connection between nerves and muscles. Animals that get this toxin, called *curarine,* in their bodies become paralyzed and can do little damage to the vine. Curarine is produced via a secondary reaction of the process of amino acid synthesis. In this case, natural selection must have favored genetic variations leading to production of not only normal amino acids but also this toxic secondary compound.

Other secondary compounds you may be familiar with are the drugs nicotine and cocaine, the poisons ricin and cyanide, and the drugs aspirin and digitalis. While humans have learned how to extract these toxins for our own uses—sometimes not so friendly ones—flowering plants have been engaged in this form of technological chemical warfare for millions of years.

## Not Quite Living: Viruses

Any accounting of biodiversity must also consider viruses, organic entities that interact with living organisms but are not quite alive themselves. Viruses are considered nonliving because they are unable to maintain homeostasis (Chapter 2) and are incapable of growth or reproduction without the assistance of another organism. Because they are not quite living, many biologists would leave viruses off of a greatest species discussion—but they are remarkable entities that have the power to take down human civilizations.

A virus typically consists simply of a strand of DNA or RNA surrounded by a protein shell, called a **capsid.** Some viruses are also enveloped in a membranous coat. Viruses are basically rogue pieces of genetic material that must use the cells of other organisms to reproduce (**FIGURE 13.17**). When a virus enters a host cell, it uses the transcription machinery of the cell to make copies of its DNA or RNA. The replicated viral genomes then use the ribosome, transfer RNA, and amino acids of the host to make new protein capsules and other polypeptides. If the virus is enveloped, it even uses the host cell's own membrane to make envelopes for the daughter viruses.

Once a cell has been infected by a virus, it cannot perform its own necessary functions and will die. But before the infected cell dies, dozens of duplicated viruses can be released and move on to infect other cells. Some of the most devastating pathogens of humans are viruses, including polio, smallpox, HIV, and influenza. HIV is especially troublesome because it destroys immune system cells, which are essential for fighting off infection. An HIV-infected individual gradually loses the ability to combat other diseases, including the HIV infection itself.

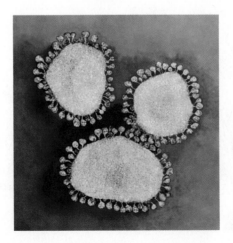

**FIGURE 13.17 Viruses.** The severe acute respiratory virus (the cause of severe acute respiratory syndrome, SARS), viewed under very high magnification. Visible here is the protein capsid, studded with receptors that bind to host cells.

# 13.3 Learning about Species

The living world is amazingly diverse, and our knowledge of it is only fragmentary. Many biologists are attracted to the study of biological diversity for just this reason: because the variety of life is remarkable, fascinating, and largely an unexplored frontier in human knowledge. Once a new species is identified, biologists seek to place it in context with other living organisms. This may be the one place where humans have a distinct advantage over nearly all other species—in the ability to use the scientific method to learn about the past and predict the future.

## Reconstructing Evolutionary History

While we can imagine many different classification schemes, most biologists argue that one reflecting evolutionary relationships, called an **evolutionary classification**, is more useful than one based on more superficial similarities.

The relationship between certain reptiles and birds can help illustrate this point. Both Komodo dragons and alligators are large reptiles, and both are ambush predators that lay in wait for their prey. Both animals walk on four legs, are covered with scales, and are cold-blooded. Birds, on the other hand, are warm-blooded, covered with feathers, and have wings (**FIGURE 13.18**). Not surprisingly, a non-evolutionary classification places alligators with Komodo dragons and other cold-blooded, scaly reptiles. However, overwhelming evidence from morphology, physiology, and genetics indicates that alligators are more closely related to birds than they are to Komodo dragons. If scientists wish to know more about the anatomy, physiology, behavior, and ecology of Komodo dragons, they should not use what is known about alligators—a much more distant relative—as a stepping-off point. Instead, they should focus on more closely related lizards.

**Developing Evolutionary Classifications.** Evolutionary classifications are based on the principle that the descendants of a common ancestor should share any biological trait that first appeared in that ancestor. For example, this principle has been used to uncover the evolutionary relationship among different species of sparrows.

All species have a two-part name—the first part of the name indicates the **genus**, or broader group of relatives to which the species belongs; the second part is specific to a particular species within that genus. In effect, a species name

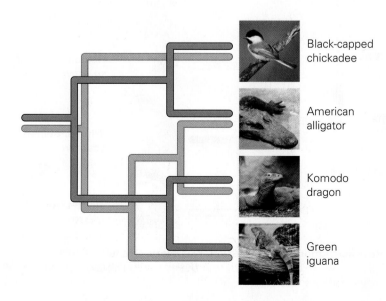

Black-capped chickadee

American alligator

Komodo dragon

Green iguana

**FIGURE 13.18 The challenge of biological classification.** Evolutionary relationships are sometimes surprising. The phylogenetic tree in red illustrates a relationship that you might expect based on the superficial similarities and differences among birds and various reptiles. However, the tree in green indicates the relationship that is best supported by the data.

is like a full name in which the "family group" is designated by the first name, and the individual in that particular family is identified by the second part of the name. Both parts of the name are always used when identifying a particular species, although if we are discussing a group of species in the same genus, the genus name may be abbreviated to just its first letter.

**FIGURE 13.19** illustrates a hypothesized **phylogeny**—the evolutionary relationship—of sparrow species in the genus *Zonotrichia*. Scientists have used a technique called **cladistic analysis,** an examination of the variation in these sparrows' traits relative to a closely related species, to determine this phylogeny. For example, if we examine just the heads of the four sparrow species compared to their relative, the dark-eyed junco, we see that three—all but the Harris's sparrow—have dark and light alternating stripes on the crown of their heads. This observation seems to indicate that crown striping evolved early in the radiation of these sparrows. Among the three species with crown stripes, two have seven stripes, while the golden-crowned sparrow has only three stripes. An increase in the number of stripes appears to have evolved after the original striped crown pattern. Finally, of the two species with seven crown stripes, only one has evolved a white throat—the aptly named white-throated sparrow. In the case of these four sparrows, it appears that every step in their radiation involved a visible change in their appearance.

Unfortunately, reconstructing evolutionary relationships is not as simple as the sparrow example suggests. Descendant species may lose a trait that evolved in their ancestor, or unrelated species may acquire identical traits via convergent evolution. You can even see convergent evolution in the phylogeny in Figure 13.19; golden-crowned sparrows, like the white-throated species, have a patch of golden feathers on their heads that appears to have evolved independently. The existence of convergent traits complicates the development of evolutionary classifications.

## Working with Data ▶

**What traits did the ancestral bird likely have?**

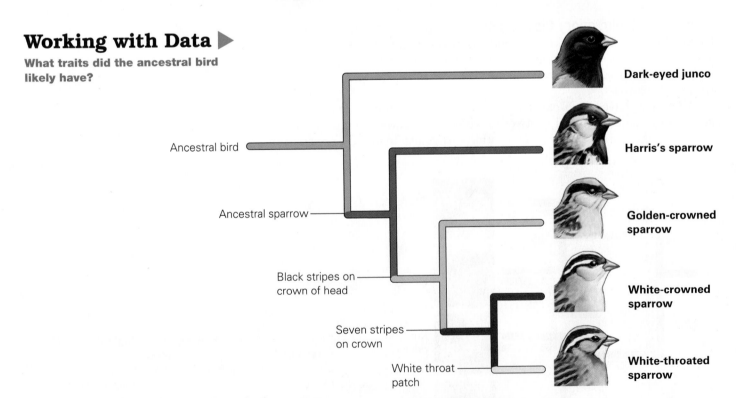

**FIGURE 13.19 Reconstruction of an evolutionary history.** The relationships among these four species of sparrow can be illustrated by their shared physical traits compared to a more distant relative, the dark-eyed junco.

**STOP & STRETCH**    Amateur bird watchers often consider vultures to be birds of prey, along with hawks, falcons, eagles, and owls. Vultures are scavengers rather than predators, however. What convergent traits do vultures share with these other birds that would cause bird watchers to classify them as birds of prey?

**Testing Evolutionary Classifications.**    An evolutionary classification is a hypothesis of the relationship among organisms. It is difficult to test this hypothesis directly—scientists have no way of observing the actual evolutionary events that gave rise to distinct organisms. However, scientists *can* test their hypotheses by using information from both fossils and living organisms.

By examining the fossils of extinct organisms, scientists can gather clues about the genealogy—the record of descent from ancestors—of various organisms. For example, fossils of birds clearly indicate that these animals evolved from crocodile-like ancestors.

Information from living organisms can provide an even finer level of detail about evolutionary relationships. As illustrated in Figure 13.4, closely related species should have similar DNA. If the pattern of DNA similarity matches a hypothesized evolutionary relationship among species, the phylogeny is strongly supported. This is the case with the relationship between birds and crocodiles; DNA sequence comparisons indicate that the DNA of crocodiles is more similar to the DNA of birds than it is to the DNA of Komodo dragons.

In contrast, DNA sequence comparisons do not support the sparrow phylogeny presented in Figure 13.19. Here, the data suggest that white-crowned birds and golden-crowned birds are closely related, and white-throated sparrows are a more distant relative. In this case, more observations are needed to discern the true evolutionary relationships among the *Zonotrichia*. (In Chapter 10, we described multiple supportive tests of another phylogeny—in that case, the evolutionary relationship among humans and apes.)

## The Greatest Species on Earth

Our survey of diversity provides only a small sampling of the remarkable life forms that have evolved on Earth. But even with a more detailed accounting, any argument about the greatest species (or group of species) of all time doesn't have a correct answer. Humans are pretty great, but we can be laid low by something as simple as a not-quite-living virus, and we depend on the attributes of many other species to survive.

Where humans clearly excel compared to all other species is in our ability to use the scientific method to understand the world and to use that knowledge to change the environment. It is clear no other species on Earth has this trait. Our understanding of the world, limited as it is, also leads us to other understandings about biodiversity. We know that thousands of other living organisms are lost every year through our destruction of natural habitats. We know that the dramatic rate of biodiversity loss not only denies humans still undiscovered biological wonders but also diminishes the ability of Earth to sustain our population.

At current rates of loss, our generation may be the last to truly enjoy the diverse wonder, beauty, and benefits of wild nature (Chapter 15 discusses the causes, consequences, and possible solutions for the current biodiversity crisis). Humans can help to reduce the rate at which biodiversity is being lost—but only if we begin to appreciate the value of the remarkable diversity that surrounds and sustains us. If we can do this, maybe we will earn the title of greatest species in the history of Earth.

# *savvy reader*

## Are Middle-Aged Humans Evolution's Greatest Accomplishment?

People in their forties and fifties might mourn their changing figure or the passing of childbearing age, but these changes are key to the success of the human species, Dr. David Bainbridge said.

Far from being over the hill, middle-aged people are arguably the "pinnacle of evolution" because they are primed to play a vital role in society which could not be filled by younger adults, he added.

…

Writing in the *New Scientist* magazine, he explained: "Middle age is a controlled and preprogrammed process not of decline but of development."

…

Many middle-aged women lament the loss of their youthful figure as fat deposits leave the breasts, hips and thighs and gather in larger quantities around the midriff.

But this happens because the body is no longer required to produce children and storing fat centrally makes it easier to carry around.

In evolutionary terms, having more fat would also have allowed our ancestors to use it for sustenance in hard times, leaving more food for younger generations to survive on.

…

Contrary to popular belief many of our prehistoric ancestors lived beyond 40, and they developed over thousands of years into skilled and experienced "super-providers," Dr. Bainbridge said.

…

Dr. Bainbridge said: "Each of us depends on culture to survive, and the main route by which culture is transmitted is by middle-aged people telling children and young adults what to do.

"Middle-aged people can do more, earn more and, in short, they run the world," he said.

1. What is the evidence provided in the excerpt that middle-aged people are "the pinnacle of evolution"?
2. Use your understanding of natural selection to describe how, according to the article, the changes in fat storage in middle-aged women came about.
3. What is the evidence provided that natural selection caused the trait described in question 2 to evolve in middle-aged women?
4. "Confirmation bias" is the tendency for people to favor information that confirms their beliefs, and the text in this excerpt displays confirmation bias for the belief that middle-aged humans are evolution's greatest accomplishment. Can you describe accurate information that would provide a counter to this belief?

http://www.telegraph.co.uk/science/science-news/9129205/Middle-aged-people-the-pinnacle-of-evolution.html?fbA

# SOUNDS RIGHT **BUT IS IT?**

English and many other languages read from left to right or top to bottom. As a result, readers often expect summaries or conclusions on the right hand side or bottom of a figure. This can lead to some mistaken conclusions. Consider the following phylogenetic tree (FIGURE 13.20) describing the relationship between a few animal phyla.

**Because humans are pictured at the top of this tree, they are the most complex animal.**

Sounds right, but it isn't.

1. All of these animals share a common ancestor, the first animal. Have all of these animals been evolving for the same amount of time?

2. What traits or abilities do some or all insects have that humans do not?

3. Which of the organisms on the tree is most closely related to humans?

4. Does switching the placement of starfish and humans change the information on the tree? Explain your answer.

5. Based on your answers to questions 1–4, explain why the statement in bold sounds right, but isn't.

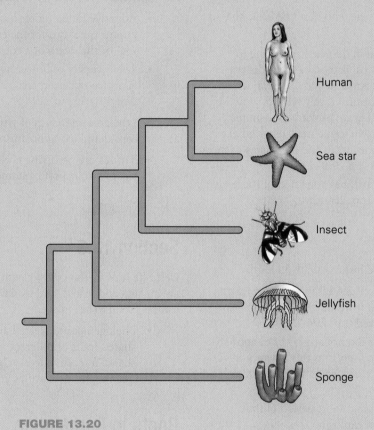

Human

Sea star

Insect

Jellyfish

Sponge

FIGURE 13.20

# Chapter Review   MasteringBiology®

Go to the Study Area in MasteringBiology® for practice quizzes, myeBook, BioFlix™ 3-D animations, MP3Tutor sessions, videos, current events, and more.

## Summary

### Section 13.1

Cite the current estimate of the number of species described by science, and explain why this number is only an estimate.

- The number of known living species is estimated to be 1.3 million, but the total number of unknown species is likely to be several times greater (pp. 288–289).

List the three major domains of life, and describe basic characteristics of each domain.

- Organisms are classified in domains according to evolutionary relationships. Bacteria share a distant common ancestor with Archaea and Eukarya, while the last two groups are more closely related. However, both Bacteria and Archaea are prokaryotic (simple single-celled organisms without a nucleus or other membrane-bound organelles), while organisms in the Eukarya are eukaryotes (pp. 289–293).
- The history of life on Earth has consisted of successive dynasties of organisms (pp. 289–290).

### Section 13.2

Provide a broad overview of the history of life on Earth.

- Life on Earth began about 3.6 billion years ago with simple prokaryotes, but it would be 1.5 billion years before eukaryotes evolved (pp. 294–295).
- Multicellular organisms did not appear until approximately 600 million years ago, and this advance in form led to the diversity of species on Earth today (p. 297).
- Animal groups evolved in a short period of time known as the Cambrian explosion (p. 299).

Summarize the endosymbiotic theory for the origin of eukaryotic cells.

- Eukaryotes, cells with nuclei and other membrane-bound organelles, probably evolved from symbioses among ancestral eukaryotes and prokaryotes (pp. 295–296).

List the major characteristics of the four major kingdoms of Eukarya: Protista, Animalia, Fungi, and Plantae.

- The kingdom Protista is a hodgepodge of organisms that are typically unicellular eukaryotes. Organisms in this group may be plant-like, animal-like, or fungus-like in lifestyle (pp. 297–298).

- Animals are motile, multicellular eukaryotes that rely on other organisms for food (pp. 297, 299–302).
- Fungi are immobile, multicellular eukaryotes that rely on other organisms for food and are made up of thin, threadlike hyphae (pp. 302–305).
- Plants are multicellular, photosynthetic eukaryotes that make their own food (p. 305).

Define *adaptive radiation,* and provide examples of where this has occurred in the history of life.

- Adaptive radiation occurs when several new species appear quickly after the evolution of a new "way of life" or the death of a competing group (pp. 306–307).
- The diversity of flowering plants may be due partly to adaptive radiation resulting from their production of defensive chemicals (pp. 307–308).

Describe the characteristics of viruses, and explain why they are not considered living organisms.

- Viruses are nonliving entities made up of genetic material in a transport container and can reproduce only by using the replication machinery inside living cells (p. 308).

### Section 13.3

Describe how evolutionary classifications of living organisms are created and tested, and explain how these classifications can be useful to scientists.

- Phylogenies are created and tested by evaluating the shared traits of different species that indicate they shared a recent ancestor, as well as examining DNA sequences for similarities and differences (pp. 310–311).

## Roots to Remember

**The following roots of words come mainly from Latin and Greek and will help you decipher terms:**

| | |
|---|---|
| **anima-** | means breath or soul. Chapter term: *animal* |
| **botan-** | means pertaining to plants. Chapter term: *botanist* |
| **eu-** | means true. Chapter terms: *eukaryote, Eukarya* |
| **in-** | means without. Chapter term: *invertebrate* |
| **-karyo-** | means kernel. Chapter terms: *prokaryote, eukaryote* |
| **myco-** | means fungal. Chapter term: *mycologist* |
| **zoo-** | means pertaining to animals. Chapter term: *zoologist* |

## Learning the Basics

1. How many different species have been identified by science? How many are estimated to exist?

2. Plants experienced an adaptive radiation after the evolution of the flower. Explain what occurred and list the hypotheses that may explain why it occurred.

3. How are hypotheses about the evolutionary relationships among living organisms tested?

4. Which of the following kingdoms or domains is a hodgepodge of different evolutionary lineages?
   **A.** Bacteria; **B.** Protista; **C.** Archaea; **D.** Plantae; **E.** Animalia

5. Comparisons of ribosomal RNA among many different modern species indicate that _____.
   **A.** there are two very divergent groups of prokaryotes; **B.** the kingdom Protista represents a conglomeration of very unrelated forms; **C.** fungi are more closely related to animals than to plants; **D.** A and B are correct; **E.** A, B, and C are correct.

6. On examining cells under a microscope, you notice that they occur singly and have no evidence of a nucleus. These cells must belong to _____.
   **A.** Domain Eukarya; **B.** Domain Bacteria; **C.** Domain Archaea; **D.** Kingdom Protista; **E.** More than one of the above could be correct.

7. The mitochondria in a eukaryotic cell _____.
   **A.** serve as the cell's power plants; **B.** probably evolved from a prokaryotic ancestor; **C.** can live independently of the eukaryotic cell; **D.** A and B are correct; **E.** A, B, and C are correct.

8. Fungi feed by _____.
   **A.** producing their own food with the help of sunlight; **B.** chasing and capturing other living organisms; **C.** growing on their food source and secreting chemicals to break it down; **D.** filtering bacteria out of their surroundings; **E.** producing spores.

9. Which of the following is/are always true?
   **A.** Viruses cannot reproduce outside a host cell; **B.** Viruses are not surrounded by a membrane; **C.** Viruses are not made up of cells; **D.** A and C are correct; **E.** A, B, and C are correct.

10. Phylogenies are created based on the principle that all species descending from a recent common ancestor _____.
    **A.** should be identical; **B.** should share characteristics that evolved in that ancestor; **C.** should be found as fossils; **D.** should have identical DNA sequences; **E.** should be no more similar than species that are less closely related.

## Analyzing and Applying the Basics

1. Unless handled properly by living systems, oxygen can be quite damaging to cells. Imagine an ancient nucleated cell that ingests an oxygen-using bacterium. In an environment where oxygen levels are increasing, why might natural selection favor a eukaryotic cell that did not digest the bacterium but instead provided a "safe haven" for it?

2. Imagine you have found an organism that has never been described by science. The organism, made up of several hundred cells, feeds by anchoring itself to a submerged rock and straining single-celled algae out of pond water. What kingdom would this organism probably belong to, and why do you think so?

3. Support for the endosymbiotic hypothesis for the evolution of eukaryotic groups includes that chloroplasts and mitochondria are surrounded by two membranes, one that derives presumably from the bacterial symbiont and one that derives from the host cell's vacuole that originally surrounded it. Interestingly, some algal phyla, for instance the Euglenophyta, contain chloroplasts composed of three membranes. Given your understanding of endosymbiosis, how do you think this extra membrane may have evolved?

## Connecting the Science

1. Scientists initially rejected the endosymbiotic theory, which is the hypothesis that eukaryotic cells evolved from a set of cooperating independent cells. Most biologists still believe that competition for resources among organisms is the primary force for evolution. Do you think our modern culture, where competition is often valued over cooperation, might make scientists less likely to search for and see the role of cooperation in evolution, or do you think cooperation leading to evolution must be truly rare?

2. Do we have an obligation to future generations to preserve as much biodiversity as possible? Why or why not? Would simply preserving the information contained in an organism's genes (in a zoo or other collection) be good enough, or do we need to preserve organisms in their natural environments?

Answers to **Stop & Stretch**, **Visualize This**, **Working with Data**, **Savvy Reader**, **Sounds Right, But Is It?**, and **Chapter Review** questions can be found in the **Answers** section at the back of the book.

CHAPTER **14**

# Is the Human Population Too Large?

**From space, Earth doesn't look too crowded.**

# Population Ecology

In its most recent estimate in 2012, the United Nations (UN) reported that the world's population is approximately 7.2 billion—more than double of what it was in 1965. The UN also predicted that the population would continue to grow for several more decades before stabilizing at around 11 billion by about 2100. Many observers greeted the report as another piece of bad news. The estimate was higher than slightly older ones; many scientists and environmentalists wonder if our planet can support 7.2 billion people for much longer, let alone 3.8 billion more.

Other commentators are skeptical of environmentalists' statements about population growth. In the middle of the twentieth century, environmentalists looked at the population size—then less than 4 billion people—and predicted food

**But Earth's human population is 7 billion—and rising.**

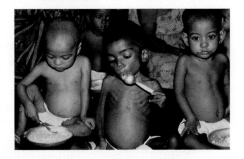

**Are these children hungry because the planet is overpopulated?**

and water shortages by the year 2000. However, most measures of human health have become more upbeat since 1970, including global declines in infant mortality rates, increase in life expectancy, and a 20% rise in per capita income. By most measures, the average person is better off today than they were 40 years ago; why should we believe doom-and-gloom predictions about our future when others have been wrong?

However, there are some indications that the large human population is rapidly reaching a real limit to growth. For example, according to the UN in the period between 2010 and 2012, 870 million people did not get enough food regularly for a healthy existence. A staggering 5.5 million deaths among children under five each year are associated with inadequate

nutrition. Despite years of international attention and billions of dollars spent to address this problem, the situation has not improved dramatically—there are only 13% fewer people suffering from malnutrition today than there were in 1990. And rapidly rising food prices around the globe pose a serious threat to future human health and survival.

So what is the truth? Is the human population larger than Earth can support for much longer? Are we headed into a global food crisis and massive famine? Or are we gradually moving toward an era when all people on Earth will be as well fed, long-lived, and affluent as the average North American?

**Can Earth support everyone at the same level as that of the average North American family?**

# Is the human population larger than Earth can support for much longer?

---

**popula-** means multitude.

---

# 14.1 Population Growth

**Ecology** is the field of biology that focuses on the interactions among organisms as well as those between organisms and environments. The relationship between organisms and their environments can be studied at many levels—from the individual, to populations of the same species, to communities of interacting species, and finally to the effects of biological activities on the nonbiological environment, such as the atmosphere.

In ecology, a **population** is defined as all of the individuals of a species within a given area. Populations exhibit a structure, which includes the spacing of individuals (that is, their distribution) and their density (abundance). Ecologists seek to explain the distribution and abundance of the individuals within populations and to understand the factors that lead to the success or failure of a population. The interactions among species make up one set of influences on distribution and abundance (Chapter 15), but another set is the internal dynamics of the population, including the relative numbers of individuals of different sexes and ages and the numbers that are born or die in a given time period.

## Population Structure

The first task of a population ecologist is to estimate the size of the population of interest. Certain populations can be counted directly, as in a census of people in the United States or a survey identifying all individuals of a particular tree species in a forest tract.

The size of more mobile or inconspicuous species can be estimated by the **mark-recapture method** (**FIGURE 14.1**). In this technique, researchers capture many individuals within a defined area, mark them in some way (for instance, with a dab of paint), and release them back into the environment. Later, the researchers capture another group of individuals in the same area and calculate the proportion of previously marked individuals in this group. This proportion can be used to estimate the size of the total population.

## Visualize This ▼

**Animals may become either "trap-happy," that is, attracted to traps, or "trap-shy," after one capture. How might each of these behaviors affect a population estimate?**

**FIGURE 14.1 The mark-recapture method.** 1. In the first step, a researcher captures, marks, and releases animals in a wild population. 2. Some time later, the researcher returns and determines what percentage of animals trapped again were previously marked. 3. In the second capture, the ratio of marked to total animals captured is approximately the ratio of the original number trapped to the whole population.

① Researcher captures 100 beetles in a trap and marks each with a dab of paint.

② After one week, a trap is set again, resulting in a captured group of marked and unmarked individuals.

③ Total population is estimated as equivalent to the percentage of marked individuals in the second trap.

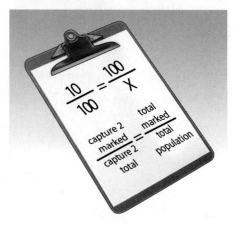

$$\frac{10}{100} = \frac{100}{X}$$

$$\frac{\text{capture 2 marked}}{\text{capture 2 total}} = \frac{\text{total marked}}{\text{total population}}$$

For an example of the mark-recapture method, imagine that a researcher captured, marked, and released 100 beetles in a small area of forest. One week later, the researcher returns to the same place and captures another group of beetles. If she finds that 10% of the beetles caught on the second round are marked, the researcher can assume that the 100 beetles captured and marked initially represented 10% of the entire beetle population. According to this mark-recapture survey, the total population is approximately 1000 beetles.

**STOP** & **STRETCH**   Imagine the same scenario described in the previous paragraph, but the researcher found that only 1% of the beetles were marked in the second round of captures. How large is the population likely to be in this case?

Another basic aspect of population structure is dispersion—that is, how organisms are distributed in space. Many species show a **clumped distribution,** with high densities of individuals in certain resource-rich areas and low densities elsewhere. Plants that require certain soil conditions and the animals that depend on these plants tend to be clumped (**FIGURE 14.2a**). On a global scale, humans show a clumped distribution, with high densities found around transportation resources, such as rivers and coastlines.

The clumped distribution of humans masks a more **uniform distribution** on a local scale; for instance, the spacing between houses in a subdivision or strangers in a classroom tends to equalize the distances among individual property owners or people. Species that show a uniform distribution are often territorial—they defend their own personal space from intruders. We can observe these same strong reactions among certain species of birds at a breeding site (**FIGURE 14.2b**).

Nonsocial species with the ability to tolerate a wide range of conditions typically show a **random distribution,** in which no compelling factor is actively bringing individuals together or pushing them apart. The distribution of seedlings of trees with windblown seeds is often random (**FIGURE 14.2c**).

The distribution and abundance of a population provide a partial snapshot of its current situation. The dispersion of the human population—and recent changes in that pattern—profoundly affects the natural environment (Chapter 16). However, to better understand how a population is responding to its environment, we need to determine how it is changing through time.

**FIGURE 14.2 Patterns of population dispersion.**   Individuals in a population may be (a) clumped, like these cattails growing only in soil with the correct water content; (b) uniformly distributed, like these territorial nesting penguins; or (c) randomly dispersed, like these seedlings grown from windblown seeds in a forest.

**(a) Clumped**

**(b) Uniform**

**(c) Random**

## Exponential Population Growth

Historians have been able to use archaeological evidence and written records to determine the size of the human population at various times during the past 10,000 years. This record, presented in **FIGURE 14.3**, dramatically illustrates the pattern of population growth.

The graph of human population growth is a striking illustration of **exponential growth**—growth that occurs in proportion to the current total. Populations growing exponentially do not add a fixed number of offspring every year; instead, the quantity of new offspring is a proportion of the size of the previous generation and thus is an ever-growing number. Exponential growth results in the J-shaped growth curve seen in Figure 14.3.

**expo-** is from the word meaning to put forth.

## Working with Data ▶

The line looks relatively flat from 8000 B.C.E. to 1500 B.C.E., though the population doubled four times in that period. Was the population growing exponentially at this time?

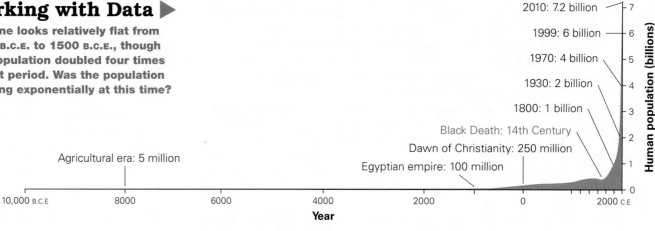

**FIGURE 14.3 Exponential growth.** The number of people on Earth grew relatively slowly until the eighteenth century. The rapid growth since then has occurred in proportion to the total, causing a J-shaped curve.

## Exponential Population Growth: A Closer Look ▼

For most of our history, the human population has remained at very low levels. At the beginning of the agricultural era, about 10,000 years ago, there were approximately 5 million humans. There were 100 million people during the Egyptian Empire (7000 years later) and about 250 million at the dawn of the Christian religion in 1 C.E. The population was growing, but at a very slow rate—approximately 0.1% per year.

Beginning around 1750, the rate at which the human population was growing jumped to about 2% per year. The human population reached 1 billion in 1800, had doubled to 2 billion by 1930, and then doubled again to 4 billion by 1970. Although the current growth rate is slower, about 1.2% per year, the rapid increase in population looks quite dramatic on a graph of human population over time.

The larger a population is, the more rapidly it grows because an increase in numbers depends on individuals reproducing in the population. So, while a growth rate of 1.2% per year may seem rather small, the number of individuals added to the 7.2-billion-strong human population every year at this rate of growth is a mind-boggling 86 million (more than the combined populations of California, Texas, and New York). Put another way, three people are added to the world population per second, and about a quarter of a million people are added every day.

What has fueled this enormous increase in human population? The annual **growth rate** of a population is the percentage change in population size over a single year. Growth rate, which is mathematically represented as $r$, is a function of the birth rate of the population (the number of births as a percentage of the population)

*A Closer Look, continued*

(a)

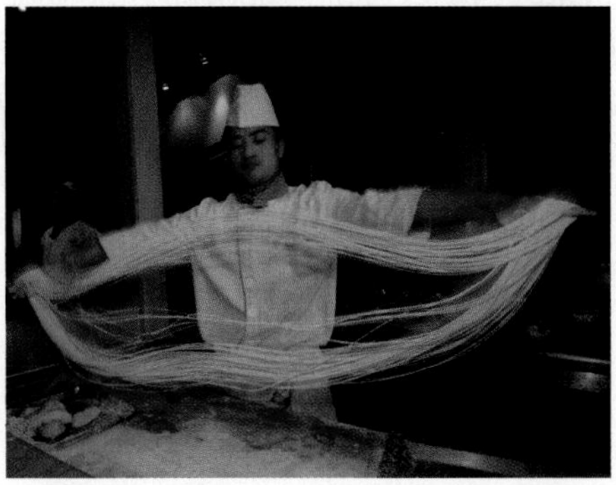

(b)

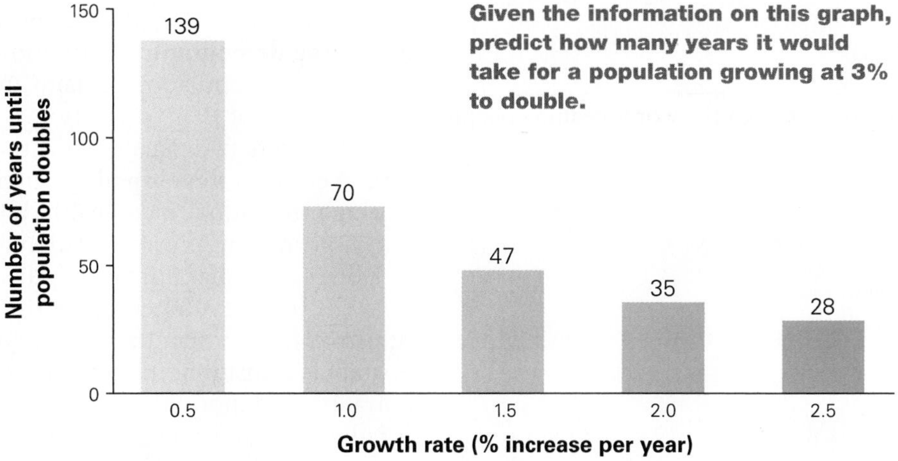


### ◀ Working with Data

**Given the information on this graph, predict how many years it would take for a population growing at 3% to double.**

**FIGURE 14.4 Growth rate and doubling time.** (a) Hand-pulled noodles are made by folding and pulling dough repeatedly, doubling the number of strands in every fold. A single thick noodle folded only 12 times will produce 4096 fine noodles (seen here), illustrating how rapidly growth occurs when it is exponential. (b) The time it takes a population to double in size depends on its growth rate. A population growing at 2% per year doubles in 35 years, while one growing at 1% takes twice as long to double.

minus the death rate (the number of deaths as a percentage of the population). For example, in the entire human population 22 babies are born per year for every 1000 people—that is, the birth rate for the population is 2.2%:

$$\frac{22}{1000} = 0.022 = 2.2\%$$

In addition, each year 10 individuals die of every 1000 people, resulting in a death rate of 1%:

$$\frac{10}{1000} = 0.01 = 1\%$$

This results in the current growth rate of 1.2%:

$$\text{Growth rate} = \text{Birth rate} - \text{Death rate}$$

$$1.2\% = 2.2\% - 1\%$$

Today's relatively high growth rate, compared to the historical average of 0.1%, is the result of a large difference between birth and death rates.

It can be easier to think of exponential growth in terms of the amount of time it takes for a population to double in size. At a growth rate of 0.1, it takes 693 years for the population to double. At a rate of 1.2%, it takes only about 58 years (**FIGURE 14.4**).

## The Demographic Transition

In human populations, the tendency has been for decreases in death rate to be followed by decreases in birth rate. The speed of this adjustment helps to determine population growth in the future. Before the Industrial Revolution, both birth and death rates were high in most human populations. Women gave birth to many children, but relatively few children lived to reach adulthood. Rapid population growth was triggered in the eighteenth century by a dramatic decline in infant mortality (the death rate of infants and children) in industrializing countries. In particular, advances in treating and preventing infectious disease has reduced the number of children who die from these illnesses.

**STOP & STRETCH** Life expectancy is the average age a newborn is expected to reach. Decline in infant mortality dramatically increases life expectancy in a country, even if adults are not living longer. Use your understanding of how averages are calculated to explain why this is the case.

demo- is from the word meaning people.

Not long after death rates declined in these countries, birth rates followed suit, lowering growth rates again. Scientists who study human population growth refer to the period when birth rates are dropping toward lowered death rates as the **demographic transition** (FIGURE 14.5). The length of time that a human population remains in the transition has an enormous effect on the size of that population. Countries that pass through the transition swiftly remain small, while those that take longer can become extremely large. Nearly all **developed countries** (those that have industrial economies and high individual incomes) have already passed through the demographic transition and have low population growth rates.

However, global human population growth rates have remained high because the **less developed countries** (countries that are early in the process of industrial development and have low individual incomes) remain in the demographic transition. In addition, several recent changes have decreased infant mortality even more dramatically. These changes include the use of pesticides to reduce rates of mosquito-borne malaria; immunization programs against cholera, diphtheria, and other fatal diseases; and the widespread availability of antibiotics. As the death rate continues to decline but the birth rate remains at historical levels in less developed countries, population growth rates soar.

The vast majority of future population growth will occur within populations in the less developed world, but these countries are also where the vast majority of food crises are occurring. Are the populations in these countries already too large to support themselves? Answering that question requires an understanding of the factors that limit population growth.

## Working with Data ▶

**Where on this graph is the growth rate of the population close to zero?**

**FIGURE 14.5 The demographic transition.** Improvements in sanitation and medical care in human populations cause a decrease in infant mortality, dropping death rates. Because birth rates remain high, growth rates soar. Eventually, people in these populations respond by decreasing the number of children they have.

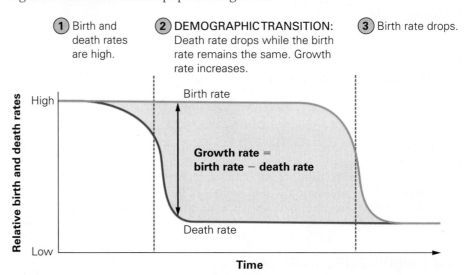

① Birth and death rates are high.

② DEMOGRAPHIC TRANSITION: Death rate drops while the birth rate remains the same. Growth rate increases.

③ Birth rate drops.

## 14.2 Limits to Population Growth

In their studies of nonhuman species, ecologists see clear limits to the size of populations. They can also observe the sometimes awful fates of individuals in populations that outgrow these limits. For this reason, many professional ecologists are gravely concerned about the rapidly growing human population.

You may know of several instances of nonhuman populations outgrowing their food supplies. The elk population in Yellowstone National Park suffered significant winter mortality in the 1990s after it grew so large that it apparently degraded its own rangeland. The massive migrations of Norway lemmings that occur every 5 to 7 years and lead to many lemming deaths result from population crowding. While these animals do not commit "mass suicide," as often assumed, the loss of high-quality food in an area as populations increase incites the lemmings to disperse, and they often meet their death in the process. Even yeast in brewing beer grow large populations that eventually use up their food source and as a result die off during the fermenting process. Let us explore what ecology can tell us about the likelihood of human populations suffering the same fate as elk, lemmings, or yeast.

## Carrying Capacity and Logistic Growth

The examples of elk in Yellowstone and Norway lemmings illustrate a basic biological principle. While populations have the capacity to grow exponentially, their growth is limited by the resources—food, water, shelter, and space—that individuals need to survive and reproduce. The maximum population that can be supported indefinitely in a given environment is known as the environment's **carrying capacity.**

A simplified graph of population size over time in resource-limited populations is S-shaped (**FIGURE 14.6**). This model shows the growth rate of a population declining to zero as it approaches the carrying capacity. In other words, birth rate and death rate become equal, and the population stabilizes at its maximum size, which is mathematically represented as $K$. Not long after ecologists first predicted this pattern of growth, called **logistic growth,** populations of organisms as diverse as flour beetles, water fleas, and single-celled protists were shown in laboratory studies to conform to this projected growth curve.

The declining growth rate near a population's carrying capacity is caused by **density-dependent factors,** which are population-limiting factors that increase in intensity as a population increases in size. Density-dependent factors include limited food supplies, increased risk of infectious disease, and an increase in toxic waste levels. Density-dependent factors cause either declines in birth rate or increases in death rate. In organisms such as fruit flies growing in laboratory culture bottles, high populations lead to increased mortality of the flies as food supplies dwindle and wastes accumulate. *Daphnia* (often known as "water fleas") living in crowded aquariums do not have enough food to support egg production, so birth rates drop. Females of white-tailed deer populations living in crowded natural habitats are less likely to have the energy reserves necessary to carry a pregnancy to term than deer in less crowded environments (**FIGURE 14.7**). Some populations of humans have experienced density-dependent

**Working with Data** ▼

Where on the graph is the number of individuals added to the population *per unit of time* highest?

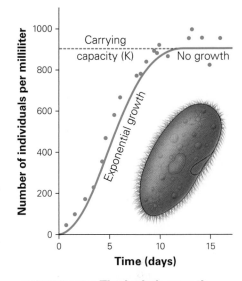

**FIGURE 14.6 The logistic growth curve.**   The graph illustrates the change in the population of *Paramecium* (a single-celled protist) over time in a laboratory culture. The S-shape of the growth curve is due to a gradual slowing of the population growth rate as it approaches the carrying capacity of the environment.

**logo-**   means proportion or ratio.

**(a) Fruit flies**

**(b) *Daphnia***

**(c) White-tailed deer**

**(d) Humans**

**FIGURE 14.7 Limits to growth.** Populations of (a) fruit flies in a laboratory culture, (b) *Daphnia* in an aquarium, and (c) white-tailed deer in the northeastern United States all experience high death rates or low birth rates as their populations approach the carrying capacity of the environment. (d) Do human populations face these same limits?

mortality as well, including the epidemic of infectious bubonic plague that occurred as a result of overcrowding in European cities in the fourteenth century.

Density-dependent factors can be contrasted with **density-independent factors** that influence population growth rates—for instance, severe droughts that increase the death rate in plant populations regardless of their density or increased temperatures that increase the birth rate in insects. The impact of climate change may be to increase the occurrence of density-independent factors as Earth warms and dries and storms become more severe (Chapter 5). However, density-independent factors do not occur in a vacuum; they can have more or less severe effects depending on the size of a population. For example, a density-independent factor such as an unusually cold winter can be deadly to individuals in a white-footed mouse population, but the likelihood of survival is also a function of how much food each individual has stored for the winter. How much food is stored by each animal depends on the density of mice competing for that food during the autumn.

**STOP & STRETCH** Why are infectious diseases considered a density-dependent factor rather than density-independent?

Are density-dependent factors beginning to reduce growth rates in the human population? That is, are humans nearing the carrying capacity of Earth for our population? If we are, will death rates increase as food resources dwindle and more people starve? Or will birth rates decline because fewer women will have enough food to support themselves and a developing baby?

## Earth's Carrying Capacity for Humans

One way to determine if the human population is reaching Earth's carrying capacity is to examine whether, and how rapidly, the growth rate is declining. As we saw in Figure 14.6, the S-shaped curve of population size over time results from a gradually declining growth rate as the population approaches carrying capacity.

Human population growth rates were at their highest in the early 1960s, at 2.1% per year, but they have since declined to the current rate of 1.1%. This steady decline is an indication that the population, though still currently growing, is nearing a stable number. Uncertainty about the future rate of growth has led the UN to produce differing estimates of this number and how soon population stability will be reached (**FIGURE 14.8**). However, the unique characteristics of humanity make it difficult to determine exactly which population size represents Earth's carrying capacity for humans.

### Signs That the Population Is Not Near Carrying Capacity.
The rates of population increase of fruit flies and *Daphnia* in the laboratory will slow as these populations near carrying capacity because their growth rates are forced down by density-dependent factors, such as lack of resources, causing increased death rates or decreased birth rates. However, this is not the case in human populations. Even as the human population has rapidly increased, death rates continue to decline—an indication that people are not running out of food. Growth rates are declining because birth rates are falling even faster than death rates. Unlike the water fleas and white-tailed deer, whose females are unable to have offspring when populations are near carrying capacity, birth rates in human populations are falling because women and families, especially those with adequate resources, are *choosing* to have fewer children.

Another way to determine if humans are near Earth's carrying capacity is to estimate the proportion of Earth's resources used by humans now. The amount

**Working with Data** ▼

**Where on this graph is the population growth rate declining toward zero?**

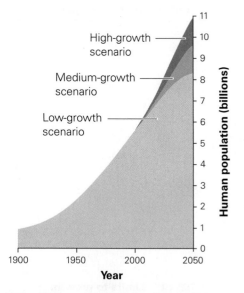

**FIGURE 14.8 Projected human population growth.** The United Nations report is based on a number of uncertainties and returns three projections: a low-growth scenario of 8.3 billion people, medium growth resulting in 9.6 billion, and a high-growth estimate of 10.9 billion.

of food energy available on the planet is referred to as the **net primary productivity (NPP)**. NPP is a measure of plant growth, typically over the course of a single year (**FIGURE 14.9**).

Several different analyses of the global extent of agriculture, forestry, and animal grazing estimate that humans use between one-quarter and one-third of the total land NPP. If we accept these rough estimates, we can approximate that the carrying capacity of Earth is three to four times the present population, or approximately 21 billion people. This theoretical maximum is the total number of humans that could be supported by all of the photosynthetic production of the planet—leaving no resources for millions of other species. Given the dependence of humans on natural systems (explored in Chapter 15), it is unlikely that our species could survive on a planet where no natural systems remained. However, even the largest population projection by the UN, 10.9 billion, falls well short of this theoretical maximum.

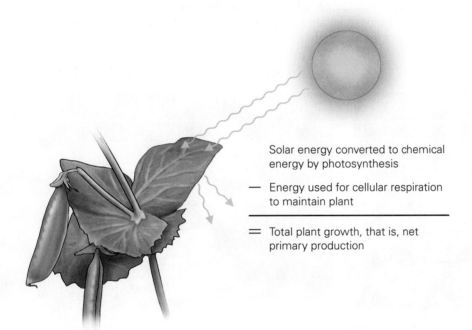

Solar energy converted to chemical energy by photosynthesis

— Energy used for cellular respiration to maintain plant

= Total plant growth, that is, net primary production

**FIGURE 14.9 Net primary production.** NPP is a measure of the total calories available on Earth's surface to support consumers.

**Signs That the Population Is Near Carrying Capacity.** Ecologists caution that the resources required to sustain a population include more than simply food, so the carrying capacity deduced from NPP estimates may be much too high. Humans also need a supply of clean water, clean air, and energy for essential tasks such as heating, food production, and food preservation.

The relationship between population size and the supply of these resources is not as straightforward as the relationship between population and food. For instance, every new person added to the population requires an equivalent amount of clean water, but every new person also introduces a certain amount of pollution to the water supply. We cannot simply divide the current supply of clean water by 10.9 billion to determine if enough will be available in the future because increased population leads to increased pollution and therefore less total clean water.

Furthermore, many essential supplies that sustain the current human population are **nonrenewable resources,** meaning that they are a one-time stock and cannot be easily replaced. The most prominent nonrenewable resource is fossil fuel, the buried remains of ancient plants transformed by heat and pressure into coal, oil, and natural gas.

The use of fossil fuel and other nonrenewable resources is a function not only of the number of people but also of average lifestyles, which vary widely

## Are we headed into a global food crisis and massive famine?

around the globe. For example, Americans make up only 5% of Earth's population but are responsible for 24% of global energy consumption. The average American uses as many resources as 2 Japanese or Spaniards, 3 Italians, 6 Mexicans, 13 Chinese, 31 Indians, 128 Bangladeshis, 307 Tanzanians, or 370 Ethiopians (**FIGURE 14.10**).

Americans also consume a total of 815 billion food calories per day—about 200 billion calories more than is required, or enough to feed an additional 80 million people. Much of modern food production relies on the energy provided by fossil fuel. When these fuels begin to run out, we might find that we need far more of Earth's NPP than we do now to sustain abundant food production. In other words, the actual carrying capacity of our planet may be much lower than our approximations.

The question posed at the beginning of this section remains unanswered; there is no agreement among scientists concerning the carrying capacity of Earth for the human population. Given that uncertainty, what can ecologists tell us about the risks facing the human population that may result from massive, rapid population growth?

American family

Malian family

**FIGURE 14.10 Population plus consumption.** Americans consume more than people in the developing world and therefore have a greater impact on resources.

# 14.3 The Future of the Human Population

Unlike nearly all other species, human populations are not simply at the mercy of environmental conditions. With its ability to transform the natural world, human ingenuity has helped populations circumvent seemingly fixed natural limits. However, ingenuity has a dark side in that it can lull people into believing that nature has an almost infinite capacity to support their ever-growing needs. Managing the growth of human populations before even the most secure of us face environmental and economic disaster requires an understanding of the risks of continued rapid growth and the strategies that help reduce it.

## A Possible Population Crash?

The use of nonrenewable resources creates a risk that the human population will overshoot a still-unknown carrying capacity. Ecologists have long known that when populations have high growth rates, they may continue to add new

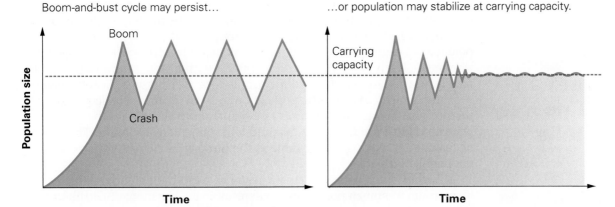

Boom-and-bust cycle may persist...                    ...or population may stabilize at carrying capacity.

**FIGURE 14.11 Overshooting and crashing.** These graphs illustrate rapid population growth followed by a population crash. Over time, the population may stay in a "boom-and-bust" cycle, or it may stabilize at its carrying capacity.

▲ **Working with Data**

**Imagine that in both populations, more resources became available, raising the carrying capacity. How would this change the graphs?**

members even as resources dwindle. This causes the population to grow larger than the carrying capacity of the environment. The members of this large population are then competing for far too few resources, and the death rate soars while the birth rate plummets. This results in a **population crash**, a steep decline in number (**FIGURE 14.11**).

For instance, in some species of *Daphnia,* healthy offspring continue to be born for several days after the food supply becomes inadequate. This occurs because females can use their fat stores to produce additional young. The size of the population continues to rise even when there is no food left for grazing; however, when the young water fleas run out of stored fat transferred from their mothers, most individuals die. For many species with high birth rates, rapid growth followed by dramatic crashes produce a **population cycle** of repeated "booms" and "busts" in number.

A population overshoot and subsequent crash affected the human population on the Pacific island of Rapa Nui (also known as Easter Island) during the eighteenth century (**FIGURE 14.12**). This 150-square-mile island is separated from other landmasses by thousands of miles of ocean; therefore, its people were limited to using only the resources on or near their island. Archaeological evidence suggests that at one time, the human population on Rapa Nui was at least 7000—apparently a number far greater than the carrying capacity of the island. By 1775, the subsequent overuse and loss of Rapa Nui's formerly lush palm forest had resulted in a rapid decline to fewer than 700 people, a population likely much lower than the initial carrying capacity of the island.

Biological populations may also overshoot carrying capacity when there is a time lag between when the population approaches carrying capacity and when it actually responds to that environmental limit. Scientists who study human populations note a lag between the time when humans reduce birth rates and when population growth actually begins to slow. They call this lag **demographic momentum.** The momentum occurs because while parents may be reducing their family size, their children will begin having children before the parents die, causing the population to continue growing. Even when families have an average of two children, just enough to replace the parents, demographic momentum causes the human population to grow for another 60 to 70 years before reaching a stable level.

**FIGURE 14.12 The crash of a human population.** On Rapa Nui, also known as Easter Island, the human inhabitants created these large statues. Soon after completely deforesting this small island, the large population suffered a severe crash.

**STOP & STRETCH** How does demographic momentum explain why species with high birth rates are more likely to experience population cycles compared to populations with low birth rates?

## Or are we gradually moving toward an era when all people on Earth will be well fed?

The demographic momentum of a human population can be estimated by looking at its **population pyramid,** a summary of the numbers and proportions of individuals of each sex and each age group. As **FIGURE 14.13** illustrates, the potential for high levels of demographic momentum occurs when the age structure most closely resembles a true pyramid, with a large proportion of young people. As this large group of young people ages, they become potential parents, with the capacity to cause the population to swell. In more stable populations, the proportion who are young is not significantly larger than the proportion who are middle aged, and the pyramid looks more like a column. The number of "parents-in-waiting" is the same as the number of current parents, keeping the population stable.

Whether our reliance on stored resources and the potential demographic momentum in human populations will result in an overshoot of Earth's carrying capacity—followed by a severe crash, as on the island of Rapa Nui—remains to be seen. But human ecologists already know what factors help to slow population growth, so a crash may become less likely.

## Avoiding Disaster

As discussed previously, when death rates drop in a human population, birth rates eventually follow. Unlike any other species known to science, humans will voluntarily limit the number of babies they produce. When more opportunities become available outside of child rearing, most women delay motherhood and

## Working with Data ▼

According to the graphs, what are the total sizes of the populations of the United States and India?

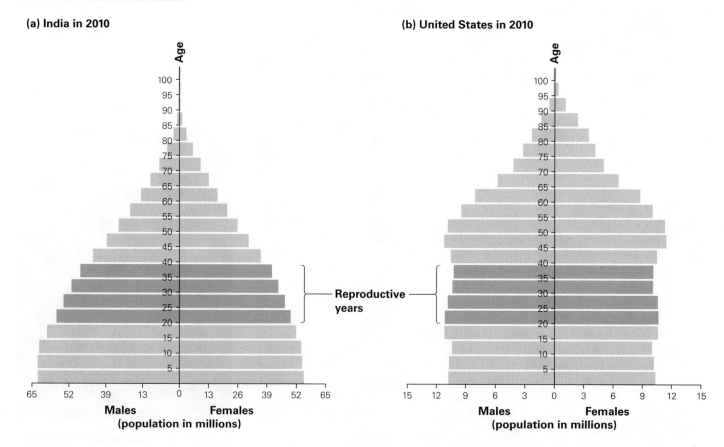

**FIGURE 14.13 Demographic momentum.**   In a rapidly growing human population like that of (a) India in 2010, most of the population is young, and the population will continue to grow as these children reach child-bearing age. In a slower-growing or stable human population, the ages are more evenly distributed, as in (b) the United States in 2010.

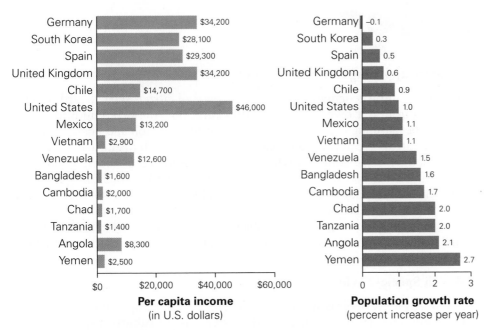

| | Per capita income (in U.S. dollars) | Population growth rate (percent increase per year) | Female literacy (percent of population) |
|---|---|---|---|
| Germany | $34,200 | -0.1 | 99 |
| South Korea | $28,100 | 0.3 | 97 |
| Spain | $29,300 | 0.5 | 97 |
| United Kingdom | $34,200 | 0.6 | 99 |
| Chile | $14,700 | 0.9 | 96 |
| United States | $46,000 | 1.0 | 99 |
| Mexico | $13,200 | 1.1 | 85 |
| Vietnam | $2,900 | 1.1 | 87 |
| Venezuela | $12,600 | 1.5 | 93 |
| Bangladesh | $1,600 | 1.6 | 41 |
| Cambodia | $2,000 | 1.7 | 64 |
| Chad | $1,700 | 2.0 | 13 |
| Tanzania | $1,400 | 2.0 | 62 |
| Angola | $8,300 | 2.1 | 54 |
| Yemen | $2,500 | 2.7 | 30 |

**FIGURE 14.14 Income, growth rate, and women's literacy.** These three graphs illustrate the relationships among income, population growth, and female literacy. Note that higher income and literacy are correlated with decreased birth rates and thus decreased population growth in most countries.

## ▲ Visualize This

**Income and female literacy do not explain all differences among nations in growth rates. Why do you suppose the United States has a relatively high population growth rate compared to other wealthy countries?**

have fewer children. In fact, birth rates are lowest in countries where income is high and women have access to education (**FIGURE 14.14**).

This information provides a clear direction for public policies attempting to decrease population growth rates: improve conditions for women, including increasing access to education, health care, and the job market, and provide them with the information and tools that allow them to regulate their fertility (**FIGURE 14.15**). Slowing growth rates before the human population reaches some environmentally imposed limit has additional benefits. Determining Earth's carrying capacity for humans as simply a function of whether food and water will be available also ignores quality-of-life issues, or what some scientists call cultural carrying capacity. An Earth that was wholly given over to the production of food for the human population would lack wild, undisturbed places and the presence of species that nurture our sense of wonder and discovery. With human populations at the limits of growth, much of our creative energy would be used for survival, taking away our ability to make and enjoy music, art, and literature. (Chapter 24 addresses issues of sustainable food production that face the human population now and which will become more pressing as the population grows.)

Limiting human population growth also leaves room for nonhuman species. Human activity is posing a direct threat to the survival of a significant percentage of Earth's biodiversity (Chapter 15)—a threat that increases in direct proportion to the size and affluence of the planet's human population.

What we have learned is that scientists cannot tell us exactly how many people Earth can support, partly because humans make unpredictable choices and partly because humans have the capacity to innovate and adjust seemingly fixed biological limits. Ultimately, the question of how many people Earth should support—and at what quality of life or including support for nonhuman species—is a question not only of science but also of values and ethics.

**FIGURE 14.15 Reducing population growth through opportunity.** The Grameen Bank in Bangladesh provides small loans to help raise people from poverty. Women make up 97% of the borrowers, and successful women borrowers are less likely to have large families.

# savvy reader

## Using Science to Urge Action

### What's Happening to Birds We Know and Love?

Audubon's unprecedented analysis of 40 years of citizen-science bird population data from our own Christmas Bird Count plus the Breeding Bird Survey reveals the alarming decline of many of our most common and beloved birds.

Since 1967, the average population of the common birds in steepest decline has fallen by 68%; some individual species nosedived as much as 80%. All 20 birds on the national Common Birds in Decline list (including northern bobwhite, whip-poor-will, and common grackle) lost at least half their populations in just four decades.

The findings point to serious problems with both local habitats and national environmental trends. Only citizen action can make a difference for the birds and the state of our future.

### Which Species? Why?

The wide variety of birds affected is reason for concern. Populations of meadowlarks and other farmland birds are diving because of suburban sprawl, industrial development, and the intensification of farming over the past 50 years.

Greater scaup and other tundra-breeding birds are succumbing to dramatic changes to their breeding habitat as the permafrost melts earlier and more temperate predators move north in a likely response to global warming. Boreal forest birds like the boreal chickadee face deforestation from increased insect outbreaks and fire, as well as excessive logging, drilling, and mining.

The one distinction these common species share is the potential to become uncommon unless we all take action to protect them and their habitat.

1. According to this article, how does the National Audubon Society know that bird species are declining?
2. What information is missing from this article that could influence how you respond to this issue?

National Audubon Society, "State of the Birds," summer 2007 update, http://stateofthebirds.audubon.org/

# SOUNDS RIGHT BUT IS IT?

The exponential growth of populations can be challenging to comprehend, in part because exponential increases are not as familiar as simple additive increases. In fact, people often grossly underestimate the effect of exponential growth. Consider the following job offer: A corporate executive needs a temporary assistant for a 30-day assignment. The executive proposes to pay you either (a) $1000/day for 30 days, or (b) to start at 1 cent per day and then double the salary on each subsequent day (such that you earn 1 cent on the first day, 2 cents on the second day, 4 cents on the third day, 8 cents on the fourth day, and so on). You take the first option, reckoning that

**$1000 per day for 30 days is a much better salary than one that starts at 0.001% of that rate and doubles every day for 30 days.**

Sounds right, but it isn't.

1. What is the definition of exponential growth?
2. Which salary proposal is an example of exponential growth?
3. What would your earnings be for the month if you choose proposal (a)?
4. What would your earnings be for the 25th day of the month if you choose proposal (b)? (You may need to make a table to calculate this, but you can also use the following equation: daily wage = $2^n$, where n = the day.)
5. Reflect on your answers to questions 1–4 above and explain why the statement bolded above sounds right, but isn't.

# Chapter Review   MasteringBiology®

Go to the Study Area in MasteringBiology® for practice quizzes, myeBook, BioFlix™ 3-D animations, MP3Tutor sessions, videos, current events, and more.

## Summary

### Section 14.1

Define *population,* and describe the aspects of populations that are typically measured by ecologists.

- A population is defined as a group of individuals of the same species living in a fixed area (p. 318).
- Ecologists measure a population by the number of individuals and their dispersion (pp. 318–319).

Describe how a mark-recapture estimate of population size is performed, and estimate the size of a population from mark-recapture data.

- Population size can be estimated by capturing a sample of the population, marking the individuals in some way, freeing them, and then capturing another sample in the same area. Determining the proportion of marked to unmarked individuals allows calculation of total population size (pp. 318–319).

Explain how the size of a population that is experiencing exponential growth changes over time.

- The human population has grown very rapidly over the last 150 years and exhibits a pattern of exponential growth, which is an increase in numbers as a function of the current population size. On a graph, exponential growth takes the form of a J-shaped curve (pp. 320–321).

Describe the demographic transition in human populations and how it contributes to population growth.

- Human population growth is spurred by decreases in death rate caused by decreased infant mortality. In most populations, this decrease is followed by a decrease in birth rates. The gap between when death rates drop and birth rates follow is called the demographic transition (pp. 321–322).

### Section 14.2

Define *carrying capacity*, and explain how the existence of a carrying capacity causes a pattern of logistic growth as populations grow.

- Nearly all populations eventually reach the carrying capacity of their environment, which is the maximum population that can be supported indefinitely given current conditions. On a graph, logistic growth takes the form of an S-shaped curve (p. 323).

Compare and contrast the effect of density-dependent and density-independent factors on population growth.

- The impact of density-dependent factors depends on the size of the population, whereas density-independent factors affect birth or death rates regardless of population size (pp. 323–324).
- Density-dependent factors include starvation and disease, and density-independent factors include weather (pp. 323–324).

List the evidence that the human population may not be near carrying capacity and the evidence that it may be near carrying capacity.

- The growth rate of the human population is declining, but not as a result of density-dependent factors. Instead, birth rates are dropping because women are choosing to have fewer children (p. 324).
- Rough calculations indicate that the energy received from the sun each year could support a population of 21 billion people, well over the largest population projections (p. 325).
- Humans' reliance on nonrenewable resources may be temporarily inflating the actual carrying capacity of Earth (pp. 325–326).

### Section 14.3

Describe the conditions (in both human and other populations) under which a population crash can occur.

- Fast-growing populations that overshoot their environment's carrying capacity may experience a crash or go through periodic booms and busts (pp. 326–327).
- It is possible that the human population will overshoot Earth's carrying capacity because of our reliance on nonrenewable resources and due to demographic momentum, the potential for future population growth comes from a population consisting primarily of young people (pp. 327–328).

# Roots to Remember

**The following roots of words come mainly from Latin and Greek and will help you decipher terms:**

| | |
|---|---|
| **demo-** | is from the word meaning people. Chapter term: *demographic* |
| **expo-** | is from the word meaning to put forth. Chapter term: *exponential* |
| **logo-** | means proportion or ratio. Chapter term: *logistic* |
| **popula-** | means multitude. Chapter term: *population* |

# Learning the Basics

**1.** What factors have led to the explosive increase in the human population over the past 150 years?

**2.** Explain why a decrease in population growth rate is expected as a nonhuman population approaches carrying capacity.

**3.** When individuals in a population are evenly spaced throughout their habitat, their dispersion is termed as _____.
**A.** clumped; **B.** uniform; **C.** random; **D.** excessive; **E.** exponential.

**4.** A population growing exponentially _____.
**A.** is stable in size; **B.** adds a fixed number of individuals every generation; **C.** adds a larger number of individuals in each successive generation; **D.** will likely expand forever; **E.** will not crash.

**5.** According to **FIGURE 14.16**, the carrying capacity for fruit flies in the environment of the culture bottle is _____.
**A.** 0 flies; **B.** 100 flies; **C.** 150 flies; **D.** between 100 and 150 flies; **E.** impossible to determine.

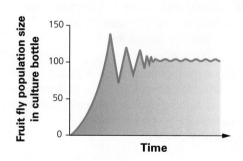

**FIGURE 14.16**

**6.** All of the following are density-dependent factors that can influence population size *except* _____.
**A.** weather; **B.** food supply; **C.** waste concentration in the environment; **D.** infectious disease; **E.** supply of suitable habitat for survival.

**7.** In contrast to nonhuman populations, human population growth rates have begun to decline due to _____.
**A.** voluntarily increasing death rates; **B.** voluntarily decreasing birth rates; **C.** involuntary increases in death rates; **D.** involuntary decreases in birth rates; **E.** density-dependent factors.

**8.** Populations that rely on stored resources are likely to overshoot the carrying capacity of the environment and consequently experience a(n) _____.
**A.** demographic momentum; **B.** cultural carrying capacity; **C.** decrease in death rates; **D.** population crash; **E.** exponential growth.

**9.** The current carrying capacity of Earth for the human population may have been inflated by _____.
**A.** demographic momentum; **B.** the tendency for women to want to control family size; **C.** an artificially low number of density-independent factors; **D.** our use of fossil fuels; **E.** recent population crashes.

**10.** Demographic momentum refers to the tendency for _____.
**A.** low population growth rates to continue to decline; **B.** high population growth rates to continue to increase; **C.** populations to continue to grow in number even when growth rates reach zero; **D.** populations to continue to grow in number even when women are reducing the number of children they bear; **E.** women to continue to have children even though they no longer wish to.

# Analyzing and Applying the Basics

**1.** A researcher captures 50 penguins, marks them with a spot of paint on their bills, and releases them. One month later, the researcher returns, captures another 50 penguins, and notes that only 1 has a previous mark. What is the likely size of the total penguin population in the researcher's study area?

**2.** Consider a population in which, for every thousand individuals in the population, 25 children are born each year and 27 individuals die. What is the growth rate of this population? Do you expect that it is near carrying capacity? Why or why not?

**3.** Review Figure 14.16. How would you expect the carrying capacity of the population to change if the flies are supplied with more food? What other factors might influence the carrying capacity in this environment?

## Connecting the Science

**1.** Review your answer to question 3 in "Analyzing and Applying the Basics." How are the factors that limit fruit fly populations in a culture bottle similar to the factors that limit human populations on Earth? How are they different?

**2.** Africa is the only continent where increases in food production have not outpaced human population growth. Many of the most severe food crises are in African countries. Should those of us in the more developed world assist African populations? How? What factors influence your thoughts on this question?

# CHAPTER 15

# Conserving Biodiversity

**The rufus red knot traverses an incredible 18,000 miles during its yearly migration.**

# Community and Ecosystem Ecology

He might be the closest any single wild bird comes to being a celebrity. Moonbird is a male rufus red knot (*Calidris canutus rufa*) tagged with leg bands by researchers in Argentina in early 1995. By the time Moonbird was spotted in Delaware in May 2014, he had traveled over 400,000 miles over the 19 years that he has annually migrated between the Arctic and Argentina. This little bird has flown the distance to the moon and back! Moonbird's story of survival has inspired many, and has spawned a book, countless news stories, and even a commemorative statue in Mispillion Harbor.

In addition to his stamina, Moonbird is remarkable for his survival. When he was banded, Moonbird was one of an estimated 150,000 birds of his subspecies. By

**One tagged red knot is known to have traversed over 400,000 miles in its lifetime—and has become famous enough to merit its own statue.**

**Moonbird survived a population crash caused when an important food source was overharvested in the 1990s.**

2013, the rufus red knot population was only about 30,000. In 2007, scientists were predicting the total extinction of the birds by 2013. That hasn't happened, but the population is today a fraction of its former size.

The dramatic decline in rufus red knots led to its addition to the list of threatened species in 2013. Already the decline in the rufus red knot population has led to restrictions on human activity. In particular, the Atlantic States Marine Fisheries Commission has limited the harvest of horseshoe crabs, the eggs of which are a primary food source for red knots during their migration. New Jersey went even further, banning horseshoe crab harvesting altogether. These restrictions seem to have made a difference—the population of horseshoe crabs in Delaware Bay has begun to rebound dramatically in the past several years, and red knot

populations, while still low, appear to have stabilized.

The ban on horseshoe crab harvesting is not without cost. The crabs are a valuable commercial species. People who harvested these crabs have lost a major source of income, thanks to concerns about a single shorebird. Said Frank Eicherley, a Delaware crab harvester, "Since all these changes, I can't make a living on the water."

Should the survival of one or a few species come before the needs of humans? Biologists say that the concept of "birds versus people" is wrong. Instead, they argue, protecting endangered species is not about pitting birds against people; it is about protecting birds—and the crabs they depend on—to ensure the survival of all. In this chapter, we explore the causes and consequences of the loss of biological diversity.

**Restrictions on crab harvests seem to have stopped the bird's population crash, but have harmed those who make a living fishing.**

# In addition to his stamina, Moonbird is remarkable for his survival.

**FIGURE 15.1 A critically endangered species.** Black-footed ferrets once numbered in the tens of thousands across the Great Plains of North America but were decimated by human activity. By 1986, they were completely gone from the wild, and only 18 remained alive in captivity.

# 15.1 The Sixth Extinction

The U.S. **Endangered Species Act (ESA)** is a law passed in 1973 to protect and encourage the population growth of threatened and endangered species. Endangered organisms are at high risk of **extinction**, defined as the complete loss of a species or subspecies, while *threatened* organisms are at high risk of becoming endangered. Rapidly declining species such as the rufus red knot are exactly the type of organisms that legislators had in mind when they enacted the ESA.

The ESA was passed because of the public's concern about the continuing erosion of **biodiversity**, the entire variety of living organisms. Whooping cranes, passenger pigeons, black-footed ferrets (**FIGURE 15.1**), and spotted skunks—once abundant species—are extinct or highly threatened in the United States as a result of human activity. The ESA was a response to the unprecedented and rapid rate of species loss at our hands.

Critics of the ESA argue that the goal of saving all species from extinction is unrealistic. After all, extinction is a natural process—the approximately 10 million species living today constitute less than 1% of the species that have ever existed. In the next section, we explore the scientific questions posed by ESA critics: How does the rate of extinction today compare to the rates in the past? Is the ESA just attempting to postpone the inevitable, natural process of extinction?

## Measuring Extinction Rates

If ESA critics are correct in stating that the current rate of species extinctions is "natural" and not a result of human activity, then the extinction rate today should be roughly equal to the rate that existed before humans evolved. The rate of extinction in the past can be estimated by examining the fossil record.

**FIGURE 15.2** illustrates what the fossil record tells scientists about the history of biodiversity as measured by the diversity of marine organisms, which have the longest continuous record. Since the rapid evolution of a wide variety

## ▼ Working with Data

**According to the graph, about how long does it take after a mass extinction for diversity to return to preextinction levels?**

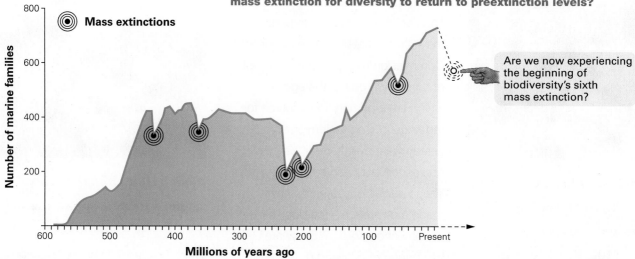

**FIGURE 15.2 Mass extinction.** The history of life on Earth is reflected by this graph plotting the change in the number of families of fossil marine organisms over time. The number of families has increased over the past 600 million years. However, this rise has been punctuated by five mass extinctions (marked here with black circles), each resulting in a global decline in biodiversity.

of animal groups approximately 580 million years ago, the number of families of organisms has generally increased. However, this increase in biodiversity has not been smooth or steady.

The history of life as pieced together by fossils has been punctuated by five **mass extinctions**—species losses that are global in scale, affect large numbers of species, and are dramatic in impact. Past mass extinctions were probably caused by massive global changes—for instance, climate fluctuations that changed sea levels, continental drift that changed ocean and land forms, or an asteroid impact that caused widespread destruction and climate change. Many scientists argue that we are now seeing biodiversity's sixth mass extinction, this one caused by human activity.

Determining whether the current rate of extinction is unusually high requires knowledge of the **background extinction rate,** the rate at which species are lost through the normal evolutionary process. Normal extinctions occur when a species lacks the variability to adapt to environmental change or when a new species arises due to the evolution of its ancestor species. The fossil record can provide clues about the background extinction rate that results from this continual process of species turnover.

The span of rock ages in which fossils of a species are found represents the life span of that species (**FIGURE 15.3**). Biologists have thus estimated that the "average" life span of a species is around 1 million years (although there is tremendous variation). We would therefore expect that the background rate of extinction is about one species per million (0.0001%) per year. Some scientists have argued that this estimate is too low because it is based on observations of fossils, a record that may be biased toward long-lived species. However, it remains the best-supported approximation of background extinction rates.

The current rate of extinction is calculated from known species disappearances. Calculating current rates is a challenge because extinctions are surprisingly difficult to document. The only way to conclude that a species no longer exists is to exhaustively search all areas where it is likely to have survived. In the absence of a complete search, most conservation organizations have adopted this standard: To be considered extinct, no individuals of a species must have been seen in the wild for 50 years.

A few searches for specific species give hints to the recent extinction rate. In Malaysia, a 4-year search for 266 known species of freshwater fish turned up only 122. In Africa's Lake Victoria, 200 of 300 native fish species have not been seen for years. On the Hawaiian island of Oahu, half of 41 native tree snail species have not been found, and in the Tennessee River, 44 of the 68 shallow-water mussel species are missing. Despite these results, few of the missing species in any of these searches are officially considered extinct.

The most complete records of documented extinction occur in groups of highly visible organisms, primarily mammals and birds. Since 1600, of an approximate 4500 identified mammal species, 83 have become extinct, while 113 of approximately 9000 known bird species have disappeared. The known extinctions of mammals and birds, spread out over the 400 years of these records, correspond to a rate of 0.005% per year. Compared to the background rate of extinctions calculated from the fossil record, the current rate of extinction is 50 times higher. If we examine the past 400 years more closely, we see that the number of mammals and birds lost, and thus the extinction rate, has actually increased over time (**FIGURE 15.4**, on the next page). In the years 1850 to 2000, the number of species lost translates to about 0.01% of the total per year, making the current rate 100 times higher than the background rate.

There are reasons to expect that the current elevated rate of extinction will continue into the future. The World Conservation Union (known by its French acronym, IUCN), a highly respected global organization composed of and funded by government agencies and nongovernmental organizations from over

▼ **Working with Data**

What would be the new estimated life span of this species if another fossil were found in rocks dated only 12.1 million years old?

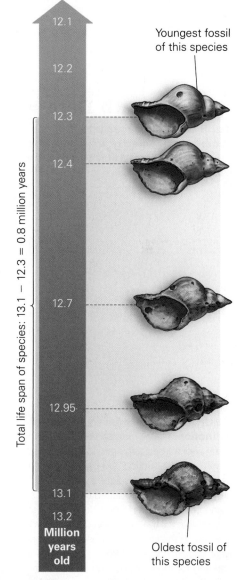

Youngest fossil of this species

Total life span of species: 13.1 − 12.3 = 0.8 million years

Oldest fossil of this species

**FIGURE 15.3 Estimating the life span of a species.** Fossils of the same species are arranged on a timeline from oldest to youngest. The difference in age between the oldest and youngest fossil of a species is an estimate of the species' life span.

**FIGURE 15.4 Modern extinctions.** This graph illustrates the number of species of mammals and birds known to have become extinct since 1600.

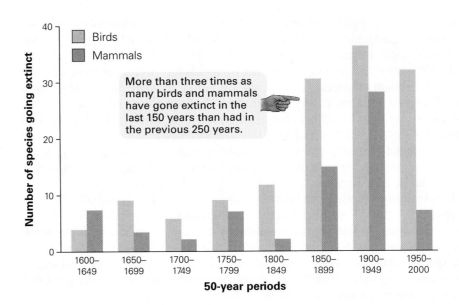

More than three times as many birds and mammals have gone extinct in the last 150 years than had in the previous 250 years.

## Working with Data ▶

**Use the graph to estimate the number of species of mammals that went extinct from 1600 to 1849 and from 1850 to 2000.**

140 countries, collects and coordinates data on threats to biodiversity. According to the IUCN's most recent assessment, 29% of amphibians, 21% of mammals, 12% of all bird species, and 3% of all plant species, including 40% of conifers, are in danger of extinction.

## Causes of Extinction

The IUCN attempts to identify all the threats that put a particular species at risk of extinction. For most, some type of human activity is to blame. The most severe threats belong to one of four general categories: loss or degradation of habitat, introduction of nonnative species, overexploitation, and effects of pollution (**FIGURE 15.5**). Of these categories, the first poses the most serious threat; the

**FIGURE 15.5 The primary causes of extinction.** (a) A road carved through formerly undisturbed rainforest is the first step toward the destruction and fragmentation of habitat for dozens of unique species. (b) The brown tree snake introduced to the Pacific island of Guam from Papua New Guinea is responsible for the extinction of native bird species on the island. (c) These tiger skins were illegally harvested from southeast Asia, primarily for the Chinese market. (d) Polar bears hunt for seals, their primary prey, from sea ice. Some bears contain high levels of persistent pollutants that appear to have disrupted their reproductive organs. Global warming has also reduced the extent of sea ice in the Arctic Ocean over the past 20 years, further threatening the bears' survival.

**(a) Habitat degradation**

**(b) Introduced species**

**(c) Overexploitation of species**

**(d) Pollution**

IUCN estimates that 83% of endangered mammals, 89% of endangered birds, and 91% of endangered plants are directly threatened by damage to or destruction of the species' **habitat**, the place where they obtain their food, water, shelter, and space.

### Habitat Destruction and Fragmentation.

Habitat destruction is a major risk to species around the globe (**FIGURE 15.6**). Rates of habitat destruction caused by agricultural, industrial, and residential development accelerated everywhere throughout the twentieth century as the human population swelled from less than 2 billion in 1900 to over 7 billion today. As the amount of natural landscape declines, the number of species supported by the habitats in these landscapes naturally decreases.

The relationship between the size of a natural area and the number of species that it can support follows a general pattern called a **species-area curve**. A species-area curve for reptiles and amphibians on a West Indies archipelago is illustrated in **FIGURE 15.7a**. Similar graphs have been generated in studies of different groups of organisms in a variety of habitats. The general pattern in all these graphs is that the number of species in an area increases rapidly as the size of the area increases, but the rate of increase slows as the area becomes very large. This rule of thumb is shown in **FIGURE 15.7b**. We can use this curve to estimate rates of extinction in regions that are rapidly being modified by human activity but difficult to survey extensively. For example, from the graph in

**FIGURE 15.6 Lost habitat = lost species.**   Lemurs are found in the wild only on the island of Madagascar, first settled by humans 1500 years ago. Of the 48 species of lemur present on the island 2000 years ago, 16 have become extinct, and 15 are at risk of extinction.

## ▼ Working with Data

**Using part (a) of this figure, calculate how many species of reptiles and amphibians you would expect to find on an island that is 15,000 square kilometers in area. Imagine that humans colonize this island and dramatically degrade 7500 square kilometers of the natural habitat. Using part (b), calculate the percentage of the species originally found on the island you would expect to become extinct.**

**(a) Species diversity increases with area.**

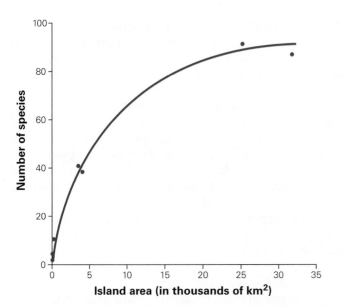

**(b) Habitat reduction is predicted to result in loss of species.**

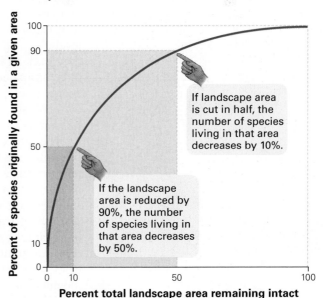

**FIGURE 15.7 Predicting extinction caused by habitat destruction.**   (a) This curve demonstrates the relationship between the size of an island in the West Indies and the number of reptile and amphibian species living there. (b) We use a generalized species-area curve to roughly predict the number of extinctions in an area experiencing habitat loss.

Figure 15.7b, we can estimate that a 90% decrease in landscape area will cut the number of species living in the remaining area by half.

Using images from satellites, scientists have estimated that approximately 20,000 square kilometers (about 7722 square miles, an area the size of Massachusetts) of rain forest are cut each year in South America's Amazon River basin. At this rate of habitat destruction, tropical rain forests will be reduced to 10% of their original size within about 35 years. According to the species-area curve, this habitat loss translates into the extinction of about 50% of species living there. If this prediction proves accurate, the extinct species in the rain forest would include about 50,000 of the 250,000 known species of plants, 1800 species of birds, and 900 species of mammals.

Of course, habitat destruction is not limited to tropical rain forests. Fresh-water lakes and streams, grasslands, and temperate forests are also experiencing high levels of modification. Many of the beaches that serve as stopover sites for the rufus red knot are impacted by human development and use, contributing to the threats this species faces. According to the IUCN, if habitat destruction around the world continues at its present rate, nearly one-fourth of *all* living species will be lost within the next 50 years.

Some critics have argued that these estimates of future extinction are too high because not all groups of species are as sensitive to habitat area as the curve in Figure 15.7b suggests. Many species may still survive and even thrive in human-modified landscapes. Other biologists contend that there are other threats to species, including **habitat fragmentation**, and therefore the rate of species loss is likely to be even higher than these estimates.

Habitat destruction rarely results in the complete loss of a habitat type. Often what results from human activity is habitat fragmentation, in which large areas of intact natural habitat are subdivided. Habitat fragmentation is especially threatening to large predators, such as grizzly bears and tigers, because of their need for large hunting areas.

Large predators require large, intact hunting areas due to a basic rule of biological systems: Energy flows in one direction within an ecological system along a **food chain,** which typically runs from the sun to **producers** (photosynthetic organisms), to the **primary consumers** that feed on them, to **secondary consumers** (predators that feed on the primary consumers), and so on. Along the way, most of the calories taken in at one **trophic level** (that is, a level of the food chain) are respired simply to support the activities of the individuals at that level. In other words, a substantial amount of the energy individuals take in is dissipated as heat. Each level of the food chain thus has significantly less energy available to it compared to the next lower level. You can see this in your own life; an average adult needs to consume between 1600 and 2400 kilocalories per day simply to maintain his or her current weight.

The flow of energy along a food chain leads to the principle of the **trophic pyramid,** the bottom-heavy relationship between the **biomass** (total weight) of populations at each level of the chain (**FIGURE 15.8**). Habitat destruction and fragmentation can cause the lower levels of the pyramid to shrink, depriving the top predators of adequate calories for survival.

Habitat fragmentation also exposes mobile species to additional dangers. For example, the American crocodile population is threatened by high levels of direct mortality, the death of individuals. Of the 143 recorded crocodile deaths in Florida between 1967 and 2007, the majority (97) were hit by cars while moving from one fragment of habitat to another.

Even the species that do survive in small fragments of habitat are made more susceptible to extinction by isolation. Habitat fragmentation often makes it impossible for individuals to follow changes in food sources or available nesting or growing sites. Isolated populations are also subject to a lack of genetic diversity resulting from inbreeding, as we discuss in detail later in the chapter.

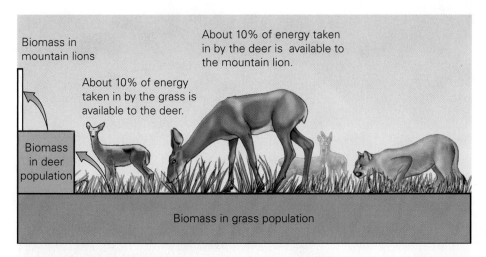

Biomass in mountain lions

About 10% of energy taken in by the deer is available to the mountain lion.

About 10% of energy taken in by the grass is available to the deer.

Biomass in deer population

Biomass in grass population

**FIGURE 15.8 A trophic pyramid.** Because most of the energy consumed by a trophic level is used within that level for maintenance, biomass decreases as position in the food chain increases.

◀ **Working with Data**
Imagine that the population of deer is 10,000 in this ecosystem. How many mountain lions would you expect if the relationship illustrated holds true?

Although habitat destruction and fragmentation are the gravest threats to endangered species, the others are not insignificant. According to the IUCN, activity unrelated to habitat modification plays a role in about 40% of all cases of endangerment. These activities include problems caused by species that actually do well in our presence.

**Introduced Species.** Introduced species include organisms brought by human activity, either accidentally or purposefully, to new environments. Introduced species are often dangerous to native species because they have not coevolved with them.

**Coevolution** occurs when pairs or groups of species adapt to each other via natural selection. For instance, many birds on oceanic islands such as Hawaii and New Zealand evolved in the absence of ground predators and have not evolved behaviors or other strategies to escape or combat these predators—therefore, introduced carnivores can rapidly deplete them via hunting. In Hawaii, the Pacific black rat, accidentally introduced from the holds of visiting ships, became very adept at raiding eggs from nests and contributed to the extinction of dozens of species of honeycreepers, birds found nowhere else on Earth. Even domestic cats, deliberately introduced by people who sought to control rodents around their houses, can take an enormous toll on wildlife. A recent study by the Smithsonian Conservation Biology Institute estimated that free-roaming domestic cats kill between 1.4 and 3.7 billion birds in the continental United States every year.

Introduced species may also compete with native species for resources, causing populations of the native species to decline. For example, zebra mussels, accidentally introduced to the Great Lakes through the ballast water in European trading ships, crowd out native mussel species as well as other organisms that filter algae from water, and the introduced vine kudzu, deliberately brought to the United States from Japan, overgrows native trees and vines in the Southeast (**FIGURE 15.9**, on the next page).

Humans continue to move species around the planet. As the global trade in agriculture and other goods continues to expand, the number of species introductions is likely to increase over the next century.

**Overexploitation.** When the rate of human use of a species outpaces its reproduction, the species experiences **overexploitation**. Overexploitation may occur when particular organisms are highly prized by humans, for example, as exotic pets or for their medicinal value.

## Visualize This ▶

Can you name any introduced species in your own region that have a negative impact on native species?

**FIGURE 15.9 An introduced species.** Kudzu is the common name for the vine *Pueraria lobata,* introduced from Japan to the southeast United States as an ornamental plant and promoted as a solution to soil erosion. In the absence of its predators and parasites, kudzu grows extremely rapidly, almost 50 centimeters per day, and effectively smothers other plants.

**FIGURE 15.10 Beneficial blood.** The blue tint of horseshoe crab blood derives from copper, which performs the same function as iron in our own blood. The blood from these animals is valuable as a screening tool to spot bacterial contamination in vaccines and medical devices.

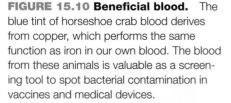

In addition to their value as bait, horseshoe crabs have a surprising medicinal use that led to their overexploitation before the restrictions were put in place. The crab's sky-blue blood contains an enzyme that reacts to certain disease-causing bacteria by forming a clot around the microbes. Biomedical companies harvest the blood and use it to screen medical devices and vaccines for the presence of bacteria (**FIGURE 15.10**). While the horseshoe crabs are returned to the ocean after their blood is collected, perhaps up to 50% do not survive the procedure.

Overexploitation is also likely when an animal competes directly with humans—as in the case of the gray wolf, which was nearly exterminated in the United States by human hunters determined to protect their livestock. The red knot and other species that cross international boundaries during migration are also susceptible to overexploitation because no single government regulates the total harvest; the same is true for species, like the horseshoe crab, that live in the oceans. The near extinction of several whale species in the nineteenth century was the result of the unregulated harvest of these animals by many nations. Stocks of cod, swordfish, and tuna are now similarly threatened.

**Pollution.** The release of poisons, excess nutrients, and other wastes into the environment—a practice otherwise known as **pollution**—poses an additional threat to biodiversity. For example, the herbicide atrazine poisons frogs and salamanders in agricultural areas of the United States, and nitrogen pollution caused by overuse of fertilizer and car and smokestack exhaust has led to drastic declines of certain plant species within native grasslands in Europe.

Excess nutrients flowing into water threaten sea life in surprising ways. The nutrients cause increased production of algae, some of which produce toxins that can accumulate in small animals and poison the wildlife and people that consume them. Thousands of rufus red knots were killed along their migration route after feeding on mussels containing high levels of one such toxin.

Algae can be harmful even if they do not produce toxins. When a large algae population dies, bacteria feeding on dead cells consume the majority of oxygen in the water. This process of oxygen-depleting **eutrophication** results in large fish kills. Eutrophication threatens animals in hundreds of waterways all over the United States. In the Gulf of Mexico, for example, fertilizer from farm fields in the midwestern United States creates a low-oxygen "dead zone" the size of New Jersey at the mouth of the Mississippi River every summer.

Perhaps the most abundant pollutant released by humans is carbon dioxide, also a principal cause of global climate change (see Chapter 5). Computer models that link predicted changes in climate to known ranges and requirements of over 1000 species of plants indicate that 15 to 37% of these plants face extinction in the next century as the climate changes. The rufus red knot is not immune from the threat of increasing carbon dioxide—from the impact of rising sea levels on the beaches it relies upon to the dramatic changes in the Arctic tundra, where it makes its nest, climate change could cause declines in the birds in several ways.

**eu-** means true or good.

**-trophic** means food, nourishment.

**STOP & STRETCH**  About 1/3 of the species of amphibians (frogs, toads, and salamanders) around the world are threatened or endangered, much higher than the rates in birds and mammals. Describe some human activities that might be affecting these animals. What information would help you determine whether these widely dispersed species were put at risk by the same cause?

**New Jersey banned horseshoe crab harvesting altogether.**

By nearly any measure, ESA critics who describe modern extinction rates as "natural" are incorrect. Over the past 400 years, humans have caused the extinction of species at a rate that far exceeds past rates. Human activities continue to threaten thousands of additional species around the world. Earth appears to be on the brink of a sixth mass extinction of biodiversity—and the massive global change responsible is human activity.

Many people feel a moral responsibility to minimize the human impact on other species and therefore support conservation. However, there is also a practical, human-centered reason to prevent the sixth extinction from occurring—the loss of biodiversity can hurt us as well.

# 15.2 The Consequences of Extinction

Concern over the loss of biodiversity is not simply a matter of an ethical interest in nonhuman life. Humans have evolved with and among the variety of species that exist on our planet, and the loss of these species often results in negative consequences for us.

## Loss of Resources

The horseshoe crab has a clear direct benefit to humans, as already detailed. Among the biological resources that are harvested directly from natural areas are wood for fuel and lumber, shellfish for protein, algae for gelatins, and herbs for medicines. The loss of any of these species affects human populations economically. One estimate places the value of wild species in the United States at $87 billion a year or about 4% of the gross domestic product.

Wild species also provide resources for humans in the form of unique biological chemicals. In fact, a number of valuable drugs, food additives, and industrial products derive from wild species. One dramatic example is the rosy periwinkle

**(a) Rosy periwinkle**

**(b) Teosinte**

**(c) Boll weevil wasp**

(*Catharanthus roseus*), which evolved on the island of Madagascar, one of the regions on Earth where biodiversity is most endangered (**FIGURE 15.11a**). Two drugs extracted from this plant, vincristine and vinblastine, have dramatically reduced the death rate from leukemia and Hodgkin's disease, two forms of cancer. If wild species go extinct before they are well studied, we will never know which ones might have provided compounds that would improve human lives.

Wild relatives of domesticated plants and animals, such as agricultural crops and cattle, are also important resources for humans. Genes and alleles that have been "bred out" of domesticated species are often still found in their wild relatives. These genetic resources are a reservoir of traits that can be reintroduced into agricultural species through breeding or genetic engineering. Agricultural scientists attempting to produce better strains of wheat, rice, and corn look to the wild relatives of these crops for genes conferring pest resistance and improved yields. For example, the Mexican teosinte species *Zea diploperennis* (**FIGURE 15.11b**), is an ancestor of modern corn. This species of teosinte is resistant to several viruses that plague cultivated corn; the genes that provide this resistance have been transferred to our domestic plants via hybridization.

By preserving wild relatives of domesticated crops in their natural habitats, scientists can also find resources that reduce pest damage and disease on the domestic crop. For example, the wasp *Catolaccus grandis* consumes boll weevils and is used to control infestations of these pests in cotton fields (**FIGURE 15.11c**). *C. grandis* was discovered in the tropical forest of southern Mexico, where it parasitizes a similar pest in wild cotton populations.

Of course, introducing an insect such as *C. grandis* into a new environment carries risk, even if the introduction is meant to reduce environmental damage. We have already noted many examples of environmental disasters caused by introduced species. Often, a less risky approach to reducing pest damage to crops is to preserve nearby habitats and the ecological interactions that persist there.

## Predation, Mutualism, and Competition

Although humans receive direct benefits from thousands of species, most threatened and endangered species are probably of little or no use to people. While birders and biologists would mourn its extinction, the loss of the rufus red knot would not likely cause direct harm to anyone.

In reality, most species are beneficial to humans because they are part of a biological **community,** consisting of all the organisms living together in a particular habitat area. Within a community, each species occupies a particular **ecological niche,** which can be thought of as the role or "job" of the species. The complex linkage among organisms inhabiting different niches in a community is often referred to as a **food web** (**FIGURE 15.12**). As with a spider's web, any disruption in one strand of the web of life is felt by other portions of the web. Some tugs on the web cause only minor changes to the community, while others can cause the entire web to collapse. Most commonly, losses of strands in the web are felt by a small number of associated species. The story of the horseshoe crab and red knot already shows us that. But some disruptions caused by the loss of seemingly insignificant species have the potential to be felt even by humans.

**FIGURE 15.11 Resources from nature.** (a) Anticancer drugs vincristine and vinblastine were first isolated from rosy periwinkle (*Catharanthus roseus*), a species of flower native to Madagascar. (b) Teosinte is the ancestor of modern corn, and is found in wild populations throughout Central America. This plant contains genes that might confer resistance to disease and drought in corn. (c) *Catolaccus grandis* was discovered preying on boll weevils on wild cotton in southern Mexico. This wasp is now released for the biological control of boll weevils on cotton crops throughout the world.

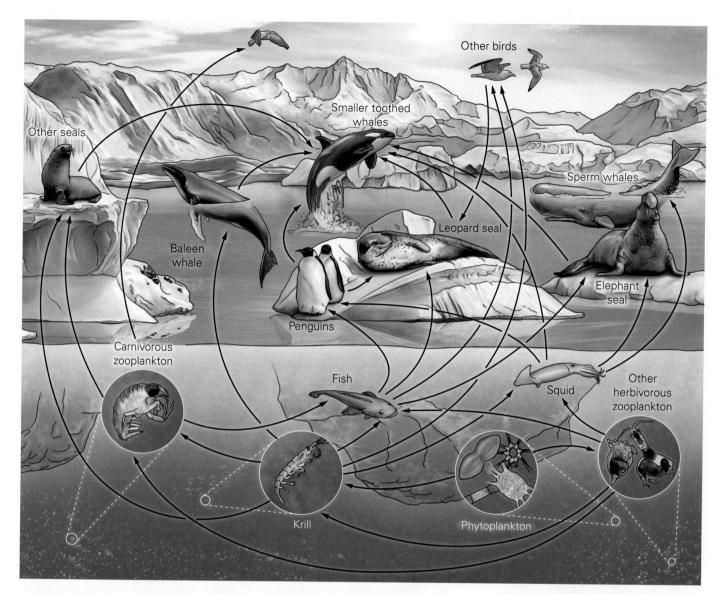

**FIGURE 15.12 The web of life.** Species are connected to other species in food chains—a network of food chains forms a food web. This drawing shows the feeding connections among species in the Antarctic Ocean. Black arrows represent feeding relationships; for example, penguins eat fish and in turn are eaten by leopard seals.

## Visualize This ▲

**Predict what would happen to the other species in this web if baleen whales went extinct.**

**Mutualism: How Bees Feed the World.** An interaction between two species that benefit each other is called **mutualism.** Mutualism can be contrasted with **commensalism**, a relationship in which one species benefits and the other is unaffected—for instance, the relationship between cattle egrets (a species of bird), and domestic cattle. The egrets follow the cattle as they graze, feeding on insects stirred up from the ground by these animals (**FIGURE 15.13**). The cattle do not appear to benefit or be harmed by the egrets' presence.

We find examples of mutualism in many environments. Cleaner fish that remove and consume parasites from the bodies of larger fish, fungi called mycorrhizae that increase the mineral absorption of plant roots while consuming the plant's sugars, and ants that find homes in the thorns of acacia trees and defend the trees from other insects are all examples of mutualism. The mutualistic interaction between plants and bees is perhaps the most important to us.

Bees occupy a very important ecological niche as the primary pollinators of many species of flowering plants. The role of pollinators is to transfer sperm, in

**FIGURE 15.13 Commensalism.**
Cattle egrets in close association with cattle. Birds follow cows, eating insects stirred up by the activity. Cattle are not affected.

the form of pollen grains, from one flower to the female reproductive structures of another flower. The flowering plant benefits from this relationship because insect pollination increases the number of seeds that the plant produces. The bee benefits by collecting excess pollen and nectar to feed itself and its relatives in the hive (**FIGURE 15.14**).

Wild bees pollinate at least 80% of all the agricultural crops in the United States, providing a net benefit of about $8 billion. In addition, populations of wild honeybees have a major and direct impact on many more billions of dollars of agricultural production around the globe.

Unfortunately, bees have suffered dramatic declines in recent years. Steady declines in wild and domesticated bee populations occurred from the 1970s through 2005; however, since 2006, the declines have reached crisis proportions, with up to 33% of captive bee colonies perishing as a result of colony collapse disorder (CCD). The exact causes of these dramatic die-offs are not known but are believed to result from an increased level of bee **parasites** (infectious organisms that cause disease or drain energy from their hosts), competition with the invading Africanized honeybees ("killer bees"), pesticide pollution, and habitat destruction. The extinction of populations of either wild or domesticated bees that are mutualists of crop plants would be extremely costly to humans.

**Predation: How Songbirds May Save Forests.** A species that survives by eating another species is typically referred to as a **predator**. The word conjures up images of some of the most dramatic animals on Earth: cheetahs, eagles, and killer whales. You might not picture wood warblers, a family of North American bird species characterized by their small size and colorful summer plumage, as predators; however, these beautiful songsters are voracious consumers of insects (**FIGURE 15.15a**). The hundreds of millions of individual warblers in the forests of North America collectively remove thousands of kilograms of insects from forest trees and shrubs every summer.

Most of the insects that warblers eat prey on plants. By reducing the number of insects in forests, warblers reduce the damage that insects inflict on forest plants. Reducing the amount of damage likely increases the growth rate of the trees. Harvesting trees for paper and lumber production fuels an industry worth over $200 billion in the United States alone. At least some of this wood was produced because warblers were controlling insects in forests (**FIGURE 15.15b**).

Many species of forest warblers are experiencing declines in abundance. The loss of warbler species has several causes, including habitat destruction in their summer habitats in North America and their winter habitats in Central and South America. Warblers also face increased predation by human-associated animals such as raccoons and housecats. Although other, less vulnerable birds may increase in number when warblers decline, these birds are typically less insect dependent. If smaller warbler populations correspond to lower forest

**FIGURE 15.14 Mutualism.** Honeybees transfer pollen, allowing a plant to "mate" with another plant some distance away.

**Benefit to bee:**
It obtains food in the form of nectar and excess pollen.

**Benefit to flower:**
Its sperm (within the pollen) is carried to the female reproductive structures of another flower, enabling cross-pollination.

**(a) Black-throated blue warbler, predator of insects**

**(b) Forests suffer when insects are unchecked by predators.**

**FIGURE 15.15 Predation.** (a) The black-throated blue warbler is one of many warbler species native to North American forests. These birds are active predators of plant-eating insects. (b) Insects can kill trees, as seen in this photo of a spruce budworm infestation. Warblers and other insect-eating birds probably reduce the number and severity of such insect outbreaks.

growth rates and higher levels of forest disease, then these tiny, beautiful birds definitely have an important effect on the human economy.

### Competition: How a Deliberately Infected Chicken Could Save a Life.
When two species of organisms both require the same resources for life, they will be in **competition** for the resources within a habitat. In general, competition limits the size of competing populations. To determine whether two species that seem to be using the same resource are competing, we remove one from an environment. If the population of the other species increases, then the two species are competitors.

We may imagine lions and hyenas fighting over a freshly killed antelope or weeds growing in our vegetable gardens as typical examples of competition, but most competitive interactions are invisible. The least visible competition occurs among microorganisms. However, microbial competition is often essential to the health of both people and ecological communities.

*Salmonella enteritidis* is a leading cause of food-borne illness in the United States. Between 2 million and 4 million people in this country are infected by *S. enteritidis* every year, experiencing fever, intestinal cramps, and diarrhea as a result. In about 10% of cases, the infection results in severe illness requiring hospitalization. Four to six hundred Americans die as a result of *S. enteritidis* infection every year.

Most *S. enteritidis* infections result from consuming undercooked poultry products, especially eggs. The U.S. Centers for Disease Control and Prevention estimate that as many as 1 in 50 consumers is exposed to eggs contaminated with *S. enteritidis* every year. Surprisingly, most of these eggs look perfectly normal and intact. These pathogens contaminate the egg when it forms inside the hen. Thus, the only way to prevent *S. enteritidis* from contaminating eggs is to keep it out of hens.

A common way to control *S. enteritidis* is to feed hens antibiotics—chemicals that kill bacteria. However, like most microbes, *S. enteritidis* strains can evolve drug resistance that makes them more difficult to kill off. But there is another way to reduce *S. enteritidis* infection in poultry—make sure another species is occupying its niche.

Most *S. enteritidis* infections originate in an animal's gut. If another bacterial species is already monopolizing the food and available space in a hen's digestive system, then *S. enteritidis* will have trouble colonizing there. Following this principle, some poultry producers now intentionally infect hens' digestive systems with harmless bacteria, a practice called **competitive exclusion,** to reduce *S. enteritidis* levels in their flocks. This technique involves feeding cultures of benign bacteria to 1-day-old birds. When the harmless bacteria become established in the niche of the chicks' intestines, the chicks will be less likely to host large *S. enteritidis* populations (**FIGURE 15.16**). There is evidence that this practice is working; *S. enteritidis* infections in chickens have dropped by nearly 50% in the United Kingdom, where competitive exclusion is common practice.

The competitive exclusion of *S. enteritidis* in hens mirrors the role of some human-associated bacteria, such as those that normally live within our intestines and genital tracts. For instance, many women who take antibiotics for a bacterial infection will then develop vaginal yeast infections because the antibiotic kills noninfectious bacteria as well, including species that normally compete with yeast. Maintaining competitive interactions between larger species can be important for humans as well. For instance, in temporary ponds, the main competitors for the algae food source are mosquitoes, tadpoles, and snails. In the absence of tadpoles and snails, mosquito populations can become quite large—potentially with severe consequences because these insects may carry deadly diseases such as malaria, West Nile virus, and yellow fever. With frogs, toads, and their tadpoles increasingly endangered, this risk is a real one.

**STOP & STRETCH**   Examine the web of relationships among organisms depicted in Figure 15.12. Which of the following species pairs are likely competitors? In each case, describe what the competition involves. *A.* leopard seals, penguins; *B.* baleen whales, squid; *C.* toothed whales, leopard seals; *D.* sperm whale, elephant seals; *E.* fish, krill. How could you test your hypothesis that these animals are in competition with each other?

## Visualize This ▼

**Why does the total number of bacteria level off over time in both graphs?**

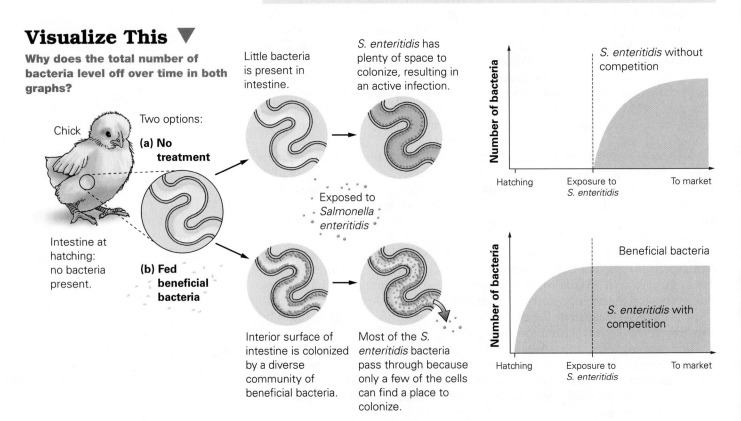

**FIGURE 15.16 Competition.**   If poultry producers feed very young chicks non-disease-causing (beneficial) bacteria, the beneficial bacteria take up the space and nutrients in the intestine that would be used by *S. enteritidis*.

**Keystone Species: How Wolves Feed Beavers.** **TABLE 15.1** summarizes the major types of ecological interactions among organisms. However, this table emphasizes the effects of each interaction on the species directly involved; it does not illustrate that many of these interactions may have multiple indirect effects.

Look again at the food web pictured in Figure 15.12. None of the species in the Antarctic Ocean's biological community is connected to only one other species—they all eat something, and most of them are eaten by something else. You can imagine that penguins, by preying on squid, have a negative effect on elephant seals, which they compete with for these squid, and a more indirect positive effect on other seabirds, which compete with squid for krill. The existence of these **indirect effects** of varying importance has led ecologists to

**TABLE 15.1** **Types of species interactions and their direct effects.**

| Interaction | Example | Effect on Species 1 | Effect on Species 2 |
|---|---|---|---|
| **Commensalism:** Association increases the growth or population size of one species and does not affect the other. | 1. Remora 2. Shark | + As the shark feeds somewhat sloppily, the remora can collect the scraps. | 0 The shark seems to suffer no negative effects from its hitchhiker. |
| **Mutualism:** Association increases the growth or population size of both species. | 1. Ants 2. Acacia tree | + The swollen thorns of the acacia provide shelter for the ants. The acacia leaves provide "protein bodies" that the ants harvest for food. | + Ants kill herbivorous insects and destroy competing vegetation, benefiting the acacia. |
| **Predation and Parasitism:** Consumption of one organism by another. | 1. Brown bear 2. Salmon | + The brown bear catches the salmon and eats it, obtaining nourishment. | − The salmon does not survive. |
| **Competition:** Association causes a decrease or limitation in population size of both species. | 1. Dandelion 2. Tomato plant | − The dandelion does not grow as well in the presence of the tomato plant. Dandelion produces fewer seeds and fewer offspring. | − The tomato plant does not grow optimally in the presence of the weed. Tomato plant produces fewer flowers and fruit. |

**FIGURE 15.17 Keystone species.**
(a) The keystone in an archway helps to stabilize and maintain the arch. (b) A keystone species, such as wolves in Yellowstone National Park, helps to stabilize and maintain other species in an ecosystem.

(a)

(b)

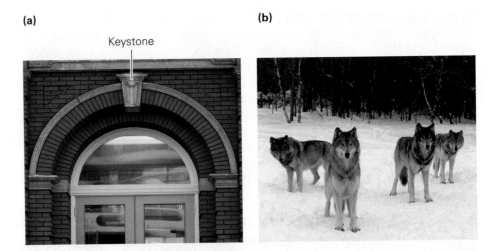

Keystone

hypothesize that, in at least some communities, the activities of a single species can play a dramatic role in determining the composition of the system's food web. These organisms are called **keystone species** because their role in a community is analogous to the role of a keystone in an archway (**FIGURE 15.17a**). Remove the keystone, and an archway collapses; remove the keystone species, and the web of life collapses. It is very difficult to predict which species in an intact ecosystem may be a keystone species, but biologists can point to several examples that became apparent after a species disappeared. One example is the population of gray wolves in Yellowstone National Park (**FIGURE 15.17b**).

Gray wolves were exterminated within Yellowstone National Park by the mid-1920s because of a systematic campaign to rid the American West of this occasional predator of livestock. However, by the 1980s, thanks to insights gained from the science of ecology and a new interest in environmental health, attitudes about the wolf had changed. A new appreciation of the role of wolves in natural systems led to renewed interest in returning wolves to their historical homeland. In the mid-1990s, 31 wolves originally trapped in Canada were released into Yellowstone National Park, and more were released in surrounding areas. Thanks to protection from hunting and the wolves' own adaptability, by the end of 2010 the number had grown to at least 1500 wolves, living both in the park and in surrounding public lands, and the animal was no longer considered endangered in the area.

During the time that wolves were extinct in Yellowstone Park, biologists noticed dramatic declines in populations of aspen, cottonwood, and willow trees. They attributed this decline to an increase in predation by elk, especially during winter when grasses become unavailable. However, just a few years after wolf reintroduction, aspen, cottonwood, and willow tree growth has rebounded in some areas of the park, even though the wolf population was still too low to make a major dent in elk populations. Besides the regions near active wolf dens, the areas of the park that saw the greatest recovery include places on the landscape where elk have limited ability to see approaching wolves or to escape. Thus, the elk will stay away from these areas to avoid wolf predation. Wolves, primarily by changing elk behavior, appear to be important to maintaining large populations of hardwood trees in Yellowstone Park.

The rebound of aspen, cottonwood, and willow populations in Yellowstone has effects on other species as well. Beaver rely on these trees for food, and their populations appear to be growing in the park after decades of decline. Warblers, insects, and even fish that depend on shelter, food, and shade from these trees are increasing in abundance as well. Wolves in Yellowstone appear to fit the profile of a classic keystone species, one whose removal had numerous and surprising effects on biodiversity.

# Energy and Chemical Flows

As the examples in the previous section illustrate, the extinction of a single species can have sometimes surprising effects on other species in a habitat. What may be even less apparent is how the loss of seemingly insignificant species can change the environmental conditions on which the entire community depends.

Ecologists define an **ecosystem** as all of the organisms in a given area, along with their nonbiological environment. The function of an ecosystem is described in terms of the rate at which energy flows through it and the rate at which nutrients are recycled within it. The loss of some species can dramatically affect both of these ecosystem properties.

**eco-** means home or habitation.

**Energy Flow.** In nearly all ecosystems, the primary energy source is the sun. Producers convert sun energy into biomass during the process of photosynthesis. The chemical energy thus captured is passed through trophic levels in the ecosystem. Biomass is partitioned among the trophic levels, with most of it residing at the bottom of the pyramid (review Figure 15.8). As we move up the pyramid, only a portion of the energy available at one level can be converted into the biomass of the next level. The amount of biomass at the producer level effectively determines how large the population of organisms at the highest level can be.

The amount of sunlight reaching the surface of Earth and the availability of water at any given location are the major determiners of both trophic pyramid structure and energy flow through it. (Chapter 16 provides a summary of how variance in sunlight and water availability leads to differences in Earth's ecosystem types.) However, the biodiversity found in an ecosystem can also have strong effects on energy flow within it.

Studies in grasslands throughout the world have provided convincing evidence that loss of species can affect energy flow. By comparing experimental prairie gardens planted with the same total number of individual plants but with different numbers of species, scientists at the University of Minnesota and elsewhere have discovered that the overall plant biomass tends to be greater in more diverse gardens. This research indicates that a decline in diversity, even without a decline in habitat, may lead to less energy being made available to organisms higher on the food chain, including people who depend on wild-caught food.

**Nutrient Cycling.** When essential mineral nutrients for plant growth pass through a food web, they are generally not lost from the environment—hence the term **nutrient cycling**. FIGURE 15.18 (on the next page) illustrates the nitrogen nutrient cycle in a natural prairie. Here, the element moves from inorganic forms in the soil, such as ammonia, into plants, where it is converted to organic forms, and then typically moves from one living organism to the next contained in food. It finally returns to its inorganic form, thanks to the activities of decomposers.

Nitrogen is a major component of protein, and abundant protein is essential for the proper growth and functioning of all living organisms. Nitrogen is, therefore, often the nutrient that places an upper limit on production in most ecosystems—more nitrogen generally leads to greater production, while areas with less available nitrogen can support fewer plants (and therefore animals).

**STOP & STRETCH** Inorganic nitrogen fertilizer applied to agricultural fields is made from an industrial process using nitrogen gas in the atmosphere. A portion of the nitrogen contained in these agricultural crops is eaten by humans and excreted in urine, which in developed countries is essentially released into nearby waterways (after waste treatment). Modify Figure 15.18 to illustrate how processing nitrogen gas and flushing wastes change the nitrogen cycle.

**Visualize This ▶**

Consider how people in cities obtain nutrition and what happens to their waste after digestion and remains after death. How do these features of modern human societies change nutrient flows from the natural cycle pictured here?

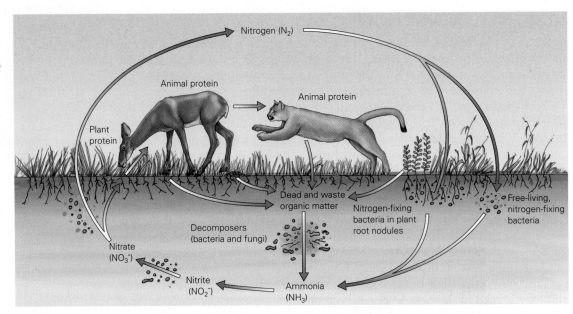

**FIGURE 15.18 Nutrient cycling.** Nutrients such as nitrogen, shown here, are recycled in an ecosystem, flowing from soil to producers to consumers and then back into the soil, where complex nutrients are decomposed into simpler forms.

**(a) Invasive earthworms absent**

**(b) Invasive earthworms present**

**FIGURE 15.19 Changes in ecosystem function.** Notice how barren of living vegetation the worm-infested forest floor appears. One reason for this dramatic change may be a disruption in the native nutrient cycle by this introduced species.

You should note as you review Figure 15.18 that plants absorb simple molecules from the soil and incorporate them into more complex molecules. These complex molecules move through the food web with relatively minor changes until they return to the soil. Here, complex molecules are broken down into simpler ones by the action of **decomposers**, typically bacteria and fungi.

Changes in the soil community can greatly affect nutrient cycling and thus the survival of certain species in ecosystems. Scientists investigating the effects of introduced earthworms, which have invaded forests throughout the northeastern United States, have observed dramatic reductions in the diversity and abundance of plants on the forest floor (**FIGURE 15.19**). The introduced earthworms have apparently caused changes to the community of native soil organisms. As a result of these changes, the nutrient cycle is disrupted and the native plant community has suffered.

The loss of biodiversity clearly can have profound effects on the health of communities and ecosystems on which humans depend. However, controversy exists over whether the current extinction may negatively affect our own psychological well-being.

## Psychological Effects

Some scientists argue that the diversity of living organisms sustains humans by satisfying a deep psychological need. One of the most prominent scientists to promote this idea is Edward O. Wilson, who calls this instinctive desire to commune with nature **biophilia.**

Wilson contends that people seek natural landscapes because our distant ancestors evolved in similar landscapes (**FIGURE 15.20**). According to this hypothesis, ancient humans who had a genetic predisposition driving them to find diverse natural landscapes were more successful than those without this predisposition because diverse areas provide a wider variety of food, shelter, and tool resources. Wilson claims that we have inherited this genetic imprint of our preagricultural past.

**FIGURE 15.20 Is our appreciation of nature innate?**   Humans evolved in a landscape much like this one in East Africa. Some scientists argue that we have an instinctive need to immerse ourselves in the natural world.

While there is no evidence of a genetic basis for biophilia, there is evidence that our experience with nature has psychological effects. A recent literature review noted 50 controlled studies in which positive benefits of interacting with nature were measured, from reduced stress levels to increased longevity. Individual experiences with pets and houseplants indicate that many people derive great pleasure from the presence of nonhuman organisms. Although not conclusive, these studies and experiences are intriguing since they suggest that a continued loss of biodiversity could make life in human society less pleasant overall.

**-philia** means affection or love.

The consequences of the loss of biodiversity are not confined to our generation. The fossil record illustrated in Figure 15.2 reveals that it takes 5 to 10 million years to recover the biological diversity lost during a mass extinction. The species that replaced those lost in previous mass extinctions were very different. For instance, after the mass extinction of the reptilian dinosaurs, mammals replaced them as the largest animals on Earth. We cannot predict what biodiversity will look like after another mass extinction. The mass extinction we may be witnessing today will have consequences felt by people in thousands of generations to come.

# 15.3  Saving Species

So far in this chapter, we have established the possibility of a modern mass extinction occurring, and we have described the potentially serious costs of this loss of biodiversity to human populations. Because the sixth extinction is largely a result of human activity, reversing the trend of species loss requires political and economic, rather than scientific, decisions. But what can science tell us about how to stop the rapid erosion of biodiversity?

**Should the survival of one species come before the needs of humans?**

## Protecting Habitat

Without knowing exactly which species are closest to extinction and where they are located, the most effective way to prevent loss of species is to preserve as many habitats as possible. The same species-area curve that is used to estimate the future rate of extinction also gives us hope for reducing this number.

According to the curve in Figure 15.7b, species diversity declines rather slowly as habitat area declines. Thus, in theory we can lose 50% of a habitat but still retain 90% of its species. This estimate is optimistic because habitat destruction is not the only threat to biodiversity, but the species-area curve tells us that if the rate of habitat destruction is slowed or stopped, extinction rates will slow as well.

**Protecting the Greatest Number of Species.** Given the growing human population, it is difficult to imagine a complete halt to habitat destruction. However, biologist Norman Myers and his collaborators have concluded that 25 biodiversity "hot spots," natural areas making up less than 2% of Earth's surface, contain up to 50% of all mammal, bird, reptile, amphibian, and plant species (**FIGURE 15.21**). Hot spots occur in areas of the globe where favorable climate conditions lead to high levels of plant production, such as rain forests, and where geological factors have resulted in the isolation of species groups, allowing them to diversify.

Stopping habitat destruction in biodiversity hot spots could greatly reduce the global extinction rate. By focusing conservation efforts on hot spot areas at the greatest risk, humans can very quickly prevent the loss of a large number of species. Of course, even with habitat protection, many species in these hot spots will likely become extinct anyway for other human-mediated reasons.

In the long term, we must find ways to preserve biodiversity while including human activity in the landscape. One option is **ecotourism,** which encourages travel to natural areas in ways that conserve the environment and improve the well-being of local people. Some hot spot countries, such as Costa Rica and Kenya, have used ecotourism to preserve natural areas and provide much-needed jobs; other countries have been less successful.

While preserving hot spots may greatly reduce the total number of extinctions, this approach has its critics, who say that by promoting a

## Visualize This ▼

Does it appear that there are more hot spots closer to the equator or closer to the poles?

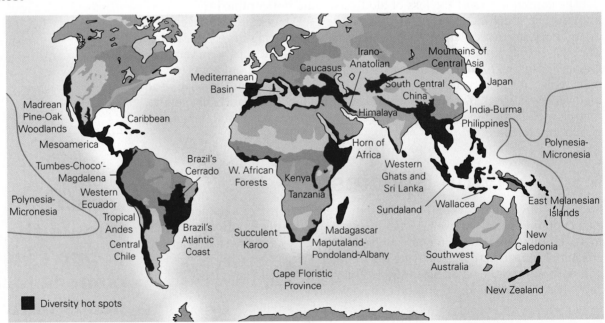

**FIGURE 15.21 Diversity "hot spots."** This map shows the locations of 25 regions around the world that have many unique species and that contain biodiversity hot spots. The hot spots themselves are the undeveloped areas within these regions. Notice how unevenly these regions of high biodiversity are distributed.

strategy that focuses intensely on small areas, we risk losing large amounts of biodiversity elsewhere. These critics promote an alternative approach—identifying and protecting a wide range of ecosystem *types*—designed to preserve the greatest range of biodiversity rather than just the largest number of species.

**Protecting Habitat for Critically Endangered Species.** Although preserving a variety of habitats ensures less extinctions, already-endangered species require a more individualized approach. The ESA requires the U.S. Department of the Interior to designate critical habitats for endangered species—that is, areas in need of protection for the survival of the species. The amount of critical habitat that becomes designated depends on political as well as biological factors.

The biological part of a critical habitat designation includes conducting a study of habitat requirements for the endangered species and setting a population goal for it. The U.S. Department of the Interior's critical habitat designation has to include enough area to support the recovery population. However, federal designation of a critical habitat results in the restriction of human activities that can take place there. The U.S. Department of Interior has the ability to exclude some habitats from protection if there are "sufficient economic benefits" for doing so—a decision that is political in nature.

**Decreasing the Rate of Habitat Destruction.** Preserving habitat is not simply the job of national governments that designate protected areas or of private conservation organizations that purchase at-risk habitats. All of us can take actions to reduce habitat destruction and stem the rate of species extinction. Conversion of land to agricultural production is a major cause of habitat destruction, so eating lower on the food chain and reducing your consumption of meat and dairy products from grain-fed animals is one of the most effective actions you can take. Reducing the use of wood and paper products and limiting consumption of these products to those harvested sustainably (that is, in a manner that preserves the long-term health of the forest) can help slow the loss of forested land.

Other measures to decrease the rate of habitat destruction require group effort. For instance, increased financial aid to developing countries may help slow the rate of habitat destruction. This would allow poor countries to invest money in technologies that decrease their use of natural resources, for example, cooking technologies that reduce the need to harvest large amounts of woody plants for fuel. Strategies that slow the rate of human population growth offer more ways to avoid mass extinction. You can participate in group conservation efforts by joining nonprofit organizations focused on these issues, writing to politicians, and educating others.

Although protecting habitat from destruction can reduce extinction rates for species on the brink of extinction, preserving habitat is not enough. Populations can become so small that they can disappear, even with adequate living space. Recovery plans for a critically endangered species may set a short-term goal of a stable population of 500 individuals or more. To understand why at least this many individuals are required to save these species from extinction, we need to review the special problems of small populations.

## Small Populations Are Vulnerable

The growth rate of an endangered species influences how rapidly that species can attain a target population size. Horseshoe crabs have relatively high growth rates and will meet their population goals quickly if the environment is ideal

(**FIGURE 15.22a**). For slower-growing species, such as the California condor (**FIGURE 15.22b**), populations may take decades to recover.

The rate of recovery is important because the longer a population remains small, the more it is at risk of experiencing a catastrophe that could eliminate it entirely. The story of the heath hen in the United States is a case study on just this point.

The heath hen was a small wild chicken that once ranged from Maine to Virginia. In the eighteenth century, the heath hen population numbered in the hundreds of thousands of individuals. Continued human settlement of the eastern seaboard resulted in the loss of habitat, causing a rapid and dramatic decline of the birds' population. By the end of the nineteenth century, the only remaining heath hens lived on Martha's Vineyard, a 100-square-mile island off the coast of Cape Cod, Massachusetts. Farming and settlement on the island further reduced the habitat for heath hen breeding. By 1907, only 50 birds were present on the island.

In response to this precipitous decline, Massachusetts established a 2.5-square-mile reserve for the remaining birds on Martha's Vineyard in 1908. The response initially seemed effective; by 1915, the population had recovered to nearly 2000 individuals. However, beginning in 1916, a series of disasters struck. First, fire destroyed much of the remaining habitat on the island. The following winter was long and cold, and an invasion of starving predatory goshawks further reduced the heath hen population. Finally, a poultry disease introduced by imported domestic turkeys wiped out much of the remaining population. By 1927, only 14 heath hens remained—almost all males. The last surviving member of the species was spotted for the last time on March 11, 1932.

The final causes of heath hen extinction were natural events—fire, harsh weather, predation, and disease. But it was the population's small size that doomed it in the face of these relatively common challenges. A population of 100,000 individuals can weather a disaster that kills 90% of its members but leaves 10,000 survivors, but a population of 1000 individuals will be nearly eliminated by the same circumstances. Even when human-caused losses to the heath hen population were halted, the species' survival was still precarious.

## Visualize This ▼

**Why is the growth rate of condors so much slower than the growth rate of horseshoe crabs?**

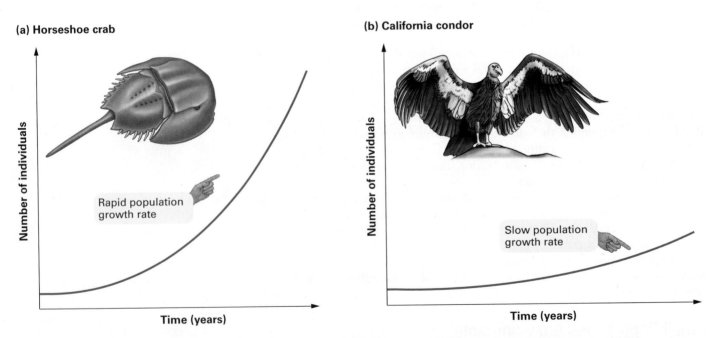

**(a) Horseshoe crab**

Number of individuals

Rapid population growth rate 👉

Time (years)

**(b) California condor**

Number of individuals

Slow population growth rate 👉

Time (years)

**FIGURE 15.22 The effect of growth rate on species recovery.** (a) This graph illustrates the rapid growth of a hypothetical population of quickly reproducing horseshoe crabs. (b) The slow growth rate of the California condor has made the recovery of this species a long process. Today, nearly 30 years after recovery efforts began, the population of wild condors is still only in the dozens.

Small populations of endangered species can be protected from the fate that befell the heath hen. Having additional populations of the species at sites other than Martha's Vineyard would have nearly eliminated the risk that *all* members of the population would be exposed to the same series of environmental disasters. This is the rationale behind placing captive populations of endangered species at several different sites. For instance, captive whooping cranes are located at the U.S. National Biological Service's Patuxent Wildlife Research Center in Maryland, the International Crane Foundation in Wisconsin, the Calgary Zoo in Canada, and the Audubon Center for Endangered Species Research in New Orleans.

Even with multiple habitats, if populations of endangered species remain small in number, they are subject to a more subtle but potentially equally devastating situation—the loss of genetic variability.

## Conservation Genetics

A species' **genetic variability** is the sum of all of the alleles and their distribution within the species. For example, the gene that determines your ABO blood type comes in three different forms. A population containing all three of these alleles contains more genetic variability than does a population with only two alleles.

The loss of genetic variability in a population is a problem for two reasons. On an individual level, low genetic variability leads to low fitness for two reasons: because homozygotes are more likely to express deleterious, mutant alleles, and because homozygotes generally have a narrower range of environments in which they can survive. On a population level, rapid loss of genetic variability may lead to extinction for two related reasons: due to the combined effects of the low fitness of individuals—a problem known as **inbreeding depression**—and because the population as a whole does not contain variants that will survive in changed environmental conditions, which are inevitable (**TABLE 15.2**).

**TABLE 15.2** **The costs of low genetic diversity.**

| Subject of Risk | Outcomes of Low Diversity | Example |
|---|---|---|
| Individuals | **Inbreeding depression:** More homozygotic genes → more deleterious alleles being expressed. | Cheetah populations with low genetic diversity exhibit poor sperm quality and low cub survival. |
| | **Loss of heterozygote advantage:** Only one allele per gene = smaller range of suitable environments possible. | Humans without one copy of the sickle-cell allele are more susceptible to malaria. |
| Populations | **Extinction vortex:** Inbreeding depression leads to a small population → vulnerable to "accidents" | Depressed heath hen population could not weather a series of unpredicted environmental challenges. |
| | **Inability to respond to environmental change:** No genetic variation = no adaptation | Human populations without type A blood allele don't evolve resistance to cholera and plague. |

## Conservation Genetics: A Closer Look ▼

### The Importance of Individual Genetic Variability

**Fitness** refers to an individual's ability to survive and reproduce in a given set of environmental conditions (Chapter 11). Low individual genetic variability decreases fitness for two reasons. As an introduction, we can use an analogy to illustrate them.

First, imagine that you could own only two jackets (**FIGURE 15.23**). If both are blazers, then you would be well prepared to meet a potential employer. However, if you had to walk across campus to your job interview in a snowstorm, you would be pretty uncomfortable. However, if you own one warm jacket and one blazer, you are ready for freezing weather as well as a job interview. In a way, individuals experience the same advantages when they carry two different functional alleles for a gene. If each allele codes for a functional protein, a heterozygous individual produces two slightly different proteins that perform essentially the same function. This phenomenon is known as **heterozygote advantage.**

The second reason high individual genetic variability increases fitness is that, in many cases, one allele for a gene is **deleterious**—that is, it produces a protein that is not very functional. In our jacket analogy, a nonfunctional, deleterious allele is equivalent to a badly torn jacket. If you have this jacket and an intact one, at least you have one warm covering (**FIGURE 15.24**). In this case, heterozygosity is valuable because a heterozygote still carries one functional allele. Often, deleterious alleles are recessive, meaning that the activity of the functional allele in a heterozygote masks the fact that a deleterious allele is present (see Chapter 7). An individual who is homozygous (carries two identical copies of a gene) for the deleterious allele will have low fitness—in our analogy, two torn jackets and nothing else. For both of these reasons, when individuals are heterozygous for many genes (or, in our analogy, have two choices for all clothing items), the cumulative effect is greater fitness relative to individuals who are homozygous for many genes.

Heterozygosity declines in small populations over time. When related individuals mate—known as **inbreeding**—the chance that their offspring will be homozygous for any allele is relatively high. In cheetahs, high levels of inbreeding have decreased fitness by causing poor sperm quality and low cub survival, both likely due to increased expression of deleterious alleles. The costs of inbreeding are seen in humans as well; the children of first cousins have higher rates of homozygosity and higher mortality rates (thus lower fitness) than children of unrelated parents. In a small population of an endangered species, inbreeding often causes low rates of survival and reproduction and can seriously hamper a species' recovery.

**Being heterozygous may confer higher fitness for responding to a changing environment.**

**Homozygote 1: Relatively low fitness**
(only one type of jacket in wardrobe)

**Homozygote 2: Relatively low fitness**
(only one type of jacket in wardrobe)

**Heterozygote: Relatively high fitness**
(two types of jackets in wardrobe)

**FIGURE 15.23 Heterozygotes can inhabit a wider range of environments.** In this analogy, each jacket represents an allele. Just as having two different jackets prepares you for a wider range of situations than having only one type, two different alleles for the same gene may allow for optimal function over a wider range of conditions.

*A Closer Look, continued*

**Being heterozygous may confer higher fitness by masking deleterious recessive alleles.**

**Homozygote 1: Relatively high fitness**
(two functional jackets in wardrobe)

**Homozygote 2: Relatively low fitness**
(two nonfunctional jackets in wardrobe)

**Heterozygote: Relatively high fitness**
(one functional jacket in wardrobe)

**FIGURE 15.24 Heterozygotes avoid the deleterious effects of recessive mutations.**   Again, each jacket represents an allele. If one type of jacket you can receive is nonfunctional, it is better to receive no more than one of them. Heterozygotes are likewise protected against the likelihood of having only nonfunctional alleles, as is present in homozygous, functional recessive individuals.

## How Variability Is Lost via Genetic Drift

Small populations lose genetic variability because of **genetic drift,** a change in the frequency of an allele that occurs simply by chance within a population. While genetic drift can be a process for causing evolutionary change (Chapter 12), in a small population, genetic drift can have detrimental consequences also.

Imagine two human populations in which the frequency of blood-type allele *A* is 1%—that is, only 1 of every 100 blood-type genes in the population is the *A* form (we use the symbol $I^A$ for this allele). In the first population of 20,000 individuals, there are 40,000 total blood group genes (that is, two copies per individual). At 1% frequency, this population contains 400 $I^A$ alleles.

In the second, much smaller population of only 200 individuals, only four copies of the allele are present. If the chance of passing on any given allele in both populations is equivalent to a coin flip, the chance that the $I^A$ allele is not passed on in the population of 20,000 is equivalent to flipping 400 "heads" in succession; in the small population, it is equivalent to only four heads in a row. Genetic drift is more severe in small populations and is much more likely to result in the complete loss of alleles (**FIGURE 15.25**).

**Visualize This ▲**

Homozygotes for the normal allele have high fitness, while homozygotes for the nonfunctional recessive allele have low fitness. Explain the difference in fitness in these two types of homozygotes.

**Working with Data ▼**

Use the graph to determine how much diversity is lost due to genetic drift in a population of 60 individuals over 100 generations.

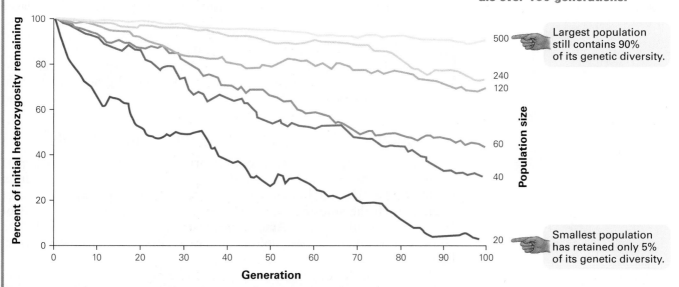

**FIGURE 15.25 Genetic drift affects small populations more than large populations.**   In this graph, each line represents the average of 25 computer simulations of genetic drift for a given population size. After 100 generations, a population of 500 individuals still contains 90% of its genetic variability. In contrast, a population of 20 individuals has less than 5% of its original genetic variability.

*(continued on the next page)*

*A Closer Look, continued*

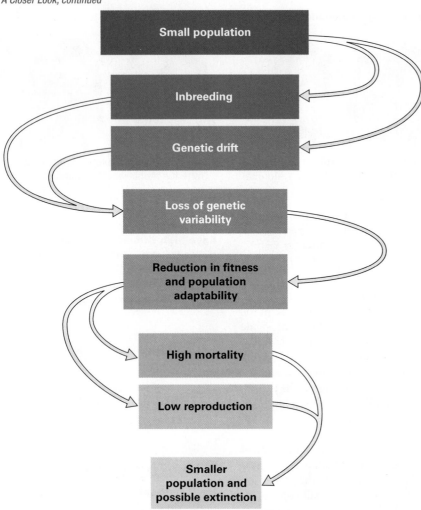

**FIGURE 15.26 The extinction vortex.** A small population can become trapped in a positive-feedback process that causes it to continue to shrink in size, eventually leading to extinction.

In most populations, alleles that are lost through genetic drift have relatively small effects on fitness at the time. However, many alleles that appear to be nearly neutral with respect to fitness in one environment may have positive fitness in another environment. When this is the case, the loss of these alleles may spell disaster for the entire species.

### The Consequences of Low Genetic Variability in a Population

Populations with low levels of genetic variability have an insecure future for two reasons. First, when alleles are lost, the level of inbreeding in a population increases, which means lower reproduction and higher death rates, leading to declining populations that are susceptible to all the other problems of small populations. This process is often referred to as the **extinction vortex** (**FIGURE 15.26**). The heath hen discussed in this chapter is an example of the extinction vortex; once the remnant population was reduced to small numbers by fire, disease, and predation, inbreeding depression prevented it from rebounding, thus dooming it to extinction.

Second, populations with low genetic variability may be at risk of extinction because they cannot evolve in response to changes in the environment. When few alleles are available for any given gene, it is possible that no individuals in a population possess an adaptation allowing them to survive an environmental challenge. For example, there is some evidence that people with type A blood are more resistant to cholera and bubonic plague than are people with type O or B blood. Loss of the $I^A$ allele would make human populations more susceptible to serious declines in the face of these diseases.

Preventing endangered species from declining to very small population levels, thus eroding genetic diversity, is critical to avoiding a similar genetic disaster even once populations recover.

## Protecting Biodiversity versus Meeting Human Needs

The ESA has been a successful tool for bringing species such as the peregrine falcon, American alligator, and bald eagle (**FIGURE 15.27**) back from the brink of extinction, but all of these successes have come with some cost to citizens. If the solution to these and other endangered species controversies is any guide, many Americans are willing to devote tax dollars to efforts that balance the needs of people and wildlife to protect our natural heritage.

As with any challenge that humans face, the best strategy is prevention. **TABLE 15.3** provides a list of actions that can help reduce the rate at which species become endangered. Meeting the challenge of preserving biodiversity requires some creativity, but it is possible to provide for the needs of people while making space for other organisms. However, it will take all of us to help preserve the balance.

**FIGURE 15.27 An endangered species success story.** The bald eagle was "delisted" from endangered status in 2007, at an event celebrated at this ceremony in Washington, DC. Its dramatic recovery from near extinction was thanks in part to government protections. However, delisting is still a relatively rare event.

## TABLE 15.3 Taking action to preserve biodiversity.

| Objective | | Why Do It? | Actions |
|---|---|---|---|
| Reduce fossil fuel use | | • Mining, drilling, and transporting fossil fuels modifies habitat and leads to pollution.<br>• Burning fossil fuels contributes to global climate change, further degrading natural habitats. | • Buy energy-efficient vehicles and appliances.<br>• Walk, bike, carpool, or ride the bus whenever possible.<br>• Choose a home near school, work, or easily accessible public transportation.<br>• Buy "clean energy" from your electric provider, if offered. |
| Reduce the impact of meat consumption | | • The primary cause of habitat destruction and modification is agriculture.<br>• Modern beef, pork, and chicken production rely on grains produced on farms. One pound of beef requires 4.8 pounds of grains or about 25 square meters of agricultural land. | • Eat one more meat-free meal per week.<br>• Make meat a "side dish" instead of the main course.<br>• Purchase grass-fed or free-range meat. |
| Reduce pollution | | Pollution kills organisms directly or can reduce their ability to survive and reproduce in an environment. | • Do not use pesticides.<br>• Buy products produced without the use of pesticides.<br>• Replace toxic cleaners with biodegradable, less-harmful chemicals.<br>• Consider the materials that make up the goods you purchase and choose the least-polluting option.<br>• Reuse or recycle materials instead of throwing them out. |
| Educate yourself and others | | Change happens most rapidly when many individuals are working for it. | • Ask manufacturers or store owners about the environmental costs of their goods.<br>• Talk to family and friends about the choices you make.<br>• Write to decision makers to urge action on effective measures to reduce human population growth and curb habitat destruction and species extinction. |

# savvy reader

## What Happens When a Species Recovers

**Gray Wolf's endangered species status appears close to end**

A new federal bill introduced in the U.S. House of Representatives would "de-list" the Gray Wolf from Endangered Species Act protections and allow individual states to design and carry out their own wolf management and recovery plans. The legislation is co-sponsored by Rep. Jim Matheson (D) from Utah, a state that is just beginning to see increases in wolf numbers, and Rep. Mike Ross (D-Arkansas), who co-chairs the Congressional Sportsman's Caucus, made up of advocates of hunting and fishing interests.

According to Rep. Matheson, "Scientists and wolf recovery advocates agree—the gray wolf is back. Since it is no longer endangered, it should be de-listed as a species, managed as other species are—by state wildlife agencies—and time, money, and effort can be focused where it's needed."

Rep. Ross argues that officials within the regions inhabited by particular endangered species should have a greater role in management decisions related to these species. "Excessive wolf populations are having a devastating impact on elk, moose, deer, and other species, and each state has its own unique set of challenges. Both the Obama and Bush administrations have already recommended the de-listing of wolves in many states and turning their management over to state wildlife agencies because it is the right thing to do to keep our nation's sensitive ecosystem in balance."

1. What evidence does the representative from Utah provide to support his statement that "the gray wolf is back"? Is that evidence convincing?

2. Given what you've learned in this chapter, do you think it is likely that the return of the gray wolf is having a "devastating impact" on elk and moose?

# SOUNDS RIGHT BUT IS IT?

Mosquitoes spend the first part of their lives in pools of stagnant water, feeding on algae, bacteria, and other microbes. In the water, they are preyed upon by dragonflies, fish, and small crustaceans. When mosquitoes hatch, both males and females feed on flower nectar—thus acting as pollinators. Once fertilized, the females feed on blood to nourish the development of their eggs. As a result, mosquitoes are possibly Earth's most annoying animals—at least to humans. Not only do they cause itchy welts when they bite and whine in your ear when you are trying to sleep, but they spread some of the most troublesome diseases—malaria, yellow fever, and dengue fever, among them. Malaria alone is responsible for hundreds of thousands of deaths every year. That mosquitoes are a plague on humankind might lead you to say:

**Earth would definitely be better if mosquitoes were extinct.**

Sounds right, but it isn't.

1. What are the ecological relationships between (a) mosquitoes and microbes in the water; (b) mosquitoes and dragonflies; (c) mosquitoes and the plants they visit; and (d) mosquitoes and humans?

2. Sketch a food web containing mosquitoes. What organisms would likely be affected by the loss of mosquitoes, and in what way?

3. Could any of the changes listed affect organisms that are not in the web you drew? Explain.

4. What is a keystone species?

5. Are mosquitoes a keystone species? Explain.

6. Given your answers to questions 1–5, explain why the statement bolded above sounds right, but isn't.

# Chapter Review MasteringBiology®

Go to the Study Area in MasteringBiology® for practice quizzes, myeBook, BioFlix™ 3-D animations, MP3Tutor sessions, videos, current events, and more.

## Summary

## Section 15.1

Describe how scientists can estimate current and historical rates of extinction, and provide the evidence that Earth is currently experiencing a mass extinction.

- Historical rates of extinction can be estimated by looking at the life span of species in the fossil record, while the rate of modern extinctions can be measured by counting the number of species known to be lost or endangered (pp. 336–337).
- The loss of biodiversity through species extinction is exceeding historical rates by 50 to 100 times (p. 337).

List the major causes of extinction, and describe how each is related to human activity.

- The loss of natural habitat caused by human activities, such as agriculture and urban development, is the primary cause of the extinctions occurring in the modern era (pp. 338–340).
- Additional threats of habitat fragmentation, the movement of introduced species to regions where they are not native, overexploitation through uncontrolled harvesting, and pollution also contribute to species extinction (pp. 340–343).

Explain how the species-area curve is used to estimate species number in an area.

- Species-area curves illustrate the relationship between the number of species and a geographic area. If the actual number of species of a particular type is not known, a species-area curve can provide some estimate of it in a given area (pp. 339–340).

Explain the principle of the trophic pyramid.

- The trophic pyramid is the relationship among biomass at different levels of a food chain. In nearly all communities, there is significantly more biomass in the producer level than at higher consumer levels (p. 340).
- Species at the top of the food chain are more susceptible to extinction because less energy is available for survival at higher trophic levels (p. 340).

## Section 15.2

Define *predation*, *mutualism*, and *competition*, and provide examples of each ecological interaction.

- Species are members of communities; they may interact with one another via mutualism, where both species are benefited by the relationship; via predation, where one benefits while the other is consumed in the relationship; or via competition, where both are struggling to obtain the same resources (pp. 344–348).

Define *food web* and *ecosystem*, and explain the role of keystone and non keystone species in both.

- A keystone species in a food web is one whose absence has a large effect on the entire community (pp. 349–350).
- A food web is the connection among species in a community, while an ecosystem includes not only the food web but all of the nonbiological aspects of the environment occupied by a community of organisms (pp. 344–345, 351).
- Species can play a role in ecosystem function, including effects on energy flow and nutrient cycling (pp. 351–352).

Provide an example of a nutrient cycle.

- Nutrients cycle through ecosystems from living organisms to inorganic forms and back again. Nitrogen is an example of a nutrient that cycles through the environment (pp. 351–352).

## Section 15.3

Explain why small population size and a loss of genetic diversity in a population are risky for both the population and individuals within the population.

- When species are already endangered, restoring larger populations is critical for preventing extinction (p. 354).
- Small populations are at higher risk for extinction due to environmental catastrophes (pp. 356–357).

- Small populations are at risk when individuals have low fitness due to inbreeding and thus are less able to increase population size (pp. 357–358).
- Genetic variability is lost in small populations because of genetic drift—the loss of alleles from a population due to chance events. Therefore, small populations may be less able to evolve in response to environmental change (pp. 359–360).

List strategies that will help reduce the number of extinctions going forward.

- Eating lower on the food chain and using sustainable methods of transportation and energy all help conserve habitat for other species (p. 361).
- The political process enables people to develop plans for helping endangered species recover from the brink of extinction while minimizing the negative effects of these actions on people (pp. 360–361).

## Roots to Remember

**The following roots of words come mainly from Latin and Greek and will help you decipher terms:**

| | |
|---|---|
| **eco-** | means home or habitation. Chapter term: *ecosystem* |
| **eu-** | means true or good. Chapter term: *eutrophication* |
| **-philia** | means affection or love. Chapter term: *biophilia* |
| **-trophic** | means food, nourishment. Chapter term: *eutrophication* |

## Learning the Basics

1. Describe how habitat fragmentation endangers certain species. Which types of species do you think are most threatened by habitat fragmentation?

2. Compare and contrast the species interactions of mutualism, predation, and competition.

3. A mass extinction _____.
A. is global in scale; B. affects many different groups of organisms; C. is caused only by human activity; D. A and B are correct; E. A, B, and C are correct.

4. Current rates of species extinction appear to be approximately _____ historical rates of extinction.
A. equal to; B. 10 times lower than; C. 10 times higher than; D. 50 to 100 times higher than; E. 1000 to 10,000 times higher than

5. According to the generalized species-area curve, when habitat is reduced to 50% of its original size, approximately _____ of the species once present there will be lost.
A. 10%; B. 25%; C. 50%; D. 90%; E. it is impossible to estimate the percentage.

6. Which cause of extinction results from humans' direct use of a species?
A. overexploitation; B. habitat fragmentation; C. pollution; D. introduction of competitors or predators; E. global warming

7. The web of life refers to the _____.
A. evolutionary relationships among living organisms; B. connections between species in an ecosystem; C. complicated nature of genetic variability; D. flow of information from parent to child; E. predatory effect of humans on the rest of the natural world.

8. Which of the following is an example of a mutualistic relationship?
A. moles catching and eating earthworms from the moles' underground tunnels; B. cattails and reed canary grass growing together in wetland soils; C. cleaner fish removing and eating parasites from the teeth of sharks; D. Colorado potato beetles consuming potato plant leaves; E. more than one of the above.

9. The risks faced by small populations include _____.
A. erosion of genetic variability through genetic drift; B. decreased fitness of individuals as a result of inbreeding; C. increased risk of experiencing natural disasters; D. A and B are correct; E. A, B, and C are correct.

10. One advantage of preserving more than one population of an endangered species at more than one location is _____.
A. a lower risk of extinction of the entire species if a catastrophe strikes one location; B. higher levels of inbreeding in each population; C. higher rates of genetic drift in each population; D. lower numbers of heterozygotes in each population; E. higher rates of habitat fragmentation in the different locations.

11. There are fewer lions in Africa's Serengeti than there are zebras. This is principally because _____.
A. zebras tend to drive off lions; B. lions compete directly with cheetahs, while zebras do not have any competitors; C. zebras have mutualists that increase their population, while lions do not; D. there is less energy available in zebras to support the lion population than there is in

grass to support the zebras; **E.** zebras are a keystone species, while lions are not.

**12.** Most of the nutrients available for plant growth in an ecosystem are _____.
**A.** deposited in rain; **B.** made available through the recycling of decomposers; **C.** maintained within that ecosystem over time; **D.** B and C are correct; **E.** A, B, and C are correct.

## Analyzing and Applying the Basics

**1.** Off the coast of the Pacific Northwest, areas dominated by large algae, called kelp, are common. These "kelp forests" provide homes for small plant-eating fishes, clams, and abalone. These animals in turn provide food for crabs and larger fishes. A major predator of kelp in these areas is sea urchins, which are preyed on by sea otters. When sea otters were hunted nearly to extinction in the early twentieth century, the kelp forest collapsed. The kelp were only found in low levels, while sea urchins proliferated on the seafloor. Use this information to construct a simple food web of the kelp forest. Using the food web, explain why sea otters are a keystone species in this system.

**2.** In which of the following situations is genetic drift more likely and why?
**A.** A population of 500 in which males compete heavily for harems of females. About 5% of the males father all of the offspring in a given generation. Females produce one offspring per season; **B.** A population of 250 in which males and females form bonded pairs that last throughout the mating season. Females produce three to four offspring per season.

**3.** The piping plover is a small shorebird that nests on beaches in North America. The plover population in the Great Lakes is endangered and consists of only about 30 breeding pairs. Imagine that you are developing a recovery plan for the piping plover in the Great Lakes. What sort of information about the bird and the risks to its survival would help you to determine the population goal for this species as well as how to reach this goal?

## Connecting the Science

**1.** From your perspective, which of the following reasons for preserving biodiversity is most convincing? **(A)** Nonhuman species have roles in ecosystems and should be preserved to protect the ecosystems that support humans; or **(B)** nonhuman species have a fundamental right to existence. Explain your choice.

**2.** If a child asks you the following question 20 or 30 years from now, what will be your answer, and why? "When it became clear that humans were causing a mass extinction, what did you do about it?"

Answers to **Stop & Stretch, Visualize This, Working with Data, Savvy Reader, Sounds Right, But Is It?,** and **Chapter Review** questions can be found in the **Answers** section at the back of the book.

# CHAPTER 16

# Where Do You Live?

**What is your "biological address"?**

# Climate and Biomes

How do you answer the question, "Where do you live?" Most of us would respond with some part of our mailing address—that is, our street, municipality, state, or country. Not too many of us would give a reply such as "the Sonoran Desert" or "the boreal forest." However, descriptions of the natural environment in which we live can be thought of as our *bioregional* addresses. Our bioregional addresses include the native flora and fauna that share our living space, as well as the climate that supports those organisms. People often have a difficult time describing the bioregion in which they live.

You may have experienced one consequence of poor bioregional awareness—local water shortages occurring when large numbers of homeowners attempt to keep their thirsty lawns green during

**Who are your neighbors?**

**It can be hard to identify a biological address in a human-designed landscape.**

summertime droughts. Other costs of lack of bioregional awareness may be more severe, including the consequences of home construction in areas where vegetation that promotes periodic fires is common.

Understanding one's own bioregion may allow people to build human settlements that are better for both humans and the natural environment. For example, in the southwestern desert regions of the United States, environmentally sensitive homeowners practice xeriscaping, a kind of landscaping that uses native, drought-tolerant plants. Xeriscaping not only prevents

the overconsumption of water but also provides a habitat for resident wildlife.

While we may pride ourselves on our ability to settle anywhere, humans remain dependent on the natural world for resources and waste disposal. If we don't understand how our human communities fit into the surrounding biological community, the results can include air and water pollution that harms our biological neighbors—and ourselves. In this chapter, we explore how the ecology of a bioregion and the biology of human habitats intersect.

**Few of us know where the resources on which we rely come from.**

**People often have a difficult time describing the bioregion in which they live.**

# 16.1 Global and Regional Climate

Why is Buffalo, New York, so much snowier than Winnipeg, Manitoba? Why does India experience monsoons? Why is the weather on Pacific islands always beautiful?

The answers to these questions require an understanding of **climate**, the average conditions of a place measured over time. Climate is different from **weather**, which is the current temperature, cloud cover, and **precipitation** (rain or snowfall). Simply put, weather information will tell you if you have to shovel snow in the morning; climate information will tell you if you need to own a snow shovel.

In general, temperatures are warmer in the tropics than in the Arctic and Antarctic, and at altitudes close to sea level. Variation in temperature throughout the year correlates to location as well—the closer one is to a pole the more variation, or seasonality, one experiences.

On a global scale, precipitation patterns are more variable than temperature patterns. There is a wide region of high rainfall in the tropics, while north and south of this point rainfall diminishes considerably and many great deserts can be found. The same pattern is repeated in the wetter temperate zones compared to drier polar regions. **FIGURE 16.1** summarizes global temperature and precipitation patterns.

The average temperature of a region is primarily determined by the amount of **solar irradiance** it receives—the total solar energy per square meter of land or water surface. Locations that receive large amounts of solar irradiance have a higher average temperature than places receiving less.

Earth's axis is roughly perpendicular to the flow of energy from the sun. The extremes of this axis are called **poles**, while the circle around the planet that is equidistant to both poles is called the **equator**. The amount of solar irradiance varies between poles and equator because of the planet's shape.

**FIGURE 16.2** shows two identical streams of solar energy flowing from the sun. One strikes Earth's surface directly at the equator, while the other strikes at an angle closer to the pole, where Earth's surface is "curving away" from the sun. The difference in geometry means that the surface area warmed by an amount of sunlight is much smaller at the equator than the surface area warmed at the poles. Because sunlight passing through the atmosphere can be scattered back into space, less light reaches the ground at the poles because it

**equ-** means equal.

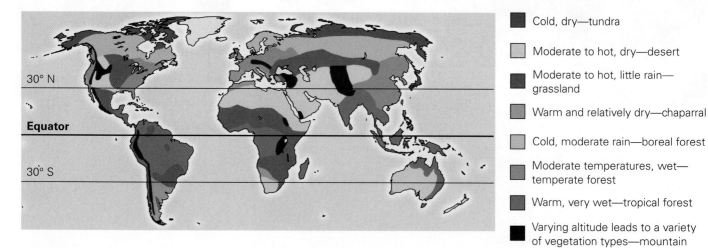

| | |
|---|---|
| ■ | Cold, dry—tundra |
| ☐ | Moderate to hot, dry—desert |
| ■ | Moderate to hot, little rain—grassland |
| ☐ | Warm and relatively dry—chaparral |
| ☐ | Cold, moderate rain—boreal forest |
| ■ | Moderate temperatures, wet—temperate forest |
| ■ | Warm, very wet—tropical forest |
| ■ | Varying altitude leads to a variety of vegetation types—mountain |

**FIGURE 16.1 Global climate patterns.** Average temperature in a region is somewhat predictable based on the region's distance from the equator. Precipitation is determined by some global patterns but also by local conditions, such as proximity to an ocean or mountain range.

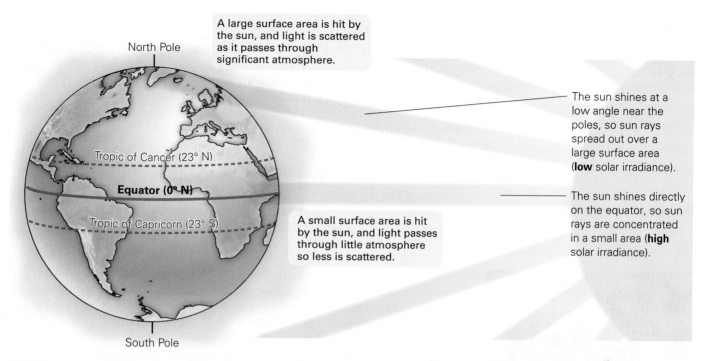

North Pole

A large surface area is hit by the sun, and light is scattered as it passes through significant atmosphere.

Tropic of Cancer (23° N)

Equator (0° N)

Tropic of Capricorn (23° S)

A small surface area is hit by the sun, and light passes through little atmosphere so less is scattered.

The sun shines at a low angle near the poles, so sun rays spread out over a large surface area (**low** solar irradiance).

The sun shines directly on the equator, so sun rays are concentrated in a small area (**high** solar irradiance).

South Pole

**FIGURE 16.2 Solar irradiance on Earth's surface.** The annual average temperature in a location on Earth's surface is most directly determined by its solar irradiance. Areas near the equator receive the greatest amount of solar energy, while areas near the poles receive the least.

## Visualize This ▲

**Explain why solar irradiance, and thus climate, differs between Florida (at 28° North of the equator) and Vermont (at 44° N).**

passes through more atmosphere at a shallow angle. In other words, the solar irradiance is greatest at the equator and lowest near the poles. This is one reason southern Florida has a warmer climate than northern Vermont.

Solar irradiance also varies in a particular location annually. This occurs because Earth's axis actually tilts approximately 23.5° from perpendicular to the sun's rays (**FIGURE 16.3**). Due to this tilt, as Earth orbits the sun, the Northern

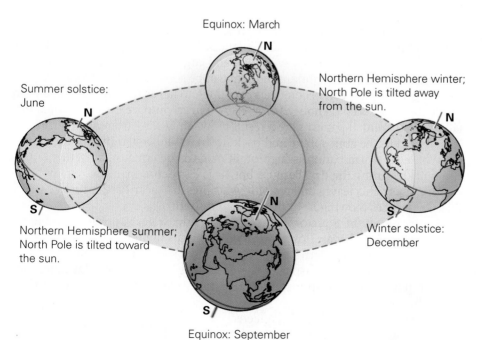

Equinox: March

Summer solstice: June

Northern Hemisphere winter; North Pole is tilted away from the sun.

Northern Hemisphere summer; North Pole is tilted toward the sun.

Winter solstice: December

Equinox: September

## ◀ Visualize This

**Why is it summer in Australia when it is winter in North America?**

**FIGURE 16.3 Earth's tilt leads to seasonality.** Because Earth's axis is 23.5° from perpendicular to the rays of the sun, as Earth orbits the sun during a year, the poles appear to move toward and away from the sun. Solar irradiance is increased at and near a pole when it tilts toward the sun and is decreased when that pole tilts away.

**sol-** means sun.

**-stice** means to stand still.

Hemisphere (north of the equator) is tilted toward the sun during the northern summer and away from the sun during the northern winter. Solar irradiance is at its annual maximum in the Northern Hemisphere during the summer **solstice**, when the sun reaches its northern maximum and the North Pole is tilted closest to the sun. The winter solstice is the point where the sun is at its minimum. Earth's tilt also helps explain why the position of the sunrise changes over the course of a year, moving from south to north as winter turns to summer (**FIGURE 16.4**).

**STOP & STRETCH**    There is evidence that Earth has "wobbled" on its axis throughout its history, such that the axis is more perpendicular to the sun at some times and tilted at a greater angle other times. How would a reduction in the tilt affect seasonality?

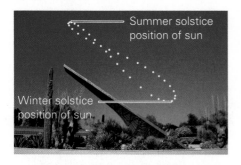

**FIGURE 16.4 The sun's travels.**    This image was produced by taking a picture of the sun at the same location and time approximately once each week throughout a year. When the sun is high in the sky, solar irradiance is high and temperatures are warm. When it is low, irradiance and thus temperatures are low.

The atmosphere of Earth plays a role in determining climate as well. The blanket of gases surrounding Earth, including water vapor and carbon dioxide, prevents the heat absorbed from the sun during the day from escaping back into space during the night. This insulating blanket makes Earth habitable. Recent human activities have changed the atmosphere of Earth. By greatly increasing carbon dioxide levels in the atmosphere via the burning of fossil fuels, the effectiveness of the blanket in retaining heat has increased. The consequences of this phenomenon are increases in global temperatures and radical changes to the climate, some of which may be unpredictable. Chapter 5 explores human-caused global climate change in depth.

## Global Temperature and Precipitation Patterns

Energy from the sun is the primary driver of precipitation—that is, rain and snowfall. To understand how sunlight causes rainfall, we must first understand some of the properties of water vapor. Condensation is the process by which molecules clump together to form liquid droplets, while evaporation occurs when molecules escape from droplets. For water molecules to remain as a vapor in the air, the rate of condensation must be less than the rate of evaporation.

The rate of evaporation depends on temperature; at high temperatures, the evaporation rate is high. The reverse is true at low temperatures. Thus, when air cools, water molecules clump into larger and larger droplets. When the droplets are large enough, concentrations of them can be seen as clouds. As clouds grow even larger, droplets can become heavy enough to fall as rain. If the temperature inside the cloud is cold enough, droplets will freeze into ice crystals, which may fall as snow. Rainfall patterns are a result of the air cycling from near Earth's surface to high in the atmosphere and back down, as illustrated in **FIGURE 16.5**.

Where solar irradiation is highest, at or near the equator, air temperatures rise quickly during the day. Because hot air is less dense than cold air, air at the equator rises. This leaves an area of low air pressure near Earth's surface that is filled by breezes blowing from the north and south.

As air rises at the equator, it cools, causing the water vapor it carries to fall as rain. This now-dry air flows in the upper atmosphere toward the poles and finally drops back to Earth's surface at about 30° north and south latitudes. This very dry falling air displaces the ground-level air at these latitudes, and that air, having picked up moisture from the surface, flows toward the poles along the surface.

The movement of air is also affected by Earth's rotation, which creates the prevailing winds in various regions of the globe. The pattern of air movement

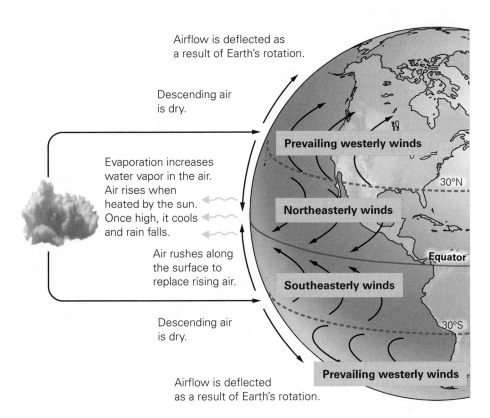

Airflow is deflected as a result of Earth's rotation.

Descending air is dry.

Evaporation increases water vapor in the air. Air rises when heated by the sun. Once high, it cools and rain falls.

Air rushes along the surface to replace rising air.

Descending air is dry.

Airflow is deflected as a result of Earth's rotation.

Prevailing westerly winds

Northeasterly winds

30°N

Equator

Southeasterly winds

Prevailing westerly winds

30°S

◀ **Visualize This**

The Doldrums and the Horse Latitudes are regions where there is little surface wind and thus where sailing ships are likely to stall. Based on the global wind patterns pictured here, where do you think these places are on the globe?

**FIGURE 16.5 Global wind patterns.** High levels of solar irradiance at the solar equator lead to high levels of evaporation and rainfall. This phenomenon drives massive movements of air near the tropics into the temperate zones.

in the atmosphere that is created by the sun's heating and Earth's rotation helps explain the band of rain forests near the equator, the great deserts found near 30° north and south of the equator, and the tendency for weather patterns in North America to come from the west.

Global rainfall patterns exhibit seasonality as well. The area of maximum solar irradiance travels from 23.5° north of the equator to 23.5° south over the course of a year due to the tilt in Earth's axis. Therefore, the regions of the tropics receiving the most solar energy also move. Rainy seasons occur north and south of the equator as this region of high irradiance shifts, and also when this shift affects the prevailing winds. The rainy monsoon seasons in India and southern Arizona are both associated with wind shifts, as breezes move over long expanses of ocean, picking up water vapor that falls on land as rain.

## Global Temperature and Precipitation Patterns: A Closer Look ▼

Both global and regional factors influence climate locally or in a particular bioregion.

### The Solstices and the Seasons

In addition to differences in solar irradiance at the solstices, the length of day is also greatest at the summer solstice and least at the winter solstice. For example, in Chicago, the time from sunrise to sunset is approximately 15 hours at the summer solstice and 9 hours at the winter solstice. The closer a region is to a pole, the greater the variance in day length over the course of a year. So, while day length in Chicago varies by 6 hours from winter to summer, in Fairbanks, Alaska, it varies by 18 hours: from less than 4 hours at the winter solstice to nearly 22 hours at the summer maximum.

*(continued on the next page)*

*A Closer Look, continued*

Although the summer solstice is the day of highest solar irradiance in the Northern Hemisphere, it is not the warmest day of the year. Instead, the warmest days of the year are about one month later. This lag occurs because Earth converts the energy from the sun to heat and releases it gradually into the atmosphere. You can think of the solar heat stored in Earth as a bank account. Heat is always dissipating, much like money being withdrawn regularly from a bank account. If more money is deposited into the bank than is withdrawn, the account balance increases. Likewise, as day length increases in a region, the amount of heat deposited there is greater than the amount withdrawn—that is, the temperature starts to rise. On the longest day of the year, the amount of heat deposited is greatest, but even after that day, the heat deposits are greater than the withdrawals, and the temperature continues to increase. Similarly, the coldest days of the year are about a month after the winter solstice—in late January or early February—because withdrawals of heat from Earth's surface remain larger than deposits for several weeks.

Because day length over the course of a year changes more dramatically near the poles than near the equator, the temperature also changes more dramatically closer to the poles. The amount of temperature variance explains why seasonal temperature swings are more pronounced closer to the poles than near the equator. The average low temperature for January in Tampa, Florida, is 10°C (50°F), and the average high in July is 32°C (90°F), a difference of 22°C (40°F). In contrast, the January low in Missoula, Montana, is –10°C (14°F), and the July high is 29°C (84°F), a difference of 39°C (70°F).

### Local Influences on Climate

Three characteristics of a location's setting have an effect on its temperatures and precipitation: (1) altitude, (2) the proximity of a large body of water, and (3) characteristics of the land's surface and vegetation (**FIGURE 16.6**).

Temperature drops as altitude increases because as gases rise over hundreds of meters, the molecules move away from each other, reducing heat content. Temperature differences due to altitude are dramatic; the summit of Mt. Everest, 8.8 km (29,035 ft) above sea level, averages –27°C (–16°F), while nearby Kathmandu, Nepal, at 1.3 km (4385 ft), averages 18°C (65°F). However, smaller differences in altitude within a region have a converse effect on air temperature. Because cold air masses are denser than warm air, pockets of cold air (for instance, air that has been in shade and not exposed to solar radiation) tend to "drain" to the lowest point on a landscape. Thus, valleys will often be colder than nearby hilltops.

Temperatures in areas near oceans, seas, and large lakes are influenced by the thermal properties of water, including great ability to store heat (Chapter 5). Water

**FIGURE 16.6 Factors that influence local climate.** Local temperature and rainfall patterns result from global climate patterns moderated by several aspects of local geography.

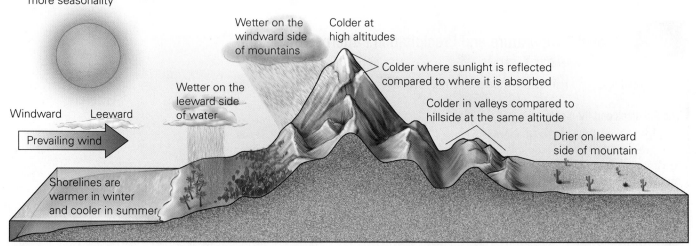

*A Closer Look, continued*

temperature rises and falls slowly in response to solar irradiation when compared to the rate of temperature change of land surfaces. Thus, air over a large body of water is comparatively cooler in summer and warmer in winter (**FIGURE 16.7**), and nearby land areas experience more moderate temperatures than do regions fur ther inland. Because of this phenomenon, the growing season on Long Island, New York, is as much as 30 days longer than areas in New Jersey just a few dozen miles inland from the Atlantic. Oceans also have an effect on local climates because heat is transferred through them via currents. The Gulf Stream is a current that carries water from the tropical Atlantic Ocean to the shores of northern Europe, producing a much milder climate in those regions than in other areas at the same distance from the equator. The warmth of the Gulf Stream makes Dublin, Ireland, as warm as San Francisco, even though Dublin is 1600 km (1000 mi) closer to the North Pole.

The amount of light absorbed or reflected by the surface of the land will also influence surrounding air temperature. A surface that reflects most light energy will have a lower nearby air temperature compared to a surface that absorbs most of that light energy, heats up, and radiates that heat into the air. Snow reflects more light than a forest does, so air over a snowpack remains cold. The low reflectance of asphalt pavement and most building materials contributes to the urban heat island effect, which is the tendency for cities to be from 0.5° to 3°C (1° to 6°F) warmer than the surrounding areas (**FIGURE 16.8**, on the next page).

Cities are also warmer because they contain little vegetation, which tends to reduce air temperature. Much of the solar energy absorbed by plants converts liquid water inside the plant to vapor, which also prevents the energy from being converted to heat.

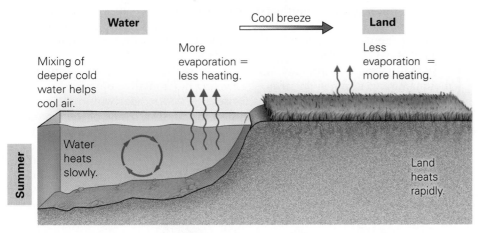

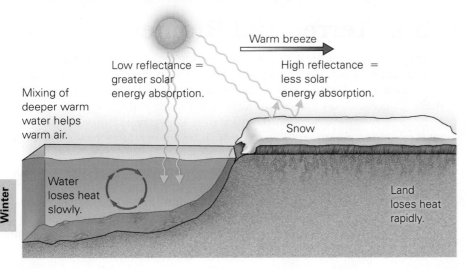

◀ **Visualize This**

**Consider this figure along with the concept of the lag of the seasons. Why is late July through mid-August traditionally the most popular time for people to visit the seashore?**

**FIGURE 16.7 The moderating influence of water.** Water heats slowly because of evaporative cooling and the mixing of warmed water with cooler deeper waters. As a result, winds blowing across water in spring and summer will cool nearby landmasses. Conversely, water absorbs heat even in winter, when the sun's rays are reflected off the snow on land, and it loses heat more slowly than land. Thus, a large body of water is a source of warmer breezes in fall and winter.

*(continued on the next page)*

*A Closer Look, continued*

Air conditioning = heat vented
into outside areas

Less vegetation = less energy used
to evaporate water, and more
energy converted to heat

Dark surfaces = solar energy
absorbed rather than reflected

Fossil fuel use = waste heat produced

**FIGURE 16.8 Urban heat island.** Urban areas are significantly warmer than nearby rural areas for a variety of reasons.

### Local Precipitation Patterns

The amount of precipitation that falls in a given land region is highly dependent on the context of that area—in particular, the proximity of the land to a large body of water. Wind blowing across warm water accumulates water vapor that condenses and falls when it reaches a cooler landmass. Communities surrounding the Great Lakes provide a dramatic example of this effect. For example, Toronto, on the northwest side of Lake Ontario, averages about 140 cm (55 in.) of snow per year, while Syracuse, New York, on the southeast side—receiving the prevailing wind after it crosses the lake—averages almost twice that—274 cm (108 in.).

Precipitation amounts are also affected by the presence of mountains or mountain ranges. When an air mass traveling horizontally approaches a mountain, it is forced upward. Cooling as it rises, the water vapor within it condenses to form clouds. Rain or snow then falls on the windward side of the mountain. Warming again as it drops down the other side, the dry air mass causes water to evaporate from land on the sheltered, or leeward, side of the mountain. The dry area that results is often referred to as the mountain's "rain shadow." The Great Basin of North America, encompassing nearly all of Nevada and parts of Utah, Oregon, and California, is a desert because of the rain shadow cast by the Sierra Nevada.

**Our bioregional addresses include the native flora and fauna that share our living space.**

# 16.2 Terrestrial Biomes

Climate plays the greatest role in determining the characteristics of a **biome**, a geographic area defined by its primary vegetation. Plants (and animals) native to a region are adapted to the water availability and temperatures experienced there. In general, the size of the vegetation is limited by water availability—large trees require large amounts of water. Water availability is a function of total precipitation, but it is also influenced by temperature; frozen water cannot be taken up by plants.

Four basic terrestrial, or land, biomes are typically recognized: forest, grassland, desert, and tundra. Each of these basic biomes may contain variants within them; for instance, a grassland may be prairie, steppe, or savanna. The relationship between climate and biome type is illustrated in **FIGURE 16.9**, and the characteristics of the terrestrial biomes are summarized in **TABLE 16.1**.

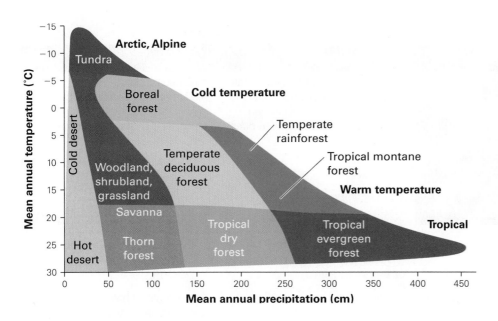

◀ **Working with Data**

**What types of biomes are possible where average annual temperature is 5°C? What factor determines exactly which biome occurs at these temperatures?**

**FIGURE 16.9 Biomes and climate.**
The primary vegetation type in a given area is determined by the region's climate.

## TABLE 16.1 Terrestrial biomes.

| Biome | Characteristics | Location |
|---|---|---|
| **Forests** | **Dominated by woody plants** | |
| Tropical | High rainfall, high average temperatures. Tall trees, relatively little understory. Great diversity of species. | Around the equator: Central and South America, central Africa, India, southeast Asia, and Indonesia |
| Temperate (Deciduous) | Sufficient rainfall, moderate with some freezing temperatures. Three distinct layers: deciduous canopy trees, shrubs, and spring-flowering plants in the understory. | Middle latitudes of the Northern Hemisphere: eastern North America, western and central Europe, and eastern China |
| Boreal | Sufficient rainfall with cold average temperatures. Dominated by evergreen coniferous trees with a very short growing season. | Northern latitudes and high altitudes: northern North America, northern Europe and Asia, and mountains in the western United States |

*continued*

**TABLE 16.1** **Terrestrial biomes.** *continued*

| Biome | | Characteristics | Location |
|---|---|---|---|
| Chapparal | | Moderate temperatures, moderate seasonal rainfall, periodic fires. Dominated by spiny evergreen shrubs and maintained by frequent fires. | Areas surrounding the Mediterranean Sea and in patches in southern California, South Africa, and southwest Australia |
| **Grasslands** | | **Dominated by nonwoody grasses with few or no shrubs or trees** | |
| Savanna | | Lower rainfall, above freezing temperatures. Scattered individual trees, maintained by periodic fires or large-animal grazing. | Tropical: about half of Africa, as well as large areas of India, South America, and Australia |
| Prairie and Steppe | | Lower rainfall, seasonally cold. Tall (prairie) or short (steppe) grasses, no woody plants. | Temperate areas in the middle of large landmasses: central North America, central Asia, parts of Australia, and southern South America |
| **Desert** | | Scant rainfall. Dry soils, very sparse vegetation. Plants often covered with spines and adapted to store water. | Near 30° north or south of the equator: northern Africa (Sahara), central Asia (Gobi), the Middle East, central Australia, and the southwestern United States |
| Tundra | | Very low average temperature. Very short growing season, permafrost. Plants are low to the ground and many animals migrate to warmer climates during the winter season. | Near the poles and at high altitudes |

# Forests and Shrublands

The major attribute of **forests** is the presence of trees—forests are found where rainfall is sufficient and average temperatures in the summer are above 10°C (50°F) and thus can support tree growth. Forests occupy approximately one-third of Earth's land surface and, when all forest-associated organisms are included, contain about 70% of the **biomass,** the total weight of living organisms, found on land. Forests are generally categorized into three groups based on their distance from the equator: (1) tropical forests at or near the equator, (2) temperate forests from 23.5° to 50° north and south of the equator, and (3) boreal forests close to the poles.

### Tropical Forests.

Extensive areas of tropical forest were once found throughout Earth's equatorial region. Tropical forests contain a large amount of biological diversity; 1 hectare (10,000 m²) may contain as many as 750 tree species.

One hypothesis regarding why tropical forests are so rich in species is the high solar irradiance they experience. Because the energy level is high, populations of many species can be supported in a relatively small area. Think of the available energy as analogous to a pizza—the larger the pizza, the greater the number of individuals who can be fed to satisfaction.

High energy and water levels also support the growth of very large trees in tropical forests. As a result, most of the sunlight is absorbed by vegetation before it hits the ground, and most living organisms are therefore adapted to survive high in the treetops. Tropical forest animals are able to fly, glide, or move freely from branch to branch, while small plants are able to obtain nutrients and water while living on the upper branches of these huge trees (**FIGURE 16.10**).

With warm temperatures and abundant water, the process of **decomposition,** the breakdown of waste and dead organisms, is rapid in tropical forests. Dense vegetation quickly reabsorbs the simple nutrients that are produced by decomposition, resulting in relatively little organic matter stored in the soil. Consequently, when vegetation is cleared and burned, the ash-fertilized soils can support crop growth for only 4 to 5 years. Once its soil is depleted of nutrients, a farm field carved out of tropical forest is abandoned, and a new field is cleared using the same method.

Among human populations in tropical forest areas, this slash-and-burn (or swidden) agricultural system is common. Tropical forests appear able to support swidden agriculture for many generations if the human population is small enough and abandoned plots have several decades to recover. However, increasing human population levels in tropical countries may be too large to allow adequate recovery time, and road building into interior forests has exposed untouched wilderness to swidden fields. The resulting **deforestation**—literally, the removal of forests—endangers thousands of organisms that depend on this biome for survival (**FIGURE 16.11**).

### Temperate Forests.

Closer to the poles, where seasonal fr[e]... tures limit tropical vegetation, forests are dominated by trees t[...] to these conditions. Temperate rain forest dominated by evergr[...] trees is found where rainfall is abundant, such as in the Paci[...] of the United States. However, in the majority of temperate for[...] abundant enough during the growing season to support the gro[...] trees, but cold temperatures during the winter limit photosynthesis[...] water in the soil.

Trees with broad leaves can grow faster than narrow-leaved tre[...] they have an advantage in the summer. However, broad leaves allow tremendous

**FIGURE 16.10 Life in the canopy.** This bromeliad, often found more than 60 m from the forest floor, is vase shaped as an adaptation to acquire water and nutrients from rainfall.

**FIGURE 16.11 Loss of tropical forest.** According to the UN, about half of Earth's tropical forests have been cleared.

**FIGURE 16.12 Springtime in a temperate deciduous forest.** Forest wildflowers are adapted to take advantage of a short period between snowmelt and tree leafing by emerging from the soil, flowering, and producing fruit rapidly.

amounts of water to evaporate from a plant, leading to potentially fatal dehydration when water supplies are limited. To balance these two seasonal challenges, most broad-leaved trees in temperate forests have evolved a **deciduous** habit, meaning that they drop their leaves every autumn. In preparation for shedding its leaves, a deciduous tree seals the connection between branch and leaf, and chlorophyll, the green pigment essential for photosynthesis, within the leaf breaks down and cannot be replaced. The colorful fall leaves that grace eastern North America in September and October result from secondary photosynthetic pigments and sugars that are left behind in the leaf after the chlorophyll disappears.

**STOP & STRETCH** Suggest an explanation for why deciduous trees reabsorb chlorophyll, but not all pigments, from their leaves without shedding them. (Hint: chlorophyll contains significant amounts of nitrogen.)

The short time lag between the onset of warm temperatures and the releafing of deciduous trees provides an opportunity for plants on the temperate forest floor to receive full sunlight. Spring in these forests triggers the blooming of wildflowers that flower, fruit, and produce seeds quickly before losing light and water to their towering companions (**FIGURE 16.12**). The thinner leaves of deciduous trees allow more sunlight through than do their tropical forest equivalents, permitting a shrub layer typically missing in the tropics. The forest floor and shrub layer are collectively known as the forest **understory.** In addition, animals in temperate forests are more evenly distributed throughout the forest rather than concentrated in the treetops.

Nearly all of the deciduous forested lands in the eastern United States were converted to farmland within 100 years of the American Revolution. However, in the late nineteenth and early twentieth centuries, farms in the eastern United States were abandoned as production switched to the south and west. The regrowth of these forests provided a unique opportunity for ecologists to examine **succession,** the progressive replacement of different suites of species over time. Succession in most forests follows a predictable pattern. The first colonizers of vacant habitat are fast-growing plants that produce abundant, easily dispersed seeds. These pioneers are replaced by plants that grow more slowly but are better competitors for light and nutrients. Eventually, a habitat patch is dominated by a set of species—called the climax community—that cannot be displaced by other species without another environmental disturbance (**FIGURE 16.13**).

Today, these resurgent forests are once again threatened—this time by expanding urban development. The World Wildlife Fund estimates that worldwide, only 5% of temperate deciduous forests remain relatively untouched by humans.

**FIGURE 16.13 Forest succession in the eastern United States.**
(a) Abandoned farms are quickly colonized by fast-growing herbs. (b) The second stage of succession consists of a dense landscape of shrubs and quickly growing trees. (c) The climax community in the northeastern United States is made up of beech and maple trees.

**(a)**

**(b)**

**(c)**

**Boreal Forests.** Coniferous plants that produce seed cones instead of flowers and fruits dominate boreal forests (**FIGURE 16.14**). In fact, boreal forests are the only land areas where flowering plants are not the dominant vegetation type.

Climate conditions for boreal forests include very cold, long, and often snowy winters and short, moist summers. The dominant conifers in these forests are evergreens, meaning that they maintain their leaves throughout the year. Coniferous tree leaves are needle shaped and coated with a thick, waxy coat—both adaptations to reduce water loss and to help shed snow during winter. Their evergreen habit likely explains conifers' dominance over flowering trees in boreal forests—the growing season is so short that the ability to begin photosynthesizing immediately after thawing gives conifers an advantage over faster-growing but slower-to-start deciduous trees.

Because they occupy regions with such cold winters and short summers, boreal forests tend to be lightly populated by humans. These areas represent some of the "wildest" landscapes on Earth—home to moose, wolves, bobcat, beaver, and a surprising diversity of summer-resident birds. Trees in these landscapes are valuable for both building materials and paper products, and logging in the boreal forest is extensive. There are increasing concerns that the boreal forest in North America is being cut at rates faster than it can be replaced.

**Chaparral.** One major biome dominated by woody plants is not a forest. This landscape is known as chaparral, and its vegetation consists mostly of spiny evergreen shrubs.

Long, dry summers and frequent fires maintain the shrubby nature of chaparral. In fact, chaparral vegetation is uniquely adapted to fire. Several species have seeds that will germinate only after experiencing high temperatures. Many chaparral plants have extensive root systems that quickly resprout after aboveground parts are damaged. Chaparral will grow into temperate forest when fire is suppressed. As a result, natural selection has favored chaparral shrubs that actually encourage fire—such as rosemary, oregano, and thyme, which contain fragrant oils that are also highly flammable.

Fire can be an important contributor to the structure and function of many biomes, including chaparral. In Southern California, the flammability of chaparral vegetation has come directly into conflict with rapid urbanization. In recent years, this area has experienced extensive and expensive fire seasons—for example, in 2009, fires near Los Angeles burned nearly 400,000 acres and cost almost $100 million. In response to these events, support has increased for a policy of suppressing fires immediately rather than letting them burn. In fact, such a strategy may contribute to the buildup of fuel and more intense fires that are destructive to both people and the wildlands. Fire policy in Southern California along with the long-term human modification of chaparral around the Mediterranean makes this biome one of the most threatened on the planet.

## Grasslands

**Grasslands** are regions dominated by nonwoody grasses and containing few or no shrubs or trees. These biomes occupy geographic regions where precipitation is too limited to support woody plants. Grasslands can be further categorized into tropical and temperate categories.

Tropical grasslands, found generally in the regions between the equatorial rain forest and the large deserts around 30° N and S latitude, are known as **savannas** and are characterized by the presence of scattered individual trees. Savannas are maintained by periodic fires or clearing. In regions where wet and dry seasons are

**FIGURE 16.14 Coniferous trees.** Coniferous trees in the boreal forest are evergreen, an adaptation that allows them to dominate this biome. These plants produce sperm and eggs on cones instead of in flowers, which may be an advantage in cold climates because they do not rely on animal pollinators but are instead wind pollinated.

## Visualize This ▼

Think about the physical differences between elephants and cattle. Why aren't domestic cattle as effective as elephants at maintaining savanna?

**FIGURE 16.15 Savanna maintenance.** Elephants severely prune acacia trees when feeding on them, keeping the size of the trees relatively small; they also actively destroy thorny myrrh bushes that they do not eat, which provides additional habitat for grass to grow.

**FIGURE 16.16 Desert plant adaptations to constant drought.** Thick, whitish, and waxy coatings reflect light and reduce evaporation. Column-like forms reduce exposure to the high-intensity midday sun. Stems store water. The modification of leaves into spines discourages predators from accessing their water or damaging their protective surface.

distinct, yearly grass fires during the dry season kill off woody plants; in regions where elephants and other large grazing animals are present, the damage these animals do to trees helps to maintain the grass expanse (**FIGURE 16.15**).

Because grazing animals eat the tops of plants, natural selection in these environments has favored plants, like grasses, that keep their growing tip at or below ground level. Grass grows from its base, so grazing (or mowing) is equivalent to a haircut—trimming back but not destroying. (Woody plants grow from the tips, so intense grazing can destroy these plants.)

Although grazing mammals are characteristic of savannas, the introduction of large numbers of cattle in these habitats has transformed the landscape from grassland to bare, sandy soil in a process known as **desertification.** In the Sahel region of central Africa, overgrazing and resulting desertification has destroyed thousands of square kilometers of savanna.

Temperate grassland biomes are generally found north and south of the great desert zones or in the rainshadow of temperate mountain ranges and include tallgrass **prairies** and shortgrass **steppes.** Generally, the height of the vegetation corresponds to the precipitation—greater precipitation can support taller grasses. These landscapes are generally flat or slightly rolling and contain no trees.

In the cooler temperate regions, decomposition is relatively slow, and the soil of prairies and steppes is rich with the remains of partially decayed plants. These soils provide an excellent base for agriculture. As a result, most native prairies and steppes have been plowed and replanted with crops. In North America, less than 1% of native prairie remains.

## Desert

Where rainfall is less than 50 cm (20 in.) per year, the biome is called **desert.** Most of these deserts are close to 30° N or 30° S of the equator, where the air masses that were "wrung" of water in the rain-forest regions around the equator fall back to Earth's surface as hot and dry winds.

Although the image of a desert is often hot and sandy, some deserts can be quite cold; most deserts have vegetation, although it can be sparse. Plants and animals in desert regions have evolutionary adaptations to retain and conserve water. For example, plants are often thickly coated with waxes to reduce evaporation, contain photosynthetic adaptations that reduce water loss through leaf pores, and may be protected from predators of their stored water by spines and poisonous compounds (**FIGURE 16.16**).

The dominant vegetation in many deserts is low, slow-growing woody plants with very deep roots. But deserts are also home to many flowering plants that complete their entire life cycle from seed to seed in a single season. In deserts, the wet season is quite short; many of these plants can germinate, flower, and produce seeds in a matter of 2 or 3 weeks. The seeds they produce are hardy and adapted to survive in the hot, dry soil for many years until the correct rain conditions return.

Some animals in these dry environments have physiological adaptations that allow them to survive with little water intake. The most amazing of these animals are the various species of kangaroo rat, which apparently never consume water directly. While these animals get some water from their foods, they also conserve water produced during the chemical reactions of metabolism and have kidneys that produce urine four times more concentrated than our own.

The appeal of the sunny, warm, and dry climate of the deserts of the southwestern United States has made this region one of the fastest-growing areas in the country. The increasing population of the desert Southwest is putting stress on water supplies, causing conflicts among water users, and depleting water sources for native animals. The Colorado River is so extensively used that often it cannot sustain a flow through its historic outlet at the Gulf of California in Mexico. As a result, up to 75% of the species that lived in the Colorado River delta are now gone.

## Tundra

The biome type where temperatures are coldest—close to Earth's poles and at high altitudes—is known as **tundra.** Here, plant growth can be sustained for only 50 to 60 days during the year, when temperatures are high enough to melt ice in the soil. High temperatures are not sustained long enough to melt all of the ice stored there. Therefore, places like the arctic tundra near the North Pole are underlain by **permafrost,** icy blocks of gravel and finer soil material. The permafrost layer impedes water drainage, and soils above permafrost are often boggy and saturated. The shallow saturated soil cannot support deep roots, and so the tundra cannot support tall trees.

Plants in tundra regions are adapted to windswept expanses and freezing temperatures, often growing in mutualistic multispecies "cushions" where all individuals are the same height and shelter one another. This low vegetation supports a remarkably large and diverse community of grazing mammals, such as caribou and musk oxen, and their predators, such as wolves.

Animals in tundra regions have evolved to survive long winters with structural adaptations, such as storing fat and producing extra fur or feathers. Other animals, such as ground squirrels, have adapted by evolving hibernation; they enter a sleep-like state and reduce their metabolism to maximize energy conservation. Grizzly bears and female polar bears also spend many of the coldest months in deep sleep; although this is not true hibernation, these bears are so lethargic that females give birth in this state without fully awakening. Other animals survive by migrating south to avoid the hardships of a long, frigid winter.

Tundra is very lightly settled by humans, but it is threatened by our dependence on fossil fuels—oil, natural gas, and coal that formed from the remains of ancient plants. Some of the largest remaining untapped oil deposits are found in tundra regions, including northern Alaska, Canada, and Siberia, and the proposed development of these resources threatens thousands of square kilometers of tundra with destruction caused by road building, oil spills, and water pollution. Fossil fuel use also contributes to global warming, which has had the greatest impact at the poles, where tundra is predominant. Winters in Alaska have warmed by 2° to 3°C (4°–6°F), whereas elsewhere they have warmed by about 1°C (2°F). As climate conditions change, areas that were once tundra have begun to support shrub and tree growth and are changing into boreal forest.

# 16.3 Aquatic Biomes

Nearly all human beings live on Earth's land surface, but most people also live near a major body of water and are both influenced by and have an influence on these **aquatic** systems. Aquatic biomes are typically classified as either freshwater or saltwater, and their characteristics are summarized in **TABLE 16.2**, on page 383.

**aqua-** means water.

## Freshwater

Freshwater is characterized as having a low concentration of salts—typically less than 0.1% total volume. Scientists usually describe three types of freshwater biomes: lakes and ponds, rivers and streams, and wetlands.

**(a) Lake in summer**

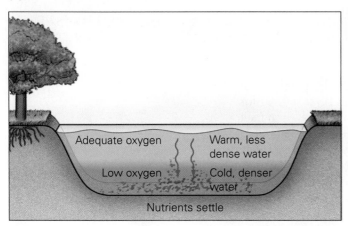

Adequate oxygen

Low oxygen

Warm, less dense water

Cold, denser water

Nutrients settle

**(b) Lake in fall and spring**

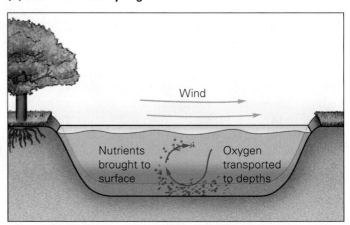

Wind

Nutrients brought to surface

Oxygen transported to depths

**FIGURE 16.17 Nutrients in lakes.** (a) A lake is stable during the summer, and oxygen levels in deeper regions diminish as nutrient levels near the surface decline. (b) Nutrients and oxygen are mixed throughout a lake during fall and spring "turnover," providing the raw materials for algae growth.

**(a)**

**(b)**

**FIGURE 16.18 Eutrophication.** (a) A boat travels through an algae "bloom" in Lake Erie. When these algae die, the bloom of bacteria that decompose them cause severe oxygen depletion, (b) leading to fish kills.

## Lakes and Ponds.

Bodies of water that are inland, meaning surrounded by land surface, are known as lakes or ponds. Typically, ponds are smaller than lakes, although there is no set guideline for the amount of surface area required for a body of water to reach lake status. Some ponds, however, are small enough that they dry up seasonally. These vernal (springtime) ponds are often crucial to the reproductive success of frogs, salamanders, and a variety of insects that spend part of their lives in water.

The aquatic environment of lakes and ponds can be divided into different zones: the surface and shore areas, which are typically warmer, brighter, and thus full of living organisms; and the deepwater areas, which are dark, low in oxygen, cold, and home to mostly decomposers. The biological productivity of lakes in temperate areas is increased by seasonal turnovers—times of the year when changes in air temperature and steady winds lead to water mixing, which redistributes nutrients from the bottom of the lake to the surface and carries fresh oxygen from the surface to deeper water (**FIGURE 16.17**).

Fertilizers applied to agricultural lands and residential lawns near lakes and ponds can increase their algae populations as these nutrients leach into the water. Ironically, having too many nutrients can lead to the "death" of these water bodies through a process known as eutrophication. Eutrophication occurs because large populations of algae lead to large populations of microbial decomposers, which consume a large fraction of oxygen in the water, thereby suffocating the native fish (**FIGURE 16.18**).

The life in freshwater lakes may also be threatened by **acid rain,** precipitation that has a high acid content resulting from high levels of air pollution. In lakes with little ability to buffer pH changes, acid rain lowers the pH and causes sensitive species to die off. Antipollution legislation in the United States has reduced the impact of acid rain in the northeastern states, but acid precipitation remains a significant problem in countries with less strict pollution regulations.

## Rivers and Streams.

Rivers and streams are bodies of flowing water moving in one direction. These waterways can be divided into zones along their lengths: the headwaters near the source, the middle reaches, and the mouth where they flow into another body of water.

**TABLE 16.2** Aquatic biomes.

| Biome | Characteristics |
|---|---|
| **Freshwater** | **Less than 0.1% total volume of salts** |
| Lakes and Ponds | Inland, some dry up seasonally. Yearly seasonal turnovers in larger lakes cause mixing of nutrients that would settle to the bottom. |
| Rivers and Streams | Flowing water moving in one direction. Different habitats along the way, from the fast-flowing, oxygen-rich regions at the headwaters to the sluggish, sediment-filled waters near the mouth. |
| Wetlands | Standing water with emergent plants. High species diversity due to high-nutrient levels at water–land interface. |
| **Marine (Saltwater)** | |
| Oceans | Contains multiple habitats, including (clockwise from top) intertidal zones, where organisms are exposed to fluctuating water levels; deep abyssal zones where no light penetrates; and open ocean. |
| Coral Reefs | Tropical areas in relatively shallow water. Reefs are made up of the minerals deposited by coral animals and provide habitat for a diverse array of species. |

*continued*

**TABLE 16.2 Aquatic biomes.** *continued*

| Biome | | Characteristics |
|---|---|---|
| Estuaries | | Where freshwater rivers flow into intertidal areas; mix of salt- and freshwater. Highly productive due to nutrients from the rivers and warmer temperatures. |

At the headwaters of a river—often by the source lake, an underground spring, or a melting snowpack—the water is clear, cold as a result of the temperature of the source, and generally flowing quickly because its volume is small; it is thus high in oxygen, providing an ideal habitat for cold-water fish such as trout. Near the middle reaches of a river, the width typically increases; thus, the flow rate slows. Also increasing is the diversity of fish, reptile, amphibian, and insect species that the river supports because the warming water provides a better habitat for their food source of photosynthetic plants and algae. At the mouth of a river, the speed of water flow is even slower, and the amount of sediment—soil and other particulates carried in the water—is high. High levels of sediment reduce the amount of light in the water and therefore the diversity of photosynthesizers that survive there. Oxygen levels are also typically lower near the mouth because the activities of decomposers increase relative to photosynthesis. Many of the fish found at the mouth of a river are bottom-feeders, such as carp and catfish, which eat the dead organic matter that flowing water has picked up along its way.

Rivers and streams are threatened by the same pollutants that damage lakes. But their habitats also face wholesale destruction with the development of dams and channels—dams that provide hydropower or reservoirs for cooling fossil-fuel-powered, electricity-generating plants, and channels that simplify and expedite boat traffic.

**Wetlands.** Areas of standing water that support emergent, or above-water, aquatic plants are called wetlands. In number of species supported, wetlands are comparable to tropical rain forests. The high biological productivity of wetlands results from high-nutrient levels occurring when soil and organic matter washed across the landscape accumulates there (**FIGURE 16.19**).

Besides their importance as biological factories, wetlands provide health and safety benefits by slowing the flow of water. Slower water flows reduce the likelihood of flooding and allow sediments and pollutants to settle before the water flows into lakes or rivers.

**STOP & STRETCH** Given the role of wetlands in slowing water flow, how might their loss contribute to eutrophication of nearby lakes and rivers?

Since the European settlement of the continental United States, over 50% of wetlands have been filled, drained, or otherwise degraded. Although legislation passed in the last few decades has greatly slowed the rate of wetland loss in the United States, about 58,000 acres of this biome type are still destroyed around the country every year.

**FIGURE 16.19 Rich wetlands.** Wetlands tend to be low spots on the landscape and thus accumulate nutrients running off from surrounding soils. These nutrients support the great diversity of plants and animals found in these biomes.

## Saltwater

About 75% of Earth's surface is covered with saltwater, or **marine**, biomes. Marine biomes can be categorized into three types: oceans, coral reefs, and estuaries.

**Oceans.** The ocean covers about two-thirds of Earth's surface but is the least well-known biome of all. In truth, it can be thought of as a shifting mosaic of different environments that vary according to temperature, nutrient availability, and depth of the ocean floor. Oceanographers typically subdivide it into three regions.

About 50% of the oxygen in Earth's atmosphere is generated by single-celled photosynthetic plankton in the open ocean (**FIGURE 16.20**). The open ocean also generates most of Earth's freshwater because water molecules evaporating from its surface condense and fall on adjacent landmasses as rain and snow, which eventually flow into lakes, ponds, rivers, and streams.

Photosynthetic plankton serve as the base of the ocean's food chain, providing energy to microscopic animals called zooplankton, which in turn feed fish, sea turtles, and even large marine mammals such as blue whales. These predators provide a source of food for yet another group of predators, including sharks and other predatory fish, as well as ocean-dwelling birds such as albatross.

Unlike lakes, oceans experience tides—regular fluctuations in water level caused by the gravitational pull of the moon. As a result of tides, ocean coasts contain unique habitats known as **intertidal zones,** which are underwater during high tide and exposed to air during low tide. Organisms in intertidal zones must be able to survive the daily fluctuations and rough wave action they experience. Adaptations such as strong anchoring structures, burrow building, and water-retaining, gelatinous outer coatings allow animals and seaweeds to take advantage of the high-nutrient environment found along the shore.

Oceans also contain a habitat known as the abyssal plain, the deep ocean floor. In these areas, sunlight never penetrates, temperatures can be quite cold, and the weight of the water above creates enormous pressure. Once thought to be lifeless because of these conditions, the abyssal plain is surprisingly rich in life, supported primarily by the nutrients that rain down from the upper layers of the ocean. In the 1970s, researchers studying the deep ocean discovered a previously unknown ecosystem supported by bacteria that use hydrogen sulfide escaping from underwater volcanic vents as an energy source. Animals in this ecosystem either use the bacteria as a food source directly or have evolved a mutualistic relationship with them, providing living space while benefiting from the metabolism of the bacteria. The abyssal plain represents the last major unexplored frontier on Earth.

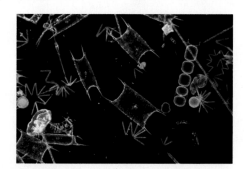

**FIGURE 16.20 The ocean's oxygen factories.** These phytoplankton are single-celled algae that are abundant in the ocean's surface waters. Their photosynthesis provides 70% of Earth's oxygen, as well as food to support an enormous diversity of life in the ocean.

**(a) Healthy coral reef**

**(b) Bleached coral reef**

**FIGURE 16.21 Threats to coral reefs.** (a) A healthy coral reef teems with life and color. (b) Stressed corals lose their photosynthetic algae and become bleached. Without healthy coral, the organisms that depend on it are lost as well.

The ocean is heavily exploited by human fishing fleets—recent estimates found that species diversity in the open ocean has declined by 50% over the last 50 years as a result of fishing pressure.

**Coral Reefs.** Coral reefs are unusual in that the structure of the habitat is not determined by a geological feature but composed of minerals excreted by the dominant organism in the habitat: coral animals. Coral animals are very simple in structure but have a unique lifestyle—they filter food items from the water but also receive nutrients from the photosynthetic algae they host inside their bodies. Up to 90% of the nutrition of a coral is provided by the algae, which receive in return a protected site for photosynthesis and easy access to the coral's nutrients and carbon dioxide. Reef-building coral live in large clones, and each individual coral secretes a limestone skeleton that protects it from other animals and from wave action.

Coral reefs are found throughout the tropics in warm and well-lit water. Their complex structure and high biological productivity make them the most diverse aquatic habitats, rivaling terrestrial tropical rain forests in species per area.

Coral reefs are sensitive to environmental conditions and prone to "bleaching," which occurs when host coral animals lose their algae companions (**FIGURE 16.21**). Without their photosynthetic mutualists, the animals may starve. Bleaching can occur for various reasons, but recent episodes seem to be associated with high ocean temperatures. Although coral can recover from bleaching, increased global temperatures due to climate change may lead to more frequent bleaching and the death of especially sensitive reef systems.

**Estuaries.** Zones where freshwater rivers drain into salty oceans are known as **estuaries.** The mixing of fresh and saltwater combined with water-level fluctuations produced by tides creates an extremely productive habitat. Some familiar and economically important estuaries are Tampa Bay, Puget Sound, and Chesapeake Bay.

Estuaries provide a habitat for the minnows of 75% of commercial fish species and 80 to 90% of recreational fish species; they are sometimes called the "nurseries of the sea." Estuaries are also rich sources of shellfish—crabs, lobsters, and clams. Vegetation surrounding estuaries, including extensive salt marshes consisting of wetland plants that can withstand the elevated saltiness compared to freshwater, provides a buffer zone that stabilizes a shoreline and prevents erosion. Unfortunately, estuaries are threatened by human activity as well, including eutrophication from increasing fertilizer pollution and outright loss as a result of housing and resort development.

# 16.4 The Human Impact

As is clear from our discussion of both aquatic and terrestrial biomes, no habitat on Earth has escaped the effects of humans. Consequently, to truly know our bioregion, we must learn how human populations use the environment. Preserving our biological neighborhoods requires being able to meet human needs while respecting the needs of other species with which we coexist.

According to the United Nations, humans have modified 50% of Earth's land surface (**FIGURE 16.22**). Most of this modification is for agriculture and forestry, but a surprisingly large amount—2 to 3% of Earth's land surface, or 3 to 4.5 million square kilometers, larger than the area of India—has been modified for human habitation. And this is only direct conversion; human activities have environmental effects far beyond their geographic boundaries, including changes to Earth's atmosphere, which receives some of our waste. In fact, it is fair to say that no natural habitat has escaped the human fingerprint.

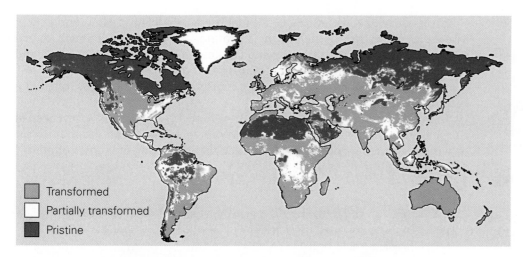

FIGURE 16.22 **Human modification of Earth's land surface.** Most of the land that is relatively untouched by humans lies in the vast boreal forest covering much of Canada and Siberia, the great deserts of Africa and Australia, and the dense Amazonian rain forest in north and central South America.

Legend:
- Transformed
- Partially transformed
- Pristine

◀ **Working with Data**

Use the map in Figure 16.1 and this figure to determine the climate conditions that are more likely to experience human modifications.

Today, half of the human population—approximately 3.5 billion people—live in cities. If current migrations from rural to urban areas continue, by the year 2050 two-thirds of the population then, or about 6 billion people, will live in urban areas. Clearly, cities are going to become increasingly important in determining the human impact on native biomes.

**STOP** & **STRETCH**   If two-thirds of the population in 2050 is estimated to be 6 billion people, what is the estimated size of the entire human population in 2050?

## Energy and Natural Resources

Cities are different from natural landscapes in a variety of ways. Consider the requirements of a forest. The only energy required for its growth is the solar irradiance striking the area, and most of the nutrients available to support this growth are already present in the soil. In addition, nearly all of the waste produced by forest organisms is processed on site—that is, it becomes part of the soil or air and is recycled in the system.

In contrast, cities rely extensively on energy imported from elsewhere and tend to be the central link in a linear flow of materials—from natural or agricultural landscapes and into waste disposed elsewhere.

**Energy Use.** In more developed countries, much of the energy required to power the activities within cities—from running buses and heating and cooling buildings to lighting traffic signals and energizing wireless phones—is derived from fossil fuels. These fuels are extracted from wells and mines around the globe, including the region surrounding the Persian Gulf, source of much of the world's oil, and the Appalachian Mountains, a rich source of coal. For most people in developed countries, a significant portion of the energy used is not associated with the bioregion in which they live. The environmental impacts of acquiring and transporting fossil fuel can be substantial, ranging from the degradation of oceans and estuaries resulting from oil spills to the wholesale dismantling of forested mountains during coal mining (**FIGURE 16.23**).

FIGURE 16.23 **The cost of fossil fuel extraction.** In the 1970s, mining companies developed a technique known as "mountaintop removal"—using explosives to blast off the rock above a coal deposit and dumping the waste rock into nearby valleys—to access coal. Since 1981, more than 500 square miles of West Virginia have been destroyed by this method of mining.

If we don't understand how our human communities fit into the surrounding biological community, the results can include pollution that harms our biological neighbors—and ourselves.

In less developed countries, fossil fuel use is still relatively low, and energy sources are more directly tied to the surrounding natural environment. A primary source of energy for heating and cooking in these countries is wood and other plant-based materials (including animal dung). As urban areas in less developed countries grow, surrounding forests are stripped of trees, often at a faster rate than they can regrow.

In both less developed and more developed countries, an awareness of cleaner sources of energy within a given bioregion—such as abundant solar or wind resources—could help to reduce the environmental costs of energy consumption.

**STOP & STRETCH** Which historical and contemporary factors do you think cause fossil fuels, rather than solar or wind power, to be the most common source of energy in the developed world today?

**Natural Resources.** In addition to energy, human settlements require materials for survival—food from agricultural production and harvesting of natural resources, metals and salts from mines, freshwater for human and industrial consumption, petroleum for asphalt and manufactured goods, and trees for processing into paper and packaging. Extracting all these raw materials changes biomes far from a city's center.

The amount of material needed to support a city can be tremendous. A year 2010 estimate calculated that the San Francisco area's **ecological footprint**— that is, the amount of land needed to support the human activity there—was 48 times the actual size of the city, equal to 70% of the land surface of the state of California, despite containing only 11% of the state's population. Clearly, the resources to support San Francisco must come from other states and countries— from the boreal forests of Oregon and British Columbia that supply wood for building and paper production to the wheat fields of Nebraska and Alberta that supply flour for the city's famous sourdough bread.

San Francisco is not exceptional in its importation of natural resources or its ecological footprint. In fact, most cities in more developed countries have similar size footprints. A city's footprint can be reduced by its citizens, especially if they reduce the amount of energy used for transportation and make sensitive consumer choices, including buying locally produced foods and less meat, which requires significant additional land to raise.

## Waste Production

The waste produced by a city presents a special problem. The density of the population is so great in cities that specialized systems are required to handle the enormous amounts of human waste generated in a small area. Human wastes can be liquid, in the form of wastewater; solid, in the form of garbage; or gaseous, in the form of air pollution.

**Wastewater.** In developed countries, cities have sewage treatment systems to handle the wastewater that drains from sinks, tubs, toilets, and industrial plants (**FIGURE 16.24**). These systems typically treat water by removing semisolid wastes through settling, using chemical treatment to kill any disease-causing microorganisms, and eventually discharging the treated water to lakes, streams, or oceans. Unfortunately, older treatment systems can be overwhelmed by storm water, causing the discharge of large volumes of wastewater directly into waterways. Untreated wastewater can cause nutrient levels to spike and algae

**FIGURE 16.24 Treatment of human waste.** In more developed countries, human waste is treated in expensive and technologically advanced sewage treatment plants. This is a great improvement over the past, when untreated waste was dumped directly in rivers, lakes, and oceans.

and bacteria growth to increase, thus leading to beaches being closed to swimming and other water-based recreation.

Disposal of semisolid wastes, called *sludge*, is a significant challenge. Sludge is often composted (allowed to decompose) and applied to land as fertilizer or trucked to landfills. Land application of composted sludge has its own problems; this material contains not only human waste—a valuable fertilizer if properly composted—but also industrial wastewater, which can contain a wide variety of toxic chemicals.

In less developed countries, safe disposal of treated sewage is often a distant dream. In many rapidly industrializing regions, the emigration of people to urban areas has overwhelmed antiquated and inadequate sewer systems. Large numbers of people in these cities live in slums where running water is scarce and untreated human waste flows in open gutters. The consequences of inadequate disposal of human waste can be severe—intestinal diseases due to contact with waste-contaminated drinking water cause the preventable deaths of over 2 million children under 5 years old every year.

**Garbage and Recycling.**  In addition to dealing with wastewater, people in urban areas must find ways to control their solid waste—that is, their garbage. In more developed countries, most of the solid waste finds its way into sanitary landfills, which are pits lined with resistant material such as plastic. Landfills have systems for collecting liquid that drains through the waste and exhaust pipes to vent dangerous gases that result from decomposition.

In most areas, landfill space is becoming more and more limited and farther removed from cities. To stave off a looming "garbage crisis," states and communities mandate recycling of paper, glass, metal, and plastic, and many have or are considering composting programs to reduce the amount of food and yard waste. Unfortunately, even where recycling rates are relatively high, household garbage production continues to increase (**FIGURE 16.25**).

In less developed countries, the problem of solid waste disposal is more severe. Many large cities in these areas have large, open dumps in which unstable, fire-prone piles of garbage provide living space for desperately poor immigrants in the city.

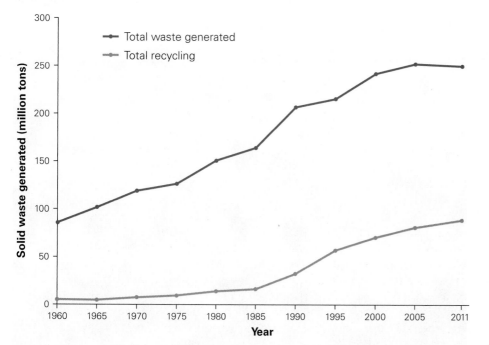

◀ **Working with Data**

How much non-recycled waste was produced in 1965, 1990, and 2011?

**FIGURE 16.25 Growing garbage.** This graph illustrates the growth in the amount of garbage produced each year in the United States since 1965. Recycling rates began to increase in the 1990s, rising from 10% in 1985 to 35% of all waste in 2011.

**Air Pollution.** Because of human dependence on fossil fuels, urban areas produce large amounts of gaseous waste. These air emissions include carbon dioxide as well as by-products of the combustion of oil, gasoline, and coal (from power plants), including nitrogen and sulfur oxides, small airborne particulates, and fuel contaminants such as mercury. Exposure to sunlight and high temperatures can cause some of these by-products to react with oxygen in the air to form ground-level ozone or smog (**FIGURE 16.26**). For individuals with asthma, heart disease, or reduced lung function, smog exposure can lead to severe illness, even death.

When gaseous pollution enters the upper reaches of Earth's atmosphere, it can be carried on air currents to less settled areas throughout the globe. Air emissions from coal-fired power plants throughout the Midwest cause severe acid rain in lightly settled regions of the northeastern United States. Airborne toxins such as benzene and PCBs (polychlorinated biphenyls), by-products of manufacturing processes, have been found in high levels in animals as a result of **bioaccumulation,** the process that causes relatively small amounts of pollutants to become concentrated to dangerous levels in animals that are high on the food chain.

Air pollution is a problem in both developed and less developed countries. In less developed countries, pollution control is weak or lacking altogether; in more developed countries, the sheer volume of fossil fuel use contributes to poor air quality. In the United States, the number of miles driven by car per household has increased by 50% since the 1970s partially due to an increase in the distance that individuals live from their workplaces.

Development of suburban settlements outside the geographical limits of cities has been termed **urban sprawl.** Urban sprawl not only contributes to an increase in fossil fuel consumption but also negatively affects wildlife through habitat destruction and fragmentation. Urban sprawl also impairs water quality via the destruction of wetlands and the increasing amount of paved surfaces, which funnel pollutants and warmed water into lakes, streams, and rivers.

**Reducing Human Impacts.** The effect of human settlements on surrounding natural biomes can be significant and severe, as we have discovered in this chapter. However, many of these impacts can be mitigated with thoughtful

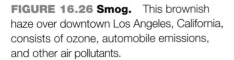

**FIGURE 16.26 Smog.** This brownish haze over downtown Los Angeles, California, consists of ozone, automobile emissions, and other air pollutants.

planning and the use of improved technology. In the United States, laws such as the Clean Air Act and the Clean Water Act—passed in 1970 and 1972, respectively—have greatly reduced air and water pollution and have contributed to the recovery of once severely impaired habitats (**FIGURE 16.27**).

Cities throughout the more developed world are supporting projects aimed at creating sustainable communities that are both economically vital and environmentally intelligent. **TABLE 16.3** poses a set of questions that will help you become more knowledgeable about your own bioregion. Perhaps by getting to know our biological neighbors and understanding how our choices affect these organisms, we can be inspired to help create communities that are safe and healthy for humans and other species.

**FIGURE 16.27 Fitting human needs into the bioregion.** Lake Erie was once so polluted that it was considered "dead" in the 1960s. Thanks to government and citizen efforts, the lake has now recovered and provides clean swimming beaches and much healthier fish populations.

**TABLE 16.3 Know your bioregion.**

| Environmental Factor | Your Bioregion |
| --- | --- |
| Climate | Describe the climate of the place where you live, including the annual average temperature, average precipitation, average number of sunny days, and names of seasons. Which global and local factors influence the climate where you live? |
| Vegetation | Describe the native vegetation of the place where you live. How much native vegetation remains, and what has taken its place? Which ecological factors (climate, fire, and so on) have influenced the native vegetation type? |
| Aquatic habitats | What are the aquatic habitats nearest to you? Can you name some of the dominant species in these habitats? What threats do these habitats face in your area? |
| Energy resources | What is the primary source for the electricity you use in your home? Is your bioregion rich in any less environmentally damaging energy resources? |
| Food resources | What agricultural products are produced in your bioregion? How easy is it to buy locally produced food? |
| Natural resources | What natural resources does your bioregion supply? |
| Wastewater and sewage | Where does your wastewater go? Does your community have any problems handling wastewater and sewage? |
| Solid waste | Where does your garbage go? What is the rate of recycling in your community, and can it be improved? |
| Air quality | How is the air quality in your community? What are the major air pollutants and their sources? |

# *savvy reader*

## Greenwashing

A clean and healthy environment is a high priority for many consumers—which is why many businesses tout their environmental credentials in their advertisements and products. While many companies have business practices that represent a true choice of placing conservation above profits, some companies have made a practice of using public relations techniques to claim that they are ecologically friendly when they have not made the same choice. This technique is called *greenwashing*.

One way to spot greenwashing, or indeed any deceptive advertising practice, is to use the same guidelines used in Chapter 1. In short, any claim made in advertising should be supported by objective evidence. Consider the following statements posted on a variety of company websites:

From two global oil and gas companies:

> "Many of the locations in which we operate present challenging environmental sensitivities, so managing our impact in these areas is always at the core of our activities."

> "We consider biodiversity early in new projects, develop biodiversity action plans, and collaborate with biodiversity experts to help protect areas with rich and delicate ecosystems."

From two multinational computer technology corporations:

> "Today's energy- and climate-related issues are at the top of our strategic agenda."

> "We know that the most important thing we can do to reduce our impact on the environment is to improve our products' environmental performance. That's why we design them to use less material, ship with smaller packaging, be free of toxic substances used by others, and be as energy efficient and recyclable as possible."

From two luxury automobile makers:

> "Every … manufacturing site has achieved Zero Landfill Status, meaning that any waste resulting from manufacture is recycled, repurposed, or used to generate additional energy."

> "Our measurement of success must go beyond the traditional bottom line and extend to recognition of the environment and the needs of individuals both inside and outside the company."

1. Which, if any, of these statements by itself contains objective information to support the claims?
2. Where would it be reasonable to expect these companies to report tests, analyses, research, studies, or other evidence to support these claims?
3. Choose one of these statements and list the evidence that you would want to see in order to determine if the claim has merit.

# SOUNDS RIGHT BUT IS IT?

In the Midwest of the United States, the sun feels so hot in July but seems weak and pale in January. The quality of the sunlight is so different that it is tempting to say:

**The sun is closer to Earth in summer than it is in winter.**

Sounds right, but it isn't.

**1.** What season (that is, winter, spring, summer, and fall) is it in January in the midwestern United States? What season is it during January in Australia?

**2.** What is "solar irradiance"?

**3.** How does the tilt of Earth's axis affect solar irradiance in January compared to July in the Midwest? How about in Australia?

**4.** Is the sun closer to Earth when it is summer in Australia?

**5.** Reflect on your answers to questions 1–4 and explain why the statement in bold sounds right, but isn't.

---

# Chapter Review MasteringBiology®

Go to the Study Area in MasteringBiology® for practice quizzes, myeBook, BioFlix™ 3-D animations, MP3 Tutor sessions, videos, current events, and more.

## Summary

## Section 16.1

Describe the factors that determine temperature and precipitation patterns on a global scale.

- Climate is the weather of a place as measured over many years. Climate in an area is determined by global temperature patterns, which are driven by solar irradiance and local factors such as proximity to large bodies of water (p. 368).

- Temperatures are warmer at the equator than at the poles because solar irradiance is greater at the equator (pp. 368–369).

- Seasonal changes are caused by the tilt of Earth's axis. For example, during summer in the Northern Hemisphere, total solar irradiance during the day is much greater than it is during winter (pp. 369–370).

- Global precipitation patterns are driven by solar energy, with areas of heavy rainfall near the solar equator and dry regions at latitudes of 30° north and south of the equator (pp. 370–371).

List factors that influence climate on a local scale, and explain how they affect local temperature and precipitation patterns.

- Local temperatures are influenced by altitude, proximity to a large body of water, nearby ocean currents, the light reflectance of the land surface, and the amount of surrounding vegetation (pp. 372–373).

- Local precipitation patterns are influenced by the presence of a large body of water, which tends to increase rainfall, and the presence of a mountain range, which tends to reduce rainfall on its leeward or sheltered side (p. 374).

## Section 16.2

Briefly describe the climate and vegetation characteristics of the terrestrial biomes, including tropical, temperate, and boreal forests; chaparral; tropical and temperate grasslands; deserts; and tundra, and give examples of animal and plant adaptations in these environments. Describe some of the threats to these biomes imposed by human activity.

- Forests are dominated by trees and categorized by distance from the equator into tropical, temperate, and boreal types. Tropical forests are dominated by large trees with little understory, while temperate forests have a diverse understory that flourishes when the trees drop their leaves. Boreal forest is dominated by evergreen conifers that conserve water during the long, cold winter (pp. 375, 377–378).

- The chaparral biome is found in slightly hotter and drier areas than forests and is dominated by highly flammable woody shrubs that regenerate quickly after fire (pp. 376, 379).

- Grasslands are found where precipitation is relatively low. The major vegetation type on grasslands is non-woody grasses. These biomes are categorized by distance from the equator into tropical and temperate types. Both tropical and temperate grasslands are maintained by periodic fires and adapted to regrow quickly (pp. 376, 379–380).

- Deserts are found where precipitation is 50 cm (20 in.) per year or less and contain mostly drought-resistant plants and animals adapted to the low-water environment (p. 380).
- Tundra occurs in areas where the growing season is 60 days or less, both near the poles and at high altitudes, and is dominated by short plants and grazing animals (p. 381).
- Forests are threatened by logging and agricultural development, while chaparral is primarily threatened by human fire-suppression activities. Grasslands are threatened by agricultural development and overgrazing, and deserts by human settlement and water depletion. Tundra is threatened by climate change and oil and gas exploration (pp. 377–381).

## Section 16.3

Briefly describe the characteristics of the aquatic biomes, including freshwater lakes, rivers, and wetlands, and saltwater oceans, coral reefs, and estuaries, and provide examples of organisms found in these biomes. Describe some of the threats to these biomes imposed by human activity.

- Freshwater biomes include lakes, areas of standing water dominated by aquatic plants and algae and fish; rivers, made up of flowing water with relatively few plants; and wetlands, which consist of waterlogged soil and emergent vegetation (pp. 381–384).
- Marine biomes include oceans, which can be divided into multiple zones depending on water temperature, light, and nutrient availability. The dominant organisms in the open ocean are photosynthetic plankton, while the great diversity of marine life is found in coral reefs and estuaries (pp. 385–386).
- Aquatic habitats are subject to pollution via acid rain and nutrient runoff from agricultural lands. Organisms in these biomes are often threatened by over harvesting (pp. 382, 384, 386).

## Section 16.4

List the ecological requirements of human communities, including food, energy, and waste disposal, and describe how these needs are met in various human societies.

- Humans have modified over 50% of Earth's land surface. Much of this modification now supports the activities of cities (pp. 386–387).
- Cities must rely on imported energy and other resources to survive; extraction of these resources carries an environmental cost felt by other bioregions (pp. 387–388).
- Waste disposal is a significant challenge for large urban areas. Sewage must be treated to avoid contaminating drinking water; garbage must be disposed of effectively,

and air emissions must be controlled. More developed countries have more resources to consume and handle the vast amounts of waste that they produce, while less developed countries struggle with inadequate systems (pp. 388–390).
- Increasing the bioregional awareness of citizens and communities can help them devise methods for supporting human activities in a more sustainable manner (pp. 390–391).

## Roots to Remember

**The following roots of words come mainly from Latin and Greek and will help you decipher terms:**

| | |
|---|---|
| **aqua-** | means water. Chapter term: *aquatic* |
| **equ-** | means equal. Chapter term: *equator* |
| **sol-** | means sun. Chapter term: *solstice* |
| **-stice** | means to stand still. Chapter term: *solstice* |

## Learning the Basics

1. Explain why the northern United States experiences a cold season in winter and a warm season in summer.

2. Why does proximity to a large body of water moderate climate on a nearby landmass?

3. Areas of low solar irradiation are _____.
   A. closer to the equator than to the poles; B. closer to the poles than the equator; C. at high altitudes; D. close to large bodies of water; E. more than one of the above is correct.

4. The solar equator, the region of Earth where the sun is directly overhead, moves from 23.5° N to 23.5° S latitudes and back over the course of a year. Why?
   A. Earth wobbles on its axis during the year; B. The position of the poles changes by this amount annually; C. Earth's axis is 23.5° from perpendicular to the rays of the sun; D. Earth moves 23.5° toward the sun in summer and 23.5° away from the sun in winter; E. Ocean currents carry heat from the tropical ocean north in summer and south in winter.

5. Which of the following biomes is most common on Earth's land surface?
   A. chaparral; B. desert; C. temperate forest; D. tundra; E. boreal forest

6. Tundra is found _____.
   A. where average temperatures are low and growing seasons are short; B. near the poles; C. at high altitudes; D. A and B are correct; E. A, B, and C are correct.

7. Which statement best describes the desert biome?
   A. It is found wherever temperatures are high; B. It contains a larger amount of biomass per unit area than any

other biome; **C.** Its dominant vegetation is adapted to conserve water; **D.** Most are located at the equator; **E.** It is not suitable for human habitation.

**8.** Which of the following biomes has a structure made up primarily of the mineral deposits secreted by its dominant organisms?

**A.** coral reefs; **B.** freshwater lakes; **C.** rivers; **D.** estuaries; **E.** oceans

**9.** An ecological footprint _____.

**A.** is the position an individual holds in the ecological food chain; **B.** estimates the total land area required to support a particular person or human population; **C.** is equal to the size of a human population; **D.** helps determine the most appropriate wastewater treatment plan for a community; **E.** is often smaller than the actual land footprint of residences in a city.

**10.** In more developed countries, wastewater containing human waste _____.

**A.** is dumped directly into waterways; **B.** is applied directly to farm fields; **C.** is piped into sanitary landfills; **D.** is treated to remove sludge and disease-causing organisms and released into nearby waterways; **E.** is a major source of infectious disease in human populations.

## Analyzing and Applying the Basics

**1.** Consider the following geographic factors and predict both the climate and biome type found in the location described. Explain the reasoning that you used to determine your answer. This small city is:

- On the coast of the Pacific Ocean
- 20° north of the equator
- 20 m above sea level
- At the base of a mountain range

**2.** One prediction of global climate change models is that significant amounts of melting ice will change the salt content of the ocean, causing the Gulf Stream current in the Atlantic Ocean to stop altogether. How will this change likely affect Europe?

**3.** What can you infer about the geographical relationship among the cities in the following table and the Cascades, the primary mountain range that influences their climate?

| City | Approximate Average Annual Rainfall in centimeters (inches) |
|---|---|
| Bend, OR | 31 (12) |
| Eugene, OR | 109 (43) |
| Portland, OR | 91 (36) |
| Tacoma, WA | 99 (39) |
| Walla Walla, WA | 53 (21) |
| Yakima, WA | 20 (8) |

## Connecting the Science

**1.** How many biomes do you rely on to supply your food? Many grocery stores label the origin of their produce. The next time you go to the grocery store, try to determine the number of different countries from which your groceries come. Could you easily change your diet and shopping habits to rely on locally produced food? Why or why not?

**2.** Consider the setting of your home. What kind of changes to the environment of your home would help it fit better into the bioregion? Are these changes feasible? Are they desirable?

Answers to **Stop & Stretch, Visualize This, Working with Data, Savvy Reader, Sounds Right, But Is It?**, and **Chapter Review** questions can be found in the **Answers** section at the back of the book.

CHAPTER **17** | # Organ Donation

**A motorcycle accident results in severe head injuries to a young man.**

# Tissues and Organs

The drama plays out in emergency rooms around the country thousands of times every year. A formerly healthy young man is rushed to the hospital in an ambulance after having been involved in a motorcycle accident. After many failed attempts to find brain activity, emergency room physicians decide that the young

**Medical personnel search his wallet for an organ donor card.**

**He is rushed to the hospital, where it is determined that his brain has stopped functioning.**

man is brain dead. In the meantime, his family has assembled in the waiting room, overcome by the loss. A physician enters the waiting room and asks the family whether they would be willing to donate their loved one's tissues and organs.

The family is understandably confused and has many questions. Is there no chance that their loved one will ever be okay—even many years from now? If they disagree, will the medical staff not work as hard to keep their loved one alive? What will

happen to the tissues and organs if they do decide to donate them? The family knows that their loved one would have liked the idea of helping others, but they have never really discussed organ donation. Even though his driver's license lists him as a donor, the family is not sure what to do.

What would you want your family to do? This chapter may help you decide.

**His family is asked whether they will donate his organs.**

# 17.1 Tissues

Groups of similar cell types that perform a common function comprise a **tissue.** Damage to tissues can sometimes prevent them from recovering their function, which may be restored by a tissue **transplant.** We will first learn about the structure and function of different tissue types and then see some situations in which transplants can help replace damaged tissues. The human body has four main types of tissues: epithelial, connective, muscle, and nervous.

**trans-** means through.

## Epithelial Tissue

**Epithelial tissue,** or **epithelium,** is tightly packed sheets of cells that cover organs and outer surfaces and line hollow organs, vessels, and body cavities (**FIGURE 17.1**).

**epi-** means on or upon.

Sheets of epithelial tissue, unlike other tissue types, are usually not attached on all surfaces. Instead, epithelial tissues are anchored on one face of the tissue but free on the other tissue face. The unattached tissue can be exposed to body fluids or the external environment. For example, the epithelial tissue that lines blood vessels is attached to the wall of the blood vessel, but its outer surface is exposed to the bloodstream. The epithelium that comprises the outer layer of skin, called the epidermis, is anchored to the underlying tissue but has a surface that is exposed to air. This is also true of the epithelia that line other parts of the body, including respiratory surfaces and the digestive, urinary, and genital (or urogenital) tracts. Epithelia can be a single layer or many layers thick.

Epithelial cells function in protection, secretion, and absorption. Epithelial cells in the skin help protect the body from injury and ultraviolet light. Some glands in the skin are formed by outgrowths of epithelial tissue. These glands

**FIGURE 17.1 Epithelial tissues.** (a) Epithelia line many organs. (b) Epithelial tissues are tightly packed cells that usually have one unattached surface, as seen in the epidermis of the skin. (c) Epithelia can be multilayered or a single layer, as is the epithelium that lines the small intestine.

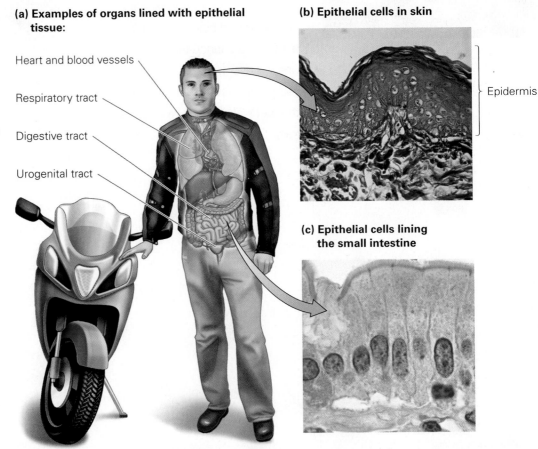

**(a) Examples of organs lined with epithelial tissue:**

Heart and blood vessels

Respiratory tract

Digestive tract

Urogenital tract

**(b) Epithelial cells in skin**

Epidermis

**(c) Epithelial cells lining the small intestine**

secrete substances such as mucus, oils, and sweat. Epithelial tissues can also function to protect the body from water loss and pathogens. Epithelial cells that line blood vessels and intestines function in absorbing nutrients.

Epithelial tissues constantly slough off cells, which are replaced by cell division. Skin cells are rubbed off by clothing, and cells that line the mouth and intestines are scraped off by food. These cells are replaced, resulting in a replacement of the epidermis every 4–6 weeks. Face transplants, in the news lately, involve replacing damaged epithelial tissues (**FIGURE 17.2**).

## Connective Tissue

Connective tissues function, as their name implies, to form connections. They usually bind organs and tissues to each other. Connective tissue is loosely organized, not tightly packed like epithelium. Our bodies contain six different types of connective tissues, all of them composed of cells embedded in a **matrix** made of protein fibers and an amorphous, gel-like ground substance. The matrix can be likened to Jell-O® with carrot shavings in it: The Jell-O is analogous to the ground substance, and the carrots are analogous to the protein fibers. The proportions of these components vary from one part of the body to another, depending on structural requirements. In some areas, the connective tissue is highly cellular; in others, the predominant feature is the protein fibers or ground substance of the matrix. The consistency of the matrix is a function of the types of fibers and ground substances present. Blood, for example, is a connective tissue with a liquid matrix. **Adipose** or fat tissue has a thick fluid, or viscous, matrix, and bone has a solid matrix. Specific compositions and functions of the six types of connective tissue follow.

**Loose Connective Tissue.**  Loose connective tissue (**FIGURE 17.3a**) is the most widespread connective tissue in animals. It connects epithelia to underlying

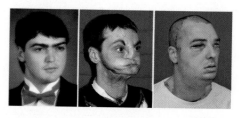

**FIGURE 17.2**  From left, Richard Norris before a shotgun accident disfigured his face, after his accident, and after his face transplant.

## Visualize This ▼

Compare the cell types and matrix components of the loose and fibrous connective tissues.

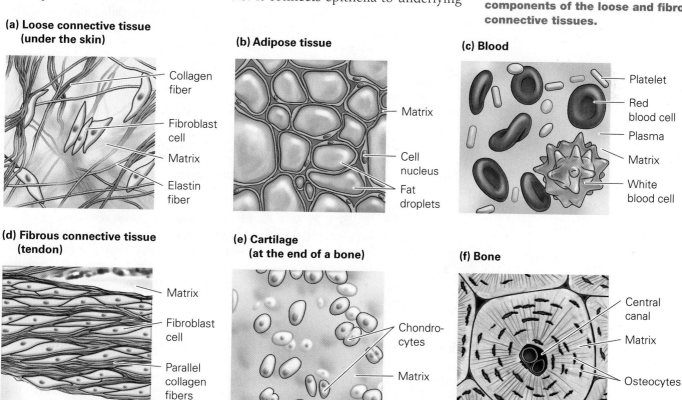

**(a) Loose connective tissue (under the skin)**
- Collagen fiber
- Fibroblast cell
- Matrix
- Elastin fiber

**(b) Adipose tissue**
- Matrix
- Cell nucleus
- Fat droplets

**(c) Blood**
- Platelet
- Red blood cell
- Plasma
- Matrix
- White blood cell

**(d) Fibrous connective tissue (tendon)**
- Matrix
- Fibroblast cell
- Parallel collagen fibers

**(e) Cartilage (at the end of a bone)**
- Chondrocytes
- Matrix

**(f) Bone**
- Central canal
- Matrix
- Osteocytes

**FIGURE 17.3 Connective tissues.**  The six different types of connective tissue are each composed of cells and a specialized matrix.

tissues, holds organs in place, and acts as padding under the skin and elsewhere. The tissue is called loose connective tissue due to the loose weave of its constituent fibers. The cells of loose connective tissue, called fibroblasts, secrete the proteins comprising the matrix. The matrix of loose connective tissue is its chief structural feature. Two proteins are important constituents of this matrix—collagen and elastin. Collagen fibers give connective tissue tremendous strength, and the elastin fibers allow connective tissue to stretch without breaking. The degradation of collagen and elastin in skin that occurs with age, exposure to sunlight, and cigarette smoke causes weakening of the connection between skin and underlying muscle, leading to wrinkles.

**Adipose Tissue.** **Adipose tissue** (**FIGURE 17.3b**) is fat tissue that connects the skin to underlying structures and insulates and protects organs. Adipose cells are specialized for the synthesis and storage of energy-rich reserves of fat. A fat droplet fills the cytoplasm of these cells and shrinks when the fat is used for energy. Cells are the predominant constituent of this connective tissue; only a small amount of matrix is associated with adipose tissue.

**adipo-** means fat.

**Blood.** **Blood** (**FIGURE 17.3c**) is a type of connective tissue that circulates throughout the body through arteries, veins, and capillaries and transports oxygen and nutrients to cells. The cellular component of blood tissue includes red cells, which carry oxygen; white cells, which help fight infection; and cell fragments called platelets, which function in clotting. Blood cells are suspended in a liquid- and protein-rich matrix called plasma.

**Fibrous Connective Tissue.** **Fibrous connective tissues** form tendons that connect muscles to bones and **ligaments** that connect bones to each other at joints. As in loose connective tissue, the cells of this tissue are called fibroblasts. The densely packed collagen fibers of the matrix, arranged in parallel, are the most conspicuous feature of fibrous connective tissue (**FIGURE 17.3d**).

**lig-** means to join or connect.

**Cartilage.** **Cartilage** is a type of connective tissue composed of cells, called chondrocytes, that secrete substances to form the dense matrix surrounding them (**FIGURE 17.3e**). The matrix is rich in collagen and other structural proteins. Cartilage connects muscles with bones, provides flexible support for the ears and nose, and allows for shock absorption. At joints, the ends of bones are covered in cartilage, which helps permit smooth gliding where joint surfaces contact each other. Cartilage has no blood vessels, so injuries heal very slowly if at all.

**STOP & STRETCH** Close to 30% of college athletes experience loss of cartilage in their joints due to injury or wear and tear. As they age, the cartilage that covers the bones can be worn down enough to allow direct contact between the heads of the bones comprising this joint; a painful condition called *osteoarthritis* results. Scientists can grow cartilage cells in the lab to provide cartilage replacements. What characteristic of cartilaginous tissues needs to be overcome for this type of transplant to succeed?

**Bone.** **Bones** that make up the skeleton are connected to each other at the joints. In addition to being a framework for the body, bones support and protect other tissues and organs. Marrow, inside the bone, produces blood cells. Bone is a rigid type of connective tissue composed of branched cells called **osteocytes.** These cells become trapped in a matrix, composed of collagen and minerals that they secrete (**FIGURE 17.3f**). Bone serves as a reservoir of calcium and minerals that the body can use if dietary levels of these substances are low. Central canals inside bones house nerves and blood vessels.

**osteo-** relates to bone.

Many connective tissues can be transplanted from tissue donors. Blood and bone marrow can be transplanted from living donors.

## Muscle Tissue

While connective tissue serves to support and connect, another important tissue is muscle tissue. In fact, muscle is the most abundant type of tissue in animals. Anything we think of as "meat" is actually muscle tissue.

Muscle is a highly specialized tissue capable of contracting. Most muscle tissue is composed of bundles of long, thin, cylindrical cells called muscle fibers. Muscle fibers contain specialized proteins, **actin** and **myosin,** that cause the cell to contract when signaled by nerve cells.

Muscle tissues are differentiated according to whether their function requires conscious thought—such as walking—and is **voluntary** or whether their function requires no conscious thought, such as the beating of the heart, and is **involuntary.** Muscle tissues are also differentiated by the presence or absence of bands that look like stripes under a microscope. Muscle that is banded is called **striated muscle.** Unbanded muscle is called **smooth muscle.** The striated appearance is due to the banding pattern formed by actin and myosin deposits in the cells. Muscle tissue that lacks these striations also contains actin and myosin deposits, just not in a banded pattern. Humans and other vertebrates have three types of muscle tissues. These contracting tissues are skeletal, cardiac, and smooth muscle.

**Skeletal Muscle.** **Skeletal muscle** is usually attached to bone and produces all the movements of body parts in relation to each other. Skeletal muscle is responsible for voluntary movements such as walking. This striated muscle tissue is shown in **FIGURE 17.4a**. Exercise causes an increase in the size of skeletal muscle cells, not an increase in the number of cells.

**(a) Skeletal muscle (biceps)**

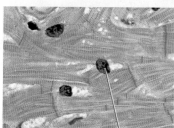

Muscle fiber   Nucleus

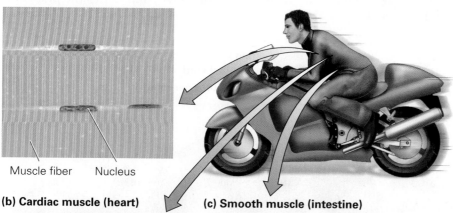

**(b) Cardiac muscle (heart)**

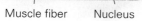

Muscle fiber   Nucleus

**(c) Smooth muscle (intestine)**

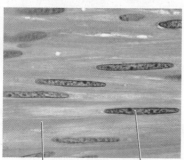

Muscle fiber   Nucleus

◀ **Visualize This**

**What causes the striped pattern seen in cardiac and skeletal muscle?**

**FIGURE 17.4 Muscle tissue.** Skeletal (a) and cardiac (b) muscles are striated, while smooth muscle (c) is not.

## Is there no chance that a brain-dead person will ever be okay?

**Cardiac Muscle.** **Cardiac muscle** is found only in heart tissue. This involuntary striated tissue undergoes rhythmic contractions to produce the heartbeat. Cardiac muscle cells are highly branched and interwoven (**FIGURE 17.4b**); this helps a contraction signal to be propagated.

**Smooth Muscle.** **Smooth muscle,** as its name implies, is not striated (**FIGURE 17.4c**). This involuntary muscle is composed of spindle-shaped cells that comprise the musculature of internal organs, blood vessels, and the digestive system. Smooth muscle contracts more slowly than skeletal muscle, but it can remain contracted for a long time. Consider the difference between the contractions of the smooth muscle lining the digestive tract to move food along as opposed to a biceps contraction. Contractions of the smooth muscles of the digestive tract are involuntary and are held for long periods of time. Contractions of the biceps are voluntary and occur more quickly. Cardiac muscle tissue can be transplanted as long as the donor heart has not been damaged. To avoid tissue breakdown in the donor heart, the transplant must take place within four hours after removal from the donor's body.

The last of the four types of tissues, and the least likely to be involved in a transplant, is nervous tissue.

### Nervous Tissue

Nervous tissue is composed mainly of cells called **neurons** (**FIGURE 17.5**) that can conduct and transmit electrical impulses. The main function of nervous tissue is to help the body sense stimuli, process the stimuli, and transmit signals from the brain out to the rest of the body and back. For example, if a car door suddenly opens in front of you while you are biking, you might react by trying to swerve out of the way. This action requires your body to recognize the threat and coordinate a response to it very quickly. It is the nervous tissue that makes this possible. Nervous tissue is found in the brain, spinal cord, and nerves. Most cells of the nervous system do not undergo cell division and therefore cannot repair themselves. This is why injuries to the brain and spinal cord are so devastating.

**neuro-** means nerve.

**FIGURE 17.5 Nervous tissue.** The brain and spinal cord are composed of nervous tissues. Nerves are composed of neurons that transmit signals to and from the brain.

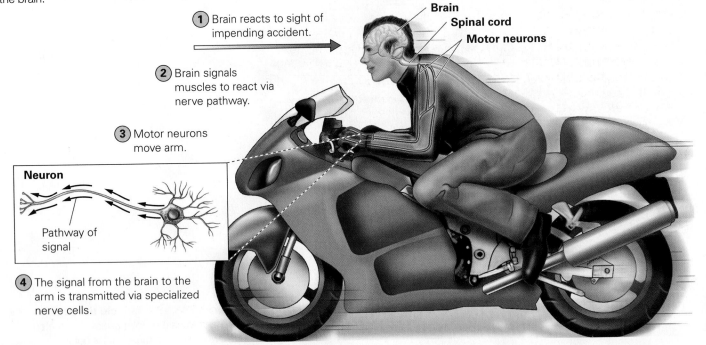

1. Brain reacts to sight of impending accident.

2. Brain signals muscles to react via nerve pathway.

3. Motor neurons move arm.

**Neuron**

Pathway of signal

4. The signal from the brain to the arm is transmitted via specialized nerve cells.

**Brain**
**Spinal cord**
**Motor neurons**

From this overview, you can now see the importance of the various tissue types. For an organism to be fully functional, damage to these tissues must be repaired or the damaged tissue must be replaced.

## Tissue Donation

If a family consents to tissue donation, tissues can be removed from a person who has been declared brain dead or from a person who has suffered cardiac death. Brain death occurs when the cerebrum and brain stem of the brain cease to function, usually the result of injury that prevents oxygen from getting to the brain. The cerebrum, the largest and most sophisticated part of the brain, is composed of two hemispheres. It controls movement and higher mental functions, such as thought, reasoning, emotion, and memory. The brain stem, located at the base of the brain, governs such automatic functions as heartbeat, respiration, and swallowing. When the brain stem ceases to function, a person has to rely on machines to perform these functions to remain alive.

Head injuries caused by motor vehicle and cycling accidents, ruptured blood vessels, and drowning are all common causes of brain death. When a person suffers this type of injury, the person's brain cells die because swelling and damage to the tissues prevent them from being supplied with oxygen and nutrients.

A person who is brain dead has a different prognosis from that of someone who is in a coma or vegetative state. A person in a coma or vegetative state might very well recover some functions, while a brain-dead person will not.

Organ donations can come from individuals who have suffered cardiac death as well. Cardiac death occurs when the heart stops functioning.

In the case of the young motorcyclist discussed at the beginning of the chapter, the family was asked whether they would donate his tissues and organs because testing showed that all activity had ceased in his cerebrum and brain stem. If the family grants permission for tissue donation, a specially trained team of surgeons will carefully remove any usable tissues that the family agrees to donate. These tissues can include bones, tendons, ligaments, cartilage, veins, skin, and corneas.

The motorcyclist's bones may be donated to someone who has had bone cancer and would require an amputation without the transplant. His tendons, ligaments, and cartilage may be used to reconstruct and stabilize the damaged joints of many different people, helping them resume normal activities after accidents and injuries. Veins from the young man's legs can be used to help recipients with circulatory problems and as bypasses for people undergoing heart bypass surgery. Donated skin can be used for burn patients as a type of biological dressing, not only to help prevent infections but also to decrease blood loss and pain due to exposed nerve endings until the patients' own skin can grow back.

The transparent cornea of the eye can be transplanted to a living person who has experienced vision loss due to corneal damage. The cornea consists of a thin layer of surface epithelium overlaying a layer of fibrous connective tissue. In a healthy eye, the cornea serves as a barrier that shields the eye from dust and germs. Damage to the cornea caused by injury, infection, or chemical burn can result in loss of vision.

Once removed, the young man's tissues will be checked for the presence of bacteria and viruses that could cause disease in the recipients. Many of his tissues will then be treated with chemicals to remove any proteins that a recipient's immune system might recognize as foreign and thus attack. Because of these treatments, tissue recipients do not normally reject tissue transplants and usually do not need to take drugs to suppress their immune system. Therefore, tissue recipients need not be genetically similar to the donor. After being tested

## ▼ **Visualize This**

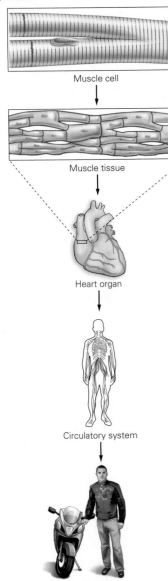

Muscle cell

Muscle tissue

Heart organ

Circulatory system

Organism

**FIGURE 17.6 Levels of organization.** Cells give rise to tissues. Tissues with a common function comprise organs. Organs working together make up an organ system, and an entire organism often has many organ systems that communicate with each other.

___

**dia-** means through.

and treated, the tissues will be frozen and sent to tissue and eye banks, where they will be catalogued and stored until they are needed. Most tissues can be stored for up to 5 years.

Tissue removal and treatment takes a few hours and can be performed during the time that the family begins to make plans for a funeral. One person's tissues can improve the lives of as many as 50 different people. In addition to improving people's lives by donating the young man's tissues, the family may save several lives by donating his organs.

# 17.2 Organs and Organ Systems

Many structures in the body are actually assemblages of different tissue types. **Organs** are structures composed of two or more tissues packaged together, working in concert to produce the organ's specific function. When many organs interact to perform a common function, these organs are said to be part of an **organ system.** All of the organ systems in a body work together to help an organism function. **TABLE 17.1** briefly outlines the functions and component organs involved in each of the twelve organ systems.

**STOP & STRETCH** The integrated functions of organs and organ systems are similar to the integrated functions of various car components and a car. Match the following car components—valves and pistons, fuel injection system, vehicle, and engine—with the corresponding biological component: organism, organ system, organ, and tissue.

Living things display increasingly complex levels of organization, from cells, to tissues, to organs, to organ systems, to an entire organism (**FIGURE 17.6**). For example, the regulation of the motorcyclist's body temperature while riding through different climates requires several levels of organization. The brain (an organ) sends signals through the cells of the nervous system (an organ system) to the epidermis (tissue) that result in changes in individual units (cells) that comprise sweat glands and blood vessels, resulting in an overall change in the amount of sweat secreted and in the dilation of blood vessels.

There are many different organs in humans and other animals, such as the heart, lungs, kidney, and brain. (Many of these organs are covered in more detail in subsequent chapters of this unit.) In the following section, we use the liver as a model to illustrate how an organ's ability to perform its required function is made possible by the actions of different tissue types. We also look at how an organ can function as part of an organ system.

## Organs: The Liver as a Model Organ

The **liver** is a large, reddish-brown organ found on the right side of the abdominal cavity below the **diaphragm.** The human liver is divided into four lobes. A greenish sac called the gallbladder is held in close contact with the liver (**FIGURE 17.7**).

Structurally, the epithelial tissue comprising the liver is covered with connective tissue that branches and extends throughout the body of the liver. This connective tissue acts as a sort of scaffolding to support the tissue and subdivides the tissue into hexagonal structures called **lobules.** In the middle of each lobule is a central vein that allows blood to reach all parts of the liver. Blood flows through the liver and is filtered by liver cells called **hepatocytes.** Filtering results in the removal of toxic materials, dead cells, pathogens, drugs, and alcohol from the bloodstream.

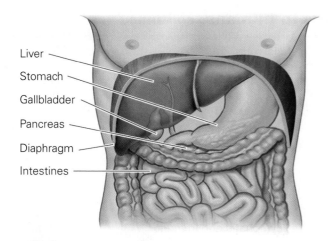

**FIGURE 17.7 The liver.** The human liver is located on the right side of the abdomen below the diaphragm. A duct connects the gallbladder to the liver.

The liver is an important part of both the circulatory and the digestive systems. As a part of the circulatory system, the liver synthesizes blood-clotting factors, detoxifies harmful substances in the blood, regulates blood volume, and destroys old blood cells. As an accessory part of the digestive system, the liver produces bile, which helps metabolize fats.

The liver also serves as a storage site for excess glucose. Glucose monomers are joined to each other to produce a polymer of glucose called glycogen. The liver removes glucose from the blood when glucose levels are high, such as after a meal, and stores them as glycogen. When a meal has not been eaten recently, blood glucose levels fall, and the liver breaks the bonds between glucose molecules in glycogen and releases glucose monomers into the bloodstream.

Liver transplants are performed when the liver is failing. Common causes of liver failure are infection with the hepatitis C virus and chronic alcohol abuse. Both of these cause a scarring of the liver called cirrhosis. Liver donors can

## Is the brain-dead accident victim an organ donor?

**TABLE 17.1 Organ systems and functions in humans.**

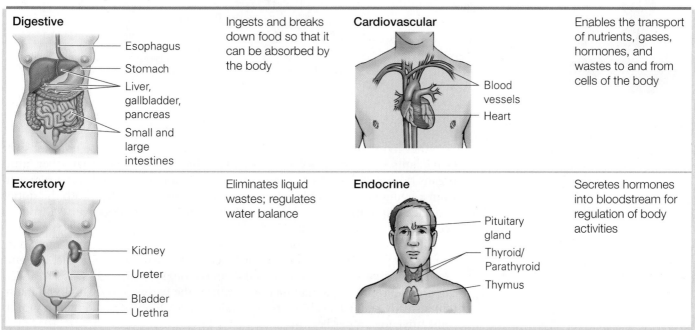

| Digestive | Ingests and breaks down food so that it can be absorbed by the body | Cardiovascular | Enables the transport of nutrients, gases, hormones, and wastes to and from cells of the body |
| Esophagus, Stomach, Liver, gallbladder, pancreas, Small and large intestines | | Blood vessels, Heart | |
| Excretory | Eliminates liquid wastes; regulates water balance | Endocrine | Secretes hormones into bloodstream for regulation of body activities |
| Kidney, Ureter, Bladder, Urethra | | Pituitary gland, Thyroid/Parathyroid, Thymus | |

*continued*

**TABLE 17.1** **Organ systems and functions in humans.** *continued*

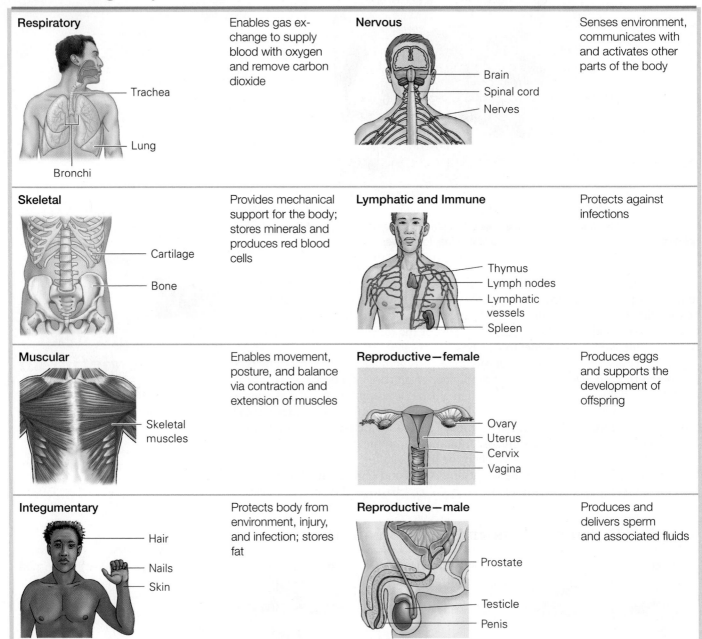

| | | | |
|---|---|---|---|
| **Respiratory** | Enables gas exchange to supply blood with oxygen and remove carbon dioxide | **Nervous** | Senses environment, communicates with and activates other parts of the body |
| Trachea, Lung, Bronchi | | Brain, Spinal cord, Nerves | |
| **Skeletal** | Provides mechanical support for the body; stores minerals and produces red blood cells | **Lymphatic and Immune** | Protects against infections |
| Cartilage, Bone | | Thymus, Lymph nodes, Lymphatic vessels, Spleen | |
| **Muscular** | Enables movement, posture, and balance via contraction and extension of muscles | **Reproductive—female** | Produces eggs and supports the development of offspring |
| Skeletal muscles | | Ovary, Uterus, Cervix, Vagina | |
| **Integumentary** | Protects body from environment, injury, and infection; stores fat | **Reproductive—male** | Produces and delivers sperm and associated fluids |
| Hair, Nails, Skin | | Prostate, Testicle, Penis | |

be either living or brain-dead individuals. Living donors can be used for liver transplants because the liver has a remarkable characteristic that other adult organs do not share—it can regenerate itself. If a piece of the liver is removed from a living donor and placed in a compatible recipient, then both the donor and the recipient's liver will grow to full size. Close to 7000 liver transplants occur every year in the United States.

Failure of an individual organ can affect an entire organ system or more than one organ system. Liver failure, for instance, would have a large impact not only on the digestive system but also on the circulatory system. In addition to affecting organ systems, organ failure compromises the body's overall ability to maintain a steady state. The ability to maintain a consistent internal environment is imperative for proper health.

# 17.3 Regulating the Internal Environment

Maintaining **homeostasis,** a consistent internal environment, requires the coordinated efforts of cells, tissues, organs, and organ systems. Each cell engages in basic metabolic activities that will ensure its own survival. Cells of a tissue perform activities that contribute to the proper functioning of organs; organs perform required functions, alone or in conjunction with organ systems, to contribute to the maintenance of a stable internal environment.

**homeo-** means the same.

Humans, for example, must keep their heart rate, blood pressure, water and mineral balance, and temperature and blood glucose levels within a narrow range for survival. Regulation of temperature, called **thermoregulation,** is an example of a bodily function that in many organisms is tightly regulated. Humans and all other mammals and birds are **endotherms**—that is, organisms with the ability to maintain a body temperature that is warmer than the surrounding environment. Endotherms derive most of their body heat from metabolism. Most invertebrates, fish, amphibians, and reptiles are **ectotherms,** or organisms that obtain their body heat primarily by absorbing it from their surroundings.

**thermo-** relates to warmth and heat.

Keeping physical and chemical properties within tolerable ranges utilizes feedback from one system to another. Feedback is information that is sent to a control center, which in turn directs a cell, tissue, or organ to respond by turning up or down a given process. Feedback can be negative or positive. In general, negative feedback negates change, and positive feedback promotes change.

## Negative Feedback

Negative feedback occurs when the product of the process inhibits the process. Thermoregulation in endotherms occurs by negative feedback in a manner very similar to the way that temperature is controlled in a house. A thermostat that functions as a sort of control center turns the heater on when the room becomes too cold. When the room becomes too hot, the heater is turned off. Therefore, the negative feedback of the temperature change regulates heating systems.

Body temperature is controlled in a similar manner. When the body temperature increases, a control center in the brain sends signals to the body to dilate blood vessels near the skin and activate sweat glands, allowing heat to escape. When the body temperature cools, blood vessels near the surface of the skin constrict, muscles shiver, and body temperature increases. These negative-feedback mechanisms keep the body temperatures of endotherms within a very narrow range.

Regulation of blood glucose levels by the liver provides another example of negative feedback (**FIGURE 17.8**, on the next page). When food is digested, the amount of glucose in the blood increases. This stimulates the pancreas to produce **insulin,** a hormone that stimulates the liver (along with fat and muscle tissues) to remove glucose from the blood and store it as glycogen.

When most of the glucose has been removed from the blood, the pancreas secretes a hormone called glucagon, which stimulates the liver to break down glycogen and secrete glucose units into the bloodstream. Consequently, negative regulation of blood glucose levels reverses the direction of diet-induced changes in glucose levels, keeping glucose levels from swinging wildly after a meal and confining glucose levels to within a very narrow range—thus promoting homeostasis.

# Visualize This ▶

Blood glucose homeostasis can be disrupted by factors other than food you ingest. What might also cause a disruption of glucose homeostasis?

**FIGURE 17.8 Feedback in the regulation of blood glucose levels.** (a) When intake of glucose is high, such as after a meal, the amount of glucose in the blood increases. The excess is stored as glycogen. (b) When blood glucose is low, glucose is secreted into the bloodstream.

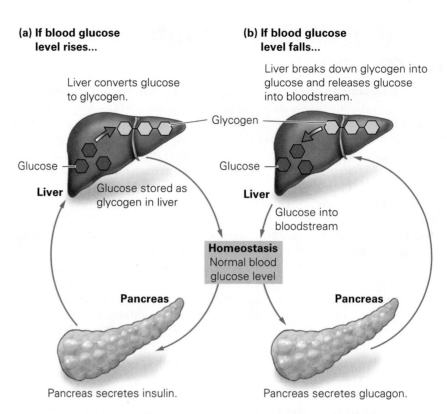

**(a) If blood glucose level rises...**

Liver converts glucose to glycogen.

Glycogen

Glucose

**Liver**

Glucose stored as glycogen in liver

**Pancreas**

Pancreas secretes insulin.

**(b) If blood glucose level falls...**

Liver breaks down glycogen into glucose and releases glucose into bloodstream.

Glucose

**Liver**

Glucose into bloodstream

**Homeostasis** Normal blood glucose level

**Pancreas**

Pancreas secretes glucagon.

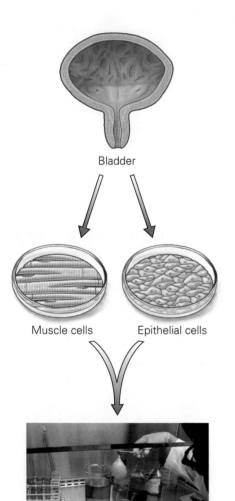

Bladder

Muscle cells    Epithelial cells

**FIGURE 17.9** The bladder is composed of epithelial and muscle tissues. Stem cells removed from the recipient can be stimulated to grow in a laboratory incubator. Layering the cells on a scaffold shaped like a bladder results in the production of a new bladder in the laboratory that can be implanted in the recipient.

## Positive Feedback

Positive feedback occurs when the product of the process intensifies the process. For example, during labor and childbirth, hormones cause the muscles of the uterus to contract, pushing the baby out of its mother's body. The pressure of the baby's head against the cervix stimulates the release of more hormones, which cause the contractions to intensify in rate and duration. This in turn causes even more hormones to be released and results in even more and longer contractions. After the cervix opens and the baby leaves the uterus, the contractions end. In animals, most homeostatic regulation is via negative feedback because positive feedback amplifies a response and can actually drive the process away from homeostasis.

Damaged and diseased organs do not respond properly to signals that help promote homeostasis. Transplanting healthy organs to replace defective ones can restore, at least temporarily, homeostasis to a diseased individual.

**STOP & STRETCH** When you walk outside into the sunlight, the pupils of your eyes constrict. When you come back in to a dark room, your pupils dilate. Is the regulation of pupil size an example of positive or negative feedback?

## Growing Replacement Organs

The immune system's job is to rid the body of foreign substances. Because transplanted tissues and organs carry surface markers unique to the donor, the recipient's immune system can attack, causing the transplant to fail. To help avoid issues with rejection and to decrease the number of people who die waiting for organ transplants, some scientists are growing organs in the lab (**FIGURE 17.9**).

Initially, scientists began working with fairly simple hollow organs such as the bladder, and windpipe or trachea, with the hope of advancing to more complex organs like lungs or the pancreas. This process, called tissue engineering, decreases the risk of rejection because cells from the recipient are used to grow the organ. Stem cells can be extracted from the organ that needs to be replaced. Once extracted, the recipient's cells are placed on a scaffold shaped like the organ requiring replacing. The scaffold is a three-dimensional replica of the natural scaffold that provides structure for most organs. It can be a synthetic biodegradable scaffold or it can be produced by removing the same organ from a donor cadaver and washing it of the donor's cells, leaving only the scaffold behind. When the scaffold is bathed in the recipient's cells and growth-supporting nutrients, the cells will fill the scaffold with new cells, ultimately producing a functional organ. To date, there have been dozens of successful bladder and trachea implants.

## Organ Donation

The young motorcyclist was a good candidate for organ donation because he died from a brain injury. People who die from cardiac death can be tissue donors, but the lack of oxygen experienced during cardiac death causes organs to deteriorate, making such people less suitable organ donors. Since many more people will die of cardiac death than brain injury, tissues can be banked for future use, but there is a shortage of organs for use in transplant operations.

If the family agrees, the young man will be kept on a ventilator so that his organs continue to be nourished with oxygen and blood until the organs can be surgically removed (**FIGURE 17.10**). While the organs are being removed,

**Even though his driver's license lists him as a donor, the family is not sure what to do.**

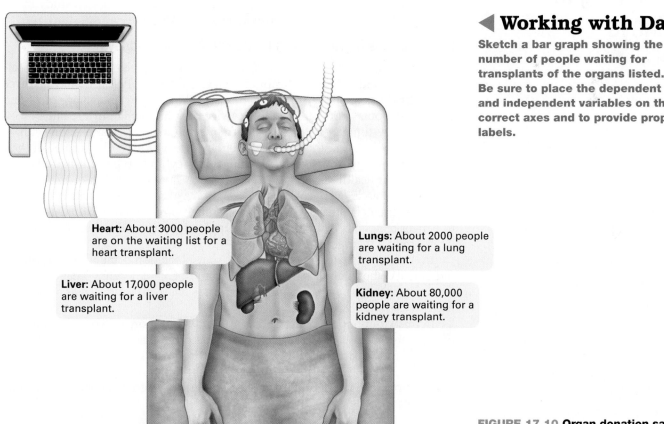

Heart: About 3000 people are on the waiting list for a heart transplant.

Liver: About 17,000 people are waiting for a liver transplant.

Lungs: About 2000 people are waiting for a lung transplant.

Kidney: About 80,000 people are waiting for a kidney transplant.

◀ **Working with Data**

Sketch a bar graph showing the number of people waiting for transplants of the organs listed. Be sure to place the dependent and independent variables on the correct axes and to provide proper labels.

**FIGURE 17.10 Organ donation saves lives.** Thousands of lives can be saved each year through organ donation.

medical personnel will attempt to find the best recipient. To select the recipient, a search of a computerized database is performed. Starting at the top of the list, medical staff will narrow down possible recipients, based in part on how long the recipient has been waiting. People at the top of the waiting list for a given organ tend to be very ill and will die if a transplant does not become available in a matter of days or weeks. Around 10% of people on waiting lists for replacement organs will die before one becomes available.

If you would like to be an organ donor, discuss your plans with your family. Even if you carry a signed organ donor card and indicate your preference to be a donor on your driver's license, your family will need to agree to donate your tissues and organs. If you should suffer an accidental death, like the young motorcyclist, your family will be forced to make many difficult decisions quickly. Having a discussion about organ donation now will relieve them of the burden of trying to make this decision for you.

# *savvy reader*

## Trafficking in Kidneys

Individuals who sell a kidney do not receive a long-term economic benefit from the sale and may have a worsening of their health, according to a study in the *Journal of the American Medical Association* (JAMA).

### Study Reveals Consequences for People Who Sell Kidneys

The investigation, led by Madhav Goyal from Geisinger Health System in State College, Pa., and Ashwini Sehgal from the CWRU School of Medicine, involved 305 individuals in southern India who had sold a kidney an average of 6 years earlier for about $1,000 apiece (about twice their annual family income).

The study compared the sellers' current economic and health status with their economic and health status before they sold a kidney. The income of the sellers declined by one-third after selling a kidney, and 86 percent had a worsening of their health status.

Because nearly every country has a shortage of organs for transplantation, providing financial incentives to donate is often proposed or justified as a way to benefit recipients by increasing the supply of organs and to benefit donors by improving their economic status.

In the United States, most organs for transplantation come from brain-dead individuals whose families have consented to donate. Providing financial incentives such as payment for funeral expenses has been proposed as a way to increase the supply of organs.

In many developing countries such as India, most organs come from living donors, and the sale of kidneys is widespread. This practice is justified as a way to both increase the supply of kidneys for transplantation and benefit the seller economically. However, critics argue that purchasing kidneys amounts to exploitation of the poor and that the poor do not overcome poverty as a result of the sale.

The researchers also sought to determine whether individuals who sell a kidney actually benefit from the sale.

"We were surprised by our findings," Goyal said. "The sellers' economic and health status appeared to worsen after selling a kidney. Most of them would not recommend that others sell a kidney."

"These findings undercut arguments in favor of financial incentives," Sehgal said. "Even modest financial incentives such as those proposed in the United States may be perceived by the general public as taking advantage of poor families. If this happens, such incentives may actually lead to fewer total organ donations."

1. What is the affiliation of the scientists who performed this study?

2. The scientists responsible for this study claimed that they were surprised by their findings. What did they do when they found that data did not turn out the way they had expected?

3. Would you find a report by an organ-trafficking agency outlining the benefits of an organ sale more or less credible than this article? Why?

*Campus News*, Case Western Reserve University, October 17, 2002 ▬

# SOUNDS RIGHT **BUT IS IT?**

Parents facing the decision to take their brain-dead daughter off life support are, understandably, overwhelmed, confused, and conflicted about doing so. They cling to the hope that their daughter may some day awaken. In fact, several years ago, a teenager from their town was in a coma for several months after a car accident and subsequently made a remarkable recovery. Besides, she does not look dead, she just appears to be asleep. They have even witnessed her moving, furthering their conviction that their daughter's brain is still working. These parents refuse to deny their daughter a chance, no matter how small, at coming back to life.

**There is always a chance that a brain-dead person will make a full recovery.**

Sounds right, but it isn't.

1. Brain-dead individuals show no electrical activity in the brain because brain cells

are not passing electrical impulses to each other. Do you think comatose people show electrical activity in the brain?

2. Brain-dead people cannot breathe on their own, because breathing is initiated in the brain stem, which has ceased to function. A mechanical ventilator can provide oxygen to their tissues, keeping nourished any tissues that are not dead. If the ventilator being used by a brain-dead person is removed, would that person be able to breathe on his or her own?

3. Is an unresponsive person who is breathing unaided in a coma or brain dead?

4. The regulation of body temperature, the ability to increase or decrease body temperature when the external environment changes, is a function of the brain. Would a brain-dead

person be able to increase his or her body temperature in response to lowering the heat?

5. If an unresponsive person can regulate body temperature, is he or she in a coma or brain dead?

6. Some movements are the product of reflexes, which involve nerves between the muscles and the spine, but not the brain. (When you jerk your hand away from a hot surface, the movement is done before you have time to use your brain to think about moving your hand.) Does the presence of these reflexive movements in an unresponsive person indicate that the brain is alive?

7. Consider your answers to questions 1–6 and explain why the original statement bolded above sounds right, but isn't.

# Chapter Review   MasteringBiology®

## Summary

### Section 17.1

Describe the structure and function of epithelial tissues.

- Epithelia line and cover organs, vessels, and body cavities. They are tightly packed tissues with one free surface. Outgrowths of epithelia form some glands. These tissues function in protection, secretion, and absorption (pp. 398–399).

Compare the structures and functions of the six connective tissues.

- Loosely packed connective tissues bind tissues and organs to each other. The six types of connective tissues are each composed of a characteristic cell type, embedded in a characteristic matrix (p. 399).
- Loose connective tissue connects epithelia to underlying tissues and holds organs in place. The cellular component is fibroblasts, and the matrix is rich in the proteins collagen and elastin, which provide this tissue with tensile strength and elasticity (pp. 399–400).
- Adipose tissue connects the skin to underlying structures and insulates and protects organs. Cells of this tissue synthesize and store fat and have very little extracellular matrix (p. 400).
- Blood is a type of connective tissue that transports oxygen and nutrients to body cells. Blood cells have a liquid matrix called plasma (p. 400).
- Fibrous connective tissue forms the tendons and ligaments. The cells of this tissue are fibroblasts, and the matrix is rich in collagen (p. 400).
- Cartilage is a flexible, shock-absorbing tissue composed of cells called chondrocytes that secrete a dense, collagenous matrix (p. 400).
- Bone tissue provides support for the body. The cells of this tissue are called osteocytes, and the matrix is rich in collagen and minerals (p. 400).

Compare the structures and functions of the three types of muscle tissues.

- Muscle tissues are composed of fibers that contract and conduct electrical impulses (p. 401).
- The three types of muscle tissue can be either voluntary or involuntary and smooth or striated (p. 401).
- Skeletal muscle is attached to bones, responsible for voluntary movements, and striated (p. 401).

- Cardiac muscle comprises the heart; it is involuntary and striated (p. 402).
- Smooth muscle comprises the musculature of internal organs and blood vessels. Smooth muscle is involuntary and not striated (p. 402).

Describe the structure and function of nervous tissues.

- Nervous tissue, found in the brain and spinal cord, is composed of neurons; it senses stimuli and transmits signals throughout the body (p. 402).

### Section 17.2

Explain what an organ is and how organs interact in an organ system.

- Organs are groups of tissues working in concert (p. 404)
- Organ systems are suites of organs working together to perform a function or functions (pp. 404–406).

List the structures and functions of the liver.

- The liver, like all organs, is composed of different tissues whose combined functions give rise to the organ's overall function (p. 404).
- The epithelial tissue of the liver is divided into lobes. Lobes are divided by connective tissue into lobules. Lobules contain a central vein that allows blood to reach the hepatocytes, which function as filters to remove toxins and pathogens (pp. 404–405).
- Other functions of the liver include the production of proteins and cholesterol, secretion of bile, and storage of vitamins and glycogen (p. 405).

### Section 17.3

Explain how positive and negative feedback mechanisms help in the maintenance of homeostasis.

- Negative feedback is one mechanism that helps facilitate homeostasis. When negative feedback occurs, the product of the process acts to turn down the process. Negative feedback negates change (p. 407).
- Positive feedback is a less commonly used homeostatic mechanism. When positive feedback occurs, the product of the process intensifies the process. Positive feedback promotes change (p. 408).

# Roots to Remember

**The following roots of words come mainly from Latin and Greek and will help you decipher terms:**

| | |
|---|---|
| **adipo-** | means fat. Chapter term: *adipose tissue* |
| **dia-** and **trans-** | mean through. Chapter terms: *diaphragm, transplant* |
| **epi-** | means on or upon. Chapter term: *epithelium* |
| **homeo-** | means the same. Chapter term: *homeostasis* |
| **lig-** | means to join or connect. Chapter term: *ligament* |
| **neuro-** | means nerve. Chapter term: *neuron* |
| **osteo-** | relates to bone. Chapter term: *osteocyte* |
| **thermo-** | relates to warmth and heat. Chapter term: *thermoregulation* |

# Learning the Basics

**1.** Describe the four tissue types found in animals.

**2.** Name the six types of connective tissues.

**3.** What are the differences between the three separate types of muscle tissue?

**4.** Epithelia _____.
A. are loosely packed tissues that hold organs in place;
B. can be free-floating tissues such as blood; C. line organs and cavities and have one surface exposed;
D. include the tissues that hold joints together; E. are tissues that conduct nerve impulses.

**5.** Connective tissues include _____.
A. cartilage; B. blood; C. fat; D. bone; E. all of the above.

**6.** Muscle tissues _____.
A. contain actin and myosin, whether or not they are striated; B. that facilitate movements requiring conscious thought are involuntary; C. increase in cell number with exercise; D. that comprise the heart are called smooth; E. all of the above.

**7.** Which of the following is not a function of the liver?
A. storing bile; B. filtering toxins; C. producing cholesterol; D. storing glycogen

**8.** When a woman is breastfeeding, the more her infant drinks, the more milk she produces. This is an example of _____.
A. negative feedback; B. positive feedback;
C. thermoregulation; D. independent regulation.

**9.** Bile _____.
A. is stored in the pancreas; B. helps break down glycogen; C. emulsifies fats; D. removes water from indigestible materials in the large intestine; E. all of the above.

**10.** Which of the following cell types is found in nervous tissue?
A. osteocyte; B. melanocyte; C. leukocyte; D. neuron;
E. brain and spinal cord

**11.** **True or False:** An organ system is a group of tissues, composed of similar cell types, with a common function.

# Analyzing and Applying the Basics

**1.** Propose a hypothesis for why shivering when cold may promote homeostasis.

**2.** How does the pancreas respond when you consume a sugary beverage?

**3.** Describe a nonbiological example of both negative and positive feedback.

# Connecting the Science

**1.** Why might families refuse to donate the tissues and organs of a loved one?

**2.** Do you think private citizens should be able to sell their own organs? Why or why not?

Answers to **Stop & Stretch, Visualize This, Working with Data, Savvy Reader, Sounds Right, But Is It?,** and **Chapter Review** questions can be found in the **Answers** section at the back of the book.

# CHAPTER 18 | Binge Drinking

A student is turning 21.

# The Digestive and Urinary Systems

It's Saturday night and Malik is hosting a surprise party to celebrate the 21st birthday of his friend Lin. Lin is several years younger than Malik. She lived with Malik's family, sharing a room with his younger sister, for 2 years when she was a high school exchange student. Now an international student attending college in the United States, Lin has had almost no experience with alcohol. Malik knows that Lin is eagerly anticipating this birthday and that she is planning to drink at least a little alcohol. Because he feels as protective of Lin as he does his little sister, Malik wants to help Lin learn how to enjoy the benefits of alcohol consumption while limiting the negative consequences that can also occur.

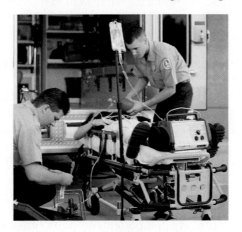

**He does not want her alcohol consumption to place her at risk of overdose …**

Some of Malik's concerns about negative consequences are based on situations he has witnessed and others on information he came across while writing a paper on alcohol abuse for a health class he took last semester.

A student who lived on Malik's dorm floor freshman year broke his ankle when he tripped while running from the police to avoid an underage consumption ticket. His chemistry lab partner broke her nose when she was riding with an intoxicated driver whose car hit a tree on a snow-covered road. While working on the paper for his health class, he came across a government website that indicated over 30,000 students required medical treatment for alcohol

poisoning last year and he does not want this to happen to Lin. He also worries about the high rate of sexual assault on college campuses. He does not want Lin to become one of the 20% of female students who will be sexually assaulted while in college.

Malik wants to develop a plan for convincing Lin, a pre-med biology major, that drinking too much is bad for her body, an argument he thinks she may find credible. Because he has heard that eating food before drinking might help absorb some of the alcohol and that alcohol consumption causes dehydration, he plans to focus his efforts on the effects of drinking on the digestive and urinary systems.

**Malik is worried about his friend Lin.**

**… or jeopardize her safety.**

**Malik wants to help Lin learn how to enjoy the benefits of alcohol consumption while limiting the negative consequences that can also occur.**

# 18.1 The Digestive System

The job of the digestive system is to break down or metabolize food so that the body can utilize the energy stored in its chemical bonds. Food is broken down into successively smaller units as it is passed through a long tube with openings at the mouth and anus. Called the **alimentary canal** or digestive tract, this tube allows food to be exposed to the actions and secretions of various organs and glands.

The breakdown of food begins in the mouth, or oral cavity, where chewing fractures food into smaller pieces (**FIGURE 18.1**). Mechanical digestion of food occurs as teeth grind down food, which increases the surface area exposed to enzymes within the mouth. Chemical digestion also begins in the mouth with the secretion of saliva from salivary glands. Saliva contains the enzyme **salivary amylase**, which breaks down sugars.

The carbohydrate breakdown that begins in the mouth is the first step in the conversion of polymers that we consume (carbohydrates, proteins, and fats) into their component subunits (monosaccharides, amino acids, and glycerol and fatty acids, respectively). This conversion is necessary because the body cannot absorb ingested polymers or utilize them to build cellular structures directly. Instead, the polymers are broken down and rebuilt according to your body's specifications.

## Mechanical and Chemical Breakdown of Food

The **tongue** has taste buds that help you taste food. This muscular structure shapes food into a ball and pushes the ball of food, called the **bolus,** to the back of the mouth. From there it is swallowed and enters the digestive

## Visualize This ▼

The third molars are the last molars to emerge. Located at the back of the mouth, they typically appear in young adults. If a person has all of these "wisdom teeth" removed, how many adult teeth would they have?

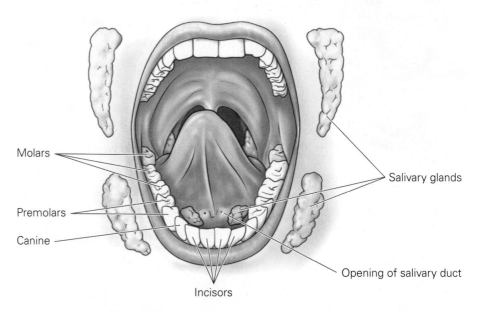

**FIGURE 18.1 The human oral cavity.** Food is ground into smaller pieces by the 32 adult teeth. Incisors are specialized to enable biting off pieces of food. Canine teeth are sharp enough to help rip food apart. Molars have broad surfaces that enhance grinding.

system (**FIGURE 18.2**). Malik knows from his research that oral cancers are more common in those who abuse alcohol than in those who do not and that alcohol has been shown to increase incidence of tooth decay, gum disease, and tooth loss.

When you swallow chewed food, it moves from your mouth to your **pharynx.** The pharynx, forming the back of your throat, branches to feed into both the trachea (which leads to the lungs) and the esophagus (which leads to the stomach). The **epiglottis** is a thin flap of cartilage below the tongue that keeps swallowed food from entering the trachea. The **esophagus** brings food to a large digestive organ called the **stomach.** At the base of the esophagus is a ring of contracting muscles called the lower esophageal sphincter, which serves to prevent backflow of stomach contents into the lower esophagus. Alcohol can weaken this sphincter, allowing acidified stomach contents to move backwards up the esophagus, causing heartburn.

The movement of a bolus of food is hastened by rhythmic waves of smooth muscle contraction down the esophagus, through a process called **peristalsis.** Once inside the stomach, food is subjected to degradation by digestive enzymes and acids secreted by specialized cells that line the stomach.

**epi-** means upon, beside, or among.

**peri-** means around.

**-stalsis** means to wrap or surround.

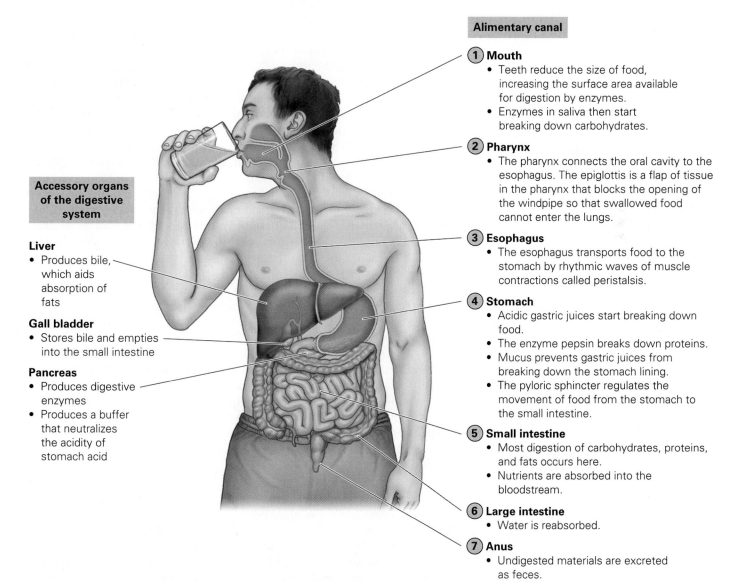

**Alimentary canal**

**1** **Mouth**
- Teeth reduce the size of food, increasing the surface area available for digestion by enzymes.
- Enzymes in saliva then start breaking down carbohydrates.

**2** **Pharynx**
- The pharynx connects the oral cavity to the esophagus. The epiglottis is a flap of tissue in the pharynx that blocks the opening of the windpipe so that swallowed food cannot enter the lungs.

**3** **Esophagus**
- The esophagus transports food to the stomach by rhythmic waves of muscle contractions called peristalsis.

**4** **Stomach**
- Acidic gastric juices start breaking down food.
- The enzyme pepsin breaks down proteins.
- Mucus prevents gastric juices from breaking down the stomach lining.
- The pyloric sphincter regulates the movement of food from the stomach to the small intestine.

**5** **Small intestine**
- Most digestion of carbohydrates, proteins, and fats occurs here.
- Nutrients are absorbed into the bloodstream.

**6** **Large intestine**
- Water is reabsorbed.

**7** **Anus**
- Undigested materials are excreted as feces.

**Accessory organs of the digestive system**

**Liver**
- Produces bile, which aids absorption of fats

**Gall bladder**
- Stores bile and empties into the small intestine

**Pancreas**
- Produces digestive enzymes
- Produces a buffer that neutralizes the acidity of stomach acid

**FIGURE 18.2 The digestive system.** The digestive system consists of accessory organs and the alimentary canal.

**Over 30,000 students required medical treatment for alcohol poisoning last year.**

Only a small percentage of alcohol is metabolized in the stomach. Epithelial cells that line the stomach secrete enzymes called alcohol dehydrogenases. These enzymes help metabolize alcohol, which means there will be less alcohol in the bloodstream to cause intoxication. Women have smaller stomachs than men and thus produce less of this enzyme. Interestingly, women also have less of this enzyme per pound of body weight than men. This means that a man and a woman of the same size and weight will metabolize alcohol at different rates. Women metabolize alcohol more slowly, thus becoming intoxicated from less alcohol consumption. Malik plans to caution Lin not to try to keep up with her male friends, even those that aren't any bigger than her.

There is one additional stomach-related factor Lin must consider when drinking alcohol. One of the alcohol dehydrogenase enzymes, abbreviated ALDH1, differs from the most common alcohol dehydrogenase by a single amino acid. This variant, common in people of East Asian descent, more slowly metabolizes one of the products of alcohol breakdown, called acetaldehyde. High concentrations of acetaldehyde cause flushing of the face, dizziness, nausea, and an irregular heartbeat. Because Lin comes from Chengdu, China, Malik will also warn her that she may experience this reaction when drinking alcohol.

Once the bolus makes its way to the stomach, peristalsis helps mix it into a slurry with digestive enzymes, becoming a substance called **chyme.** When chyme contains a high concentration of alcohol, the stomach lining is irritated, which can trigger the vomiting reflex. At the base of the stomach is another sphincter. The pyloric sphincter regulates the secretion of chyme into the **small intestine,** a long tube (around 20 feet long in adult humans) that serves as the major site of chemical digestion and the absorption of nutrients into the bloodstream via its epithelium. It is in the small intestine that the majority of food and alcohol is broken down and absorbed across the intestinal wall and into the bloodstream. When alcohol relaxes muscles involved with peristalsis, food spends more time in the digestive tract than normal and this increased exposure to digestive enzymes can cause diarrhea.

Malik has heard that it is good to eat a large meal before drinking. This is because the presence of food in the stomach causes the pyloric sphincter to remain closed. Since the stomach does not absorb alcohol as readily as the small intestine, preventing the alcohol from reaching the small intestine can slow the rate at which it reaches the bloodstream. Therefore, Malik plans to take Lin out to eat before the birthday party.

Many of the digestive enzymes used in the small intestine are produced by an organ called the **pancreas.** Secretions from the pancreas neutralize stomach acids that enter the small intestine and contain enzymes that break down carbohydrates, fats, proteins, and nucleic acids. While in the small intestine, chyme is exposed to a substance called bile, which is synthesized by the liver.

The liver, pancreas, and gallbladder are considered **accessory** organs. In a sense, they are accessories to the alimentary canal because they are outside the tube but produce or secrete substances required for digestion. The **gallbladder** stores and concentrates bile, which will be released into the small intestine to help dissolve fats. Concentrating bile involves removing water from it. When the gallbladder removes too much water, the bile can crystallize, causing gallstones.

The liver and pancreas are very susceptible to the effects of alcohol, in the form of both long-term alcohol abuse and **binge drinking,** which is typically defined as the consumption of more than four drinks in a 2-hour time period.

One of the functions of the liver is to help metabolize toxins, including some drugs and alcohol. Heavy drinking can damage the liver, in some cases causing healthy liver tissue to be replaced by scar tissue, a progressive and irreversible process called cirrhosis. Scar tissue prevents proper blood flow through the liver and can result in liver failure.

Small intestine                    One villus                    Micrograph

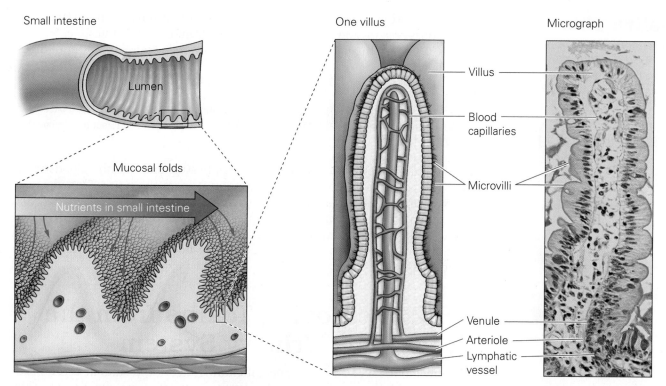

**FIGURE 18.3 Absorption of nutrients in the small intestine.** The foldings of the small intestine to produce finger-like projections called villi and microvilli increase the surface area across which nutrients can enter the bloodstream. Once in the bloodstream, nutrients can be transported to cells throughout the body.

Inflammation of the pancreas, called **pancreatitis,** can be caused by excessive alcohol consumption. Pancreatitis prevents the pancreas from secreting digestive enzymes and thus disrupts digestion, which can lead to life-threatening complications.

-----

**-itis** refers to inflammation.

-----

## Absorption of Digested Foods

Substances that move from the stomach to the small intestine are absorbed into the bloodstream through finger-like membranous projections of the small intestine called **villi.** An individual villus contains both a blood and a lymphatic vessel. Surface area is further increased by the presence of even smaller **microvilli** that transport nutrients into the blood vessels inside each villus (**FIGURE 18.3**). Together, the villi and microvilli have as much surface area as does a tennis court! Alcohol can interfere with the absorption of nutrients by damaging intestinal villi.

After traversing the small intestine, most polymers have been converted into monomers and can move across the small intestine into the bloodstream, where they are transported to individual cells. Materials that are not absorbed are passed through the **large intestine** (also known as the **colon**) through the rectum to the anus, where it will exit the body as feces. Fecal matter consists largely of indigestible plant fibers.

## Regulation of Digestive Secretions

The secretion of digestive juices is regulated hormonally (**FIGURE 18.4**). After a meal, the stomach produces the hormone **gastrin,** which stimulates the upper part of the stomach to produce acidic gastric juices,

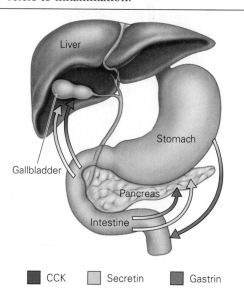

■ CCK   ▨ Secretin   ■ Gastrin

**FIGURE 18.4 Hormonal control of the digestive system.** The presence of food in the stomach causes the release of the hormone gastrin, which results in the intestinal release of the hormones secretin and CCK, which act on the gallbladder and pancreas to increase their output of bile and digestive enzymes.

## Visualize This ▼

An untreated bladder infection can cause kidney damage. What path would the bacteria take to move from the bladder to the kidney?

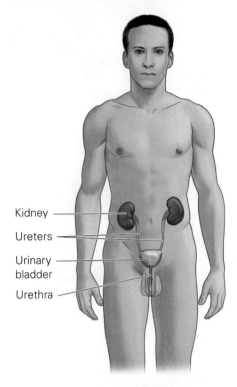

Kidney

Ureters

Urinary bladder

Urethra

**FIGURE 18.5 The human urinary system.** Kidneys filter the blood and transfer waste urine to the urinary bladder, where it is held until voluntarily released through the urethra.

---

**renal** comes from the Greek word for kidney.

---

thus facilitating digestion. The secretion of gastrin also results in release of the hormones *secretin* and *cholecystokinin* (CCK) from the small intestine. Secretin and CCK cause the pancreas and gallbladder to increase their output of digestive juices.

> **STOP & STRETCH** Gastrin release is inhibited by high concentrations of stomach acid (HCl). How might this negative feedback loop protect the stomach?

After alcohol makes its way out of the intestine and into the bloodstream, the body begins the process of removing it from the body. This is necessary to prevent damage such as the decrease in testosterone production that occurs in males after binge drinking. This hormonal decline can result in erectile dysfunction and lower sperm production while alcohol remains in the system. Removing such toxins from the body is one of the jobs of the urinary system.

# 18.2 Removing Toxins from the Body: The Urinary System

The metabolism of any substance, including alcohol, results in waste products that circulate through the bloodstream until they are expelled from the body. The **urinary system** has the task of efficiently removing wastes while retaining valuable materials that can be reused and recycled. In humans, the major organs of the urinary system are the **kidneys**, which filter and cleanse circulating blood before sending the waste through **ureters** to the bladder. Urine is stored in the **urinary bladder** until it is expelled via the **urethra** (**FIGURE 18.5**).

## Kidney Structure and Function

The kidneys are paired, approximately fist-sized organs that sit behind the liver and stomach in the upper abdominal cavity. Each kidney is densely packed with looped tubules called **nephrons.** There are close to 1,250,000 nephrons in each kidney, with a combined length in an adult human of about 145 kilometers (85 miles). Networks of capillaries surround the nephrons, allowing wastes to diffuse out of the blood and into these tubules for excretion. The entire volume of blood in the circulatory system passes through the kidneys hundreds of times per day, such that each kidney filters about 1000 liters of blood every 24 hours.

Blood is brought to the kidneys by large vessels called the **renal arteries.** The processing of waste in the kidneys has four distinct phases, which we can follow on a diagram of a single, loop-shaped nephron embedded in the body of a kidney (**FIGURE 18.6**). The first step, **filtration,** occurs within the Bowman's capsule, a structure at the head of a nephron. The Bowman's capsule encloses the glomerulus, a compact ball of blood vessels. The capillaries in the glomerulus contain tiny pores, and blood pressure forces the plasma portion of the blood through these pores and into the upstream end of the nephron. When blood reaches the glomerulus, this filter allows water and small molecules through, but retains large proteins in the plasma. The fluid that enters the interior of the nephron via this process is called filtrate.

One way that kidney health is assessed is by measuring the glomerular filtration rate. Preliminary studies suggest that those who drink alcohol in

# Visualize This ▼

**Where does urine go after it makes it to the collecting duct?**

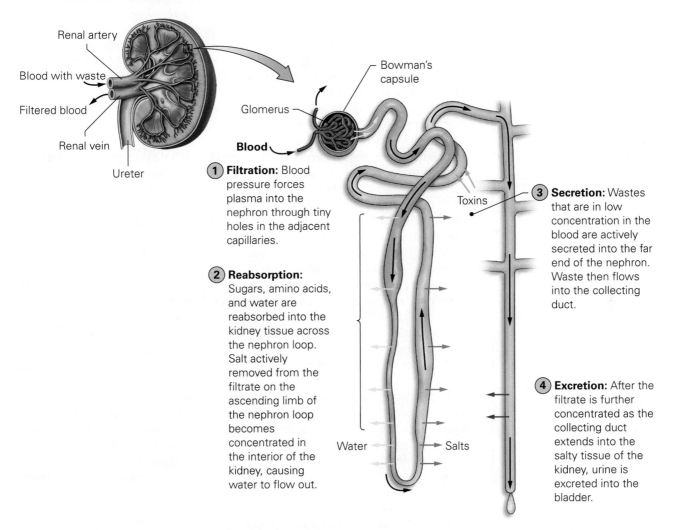

Renal artery

Blood with waste

Filtered blood

Renal vein

Ureter

Bowman's capsule

Glomerus

Blood

**1** **Filtration:** Blood pressure forces plasma into the nephron through tiny holes in the adjacent capillaries.

**2** **Reabsorption:** Sugars, amino acids, and water are reabsorbed into the kidney tissue across the nephron loop. Salt actively removed from the filtrate on the ascending limb of the nephron loop becomes concentrated in the interior of the kidney, causing water to flow out.

Toxins

Water

Salts

**3** **Secretion:** Wastes that are in low concentration in the blood are actively secreted into the far end of the nephron. Waste then flows into the collecting duct.

**4** **Excretion:** After the filtrate is further concentrated as the collecting duct extends into the salty tissue of the kidney, urine is excreted into the bladder.

**FIGURE 18.6 Nephron function.** The nephron, the functional unit of a kidney, controls the excretion of nongaseous waste from the blood.

moderation have slightly better filtration rates than those who don't. Drinking in moderation is usually defined as the consumption of one to two alcoholic drinks per day for men and one alcoholic drink per day for women. If you don't drink alcohol, however, this slight potential benefit does not make it advisable to start drinking.

Because filtrate contains both wastes and valuable substances, such as sugars, amino acids, and water, the next step of waste processing is **reabsorption** of these materials across the walls of each nephron. Water moves by osmosis out of the nephron and into the kidney interior as it descends into this salty environment. On the ascending limb, the nephron walls are impermeable to water but actively transport salt into the kidney. The structure of the nephron allows for a gradient of salt concentration to be maintained, from relatively low at the top of the loop to relatively high at the base. What remains in the filtrate after the reabsorption phase is water containing a high concentration of urea, the waste product of the breakdown of amino acids.

The last segment of the nephron permits the **secretion** into the filtrate, via active transport, of certain wastes that are in low concentration in the plasma. The filtrate then moves from the nephron into a **collecting duct** that leads to the center of the kidney, called the renal pelvis. As the collecting duct crosses the salty kidney interior, more water moves by osmosis from the filtrate into the kidney tissue. Water and other materials removed from the filtrate are drawn into the capillaries, returned to the bloodstream, and eventually leave the kidney via the renal vein. The fluid that collects in the renal pelvis is called **urine** and is made up of water and organic wastes including urea and various ions. During urine **excretion,** urine leaves the kidneys and flows to the bladder.

Alcohol is a diuretic, which means that it promotes the formation of urine and increases the volume of urine that is released from the bladder, a process called **micturition.** Coupling the increased volume of urine produced with the deadening of awareness of the need to urinate that goes with intoxication can result in a very full bladder. Even though micturition is typically under conscious control, an intoxicated person that passes out before emptying the bladder may end up urinating on himself. In this case, the body overrides the conscious control of micturition to prevent a potentially lethal bladder rupture.

Alcohol is a depressant, slowing down brain function and altering perceptions, reflexes, and balance, and causing slurred speech. In an attempt to prevent the depressant effects of intoxication, some of Malik's friends mix alcohol with energy drinks. Malik will recommend to Lin that she does not do this because Lin should develop an awareness of when to stop drinking. This is harder to do if the depressant effects of intoxication are, in part, masked by the stimulant effects of the energy drink.

In addition to managing wastes, the urinary system also plays an important role in regulating blood volume, acidity, and salt balance. The kidneys regulate acidity and salt balance by actively excreting acids and reabsorbing salts during the secretion and reabsorption steps of urine formation. The kidneys help maintain blood pressure by regulating water excretion. When blood pressure is low, antidiuretic hormone (ADH) released by the pituitary gland increases the permeability of the collecting duct to water, allowing more water from the filtrate to return to the bloodstream. When blood pressure is high, ADH release is curtailed, and more water is excreted.

**ex-** means out or out of.

**STOP & STRETCH** High blood pressure damages cells in the filtering structure of the nephron, making the pores between cells much larger. How is the urine of someone with damage to these structures likely to be different from someone with an undamaged filter?

Alcohol acts on the pituitary to lessen ADH secretion. As ADH levels drop, the kidneys reabsorb less water and thus produce even more urine. Therefore, you actually urinate a higher volume of liquid than you consume when drinking alcohol. This is, in part, the cause of the dehydration experienced after an episode of heavy drinking. Dehydration can also cause nausea and headaches. For this reason, Malik will suggest that Lin alternate drinking alcohol with drinking water.

## Engaging Safely with Alcohol

Malik's research has also made it clear that moderate alcohol consumption can have some positive physiological effects as well. He will discuss the evidence suggesting that moderate drinkers experience lower rates of cardiovascular

**Malik does not want Lin to become one of the 20% of female students who will be sexually assaulted while in college.**

disease and stroke with Lin. This may be because small amounts of alcohol can relax the heart muscle, thereby lowering the rate the heart beats and lowering blood pressure. This is true whether the alcohol consumed is wine, beer, or distilled spirits, but these benefits occur only with *moderate* drinking. In fact, larger doses of alcohol have the opposite effect, increasing risk of heart disease and stroke when blood pressure increases in the body's attempt to rid itself of the toxin via increased breathing, sweating, and urination. He hopes it will be clear to Lin that when it comes to heart health, it is better not to drink and to perform cardiovascular exercise than it is to drink moderately and not exercise, or to drink to excess. **TABLE 18.1** gives other examples of how to more safely engage with alcohol.

Alcohol in the circulatory system will also make its way to the lungs, where it evaporates from the lung surface and is expelled in each breath. Because the concentration of alcohol in the air breathed out is proportional to the amount of alcohol in the blood, testing devices like the breathalyzer can be used to determine the blood alcohol concentration (BAC). You can also approximate your BAC using your weight, gender, and rate of consumption (**TABLE 18.2**, on the next page).

The effects of alcohol in the blood are progressive. A BAC of 0.03 to 0.04 results in relaxation, mild euphoria, and decreased inhibition. Concentration is slightly impaired. By 0.05, impairment increases, affecting reasoning and motor skills. By 0.08, a person is considered legally intoxicated and should not drive. Above 0.1, emotions can become uncontrolled and behavior boisterous; reflexes, reaction time, and speech are impaired. By 0.2, a person can become unconscious and experience blackouts. A BAC of 0.3 or higher results in central nervous system depression, causing impaired breathing and heart rate. Death can result. In a 140 lb female or male, ingesting three drinks in rapid succession results in a BAC of 0.1 and 0.08, respectively.

The brain is particularly sensitive to toxins during times of rapid development, most of which occurs before a person reaches their middle twenties. During these years, lifelong traits such as the ability to reason and to critically evaluate information are developing. Binge drinking and excess alcohol consumption have been shown to disrupt brain activity and impair memory, decreasing the ability to develop these traits.

**TABLE 18.1** **Safer engagement with alcohol.**

| Strategy | Reason |
|---|---|
| Eat before consuming alcohol. | Alcohol will pass into the intestine more slowly and thus enter the circulatory system more slowly. |
| Pay attention to how your body responds to alcohol. | While most people can metabolize around one drink per hour, your metabolic rate could be slower. |
| Alternate drinking water with drinking alcohol. | Can lessen the severity of the dehydration caused by lowered ADH secretion. |
| Understand Medical Amnesty laws in your state. | Some students delay calling 911 when concerned about an intoxicated friend for fear of prosecution. Medical Amnesty laws prohibit prosecution of both the person who calls for help and the victim. |
| Never have sex with an intoxicated person. | It is a felony offense to have sex with a drunk person. An intoxicated person is not legally capable of giving consent, even if he or she expresses interest in having sex. |

**TABLE 18.2** **Blood alcohol content in women (W) and men (M).**
Subtract 0.01% for every 40 minutes of drinking.

| Drinks | 90 | 100 | 120 | 140 | 160 | 180 | 200 | 220 | 240 | Condition |
|---|---|---|---|---|---|---|---|---|---|---|
| | | | | Body Weight in Pounds | | | | | | |
| 0 | .00 | .00 | .00 | .00 | .00 | .00 | .00 | .00 | .00 | Only Safe Driving Limit |
| 1 (W) | .05 | .05 | .04 | .03 | .03 | .03 | .02 | .02 | .02 | |
| 1 (M) | | .04 | .03 | .03 | .02 | .02 | .02 | .02 | .02 | Driving Skills Significantly Affected |
| 2 (W) | .10 | .09 | .08 | .07 | .06 | .05 | .05 | .04 | .04 | |
| 2 (M) | | .08 | .06 | .05 | .05 | .04 | .04 | .03 | .03 | |
| 3 (W) | .15 | .14 | .11 | .10 | .09 | .08 | .07 | .06 | .06 | |
| 3 (M) | | .11 | .09 | .08 | .07 | .06 | .06 | .05 | .05 | |
| 4 (W) | .20 | .18 | .15 | .13 | .11 | .10 | .09 | .08 | .08 | Possible Criminal Penalties |
| 4 (M) | | .15 | .12 | .11 | .09 | .08 | .08 | .07 | .06 | |
| 5 (W) | .25 | .23 | .19 | .16 | .14 | .13 | .11 | .10 | .09 | |
| 5 (M) | | .19 | .16 | .13 | .12 | .11 | .09 | .09 | .08 | |
| 6 (W) | .30 | .27 | .23 | .19 | .17 | .15 | .14 | .12 | .11 | Legally Intoxicated |
| 6 (M) | | .23 | .19 | .16 | .14 | .13 | .11 | .10 | .09 | |
| 7 (W) | .35 | .32 | .27 | .23 | .20 | .18 | .16 | .14 | .13 | |
| 7 (M) | | .26 | .22 | .19 | .16 | .15 | .13 | .12 | .11 | |
| 8 (W) | .40 | .36 | .30 | .26 | .23 | .20 | .18 | .17 | .15 | Criminal Penalties |
| 8 (M) | | .30 | .25 | .21 | .19 | .17 | .15 | .14 | .13 | |
| 9 (W) | .45 | .41 | .34 | .29 | .26 | .23 | .20 | .19 | .17 | |
| 9 (M) | | .34 | .28 | .24 | .21 | .19 | .17 | .15 | .14 | |
| 10 (W) | .51 | .45 | .38 | .32 | .28 | .25 | .23 | .21 | .19 | |
| 10 (M) | | .38 | .31 | .27 | .23 | .21 | .19 | .17 | .16 | Death Possible |

## ▲ Working with Data

**What is the maximum number of drinks a 120 lb woman could consume over a 2-hour period to avoid becoming legally intoxicated? How many drinks can you have in a 4-hour period to avoid becoming legally intoxicated?**

After explaining the biological effects of alcohol on the body to Lin, Malik will plan to present her with some data, because he knows biologists, including Lin, like to make evidence-based decisions. He will show her study after study showing that students who do not drink, or only drink in moderation, have higher grades, better social relationships, and less involvement with law enforcement agencies; that they sleep better, suffer fewer injuries, and are less likely to engage in unplanned or unprotected sexual activity. Once she understands the biological basis of the damage caused by excess consumption of alcohol and the negative effects that can ensue (**FIGURE 18.7**), he hopes she will choose to engage safely with alcohol, if at all.

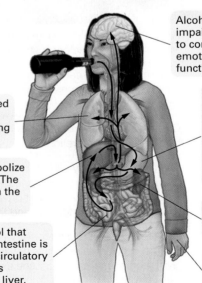

Alcohol progressively impairs the brain's ability to control behavior, emotions, and body functions.

Only a small percentage of alcohol is digested in the stomach. Most moves to the liver. Alcohol in the stomach causes increased gastric secretions that can irritate the stomach lining, triggering the vomiting reflex.

Some unmetabolized alcohol is removed from the body during exhalation.

The liver can metabolize only ~1 drink/hour. The rest moves through the circulatory system.

Some of the alcohol that reaches the small intestine is absorbed into the circulatory system and some is metabolized by the liver.

Alcohol causes increased urine formation, leading to dehydration.

Some unmetabolized alcohol is removed from the body in sweat.

**FIGURE 18.7 The destructive path of alcohol.** Alcohol crosses cell membranes readily, leaving the stomach and small intestine and entering the liver and bloodstream before it can be metabolized. When a person drinks more rapidly than the liver can process the alcohol, the excess alcohol is circulated throughout the entire body.

# savvy reader

## Sexual Assault on College Campuses

Glossy college brochures mailed to prospective students extol not only the educational experiences students can have, but also provide reassurance about the safety of the campus. These statements were excerpted from admission brochures sent out by different colleges across the United States and Canada.

- "Nothing is more important to this College than your safety."
- "This University does everything in its power to protect you."
- "The campus police department provides around-the-clock police and patrol protection."
- "The college-wide safe walk program provides male and female escorts to walk you home from night classes and on-campus activities."

But college students are not safe from sexual assault. Approximately 1 in 5 women and 1 in 70 men will be sexually assaulted during their 4-year college career. In spite of this, most people believe that campuses are safe places for students. Recently, a television journalist came under fire after asking "When was the last time you heard about rape on a college campus?"

More often than not, rape on college campuses is committed by a student at the same university, and alcohol is involved.

1. Federal law requires that colleges and universities disclose information about crime on and around their campuses. Only rapes that are reported to the police are disclosed, and the vast majority of rapes go unreported. Do you think disclosure laws could lead to university personnel pressuring rape victims not to press charges? Why or why not?

2. Most rapes on college campuses involve alcohol. In fact, some rapists will deliberately get the victim intoxicated, so it is easier to commit the assault. Women who are victims of sexual assault often do not report the crime. Do you think there is a difference in how we view a male student who is drunk and gets robbed and beaten up, and a female who is sexually assaulted while drunk? Why or why not?

3. What kinds of actions by colleges and universities might back up the safety claims they make on brochures?

4. When you plan to go out, before drinking, have a conversation with your friends about what you would like them to do if you get intoxicated. Clearly state whether you want them to leave you at the party or make sure you come home with them. Can you think of other strategies for preventing sexual assault from occurring?

# SOUNDS RIGHT BUT IS IT?

After a night of drinking at a house party, your roommate appears to have passed out. He is seated at the end of a couch, slumped sideways on the armrest. The fingers of one hand are wrapped around the neck of a half-empty beer bottle, which also leans against the armrest. It's 2 a.m. You are tired and want to go home. You gently shake him to see if he will awaken but he does not. You are a little relieved because he was getting pretty obnoxious before he passed out, and it would likely be a struggle to get him home. You ask the people you came with—some sober, some not—for help. The consensus of the group is that because he's breathing, it is probably okay to leave him on the couch for the night.

While your friends grab him under the arms and lay him on his side on the couch in case he vomits, you find a pillow and blanket for him. You alert the few remaining partygoers, all people you know fairly well, that you will return for him in the morning.

**A passed-out drunk is safe as long as he is breathing and placed on his side.**

Sounds right, but it isn't.

1. If your friend had been drinking steadily all night, then had two shots of alcohol right before passing out, what is the minimum amount of time it will take before his blood alcohol concentration peaks?

2. If your friend is breathing now, is there any guarantee he will be when his blood alcohol concentration peaks?

3. If you are trying to determine whether an intoxicated person is responsive, what might you do, aside from gentle shaking, to get a response?

4. If an intoxicated person is unresponsive, what should you do?

5. Consider your answers to questions 1–4 and explain why the original statement bolded above sounds right, but isn't.

# Chapter Review MasteringBiology®

Go to the Study Area in MasteringBiology® for practice quizzes, myeBook, BioFlix™ 3-D animations, MP3Tutor sessions, videos, current events, and more.

## Summary

### Section 18.1

Describe the structures and functions of the digestive system.

- The digestive system is a group of organs and glands working together to break foods into their component parts for reassembly into forms that the body can use or for use in generating energy (p. 416).

- Food moves from the mouth to the pharynx and through the esophagus to the stomach to the small intestine. Digested nutrients are absorbed into the bloodstream across the small intestine and brought to cells (pp. 416–417).

List the accessory organs of the digestive system, and outline their roles in digestive processes.

- The pancreas, liver, and gallbladder are accessory organs that secrete substances that aid in digestion (p. 418).

- Digestive enzymes produced by the pancreas help break down most food molecules. Bile produced by the liver facilitates the breakdown of fats and the gallbladder stores and concentrates bile before its release into the small intestine (p. 418).

### Section 18.2

List the structures composing the mammalian urinary system.

- The urinary system consists of the kidneys, bladder, ureters, and urethra (p. 420).

Describe the steps in the process of urine excretion.

- In the nephrons of the kidney, the process of filtration forces most liquid, but not cells or larger molecules, from the plasma into the kidney tubules. As the filtrate travels through the nephron, water, glucose, and other valuable molecules are reabsorbed through both active and passive mechanisms. Secretion occurs when waste materials that did not leave with the filtrate are actively brought into the nephron tubules from surrounding capillaries (pp. 420–422).

- Hormones that regulate blood pressure control the concentration of water in urine. Urine collects in the renal pelvis, then travels down the ureters to be stored in the bladder. During urination, urine is released from the bladder through the urethra and exits the body (p. 422).

## Roots to Remember

**The following roots of words come mainly from Latin and Greek and will help you decipher terms:**

| | |
|---|---|
| **epi-** | means upon, beside, or among. Chapter term: *epiglottis* |
| **ex-** | means out or out of. Chapter term: *excretion* |
| **-itis** | refers to inflammation. Chapter term: *pancreatitis* |
| **peri-** | means around. Chapter term: *peristalsis* |
| **renal** | comes from the Greek word for kidney. Chapter term: *renal arteries* |
| **-stalsis** | means to wrap or surround. Chapter term: *peristalsis* |

## Learning the Basics

1. Describe the hollow organs a piece of apple would move through as it makes its way through the digestive system, from its ingestion in the mouth to its excretion by the anus.

2. List the accessory organs in the digestive system.

3. Describe the main function of the kidney.

4. The pharynx _____.
A. forms the connection between the small and large intestine; B. keeps swallowed food from entering the epiglottis; C. connects the esophagus to the stomach; D. branches to feed into the trachea and esophagus.

5. Which of the following lists digestive processes in the correct order?
A. ingestion, absorption, peristalsis; B. peristalsis, absorption, ingestion; C. peristalsis, ingestion, absorption; D. ingestion, peristalsis, absorption

6. The pancreas _____.
A. secretes bile; B. produces stomach acid; C. secretes digestive enzymes; D. all of the above.

7. True/False: Most of the digestion of nutrients occurs across the membrane that lines the interior stomach wall.

8. The villi that help absorb nutrients line the _____.
A. small intestine; B. large intestine; C. stomach; D. pharynx.

9. Blood enters the kidneys from the _____.
A. duodenal vein; B. nephron; C. renal artery; D. gall bladder.

10. True/False: The kidneys help regulate blood pressure.

## Analyzing and Applying the Basics

1. Studies show that aspirin inhibits the alcohol dehydrogenase family of enzymes. What advice would you give a friend who thinks he should take aspirin before drinking alcohol?

2. When food goes "down the wrong tube" after swallowing, what path does it take?

3. When food is in the alimentary canal, is it actually part of the body? Why or why not?

## Connecting the Science

1. Based on your understanding of the movement of water into and out of cells, which is more likely to help mitigate the effects of dehydration: tap water or a sugar- and electrolyte-containing sports drink? Why?

2. What would you say to a friend who believes that limiting her drinking to 1 night a week, but on that night drinks six drinks in 2 hours, will not affect her studies? What would be a better strategy for limiting the negative consequences of drinking on her studies?

Answers to **Stop & Stretch**, **Visualize This**, **Working with Data**, **Savvy Reader**, **Sounds Right, But Is It?**, and **Chapter Review** questions can be found in the **Answers** section at the back of the book.

# CHAPTER 19 | Clearing the Air

**Many communities have banned smoking in all public places, including restaurants and bars.**

# Respiratory and Cardiovascular Systems

In 2006, the U.S. Surgeon General released a report that summarized a medical and scientific consensus—that secondhand smoke from the lit end of cigarettes, pipes, and cigars was dangerous to nonsmokers in the same environment. This announcement accelerated a decades-old trend in the enactment of local and state laws banning smoking in all public indoor environments. In 1993, only two such laws could be found nationwide; by 2013, the number was 628, including 24 states and dozens of cities, towns, and counties. Today, over 80% of the U.S. population lives in regions with restrictions on smoking in most public indoor spaces. The success of the campaign to ban public exposure to

**Supporters of these bans say that secondhand tobacco smoke is a public health threat.**

**Opponents include smokers, who feel discriminated against while engaging in a lawful activity ...**

secondhand smoke is clear from the latest trend in state and local laws—bans of smoking in outdoor spaces, such as parks and playgrounds.

The remaining states and municipalities with few restrictions on smoking now face organized pressure to enact laws to impose limits. Even with the Surgeon General's report, the passage of these laws is not without controversy. While nonsmokers demand a right to smoke-free air, smokers claim a right to engage in a lawful activity where and when they wish. While restaurant employees advocate for work environments that are safe from secondhand smoke, bar owners worry about the effects of a ban on their bottom line. The rising popularity of vaping—that is, use of

electronic e-cigarettes—has added another layer to the issue, leading municipalities with smoking bans in place to consider vaping bans as well. Citizens in every community that considers smoking or e-cigarette bans must try to find the right balance between protecting public health and respecting the rights of private individuals and property owners.

Is there a public place for smoking and vaping, or do the risks of both demand that we protect all nonconsumers from exposure to the smoke and vapor from these products? In this chapter, we examine the pathway of secondhand smoke and vapors into a nonsmoker's lungs and body and investigate the evidence that links exposure to these gases to preventable diseases that affect the lungs, heart, and other organs.

**... as well as bar owners, who worry that smoking bans will cost them business.**

# 19.1 Effects of Smoke on the Respiratory System

The secondhand smoke emitted by the lit end of a cigarette, combined with the smoke **exhaled** by active smokers, is more accurately referred to as **environmental tobacco smoke,** or **ETS.** This smoke affects not only the **active smoker,** who is holding and **inhaling** from the cigarette, but also nonsmoking individuals in the environment. A nonsmoker in an environment high in ETS is called a **passive smoker.** ETS is similar to the smoke inhaled by an active smoker, but it has some key differences.

According to the National Cancer Institute, more than 4500 chemicals have been definitively identified in tobacco smoke. Passive smokers are exposed to the same mix of chemicals as active smokers, although typically in much lower concentrations. Surprisingly, certain chemical compounds are actually higher in concentration in ETS than in actively inhaled smoke. These compounds are produced when the tobacco burns incompletely on the end of a cigarette or in the bowl of a pipe in between puffs. Just like when someone blows on a smoldering fire, when a smoker takes a drag on a cigarette, oxygen is provided to the smoldering tobacco leaves, causing them to burn more completely.

Carbon monoxide is a by-product of incomplete combustion. It is also the most abundant gas in ETS. In fact, carbon monoxide is approximately five times more abundant in ETS than it is in the smoke inhaled by active smokers. You may have heard of carbon monoxide as a poisonous air pollutant that occurs inside homes with faulty furnaces or other gas-burning appliances. While the levels of carbon monoxide in ETS alone are not deadly to passive smokers, effects of exposure to even low levels of carbon monoxide over prolonged periods can lead to serious consequences.

Also as a result of incomplete combustion, ETS contains high concentrations of airborne **particulates,** particles with a diameter less than half the width of a human hair. These particles make up the visible portion of ETS and are commonly known to cigarette smokers as tar. Because most active smokers inhale cigarette smoke through a filter tip, the amount of particles acquired by primary smoking is reduced. However, because neither active nor passive smokers breathe through a filter otherwise, both are exposed to the full concentration of particles present in ETS. Chronic exposure to airborne particulates can lead to a number of negative health effects, including emphysema and cancer.

The vapors produced by e-cigarette use are significantly different from ETS. Most importantly, there is no combustion in these devices, so all of the secondhand vapor is what is exhaled by the user. This vapor contains a number of different chemicals, notably nicotine, but none of the products of incomplete combustion, including carbon monoxide. The particulates in vapor are liquids, rather than the tiny solids found in cigarette smoke.

Airborne particulates enter the human body via the mouth and nose; pass through the **pharynx** (throat) and **larynx** (voice box) to the distribution network of the **trachea** (windpipe), **bronchi,** and **bronchioles;** and ultimately settle in the **lungs,** the site of gas exchange (**FIGURE 19.1**), where they cause some of their most severe consequences. However, some of the chemicals in these particles, as well as carbon monoxide gas, have the ability to cross from the lungs into the bloodstream and thus can be carried throughout the body. To understand how chemicals pass from the air into our bodies, first we must understand the structure and function of lungs.

**-hale** means to breathe.

**The secondhand smoke from the lit end of cigarettes, pipes, and cigars is dangerous to non-smokers in the same environment.**

## What Happens When You Take a Breath?

You take a breath approximately 12 times per minute, or over 6 million times per year, without thinking much about it. In each normal breath at rest, you

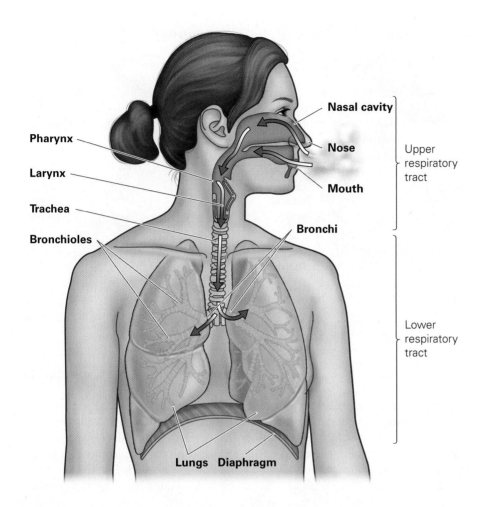

Pharynx

Larynx

Trachea

Bronchioles

Nasal cavity

Nose

Mouth

Bronchi

Upper respiratory tract

Lower respiratory tract

Lungs   Diaphragm

**FIGURE 19.1 The human respiratory system.**   Smoke enters the upper respiratory system through the nasal cavity or mouth, then passes through the pharynx and larynx into the lower respiratory system. Here, it flows into a series of smaller and smaller tubes within the lungs until it comes into close contact with circulating blood.

take in about 500 milliliters—approximately 2 cupfuls—of air and release the same amount. Over the course of a single minute, about 6 liters of air come into direct contact with the surfaces of your lungs.

**Diaphragm.**  In contrast to many other animals, respiration in humans and other mammals is an active process. The human respiratory system is separated from the organs that make up the bulk of the digestive and reproductive systems by a strong, dome-shaped muscle called the **diaphragm.** When we inhale at rest, the diaphragm contracts and flattens out, increasing the volume of our chest cavity (**FIGURE 19.2a**, on the next page). At the same time that the diaphragm is contracting, the rib cage is lifted up and outward by muscles located between the ribs, further increasing the volume of the cavity. Increasing the volume decreases the air pressure inside the chest, causing air to flow into the lungs.

The change in pressure in our chest cavity during inhalation is similar to the change in the air pressure inside a syringe when the plunger is pulled back. If the syringe is in a liquid solution, we can see the liquid flow toward the lower pressure inside the syringe, just as air flows to the lower-pressure environment inside the lungs.

When we are resting, exhalation requires no muscle contraction. The diaphragm relaxes, and the volume of the chest cavity decreases (**FIGURE 19.2b**). The air that had rushed into the lungs is squeezed into a smaller volume, increasing the pressure and causing air to flow back out of the trachea and the nasal cavity. Passive exhalation is similar to the release of air from an expanded balloon. Because the elastic walls of the balloon exert pressure on the

# Visualize This ▶

**How does a hiccup, an involuntary contraction of the diaphragm, affect normal breathing?**

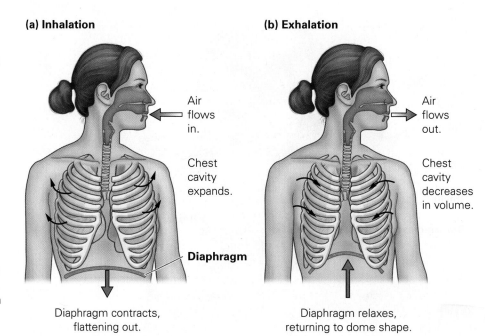

**(a) Inhalation**

Air flows in.

Chest cavity expands.

**Diaphragm**

Diaphragm contracts, flattening out.

**(b) Exhalation**

Air flows out.

Chest cavity decreases in volume.

Diaphragm relaxes, returning to dome shape.

**FIGURE 19.2 How the diaphragm works.** (a) Contraction of the dome-shaped diaphragm flattens it out, increasing the volume of the chest cavity and lowering air pressure. (b) Relaxation of the diaphragm causes the chest cavity to lose volume, increasing air pressure.

contents, air rushes out of an open nozzle even without actively squeezing the balloon.

Exhalation becomes an active process during intense aerobic exercise, when the body must actively rid itself of carbon dioxide. In this case, muscles in the abdomen contract, forcing the diaphragm upward, and muscles in the chest wall tighten, squeezing the chest cavity further. Forcing liquid out of a syringe by depressing the plunger and reducing the volume is much like the process of active exhalation. The graph in **FIGURE 19.3** illustrates how the mechanics of breathing correspond to changes in lung volume, and thus the amount of air in the lungs.

**STOP & STRETCH** The Heimlich maneuver is a strategy for forcing an obstruction out of the trachea of a person who is choking by forcibly squeezing the abdomen directly under the rib cage. Use your understanding of inhalation and exhalation to explain why this maneuver is effective.

# Working with Data ▶

**According to the graph, how much air is exchanged during a single inhalation/exhalation at rest?**

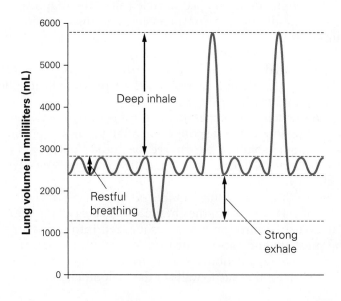

Lung volume in milliliters (mL)

Deep inhale

Restful breathing

Strong exhale

**FIGURE 19.3 Lung capacity and breathing.** The muscular movements of the chest and abdomen walls change the lung volume. A deeper inhale results from a strong contraction of the diaphragm and intercostal muscles, leading to a larger lung volume.

Our rate of breathing is regulated by control centers in the most ancient part of the brain, called the brain stem. The brain stem signals the diaphragm to contract in response to carbon dioxide levels in the blood. The breathing rate increases when carbon dioxide levels in the body increase, such as during exercise. Rapid breathing moves carbon dioxide out of the body quickly while also providing oxygen for optimal ATP production.

Contraction of the diaphragm can also be controlled voluntarily. Holding your breath requires consciously overriding the signals from the brain stem—at least for a time. Taking a drag on a cigarette or e-cigarette requires stronger contraction of the diaphragm than normal breathing does to pull air through the device and into the mouth and lungs.

Speaking or making any vocal noises requires active exhalation as air forced out past the vocal cords in the larynx causes these structures to vibrate and produce sound. The amount of air forced past the vocal cords determines the volume of our speech, while muscles that control the length of the vocal cords help to determine the pitch of our speech. The shape of our mouth, lips, and tongue and the position of our teeth determine the actual sound that is produced. Sustained exposure to tobacco smoke can cause parts of the larynx to become covered with scar tissue, often making long-time smokers sound quite hoarse. Nonsmokers regularly exposed to environmental tobacco smoke demonstrate some changes to their vocal cords relative to other nonsmokers, although it is not clear that these changes are enough to affect the voice.

The diaphragm also may contract involuntarily for other reasons, for instance, when we inhale deeply before a cough. Coughing is a reflex that helps remove irritants from the trachea. Keeping air passageways into the lungs clear is crucial for maintaining adequate oxygen uptake.

**Lungs.**   Healthy lungs are pink, rounded, and very spongy in texture (**FIGURE 19.4a**). Even at rest, these organs almost completely fill the chest cavity. The external surface of the lungs is linked to the lining of the chest cavity by two thin membranes, one covering the lungs and one lining the chest wall. These membranes stick together because they are moist, much like two wet pieces of plastic wrap stick together. This ensures that when the chest cavity enlarges, the lungs expand. If the two membranes become separated—for instance, as a result of injury to the chest wall—then air will enter the space between them, sometimes causing the underlying lung to collapse.

Once air enters the lungs, it flows through the bronchi and bronchioles, each of which dead-ends at structures called **alveoli**. The alveoli are 300 million tiny sacs within our lungs that contain the **respiratory surface** of our

**(a) Human lungs**

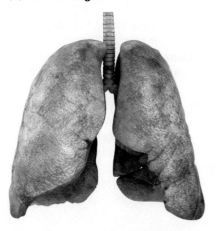

**(b) Tennis court**

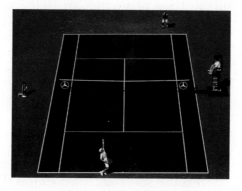

**FIGURE 19.4 The respiratory surface of the human body.**   (a) Lungs contain the surfaces over which gases are exchanged between the body and the environment. (b) The total respiratory surface area contained within a pair of healthy human lungs is approximately the size of a tennis court.

**FIGURE 19.5 Alveoli.** (a) Alveoli occur in clusters at the end of bronchi. The clusters are enmeshed in a dense network of thin-walled capillaries, which bring blood into close proximity with the gas in the alveoli. (b) An electron micrograph of alveoli in lungs.

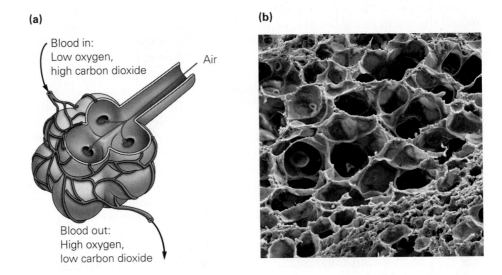

(a)

Blood in:
Low oxygen,
high carbon dioxide

Air

Blood out:
High oxygen,
low carbon dioxide

(b)

bodies, the tissue across which gases from inside the body are exchanged with gases in the air. The area of this respiratory surface in humans is approximately 160 square meters (1725 square feet), about the size of a tennis court (**FIGURE 19.4b**). A large surface area is required to supply oxygen and remove carbon dioxide from large-bodied, active mammals. Each grape-like cluster of alveoli is surrounded by a net of tiny blood vessels, called **capillaries**, connecting the gases exchanged in these structures with the entire body (**FIGURE 19.5**).

Exposure to smoke can damage and eventually destroy the respiratory surfaces within alveoli. A reduction in the total amount of respiratory surface results in shortness of breath, wheezing, and an inability to participate in even moderately vigorous activity. These symptoms are all caused by the reduced capacity for gas exchange in the lung.

## Gas Exchange

The primary function of the lungs is **gas exchange**, the process that allows us to acquire oxygen for cellular respiration and to expel carbon dioxide, one of the waste products of respiration. Exchange of gases occurs between the air spaces of the alveoli and the blood contained in neighboring capillaries. The exchange works by simple diffusion across the thin membranes surrounding the blood vessels and the alveoli walls. Recall that diffusion is the passive movement of substances from where they are in high concentration to areas where they are in low concentration. Carbon dioxide in high concentration within the capillaries passes via diffusion from these structures into the alveoli, where it is maintained in low concentration by exhalation. Oxygen in high concentrations, maintained by inhalation, passes by diffusion from the alveoli to the deoxygenated blood in the capillaries (**FIGURE 19.6**).

Carbon dioxide and oxygen gas dissolve in the fluid that coats the respiratory surface prior to diffusing across the cell membrane. Because the size of the alveoli increases and decreases over the course of a single breath, this layer of fluid must be somewhat slippery to maintain constant coverage of the surface and to prevent the alveoli walls from actually sticking together. A soap-like substance called surfactant provides this slippery consistency. The composition of surfactant, and thus the flexibility of the lung tissue itself, can be negatively affected by tobacco smoke.

# Visualize This ▼

**Oxygen levels in the atmosphere are low at high altitudes. How do you think this affects the rate of gas exchange in the lungs?**

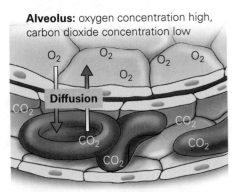

**Alveolus:** oxygen concentration high, carbon dioxide concentration low

$O_2$ $O_2$ $O_2$ $O_2$

$O_2$

**Diffusion**

$CO_2$ $CO_2$ $CO_2$ $CO_2$

$CO_2$

**Capillary:** oxygen concentration low, carbon dioxide concentration high

**FIGURE 19.6 Gas exchange in the lungs.** The movement of oxygen and carbon dioxide gas at the respiratory surface occurs via simple diffusion.

## The Role of Hemoglobin in Gas Exchange

Carbon dioxide is in relatively low concentrations in the atmosphere, so it readily moves from the blood into the alveoli and out into the atmosphere. However, oxygen does not dissolve as readily into the blood from the air; thus, most land animals (and even some plants) produce specialized proteins that can bind oxygen.

**Hemoglobin** is the protein that acquires and transports oxygen in humans and nearly all other vertebrates. Hemoglobin is actually made up of four separate protein chains, each containing a single iron atom that can bind to a single oxygen molecule (**FIGURE 19.7**). Each of our red blood cells contains approximately 250 million hemoglobin molecules; therefore, each cell can carry about 1 billion oxygen atoms.

In areas of the body where oxygen levels are low and carbon dioxide production is high, hemoglobin releases oxygen molecules. In this way, red blood cells are loaded with oxygen in the lungs and will drop their load in areas of the body where high levels of cellular respiration are occurring. Hemoglobin picks up oxygen effectively, but surprisingly, it binds to the smoke by-product carbon monoxide about 200 times more strongly. Carbon monoxide inhaled by breathing tobacco smoke is preferentially loaded into red blood cells at the respiratory surface. Because the binding of carbon monoxide is so strong, hemoglobin is slow to release it. As a result, fewer hemoglobin molecules are available to transport oxygen. Even small amounts of carbon monoxide can tie up large amounts of hemoglobin in the body, causing severe oxygen shortages in body tissues. While the amounts of carbon monoxide in ETS are not high enough to cause death from lack of oxygen, chronic low levels of oxygen deprivation can damage tissues and organs.

Carbon monoxide is especially damaging to developing embryos and fetuses because they must acquire the oxygen they need through exchange with their mother's blood supply. The lower-than-average birth weights of babies born to mothers who are active smokers may be due to their relative oxygen deprivation. There is some evidence from animal studies that long-term exposure to the carbon monoxide levels found in ETS may contribute to diminished brain function in infants and children for the same reason—oxygen deprivation.

The biggest threat ETS poses to respiratory function does not stem from the interaction between hemoglobin and carbon monoxide. Instead, the threat comes from damage caused to lungs by microscopic particles of smoke.

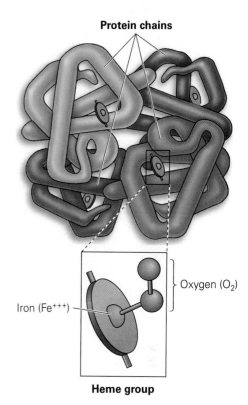

**FIGURE 19.7 Hemoglobin, the blood's oxygen shuttle.** The four protein chains in a single hemoglobin molecule are colored differently here, but function identically. The heme group in each chain is the oxygen-binding site.

## Smoke Particles and Lung Function

The particulates in ETS are a significant hazard to infants and children as well as to anyone with impaired lung function, including individuals who have been chronically exposed to particulate-laden air. While the liquid particulates from e-cigarettes appear less damaging than the solids produced by tobacco smoke, research on the effect of vapor particulates on passive "vapers" is still lacking.

**Bronchitis and Asthma.** Our lungs have various mechanisms for expelling and removing foreign objects. Coughing is our first response to larger particles or droplets that enter the trachea. Particles too small to trigger a cough can settle in the upper respiratory tract, trachea, or larger bronchi. These particles become trapped in mucus produced by the tissue lining these structures. The mucus and particles are swept upward toward the mouth and nose by the actions of tiny hairs called cilia on the tissue surface. When a glob of mucus reaches the tip of the larynx, it is coughed up, expelled out of the nose or mouth, or swallowed.

Airborne particles in tobacco smoke not only increase the production of mucus but also damage the cilia lining the bronchi, making it more difficult to expel these particles. This damage in turn can lead to inflammation of the bronchi, known as **bronchitis.** Active smokers may develop chronic bronchitis, which often manifests itself as a lasting cough, as abundant mucus is produced by inflammation. Particulates, even the liquid droplets in e-cigarette vapor, are also known to worsen **asthma,** an allergic response that results in the muscular constriction of bronchial walls as well as an overproduction of mucus. An acute asthma episode is often characterized by a wheezing sound during breathing, as the greatly restricted airflow creates sound oscillations. According to the U.S. Environmental Protection Agency (EPA), the number of additional cases of asthma caused by ETS among children in the United States is likely to be 26,000 per year. It is not clear whether exposure to e-cigarette vapor will cause additional asthma cases.

**-itis** is used to describe inflammation (of an organ).

**Emphysema.** The physical consequences of chronic bronchitis and asthma can lead to the formation of scar tissue in the lungs, which permanently blocks bronchi. The inflammation also causes damage to alveoli walls, causing the many small alveoli to merge into fewer, larger sacs (**FIGURE 19.8**). This merging of alveoli reduces the overall area of the respiratory surface, interfering with gas exchange. The sacs themselves become surrounded by thick scar tissue, which interferes with the passage of oxygen into the capillaries. The buildup of scar tissue also makes the lungs less elastic, meaning that the passive process of exhalation is ineffective. As more "dead air" remains in the lungs, they become overinflated, gradually increasing the size of the chest. A barrel-shaped chest can be a physical sign of the underlying disease, called **emphysema.**

Because alveoli cannot be regenerated, the lung damage that results in emphysema is permanent and irreversible. Individuals with emphysema are chronically short of breath and unable to participate in vigorous activity. The most severely affected individuals require supplemental oxygen simply to engage in the activities of daily living. The EPA estimates that adult nonsmokers exposed to ETS have a 30 to 60% higher occurrence of emphysema, asthma, and bronchitis when compared to unexposed nonsmokers.

## Visualize This ▼

Smaller and more numerous alveoli would increase the respiratory surface area even more. What do you think are disadvantages to more, smaller alveoli?

**(a) Alveoli in a nonsmoker's lung**

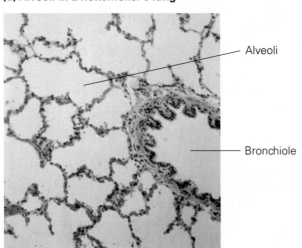

Alveoli

Bronchiole

**(b) Alveoli in a smoker's lung**

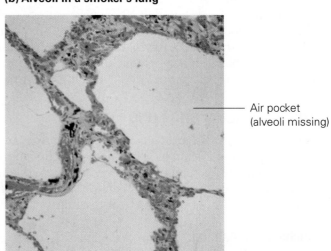

Air pocket (alveoli missing)

**FIGURE 19.8 The effect of chronic smoke exposure on alveoli.** (a) In this cross-section of a normal lung, the alveoli are tiny. The walls surrounding each air pocket are areas where gas exchange occurs. (b) In a smoker's lung, many of the alveoli walls have been destroyed, leading to larger air pockets and less surface for gas exchange. Both of these images are photographed at the same magnification.

**STOP & STRETCH**   Consider the nature of the damage that causes emphysema. Is there any potential cure for this condition?

**Lung Cancer.** The tiniest particulates in ETS can be drawn deeply into the lungs, even into the alveoli. Because alveoli lack cilia, the movement of foreign materials out of these structures is much more limited. Small particles can remain in the alveoli for long periods. We can see the accumulation of particles in the lungs of a long-time smoker, which are often black with tar trapped inside the alveoli (**FIGURE 19.9**). The accumulated particles promote inflammation and fluid accumulation and generally reduce gas exchange. Even more seriously, because many components of tobacco smoke are carcinogens known to cause mutations, lung cells are susceptible to transforming into cancer cells long after smoke inhalation has stopped.

About 170,000 new cases of lung cancer are diagnosed in the United States each year, 90% in current or former smokers. However, for the 17,000 cases of lung cancer not due to active smoking, epidemiologists at the National Institutes of Health (NIH) have estimated that 3000 cases are directly attributable to exposure to ETS, and that passive smokers have a 20 to 30% higher risk of lung cancer than unexposed nonsmokers.

Epidemiological studies, such as the one that linked ETS exposure to lung cancer risk, have the same problem as all other correlations—it is impossible to eliminate the chance that some other difference between individuals in the study is, in reality, causing a difference between groups (Chapter 1). In the case of ETS and lung cancer, the hypothesized link is also supported by the presence of known carcinogens in secondhand smoke, by experiments using animals that showed higher rates of cancer in ETS-exposed individuals, and by a link established by hundreds of studies between active smoking and lung cancer.

Because e-cigarettes are so new, scientists have much less data about the relationship between secondhand vapor exposure and any of these lung conditions. The relationship between exposure to secondhand smoke and diseases in other organs and organ systems is also less clear from the data collected. However, given the connection between the lungs and the bloodstream, one would expect that passive smokers have a higher risk than unexposed individuals of all the diseases that are more common in active smokers. To understand how smoke and vapor breathed into the lungs can affect the entire body, we must understand first how blood and its components are moved around via the bloodstream.

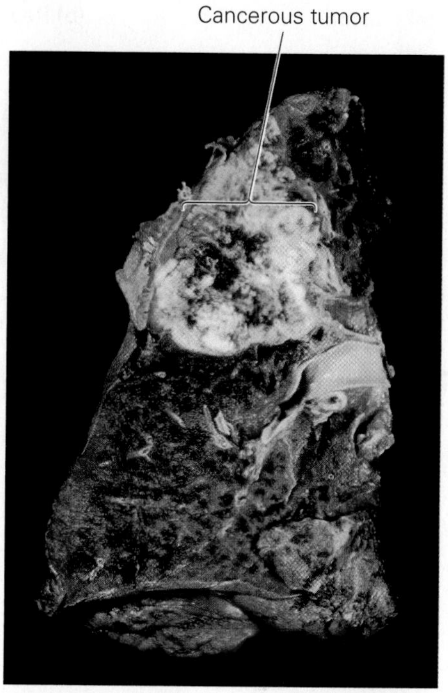

Cancerous tumor

**FIGURE 19.9 Tobacco tar trapped in lungs.** These lungs show the dark staining caused by the accumulation of particulates. The lighter-colored mass is a cancerous tumor.

# 19.2 Spreading the Effects of Smoke: The Cardiovascular System

The risks of active smoking, in addition to lung and airway damage, include increased rates of throat, bladder, and pancreatic cancer; higher rates of heart attack, stroke, and high blood pressure; and even premature aging of skin. All of these effects occur because many of the components of tobacco smoke can cross the thin walls of alveoli into the bloodstream and move throughout the body.

## Structure of the Cardiovascular System

Gases and other materials are distributed around the body of most animals by a **cardiovascular** system, which consists of three major components: a circulating fluid (often called blood), a pump (in most cases, this is called a heart), and a vascular system (blood vessels and capillaries).

**card-** and **cardio-** relate to the heart.

**vascul-** means vessel.

**(a) Plasma**

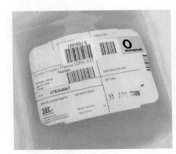

**(b) Red blood cells**

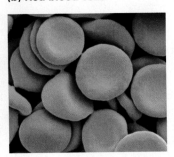

**(c) White blood cells**

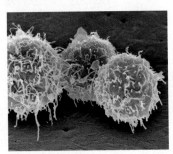

**(d) Platelets**

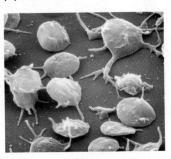

FIGURE 19.10 **Components of blood.** Blood consists of the fluid plasma as well as large amounts of red blood cells and lesser amounts of white blood cells and platelets.

**Blood.** The 5 liters (11 pints) of blood in the vascular system of an adult human are made up of both liquid and solid (that is, cellular) portions. The liquid portion, called plasma, consists of water and dissolved proteins, salts, and gases. The cellular, or solid, portion is primarily made up of red blood cells (giving blood its reddish color), with much smaller numbers of white blood cells and platelets (**FIGURE 19.10**). In adults, all of the cellular components of blood are produced by stem cells in the bone marrow, tissue found in the cavities of certain bones. Stem cells are unique among cells in adults—instead of being completely differentiated into a single cell type, such as nerve or skin, they have the capacity to produce descendants of a variety of cell types. Blood stem cells give rise to two classes of more specialized stem cells that together can generate nine different blood cell types (**FIGURE 19.11**).

# Visualize This ▶

**Leukemia is a form of cancer caused by the uncontrolled division of blood stem cells. How do you think leukemia primarily manifests itself in the blood?**

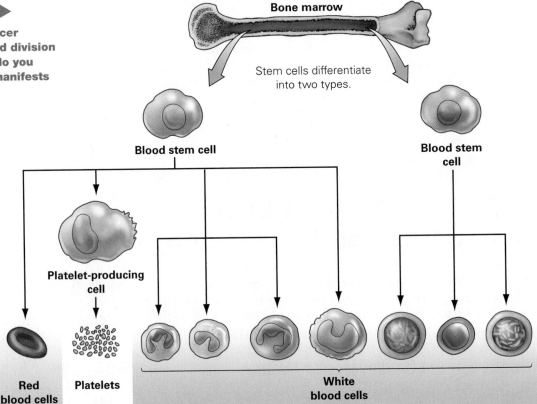

FIGURE 19.11 **Blood stem cells.** Blood cells are produced by stem cells found in the marrow of larger bones. Notice all cellular elements in the blood derive from these stem cells.

**Red blood cells** are uniquely adapted to their primary task—shuttling oxygen from the lungs to the rest of the body. These cells are packed with hemoglobin molecules and lack a nucleus and other organelles. Red blood cells are also small and pinched in the middle, which makes their surface area large relative to their volume, ensuring rapid diffusion of oxygen into and out of the cells. **White blood cells** come in many different varieties and are the essential component of immune system function, attacking invading organisms as well as removing toxins, wastes, and damaged cells throughout the body. The principal function of **platelets** is to prevent blood loss. These cell-like structures are actually membrane-bound fragments of larger cells.

Platelets work with proteins in the blood in **blood clotting,** the process that stems the flow of blood out of damaged blood vessels. Immediately after injury occurs, muscles in the walls of damaged blood vessels contract to restrict the flow of blood. Within seconds, sticky platelets become attached to any breaches in the vessels, forming a temporary plug to help restrict blood flow. At the same time, chemical signals released by the damaged tissue initiate the complex clotting process. A blood clot consists of a net made up of the protein fibrin, which forms a patch over the damaged area (**FIGURE 19.12**). Cell division in the walls of the damaged vessel and surrounding tissues eventually seals the cut, and the patch dissolves.

A substance in tobacco smoke, perhaps nicotine, increases the stickiness of platelets and promotes production of fibrinogen, the precursor to fibrin. Both of these effects result in the increased risk of blood clots. When a clot forms within a blood vessel, it can block blood flow where it forms (a condition called a *thrombosis*) or break free and travel throughout the bloodstream until it becomes lodged in another blood vessel (in which case it is called an *embolism*). When a thrombus or embolus becomes lodged within the blood vessels of the heart or any other organ, it can restrict the flow of oxygen to nearby cells, killing them and causing severe damage to the organ.

**Heart.** The fist-sized heart in a human consists of two muscular pumps that are coordinated but also somewhat independent. One pump on the right side of the heart receives oxygen-poor blood from the body and sends it to the lungs, while the left side receives oxygen-rich blood from the lungs and sends it into general circulation within the body. The two sides are each divided into two chambers, a relatively thin-walled atrium and a thick-walled ventricle. Each side also contains two valves to help control blood movement: an **AV (atrioventricular) valve** between the atrium and ventricle and a **semilunar valve** between the ventricle and the major artery that it supplies (**FIGURE 19.13**).

The heart muscle contracts with an intrinsic rhythm determined by a small patch of muscle tissue in the wall of the right atrium. This patch,

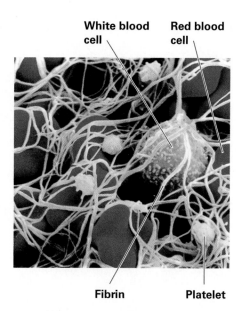

White blood cell    Red blood cell

Fibrin    Platelet

**FIGURE 19.12 A blood clot.** A blood clot consists of a net of protein fibers that trap red blood cells, forming a temporary patch over a damaged blood vessel.

## Visualize This ▼

The left and right ventricles move the same amount of blood per heartbeat, but the left ventricle walls are thicker and more powerful. Why is this necessary?

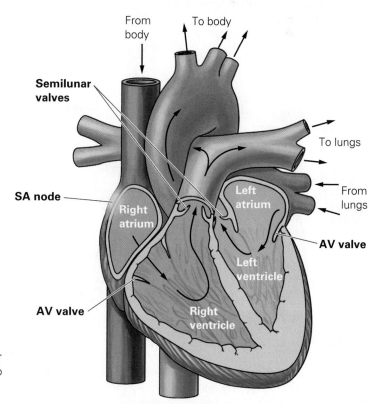

From body

To body

Semilunar valves

To lungs

From lungs

SA node

Left atrium

Right atrium

AV valve

Left ventricle

AV valve

Right ventricle

**FIGURE 19.13 The human heart.** The heart consists of four chambers making up two mostly independent pumps. One pump sends deoxygenated blood to the lungs, and the other sends blood returning from the lungs to the rest of the body.

called the **sinoatrial (SA) node,** sends out electrical signals that first cause both left and right atria to contract and then, one-tenth of a second later, cause both ventricles to contract. The complete sequence within the heart of filling with blood and then pumping is called the **cardiac cycle,** alternating between a relaxed period (called **diastole**) and a contraction phase (called **systole**).

During systole, the AV valves close as the ventricle contracts (making the first heart sound, "lubb"). At the beginning of diastole, the semilunar valves close as the ventricle relaxes ("dupp"). This cycle of contraction and relaxation leads to that distinctive heartbeat sound ("lubb-dupp, lubb-dupp").

The speed of the cardiac cycle, also known as the heart rate, is initially controlled by the SA node, which functions as an internal pacemaker (although its function can be replaced by an implanted artificial pacemaker). The speed of the pacemaker can also be affected by signals from the brain stem and spinal cord at rest, and from other parts of the brain in response to experiencing danger or intense emotion. Certain drugs, such as nicotine, may also affect heart rate.

**STOP & STRETCH** In what organ besides the heart would you expect a blood clot, which restricts oxygen delivery, to have the most severe effect on health and survival?

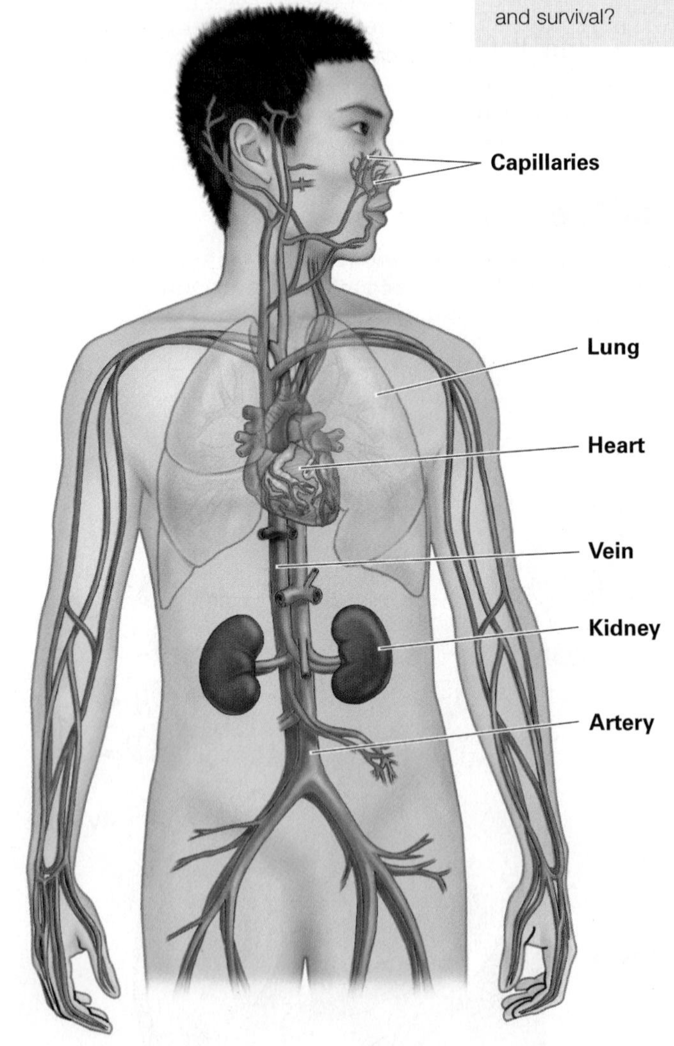

FIGURE 19.14 labels: Capillaries, Lung, Heart, Vein, Kidney, Artery

**FIGURE 19.14 The vascular system.** Arteries carry blood away from the heart, while veins carry blood to the heart. The two types of blood vessels are connected by nets of capillaries.

**Blood Vessels.** The system of tubes that carries blood to and from the heart is generally referred to as the vascular system. Components of the vascular system, broadly termed the *blood vessels,* include arteries, veins, and capillaries (**FIGURE 19.14**). Arteries are the branching blood vessels that carry blood from the heart, and veins are the converging vessels that bring it back. As in the lungs, capillaries throughout the body are tiny, thin-walled blood vessels that create a net of channels between the smallest arteries (arterioles) and the smallest veins (venules).

Arteries have thick, elastic walls that balloon out as contraction of the ventricles causes a mass of blood to flow into the system and that snap back to resting size once the blood passes by. The wave of blood is called a **pulse;** you can measure your heart rate by feeling the pulse as it passes through an artery close to the surface of the skin. The muscular walls of arteries also help regulate the rate of blood flow—when arteries are constricted, blood within them is under pressure and moves faster; blood flow slows down when they are relaxed.

Exchange of gases and other materials between the circulating fluid and the body's tissues and organs occurs across the porous walls of the capillaries. Capillaries are extremely diffuse and abundant; an adult human body contains an estimated 100,000 kilometers (60,000 miles) of capillaries, and few living cells are more than 0.1 mm (about the thickness of a sheet of paper) away from a capillary.

Liquid and materials in these vessels are forced out of capillaries due to higher blood pressure near the arterial end of a net of capillaries, called a **capillary bed.** Materials and liquid from the body tissues then flow back into the capillaries as a result of concentration differences at the venous (vein) end of the capillary bed (**FIGURE 19.15**). Muscles surrounding the arterial ends of capillary beds can contract to cut off blood flow to less needy organs to effectively deliver blood and nutrients to more essential regions. Capillaries that nourish the skin often become constricted in response to a crisis in another organ, which is why a common symptom of illness is paler skin (most apparent in the tissues lining the eyes and inside the mouth). Skin capillaries also play an essential role in the regulation of body temperature, opening when temperature rises to allow body heat to dissipate and closing to conserve heat in cold temperatures.

Veins have much thinner, less elastic walls than arteries, and the pressure of the blood is much lower once it reaches these vessels. Blood tends to pool within veins, a fact you can easily see by lowering your hand to your side. Blood pooling in the veins on the back of the hand will make these veins become distended and stand out. Movement of blood from the veins back to the heart is facilitated by the contraction of skeletal muscles, which compress the veins and squeeze the blood through them. The blood flows in only one direction, toward the heart, due to the presence of one-way valves within the veins (**FIGURE 19.16**).

The efficient flow of oxygen and materials to the body's tissues requires that the blood is maintained under enough

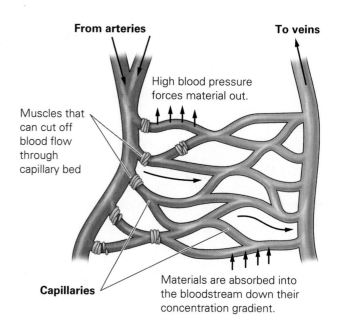

**FIGURE 19.15 Capillary function.** Liquid forced out of the upstream end of a capillary bed carries oxygen and nutrients to the tissues. Wastes and other materials are absorbed into the blood at the downstream end. Muscles on the upstream side can contract to restrict flow through the bed.

# Visualize This ▼

**Many athletes have taken to wearing compression socks during sporting events. These socks compress the calves and shin muscles. How might compression socks affect blood circulation and oxygen delivery to the leg muscles?**

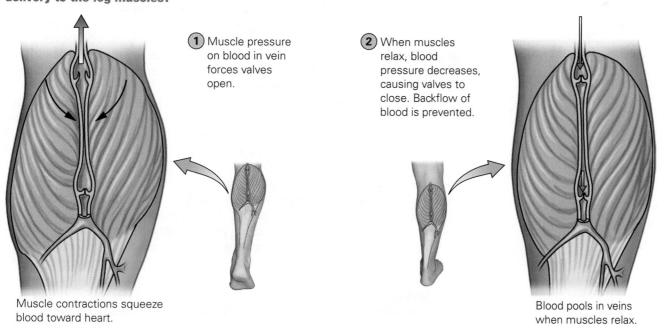

**①** Muscle pressure on blood in vein forces valves open.

**②** When muscles relax, blood pressure decreases, causing valves to close. Backflow of blood is prevented.

Muscle contractions squeeze blood toward heart.

Blood pools in veins when muscles relax.

**FIGURE 19.16 Flow of blood in veins.** Blood in the veins is under low pressure and returns to the heart due to the contraction of skeletal muscle. The blood flows in one direction, thanks to the presence of one-way valves.

pressure to move it throughout the vascular system. **Blood pressure,** the force of the blood against blood vessel walls, is created in part by the pulse of blood from the contracting heart and in part by the diameter of the arteries. Blood pressure rises when arteries become constricted by the action of muscles in the vessel walls. For instance, this happens when release of the stress hormones epinephrine and norepinephrine indicates that muscles need oxygen delivered quickly. In a healthy individual, blood pressure will rise in response to vigorous activity or excitement but return to normal under restful conditions.

Chronic high blood pressure, called **hypertension,** may be caused by arteries that are narrowed by constant psychological stress or by the accumulation of fatty material within the walls of the arteries. The latter condition is called **atherosclerosis** and occurs as a result of aging and a high-fat diet, as well as exposure to tobacco smoke. Hypertension can lead to damage in the walls of blood vessels, and it causes the heart to work much harder to push blood through the arteries.

## Movement of Materials through the Cardiovascular System

The components of tobacco smoke can exert effects on all organ systems because materials crossing the respiratory surface enter the bloodstream, which then circulates these materials throughout the body.

In humans, lungs and the cardiovascular system are connected by a double circulation system in which blood flows in two distinct but related circuits. The **pulmonary circuit** circulates the blood into the lungs, where oxygen and other components inhaled in the air are picked up and carbon dioxide released. The pulmonary circuit then returns this blood to the heart, where it enters a second, separate circuit. This **systemic circuit** pumps blood to the rest of the body, where the blood drops off its oxygen and picks up carbon dioxide. Blood flowing through the systemic circuit also picks up and distributes materials from the digestive system and filters nongaseous wastes via the kidneys.

The path of circulation between the heart, lungs, and body is detailed in **FIGURE 19.17**. Here you can see that in the pulmonary circuit, blood from the right side of the heart is pumped to the lungs. While in the lungs, the blood drops off carbon dioxide and picks up oxygen, as well as carbon monoxide and other chemicals in smoke and inhaled vapors. The blood then returns to the heart via a pair of large veins, which empty into the left side of the heart. When the heart first contracts, the initial effect is to force the remaining blood in the atria into the ventricles. A slight lag before the ventricles contract allows them to fill completely with blood, increasing the efficiency of the heartbeat.

Once the left ventricle contracts, the blood is forced at high pressure into the systemic circulation, first to the arteries, which repeatedly branch out and become smaller in diameter, and then into capillary beds. In the capillaries, oxygen and other components in high concentration within the blood diffuse out, and carbon dioxide and wastes diffuse in. The deoxygenated blood then travels to the systemic veins, which empty into the right side of the heart. Contraction of the right ventricle sends blood back into the arteries and capillary beds of the pulmonary circuit.

## Visualize This ▼

Some children are born with a "hole in the heart," which occurs when a passageway exists between the left and right atria. Referencing the path of blood flow on this figure, explain how a hole in the heart might reduce oxygen delivery to the body tissues.

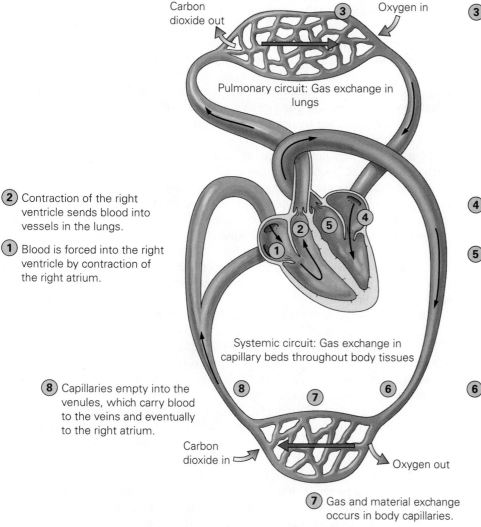

Carbon dioxide out

Oxygen in

Pulmonary circuit: Gas exchange in lungs

**3** Oxygenated blood from the lungs travels to the left atrium of the heart.

**2** Contraction of the right ventricle sends blood into vessels in the lungs.

**1** Blood is forced into the right ventricle by contraction of the right atrium.

**4** Contraction of the heart muscle forces blood into the left ventricle.

**5** Contraction of the left ventricle forces blood into body arteries at high pressure.

Systemic circuit: Gas exchange in capillary beds throughout body tissues

**8** Capillaries empty into the venules, which carry blood to the veins and eventually to the right atrium.

**6** Blood flows through the arteries and arterioles into capillaries.

Carbon dioxide in

Oxygen out

**7** Gas and material exchange occurs in body capillaries.

**FIGURE 19.17 Connecting the respiratory and circulatory system.** The lungs are connected to the other organ systems by the heart and blood vessels.

A single red blood cell carrying a load of oxygen—or carbon monoxide—can travel through the entire circulatory system in approximately 1 minute. Thus, the chemicals present in ETS have almost immediate effects on the body. These quickly occurring effects can cause more severe long-term damage as well.

## Smoke and Cardiovascular Disease

Most people believe that lung cancer and other lung diseases are the primary risk associated with exposure to tobacco smoke. In reality, most of the deaths due to smoking result from heart and blood vessel damage, or **cardiovascular disease.**

Opponents of e-cigarettes point to evidence that some of the cardio-vascular damage that results from tobacco smoke appears to be caused by

# Is there a public place for smoking and vaping?

## Visualize This ▼

Angioplasty is a procedure that threads a tiny balloon into a narrowed artery and inflates it near a plaque. The balloon is then removed. How would this improve the condition of the artery?

**(a) Normal artery**

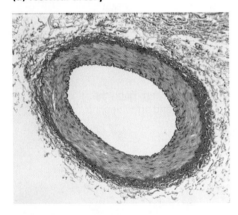

**(b) Atherosclerotic artery**

Atherosclerotic plaque

**FIGURE 19.18 Atherosclerosis.** Fat and cholesterol accumulate in the walls of arteries, reducing their diameter and their ability to carry blood.

**nicotine,** the primary active ingredient in tobacco and the drug that is delivered by e-cigarettes. Nicotine is a compound produced by tobacco plants as a natural pesticide—in fact, nicotine sulfate spray is sold commercially for use by organic gardeners to combat aphids and other insects. The nicotine in tobacco smoke and e-cigarette vapor is readily absorbed across the alveoli walls and enters the bloodstream almost immediately after it is drawn into the lungs.

In high doses, nicotine is toxic to humans and other mammals. In low doses, nicotine interacts with cells in the brain to stimulate the release of epinephrine, which in turn increases heart rate and blood pressure. Some of the brain cells activated by nicotine include those in the brain's "reward pathways," regions with activity that is associated with feelings of well-being and happiness. The reward pathways of the brain are an adaptation that reinforces behaviors important for our survival and reproduction; for instance, these pathways are activated in response to tasting delicious food or engaging in sexual activity.

By hijacking this pathway, nicotine and other drugs such as cocaine and methamphetamine become addictive, driving the user to seek out these drugs again and again to get the same response. Nicotine addiction is part of what keeps smokers smoking, despite the extensive financial and physical costs associated with tobacco use. Nicotine-containing patches, chewing gum, and nasal spray provide low doses of nicotine that can help dull the drug cravings that occur when a smoker quits. In fact, e-cigarettes were first introduced as a safer alternative to smoking. One concern about widespread acceptance of e-cigarettes is that it will entice new nicotine users who would not have started smoking to begin with.

Nicotine has effects on other parts of the body besides the brain. It appears to increase production of LDL (low-density lipoprotein), "bad cholesterol," and reduce production of HDL (high-density lipoprotein) or "good cholesterol" (Chapter 3). As a result, individuals exposed to nicotine have higher levels of circulating lipids and thus a higher risk of atherosclerosis, the accumulation of fats and other debris on the interior walls of arteries (**FIGURE 19.18**). Recall that nicotine also may stimulate blood clot formation. A clot can cut off blood flow in arteries narrowed by both atherosclerosis and nicotine-induced hypertension. When a blockage occurs in an artery that brings blood to the brain, it results in a stroke, which causes the death of brain tissue. When a blockage stops blood flow to part of the heart muscle, it can result in the death of that muscle, that is, a **heart attack.** Stroke and heart attack are the primary cardiovascular diseases that cause the death of affected individuals.

Combining the artery-damaging and clot-inducing results of nicotine consumption with the oxygen-robbing increase in inhaled carbon monoxide, exposure to tobacco smoke represents a serious risk to heart function. According to the U.S. Centers for Disease Control and Prevention (CDC), heart disease accounts for about 147,000 deaths per year among active smokers. A smoker has a 200 to 300% greater chance of dying from heart disease than does a nonsmoker. However, the CDC also estimates that 35,000 deaths per year from heart disease among nonsmokers result from exposure to carbon monoxide in environmental tobacco smoke. There is good reason to suspect that the artery-damaging effects of smoke exposure also occur in passive smokers, but the epidemiological studies have not convincingly demonstrated this link. And research is still minimal on the effects of nicotine delivered via e-cigarettes.

Unlike the permanent lung changes caused by tobacco smoke, the effects of both nicotine and carbon monoxide on cardiovascular health are much more reversible. According to the World Health Organization (WHO), the risk of heart disease decreases by 50% in ex-smokers within 1 year of quitting, and by 15 years after quitting, the risk to ex-smokers is the same as that for people who have never smoked.

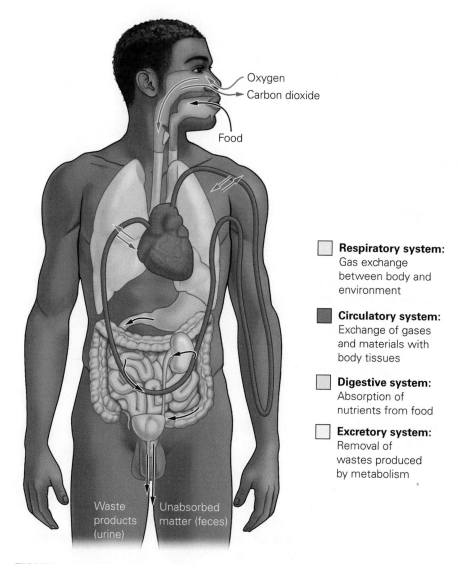

Oxygen
Carbon dioxide
Food

**Respiratory system:**
Gas exchange
between body and
environment

**Circulatory system:**
Exchange of gases
and materials with
body tissues

**Digestive system:**
Absorption of
nutrients from food

**Excretory system:**
Removal of
wastes produced
by metabolism

Waste
products
(urine)

Unabsorbed
matter (feces)

**FIGURE 19.19 Digestion, respiration, circulation, and excretion.** This figure illustrates the relationship between the four functions described in the last two chapters by simplifying the actual structure of their associated organs. These organ systems provide the primary routes of materials into, out of, and throughout the body.

The links between our respiratory and cardiovascular systems provide a mechanism by which what we inhale affects all the organs of our bodies. These systems in combination with the urinary and digestive systems discussed in Chapter 18 compromise the major pathways by which our bodies interact with the environment (**FIGURE 19.19**). While many of the links between environmental tobacco smoke and disease have not been "proven beyond reasonable doubt," we know from our own experience that our choices about what we put into our bodies can have a direct effect on our health. The role of the respiratory system in bringing materials into the body should make environmental exposure to tobacco smoke, e-cigarette vapor, and other air pollutants concern us. Accumulating evidence from both epidemiology and the known risks of active smoking provide additional support for the hypothesis that ETS is dangerous and makes public health advocates wary of the possible costs of secondhand vapor exposure.

As citizens, we are often called on to make decisions in the absence of perfect knowledge. In most of the communities that have considered imposing smoking bans, voters have decided that the risk of ETS is large enough to justify restricting smokers. Now that these bans have been in place for many

**What is the right balance between protecting public health and respecting the rights of private individuals?**

years, it appears that clearing the air in formerly smoky eating establishments and entertainment venues has been mostly good for business. Given this experience with smoking laws, many communities are making the decision that e-cigarette bans are a wise precaution. In the case of smoking bans at least, resolution of the controversy has been a win for both the business owners and the public's health.

# *savvy reader*

## Opinion Polling

News organizations in communities considering e-cigarette bans may choose to survey opinions on their own websites. Here is a sampling:

From the *Staten Island Advance*
Should [New York] city ban electronic-cigarette smoking in public places?
No. E-cigarettes produce vapor, not smoke, and should not be in the same class as cigarettes. 72.55%
Yes. E-cigarettes make the smoking ban difficult to enforce and send a message that smoking is safe. 27.45%

From the *Oregonian*
Should Oregon ban the sale of e-cigarettes to minors?
Yes 69.88%
No 29.73%

From the *Syracuse Post Standard*
Do you think electronic cigarettes are safe?
Yes 41.41%
No 34.21%
Don't know 24%

1. Summarize these results. Does it appear that the public in NYC supports an e-cigarette ban? Should Oregon ban sales to minors? Do the citizens of Syracuse believe these devices are dangerous?
2. Participation in each of these surveys is voluntary. How does this affect your interpretation of the survey results? How does permitting people to self-select whether to participate in a survey about smoking affect the sample?
3. How could these surveys be improved to provide a better understanding of opinions on this topic?

# SOUNDS RIGHT **BUT IS IT?**

Thanks to sustained public health campaigns, most people understand that smoking carries significant risks. As a result, rates of smoking have declined over the last few decades. However, rates of smokeless tobacco (that is, chewing tobacco and snuff) have remained the same. This phenomenon is due to a widespread belief that:

**Smokeless tobacco is safer than smoking.**

Sounds right, but it isn't.

Answer the following questions to understand why.

1. What dangerous components of tobacco are present even when it is not burned?
2. Smokeless tobacco contains nicotine. What are the effects of nicotine on the circulatory system?

3. The components of tobacco get into the bloodstream even if they are not inhaled. Think back to Chapter 18. How do materials in the organs of the digestive system, such as the mouth, stomach, and intestines, get into the blood?
4. Reflect on your answers to questions 1–3 and explain why the statement in bold above sounds right, but isn't.

# Chapter Review MasteringBiology®

Go to the Study Area in MasteringBiology® for practice quizzes, myeBook, BioFlix™ 3-D animations, MP3Tutor sessions, videos, current events, and more.

## Summary

### Section 19.1

List the components of the mammalian respiratory system, and describe the path of air into the body.

- Air flows into the body via the mouth and nose and enters the respiratory system via the pharynx and trachea. Once in the lungs, air flows through bronchi and into alveoli, where gas exchange occurs (pp. 430, 433–434).

Describe the muscles involved in breathing, including the diaphragm, and explain how their movements facilitate air movement into and out of the lungs.

- The movement of air into and out of the lungs depends largely on the action of the diaphragm, a dome of muscle that sits directly below the lungs (p. 431).
- Contraction of the diaphragm increases the volume of the chest cavity, decreasing air pressure and allowing air to flow in. Relaxation of the diaphragm causes the opposite to occur. Muscles surrounding the rib cage and in the abdomen can also contract or relax to cause changes in the chest cavity volume (pp. 431–432).

Explain the role of hemoglobin in gas exchange.

- As blood flows through the lungs, hemoglobin reversibly binds to oxygen molecules in high concentrations there. In the body tissues, the hemoglobin releases some of its oxygen load to supply active tissues (p. 435).

Describe the effect of smoking on the respiratory system.

- Small particles in tobacco smoke enter the lungs, causing cell damage that eventually leads to chronic bronchitis. Chronic bronchitis in turn can lead to emphysema. The tiniest smoke particles are drawn into the alveoli, where they may remain for long periods, exposing the alveolar cells to carcinogens within the smoke particles (pp. 435–437).

### Section 19.2

List the organs and tissues of the circulatory system, and describe the function of each.

- Circulatory systems in animals consist of a fluid for gas and material exchange (for example, blood), "tubes" for carrying the fluid throughout the body (for example, veins and arteries), and a pump to facilitate fluid flow (the heart) (p. 437).

Describe how blood moves through the double circulation system of the heart.

- Blood is made up of a liquid portion, called plasma, and a cellular portion, consisting of red blood cells, white blood cells, and platelets (p. 438).
- The heart is a double pump consisting of four chambers—two atria and two ventricles. The right-side pump sends oxygen-poor blood to the lungs; the left-side pump sends oxygen-rich blood to the body (p. 439).
- Blood from the lungs flows to the heart, where it is pumped into the systemic circulation. After dropping off a load of oxygen and picking up carbon dioxide in the body tissues, the blood returns to the heart and is pumped back to the lungs (pp. 442–443).

Explain the steps of the cardiac cycle, including the role of the heart's electrical system.

- The heart can generate its own beat, via the electrical activities of the SA node. Signals from the SA node are transmitted to the atria, causing these chambers to contract and forcing blood into the ventricles. The signal is carried by conductive fibers to the ventricles, which then contract to force blood out of the heart and into circulation (pp. 439–440).

Describe the effects of smoking on the cardiovascular system.

- The nicotine in smoke increases heart rate and blood pressure, putting strain on the heart muscle. Nicotine also increases the production of LDL ("bad cholesterol"), causing atherosclerosis, and increases the likelihood of blood clot formation, causing blockages in blood flow (pp. 443–444).

## Roots to Remember

**The following roots of words come mainly from Latin and Greek and will help you decipher terms:**

| | |
|---|---|
| **card-** and **cardio-** | relate to the heart. Chapter term: *cardiovascular* |
| **-hale** | means to breathe. Chapter terms: *inhale, exhale* |
| **-itis** | is used to describe inflammation (of an organ). Chapter term: *bronchitis* |
| **vascul-** | means vessel. Chapter term: *vascular* |

## Learning the Basics

**1.** Describe how contraction of the diaphragm allows the lungs to fill and how oxygen from inhaled air enters the lungs.

**2.** Describe the movement of blood through the four chambers of the heart. Where does the blood come from? Where does it go?

**3.** Alveoli are _____.
A. small air sacs at the ends of bronchi in the lungs; B. the respiratory surface in mammals; C. surrounded by a net of capillaries; D. subject to damage because of exposure to tobacco smoke; E. more than one of the above is correct.

**4.** All of the following statements about hemoglobin are true, except _____.
A. it is a protein that can bind oxygen; B. it is carried by red blood cells; C. it can bind carbon monoxide; D. it picks up oxygen in the body tissues and releases it in the lungs; E. it contains iron, which is responsible for its red color.

**5.** The blood vessels that carry blood from the heart are called _____.
A. arteries; B. veins; C. atherosclerosis; D. capillaries; E. ventricles.

**6.** The sound of a heartbeat as heard through a stethoscope is produced by _____.
A. electrical signals from the SA node; B. rush of blood into the ventricles and out of the atria; C. the ribs expanding as the diaphragm contracts; D. the closing of valves between heart chambers and vessels; E. the pulse of blood flowing through the arteries.

**7.** Blood flowing from body tissues to the heart _____.
A. is next pumped via veins to the rest of the body; B. is next pumped via arteries to the lungs; C. is under relatively high pressure compared to blood leaving the heart; D. returns via millions of capillaries tied directly to the heart's two atria; E. has used up all of its hemoglobin.

**8.** Deoxygenated blood from the body first enters the _____ of the heart, and oxygenated blood from the lungs is pumped to the body by the _____.
A. right atrium, right ventricle; B. right atrium, left atrium; C. left atrium, right ventricle; D. right atrium, left ventricle; E. right ventricle, left atrium

**9.** Heart attacks _____.
A. are typically caused by a blockage in blood flow; B. result in the death of heart tissue; C. are as common in smokers who quit 15 years ago as in people who never smoked; D. A and B are correct; E. A, B, and C are correct.

**10.** Which of the following diseases is associated with exposure to tobacco smoke?
A. heart disease; B. lung cancer; C. stroke; D. asthma; E. all of the above

## Analyzing and Applying the Basics

**1.** Some endurance athletes, such as cyclists, engage in the practice of "blood doping," in which packed red blood cells are transfused into their bloodstream immediately before an event. How might increasing the volume of red blood cells provide an advantage to an athlete? What do you think could be the risks of this practice?

**2.** Congestive heart failure occurs when damage to the heart results in a severely weakened heart muscle. Given your understanding of heart function, describe some likely symptoms of this condition.

**3.** As tobacco plants grow, radioactive minerals found in soil stick to the plants' leaves. Minerals found in phosphate fertilizer, such as radium, lead-210, and polonium-210, can also accumulate on the tobacco plant. Radioactive substances on tobacco are not removed as the tobacco is processed to make cigarettes. Therefore, each cigarette delivers a dose of radiation along with its dose of nicotine. Over the course of a year, someone who smokes about 1.5 packs per day is exposed to a radiation dose equivalent to 300 chest X-rays. How could radioactive minerals in cigarettes produce disease throughout the human body?

# Connecting the Science

**1.** Exposure to tobacco smoke causes early death, but it also causes years or even decades of disabling, chronic health problems such as emphysema and lung cancer. Some people have argued that the best way to reduce the number of smokers is to make smoking, and the health care costs associated with it, more expensive. This strategy entails placing high taxes on cigarettes and other tobacco products and restricting current and former smokers' access to subsidized medical care. What do you think of these strategies? Do you think they would be successful? Do you think they might be unfair?

**2.** Although tobacco causes more cases of long-term disability, alcohol is more deadly to young men and women than cigarettes because of automobile and other accidents. Do you think the success of tobacco bans means that alcohol will become more strongly restricted? Do you think this is a good policy? Why or why not?

Answers to **Stop & Stretch, Visualize This, Working with Data, Savvy Reader, Sounds Right, But Is It?**, and **Chapter Review** questions can be found in the **Answers** section at the back of the book.

CHAPTER **20** | # Vaccination: Protection and Prevention or Peril?

**People living in close proximity can transmit infectious diseases to each other.**

# Immune System, Bacteria, Viruses, and Other Pathogens

Deciding whether to be vaccinated against the flu is often one of the first health-care decisions new college students will have to make without parental input. Beginning in the late fall months, university health services personnel attempt to vaccinate as many students as possible. Many students, hoping that the few minutes and the mild discomfort spent on prevention will save them from losing several days to this illness, do get vaccinated. Some students,

**There are vaccines against some diseases.**

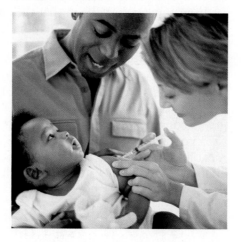

**Which diseases did your parents vaccinate you against?**

however, are reluctant to comply. They may not want to take the time to get vaccinated, or may choose to take their chances, hoping they won't get the flu. Other students refuse to get vaccinated because they believe the flu vaccine might make them sick or otherwise harm their health.

To help you make an informed decision about the flu vaccine, and

about other recommended vaccinations, we will first learn about the causes of diseases that vaccines attempt to prevent. If you do decide to get vaccinated, the flu vaccine will be very likely to help prevent this disease. Unfortunately, this is not the case for all diseases, as evidenced by the deadly outbreak of the Ebola virus, for which there is not currently an effective vaccine.

**What happens when a disease begins to spread among the unvaccinated?**

# 20.1 Infectious Agents

Tissues can be invaded and undergo damage induced by infectious agents. Once acquired, some infections can be transmitted from one individual to another, spreading disease. Such communicable or transmissible diseases differ from genetic diseases in that they are caused by organisms rather than by malfunctioning genes—although malfunctioning genes can make an organism more susceptible to infection.

Disease-causing organisms are called **pathogens**. There are many different types of pathogens, including viruses and bacteria, as well as numerous eukaryotic pathogens. When a pathogen can be spread from one organism to another, it is said to be **contagious**. Organisms that depend on other organisms to obtain nutrients and shelter required for growth and development while contributing nothing to the survival of the host organism are **parasites**. Most infectious agents are parasites. Some organisms can cause an infection in an individual, but they are not contagious if the infected individual cannot pass the infection to other individuals.

-**path-** or **patho-** relates to disease.

**STOP & STRETCH** Lyme disease comes from the bite of an infected tick but cannot be passed directly from one person to another. Is this disease infectious, contagious, or both?

Organisms that can be seen only when viewed through a microscope are called *microscopic organisms*, or **microbes**. The most common infectious microbes are bacteria and viruses.

## Bacteria

**Bacteria** are a diverse group of tiny, single-celled prokaryotic organisms. There are more bacteria in your mouth than there are humans on Earth. Bacteria are commonly rod shaped (bacilli), spherical (cocci), or spiral (spirochetes) (**FIGURE 20.1**). While we often think of bacteria only in their role as disease-causing agents, many different types of bacteria are beneficial to living organisms and the environment.

### Visualize This ▼

Note that the bacterial cell does not contain any membrane-bound organelles such as a Golgi apparatus or an endoplasmic reticulum. Would you expect that this prokaryotic cell could have ribosomes? Why or why not?

**FIGURE 20.1 Bacteria.** The drawing shows the structure of a typical bacterium. The photos show the three characteristic bacterial shapes.

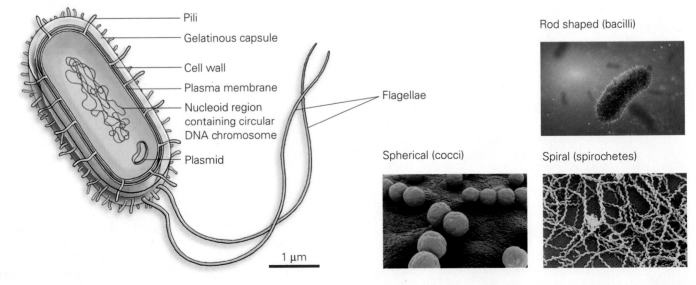

Pili
Gelatinous capsule
Cell wall
Plasma membrane
Nucleoid region containing circular DNA chromosome
Plasmid
Flagellae

1 µm

Rod shaped (bacilli)
Spherical (cocci)
Spiral (spirochetes)

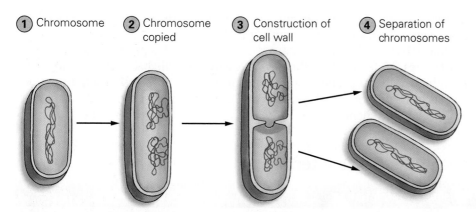

**1** Chromosome   **2** Chromosome copied   **3** Construction of cell wall   **4** Separation of chromosomes

**FIGURE 20.2 Binary fission.**   (1) The bacterial cell begins with one copy of the circular DNA chromosome that is wound around itself. (2) The chromosome is copied, and each copy is attached to the plasma membrane. (3) Continued growth separates the two chromosomes. The plasma membrane pulls inward in the middle, and a new cell wall is constructed. (4) Daughter cells separate.

Like other prokaryotes, bacteria lack a nucleus. Instead, they have a **nucleoid region** containing a double-stranded, circular DNA chromosome. In addition to the large DNA chromosome, bacteria may also contain small, circular extrachromosomal DNA (DNA that is separate from the chromosome) molecules called **plasmids.** Plasmids carry a few genes that are helpful to the survival of the bacteria and can be passed from one bacterium to another.

Most bacterial cells are surrounded by a **cell wall** that provides rigidity and protection; it is composed of carbohydrate and protein molecules. The cell walls of many bacteria are surrounded by a gelatinous **capsule,** which helps bacteria attach to cells within tissues they infect. This capsule also allows bacteria to escape destruction by cells of the immune system. Bacteria also may have one or more external **flagella** to aid in movement, and **pili,** which help some bacterial cells attach to each other and pass genes.

Bacteria reproduce by a process called **binary fission** (**FIGURE 20.2**), in which a single bacterial parent cell gives rise to two genetically identical daughter cells. When a bacterial cell divides by binary fission, the single circular chromosome that is attached to the plasma membrane inside the cell wall is copied. The copy is then attached to another site on the plasma membrane, and the membrane between the attachment sites grows and separates the two chromosomes, eventually producing two separate, genetically identical daughter cells.

Bacteria can reproduce rapidly under favorable conditions, some doubling their population every 20 minutes or so. This exponential growth can result in the production of millions of bacterial cells after 8 hours at room temperature (**FIGURE 20.3**). Fortunately, refrigeration slows the rate at which bacteria divide.

**Bacterial Infections.**   When a bacterial infection occurs, the host cell's nutrients are used by the bacteria to allow for rapid bacterial multiplication, thus preventing host cells from functioning properly. The actual symptoms of bacterial infection result from more than just the large numbers of bacteria in the body. The symptoms arise due to the effects of molecules called **toxins** that are secreted

# Working with Data ▼

How many bacteria would be present at 1:00? Sketch a line graph approximating the growth rate of bacteria over 8 hours. Add a second line to the graph approximating the growth of bacteria, when refrigerated, over the same time period.

Chicken salad sandwich

Noon
**2 bacteria**

8:00 p.m.
**Over 33 million bacteria**

**FIGURE 20.3 Exponential growth of bacteria.**   One kind of bacterium that can cause food poisoning reproduces every 20 minutes. If a sandwich with 2 bacteria is left out for 20 minutes, each bacterial cell will make a copy of itself, yielding 4 bacteria. Within 8 hours, there will be 33 million bacteria.

by the bacterial cells, which can result in cell death. Several examples of diseases caused by bacteria are listed in **TABLE 20.1**.

Many bacterial infections are routinely treated with medications called **antibiotics**. Unfortunately, the days of easily treatable bacterial infections may soon be over because the bacteria that cause many diseases—including tuberculosis, ear infections, and gonorrhea—have become resistant to the antibiotics that used to kill them.

**TABLE 20.1 Examples of diseases caused by bacteria.**

| Disease | Disease Information |
|---|---|
| Anthrax  | *Bacillus anthracis* lives mainly in soil and, as a part of its life cycle, can exist as a tough, resistant structure called a spore. Anthrax spores can be lethal if they are inhaled. Anthrax is infectious but not contagious. |
| Botulism  | *Clostridium botulinum* produces a powerful toxin that acts on the nervous system, causing paralysis of respiratory and facial muscles. Botox—injections of the toxin produced by this bacterium—temporarily paralyzes facial muscles to prevent drooping and the appearance of wrinkles. This soil-dwelling bacterium can adhere to vegetables. Improper sterilization allows the bacteria to proliferate and produce a gas that causes telltale bulges in canned foods. |
| *Escherichia coli* infection  | *E. coli* bacteria live in the intestines of humans and other organisms. Disease can occur when human sewage or animal waste contaminates water supplies. *E. coli* infection can also occur from eating undercooked, contaminated beef or drinking contaminated, unpasteurized milk. |
| *Staphylococcus* (commonly known as staph infection)  | Staphylococcus bacteria that reside on human skin and mucous membranes can invade and destroy tissue through an opening in the skin. *S. aureus* can cause toxic shock syndrome when tampons are left in place long enough to allow bacterial growth. Methicillin-resistant *S. aureus* (*MRSA*) infection is deadly because resistances to antibiotics have evolved in these bacteria. Healthy young athletes can succumb when resistant bacteria are present on locker room floors or shared equipment enter the skin through cuts or scrapes. |
| Tuberculosis 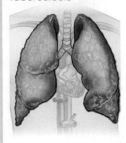 | Tuberculosis is caused by the bacterium *Mycobacterium tuberculosis*. This infectious agent is spread when a person with the infection coughs or sneezes. Once inside the body, the bacteria take up residence in the lungs. The immune system responds by walling off the bacteria in a manner much like the formation of a scab. The bacteria can live within the walls for years, during which time the infection is not considered to be active, and it is not transmissible. If the bacteria escape the immune system's attempts at control, the infection becomes active. |

Antibiotic resistance develops as a result of natural selection in a bacterial population. Variation exists within any population, including within a population of bacterial cells. Some bacteria, even before exposure to an antibiotic, carry genes that enable them to resist the antibiotic. You can develop a drug-resistant infection if the resistant bacteria are selected for in your body or when you contract the resistant bacteria from someone else.

**Vaccines against Bacterial Diseases.** Immunizations against many bacterial diseases are recommended by the U.S. Department of Health and Human Services Centers for Disease Control and Prevention (CDC). In fact, children are supposed to have proof of vaccination to attend school. Young children receive a series of injections known as the DPT vaccine, containing vaccines against three bacterial diseases. The first letter in the name of this vaccine is for diphtheria, a respiratory infection that has been eradicated in the United States but has a 10% death rate in countries that don't vaccinate. Continued vaccinations are necessary in the United States, in spite of the eradication of this disease, because travelers can bring the disease back into the country. The DPT vaccine also protects against pertussis, another respiratory tract infection. This potentially deadly infection results in violent coughing that makes breathing extremely difficult. The whooping sound emitted during inhalation led to the disease also being called whooping cough. The T in DPT stands for tetanus, a disease caused by a toxin produced by the *Clostridium tetani* bacterium. This bacterium can persist as a resistant spore in the soil. Puncture wounds produced by objects carrying contaminated soil allow the bacteria into tissues where it can grow, producing a toxin that has devastating, often fatal, effects on the nervous and muscular systems.

There is also a vaccine against bacterial meningitis, a disease caused by a bacterium that can cause inflammation of the membranes surrounding the brain. Around 10% of people who have meningitis will die, and many more will suffer permanent disability. Because this disease can be spread when an infected person sneezes or coughs, outbreaks of meningitis are most common in places where many people live in close quarters, like university dormitories. For this reason, the CDC recommends that the vaccine against meningitis be given to adolescents. If you were not vaccinated against meningitis, you may want to discuss this with your doctor.

## Viruses

Many of the CDC-recommended vaccines protect against viral diseases. **Viruses** (**FIGURE 20.4**) are little more than packets of nucleic acid (DNA or RNA) surrounded by a protein coat. Viruses are not considered to be living organisms for two reasons: (1) They cannot replicate without the aid of a host cell and (2) they are not composed of cells. Viruses lack the enzymes for metabolism and contain no ribosomes. Therefore, they cannot make their own proteins. They also lack cytoplasm and membrane-bound organelles. They cannot produce toxins like bacteria can.

The genetic material, or **genome,** of a virus can be DNA or RNA; it can be double-stranded or single-stranded, and it can be linear or circular. For example, the herpes virus has a double-stranded DNA genome, and the poliovirus has a single-stranded RNA genome. The genes of a virus can code for the production of all the proteins required to produce more viruses inside a host cell. RNA viruses called **retroviruses** are packaged with the enzyme reverse transcriptase, which synthesizes DNA from the RNA genome rather than the reverse as in "normal" genetic transcription, where RNA is copied from DNA.

The protein coat surrounding a virus is called its **capsid.** Many of the viruses that infect animals have an additional structure outside the capsid called the **viral envelope.** The envelope is derived from the cell membrane of the host cell and may contain additional proteins encoded by the viral genome. Viruses that are surrounded by an envelope are called **enveloped viruses.**

# Flu vaccines are offered on most college campuses beginning in the late fall.

## Visualize This ▼

**Based only on structures shown in this figure, can you guess which parts of the virus are most likely to help it attach to a cell?**

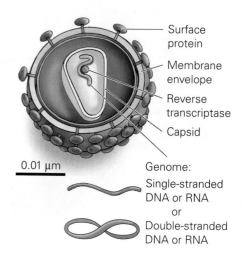

Surface protein

Membrane envelope

Reverse transcriptase

Capsid

0.01 μm

Genome:

Single-stranded DNA or RNA

or

Double-stranded DNA or RNA

**FIGURE 20.4 Viral structure.** Viruses are composed of genetic material surrounded by a protein coat. Some viruses, including the one shown, are also surrounded by an envelope.

**Before you came to college, your parents made decisions about whether you should be vaccinated against 15 different diseases.**

**Viral Infection.** Infection by an enveloped virus occurs when the virus gains access to the cell by fusing its envelope with the host's cell membrane (**FIGURE 20.5**). An unenveloped virus uses its capsid proteins to bind to receptor proteins in the plasma membrane of a host cell. Some capsid proteins function as enzymes that digest holes in the plasma membrane, thereby allowing the viral genome to enter cells. Once inside the host cell, the capsid is removed.

Regardless of how it enters, after the genome enters the host cell, the infection continues when the virus replicates itself. First, the genome is copied, then the virus uses the host-cell ribosomes and amino acids to make viral proteins for building new capsids and synthesizing some of the envelope proteins. Once assembled, the new viruses exit the cell, leaving behind some viral proteins in the host's cell membrane. The newly constructed viruses then move to other cells, spreading the infection.

Some viruses, called **latent viruses**, enter a state of dormancy in the body. During that time, the virus is not replicating. Herpes viruses, for example, which cause outbreaks of painful sores and blisters on the mouth and genitals, can undergo long periods of dormancy, during which time they are present in the body but do not cause symptoms. Herpes simplex is sexually transmitted, as are many viruses. Examples of diseases caused by viruses are listed in **TABLE 20.2**.

**Vaccines against Viral Diseases.** Newborns are vaccinated against hepatitis B, a sexually transmitted virus, because this disease is easily passed from mothers who do not realize they have the virus to their babies during delivery. The potentially fatal hepatitis virus attacks the liver and prevents it from functioning properly. Infants are also vaccinated against rotavirus, which causes severe diarrhea, fever, abdominal pain, and dehydration, and results in half a million deaths worldwide in children under the age of 5. Young children are vaccinated against polio, a disease that can cause paralysis and death. They are also given the MMR triple vaccine, a vaccine against measles, a severe and sometimes deadly rash and fever disease still common in many parts of the world; mumps, a swelling of the salivary glands that can, though rarely, cause deafness or reproductive difficulties; and rubella, a disease that causes fever and rash and can cause birth defects if a pregnant woman acquires the infection.

The MMR triple vaccine is given at around 15 months of age, which happens to coincide with the age when the first symptoms of autism appear. This correlation has led many parents of autistic children to assume that the vaccine can

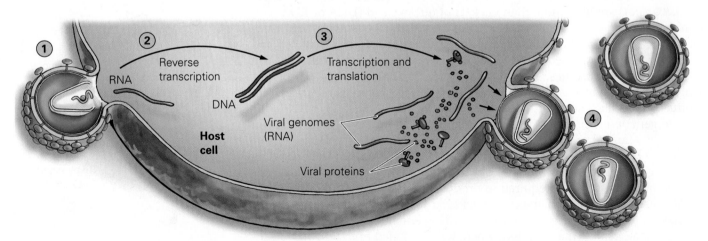

**FIGURE 20.5 Viral replication.** (1) Viral membrane fuses with the host cell; the capsid is removed, and the genome enters. (2) In the case of some RNA viruses, such as HIV, the virus that causes AIDS, the viral genome is used to synthesize many copies of DNA by the reverse transcriptase enzyme. (3) The DNA is transcribed and translated by the host cell. More viral proteins and copies of the viral genome are produced. (4) Once assembled, new viruses leave the host cell and infect other cells.

**TABLE 20.2 Examples of diseases caused by viruses.**

| Disease | Disease Information |
|---|---|
| AIDS 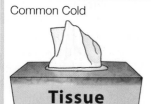 | **AIDS** is caused by the human immunodeficiency virus or HIV. HIV is a fragile virus that is transmitted only through direct contact with bodily fluids, primarily blood, semen, or vaginal fluid. There is no evidence that the virus is spread by tears, sweat, coughing, or sneezing. It is not spread by contact with an infected person's clothes, phone, or toilet seat, nor can it be transmitted by an insect bite. And it is unlikely to be transmitted by kissing, unless both individuals have lesions or cuts. HIV is frequently spread through needle sharing among injection drug users, but the primary mode of HIV transmission is via unprotected sex, including oral sex, with an infected partner. |
| Common Cold  | As many as 200 different viruses cause colds, with rhinovirus being the most common cause. Cold viruses spread easily in droplets from sneezes or coughs or on objects such as doorknobs and utensils that an infected person contaminated. |
| Ebola  | Humans become infected when they handle infected wild animals or are exposed to the secretions of such animals. Human to human transmission is via direct contact with bodily fluids of an infected person, even if that person is recently deceased. Infection also occurs by exposure to bodily fluids present on medical supplies used to treat an infected person. |
| Mononucleosis  | Mononucleosis is caused by the Epstein-Barr virus, a disease that commonly affects young adults. The symptoms—fatigue, weakness, fever, headache, and sore throat—can be severe and may last a month or two. The virus can be transmitted in saliva when kissing, and by coughing and sneezing or by touching an object contaminated by an infected person. |
| Rabies  | Rabies is a viral disease that typically spreads to humans through a bite from an infected animal. The initial signs of infection include flu-like symptoms but progress to more serious symptoms, including paralysis, convulsions, and hallucinations. Most domesticated animals are vaccinated against rabies, so transmission is most common from bites by wild animals. |
| West Nile  | The West Nile virus is transmitted to humans via the bite of an infected mosquito. In the United States the virus first emerged in New York and quickly spread west and south throughout the country. In infected persons, the brain and spinal cord can become inflamed, resulting in paralysis and death. |

cause autism. However, many scientific studies have been performed to test this hypothesis, and no evidence has emerged to support it. One study published in 1998 used falsified data to show a fraudulent correlation between the vaccines and autism. This study has been retracted by the journal that published it, and the study's author has been banned from practicing medicine.

More recently, the CDC began recommending childhood vaccination against chicken pox, a skin rash of blisterlike lesions that usually resolves itself but can cause swelling of the brain, or **encephalitis**, and pneumonia, typically when older children come down with the disease.

Of particular interest to college students is a vaccine called Gardasil that provides protection against some strains of human papillomavirus (HPV), known to cause a reproductive cancer in women. The lower third of the uterus, called the cervix, is very susceptible to infection with HPV, even when condoms are used. Even with good medical care, 37% of women with cervical cancer die from the disease. In poor countries, the toll is even more dramatic—cervical cancer is the leading cause of cancer deaths in the developing world, killing over 300,000 women every year. If a larger percentage of the population were vaccinated against HPV, most of these cervical cancer deaths could be prevented.

The success of vaccination programs lies not with providing immunity to every susceptible individual, but with reducing the total number of susceptible individuals so that the disease cannot easily spread (**FIGURE 20.6**). Effectively reducing the toll of cervical cancer requires that as many potential carriers of HPV be immunized as possible. In other words, men as well as women should receive Gardasil, because this will help lower the population's infection rate. Gardasil also prevents infection by strains of the virus that cause genital warts, which may make it more attractive to men. If you have not had this vaccination, and are sexually active or plan to be sexually active in the future, you should discuss this vaccination with your physician.

The CDC also recommends annual vaccination against the flu for children and adults. Composed of three or four viruses, the flu shot is designed to stimulate your body's immune system, giving it a head start against your next exposure to influenza.

**enceph-** and **encephalo-** mean brain.

**STOP & STRETCH** Many people cannot get vaccinated. For instance, newborns are not vaccinated against most diseases, and people undergoing cancer treatments are not vaccinated against the flu. Children undergoing treatment for cancer do not receive vaccines against childhood diseases. In addition, no vaccine is 100% effective. What is likely to happen to infants and people with compromised health if they are exposed to someone who has refused to be vaccinated and is sick with disease?

## Visualize This ▶

If large numbers of people refuse vaccinations, what effect may that have on the likelihood of a more widespread occurrence, or epidemic, occurring?

**FIGURE 20.6 Disease eradication.**
Not all members of a population need to be immunized for a disease to die out. As long as enough individuals are protected, a person with the infection cannot transmit it effectively to someone else, so the pathogen cannot reproduce.

**(a) Population where no one has been immunized**

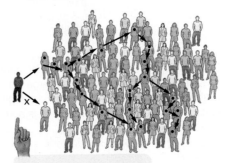

The infection can spread rapidly in the population.

**(b) Population where most people have been immunized**

With few susceptible people, the disease cannot spread.

■ Infected with disease    ■ Susceptible to disease    ■ Vaccinated

■ Genetically resistant to disease    ● Newly infected

## Eukaryotic Pathogens

While many infectious diseases are caused by bacteria or viruses, some are also caused by eukaryotic organisms. Eukaryotic pathogens such as single-celled protozoans, along with some worms and fungi, cause tremendous human suffering and death worldwide. Protozoans are often spread by water and food contaminated with animal feces. Most worms gain access to the body because of poor sanitation or by consumption of contaminated meat or fish. Worms that take up residence in intestines or other organs can cause extensive damage to internal tissues. Fungi cause diseases of the skin and internal organs. They damage tissue by secreting digestive enzymes into it and absorbing the products of this digestion. Examples of diseases caused by eukaryotic pathogens are found in **TABLE 20.3**. There are no vaccines routinely given in the United States against diseases caused by these pathogens.

**TABLE 20.3** **Examples of diseases caused by eukaryotic pathogens.**

| Disease | Disease Information |
|---|---|
| Giardiasis 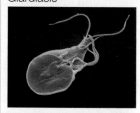 | Giardiasis is caused by the waterborne *Giardia lamblia*. During one part of its life cycle, *Giardia* exists as a tough, resistant structure called a cyst. *Giardia* cysts enter bodies of water or food supplies in contaminated feces. The organism completes its life cycle in the intestines of an infected person, causing severe diarrhea and gas. |
| Malaria 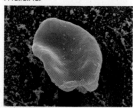 | Malaria is caused by protozoans from the genus *Plasmodium*. These protozoans are transmitted when infected mosquitoes bite humans. This disease kills close to 2 million people annually. |
| Schistosomiasis  | Freshwater snails carry this parasitic disease, caused by flatworms. Humans affected by this disease live around contaminated water, commonly found in Asia, Africa, and South America. |
| Tapeworm  | Tapeworm infections are usually caused by ingesting infected raw or undercooked pork or beef. This worm lives in the intestines and can cause abdominal pain and diarrhea. |
| Athlete's foot and jock itch 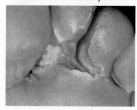 | Athlete's foot and jock itch are fungal infections that affect the skin between the toes and on genitals and inner thighs respectively. Fungi spread easily in public places such as locker rooms and the floors of communal showers. Excessive perspiration can wash away naturally occurring fungus-killing oils present on skin. Due to their routine exposure to locker room floors and communal showers, and their tendency to sweat during workouts, athletes are more prone to these infections. |

**FIGURE 20.7 Three lines of defense.**
(1) The skin and mucous membranes serve as a nonspecific barrier to infection. (2) Macrophages attack pathogens that gain access to the body and cause an immune response. (3) Lymphocytes initiate a specific response by targeting specific pathogens and attempting to contain them.

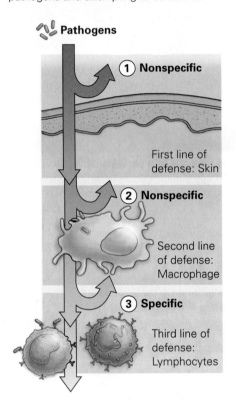

# 20.2 The Body's Response to Infection: The Immune System

In humans and most vertebrates, it is the job of the immune system to protect against infection. The immune system consists of three different lines of defense (**FIGURE 20.7**). The first line of defense helps prevent access to the body, while the remaining two lines of defense operate if the pathogen gets past the first line.

## First Line of Defense: Skin and Mucous Membranes

The skin and mucous membrane secretions comprise the first line of defense against pathogens. These external physical and chemical barriers are nonspecific defenses; that is, they do not distinguish one pathogen from another.

Organisms living on the body's surface are kept out by the skin. In addition to being a physical barrier, the skin sheds, taking pathogens with it, and has a low pH that can help to repel microorganisms. Likewise, glands in the skin secrete chemicals that slow the growth of bacteria. For example, tears and saliva contain enzymes that break down bacterial cells, and earwax traps microorganisms.

Mucous membranes lining the respiratory, digestive, urinary, and reproductive tracts also secrete mucus that traps pathogens, which are coughed or sneezed away from the body, destroyed in the stomach, or excreted in the urine or feces. Digestive secretions, including acids, kill many microorganisms that gain access to the stomach. Vomiting can also rid the body of toxins or infectious agents. Pathogens that are able to evade the first line of defense next encounter an internal, second line of defense.

## Second Line of Defense: Phagocytes and Macrophages, Inflammation, Defensive Proteins, and Fever

If a pathogen is able to get past the first line of defense, the internal second line of defense can often stop the infection. Participants in the second line of defense are nonspecific in that they do not target particular pathogens.

**Phagocytes and Macrophages.** Phagocytes are white blood cells that indiscriminately attack and ingest invaders by engulfing and digesting the invader. Macrophages are one type of phagocytic white blood cell that move throughout the lymphatic fluid, engulfing dead and damaged cells (**FIGURE 20.8**). Enzymes inside the macrophage help break the invader apart. Macrophages can also clean up old blood cells, dead tissue fragments, and other cellular debris. They release chemicals that stimulate the production of more white blood cells. Much of the destruction of the offending cells occurs in the lymph nodes.

When an invader is too big to be engulfed (protozoans and worms, for example), other white blood cells cluster around the invader and secrete digestive enzymes that irritate or may even destroy the organism.

Additional white blood cells that circulate through the blood and lymph destroying invaders are called **natural killer cells.** These nonspecific cells attack tumor cells and virus-invaded body cells on first exposure. They release chemicals that break apart plasma membranes of their target cells, causing them to burst. Fluid, dead cells, and microorganisms accumulate at the site of the infection, producing pus. If pus cannot drain, the body may wall it off with connective tissue, producing an abscess. Another component of the second line of defense is the inflammatory response.

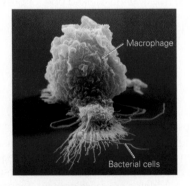

**FIGURE 20.8 Phagocytosis.**
A macrophage attacks bacterial cells.

**Inflammation.**  Whenever a tissue injury occurs, an inflammatory response begins. Damaged cells release chemicals that stimulate specialized cells to release **histamine.** Histamine promotes increased size of blood vessels, or vasodilation, near the injury. As more blood and cells arrive at the site of infection to speed cleanup and repair, redness, warmth, and swelling occur. The extra blood flow also brings oxygen and nutrients required for tissue healing.

**Defensive Proteins.**  In addition to white blood cells, some proteins act as nonspecific defenders. **Interferons** are proteins produced by virus-infected body cells to help uninfected cells resist infection. When infected cells die, they release interferons that bind to receptors on uninfected cells and stimulate the healthy cells to produce proteins that inhibit viral reproduction.

The over 20 different proteins comprising the complement system are another class of defensive proteins. **Complement proteins** circulate in the blood and help, or complement, other defense mechanisms. These proteins coat the surfaces of microbes, making them easier for macrophages to engulf. Complement proteins can also poke holes in the membranes surrounding microbes, causing them to break apart (**FIGURE 20.9**) and serve to increase the inflammatory response.

**Fever.**  A body temperature above the normal range of 97–99°F is called fever. Macrophages release chemicals called pyrogens that cause body temperature to increase. A slightly higher-than-normal temperature decreases bacterial growth and increases the metabolic rate of healthy body cells. Both of these processes help to fight infection by slowing pathogen reproduction and allowing repair of tissue to occur more quickly. When the infection is controlled, macrophages stop releasing pyrogens, and body temperature returns to normal.

## Third Line of Defense: Lymphocytes

If a pathogen makes it past the first two nonspecific defense systems, the immune system presents a third line of defense. Cells of the immune system identify and attack specific microorganisms that are recognized as foreign. This **specific defense** system consists of millions of white blood cells, called **lymphocytes.** Lymphocytes travel throughout the body by moving through spaces between cells and tissues or being transported via the blood and lymphatic systems (**FIGURE 20.10**).

The specific response generated by the immune system is triggered by proteins and carbohydrates on the surface of pathogens or on cells that

**FIGURE 20.10 The lymphatic system.** Pathogens trigger a response by the lymphatic system. The various organs of this system work to eliminate infections.

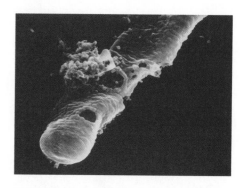

**FIGURE 20.9 A bacterial cell after complement exposure.**  Complement proteins cause holes to form in the bacterial cell, penetrating the cell wall and cell membrane and causing the bacterium to fill with fluids and burst.

## Visualize This ▼

**If an infection is suspected, a blood sample can be analyzed. Because it is not possible to test for every pathogen, the blood is tested for cells that increase in number during an infection. What type of cell proliferates during infection?**

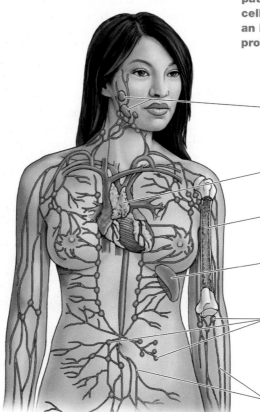

**Tonsils and adenoid:** Type of lymph node

**Thymus:** Where some lymphatic cells go to mature

**Bone marrow:** Produces some lymphatic cells

**Spleen:** Stores and purifies blood; contains high concentration of lymphocytes

**Lymph nodes:** Store cells and filter out bacteria and other unwanted substances to purify the lymphatic fluid; become swollen and painful when infection occurs

**Lymphatic vessels:** Transport fluid from tissues to lymph nodes

have been infected by pathogens. Molecules that are foreign to the host and stimulate the immune system to react are called **antigens**. Examples of antigens are found on invading viruses, bacteria, fungi, protozoans, and worms, as well as on the surface of foreign substances such as dust, pollen, or transplanted tissues.

Regardless of the source, when an antigen is present in the body, the production of two types of lymphocytes is enhanced: the **B lymphocytes (B cells)** and **T lymphocytes (T cells)**. Like macrophages, these lymphocytes move throughout the circulatory and lymphatic systems and are concentrated in the spleen and lymph nodes. Lymphocytes display **specificity** because they recognize specific antigens.

The specificity of lymphocytes is based on the presence of proteins, called **antigen receptors,** whose shape fits perfectly to a portion of the foreign molecule. The receptor is able to bind to the antigen much the way that a key fits into a lock. These receptors are either attached to the surface of the lymphocyte or secreted by the lymphocyte. B and T cells recognize different types of antigens: B cells recognize and react to small, free-living microorganisms such as bacteria and the toxins they produce (**FIGURE 20.11a**). T cells recognize and respond to body cells that have gone awry, such as cancer cells or cells that have been invaded by viruses. T cells also respond to transplanted tissues and larger organisms such as fungi and parasitic worms (**FIGURE 20.11b**).

B and T cells both recognize and help eliminate antigens, but they do so in different ways. B cells secrete **antibodies**, proteins that bind to and inactivate antigens. T cells, in contrast, do not produce antibodies but instead directly attack invaders.

Antibodies are found in the blood, lymph, intestines, and tissue fluids. They are also found in breast milk. When a baby breast-feeds, some antibodies are passed from the mother to the baby, conferring short-term **passive immunity** to the child. Because the child did not make the antibodies himself or herself, this immunity lasts only as long as the antibody lasts in his or her blood. An infant's passive immunity is different from the immunity conferred by exposure to an antigen. Exposures to antigens cause the production of antibodies to combat the infection for the individual's lifetime, which is called **active immunity.**

**Allergy.** An **allergy** is an immune response that occurs even though no pathogen is present. The body simply reacts to a nonharmful substance as though a pathogen were present. For example, some people have allergic reactions to peanuts or ragweed pollen. Asthma may also be the result of an allergic reaction.

The fact that infants today receive about twice as many immunizations as they did 30 years ago has led many people to speculate that the increased rate of immunization may be the cause of increasing rates of allergies and asthma. Well-controlled scientific studies, many of which were performed by independent nonprofit organizations such as the Institute of Medicine

**FIGURE 20.11 B cells and T cells.**
(a) The receptors produced by B cells function as cell-surface antigen receptors or as secreted antibodies that are present in bodily fluids. (b) T cells have surface antigen receptors only. They do not secrete antibodies. B cells and T cells differ with respect to the types of cells and organisms they respond to.

**(a) B lymphocyte**

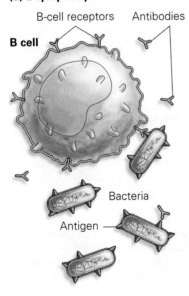

B-cell receptors    Antibodies
B cell
Bacteria
Antigen

**(b) T lymphocyte**

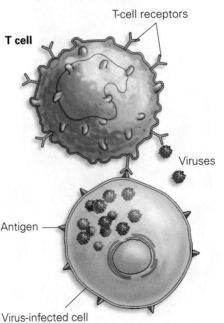

T-cell receptors
T cell
Viruses
Antigen
Virus-infected cell

(IOM), show no connection between the rates of allergy and asthma and vaccination.

**Anticipating Infection** The ability of the body's lymphocytes—B and T cells—to respond to specific antigens begins before our birth. Thus, we are able to respond to infectious agents the very first time we are exposed. This ability continues into adulthood because these cells are manufactured, at the rate of about 100 million per day, throughout our lives. The ability to respond to an infection, the **immune response**, ultimately results from the increased production of B and T cells.

Lymphocytes are produced from special cells called **stem cells**, immature cells that can differentiate into adult cell types. Many parts of the body, including the bone marrow, retain a supply of stem cells that can develop into more specialized cells. Bone marrow stem cells enable the bone marrow to produce blood cells throughout a person's lifetime. Lymphocytes are produced from the stem cells of bone marrow and released into the bloodstream.

Some lymphocytes continue their development in the bone marrow and become B cells. Others take up residence in the **thymus** gland. The thymus gland, located behind the top of the sternum, stimulates T cells to develop. When immune cells finish their development in the bone marrow, they are called B cells; when they mature in the thymus, they are called T cells (**FIGURE 20.12**).

Lymphocytes can recognize trillions of different antigens. This extraordinary ability results from the trillions of different antigen receptors that are produced by B and T cells. How is it that such an astonishing diversity of receptors can be produced?

**Antigen Diversity.** If every antigen receptor were encoded by its own gene, trillions of genes would have to code immune system proteins alone. Yet the entire human genome contains only approximately 20,000 genes, so this cannot be the case. Instead, a mechanism has evolved in B and T cells that generates an enormous variety of receptors from a limited number of genes.

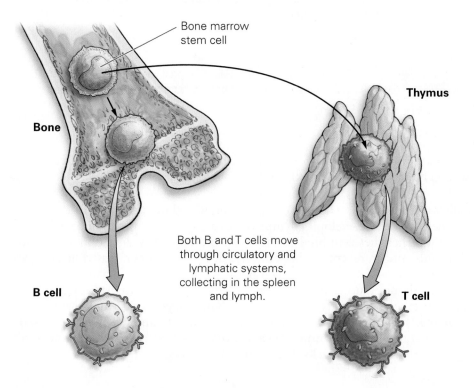

Bone marrow stem cell

Thymus

Bone

Both B and T cells move through circulatory and lymphatic systems, collecting in the spleen and lymph.

B cell

T cell

**FIGURE 20.12 Lymphocyte development.** Lymphocytes develop from cells in the bone marrow. Those that continue their development in the bone become B cells, while those that move to the thymus gland (located in the neck) to continue their development become T cells.

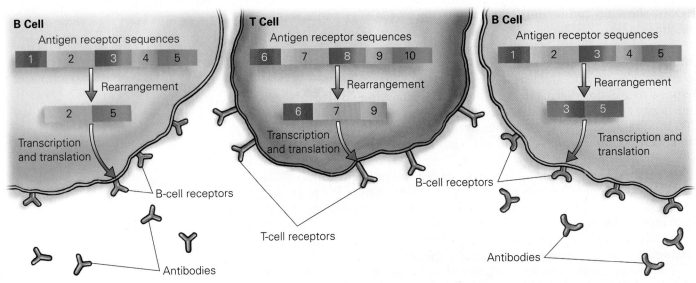

**FIGURE 20.13 Genetic rearrangements allow for the production of millions of different antigen receptors.** The genetic information encoding antibody structures is stored in bits and pieces. These bits and pieces are arranged in various orders during development of the antibody-producing cells. Removal of various DNA segments produces a wide variety of antigen receptors and increases the number of antigens to which each of us can respond.

As B and T cells develop in a fetus, each cell's DNA segments that code for the production of antibodies do something very unusual: They rearrange themselves. Some portions are cut out, and the remaining DNA segments are joined together (**FIGURE 20.13**). Each unique arrangement of DNA produced by a given cell encodes a different receptor protein. Once synthesized, the proteins move to the surface of the B or T cell and act as antigen receptors. Thus, from a pool of only a few hundred genes, a virtually unlimited variety of diverse antigen receptors can be created.

**FIGURE 20.14 Testing lymphocytes for self-proteins.** Cells have proteins and molecules on their surfaces that function as antigens. Developing lymphocytes are tested in the thymus to determine whether they are self or nonself. Lymphocytes that bind to antigens are destroyed; those that do not develop.

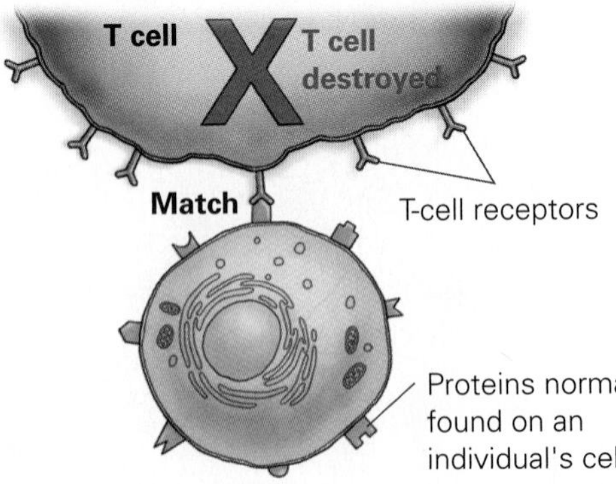

**STOP & STRETCH** Based on what you know about antigenic diversity, explain why a particular individual might become ill from an infection that most other members of the population are able to fight off.

Another important facet of the immune system is its ability to determine whether a molecule is part of the host cell or foreign.

**Self versus Nonself.** The cells of a given individual have characteristic proteins on their surfaces. Developing lymphocytes are tested, in the thymus, to determine whether they will bind to these self-proteins. Any developing lymphocyte with antigen receptors that bind to self-proteins is eliminated, making an immune response against one's own body less likely. Lymphocytes with receptors that do not bind are then allowed to develop to maturity (**FIGURE 20.14**). Thus, the body normally has no mature lymphocytes that react against self-proteins, and the immune system exhibits self-tolerance. Without this self-tolerance **autoimmune diseases**, diseases that result when a person's immune system attacks their own cells, can occur.

**auto-** means self.

**Multiple sclerosis** is an autoimmune disease that occurs when T cells specific to a protein on nerve cells attack these cells in the brain. In **insulin dependent diabetes**, T and B cells attack cells in the pancreas that produce the hormone insulin.

Even though we each have a single immune system, it is diversified into two subsystems so that we can combat the multitude of infectious agents encountered in our lifetime. This diversification is a result of the differing approaches of B and T cells to ridding the body of infectious agents once they are found.

## Cell-Mediated and Antibody-Mediated Immunity

T cells provide immunity that depends on the involvement of cells rather than antibodies and is called **cell-mediated immunity.** The protection afforded by B cells is called **antibody-mediated immunity.**

**Cell-Mediated Immunity.** T cells respond to infection by undergoing rapid cell division to produce memory cells, which then become specialized cells (**FIGURE 20.15**) that directly attack other cells. Two of these attacking cell types are the **cytotoxic T cells** and **helper T cells.** Cytotoxic T cells attack and kill body cells that have become infected with a virus. When a virus infects a body cell, viral proteins are placed on the surface of the host cell. Cytotoxic T cells

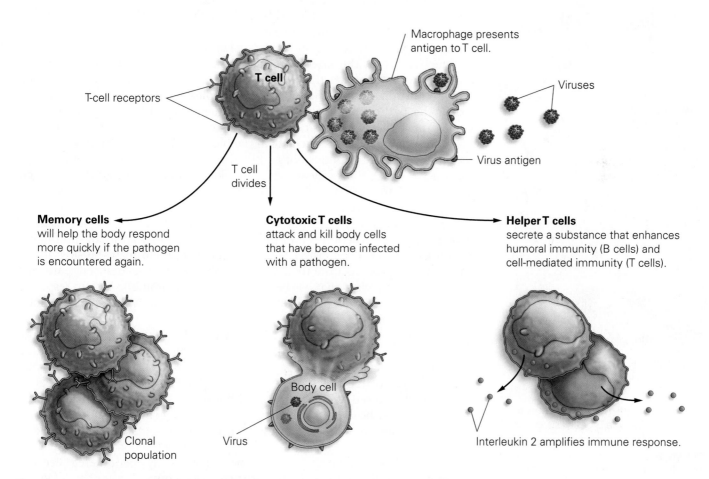

**FIGURE 20.15 Cell-mediated immunity.** T lymphocytes divide to produce different populations of cells: Memory cells carry the specific antigen receptor. Cytotoxic T cells attack and dismantle infected bodily cells. Helper T cells boost the immune response.

# There is not currently an effective vaccine against the virus that causes Ebola.

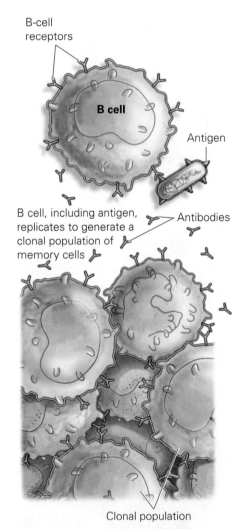

B-cell receptors

B cell

Antigen

B cell, including antigen, replicates to generate a clonal population of memory cells

Antibodies

Clonal population

**FIGURE 20.16 Antibody-mediated immunity.** When a B cell binds an antigen, the B cell makes memory cells, each copy carrying the same antibody. The clonal population of such identical cells can aid in overcoming the infection. After the infection, some memory cells remain and are able to recognize the antigen in the future.

recognize these proteins as foreign, bind to them, and destroy the entire cell. By releasing a chemical that causes the plasma membranes of the target cell to leak, they break down the cell before the virus has had time to replicate.

**Helper T cells,** also called T4 cells, can be thought of as boosters of the immune response. These cells detect invaders and alert both the B and T cells that infection is occurring. Without helper T cells, there can be almost no immune response. HIV infects and destroys helper T cells, thus crippling the body's ability to respond to any infection. Intensive efforts have been made to produce a vaccine against human immunodeficiency virus (HIV), to little avail. HIV causes **acquired immunodeficiency syndrome,** or AIDS.

**Antibody-Mediated Immunity.**   Unlike T cells, B cells do not directly kill cells bearing antigens. Instead, they make and secrete antibodies that help rid the body of antigenic cells. When a B cell responds to a specific antigen, it immediately makes copies of itself, resulting in a population of identical cells, called **memory cells,** able to help fight the infection (**FIGURE 20.16**). This population of cells is called a **clonal population.** The sheer number of cells in the clonal population strengthens the immune system's ability to rid the body of the infectious agent.

The entire clonal population has the same DNA arrangement. Therefore, all the cells in a clonal population carry copies of the same antigen receptor on their membrane and secrete the same antibody. Some cells of the clonal population, called memory cells, will help the body respond more quickly if the infectious agent is encountered again. Should subsequent infection occur, the presence of memory cells facilitates a quicker immune response.

Inactivation of the infectious agent occurs when antibodies encounter a pathogen that matches the variable region. The antibody then binds to the antigen, forming the antibody-antigen complex. Formation of this complex marks the pathogen for phagocytosis or degradation by complement proteins. Complement proteins cause destruction of the foreign and infected cells. The complement proteins circulate in the blood and lymph in an inactive state until they come into contact with antibodies bound to the surface of microorganisms, at which time they are activated. Some antibodies cause the pathogen and attached antibodies to clump together or agglutinate, rendering them unable to infect other cells.

The vaccinations you will be offered while in college, like all vaccinations, take advantage of the long-term protection provided by antibody-producing memory cells. Usually consisting of components of the disease-causing organisms, such as proteins from the plasma membrane of a bacterial cell, parts of a virus, or a whole virus that has been inactivated, the immune system responds by producing the clonal population of memory cells that will be prepared for a real infection should it happen.

The Gardasil vaccine against human papilloma virus and the vaccine against bacterial meningitis both require multiple doses, sometimes called booster shots, before a sufficient response is generated. Even if you were vaccinated in junior high or high school, you should check with your healthcare provider to see whether you need booster shots. Flu vaccines must be given every year because the flu virus rearranges its genetic sequences swiftly. This shifting genetic sequence results in different proteins being encoded and placed on the surface of the virus, thereby preventing the body's existing memory cells from recognizing a virus that they have already encountered. Because the proteins on the surface of a flu virus change so quickly, a vaccine that was prepared to protect you from last year's flu virus will not likely work against this year's flu virus.

# *savvy reader*

## HIV and AIDS

Thabo Mbeki was the president of South Africa from 1999 to 2008. During his tenure, he became convinced that AIDS was not caused by a virus and that antiviral medicines should not be used to treat this disease.

In a speech given at the 13th International AIDS Conference in the South African city of Durban, he stated that "The world's biggest killer and the greatest cause of ill health and suffering across the globe, including South Africa, is extreme poverty." Mbeki also stated that AIDS was caused by a "collapse of the immune system" due to poverty. Because Mbeki did not believe AIDS was caused by a virus, he turned down free anti-retroviral medications offered by health organizations.

In a 2008 study published in the *Journal of Acquired Immune Deficiency Syndrome*, researchers at Harvard University estimate that Mbeki's policies on AIDS led to the preventable deaths of over 300,000 South Africans. Many of the dead were babies who were infected by their mothers during childbirth, a form of transmission that can be prevented with the proper medications.

1. Is it possible that a comprehensive review of the scientific literature can lead to different interpretations of the science being presented? Is this likely what informed Mbeki's opinion about HIV not causing AIDS?

2. Do you think government officials have an obligation to accept prevailing scientific opinions when making decisions that affect the health of their constituents?

3. Other scientific claims are also denied by governments. For example, climate change is denied by many U.S. politicians. When these differing views occur, should sound scientific evidence trump the opinions of nonscientists? Why or why not?

# SOUNDS RIGHT BUT IS IT?

Some good friends are having their first child in a few months. In addition to making decisions about breast-feeding, co-sleeping, and whether to let the baby use a pacifier, they are also concerned about which vaccinations to give their child. While many of their vaccine-related questions have been answered by speaking with their doctor or looking at reliable medical Internet sources, there is one question that still bothers them. The couple came across information showing that the number of vaccines given to children today is much greater than the number given in the past. They also read on an anti-vaccine movement website that the developing immune system of a newborn baby or young child is not equipped to deal with this number of vaccines.

**The number of vaccinations given to modern children is too much for the average immune system to handle.**

Sounds right, but it isn't.

1. We know that the human brain undergoes rapid development during infancy and early childhood. In fact, newborns have a membrane-covered gap between their skull bones, called the fontanelle or soft spot, that accommodates a brain growing faster than new bones can form. Does the immune system have structures that change after birth or is it fully formed at birth?

2. How does the immune system get better at fighting disease, with age or with exposure to substances that cause an immune response?

3. Vaccines given a few decades ago contained whole, intact viruses that had been exposed to heat or chemicals that rendered them inactive. Do most modern vaccines contain entire viruses or small parts of a virus like a cell membrane protein?

4. Which would cause the generation of more kinds of antibodies, the whole viruses present in vaccines of old or the parts of viruses present in modern vaccines?

5. Consider your answers to questions 1–4 and explain why the original statement bolded above sounds right, but isn't.

# Chapter Review MasteringBiology®

## Summary

## Section 20.1

Compare the structure of bacteria to that of viruses.

- Bacteria are single-celled organisms with no nucleus or membrane-bound organelles. They are surrounded by a cell wall (pp. 452–453).
- Viruses are composed of nucleic acids encased in a capsid and sometimes an envelope (p. 455).

Contrast bacterial replication with viral reproduction.

- Bacteria replicate by binary fission (p. 453).
- Viruses replicate their genome inside host cells. The viral genome then directs the synthesis of viral components (pp. 455–456).

## Section 20.2

List the structures involved in the first line of defense against infection, and describe how they function in protecting the body against pathogens.

- The skin and mucous membrane secretions are non-specific defenses that comprise the first line of defense against infection (p. 460).

List the participants in the second line of defense against infection, and describe their functions.

- The second line of defense consists of nonspecific internal defenses, including white blood cells such as phagocytic macrophages, which engulf and digest foreign cells, and natural killer cells, which release chemicals that disintegrate cell membranes of tumor cells and virus-infected cells (p. 460).
- Inflammation attracts phagocytes and promotes tissue healing (pp. 460–461).
- Defensive proteins, including interferon, help protect uninfected cells from becoming infected. Complement proteins coat microbes and make them easier for macrophages to ingest (p. 461).

List the two different cell types involved in the third line of defense against infection, and compare how they recognize and eliminate antigens.

- Lymphocytes help comprise the third specific line of defense in response to antigens on the surface of pathogens. Exposure to antigens causes increased production of B and T lymphocytes (pp. 461–462).

- B cells secrete antibodies against pathogens, and T cells attack invaders (p. 462).

Explain the mechanism by which allergic reactions occur.

- Allergic reactions are the immune system responding to antigenic substances that are not normally pathogens (p. 462).

Explain the process immune cells undergo to be able to respond to millions of different antigens.

- Lymphocytes are produced during fetal development and allow the immune system to respond to trillions of different antigens (pp. 463–464).
- Antigen receptors on the surface of immune cells are produced by rearrangements of gene segments (p. 464).

Describe the process used to determine whether a cell is foreign or native to the body and what happens when this process fails.

- The antigen receptors of B and T cells are tested for self-reactivity, and those that react against self are eliminated (p. 464).

Explain the role of memory cells in helping to protect against infection.

- Memory cells carry copies of the antigen receptor and secrete antibodies against a previously encountered pathogen (pp. 465–466).

Describe the mechanism by which vaccines help confer immunity.

- Vaccines contain parts of disease-causing organisms or entire organisms that have been inactivated. The immune system produces memory cells in response to vaccination, thereby affording protection in the case of actual infection (p. 466).

## Roots to Remember

**The following roots of words come mainly from Latin and Greek and will help you decipher terms:**

| | |
|---|---|
| **auto-** | means self. Chapter term: *autoimmune* |
| **enceph-** and **encephalo-** | mean brain. Chapter term: *encephalitis* |
| **-path-** or **patho-** | relates to disease. Chapter term: *pathogen* |

# Learning the Basics

1. Describe the structure of a typical bacterium and a typical virus.

2. How do bodily secretions help protect against infection?

3. What roles do B cells and T cells play in the immune response?

4. Binary fission _____.
   A. allows bacterial cells to produce more plasmids; B. is a type of sexual cell division that viruses undergo; C. allows immune cells to replicate; D. is the method by which bacteria replicate.

5. Viral replication _____.
   A. occurs only inside living cells; B. requires that copies of the viral genome be produced; C. requires that copies of viral proteins be synthesized; D. all of the above.

6. The immune system _____.
   A. has a gene for each antigen; B. undergoes genetic rearrangement in response to different antigens; C. is able to make many antigen receptors by rearranging the DNA of immune cells; D. can always devise an antibody that will bind to an antigen.

7. Autoimmune diseases result when _____.
   A. a person's endocrine system malfunctions; B. liver enzymes malfunction; C. B cells attack T cells; D. the immune system fails to differentiate between self and nonself cells.

8. Which of the following cell types divides to produce cells that make antibodies?
   A. helper T cells; B. B cells; C. cytotoxic T cells; D. all of the above

9. Helper T cells secrete substances that _____.
   A. help prevent leukemia; B. prevent bacteria from entering cells; C. boost B cell and cytotoxic T cell response; D. inhibit reverse transcriptase; E. stimulate the thymus to make more B cells.

10. Your immune system can recognize a virus you have been exposed to once because _____.
    A. you harbor the virus for many years; B. you have genes to combat every type of virus; C. a cell that makes receptors to a virus multiplies on exposure to it and produces memory cells; D. a copy of the viral genome is inserted into a memory cell.

11. Which of the following is not involved in the second line of defense?
    A. natural killer cells; B. inflammation; C. complement proteins; D. macrophages; E. T lymphocytes

12. **True/False:** Allergic reactions occur when the body can't make antibodies against an antigen.

# Analyzing and Applying the Basics

1. Why are viruses not considered to be living organisms?

2. Why do you need a flu shot every year but only one inoculation against some other diseases?

3. Why is your immune system usually more effective at fighting off infection the second time you are exposed to a pathogen?

# Connecting the Science

1. Many states allow parents to opt out of the vaccinations required to start school. Should parents be allowed to opt out of vaccinating their children? Defend your answer.

2. What would you say to a friend who refuses the flu vaccine because he did not get vaccinated last year and did not get sick?

Answers to **Stop & Stretch, Visualize This, Working with Data, Savvy Reader, Sounds Right, But Is It?,** and **Chapter Review** questions can be found in the **Answers** section at the back of the book.

# CHAPTER 21 | Human Sex Differences

**Why were women prevented from Olympic ski jumping until 2014?**

# Endocrine, Skeletal, and Muscular Systems

Women's ski jumping made its Olympic debut in the 2014 Winter Olympics, making the Sochi Olympics the first winter games where men and women competed in every sport. Men have been competing in Olympic ski jumping since 1924. Why did it take so long for women to be allowed to compete? One reason was that some of those governing and coaching the sport believed that the repeated impacts of a sport like ski jumping could damage women's reproductive organs, a contention that is entirely without medical support, yet prevented women from

**Does alcohol affect the sexes differently?**

being able to compete at the Olympic level for decades.

Not all ideas about sex differences, or the average differences between females and males, are so baseless. For example, there are differences in the way the sexes metabolize alcohol and in life expectancy. Some sex differences, like differences in reproductive anatomy, are the result of biology only. Other sex differences result from biological differences that are exaggerated by

cultural expectations. As an example, consider the higher rate of knee injury seen in female athletes than male athletes. This sex difference is thought to be caused, in part, by biological differences in skeletal structure. But cultural expectations can also affect this sex difference. Until the last decade or so, women were not likely to be found in weight rooms performing exercises to strengthen the leg muscles that help stabilize the knee joint, another factor thought to play a role in the sex difference in the rate of knee injury. This sex difference, like many we will outline in this chapter, originated and is maintained by the actions of the endocrine system.

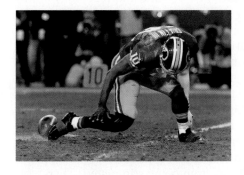

**Are there injuries that are more common in one sex than the other?**

**Why would one sex live longer than the other?**

# 21.1 **The Endocrine System**

The **endocrine system** is an internal system of regulation and communication involving hormones, the glands that secrete them, and the particular cells that respond to the hormones. The endocrine system oversees the regulation of many different processes, including reproduction and development.

## Hormones

**Hormones** are chemicals that travel through the circulatory system and act as signals to elicit a response in **targeted cells.** There are two general mechanisms by which hormones elicit a specific response: Hormones can either bind to receptors on the surface of the target cell and trigger a change inside the cell, or they can diffuse across the cell membrane and bind to receptors inside the cell to trigger the response.

Protein hormones (**FIGURE 21.1a**) cannot cross cell membranes. Typically, they bind to a receptor on the surface of the target cell, and that receptor stimulates a series of other proteins to become involved in relaying the message to the inside of the cell, eliciting a specific cellular response. The chain reaction that relays, or transduces, a signal from the outside of the cell to the inside is called **signal transduction.**

**Steroid** hormones are fat-soluble hormones that can cross cell membranes and bind to receptors inside the target cell, causing cells to turn specific genes on or off (**FIGURE 21.1b**). The steroid hormones most involved in producing anatomical sex differences are called **sex hormones.** These include the male hormone **testosterone** and the female hormone **estrogen.** Once a sex hormone passes into the cell and binds to a receptor, the hormone and receptor bind to the DNA in the nucleus of the cell to regulate the expression of different genes (**FIGURE 21.2**).

Males and females synthesize and utilize both of these sex hormones, but testosterone is the predominant hormone in males as is estrogen in females. For many sex-typical traits, it is the balance of these two hormones that maintains the sex difference. As an example, let's consider body hair. Before puberty, levels of sex hormones are low in both boys and girls. Thus, children have fine and light body hair. Once a hair follicle has been exposed to testosterone and is stimulated to produce thicker, coarser hair, all subsequent hairs produced by the follicle will be the same type. When estrogen predominates, it opposes the actions of testosterone and stops follicles from being stimulated. Therefore, the faces, chests, and abdomens of women are usually covered in fine, thin hair. At menopause, when estrogen declines and the actions of testosterone become more pronounced, some women will experience what appears to be an increase in facial hair, which is really just hair follicles responding to the effects of unopposed testosterone.

Differences in the length, thickness, and location of human hair result from both biological and cultural phenomena. Hair grows from follicles present in equal numbers in the skin of females and males. These differences are exaggerated by cultural forces that influence grooming practices. The growth of a full beard in males is considered, by some, to be manly. Likewise, many women shave, wax, and pluck body hair that they consider to be unsightly.

## Endocrine Glands

**Endocrine glands** are groups of cells or organs that secrete hormones. There are many organs in the endocrine system that don't directly affect sex differences. **FIGURE 21.3** shows the locations of the most well-known endocrine organs.

**(a)**

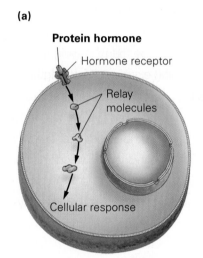

**Protein hormone**
Hormone receptor
Relay molecules
Cellular response

**(b)**

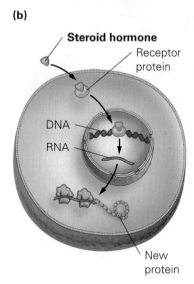

**Steroid hormone**
Receptor protein
DNA
RNA
New protein

**FIGURE 21.1 Hormone actions.**   (a) A protein hormone binds to a receptor on the cell membrane of its target cell. This binding can induce a shape change in the receptor molecule, serving as a way of transmitting a signal to the inside of the cell that some response is required. (b) Steroid hormones diffuse across the membrane and bind to receptors, which in turn regulate gene expression.

**(a) Breast tissue can respond to estrogen.**

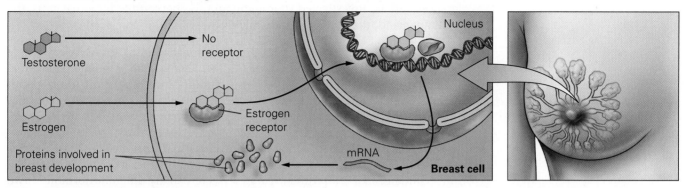

**(b) The larynx responds to testosterone.**

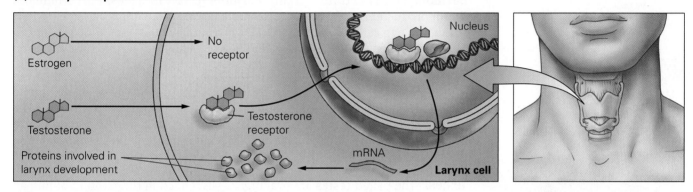

**FIGURE 21.2 Sex hormones.**   (a) Estrogen produced by females enters the cells that compose the breast tissue, where it binds to its receptor. The estrogen and its receptor together bind to DNA to increase transcription of genes required for the growth and development of the breast at puberty. (b) In males, testosterone enters the cells of the larynx and binds receptors that increase the production of proteins that change the structure of the larynx, altering the sounds it produces and resulting in voice deepening.

## Visualize This ▲

How would the breast and larynx tissue of a female respond to the presence of testosterone?

**(a) Some examples of endocrine organs**

**Thyroid** secretes calcitonin to lower blood calcium levels.

**Parathyroids** secrete parathyroid hormone to raise blood calcium levels.

**Thymus** secretes thymosin, which stimulates T cells of the immune system.

**Pancreas** secretes insulin and glucagon to regulate blood glucose levels.

**FIGURE 21.3 Endocrine glands and sex differences.**   The locations and functions of some common endocrine glands are shown. Those involved in producing sex differences are listed on the right.

**(b) Endocrine organs involved in producing sex differences**

**Hypothalamus** secretes gonadotropin-releasing hormone (GnRH).

**Pituitary gland** responds to GnRH by secreting the pituitary gonodatropins—follicle-stimulating hormone (FSH) and luteinizing hormone (LH).

**Adrenal glands** secrete adrenaline, testosterone (masculinizing hormone), and estrogen (feminizing hormone).

**Ovaries** respond to FSH and LH by secreting **estrogen**, which regulates menstruation, maturation of egg cells, breast development, pregnancy, and menopause.

**Testes** respond to FSH and LH by secreting **testosterone**, which aids in sperm production, increased muscle mass, and voice deepening.

Five endocrine organs are involved in the production of biological sex differences and are investigated thoroughly in this chapter: the hypothalamus, the pituitary gland, the adrenal glands, the ovaries, and the testes.

**Hypothalamus and Pituitary Gland.** The **hypothalamus,** located deep inside the brain, regulates body temperature and behaviors such as hunger and thirst, in addition to reproduction. Beneath the hypothalamus lies the **pituitary gland,** which secretes growth hormone, which acts on many organs, including bones that help children reach adult height.

To help regulate the activities of the reproductive system, the hypothalamus secretes a hormone called **gonadotropin-releasing hormone (GnRH).** The target cells of GnRH are found in the pituitary gland that lies below the hypothalamus. GnRH causes the pituitary gland to secrete many different hormones: two in particular are involved with producing sex differences: **follicle-stimulating hormone (FSH)** and **luteinizing hormone (LH).** In males, FSH stimulates sperm production, and LH stimulates testosterone production. In females, FSH stimulates egg cell development and LH stimulates the release of an egg cell during ovulation (**FIGURE 21.4**). GnRH gets its name, in part because it acts on the ovaries and testes, collectively called the gonads.

**Adrenal Glands.** An **adrenal gland** sits atop each kidney. These glands secrete the hormone adrenaline (epinephrine) in response to stress or excitement. The adrenal glands of both males and females also secrete a small amount of both of the sex hormones, but the majority of sex hormone synthesis and secretion occurs in the gonads.

**Testes.** The paired **testes** are oval organs suspended in the scrotum of human and other mammalian males. Testes secrete testosterone, an **androgen** or masculinizing hormone. Testosterone causes sperm production, increased muscle mass, and voice deepening. Sperm production is most efficient at temperatures that are lower than body temperature. Therefore, the testes are situated outside the body cavity, in the scrotum.

**Ovaries.** The paired **ovaries** are about the size and shape of almonds in the shell. They produce and secrete estrogen. Estrogen regulates many functions in the female body, including menstruation, the maturation of egg cells, breast development, pregnancy, and the cessation of menstruation after reproductive age, called menopause. Inside the ovaries are all of the cells that can mature into the egg cells for ovulation.

**Puberty. Puberty** marks the beginning of sperm production in males and the beginning of egg cell maturation and menstruation in females. Boys typically begin puberty at the age of 9 to 14 years, with the average being 13 years. Puberty in males includes enlargement of the penis and testes; an overall growth spurt; growth of muscles and the skeleton, resulting in wide shoulders and narrow hips; and changes in hair growth, including the production of pubic, underarm, chest, and facial hair. In addition, the larynx enlarges and vocal cords lengthen to produce a deeper voice.

For girls, the first signs of puberty occur between ages 8 and 13, most commonly around age 11. These signs include breast development, increased fat deposition, a growth spurt, pubic and underarm hair growth, and the commencement of menstruation.

**STOP & STRETCH** Androgen insensitivity syndrome occurs when cells cannot respond to the presence of testosterone. A person with X and Y sex chromosomes with complete androgen insensitivity has the outward appearance of a normal female. Propose a genetic hypothesis for how this condition could arise.

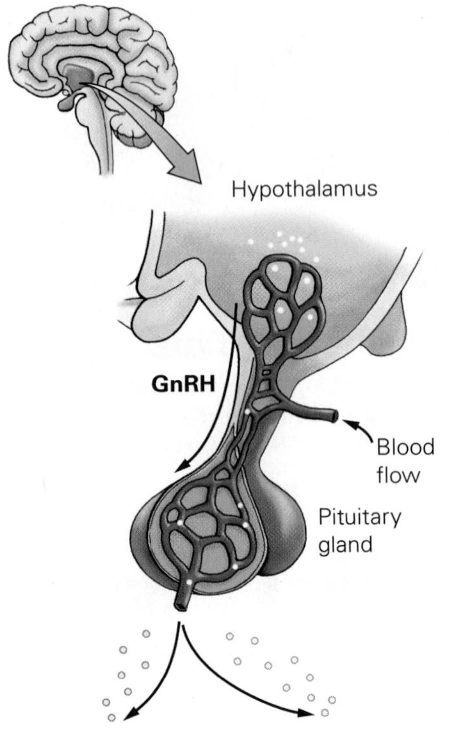


| | FSH | LH |
|---|---|---|
| ♂ | Stimulates sperm production | Stimulates testosterone production |
| ♀ | Stimulates egg cell development | Stimulates ovulation |

**FIGURE 21.4 FSH and LH secretion.** The hypothalamus secretes GnRH, causing the pituitary gland to secrete FSH and LH. FSH and LH are secreted by the brain of both males and females but stimulate a different response in each sex.

It is at puberty that biologically induced sex differences begin to emerge, in part because the hormones produced by testes and ovaries begin to exert their effects on the skeletal system.

## 21.2 The Skeletal System

The skeletal system provides support for the body, protects internal organs, aids in movement, and stores minerals. This internal framework is composed of bones and cartilage. Bones are rigid body tissues that function to provide support for the actions of muscles and protection for soft parts of the body—the skull for example, which protects the brain. Bones also help produce blood cells within the bone marrow and serve as a reservoir for minerals like calcium that circulate in the body. The ends of bones are covered with cartilage, which protects against the degradation caused when two bones in a joint rub against each other.

There are 206 bones in the human skeleton, which is organized into two basic units, the axial and the appendicular skeleton (**FIGURE 21.5**). The **axial**

**Differences in skeletal structure may account for the slightly higher rate of knee injury in women than men, but men are still susceptible.**

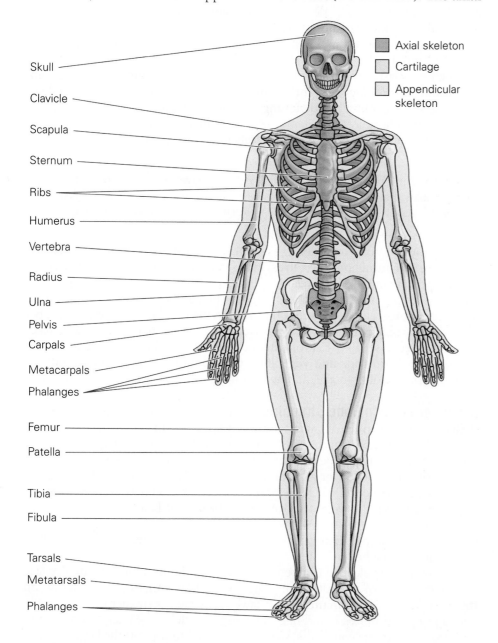

Skull
Clavicle
Scapula
Sternum
Ribs
Humerus
Vertebra
Radius
Ulna
Pelvis
Carpals
Metacarpals
Phalanges
Femur
Patella
Tibia
Fibula
Tarsals
Metatarsals
Phalanges

- ■ Axial skeleton
- □ Cartilage
- □ Appendicular skeleton

**FIGURE 21.5 The human skeleton.**
The human skeleton can be divided into those bones that support the trunk, called the axial skeleton, and those that compose the limbs, called the appendicular skeleton. Cartilage cushions joint surfaces.

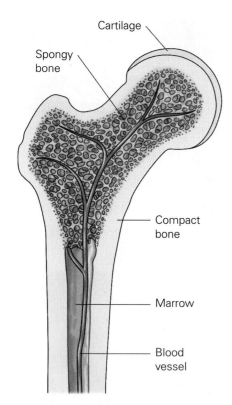

Cartilage

Spongy bone

Compact bone

Marrow

Blood vessel

**FIGURE 21.6 Bone structure.** Bone is composed of compact and spongy bone. There are blood vessels running through bone, and the marrow inside bones helps produce blood cells.

osteo- relates to bones (the Latin form is os- or ossi-).

-blast refers to an early formation or to a place where something forms.

-clast is from the Greek verb to break or to break down.

skeleton supports the trunk of the body and consists largely of the bones making up the vertebral column (or spine) and the skull. The **appendicular skeleton** is composed of the bones of the hip, shoulder, and limbs. Bones are kept in place by ligaments and moved by muscles.

## Bone Structure and Remodeling

Bones are living tissues composed of cells that use the mineral calcium to actively remodel the skeleton. The tissue composing bones can be tightly or loosely packed. **Compact bone** is the tightly packed hard outer shell of bones, and **spongy bone** is the loosely packed, porous, honeycomb-like inner bone (**FIGURE 21.6**). Because bone is a living tissue, it has blood vessels throughout. The interior of bone contains **bone marrow**, which helps to produce blood cells.

**Osteoblasts, Osteoclasts, and Calcium.** The cells that make up bone are called **osteoblasts** and **osteoclasts**. Osteoblasts help bone tissue regenerate itself, by a process called bone deposition. Osteoclasts are involved in breaking down or re-absorbing bone tissue. The delicate balance between deposition and reabsorption is integral to the maintenance of bones.

When calcium levels in the blood are high, osteoblasts remove calcium from the blood and use it to make new bone. When calcium levels in the bloodstream are low, osteoclasts break down bone and release calcium into the bloodstream. The maintenance of bloodstream calcium levels is important in many physiological processes, including blood clotting, muscle contraction, production of nerve impulses, and the activity of many enzymes. When bone reabsorption outpaces bone deposition, **osteoporosis**, bone weakening, can result (**FIGURE 21.7**).

**STOP & STRETCH** The declining estrogen levels associated with menopause lead to increased rates of osteoporosis in older women. Based on your understanding of bone remodeling, predict which type of bone cell is inhibited by estrogen.

Even though bone remodeling is mediated by estrogen, males can also get osteoporosis. Throughout their lives, males convert testosterone to estrogen, in part to maintain bone health. Close to 20% of aging men who produce less testosterone will get osteoporosis. The number of women who will get osteoporosis is several times higher than the number of males, in part because women outlive men by an average of 5 years. Both males and females can decrease their risk of osteoporosis by exercising, eating a healthy diet rich in fruits and vegetables that contain calcium and vitamin D, limiting alcohol intake, and not smoking.

**Sex Differences in Bone Structure.** Average differences in male and female skeletons arise at puberty when a growth spurt occurs. During the growth spurt, a larger proportion of growth occurs in the arms and legs than in the torso. In part because puberty starts later and lasts longer in boys, the average adult man has longer arms and legs. Because men generally have shorter torsos relative to their leg length, their center of gravity, the point on the body where the weight above equals the weight below, is higher than in women (**FIGURE 21.8a**). Women stop growing after puberty while men continue growing into their early twenties resulting in an average height difference of about 6 inches.

The remaining skeletal differences between males and females involve size differences in six of the 206 bones. Four of the bones that differ are found on the head. These are the **mandible**, or jawbone, which is smaller in females; the paired **temporal bones** found near the temple, which have a larger opening in males to allow for the connection of thicker muscles to support the large jaw; and

FIGURE 21.7 **Bone remodeling.** In normal bone (a), old bone is removed by osteoclasts and new bone is laid down by osteoblasts. Osteoporosis (b) occurs when bone is broken down more quickly than it is replaced.

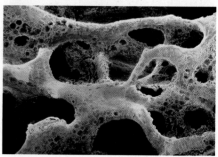

the **frontal bone** of the cranium, or forehead, which is generally more rounded in females and has a less-pronounced ridge above the eyes (**FIGURE 21.8b**).

The other two bones that show general differences between men and women are the two **ossa coxae**, or hip bones, which form the bony pelvis. The round pelvic inlet that is typical of most women is produced by a bony pelvis that is flatter and broader than a man's, which has an inlet that resembles an elongated oval. A flatter pelvis also requires that the bony pelvis be tipped forward to bring the hip bones to the front of the body; this tipping forward is maintained by the curvature of the lower spine. The spinal curvature and greater pelvic tilt in individuals with broad pelvises elevate the buttocks and give a curvy appearance to the profile of women. Conversely, the male pelvis lowers the buttocks and gives a flat appearance to a man's profile (**FIGURE 21.8c**).

The differences between the pelvises of men and women evolved in response to different selective pressures. The evolution of upright walking in early

## Visualize This ▼

List the bones involved in determining the Q angle.

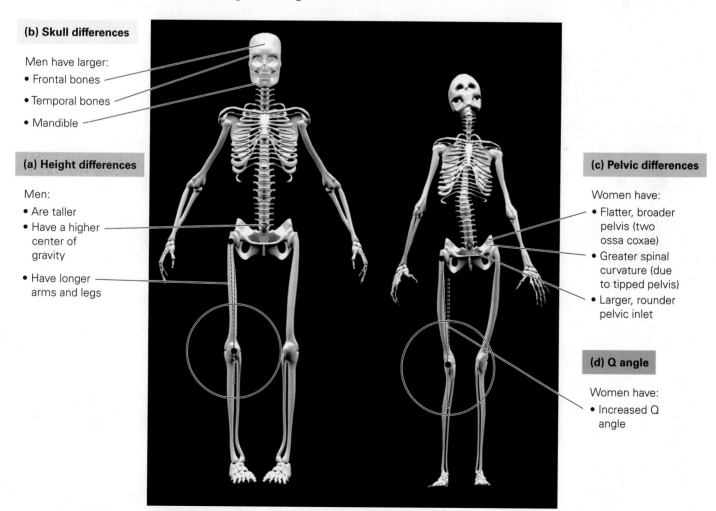

**(b) Skull differences**

Men have larger:
- Frontal bones
- Temporal bones
- Mandible

**(a) Height differences**

Men:
- Are taller
- Have a higher center of gravity
- Have longer arms and legs

**(c) Pelvic differences**

Women have:
- Flatter, broader pelvis (two ossa coxae)
- Greater spinal curvature (due to tipped pelvis)
- Larger, rounder pelvic inlet

**(d) Q angle**

Women have:
- Increased Q angle

**FIGURE 21.8 Sex differences in the skeleton.** While human male and female skeletons are difficult to distinguish, consistent differences can be found in overall size (a), the size of four skull bones (b), the size and shape of the pelvic bones (c), and the Q angle (d).

**Males metabolize alcohol more readily than females.**

humans required that these changes take place. However, the changes to the ossa coxae, combined with our large head size, resulted in one of the most difficult birth processes of any mammal. Early human females with rounder pelvic inlets may have had a selective advantage because they were less likely to die during childbirth, since rounder inlets generally permit the easier passage of a child's head through the birth canal.

One last sex difference involving the skeleton has to do with the angle formed between the kneecap, femur, and the line of the tendon from the kneecap to the shinbone, called the **Q angle** (**FIGURE 21.8d**). When you stand with your feet together, your femurs, or thighbones, extend diagonally to the knees from where they attach at the hips. A broader pelvis means that the femurs are farther away from each other at the point of attachment than are femurs attached to a narrower pelvis. Women generally have greater Q angles than men do and there is some evidence that the larger Q angle in females may be partly responsible for their higher rates of knee injury.

## 21.3 The Muscular System

The job of the muscular system, composed primarily of skeletal muscle, is to aid in movement. Skeletal muscle attaches to bones at tendons and interacts with the skeleton to allow movement of bones. Skeletal muscle is nourished by blood vessels and controlled by nerves that help signal muscles to contract.

### Muscle Structure and Contraction

Muscle tissue is composed of smaller and smaller parallel arrays of filaments (**FIGURE 21.9a**). A muscle contains a sheathed bundle of parallel **muscle fibers.** Each muscle fiber, composed of a single cell, contains parallel arrays of threadlike

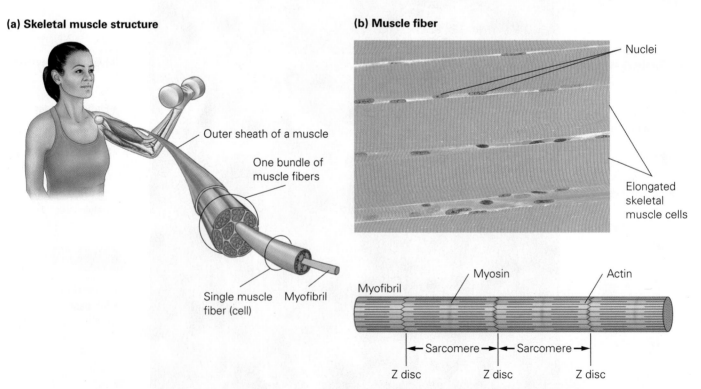

**(a) Skeletal muscle structure**

Outer sheath of a muscle

One bundle of muscle fibers

Single muscle fiber (cell)  Myofibril

**(b) Muscle fiber**

Nuclei

Elongated skeletal muscle cells

Myofibril  Myosin  Actin

←— Sarcomere —→←— Sarcomere —→

Z disc  Z disc  Z disc

**FIGURE 21.9 Skeletal muscle structure.** (a) Muscle cells (fibers) are arranged in parallel bundles inside the muscle's outer sheath. (b) A skeletal muscle cell, with its myofibrils in register, produces the characteristic banded or striated pattern. The myofibrils inside a muscle cell are linear arrangements of sarcomeres. Sarcomeres are bounded by Z discs.

filaments called **myofibrils.** Bands of all myofibrils in a cell align in register, like pipes of the same length stacked with their ends aligned (**FIGURE 21.9b**). This gives the skeletal muscle its characteristic striated or striped appearance. Each myofibril is a linear arrangement of **sarcomeres,** the unit of contraction of a muscle fiber. A sarcomere is composed of thin filaments of the protein **actin** and thick filaments of the protein **myosin.** The sarcomere is the region between two dark lines, called **Z discs,** in the myofibril.

To contract the muscles, the sarcomeres must shorten via an elaborate mechanism involving actin and myosin filaments.

**sarc-** or **sarco-** means body.

**my-** or **myo-** means muscle.

### Sliding-Filament Model of Muscle Contraction.
Sarcomere shortening involves coordinated sliding and pulling motions within the sarcomere (**FIGURE 21.10**). Actin filaments are anchored at the Z discs at the ends of a sarcomere. They

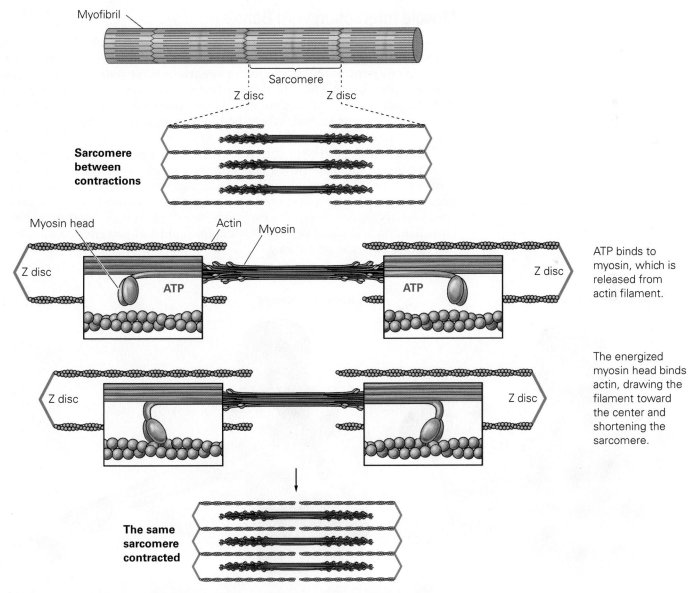

**FIGURE 21.10 Muscle contraction.** The sarcomere of a myofibril consists of actin molecules attached to the Z disc and myosin molecules. Using energy from ATP, the myosin head binds to actin and pulls it toward the center. Because this is happening at both ends of the sarcomere, the sarcomere shortens, allowing contraction of the muscle cell and movement of the muscle.

overlap a parallel, stationary set of myosin molecules that are not attached to Z discs. During contraction, actin filaments slide over the fixed myosin filaments, pulling the ends of the sarcomere with them. The sarcomere ends are brought toward the center when the head of the myosin molecule attaches to binding sites on the actin molecules and pulls toward the center.

The sliding-filament mechanism proceeds as the sarcomere shortens in a stepwise manner. First, the high-energy ATP molecule binds to the myosin head, causing the myosin head to detach from the binding site on the actin filament. Second, the breakdown of adenosine triphosphate (ATP) provides energy to the myosin head, which changes shape or "cocks." Third, the myosin head binds to another actin-binding site; pulls in short, powerful strokes; releases the actin; and reattaches farther along. This process serves to pull the Z discs inward, shortening the sarcomere.

## Muscle Interaction with Bones

When muscles exert a pulling force on bone, the use of one muscle to perform a task, like lifting a child, is followed by the use of another muscle to return the arm to its original position. Such **antagonistic muscle pairs**, in which each muscle is paired with a muscle of the opposite effect, are a common feature of the muscular system. Contraction of the biceps muscle shortens it, while the triceps muscle remains relaxed. When the biceps muscle is subsequently relaxed, the triceps muscle contracts (**FIGURE 21.11**).

Skeletal muscles can be of different varieties based on how quickly they produce a contraction. Slow-twitch fibers contract slowly, while fast-twitch fibers contract quickly. Slow-twitch fibers contain many mitochondria, are well supplied with blood vessels, and can therefore use ATP as quickly as it is synthesized and oxygen as quickly as it can be delivered. Fast-twitch fibers have fewer mitochondria and blood vessels allowing for rapid and powerful contractions that cannot be sustained for long.

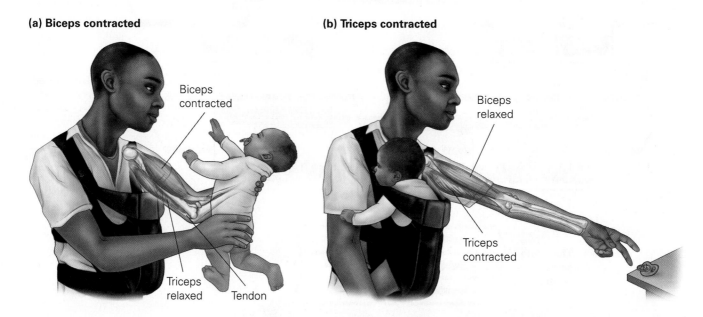

**(a) Biceps contracted**

Biceps contracted

Triceps relaxed   Tendon

**(b) Triceps contracted**

Biceps relaxed

Triceps contracted

**FIGURE 21.11 Antagonistic muscle pairs.** Movements are often produced through the actions of two opposing muscles. (a) Contraction of the biceps raises the forearm and (b) contraction of the triceps lowers the forearm.

## Sex Differences in Muscle

Sex differences in strength occur because testosterone released by the testes at puberty increases the size, but not the number, of muscle fibers. Muscle fibers contain testosterone receptors in their cytoplasm, and the presence of testosterone stimulates the cells to increase in mass. This is accomplished, in part, when males retain dietary nitrogen, which is used to produce more muscle proteins.

Although the ovaries and adrenal glands of both sexes release small quantities of testosterone, males have more of this hormone than do females, and the subsequent difference in muscle mass accounts for an overall difference in strength between males and females.

**STOP & STRETCH**   If a male and a female were the same height and had the same amount of total muscle mass in each bicep, would you expect the male to be able to curl a heavier dumbbell than the female? Why or why not?

# 21.4   Other Biology-Based Sex Differences

Differences in how fat is metabolized and stored are mediated by the endocrine system. Androgens, found in higher concentrations in males, stimulate the production of proteins that help break down fats. This means that if a man and a woman of the same height and weight both eat an ice cream cone, the man is likely to burn more of the calories and the woman is likely to store more of the calories on her body as fat. In this same-sized pair, the man is likely to have more muscle mass, which requires substantial energy to maintain, than the woman who is likely to carry more body fat. Since fat is not a very active tissue, it does not undergo contraction and flexion like muscle tissue; it costs less energy to maintain. Therefore, if this same couple were to walk while eating their ice cream cones, the man will burn calories at an even higher rate than the woman.

This more efficient breakdown of fats by males is also partly responsible for their higher rates of cardiovascular disease. Because fat is sent through the bloodstream to be metabolized by tissues instead of being stored on the body, men have higher levels of fat circulating in their bloodstreams. Fat in the bloodstreams can adhere to blood vessel walls, causing a dangerous buildup and decreasing the diameter of blood vessels. This causes blood pressure to increase and escalates risk for other cardiovascular diseases.

Fat is also stored in different locations in females and males. Women tend to have more adipocytes in their hips, thighs, and buttocks. Conversely, when testosterone is high, fat is more likely to be stored in the abdomen.

**STOP & STRETCH**   As women age, they begin to store more body fat in their abdomen and less on their extremities. Propose an endocrine-based hypothesis for why this change in the location of body fat storage might occur.

**There are differences between the sexes in life expectancy.**

Because women store more fat than men, women, on average, have an 8% thicker layer of subcutaneous fat covering muscle tissues. This means that even in a man and a woman with the same amount of muscle mass, individual muscles in the woman will appear to be more smoothed out. In order to maintain their fertility, women require around 10% more body fat than males (**FIGURE 21.12**). When a female does not have enough stored body fat, a hormone called leptin signals the brain that she will not be able to sustain a pregnancy and the menstrual cycle is prevented from occurring, which can cause permanent sterility, even if she increases her body fat levels.

Because fat is not very metabolically active, it does not require many blood vessels to deliver nutrients and remove wastes. Women have fewer blood vessels for their size and have less blood volume per body weight. If a man and woman of the same size both drink the same volume of the same alcoholic beverage, the alcohol will be more concentrated in her bloodstream and the effects of the alcohol will be more pronounced. And, as you learned in Chapter 18, women are less able to begin the process of alcohol metabolism in the stomach. These factors mean that it isn't a good idea for women to pace their drinking with that of male friends, even same-sized ones.

Another sex difference in humans is in average life expectancy. A man and woman born on the same day have different life expectancies. According to the World Bank, a woman born in the United States between 1999 and 2003 has a life expectancy of 81 years while a male born in the United States during that time has a life expectancy closer to 76.

Among the reasons for the sex difference in life expectancy are that men are more likely to be murdered, commit suicide, and die in automobile accidents than are women. These reasons alone do not account for the entire difference in life expectancy. A hypothesis that might help explain more of the difference is based on the average age of onset of cardiovascular disease. Women most often experience heart attacks and strokes in their 70s and 80s, while men most often experience them a decade or so earlier. This hypothesis posits that blood loss during menstruation has a protective effect against heart disease by lowering women's overall exposure to iron. The oxygen-carrying component of blood, hemoglobin, contains iron as part of its structure. Iron has been shown to produce free radicals that damage cells and their DNA, which could increase the risk of heart disease. If this turns out to be true, males and females would be advised to limit their consumption of foods, like red meat, rich in iron as well as decrease exposure to other free radical–generating substances like cigarette smoke and alcohol.

## Working with Data ▶

(a) Is there more variation in healthy body fat percentages between or within the sexes? (b) If a person has a 18% body fat, can we know whether the person is male or female?

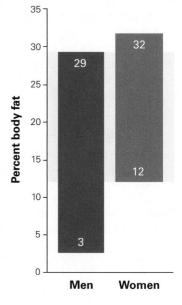

**FIGURE 21.12 Body fat.** This bar graph shows the range of percentages of body fat in healthy males and females.

# *savvy reader*

## Anabolic Steroids Use

A peer-reviewed article in the journal *Drug and Alcohol Dependence* (2007) reports the results of a study on the non-medical use of anabolic steroids among U.S. college students from 1993-2001.

Non-medical means the illegal use of testosterone-mimicking drugs to gain muscle mass, instead of physician-prescribed use for various medical conditions. Use of anabolic steroids can lead to cancer, high blood pressure, depression, irritability, insomnia, and mood swings in both sexes. In males, the use of steroids may cause shrunken testes and decreased sperm production, which can lead to infertility. Hair loss and breast development may also occur. In females, steroids can disrupt the menstrual cycle, increase facial hair growth, and deepen the voice.

Surveys were sent to over ten thousand randomly chosen college students over a 10-year period. The students queried were all from the same suite of colleges. The scientists performing this study found that non-medical use of steroids occurred in around 1% of those surveyed and that steroid use was increasing among male college students. Statistical analysis showed that male student athletes were the most likely users. There were also correlations between non-medical steroid use and the use of other illegal drugs, alcohol and cigarette use, as well as with drinking and driving.

1. List two controls used in this study.

2. The scientists who performed this study also found that users of anabolic steroids were more likely to use drugs and alcohol. How might the symptoms of steroid abuse lead to increased drug use?

3. Propose a hypothesis to explain why non-medical use of anabolic steroids increased among men during the course of this study.

http://www.sciencedirect.com/science/article/pii/S0376871607001585

# SOUNDS RIGHT BUT IS IT?

The normal cycling of estrogen that occurs in a fertile female involves an increase in estrogen level before the release of an egg cell from the ovary. The release of the egg cell, a process called ovulation, occurs around the 14th day of the 28-day menstrual cycle. After ovulation, estrogen levels decrease back to a basal level and then slowly increase again in a continuous cycle until they peak before ovulation.

When a woman takes the birth control pill, this cycling of estrogen is prevented. Instead, her body is exposed to one continuous low dose of estrogen.

Low estrogen levels can lead to increased risk of osteoporosis.

**Taking the birth control pill leads to increased risk of osteoporosis.**

Sounds right, but it isn't.

1. Assume the ovaries of a fertile woman produce a total of 49 mg of estrogen every cycle. Assume that the basal level of estrogen produced is 1 mg/day except during the increase, plateau, and decrease that occur on days 11, 12, and 13, respectively. How many days during the cycle is a woman producing a basal level of estrogen?

2. What is the total combined mg amount of estrogen that this woman produces during all the days she is producing estrogen at the basal level?

3. How much estrogen must she produce in the 3 days before ovulation (days 11, 12, and 13)?

4. Pharmaceutical companies design birth control pills that mimic the bone-protecting effects of the normal female menstrual cycle. Assuming that a woman takes the birth control pill every day of the month (which is not quite true; see the next chapter), how many mg of estrogen should be found in each pill?

5. Consider your answers to questions 1–4 and explain why the original statement bolded above sounds right, but isn't.

# Chapter Review MasteringBiology®

Go to the Study Area in MasteringBiology® for practice quizzes, myeBook, BioFlix™ 3-D animations, MP3 Tutor secessions, videos, current events, and more.

## Summary

## Section 21.1

Differentiate between protein and steroid hormones.

- Protein hormones trigger a response by binding to receptors on the surface of the target cell, which relay the signal to the inside of the cell (p. 472).
- Steroid hormones diffuse into the cell and act on receptors there (p. 472).

List the major endocrine glands, the hormones they secrete, and the effects these hormones have on their target organs.

- The thyroid gland secretes hormones that stimulate metabolism; along with the parathyroid gland, it regulates calcium levels (p. 473).
- The thymus helps the immune system form, and the pancreas helps to regulate blood glucose levels (p. 473).
- The hypothalamus regulates body temperature, hunger, thirst, and reproduction. The hypothalamus secretes GnRH, which stimulates the pituitary gland to release FSH and LH (pp. 473–474).
- Adrenal glands secrete adrenaline, androgens, and estrogens (pp. 473–474).
- Testes secrete testosterone, which affects the development of male characteristics. Ovaries secrete estrogen, which fosters the development of female characteristics (pp. 473–474).

## Section 21.2

Describe the main structural components of the skeleton and their functions.

- The skeletal system is composed of bones and cartilage. It provides support, protection, movement, and mineral storage (p. 475).
- Joints are regions where bones come together (p. 475).
- Ligaments hold bones together at joints (p. 476).

Discuss the process of bone remodeling.

- Bone remodeling is under the control of the endocrine system. Hormones stimulate the osteoblasts to use calcium to build bone, and the osteoclasts to break down bone and release calcium (p. 476).

## Section 21.3

Describe the main structural components of the muscular system and their functions.

- The muscular system functions in movement, which is coordinated by the combined actions of antagonistic pairs of muscles (p. 480).

Outline the structure of a muscle fiber, and describe how muscles contract.

- Muscles are bundles of parallel fibers. Muscle fibers are bundles of parallel myofibrils. Myofibrils are linear arrays of sarcomeres, which are the unit of muscle contraction (pp. 478–479).
- Sarcomeres are composed of actin and myosin. Actin filaments attached to the ends of a sarcomere are pulled toward the middle of the sarcomere by the movement of myosin heads. These myosin heads then attach to the filaments, pull them toward the sarcomere, release the filaments, and reattach farther down the filament. This process is called the sliding-filament model of muscle contraction (pp. 479–480).

## Section 21.4

Describe average sex differences in body fat utilization, prevalence, and location.

- Men utilize fat more readily than women (p. 481).
- In order to maintain their fertility, women require more body fat than do men (p. 482).
- Women store fat on their hips and thighs; men store fat in their abdomen (p. 481).

Describe the sex differences in alcohol metabolism and life expectancy.

- Men metabolize alcohol more quickly than women (p. 482).
- Women live, on average, at least 5 years longer than men (p. 482).

# Roots to Remember

**The following roots of words come mainly from Latin and Greek and will help you decipher terms:**

| | |
|---|---|
| **-blast** | refers to an early formation or to a place where something forms. Chapter term: *osteoblast* |
| **-clast** | is from the Greek verb to break or to break down. Chapter term: *osteoclast* |
| **my-** or **myo-** | means muscle. Chapter terms: *myosin, myofibril* |
| **osteo-** | relates to bones (the Latin form is os- or ossi-). Chapter terms: *osteoblasts, osteoclasts, osteoporosis* |
| **sarc-** or **sarco-** | means body. Chapter term: *sarcomere* |

# Learning the Basics

**1.** List the endocrine organs involved in producing sex differences and describe their functions.

**2.** Describe bone structure.

**3.** Describe muscle structure.

**4.** Steroid hormones act by _____.
**A.** causing neurons to fire; **B.** increasing muscle contraction; **C.** diffusing across membranes and regulating the expression of genes; **D.** binding to receptors on the cell surface and transducing a response without entering the cell.

**5.** The endocrine organ that sits atop a kidney is _____.
**A.** the pituitary gland; **B.** the hypothalamus; **C.** the ovary; **D.** the adrenal gland; **E.** the testicle.

**6.** Bones of the limbs form part of the _____.
**A.** appendicular skeleton; **B.** axial skeleton; **C.** spongy bone; **D.** compact bone.

**7.** Osteoblasts _____.
**A.** are found in the marrow of bone; **B.** regulate bone deposition; **C.** regulate bone reabsorption; **D.** are more active when calcium is low.

**8.** The antagonistic pairs of arm and leg muscles _____.
**A.** allow healthy muscles to compensate for injured ones; **B.** allow muscles to produce opposing movements; **C.** allow different types of rotations of joints; **D.** allow myofibrils to fire in response to different stressors.

**9.** During muscle contraction, _____.
**A.** the actin heads pull the sarcomere closed; **B.** myosin attaches to actin, pulls it toward the center of the sarcomere, releases it, and reattaches farther along; **C.** myofibrils shrink due to the actions of testosterone; **D.** muscle fibers contract under the actions of testosterone.

**10.** Which of the following is a true statement regarding muscle contraction?
**A.** Actin filaments are stationary, and myosin heads can move; **B.** Z discs are pulled toward the center of the sarcomere; **C.** Actin is used up during this process; **D.** No ATP is required for movement to occur.

# Analyzing and Applying the Basics

**1.** When it comes to sex differences, the overlap between genders, even in terms of anatomical differences, can be quite significant. For instance, the female and male pelvises have different shapes, which can affect how the legs are attached. Yet according to the classic study on pelvises, only 40 to 50% of actual women have a female pelvis as defined in anatomy books; 33% have an "android" or male-type pelvis. How is our ability to make generalizations about the sexes impacted by such differences?

**2.** Males experiencing infertility are often advised to wear loose-fitting pants and underwear. How might this practice help increase sperm production?

**3.** Some studies have shown that chronic marijuana use can lower testosterone levels in males. This can lead to breast development in males, a condition called gynomastia. Why might this occur?

# Connecting the Science

**1.** The use of anabolic steroids to build muscle is illegal, but the practice has become widespread in certain professional sports. Do you agree that using steroids is a form of cheating? If so, how is it different from other forms of physical enhancement (for instance, special diets, vision correction, even drugs that reduce asthma symptoms)? If not, explain why.

**2.** Human beings exaggerate sex differences in hair composition by shaving, waxing, and plucking. However, when it comes to sex differences in body fat, the opposite seems to be true, with many females feeling like they need to override their body's biological requirements for body fat by compulsively dieting. Why do you think we exaggerate some sex differences but not others?

Answers to **Stop & Stretch, Visualize This, Working with Data, Savvy Reader, Sounds Right, But Is It?**, and **Chapter Review** questions can be found in the **Answers** section at the back of the book.

CHAPTER **22** | # Is There Something in the Water?

Frogs with extra limbs ...

# Reproductive and Developmental Biology

In addition to concerns over a decline in the overall number of different species living in our lakes and streams, biologists are troubled by an increasing number of aquatic organisms with physical malformations. Some of these deformities are external, like extra

**Fish that are all female ...**

limbs, while others are hidden inside the bodies of affected organisms. Male frogs that develop female reproductive organs, aquatic insects that are unable to reproduce, and entire schools of fish that have both male and female reproductive organs are no longer anomalies. Changes like these can have profound implications for the reproduction of individual organisms and the survival of species.

Is there something in our waters that is altering the development of normal reproductive structures in the organisms that inhabit them? Scientific studies seem to implicate endocrine

disruptors, a class of chemicals that alter the actions of the hormone-producing endocrine system. Evidence of endocrine disruption in aquatic organisms has led to concerns about the water that humans are drinking. Are these chemicals present in the water supply, and if so, are they affecting human reproductive health? To answer these questions, a thorough understanding of reproduction and development is required. We begin by looking at basic reproductive processes.

**Aquatic insects that can't swim or reproduce ...**

**Is there something in the water altering development and reproduction?**

**Biologists are troubled by an increasing number of aquatic organisms with physical malformations.**

# 22.1 Principles of Animal Reproduction

Reproduction is the process by which organisms produce offspring. When this process is disrupted, populations decline and species can become extinct. This is why a clue to the presence of a biologically active pollutant in an area is often a decline in the survival and health of offspring produced in that environment.

Most animals reproduce via either asexual or sexual processes.

## Asexual Reproduction

**Asexual reproduction** occurs when a single organism reproduces by itself, without mating with another organism. The offspring of asexual reproduction are genetically identical to the parent and to each other. Cells that divide by mitosis undergo asexual reproduction, as do bacterial cells that divide by **binary fission**, a process that begins with the duplication of the bacterial chromosome, followed by the splitting of the single cell into two identical daughter cells. Sometimes, multiple fissions can occur, fragmenting one individual into many individual daughter cells that can develop into adults. Sponges are multicellular animals that can reproduce by fission, during which masses of cells break away from the parent to form offspring (**FIGURE 22.1a**).

**Budding** is a form of asexual reproduction that is similar to fission. A yeast cell budding in bread dough undergoes cell division to produce a daughter cell that remains attached to the parent until it achieves sufficient growth to break away. Hydra are simple animals that can also reproduce by budding (**FIGURE 22.1b**).

Asexual reproduction has advantages and disadvantages. There is no need to find a partner to reproduce, which is particularly advantageous for organisms that do not move or are isolated from other organisms. When conditions are favorable, asexual reproduction allows for the rapid production of large numbers of offspring. However, because all offspring are genetically identical to the parent, they lack the ability to adapt and a single environmental change could decimate the entire population.

## Sexual Reproduction

Organisms subject to changing environmental conditions often undergo **sexual reproduction** to produce offspring that are genetically diverse. Populations of individuals that are not identical will likely have a few members with genetic makeup that will allow survival and reproduction when conditions change.

Creation of offspring by the fusion of sex cells from two different individuals—for example, a mating between one male and one female parent—is required

## Visualize This ▶

The white cloud at the top of this photograph consists of cells that are breaking away from the parent. What is the fate of these cells?

**(a) Sponge**

**(b) Hydra**

**FIGURE 22.1 Asexual reproduction in animals.** (a) Sponges can reproduce by fission. (b) Hydra can reproduce by budding. The newly formed hydra on the left will break away from its parent.

for sexual reproduction. Offspring are produced after the gametes from two different individuals combine genetic information at fertilization. Gamete-producing structures are called **gonads.** In males, gametes are called sperm, and the gonads are **testes,** which produce sperm. In females, eggs are the gametes, and the gonads are **ovaries,** which produce the eggs.

> **gon-, gono-,** and **gonado-** mean seed or generation.

> **STOP & STRETCH** There is some concern that estrogen secreted in the urine of women using birth control pills can make its way into the waterways and affect aquatic organisms. Canadian scientists placed estrogen in a remote lake, which would not be exposed to estrogen from municipal waterways. When they compared fish exposed to estrogen with those from the same lake before estrogen treatment, they found significant increases in the numbers of male fish that were producing eggs. They found no increase in such fish in two nearby untreated remote lakes. Describe the controls used in this study and explain whether the results indicate that estrogen secreted in the urine of birth control pill users can affect aquatic organisms.

Because two different individuals contribute genetic information to their gametes, the offspring will be a genetic mixture of both parents. Each gamete is haploid (n) and contains half the genetic information of the individual that produced it. Fusion of two haploid gametes produces a diploid (2n) **zygote,** or fertilized egg cell. Because a mixing of parental genetic information occurs in the offspring, sexual reproduction produces an almost infinite variety of genetically unique individuals. Most commonly, an organism will produce either male or female gametes but not both. However, there are exceptions. **Hermaphrodites** are animals that have both male and female reproductive systems. Earthworms, for example, can both donate and receive sperm during mating. Because the production of male and female gametes by one individual is very common in the plant kingdom, the term *hermaphrodite* applies only to animals.

Fertilization can occur inside or outside the body of sexually reproducing organisms. Depositing sperm in or near the female reproductive tract through **copulation,** or sexual intercourse, allows **internal fertilization** to occur. Sharks, reptiles, birds, and mammals reproduce via internal fertilization. Many aquatic invertebrates, most fish, and some amphibians use **external fertilization,** during which the parents need not have physical contact with each other. The female lays eggs in the water, and the male releases sperm over the eggs. Because these eggs are fertilized and develop outside the body, in water, they are very susceptible to aquatic contaminants.

Frogs are sexually reproducing organisms that use external fertilization. The fertilized eggs of frogs give rise to tadpoles that develop into adults in the water. Because so much of a frog's development occurs in the water, frogs in early developmental stages are highly sensitive to environmental contaminants. Because of this, frogs can be used as environmental early-warning signals in the same way that canaries were once used to warn miners of high carbon monoxide concentrations in mines. Miners knew that a distressed or dying canary meant they should get out of the mine quickly to avoid carbon monoxide poisoning.

When male frogs are exposed to the herbicide atrazine, the most commonly used herbicide in the United States, they develop female reproductive structures. Is this an early warning sign that humans should be concerned about?

# 22.2 **Human Reproduction**

To understand how human reproduction might be affected by environmental contaminants, we will first consider the anatomical structures that comprise the human male and female reproductive systems.

## Reproductive Systems

The reproductive systems of males and females consist of external and internal structures. These structures allow the production and maturation of gametes, signal the synthesis and secretion of substances required for reproductive function, and provide a route to deliver the gametes through ducts.

**Male Reproductive Anatomy.** The external and internal components of the male reproductive system are shown in **FIGURE 22.2**.

The **penis** delivers sperm to the female reproductive tract during sexual intercourse. It is composed of spongy erectile tissue. During sexual arousal, this tissue fills with blood. Pressure from the increased volume of blood in the penis seals off the veins that drain it of blood. This causes the penis to become engorged, keeping it erect. An erection is essential for insertion of the penis into the vagina, which facilitates delivery of sperm to the egg. A tube inside the penis, the **urethra**, provides passage out of the body for both sperm and urine. The head, or **glans penis**, consists of highly sensitive skin that is covered by a fold of skin called the foreskin. In circumcised males, the foreskin has been removed.

The question of whether to circumcise newborn males is a difficult one. Some parents forgo the procedure, arguing that it is unnecessary, while others choose it for religious or cultural reasons. Emerging science suggests that circumcision lowers men's chances of developing and transmitting some sexually transmitted infections, but whether the decreased risk warrants routine circumcision is still a matter of debate.

The **scrotum** is a pouch below the penis that contains the testes, or **testicles**. The skin of the scrotum is thin and devoid of fatty tissue. It is folded or wrinkled with little hair. Below the skin is a layer of involuntary smooth muscle that regulates the position of the testes relative to the body to keep them at a temperature that maximizes sperm production. The scrotum contracts in the cold to keep the testes closer to the body.

In addition, testes also produce hormones called **androgens**. Inside each testis are many highly coiled tubes, called **seminiferous tubules**, where sperm form. The seminiferous tubules are held in place by connective tissues that are rich in hormone-producing cells called **Leydig cells**. The process of sperm formation takes about 60 days and begins in the seminiferous tubules. From there, sperm pass through the coiled **epididymis**, an approximately 6-meter (20-foot) tube that rests atop each testis. Sperm become motile during the approximately 20 days that they pass through the epididymis.

**FIGURE 22.2 Male reproductive anatomy.** (a) Side view and (b) front view of the male reproductive system.

**(a) Side view**

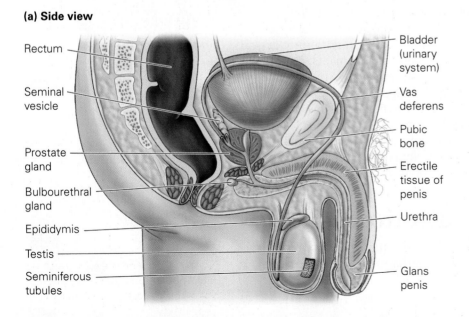

Rectum

Seminal vesicle

Prostate gland

Bulbourethral gland

Epididymis

Testis

Seminiferous tubules

Bladder (urinary system)

Vas deferens

Pubic bone

Erectile tissue of penis

Urethra

Glans penis

**(b) Frontal view**

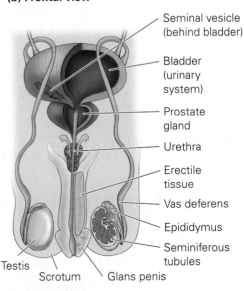

Seminal vesicle (behind bladder)

Bladder (urinary system)

Prostate gland

Urethra

Erectile tissue

Vas deferens

Epididymus

Seminiferous tubules

Testis

Scrotum

Glans penis

During ejaculation, sperm are propelled from the epididymis through ducts called the **vas deferens.** Each duct is surrounded by smooth muscle and is capable of wave-like peristaltic contractions that propel sperm through the duct.

Several glands add secretions to the developing sperm as they make their way through the reproductive system. **Seminal vesicles** are glands that secrete mucus and sugars for the sperm to use as energy. The **prostate** gland secretes a thin, milky white, nutrient-rich fluid into the urethra. The **bulbourethral glands** lie below the urethra between the prostate and the penis. Before ejaculation takes place, these glands secrete clear mucus that helps neutralize any acidic urine present in the urethra. Sperm and these secretions make up **semen,** which can survive for up to five days in the female reproductive tract.

### Female Reproductive Anatomy.   **FIGURE 22.3** shows female reproductive anatomy. The most obvious feature of the external genitalia is the **vulva.** The vulva consists of two sets of lips, or **labia:** the outer **labia majora,** which are fatty and have a hairy external surface, and the inner **labia minora,** which contain neither fat nor hair. At the front of the vulva, the labia minora divide around the **clitoris,** an important organ for female sexual arousal.

Between the folds of the labia, there is an opening for the urethra, which serves as a passageway for urine from the bladder. In women, it is short (4 centimeters on average) compared to the urethra in men (18 centimeters on average). Partially due to this difference, women are more susceptible to bladder infections because bacteria have a shorter distance to travel from outside the body. A few centimeters below the urinary opening is the vaginal opening.

The organs that make up the internal genitalia are the ovaries, vagina, uterus, and oviducts. The ovaries produce the gametes and sex hormones that circulate in a woman's body during her reproductive years. Ovaries are about the size of a walnut, but at birth they contain about 2 million ova (eggs) each. During the menstrual cycle, one or more eggs mature and are released from a structure called a **follicle.** The follicle, a fluid-filled sac containing the developing egg, secretes the hormone estrogen. During ovulation the follicle ruptures, much like a blister popping, to release the egg. The remnant of the follicle, called the **corpus luteum,** secretes both estrogen and progesterone.

The **vagina** is a muscular organ that acts as a passageway into and out of the uterus. The **uterus** is about the size of a fist. The wall is thick (about 1 centimeter) and is composed of some of the most powerful muscles in the human body. These muscular walls contract rhythmically during labor, childbirth, and orgasm. The internal surface of the uterine wall is called the **endometrium,** which changes in thickness during the course of the menstrual cycle. The lower third of the uterus is narrower than the upper portion and is called the **cervix.**

The **oviducts** are actually an extension of the top surface of the uterus. These tubes extend from the body of the uterus toward the

## Visualize This ▼

Women are more susceptible to pelvic infections than men. Trace the path bacteria could follow from the exterior to the opening to the pelvic cavity via the oviduct.

**(a) Side view**

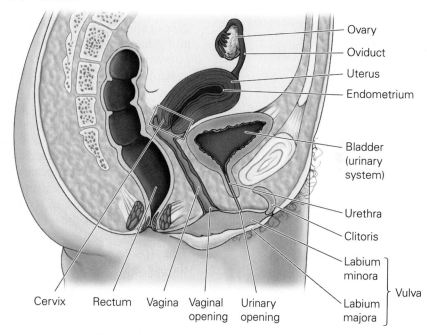

**(b) Frontal view**

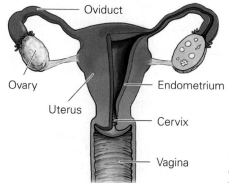

**FIGURE 22.3 Female reproductive anatomy.**   (a) Side view and (b) front view of the female reproductive system.

ovaries, which are suspended within the abdominal cavity. The oviducts are not attached to the ovaries directly. Instead, they end in brushy structures that move over the surface of the ovary. These movements, along with suction inside the oviducts, direct the eggs released by the ovary into the oviducts.

A well-studied example of an endocrine disruptor that affects human fertility is the chemical called diethylstilbestrol, or DES. Between 1938 and 1971, many pregnant women were given this synthetic drug to help prevent miscarriage. However, the physicians who prescribed this drug did not know that it behaved as a mimic of the female hormone estrogen and altered the development of the uterus in the developing baby, resulting in a misshapen uterus less able to sustain a pregnancy. Approximately 24% of the daughters of women who took this drug are infertile.

In males living in developed countries, sperm counts are declining. Could endocrine disruptors affect the number and quality of gametes humans produce?

## Gametogenesis

The production of sex cells, or gametes, is called **gametogenesis.** In human males and females, gametogenesis involves the process of meiosis, a type of cell division that reduces the chromosome number by one-half. Because human body cells contain 46 chromosomes, meiosis in humans produces gametes that contain 23 chromosomes. Human body cells with 46 chromosomes (23 pairs) are diploid (2n), and the gametes produced after meiosis are haploid (n).

Meiosis alone is not enough to produce a functional gamete. Other changes to sex cells occur as they go through gametogenesis to enable them to mature into gametes capable of being involved in fertilization. For instance, sperm cells gain motility, and egg cells increase in size and nutrient content to allow for the development of the early embryo.

**Spermatogenesis.** The production of sperm, called **spermatogenesis** (**FIGURE 22.4a**), begins to occur in the testes at puberty. Cells lining the walls of the seminiferous first duplicate by mitosis, then one of the two daughter cells produced undergoes meiosis. Cells that aid the developing sperm, called **Sertoli cells,** are also located in the seminiferous tubules. These cells secrete the substances required for further sperm development. A mature sperm is composed of a small head containing DNA, a midpiece that has mitochondria to provide energy for the journey to the oviduct, and a tail (or flagellum) to propel the sperm (**FIGURE 22.4b**). At the tip of the sperm's head lies the **acrosome.** This structure contains digestive enzymes that help a sperm cell gain access to the egg cell.

**Oogenesis.** The formation and development of female gametes, called **oogenesis,** occurs in the ovaries and results in the production of egg cells (**FIGURE 22.5a**). A small percentage of these egg cells will be released over a woman's reproductive life span, and an even smaller percentage will be fertilized.

**oo-** is from the Greek word for egg.

**-genesis** means generation of or birth of.

While spermatogenesis begins at puberty, oogenesis begins while a female is still in her mother's uterus and then pauses until puberty. A female produces around 2 million potential egg cells prior to her birth, but some of these cells degenerate, from about 700,000 at birth to around 350,000 at puberty. At puberty, development of preexisting eggs continues each month until the cessation of menstruation, or **menopause.**

The ovary contains structures called **follicle cells,** which are composed of estrogen-secreting cells surrounding the immature egg cell or **oocyte.** Follicle cells nourish and protect the developing oocyte until the follicle bursts open, releasing the egg into the oviduct, a process called **ovulation** (**FIGURE 22.5b**).

**(a) Spermatogenesis**

**Testicle**

**Cross section of seminiferous tubule**

Vas deferens

Epididymis

Seminiferous tubules

Spermatogonia

Sertoli cells within the seminiferous tubules secrete substances that nourish sperm.

Leydig cells scattered between seminiferous tubules produce testosterone and other androgens.

**FIGURE 22.4 Spermatogenesis.** (a) Sperm are produced in the seminiferous tubules of each testicle. (b) The head of the sperm contains the DNA. The midpiece contains mitochondria to provide energy for the flagellum or tail.

**(b) Sperm**

Head

Midpiece

Tail

**(a) Follicle cycle**

Uterus    Oviduct    Ovary

**1** A follicle contains an oocyte and begins producing the sex hormone estrogen.

**Ovary**

Corpus luteum

**4** Unless a pregnancy has occurred, the corpus luteum degenerates.

**2** The follicle continues to support the development of the oocyte until it is ovulated.

**3** After the egg cell is ovulated, what is left of the follicle, called the corpus luteum, secretes progesterone.

**(b) Ovulation**

Egg

Ovary

## Visualize This ▲

In this figure, it appears that the development of the follicle takes place as it moves clockwise inside the ovary. Do you think this is how development actually occurs or is it depicted this way to show the changes that occur over time?

**FIGURE 22.5 Oogenesis.** (a) Over the course of one menstrual cycle, one follicle grows and develops in preparation for ovulation. (b) During ovulation, the egg cell bursts from the ovary.

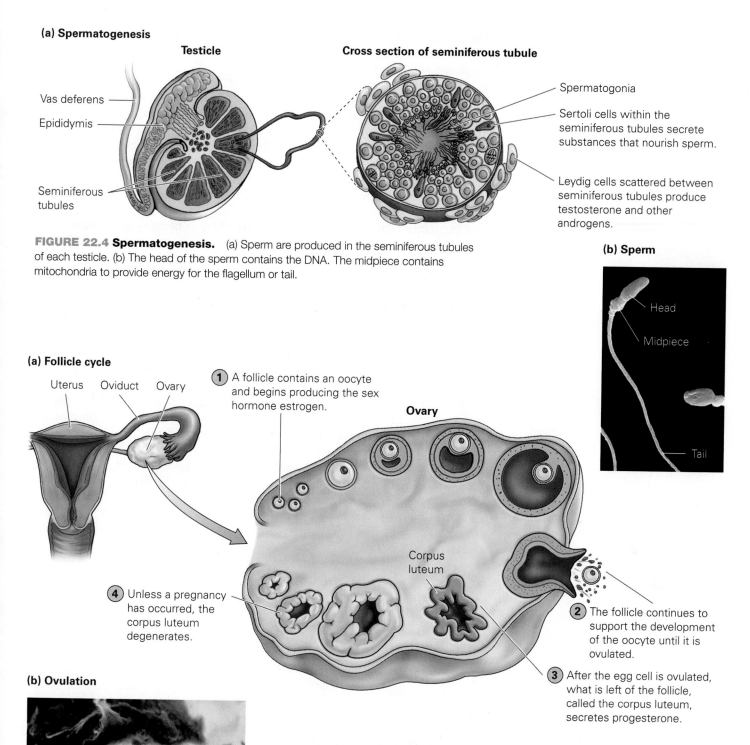

The remnant of follicle (minus the oocyte that was ovulated), the **corpus luteum**, secretes reproductive hormones but degenerates after about 10 days if fertilization does not occur.

After ovulation, the egg cell moves into the oviduct where, if there are sperm present, fertilization can occur. If not fertilized, it takes about 3 days for the secondary oocyte—an object about the size of a pinhead—to move from the ovary through the oviduct, uterus, and cervix to be shed with menstrual fluid.

Women are fertile, or able to produce an unfertilized egg cell, a few days per month from puberty until menopause. Men are fertile throughout their lifetimes, with a slow but progressive decrease in sperm count as they age. Many different chemicals, including endocrine disruptors, have been shown to hasten this decline in sperm counts.

One study that compared reproductive health of men from around the United States showed that sperm counts of men from rural Missouri were lower than they were for men from many large cities across the country. Scientists who performed this study found higher levels of three commonly used agricultural pesticides in the bodies of men with lower sperm counts than were present in the bodies of men with higher sperm counts.

Even if a causal link can be found between pesticide exposure and lower sperm counts, it would also need to be established that the lower sperm counts are actually having an effect on fertility. It might be the case that these pesticide exposures don't lower sperm counts enough to have any effect on how many children these men have.

Two very common sexually transmitted infections, chlamydia and gonorrhea (**FIGURE 22.6**), are associated with lower fertility. It is not uncommon for sexually active people, especially women, to be unaware that they have been infected with the bacteria that cause either of these diseases, due to the lack of any noticeable symptoms. This may lead to delays in treatment that allow bacteria time to ascend through the cervix and upper uterus to the oviducts, setting up a tissue-damaging infection that leads to permanent scarring and blockages. This condition, called pelvic inflammatory disease, results in permanent infertility in over 10% of cases. For information on other sexually transmitted infections, see **TABLE 22.1**.

## Working with Data ▶

**Which age group is most at risk for each of these diseases?**

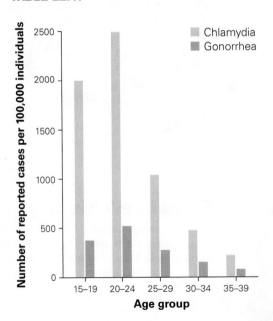

**FIGURE 22.6 Chlamydia and gonorrhea infection rates in 2012.** Data from the Centers for Disease Control and Prevention showing the number of reported cases of chlamydia and gonorrhea infections per 100,000 people in the United States.

**TABLE 22.1 Sexually transmitted infections.**
The causes, symptoms, and treatment or prevention of common STIs.

| Bacterial Pathogens | General Information | Symptoms | Treatment and Prevention |
|---|---|---|---|
| **Chlamydia is caused by _Chlamydia trachomatis_.** | Infects the urethra, cervix, and oviducts of women and the urethra of men. | Pelvic pain, fluid discharge | • Treat with antibiotics<br>• Reduced transmission with condom use |
| **Gonorrhea is caused by _Neisseria gonorrhoeae_.** | Known as "the clap"; can be spread from infected mothers to babies during birth. | Thick discharge from penis or vagina | • Treat with antibiotics<br>• Reduced transmission with condom use |
| **Syphilis is caused by _Treponema pallidum_.** | Infectious sores, or chancres, can be found on the vagina, penis, anus, rectum, lips, and mouth. | If untreated, neurological problems, paralysis, and death can occur. | Can be treated with antibiotics if caught early. |

| Viral Pathogens | General Information | Symptoms | Treatment and Prevention |
|---|---|---|---|
| **AIDS is caused by human immunodeficiency virus (HIV).** | Transmitted via oral, anal, and vaginal sex, as well as via blood transfusion. | Over time weakens the immune system; eventually severe tissue and organ damage | • Combination-drug therapies to halt progression; no cure<br>• Reduced transmission with condom use |
| **Genital warts are caused by human papilloma virus (HPV).** | Can be transmitted by oral, anal, and vaginal sex. | • Growths or bumps on the pubic area, penis, vulva, or vagina<br>• Genital warts (from some types of HPV); cervical cancer (from certain types of HPV) | • Vaccination to prevent infection by many strains<br>• Reduced transmission with condom use |

_continued_

**TABLE 22.1** **Sexually transmitted infections.** *continued*

| Viral Pathogens | General Information | Symptoms | Treatment and Prevention |
|---|---|---|---|
| **Hepatitis B is caused by hepatitis B virus (HBV).** | • Transmitted via blood and other body fluids.<br>• An infant can acquire infection from the mother during delivery. | Inflammation and scarring of the liver (cirrhosis), which may be fatal | • Vaccination for prevention<br>• Reduced transmission with condom use |
| **Herpes is caused by two types of herpes simplex viruses.** | Transmitted via direct skin-to-skin contact or kissing, or by oral, anal, or vaginal sex. | • Cold sores or fever blisters on mouth or face<br>• Genital herpes is sores in the genital area | • Antiviral medications to lessen symptoms; no cure<br>• Reduced transmission with condom use |

| Insect, Protozoan, and Fungal Pathogens | General Information | Symptoms | Treatment and Prevention |
|---|---|---|---|
| **Pubic lice are caused by *Pediculus pubis* (insect).** | Also known as "crabs"; transmitted via skin-to-skin contact or contact with an infected towel or clothing. | Itching of pubic area around 5 days after initial infection | Cured by washing the affected area with a delousing agent |
| **Trichomoniasis is caused by *Trichomonas vaginalis* (protozoan).** | Also known as "trich"; transmitted by oral, anal, and vaginal sex. | • In women, vaginal itching with a frothy yellow-green vaginal discharge<br>• In some men, irritation in urethra after urination or ejaculation | • Treated with antibiotics<br>• Reduced transmission with condom use |
| **Yeast infections are caused by *Candida albicans* (fungi).** | • Normal inhabitants of the female reproductive tract.<br>• Antibiotic use may deplete normal vaginal flora allowing yeast to proliferate. | In women, thick whitish discharge from vagina and vaginal itching | Cured by antifungal medicines |

Chlamydia and gonorrhea, while not related to exposures to endocrine disruptors, are the most preventable causes of infertility. A cause of infertility that may be affected by exposures to endocrine disruptors involves the menstrual cycle.

## The Menstrual Cycle

The term **menstrual cycle** refers specifically to cyclic changes that occur in the uterus. During the course of a single menstrual cycle, a woman's body prepares an egg for potential fertilization and her uterus for a potential pregnancy. If no pregnancy has occurred, the uterine lining is excreted during **menstruation**, and a new cycle begins.

**FIGURE 22.7** illustrates the changes in hormone levels, ovaries, and condition of the endometrium that occur throughout the 28-day cycle. Most women's

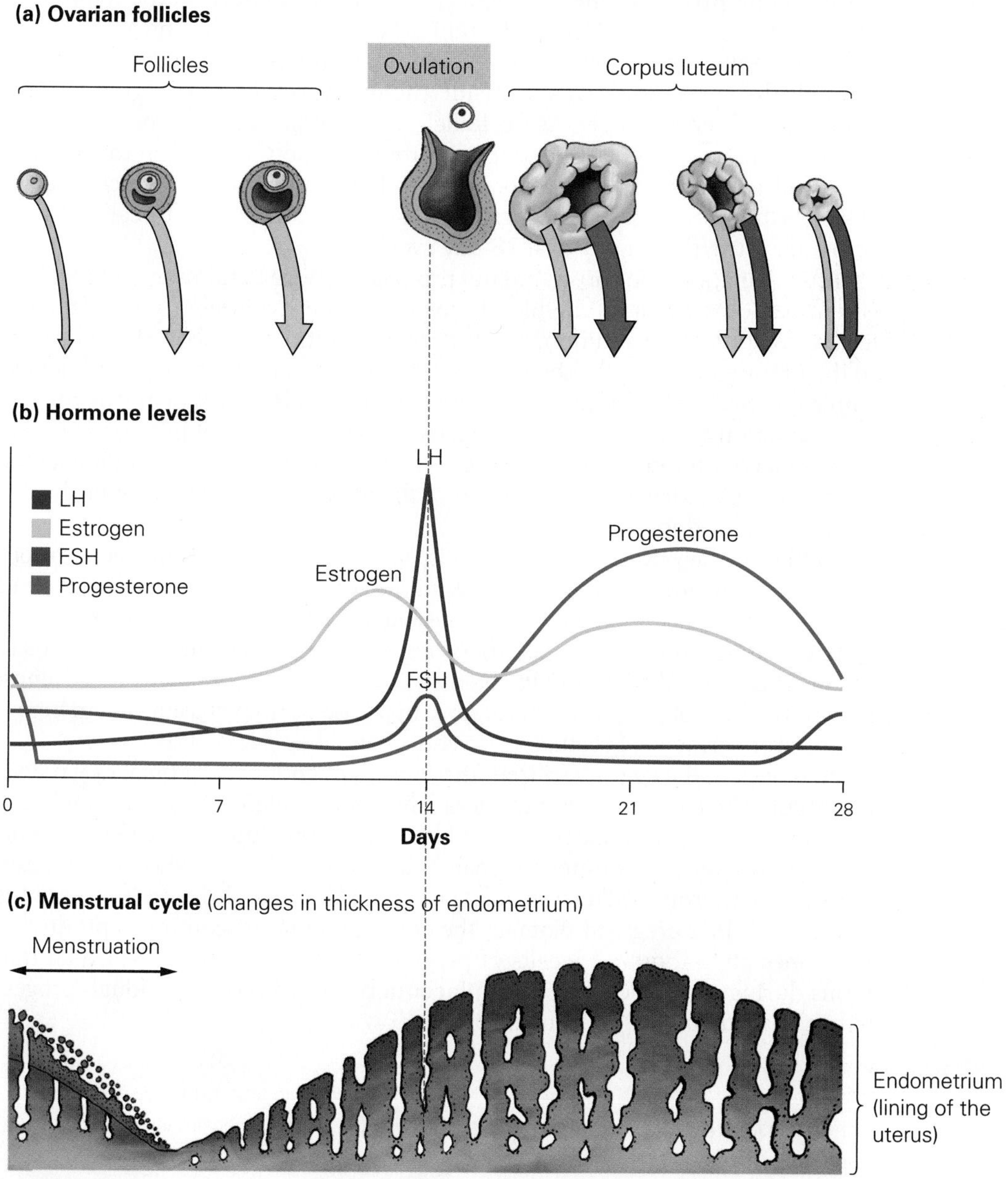

**(a) Ovarian follicles**

**(b) Hormone levels**

**(c) Menstrual cycle** (changes in thickness of endometrium)

**FIGURE 22.7 The menstrual cycle.** Changes in hormone levels are linked to (a) the state of ovarian follicles, (b) hormone levels, and (c) uterine condition over the course of the menstrual cycle.

bodies do not adhere precisely to a 28-day cycle—some women have longer cycles, and others have shorter cycles. The first day of a menstrual cycle is considered to be the first day of actual bleeding.

You learned in Chapter 21 that the hypothalamus, located in the brain, secretes gonadotropin-releasing hormone (GnRH), which stimulates another endocrine structure in the brain—the pituitary gland—to release follicle-stimulating hormone (FSH) and luteinizing hormone (LH). In males, FSH and LH are involved in the production of sperm; in females, these hormones help regulate the development of egg cells and ovulation.

Like many other biological processes, the menstrual cycle is self-regulating, operating by responding to feedback. High levels of estrogen provide positive feedback to the hypothalamus. The hypothalamus secretes GnRH, which acts on the pituitary gland to increase the secretion of FSH and LH. In this manner, FSH and LH progressively increase in concentration. Conversely, high levels of the hormone **progesterone** have a negative-feedback effect on the hypothalamus; GnRH secretion is decreased, and FSH and LH levels decline.

When a woman menstruates, the endometrial tissue is shed and released through the cervix and vagina. Inside the ovary, the follicle begins to grow under the influence of FSH. As the follicle grows, it produces estrogen.

When the follicle is large enough, it produces enough estrogen to stimulate GnRH release. This leads to a spike in both FSH and LH levels, which lasts for about 24 hours. Ovulation occurs 10 to 12 hours after the LH peak, 14 days before menstruation in a typical 28-day cycle.

After ovulation, the corpus luteum produces estrogen and progesterone. Progesterone helps maintain the blood flow to the uterine lining to support early fetal development. In addition, progesterone inhibits continued LH production, so the LH surge is inhibited by the ovarian hormones that it stimulates—an example of negative feedback. Some hormonal methods of birth control circumvent ovulation by preventing the LH surge. The birth control pill, for example, supplies a continuous dose of estrogen that is not high enough to stimulate the secretion of LH. Mechanisms of action of the most common forms of birth control are discussed in **TABLE 22.2**.

In a hormonally cycling woman (that is, women not using a hormonal birth control method), if fertilization does not occur, the corpus luteum degenerates around 10 days after ovulation. Because the corpus luteum is no longer secreting them, progesterone and estrogen levels fall, triggering the arteries that supply the uterus to spasm. This allows the lining of the uterus to be shed and causes menstrual cramps. Decreasing levels of progesterone also serve to release the hypothalamus from inhibitory control. Therefore, LH and FSH levels rise, and the cycle starts over.

If, however, the egg is fertilized, the process of endometrial breakdown does not occur. The early embryo produces a hormone called human chorionic gonadotropin (hCG) that extends the life of the corpus luteum. hCG is the hormone most over-the-counter pregnancy kits test for. With the corpus luteum intact, progesterone and estrogen levels remain high, and the endometrium is maintained. In a pregnant woman, the corpus luteum discontinues producing hormones after about 12 weeks of pregnancy, when an endocrine organ that forms during pregnancy called the **placenta** begins to produce enough progesterone to maintain the uterine lining on its own.

**STOP & STRETCH** Women who live together often report that their menstrual cycles are synchronized. In other words, they start menstruating at around the same time each month. Although scientists have tried to determine whether this is actually happening and if so why, results have been inconclusive. Assume that four fertile women move in together. Design an experiment, with appropriate controls, that would measure whether their menstrual cycles were synchronizing over time.

**TABLE 22.2** **Birth control methods.** The mechanism, risk, and efficacy of various birth control methods. Unintended pregnancy rates assume the method is used correctly and consistently as directed. The percentages indicate the number out of every 100 women who had an unintended pregnancy every year.

| | Method | | Mode of Action | Risk and Efficacy |
|---|---|---|---|---|
| **Abstinence** | Abstinence | | Sperm and egg never have contact. | • No risks<br>• Unintended pregnancy rate of 0% |
| **Hormonal methods** | Combination birth control pill | | Synthetic estrogen and progesterone given at continuous doses prevent ovulation. | • Increased risk of heart disease and fatal blood clots for women over age 35 who smoke; slightly increased risk of breast cancer; does not protect against STIs<br>• Unintended pregnancy rate of 6% |
| | Minipill | | Progesterone-only pill thickens cervical mucus, impedes sperm ascent, and does not allow the uterine to support a pregnancy. | • Increased risk of ovarian cysts<br>• Does not protect against STIs<br>• Unintended pregnancy rate of 6% |
| | Emergency birth control pills | | High doses of hormones prevent sperm from reaching egg or prevent fertilized egg from attaching to wall of uterus, or prevent ovulation. "Morning after pills" are not the same as the mifepristone abortion pill. These pills prevent pregnancy, while mifepristone terminates an established pregnancy. | • Does not protect against STIs<br>• Nausea, vomiting, abdominal pain, fatigue, headache<br>• Unintended pregnancy rate of 20% |
| | Patch | | When applied to skin, the patch delivers progesterone and estrogen to prevent ovulation and thicken cervical mucus. | • Same as combination pill<br>• Does not protect against STIs<br>• Unintended pregnancy rate of 9% |
| | Vaginal ring | | When inserted in vagina, the vaginal ring delivers progesterone and estrogen to prevent ovulation and thicken cervical mucus. | • Same as combination pill<br>• Does not protect against STIs<br>• Unintended pregnancy rate of 9% |
| | Injectables | | This method requires the user to have progesterone injections every 3 months, and the mode of action is the same as the minipill. | • Irregular vaginal bleeding<br>• Does not protect against STIs<br>• Unintended pregnancy rate of 6% |
| | Implantables | | Implantables, such as Implanon, secrete progesterone to prevent ovulation and to thicken cervical mucus. | • Irregular vaginal bleeding<br>• Does not protect against STIs<br>• Unintended pregnancy rate of 1% |

*continued*

**TABLE 22.2 Birth control methods.** *continued*

| | Method | | Mode of Action | Risk and Efficacy |
|---|---|---|---|---|
| **Barrier methods** | Cervical cap | | When inserted against cervix before intercourse, the cervical cap prevents sperm and egg contact. Usually used in concert with spermicidal agent | • No known risks<br>• Does not protect against STIs<br>• Unintended pregnancy rate of 15% |
| | Diaphragm | | When inserted into vagina before intercourse, the diaphragm prevents sperm and egg contact. | • No known risks<br>• Does not protect against STIs<br>• Unintended pregnancy rate of 12% |
| | Sponge | | The spermicide-soaked sponge fits against the cervix to prevent sperm and egg contact. | Unintended pregnancy rate of 12% in women who have not had children and 24% in women who have had children |
| | Female condom | | The female condom is held against the cervix by a flexible ring to prevent sperm and egg contact. | • No known risks<br>• Unintended pregnancy rate of 21% |
| | Male condom | | The male condom fits over the penis to prevent sperm and egg contact. | • No known risks<br>• Unintended pregnancy rate of 18% |
| **Other methods** | Spermicides | | When inserted into vagina 1 hour before intercourse, sperm are killed. | • No known risks<br>• Does not protect against STIs<br>• Unintended pregnancy rate of 28% |
| | Fertility awareness | | Fertility awareness requires abstinence for the 4 days before ovulation, the day of ovulation, and several days after ovulation. | • Does not protect against STIs<br>• Unintended pregnancy rate of 24% |
| | Intrauterine device | | This small plastic device inserted into the uterus by a physician prevents fertilization and prevents the uterus from supporting a pregnancy for many years, until its removal. | • May increase risk of pelvic inflammatory disease<br>• If pregnancy does occur, increased risk of miscarriage or ectopic pregnancy<br>• Unintended pregnancy rate of 1% per year |
| | Female sterilization | | Female sterilization permanently blocks oviducts with a tightly coiled tube. | • Does not protect against STIs<br>• Unintended pregnancy rate of 0.5% per year |
| | Male sterilization (vasectomy) | | Each vas deferens is cut or blocked to prevent sperm from being ejaculated in the semen. | • Does not protect against STIs<br>• Unintended pregnancy rate of 0.15% per year |

Female fertility can be lowered by a condition called *endometriosis,* which occurs when menstrual tissues that normally line the uterus begin to grow in other areas, such as the oviducts and ovaries. These tissues continue to respond to the hormonal cycle, growing and shedding in sync with the lining of the uterus. In addition to being very painful, endometriosis can lead to scarring and inflammation of the oviducts and disruption of ovulation. *Polycystic ovarian syndrome,* another fertility-lowering condition, occurs when fluid-filled cysts develop in the ovaries and ovulation is disrupted. Preliminary evidence, mostly in nonhumans, seems to show a correlation between exposure to chemicals that are added to plastics to keep them soft and pliable, called plasticizers, and these conditions. While there is no scientific consensus on the role of plasticizers in these conditions, some people worry that drinking water from plastic bottles or heating food in plastic containers might increase an individual's exposure to plasticizers.

Reproductive problems can arise because of exposures to toxins that a person experiences during his or her lifetime. Such problems also can result from exposures experienced in the womb during development, when reproductive and other structures are forming.

# 22.3  Human Development

**Development** is the series of events that take place after fertilization and leads to the formation of a new multicellular organism.

## Fertilization

Of the roughly 300 million sperm contained in each ejaculate, only about 200 reach the site of fertilization in the oviduct. Initially, sperm must survive the acidic (low pH) conditions of the upper vagina. The natural acidity of the upper vagina works to help prevent bacterial infections in women but also can kill sperm. Some sperm get stuck in folds of the cervix and in the uterine wall, and only half of the sperm that make it to the uterus enter the oviduct that contains the ovulated egg.

The oviducts themselves undergo muscular contractions during the time of ovulation, and these contractions allow the sperm to move toward the upper portion of the oviduct to wait for an egg cell to be ovulated. It takes about half an hour to move a sperm from the upper vagina to the oviduct.

### How Fertilization Takes Place.
When both sperm and egg are present in the oviduct, fertilization can occur (**FIGURE 22.8**). First, the sperm cells must move through the follicle cells, and then they must pass through a translucent covering on the egg cell called the *zona pellucida.* This

## Visualize This ▼

At what stage would the process be halted if the sperm and egg were from different species?

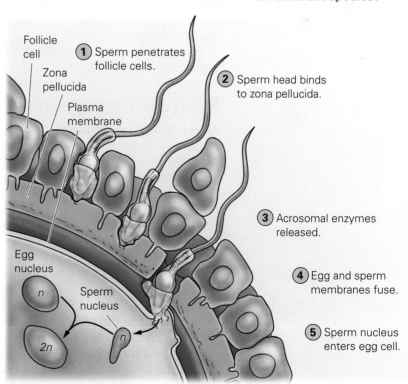

**FIGURE 22.8 Fertilization.** The sperm cell must traverse follicle cells and the zona pellucida before fusing with the plasma membrane and depositing its chromosomes.

protective layer acts as a species-specific barrier. Only sperm produced by a male of the same species as the female that produced the egg can pass through the zona pellucida. This is because getting through the zona pellucida requires binding of a specific receptor on the sperm head. This binding of the sperm head to the zona pellucida triggers the release of enzymes present in the acrosome of a sperm. The acrosomal enzymes interact with the egg cell's zona pellucida to allow the formation of a tunnel through the zona pellucida toward the egg cell's plasma membrane. Once this sperm cell makes it through the zona pellucida, its plasma membrane and the egg's plasma membrane fuse, and the egg cell draws the sperm cell's nucleus inward. The fertilized egg is now ready to begin the divisions of early development.

Chemicals in the environment can affect the ability of sperm to bind to the egg during fertilization. Trichloroethylene (TCE) is an industrial solvent that is used to remove grease from metal and is found in many household products. Large amounts of this chemical can get into the water supply via air emissions from metal-degreasing plants and via wastewater from factories that produce metals, paints, cleaning products, electronics, and rubber products. Studies have shown that males exposed to TCE while working as mechanics or dry cleaners have a larger number of abnormally shaped sperm than men not exposed to TCE. Abnormally shaped sperm are less likely to be able to fertilize egg cells.

**STOP** & **STRETCH**   Propose an alternate hypothesis, one that does not involve endocrine-disrupting, that might explain the correlation between lowered sperm counts and pesticide exposure in rural males.

While correlational studies do not provide definitive evidence that endocrine-disrupting chemicals in the water supply are compromising human health, there still may be cause for concern and caution. In the next section, we describe something scientists are surer of—that minute chemical exposures during critical windows of development can have tremendous impacts on a developing baby.

## Embryonic Development

After fertilization, the fertilized egg cell, or zygote, undergoes a series of changes to produce a multicellular structure capable of producing adult tissues (**FIGURE 22.9**). In humans, the term **embryo** is used to describe the stage of development from the first divisions of the zygote until body structures begin to appear at about the ninth week. After that time, the developing human is referred to as a **fetus.** Early development of the embryo involves a series of rapid cell divisions, called **cleavage,** that begin while the zygote is still in the oviduct. These divisions continue until the now-multicellular embryo is propelled down the oviduct toward the uterus.

When the embryo reaches the uterus, a few days after fertilization, it is in the form of a hollow ball of cells known as the **blastocyst.** Inside the blastocyst, a cluster of cells called the **inner cell mass** forms. Some cells of the inner cell mass give rise to the embryo. The outer cells of the blastocyst form part of the placenta. The inner cell mass undergoes a dramatic reshuffling of cells to produce a structure with three layers of cells, called the **gastrula.** The cells comprising the three layers of the gastrula will begin to specialize, or **differentiate,** into the adult cells and tissues with their specific functions.

**Cleavage**                                                                           **Gastrulation**

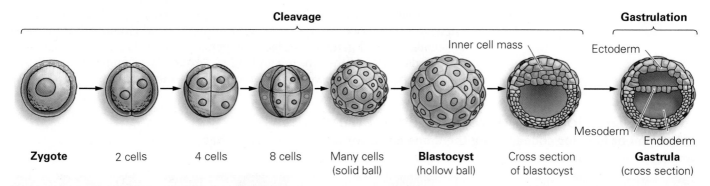

**Zygote** | 2 cells | 4 cells | 8 cells | Many cells (solid ball) | **Blastocyst** (hollow ball) | Cross section of blastocyst | **Gastrula** (cross section)

FIGURE 22.9 **Development of the early embryo.** The human zygote undergoes cleavage divisions to produce the blastocyst, which implants in the uterus. Continued cell divisions along with specialization result in the formation of the three-layered gastrula.

These three layers of tissue include an outer, middle, and inner layer. The outer layer, or **ectoderm**, gives rise to the skin, nervous system, and sense organs. The middle layer, or **mesoderm**, gives rise to muscles, excretory organs, circulatory organs, gonads, and the skeleton. The innermost layer, or **endoderm**, lines the digestive and respiratory organs.

This differentiation occurs in response to chemical signals that pass between the different cell layers and adjacent cells, turning on some genes in a given cell and leaving others off. In this manner, the specialized cells that comprise the adult body all arise from the single fertilized egg cell that undergoes mitotic cell division and cell differentiation.

Exposures to toxins of any kind very early in development tend to have an all-or-nothing impact. Exposures can be so damaging that they kill many cells, and the early embryo is spontaneously aborted—often without a woman ever realizing that she was pregnant. Conversely, exposures can have very little effect since cells are dividing so rapidly that if a few cells die due to exposure, they are quickly replaced. Exposures of greatest concern occur while organs are developing. The development of reproductive organs in particular can be derailed by exposures to endocrine-disrupting chemicals.

**Development of Human Reproductive Organs.** Prior to week 7 of a pregnancy, the reproductive organs of male and female embryos are indistinguishable except at the level of their chromosomes.

In the early embryo, the structures that will become the gonads are located in the abdomen. The same starting material can be used to produce either the paired ovaries of a female or the paired testes of a male. Differentiation of the embryonic gonads into ovaries or testes is determined by the expression of sex-specific genes. Once differentiated, the male gonads produce androgens; but the female gonads do not start to produce estrogen until puberty.

The cells of the embryonic testes make the androgen testosterone, which directs the development of internal, gamete-carrying, ductal structures. Prior to differentiation, two different ductal structures exist side by side in both male and female embryos. In each sex, one embryonic duct system persists and the other regresses, based in part on the presence or absence of androgens. In males, the ductal structure that persists develops into the sperm-carrying vas deferens, epididymis, and urethra. In female embryos, the ductal structure that persists develops into the oviducts, uterus, cervix, and vagina.

**Entire schools of fish that have both male and female reproductive organs are no longer anomalies.**

Androgens also affect the development of external genitalia. In male and female embryos, one embryonic structure can be molded into either the male or female external genitalia. In fact, the penis and clitoris arise from the same starting tissue, as do the scrotum and vulva. **TABLE 22.3** summarizes the developmental pathways followed by embryonic males and females.

**TABLE 22.3 Development of reproductive structures.**
Development is color coded, using the same color for adult structures that develop from a common embryonic structure.

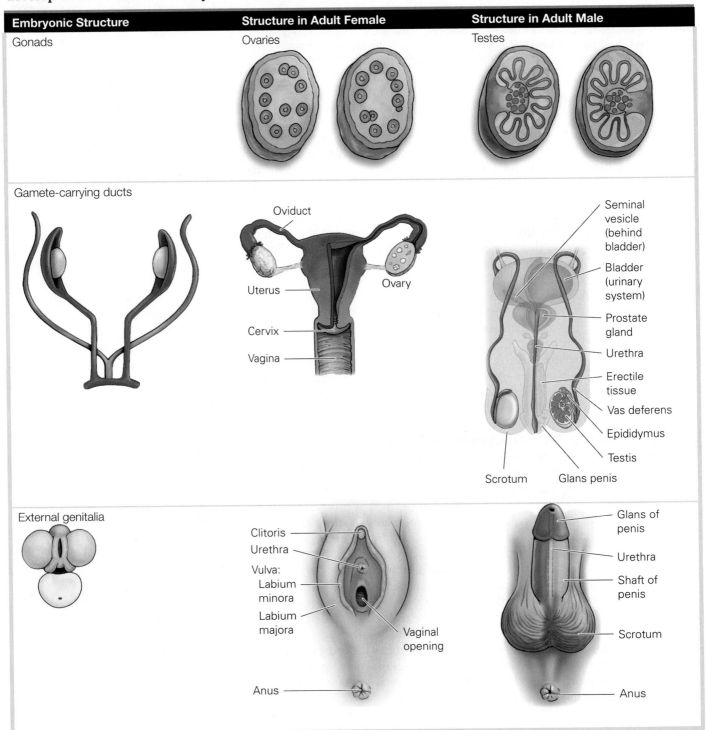

| Embryonic Structure | Structure in Adult Female | Structure in Adult Male |
|---|---|---|
| Gonads | Ovaries | Testes |
| Gamete-carrying ducts | Oviduct, Uterus, Ovary, Cervix, Vagina | Seminal vesicle (behind bladder), Bladder (urinary system), Prostate gland, Urethra, Erectile tissue, Vas deferens, Epididymus, Testis, Scrotum, Glans penis |
| External genitalia | Clitoris, Urethra, Vulva: Labium minora, Labium majora, Vaginal opening, Anus | Glans of penis, Urethra, Shaft of penis, Scrotum, Anus |

Earlier in the chapter, we related the story of DES—the synthetic estrogen taken by women in the 1960s and 1970s that caused some of their daughters to have deformed uteruses. The development of male reproductive structures also appears to be vulnerable to the effects of environmental contamination.

Several nonhuman animal studies have shown that exposure to commonly used fungicides and herbicides results in a higher-than-average rate of undescended testes, a condition called **cryptorchidism.** This anomaly decreases reproductive success because the sperm of males with this condition are held inside the abdomen, where the temperature is higher than optimum for the development of sperm. It is thought that these fungicides and herbicides function as endocrine disruptors by blocking androgen receptors.

Cryptorchidism occurs in 2 to 4% of human male newborns and is usually surgically corrected. Endocrine disruptors have been proposed to be one of several causes of cryptorchidism, although no particular chemical exposure is known to cause the condition.

Deformities in reproductive or other structures in humans occur when chemicals present in the mother's bloodstream gain access to the developing offspring during pregnancy. This is why pregnant women must be very careful to limit their exposure to toxins.

## Pregnancy

Pregnancy or gestation involves carrying a developing baby within the female reproductive tract. Gestation encompasses the period of time between fertilization and childbirth; in humans, this lasts about 38 weeks.

To sustain a pregnancy, the placenta must form inside the uterus. The placenta is produced by the cooperative actions of the embryo and the mother's uterus. By the time the embryo has made its way to the uterus, it is composed of the inner cell mass, which will become the fetus, and an outer ring of cells called the **trophoblast,** which becomes part of the placenta. Cells of the trophoblast secrete enzymes that enable the embryo to implant in the wall of the uterus.

Eventually, the implanted trophoblast begins to infiltrate the uterine lining, forming fingerlike projections that are able to carry blood. Maternal uterine blood vessels erode to form pools of blood surrounding the projections that carry fetal blood. This close positioning of fetal and maternal blood supplies allows the exchange of nutrients and wastes to occur. Blood cells and bacteria do not normally pass between fetal and maternal blood supplies. Substances that can be freely exchanged between the fetus and the mother include oxygen, carbon dioxide, water, salts, hormones, viruses, and many drugs.

The cells of the early blastocyst synthesize hCG, which serves to maintain the corpus luteum, the ruptured follicle remaining in the ovary after the ovulation of its egg cell. Continued activity of the corpus luteum results in a steady increase in the levels of progesterone and estrogen during the first 5 weeks of pregnancy. In humans and many other mammals, once the placenta has established its own hormone production, the corpus luteum is no longer necessary. Consequently, hCG is secreted only during the first months of pregnancy.

An additional hormone synthesized by the placenta is called **somatomammotropin,** which works with estrogen, progesterone, and another hormone, **prolactin,** to stimulate development of the mother's mammary glands. Near term, the mother's estrogen level drops, and prolactin is produced by the pituitary gland of the brain. Prolactin continues to stimulate the production and secretion of breast milk during breastfeeding or **lactation.**

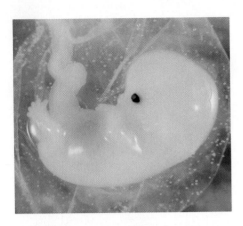

**FIGURE 22.10 Human embryo at 1 month.**

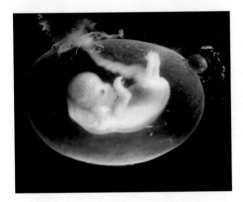

**FIGURE 22.11 Fetus at 9 weeks.**

**Evidence of endocrine disruption in aquatic organisms has led to concerns about the water that humans are drinking.**

After an embryo implants in the uterus, it undergoes remarkable growth and development. A month-old embryo is about 7 millimeters long and is beginning to develop a brain and spinal cord. It has buds that will give rise to limbs (**FIGURE 22.10**). A 9-week-old developing fetus has all of its organs and limbs in place (**FIGURE 22.11**). It can move its arms and legs and turn its head. For the remainder of the pregnancy, the baby increases in size; its features become more refined, with developments such as eyebrows, eyelashes, fingernails, and toenails. During the last stages of development, the fetal circulatory and respiratory systems undergo changes that will allow the fetus to breathe air.

During pregnancy a woman must be very careful about exposing a developing fetus to many types of chemicals. It has been clearly established that environmental chemicals can have negative effects on fetal development. Exposure to secondhand cigarette smoke is known to cause decreased birth weight. Low-birth-weight babies weigh less than 2500 grams (5 pounds, 8 ounces) at birth and are at increased risk of respiratory and heart problems at birth as well as of cerebral palsy and learning problems as they grow. Likewise, exposure to alcohol can cause severe developmental problems. Exposure to endocrine disruptors may also have negative effects. Now banned in the United States, polychlorinated biphenyls (PCBs) were used in the manufacture of many consumer goods, including plastics and rubber products, paints, and dyes. Fetal exposure to PCBs is correlated with low birth weight. Because labor is hormonally controlled, there is some concern that exposure to PCBs, which persist long after their manufacture, might also increase the risk of premature birth.

## Childbirth

Giving birth involves both labor and delivery. **Labor** involves strong, rhythmic contractions of the uterus that result in the birth. Induction of labor is hormonally controlled (**FIGURE 22.12**). Maternal estrogen levels increase toward the end of pregnancy, triggering the formation of receptors on the uterus for a hormone called **oxytocin** that serves to stimulate the smooth muscles lining the wall of the uterus to contract. Oxytocin is produced by fetal cells and by the mother's pituitary gland. During the induction of labor, oxytocin and estrogen are regulated by positive feedback. Oxytocin causes uterine contractions. Uterine contractions stimulate the release of more oxytocin, resulting in progressively intensifying muscle contractions that ultimately expel the baby from the uterus. Oxytocin also acts on the brain to increase feelings of pleasure, generosity, and bonding.

Uterine contractions normally signal the beginning of labor. Passage of the thick mucus plug blocking the cervix during pregnancy is another sign that labor is imminent. When the cervix begins to dilate, the plug loosens and is passed through the vagina. The beginning of labor can also be signaled by the rupture of a fluid-filled sac, called the **amnion**, which surrounds and protects the fetus. This is followed by the loss of fluid, occurring as either a gush of fluid or a slow trickle. It is often referred to as a woman's *water breaking* prior to delivery.

**Stages of Labor.** Labor can be thought of as occurring in three stages (**FIGURE 22.13**). During the first stage of labor, the cervical opening enlarges or dilates, going from a tiny, pencil-lead-sized opening to 10 centimeters. Contractions of the uterine muscle tissues help to dilate the cervix because they

# Visualize This ▼

**To artificially induce labor in their pregnant patients, physicians will provide an oxytocin-like drug. What effect would that have?**

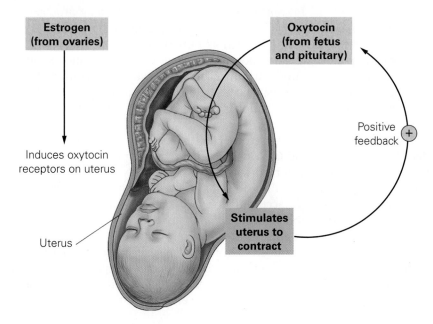

**FIGURE 22.12 Induction of labor.**   Estrogen and oxytocin are involved in a feedback loop that stimulates labor.

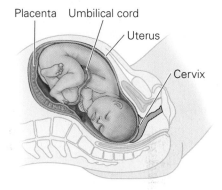

**Stage 1:** Dilation

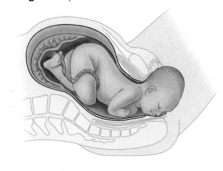

**Stage 2:** Expulsion

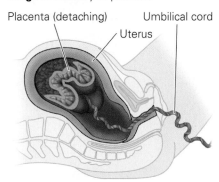

**Stage 3:** Delivery of placenta

**FIGURE 22.13 The three stages of labor.**   During labor (1) the cervix dilates, (2) the baby is expelled, and (3) the placenta is delivered.

apply pressure to the baby. The baby is forced against the cervical opening, which forces the opening to enlarge. The second stage of labor involves delivery of the baby through the narrow pelvic opening and birth canal, often aided by the mother's active "pushing" (contractions of the abdominal muscles); the third stage of labor is delivery of the placenta.

As any new parent knows, concern about exposure to toxic chemicals does not end when the baby is born. Newborns must be protected against toxic exposures as well, and a lot of this concern has centered on plasticizers. Unease about materials leaching out of plastics has led the Consumer Product Safety Commission to recommend that parents dispose of any soft vinyl baby products such as pacifiers and teething rings that contain the plasticizer DEHP. Even though DEHP has not definitively been shown to act as an endocrine disruptor in humans, negative effects of DEHP exposure have not been ruled out. Parents have also been cautioned about feeding babies from plastic bottles and feeding cups containing the chemical bisphenol A (BPA). It has been shown that this chemical can leach out of hard plastic bottles and be ingested. While toxicological studies on humans ingesting low levels of BPA have so far shown the chemical to be safe, newer studies, able to test for more subtle effects, have led to concern about the effects of BPA on the brains and behavioral development of infants and young children.

While there is no conclusive evidence that endocrine disruptors in the water supply are harming human reproductive health, neither is there any reason to disregard this hypothesis. Further research is necessary before the role of endocrine disruptors in human reproductive health can be firmly established.

# savvy reader

## Endocrine Disruption and Early Puberty

*What's a mom to do? Preventing early puberty and hormone problems in our daughters—here's the why and how.* Feb. 12, 2014, Aviva Romm, MD

Early puberty in girls is defined as puberty that starts before age 8. The signs of early puberty are the same as in normal puberty and include increased growth and pubic and underarm hair, breast growth, and menstruation. This condition can be caused by genetic mutations, hormone disorders, or hormone-secreting tumors. In a large number of cases, however, the cause of early puberty is unknown.

In an essay on her Own Your Health, Change the World, Integrative Medicine for Women and Children website, Aviva Romm, MD, states that "For years scientists have disagreed whether early puberty was really an emerging phenomenon. Now there's no doubt. Girls are getting their periods earlier. Many about a year earlier, according to a 2007 article in the *Journal of Adolescent Health*. But a study published in *Pediatrics* in 2011 found that in the United States, 15% of American girls begin puberty by age 7."

To help prevent early puberty Dr. Romm suggests, among other things, that young girls avoid drinking from plastic bottles and eating plastic-wrapped foods or foods that have been reheated in plastic containers. The reason she states for this concern is that plastics "can mimic the effects of estrogens in our bodies."

1. Dr. Romm states that in the United States, 15% of girls begin puberty by the age of 7. From the information given, is it possible to determine whether early puberty is on the rise? If not, what information would you need to determine whether early puberty is on the rise?

2. There is evidence that the age of onset of puberty has decreased by approximately 4 months every decade between 1840 and 1950. If this is true, and the author's claim that endocrine-disrupting chemicals like plastics are part of the cause, what should be true of the rate of plastic usage during this time period?

3. If you were interested in finding out more about the link between exposure to plastics and early puberty, what could you do?

4. This article was written by a medical doctor whose website also serves as a way to solicit patients. Does the fact that Dr. Romm is soliciting patients make her claims about plastics and early puberty less credible? Why or why not?

http://avivaromm.com/preventing-early-puberty-and-hormone-problems-in-our-daughters-heres-the-why-and-how

# SOUNDS RIGHT **BUT IS IT?**

Lexi and Jonah are newlyweds, both beginning graduate school programs at the same university. While they are excited to start a family, they would like to wait to have children until both of them have paying jobs. Lexi has never liked hormonal birth control, and disposable methods like sponges and condoms are out of their price range. In doing some research, she came across several high-quality science resources stating that an egg cell is only capable of surviving for 12 to 24 hours after ovulation. She also learned that cervical secretions change in response to hormonal conditions. When progesterone is high, the mucus thickens and is less abundant and slows the ascent of sperm. But when estrogen is high, as it is right before ovulation, the cervical mucus is abundant, stringy, has the consistency of raw egg white, and aids sperm in traveling through the reproductive tract toward a fertilized egg. Based on the information above, all of which is correct, she and Jonah come to the conclusion that they can avoid a pregnancy if Lexi pays close attention to her cervical secretions and that they abstain from sexual intercourse, to be safe, for the 24 hours after her cervical mucus changes to the type that facilitates sperm travel to the oviducts.

**A woman can become pregnant during only one 24-hour period every month.**

Sounds right, but it isn't.

**1.** Look back at the figure showing the menstrual cycle. What day of the month does estrogen peak?

**2.** If estrogen causes the cervical mucus to change to the slippery, fertility-enhancing mucus, when would this occur?

**3.** What day in an idealized cycle does ovulation occur?

**4.** Recall that sperm can survive in the female reproductive tract for up to 5 days. If a couple had unprotected sex on day 12 of the woman's menstrual cycle and ovulation occurred on day 14, could the woman become pregnant?

**5.** Consider your answers to questions 1–4 and explain why the original statement bolded above sounds right, but isn't.

# Chapter Review MasteringBiology®

Go to the Study Area in MasteringBiology® for practice quizzes, myeBook, BioFlix™ 3-D animations, MP3Tutor sessions, videos, current events, and more.

## Summary

### Section 22.1

Compare and contrast asexual and sexual reproduction.

- Asexual reproduction is a method of reproduction by which one parent produces offspring that are genetically identical. Large numbers of offspring can be reproduced quickly during asexual reproduction, but the absence of any genetic diversity makes the offspring less able to adapt to changes in the environment (p. 488).

- Sexual reproduction requires a mating between two parents. Males and females produce gametes in structures called gonads. Gametes unite at fertilization to produce genetically distinct offspring (pp. 488–489).

### Section 22.2

List the male reproductive structures and their functions.

- The penis of the male reproductive system is involved in sperm delivery; the urethra delivers both sperm and urine, the scrotum houses the androgen producing testes. Sperm are produced in the seminiferous tubules inside each testis and are stored in the epididymis before traveling through the vas deferens to the urethra. The seminal vesicles, prostate, and bulbourethral glands add secretions to sperm that help them develop and provide a source of energy. Semen is composed of the secretions from these glands combined with sperm (pp. 490–491).

List the female reproductive structures and their functions.

- The female reproductive system consists of the external vulva and clitoris, the internal vaginal passageway, the cervix at the base of the uterus that opens during childbirth, the uterus that houses the developing fetus, oviducts for the passage of gametes, and ovaries that produce egg cells and hormones (pp. 491–492).

Compare the processes of spermatogenesis and oogenesis.

- Spermatogenesis occurs in the seminiferous tubules of the testes and begins at puberty. Mature sperm are composed of a small head containing the DNA, a midpiece that has mitochondria to provide energy for the journey to the oviduct, and a flagellum. An acrosome at the sperm's head contains enzymes that help the sperm gain access to the egg (pp. 492–493).

- Oogenesis occurs in the ovaries and results in the production of egg cells. This process begins while the female is still in her mother's uterus, then pauses until puberty. At puberty, development of preexisting eggs continues each month until menopause (pp. 492–493).

Describe the events that occur during the menstrual cycle and how they are hormonally regulated.

- During the menstrual cycle, levels of estrogen, produced by the follicle, begin to rise, eventually stimulating GnRH release and causing a spike in FSH and LH levels. LH causes ovulation to occur (pp. 497–498).
- The remnant of the follicle that stays in the ovary after ovulation is the corpus luteum, which secretes progesterone and estrogen. If fertilization does not occur, the corpus luteum degenerates, progesterone and estrogen levels fall and menstruation occurs. Then LH and FSH levels rise, and the cycle starts over. If fertilization does occur, the embryo implants itself within the uterus and secretes hCG, which extends the life of the corpus luteum. Progesterone and estrogen levels remain high. Eventually, the placenta begins to produce these hormones (pp. 494, 497–498).

## Section 22.3

Outline the path sperm and egg must follow for fertilization to occur.

- Once inside the oviduct, sperm slide through the follicular cells that surround the egg to reach the zona pellucida, triggering the acrosomal enzymes, which break down the zona pellucida and allow the sperm to reach the egg cell's plasma membrane. After fusion of the egg and sperm plasma membranes, the sperm nucleus enters the egg cell, and the zygote can begin to develop (pp. 501–502).

Describe the stages of early development from the rapid, early divisions through early tissue differentiation.

- The zygote undergoes rapid cleavage divisions to produce the blastocyst. The inner cell mass of the blastocyst becomes the embryo and eventually the fetus. The blastocyst gives rise to the three-layered gastrula. Each layer of the gastrula (ectoderm, mesoderm, and endoderm) differentiates into specific tissues (pp. 502–503).

Compare the events leading to differentiation of male and female reproductive organs.

- Embryonic gonads become either testes or ovaries. Ductal structures exist side by side in male and female embryos. In each sex, one structure regresses. Male and female external genitalia are fashioned from the same embryonic material. The expression of sex-specific genes and hormones determines whether a particular embryo will become male or female (pp. 503–504).

Outline the hormonal controls regulating pregnancy and birthing.

- Pregnancy hormones are produced by the placenta, which forms in the uterus from the trophoblast of the blastocyst. These hormones help maintain the uterus, prevent ovulation, and prepare the breasts for lactation. Childbirth involves both labor and delivery. Labor involves contractions that help expel the baby from the uterus. During labor, the cervix dilates to allow the baby's passage out of the uterus. After her child's birth, the mother delivers the placenta (pp. 505–507).

## Roots to Remember

**The following roots of words come mainly from Latin and Greek and will help you decipher terms:**

| | |
|---|---|
| **-genesis** | means generation of or birth of. Chapter terms: *oogenesis, spermatogenesis* |
| **gon-, gono-,** and **gonado-** | mean seed or generation. Chapter term: *gonad* |
| **oo-** | is from the Greek word for egg. Chapter terms: *oocyte, oogenesis* |

## Learning the Basics

1. In what ways do asexual and sexual reproduction differ?

2. Describe the male and female reproductive structures.

3. Describe gametogenesis in males and females.

4. A sperm cell follows which path?
   **A.** seminiferous tubules, epididymis, vas deferens, urethra; **B.** urethra, vas deferens, seminiferous tubules, epididymis; **C.** seminiferous tubules, vas deferens, epididymis, urethra; **D.** epididymis, seminiferous tubules, vas deferens, urethra; **E.** epididymis, vas deferens, seminiferous tubules, urethra

5. An egg cell that is not fertilized follows which path?
   **A.** ovary, oviduct, uterus, cervix; **B.** ovary, uterus, oviduct, cervix, vagina; **C.** oviduct, ovary, cervix, uterus; **D.** oviduct, ovary, uterus, cervix; **E.** ovary, oviduct, cervix, uterus

6. Which of the following is mismatched?
   **A.** urethra: sperm passage; **B.** Leydig cells: androgen production; **C.** vas deferens: semen production; **D.** seminiferous tubules: sperm production

7. Gametogenesis _____.
   **A.** begins at puberty in males and females; **B.** requires that the Leydig cells of males produce semen; **C.** results

in the production of diploid cells from haploid cells;
**D.** begins at puberty in females; **E.** requires that meiosis halve the chromosome number so that sperm and eggs carry half the number of chromosomes as non-gametes.

**8.** The release of the oocyte from the follicle is caused by _____.

**A.** a decrease in estrogen; **B.** an increase in FSH; **C.** an increase in LH; **D.** an increase in progesterone.

**9.** Menstruation is regulated such that _____.

**A.** increasing estrogen levels have a positive feedback effect on FSH and LH; **B.** increasing FSH levels lead to ovulation; **C.** as progesterone levels increase, so do FSH and LH levels; **D.** ovulation occurs on the fifth day of the cycle; **E.** the placenta produces FSH, which stimulates ovulation.

**10.** The path of a sperm during fertilization is _____.
**A.** follicle cells, egg plasma membrane, zona pellucida;
**B.** egg plasma membrane, zona pellucida, follicle cells;
**C.** egg plasma membrane, follicle cells, zona pellucida;
**D.** zona pellucida, follicle cells, egg plasma membrane;
**E.** follicle cells, zona pellucida, egg plasma membrane.

## Analyzing and Applying the Basics

**1.** Some plants must be propagated from seeds but others, like the poinsettia, can be propagated by cutting a section of stem from one plant and allowing that to grow into a new plant. What kind of reproduction is occurring?

**2.** The external female genitalia is often mistakenly referred to as the vagina, which is an internal passageway that abuts the cervix. What is the correct term for the female external genitalia?

**3.** A typical birth control pill pack is composed of 21 hormone-containing pills and 7 inactive placebo pills. Bleeding occurs a few days after a woman begins taking the placebos. Some women choose to delay their period by not taking the placebos and instead starting a new pack of pills. Why would skipping the placebo pills prevent bleeding from occurring?

## Connecting the Science

**1.** Most birth control methods are designed for and marketed to women. Why do you think this is the case? List some advantages of creating more birth control methods for men.

**2.** Scientists use different terms to describe the stages of in utero development, from zygote to early embryo to fetus to baby. At what point in this developmental progression do you think the developing organism becomes "human"? Justify your answer.

Answers to **Stop & Stretch, Visualize This, Working with Data, Savvy Reader, Sounds Right, But Is It?**, and **Chapter Review** questions can be found in the **Answers** section at the back of the book.

CHAPTER **23** | # Study Drugs: Brain Boost or Brain Drain?

**Not many college students use heroin ...**

# Brain Structure and Function

How dangerous is it to use your friend's attention deficient disorder (ADD) medication as a study aid, or as a way to stay up later when partying? How would you rate the risk of abuse and addiction of non-prescribed ADD medication relative to the use of heroine, methamphetamine and cocaine, anabolic steroids, and highly caffeinated energy drinks? If you are like most college students, you would rank the nonmedical use of these prescription drugs as more dangerous than drinking energy drinks, but less dangerous than using steroids and far less dangerous than using methamphetamine, cocaine, or heroin. According to the United

... or to stay awake to party longer.

States Drug Enforcement Administration (DEA), you should have ranked the use of nonprescribed ADD medications as about likely to result in abuse and addiction as using methamphetamine and cocaine, and only slightly less risky than using heroin.

The use of ADD medications, like Adderall™, Vyvance, Ritalin, and Concerta, is legal for those diagnosed with ADD or ADHD (attention deficit hyperactivity disorder). When used properly, these medications can help people function properly in spite of a propensity for forgetfulness, impulsivity, distractibility, fidgeting, impatience, and restlessness.

However, more and more students are taking these so-called study drugs in order to increase their drive, focus, and productivity or just

to help them stay awake longer. While this practice seems to have started in the top-tier highly competitive universities, it now occurs at every type of post-secondary institution from community colleges, to state schools, to private colleges, and flagship universities. Whether these pills are obtained from a friend with a prescription or by making false claims to a health-care provider to obtain a prescription, studies indicate that at least one-third of U.S. college students have abused these medications.

Is the potential gain of taking these study drugs when you don't have ADD worth the potential costs? To understand this, we must first learn about the biology of the human nervous system.

All of these practices are illegal, and have surprisingly similar negative health consequences.

But many do use nonprescribed ADD medication to pull an all nighter...

# 23.1  The Nervous System

Every second, millions of signals make their way to your brain and tell it what your body is doing and feeling. Your **nervous system** receives and interprets these messages and decides how to respond. This requires the actions of specialized cells called **neurons.** Neurons, or nerve cells, carry electrical and chemical messages between your brain and other parts of your body. These electrochemical message-carrying cells can be bundled together, producing structures called **nerves.**

## Central and Peripheral Nervous Systems

The brain and spinal cord together form the **central nervous system (CNS).** Neurons in your eyes, ears, or skin transmit information to your CNS. There the information is processed, and the appropriate response is relayed back to your body by other neurons out of the CNS (**FIGURE 23.1**).

The CNS is the seat of intelligence, learning, memory, and emotion. The components of the CNS work with a second subdivision of the nervous system, the **peripheral nervous system (PNS),** which includes the network of nerves that radiates out from the brain and spinal cord, gathering information to relay back to the CNS.

To carry information, the cells of the nervous system pass electrical and chemical signals to each other. Signals are transmitted from one end of a neuron to the other, between and from neighboring neurons, and to the cells of tissues, organs, or glands that respond to nerve signals. These are called **effectors.** Information is carried along nerves by electrical changes called **nerve impulses** and transmitted by changes in ion concentration or by chemicals, called **neurotransmitters,** released from nerve cells.

Neurons can be grouped into three categories (**FIGURE 23.2**): (1) **sensory neurons,** which carry information toward the CNS; (2) **motor neurons,** which carry information away from the CNS toward effector tissues; and (3) **interneurons,** which are located between sensory and motor neurons within the brain or spinal cord. Most actions involve input from all sources; sensory activity is followed by integration and motor output.

Sensory input is detected by **sensory receptors.** They detect changes in conditions inside or outside the body. When receptors are stimulated, signals are generated and carried to the brain.

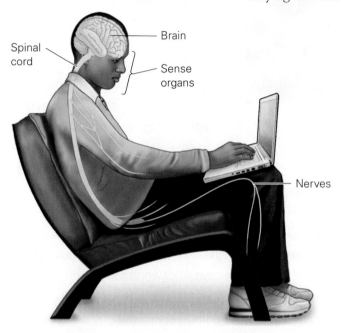

**FIGURE 23.1 The nervous system.** The nervous system includes the brain, spinal cord, sense organs, and the nerves that interconnect those organs.

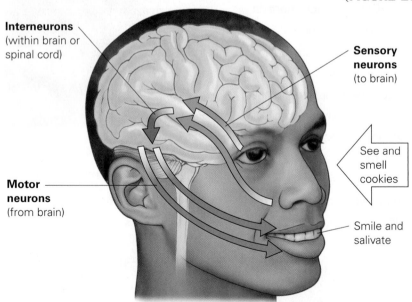

**FIGURE 23.2 Neurons.** The nervous system's three main functions are sensory input, integration, and motor output. Most actions, such as being tempted to eat cookies, involve input from sensory neurons, followed by integration via interneurons and motor output from motor neurons.

## The Senses

The **general senses** are temperature, pain, touch, pressure, and body position (proprioception). Sensory receptors for the general senses are scattered throughout the body. The **special senses** are smell, taste, equilibrium, hearing, and vision. The sensory receptors for these five special senses are found in complex sense organs in the head (**TABLE 23.1**).

**TABLE 23.1** **The senses.** There are two groups of senses: general and special. Specialized receptors from all of these senses relay information to the brain.

| Sense and Its Location | Sensory Receptors | Description |
|---|---|---|
| **General senses—throughout body** | | |
| Temperature, pain, pressure, touch<br><br>Skin | | Sensory nerve endings in the skin can moderate their response based on the intensity of the stimulus, e.g., cold temperature versus freezing temperature. |
| Proprioception<br><br>Muscle | | Specialized neurons sense joint position, tension in joints and ligaments, and muscle contraction. |
| **Special senses—found in complex sense organs** | | |
| Smell<br>Receptor in nasal cavity | | Hair-like projections called cilia stimulate olfactory organs with specialized receptors in the nasal cavity. |
| Taste<br><br>Taste bud | | Taste buds on the surface of the tongue are specialized to respond to one of five primary taste sensations: sweet, salty, sour, bitter, or umami (savory or meaty). |
| Vision<br>Retina | | Receptors in the retina of the eye allow for sight and mediate responses to changes in the amount of light. |
| Equilibrium and hearing<br>Receptors in inner ear | | These receptors, located on cilia in the inner ear, determine the position of the body in space and detect and interpret sound waves. |

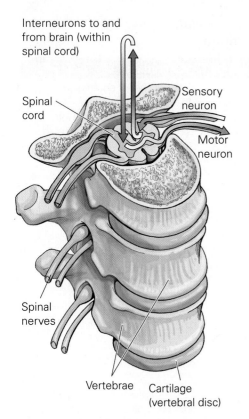

Interneurons to and from brain (within spinal cord)

Spinal cord

Sensory neuron

Motor neuron

Spinal nerves

Vertebrae

Cartilage (vertebral disc)

**FIGURE 23.3 Spinal nerves and vertebrae.** Spinal nerves branch out between the vertebrae and go to all parts of the body.

As sensory information is passed to your brain, it travels through the main nerve pathway, the **spinal cord.** Your spine, which protects your spinal cord from injury, is made up of bones called **vertebrae.** Spinal nerves branch out between the vertebrae and go to every part of the body (**FIGURE 23.3**).

In addition to transmitting messages to and from the brain, the spinal cord serves as a reflex center. **Reflexes** are automatic responses to a stimulus. They are prewired in a circuit of neurons called a **reflex arc,** which often consists of a sensory neuron that receives information from a sensory receptor, an interneuron that passes the information along, and a motor neuron that sends a message to the muscle that needs to respond.

Reflexes allow a person to react quickly to dangerous stimuli; for instance, the withdrawal reflex occurs when you come in contact with a dangerous stimulus such as something hot (**FIGURE 23.4**). When you touch something hot, sensory neurons from touch receptors send the message to your spinal cord. Within the spinal cord, interneurons send the message to motor neurons to withdraw your hand from the hot surface. While the spinal reflexes are removing your hand from the source of the heat, pain messages are also being sent through your spinal cord to your brain.

**STOP & STRETCH** When you touch something hot, you will often remove your hand before actually feeling the pain. Why might this be?

## Amphetamines Act on the Nervous System

Many study drugs are amphetamines. Amphetamine use heightens the CNS response, causing increased alertness, mental focus, and even euphoria, but it can also lead to permanent changes in the functions the CNS controls, resulting in increased rates of depression and anxiety long after use ends.

One of the jobs of the PNS is to recognize dangerous or stressful situations and alert the CNS. This can involve the so-called fight-or-flight response. When in this state of hyperarousal, whether caused by a real threat or by amphetamine use, digestion is slowed (hence a lack of appetite), and blood pressure and heart rate increase. These physiologic changes cause the anxiety, jitters, insomnia, irritability, hostility, and paranoia experienced by amphetamine users.

Some students will drink coffee or energy drinks in an effort to maximize the effects of study drugs on focus and drive, a practice that can lead to dangerous increases in heart rate even leading, in rare circumstances, to heart attack or stroke. Students may also mix study drugs with alcohol, hoping to be able to stay up later and party longer. Mixing amphetamines with alcohol can make the user less aware of the physiologic changes of both alcohol poisoning and amphetamine overdose.

Long-term use of amphetamines and alcohol can lead to **psychosis,** a mental disorder that includes the loss of contact with reality.

It is important to note that those diagnosed with ADD are far less likely to experience these side effects than are their non-ADD peers. This may be due to the manner in which the medications are taken. Physicians typically initially prescribe a low dose and slowly work up to larger doses, giving the nervous system time to adjust and giving the physician the option of dialing back if side effects are severe.

It may also be true that study drugs act differently in people with ADD because their brains differ structurally from those of their non-ADD peers. In the next section, we take a look at the structure and function of the brain and learn what science tells us about brain differences between people with and without ADD.

# 23.2 The Human Brain

The brain is the region of the body where decisions are reached and where bodily activities are directed and coordinated. Housed inside the skull, the brain sits in a liquid bath called **cerebrospinal fluid** that protects and cushions it.

The functions of the brain are divided between the left and right hemispheres. Because many nerve fibers cross over each other to the opposite side, the brain's left hemisphere controls the right half of the body and vice versa. The areas that control speech, reading, and the ability to solve mathematical problems are located in the left hemisphere, while areas that govern spatial perceptions (the ability to understand shape and form) and the centers of musical and artistic creation reside in the right hemisphere.

In addition to housing 100–200 billion neurons, the brain is composed of other cells called **glial cells.** There are 10 times as many glial cells in the brain as there are neurons. In contrast to neurons, glial cells do not carry messages. Instead, they support the neurons by supplying nutrients, helping to repair the brain after injury, and attacking invading bacteria. Structurally, the brain is subdivided into many important anatomical regions, including the cerebrum, thalamus, hypothalamus, cerebellum, and brain stem (**FIGURE 23.5**).

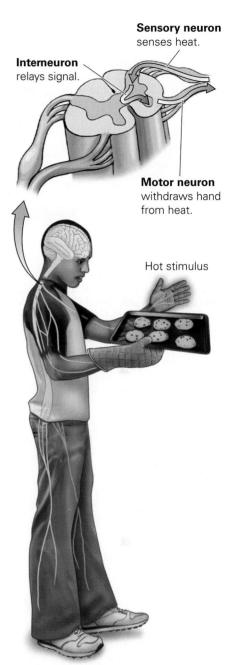

**Sensory neuron** senses heat.

**Interneuron** relays signal.

**Motor neuron** withdraws hand from heat.

Hot stimulus

**(a)**

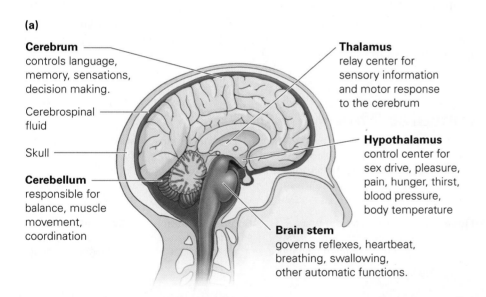

**Cerebrum**
controls language, memory, sensations, decision making.

Cerebrospinal fluid

Skull

**Cerebellum**
responsible for balance, muscle movement, coordination

**Thalamus**
relay center for sensory information and motor response to the cerebrum

**Hypothalamus**
control center for sex drive, pleasure, pain, hunger, thirst, blood pressure, body temperature

**Brain stem**
governs reflexes, heartbeat, breathing, swallowing, other automatic functions.

**(b)**

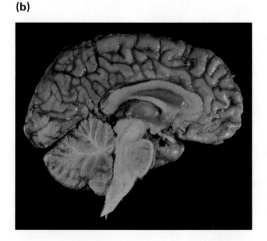

**FIGURE 23.4 A reflex arc.** A reflex arc can consist of a sensory receptor, a sensory neuron, an interneuron, a motor neuron, and an effector. Touching a hot baking sheet evokes the withdrawal reflex.

◀ **Visualize This**

**Find the locations of the cerebrum, thalamus, hypothalamus, brain stem, and cerebellum in this photo.**

**FIGURE 23.5 Anatomy of the human brain.** (a) The location and function of the cerebrum, thalamus, hypothalamus, brain stem, and cerebellum are shown. (b) Photo of a cross-sectioned human brain.

## Cerebrum

The **cerebrum** fills the whole upper part of the skull. This part of the brain controls language, memory, voluntary movements, sensations, and decision making. The cerebrum has two hemispheres, each divided into four lobes (**FIGURE 23.6**):

**cereb-** is from the Latin word for the brain.

- The **parietal lobe** processes information about touch and is involved in self-awareness.
- The **frontal lobe** processes voluntary muscle movements and is involved in planning, working memory, and impulse control.
- The **temporal lobe** is involved in processing auditory and olfactory information and is important in memory and emotion.
- The **occipital lobe** processes visual information from the eyes.

The deeply wrinkled outer surface of the cerebrum is called the **cerebral cortex**. In humans, the cerebral cortex, if unfolded, would be the size of a large (16-inch) pizza. The folding of the cortex increases the surface area and allows it to fit inside the skull. The cortex contains areas for understanding and generating speech, areas that receive input from the eyes, and areas that receive other sensory information from the body. It also contains areas that allow planning.

**cortex** is the Latin word for the bark of a tree.

The cerebrum and its cortex are divided from front to back into two halves—the right and left cerebral hemispheres—by a deep groove or fissure. At the base of this fissure lies a thick bundle of nerve fibers, the **corpus callosum**, which connects the hemispheres to each other and allows communication from one side of the brain to the other.

## Visualize This ▼

**Impulsivity, or acting before thinking, is a symptom associated with ADD. Based on the functions listed, an abnormality in which structure do you think might lead to impulsivity?**

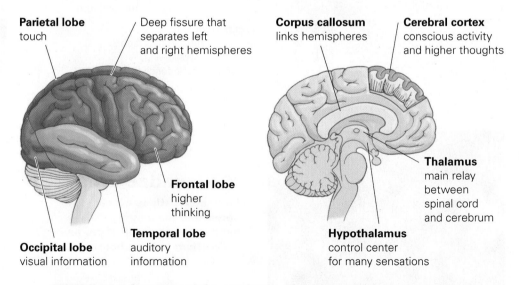

**Parietal lobe**
touch

Deep fissure that separates left and right hemispheres

**Corpus callosum**
links hemispheres

**Cerebral cortex**
conscious activity and higher thoughts

**Frontal lobe**
higher thinking

**Thalamus**
main relay between spinal cord and cerebrum

**Temporal lobe**
auditory information

**Hypothalamus**
control center for many sensations

**Occipital lobe**
visual information

**FIGURE 23.6 Structure of the cerebrum.** The right and left hemispheres of the cerebrum are each divided into four lobes and connected to each other by the corpus callosum. Deep inside and between the two cerebral hemispheres are the thalamus and the hypothalamus.

**STOP & STRETCH**   The frontal lobes of the brain are thought not to be fully developed until around age 25. Based on your understanding of the functions of the frontal lobe, propose an explanation for why people under age 25 might be more willing to use drugs than those with fully developed brains.

**How dangerous is it to use a friend's ADD medication to help you stay alert and focused?**

## Thalamus and Hypothalamus

Deep inside the brain, lying between the two cerebral hemispheres, are the thalamus and the hypothalamus. The **thalamus** relays information between the spinal cord and the cerebrum. The thalamus is the first region of the brain to receive messages signaling such sensations as pain, pressure, and temperature. The thalamus suppresses some signals and enhances others, which are then relayed to the cerebrum. The cerebrum processes these messages and sends signals to the spinal cord and to neurons in muscles when action is necessary. The **hypothalamus**, located just under the thalamus and about the size of a kidney bean, is the control center for sex drive, pleasure, pain, hunger, thirst, blood pressure, and body temperature. The hypothalamus also releases hormones, including those that regulate the production of sperm and egg cells as well as the menstrual cycle.

**hypo-** means under or below.

## Cerebellum and Brain Stem

The **cerebellum** controls balance, muscle movement, and coordination (**FIGURE 23.7**). Since this region of the brain ensures that muscles contract and relax smoothly, damage to the cerebellum can result in rigidity and, in severe cases, jerky motions. The cerebellum looks like a smaller version of the cerebrum. It is tucked beneath the cerebral hemispheres and, like the cerebrum, has two hemispheres connected to each other by a thick band of nerves.

The **brain stem** lies below the thalamus and hypothalamus. It governs reflexes and some spontaneous functions, such as heartbeat, respiration, swallowing, and coughing.

The brain stem is composed of the midbrain, pons, and medulla oblongata. Highest on the brain stem is the **midbrain,** which adjusts the sensitivity of your eyes to light and of your ears to sound. Below the midbrain is the **pons.** The pons functions as a bridge, allowing messages to travel between the brain and the spinal cord. The **medulla oblongata** is the lower part of the brain stem. It helps control heart rate and conveys information between the spinal cord and other parts of the brain.

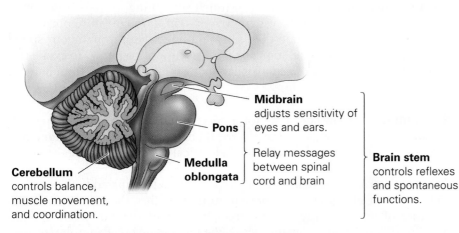

**FIGURE 23.7 The cerebellum and brain stem.** The cerebellum lies below the cerebrum. The brain stem consists of the midbrain, pons, and medulla oblongata.

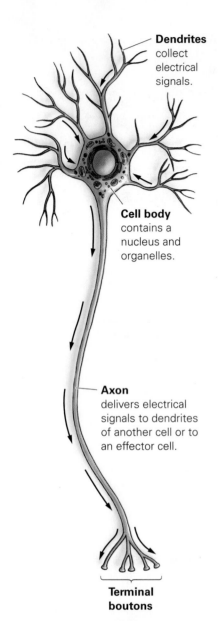

**Dendrites** collect electrical signals.

**Cell body** contains a nucleus and organelles.

**Axon** delivers electrical signals to dendrites of another cell or to an effector cell.

**Terminal boutons**

**FIGURE 23.8 The structure of a generalized neuron.** A neuron consists of the branching dendrites, the cell body, the axon, and terminal boutons. Nerve impulses are propagated in the direction of the arrows.

Some scientific studies seem to implicate differences in the frontal cortex between those with and without ADD, with the frontal cortex being smaller in those with ADD. This may be part of the reason why ADD medications help those with ADD feel normal but give a heightened response to those using it for nonmedical reasons.

Underlying differences in brain structure are differences in the ability neurons to function properly.

# 23.3 Neurons

Neurons are highly specialized cells that usually do not divide. Therefore, damage to a neuron cannot be repaired by cell division and often results in permanent impairment. For example, damage to spinal motor neurons results in lifelong paralysis because messages can no longer be transmitted from the CNS to muscles. Likewise, injury to the brain can result in permanent brain damage if neurons in the brain are harmed.

## Neuron Structure and Function

Neurons are composed of branching **dendrites** that radiate from a bulging **cell body**, which houses the nucleus and organelles, and a long, wirelike **axon**, terminating in knobby structures called the **terminal boutons** (**FIGURE 23.8**). A nerve impulse usually travels down the axon of one neuron and is transmitted to the dendrites of another neuron.

The axons of many neurons are coated with a protective fatty layer called the **myelin sheath** (**FIGURE 23.9a**). The myelin sheath is composed of lipid-rich, neuron-supporting cells called **Schwann cells** that wrap around an axon and give tissues coated in myelin a glistening white color called **white matter.** The myelin sheath functions like insulation on a wire by preventing sideways message transmission, thus increasing the speed at which the electrochemical impulse travels down the axon.

Long stretches of the myelin sheath are separated by small interruptions leaving tiny, unmyelinated patches called the **nodes of Ranvier.** Nerve impulses "jump" successively from one node of Ranvier to the next (**FIGURE 23.9b**), which is a much faster transmission than takes place on unmyelinated neurons. Unmyelinated axons, combined with dendrites and the cell bodies of other neurons, look gray in cross sections and thus are called **gray matter.**

Neurons transmit impulses from one part of the body to another. Many kinds of stimuli, including touch, sound, light, taste, temperature, and smell, cause a neuronal response. When you touch something, signals from touch sensors travel along sensory nerves from your skin, through your spinal cord, and into your brain. Your brain then sends out messages through your spinal cord to the motor neurons, telling your muscles how to respond. To evoke this response, nerve cells must transmit signals along their length and from one cell to the next.

**Action Potential.** Another name for a nerve impulse is an action potential. An action is a brief reversal of the electrical charge across the membrane of a nerve cell. It is propagated as a wave of electrical current down the length of the axon.

The inside of a resting neuron is negatively charged compared to the outside. Because opposite charges tend to move toward each other, the membrane serves as a source of stored energy, like a battery, by keeping these opposite charges apart. This difference in the charge across the membrane is called **polarization.**

**(a) Axons covered in myelin sheaths**     **(b) Nerve impulses travel more quickly on myelinated than on unmyelinated neurons.**

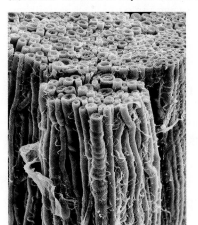

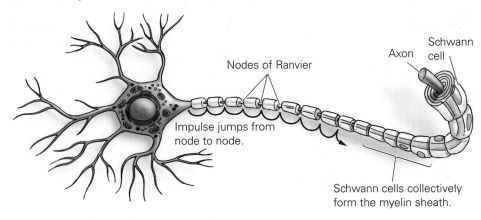

FIGURE 23.9 **Myelination.** (a) A cut bundle of nerve fibers. Each nerve consists of an internal axon (greenish) covered in an insulating myelin layer (whitish). (b) By jumping from one node of Ranvier to the next, nerve impulses travel quickly on myelinated nerves than on unmyelinated.

# Working with Data ▲

Nerve impulses move 100 meters per second along myelinated axons and 1 meter per second along unmyelinated axons. If a person is 6'6" how long would it take for a nerve impulse to travel from his head to his feet on (a) myelinated neurons (1 meter is about 3.3 feet) and (b) unmyelinated neurons?

When signaled, gated protein channels along the length of the axon open to allow positively charged sodium and potassium ions to move across the membrane. The influx of positively charged sodium ions depolarizes the nerve cell. A domino effect occurs as the ions spread charges toward each successive voltage-sensitive protein channel, causing each to open and let in more sodium, propagating the depolarization along the length of the neuron. This impulse travels only in one direction, toward the terminal boutons (**FIGURE 23.10**). Repolarization results from diffusion of potassium ions out of the cell, causing the inside of the cell to become more negative than the outside.

**STOP & STRETCH** A neurotoxin carried by pufferfish binds to sodium channels and prevents passage of ions. What effect would this neurotoxin have on the nervous system of someone who ingested it?

**Synaptic Transmission.** Once a signal has traveled along the length of the axon, it must pass to the next neuron or to the effector organ, muscle, or gland. Most neurons are not directly connected to each other; the signal must be transmitted to the next neuron across a gap between the

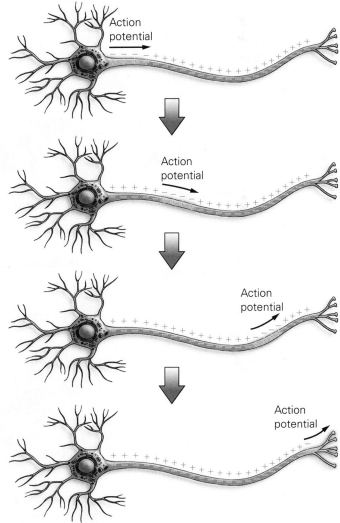

FIGURE 23.10 **Action potential.** The short-lived rise and fall in membrane potential travels the length of the axon, then relays the signal to the effector organ, muscle or gland, or to the next neuron.

**Close to 1/3 of students have taken these so-called study drugs in order to increase their drive, focus, and productivity or just to help them stay awake longer.**

two called the **synapse** (**FIGURE 23.11a**). The synapse consists of the terminal boutons of the presynaptic neuron, the space between the two adjacent neurons, and the cell body or dendrite of the postsynaptic neuron. Structures called vesicles are found in the terminal boutons of the presynaptic neuron. Each is filled with a specific chemical neurotransmitter. When an electrical impulse arrives at the terminal bouton of a nerve cell, neurotransmitters are released and they diffuse across the synapse, binding to specific receptors on the cell membrane of the postsynaptic neuron. The binding stimulates a rapid change in the uptake of sodium ions by the postsynaptic neuron. The sodium channels open, and sodium ions flow in, causing another depolarization and generating another action potential. The nerve impulse is propagated from one neuron to the next until the signal reaches a tissue.

After a neurotransmitter evokes a response, it is removed from the synapse. Some are broken down by enzymes in the synapse; others are reabsorbed by the neuron via a process called **reuptake** (**FIGURE 23.11b**). Reuptake occurs when neurotransmitter receptors on the presynaptic cell permit the neurotransmitter to re-enter. This allows the neuron to use it again. Both breakdown by enzymes and reuptake by receptors prevent continued stimulation of the postsynaptic cell.

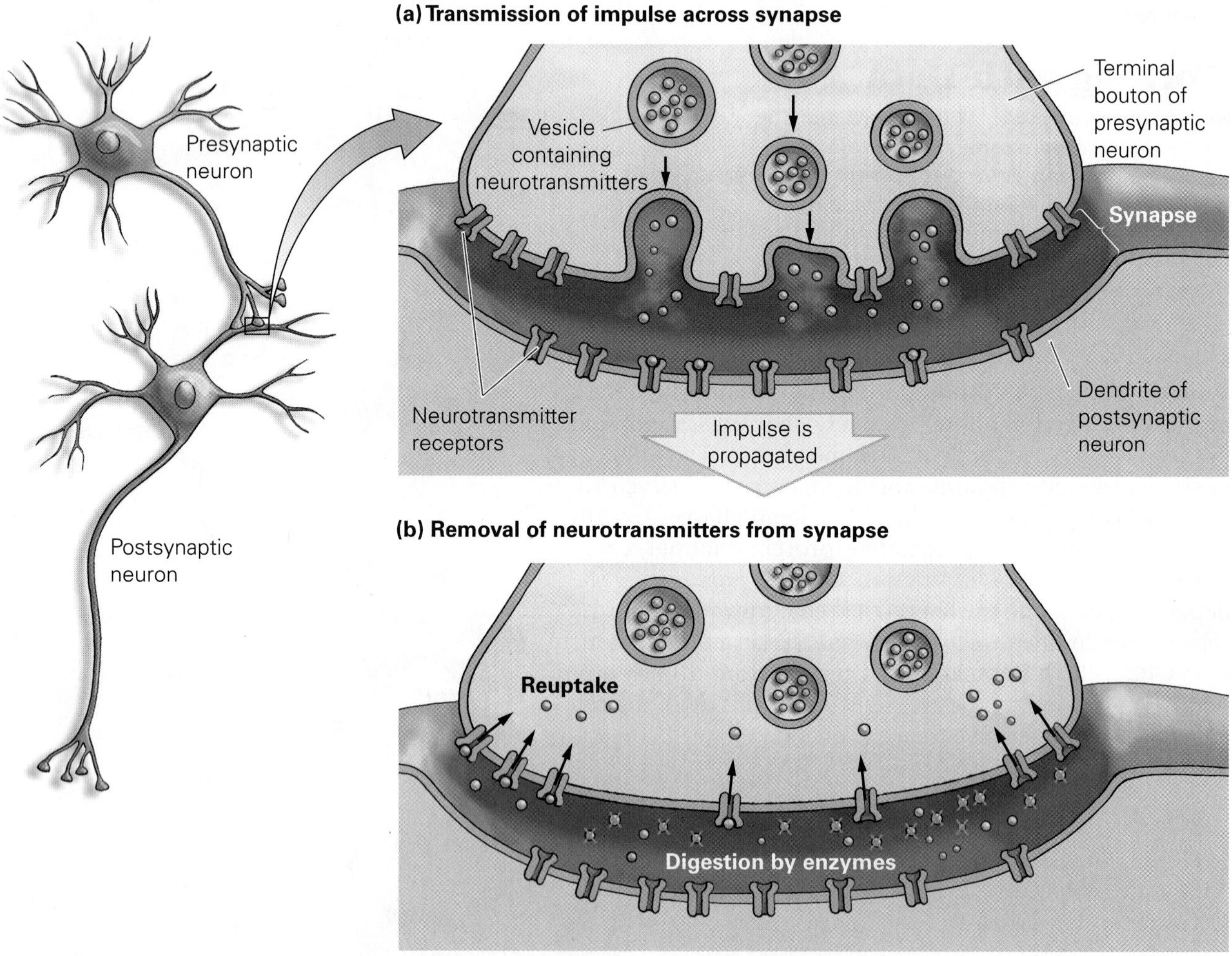

**(a) Transmission of impulse across synapse**

Presynaptic neuron

Vesicle containing neurotransmitters

Terminal bouton of presynaptic neuron

Synapse

Neurotransmitter receptors

Impulse is propagated

Dendrite of postsynaptic neuron

Postsynaptic neuron

**(b) Removal of neurotransmitters from synapse**

Reuptake

Digestion by enzymes

**FIGURE 23.11 Propagating the nerve impulse between neurons.** (a) The nerve impulse can be transmitted from the terminal bouton of one neuron to the dendrite of the next. Neurotransmitters are released from the presynaptic neuron, travel across the synapse to the postsynaptic neuron, and bind to receptors on dendrites of the postsynaptic neuron. (b) After the neurotransmitter evokes a response, it is removed from the synapse by enzymes present in the synapse that break the neurotransmitter apart or by reuptake via the presynaptic cell.

There is evidence linking disturbances in neurotransmission to various diseases. For example, **Alzheimer's disease**, a progressive mental deterioration which causes memory loss along with the loss of control of bodily functions that eventually results in death, is thought to involve impaired function of the neurotransmitter **acetylcholine** in some neurons. **Depression** is a disease that involves feelings of helplessness, despair, and thoughts of suicide. Three neurotransmitters, serotonin, dopamine, and norepinephrine, seem to play a role in depression. **Parkinson's disease**, thought to be caused by low levels of dopamine, causes tremors, rigidity, and slowed movements.

Abnormal levels of the neurotransmitters **dopamine** and **norepinephrine** seem to be involved in producing the symptoms of ADD. Dopamine controls emotions as well as complex movements while norepinephrine helps regulate response to stress and regulates attention and focus. Therefore, someone with a low concentration of dopamine may respond impulsively to situations in which pausing to process the input would be more effective. Those with low norepinephrine may have trouble with focus and concentration.

While no one knows for sure what causes the lower levels of these neurotransmitters in those with ADD, drugs used to treat symptoms of ADD work to decrease the impact of these deficiencies by increasing the level of neurotransmitter in the synapse. This can be done by causing the release of more neurotransmitter (**FIGURE 23.12a**) or by blocking the actions of the reuptake receptors (**FIGURE 23.12b**).

**(a) ADD drugs can block dopamine reuptake.**

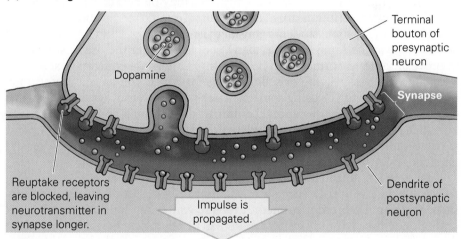

**(b) ADD drugs can increase the amount of dopamine released into the synapse.**

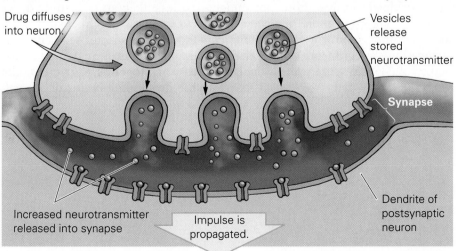

**FIGURE 23.12 Mechanism of ADD prescription drug action.** Some ADD drugs work by (a) increasing the release of neurotransmitters or (b) preventing the reuptake of neurotransmitters, allowing longer stimulation of postsynaptic neuron.

**Does the illegal use of ADD medications really help you learn? Or do they cause exhaustion and brain drain?**

## Nonmedical Use of ADD Medications

When a user takes ADD medications for nonmedical reasons, the brain is rapidly drained of neurotransmitter that would otherwise have been utilized slowly over the course of several days. The synthesis of new neurotransmitter occurs at a rate that only slowly compensates for the loss. During the time period after the neurotransmitter is used but before the body can restock, the user will experience symptoms sometimes known as the "crash." These symptoms can include exhaustion, depression, inability to concentrate, anger, aggression, anxiety, tearfulness, and many more.

For some users, the strategy of taking study drugs immediately backfires when the effects of the crash are experienced during an exam. For others, one crash is enough to dissuade them from further use. The worst outcomes occur when users attempt to immediately re-attain or intensify the desired symptoms or alleviate crash symptoms by using more of the drug. One of the ways the body responds to the increased amount of neurotransmitter present is to increase the number of receptors on the postsynaptic neuron, thus limiting the drug's effects. As this tolerance for the drug builds, the amount of drug required to achieve the initial effects continues to escalate, leading to a higher risk of overdose. With prolonged use, the body will respond by permanently lowering the amount of neurotransmitter produced. Now the user needs to use more just to feel normal.

While around one-third of students use ADD medications for nonmedical reasons, the majority of students do not. This may be because of a choice not to use any nonprescription drugs or may stem from concerns about the health risks. In addition, some students consider the practice of taking ADD drugs to be cheating, in the same way that many people consider the use of steroids to improve athletic performance to be cheating. Others are concerned about the legal consequences. Giving or selling drugs for which you have a prescription is illegal, as is buying or using someone else's prescription drugs. Most students choose to avoid using substances that impair their driving abilities so as not to hurt themselves or others. Most students also understand that developing time management skills and discipline now, instead of popping a pill, will be very helpful to them when they get out of college.

The time lost to the misery of crashing, coupled with the very real health risks associated with study drug use, will hopefully lead students using or considering using nonprescribed medications to conclude that they are better off setting aside time to study every day instead of relying on illegal drugs to help them pull a last minute all-nighter. Universities, too, can help by understanding the scope of the problem and by helping students find strategies to manage their workload. Studies show that students are less likely to take study drugs when they are given more low-stakes exams versus a few high-stakes exams, and when they submit written assignments in multiple shorter segments versus one long assignment. Whatever the mechanism, it is clear that students, faculty, and health-care professionals need to work together to decrease the consequences of drug use (**TABLE 23.2**) on the developing brains of our college students.

**TABLE 23.2 Recreational drugs and the nervous system.**

| Drug | Mechanism of Action | Desired Mental Effect | Side Effect |
|---|---|---|---|
| Alcohol is a by-product of fermentation. | • A depressant that diffuses easily across cell membranes <br> • Inhibits neurotransmission and interferes with the activity of neurons throughout the brain | Reduced anxiety and a sense of well-being, loss of concern for social constraints | • Impaired judgment, slurred speech, unsteady gait, slower reaction times, uncontrollable emotions <br> • Chronic alcohol abuse damages nerve cells in frontal lobes, leading to impaired intellect, judgment, thought, and reasoning. Liver damage is also likely. |
| Amphetamines are used legally to treat obesity, asthma, ADD, and narcolepsy. Methamphetamines are illegal amphetamines. | • Increase dopamine and norepinephrine release <br> • Block reuptake and inhibit breakdown of neurotransmitters | Small doses make a person feel more energetic, alert, focused and confident. | • Effects wear off quickly, causing sudden "crashes" from depleted neurotransmitter stores and resulting in depression and fatigue. <br> • Prolonged use results in aggressiveness, delusions, hallucinations, violent behaviors, and death. |
| Caffeine is a naturally occurring chemical found in plants such as coffee, tea, and cocoa. | A general stimulant of all cells, not just those of the central nervous system | Mental alertness, increased energy | Insomnia, anxiety, irritability, and increased heart rate |
| Cocaine is extracted from the leaves of the coca plant of South America. | A stimulant that increases levels of dopamine and norepinephrine by decreasing reuptake | A rush of intense pleasure, increased self-confidence, and increased physical vigor | • Increased heart rate and blood pressure <br> • Crash involves deep depression, anxiety, and fatigue. <br> • Abuse may change neurons to prevent experience of positive feelings without the drug. |

*continued*

**TABLE 23.2 Recreational drugs and the nervous system.** *continued*

| Drug | Mechanism of Action | Desired Mental Effect | Side Effect |
|---|---|---|---|
| Ecstasy is MDMA in pill form; Molly is MDMA in crystal or powder form. | A stimulant and hallucinogenic that acts to prevent serotonin reuptake while flooding neurons with other neurotransmitters | Euphoria, enhanced emotional and mental clarity, increased energy and sexual response, heightened sensitivity to touch | • Severe dehydration, confusion, anxiety, paranoia, depression, and sleeplessness lasting several days<br>• Permanent memory damage |
| Lysergic acid diethylamide (LSD) and mushrooms: LSD is a derivative of the fungus *Claviceps purpurea*, which grows on rye. | Hallucinogen that binds to serotonin receptors in the brain, increasing the normal response to serotonin | Heightened sensory perception and bizarre changes in thought and emotion, hallucinations | • Hallucinations can lead users to dangerous actions.<br>• Heavy use leads to permanent brain damage, including losses of memory, decreased attention span, and psychosis. |
| Marijuana is derived from the *Cannabis sativa* plant and contains delta-9-tetrahydrocannabinol (THC). | • Receptors for THC are in the areas of the brain that influence mood, pleasure, memory, pain, and appetite.<br>• THC is thought to work by increasing dopamine release. | • Altered sense of time, enhanced feeling of closeness to others, increased sensitivity to stimuli<br>• Large doses can cause hallucinations. | • Slowed reaction time and decreased coordination<br>• Permanently impairs short-term memory, learning, attention span, ability to store and acquire information<br>• Decreased testosterone production and disrupted menstrual cycle |
| Nicotine is found in tobacco plants. | Stimulant that triggers neurons of cerebral cortex to produce acetylcholine, epinephrine, and norepinephrine | Increased alertness and awareness, appetite suppression, relaxation | • Increased odds of most cancers<br>• Causes increased heart rate and blood pressure |
| Opiates—including heroin, morphine, and codeine—are derived from the opium poppy. | Like the endorphins produced by exercise, opiates bind receptors in neurons that control feelings of pleasure. | A quick, intense feeling of pleasure, followed by a sense of well-being, then drowsiness | • Poor motor coordination, depression, can cause coma and death<br>• May change the ability of the brain to respond to normal pleasures |

# *savvy reader*

## Prescription Drug Abuse

*Nonmedical Use of Prescription Stimulants among U.S. College Students: Prevalence and Correlates from a National Survey*

This peer-reviewed article reports the results of a study on the illegal use of prescription attention-deficit medications, such as Ritalin™ and Adderall™, by U.S. college students who have not been prescribed the medication. Surveys were mailed to over ten thousand randomly selected students at over 100 four-year colleges in the United States. Statistical analysis showed that use was highest among white male fraternity members and that schools with more rigorous admissions standards had more students who used these drugs.

The authors of this article also present evidence that students who used these stimulants are more likely to use other legal and illegal drugs.

1. List several controls used by the authors of this peer-reviewed study.
2. Why is it important that statistical analyses were applied?
3. Did the authors speculate about anything other than the data they present?

McCabe SE, Knight JR, Teter CJ, Wechsler H. *Addiction 2005,* Jan; 100(1):96–106

# SOUNDS RIGHT BUT IS IT?

You are flipping through a magazine and come across a quiz designed to help you determine whether you are right brained or left brained. The preamble to the quiz states that the human brain exhibits a phenomenon called lateralization where each side, or hemisphere, of the brain specializes in particular types of thinking. According to this theory, people who are right brained are more creative, intuitive, thoughtful, and better at reading and expressing emotions, because the right side of their brain is dominant. Left-brain dominance yields people who are more analytical, logical, and better at math, reasoning, and critical thinking.

**Traits like being creative or logical occur when one hemisphere of the brain dominates over the other hemisphere.**

Sounds right, but it isn't.

1. Are there structures in one hemisphere that are not in the other hemisphere?
2. How do the two hemispheres of the brain communicate with each other?
3. Brain imaging scans taken while people are solving math problems show that both sides of the brain are active. Does this provide support for the right brain–left brain hypothesis? Why or why not?
4. Evidence from neurobiological studies of people listening to spoken language suggests that the left hemisphere determines which sounds form words while the right hemisphere processes whether the words are meant to convey emotion or stress. How would you describe the relationship between the left and right hemispheres when it comes to language?
5. Does the brain lateralization hypothesis account for people who are very creative and very good at math?
6. Consider your answers to questions 1–5 and explain why the statement bolded above sounds right, but isn't.

# Chapter Review MasteringBiology®

Go to the Study Area in MasteringBiology® for practice quizzes, myeBook, BioFlix™ 3-D animations, MP3Tutor sessions, videos, current events, and more.

## Summary

## Section 23.1

List the components of the nervous system, and differentiate between the CNS and PNS.

- The nervous system consists of the brain and spinal cord of the central nervous system as well as the nerves of the peripheral nervous system that carry information to and from the brain (p. 514).

Describe the roles that sensory, motor, and interneurons play in the nervous system.

- Neurons are specialized cells of the nervous system that carry chemical and electrical messages (p. 514).
- Sensory neurons carry information toward the brain; motor neurons carry information away from the brain (p. 514).
- Interneurons are located between sensory and motor neurons (p. 514).

Compare the general and special senses.

- Sensory receptors for general senses, found throughout the body, relay information to the brain. Receptors for the special senses are found in more complex sense organs (pp. 514–515).

Explain how reflexes work.

- Reflexes are prewired responses that carry messages from sensory receptors to interneurons to a motor neuron (p. 516).

## Section 23.2

Outline the structure and function of the cerebrum, thalamus, hypothalamus, cerebellum, and brain stem.

- The cerebrum is where most thinking occurs. The two hemispheres of the cerebrum consist of four lobes— temporal, occipital, parietal, and frontal. The outer surface of the cerebrum is the cortex (p. 518).
- The thalamus and hypothalamus are located between the two cerebral hemispheres. The thalamus relays information between the spinal cord and the cerebrum, and the hypothalamus regulates many vital bodily functions (pp. 518–519).
- The cerebellum is located at the base of the brain, beneath the cerebral hemispheres. It regulates balance and coordination (p. 519).
- The brain stem is located below the thalamus and hypothalamus. It controls many unconscious functions (p. 519).

## Section 23.3

Describe the structure of a neuron.

- Neurons are specialized cells with a structure that consists of the branching dendrites, the cell body, the axon, and terminal boutons (p. 520).

Differentiate between neurons comprising white and gray matter.

- Axons that are covered in myelin conduct impulses faster and comprise the white matter (p. 520).
- Unmyelinated axons conduct impulses more slowly and comprise the gray matter (p. 520).

Describe the events involved in generating an action potential.

- Nerve impulses, or action potentials, are generated when depolarization of the cell membrane occurs (pp. 520–521).
- A resting neuron is polarized in that it is negatively charged in comparison to the outside of the cell. Stimulation of a neuron opens the gates on the sodium channel proteins in the membrane, allowing sodium to diffuse into the cell and depolarizing the cell (pp. 520–521).
- The depolarization moves in a wave down the cell, activating sodium channels in adjacent parts of the membrane and thereby moving the impulse along the length of the neuron (p. 521).

Describe how an action potential is relayed across a synapse.

- The electrical impulse is propagated to the ends of the axon, which house chemical neurotransmitters (pp. 521–522).
- Nerve impulses cause neurotransmitters to be released into the synapse, from which they can bind to receptors on the next neuron (p. 522).

## Roots to Remember

**The following roots of words come mainly from Latin and Greek and will help you decipher terms:**

| | |
|---|---|
| **cereb-** | is from the Latin word for the brain. Chapter term: *cerebrum* |
| **cortex** | is the Latin word for the bark of a tree. Chapter term: *cerebral cortex* |
| **hypo-** | means under or below. Chapter term: *hypothalamus* |

# Learning the Basics

**1.** How do the CNS and PNS differ?

**2.** How do sensory, motor, and interneurons differ?

**3.** What are the general senses?

**4.** Which brain structure controls language, memory, sensations, and decision making?
**A.** cerebrum; **B.** thalamus; **C.** hypothalamus; **D.** cerebellum; **E.** brain stem

**5.** Where in the brain is the thalamus located?
**A.** in the pons; **B.** in the medulla; **C.** in the cerebral cortex; **D.** between the cerebral hemispheres; **E.** in the occipital lobe

**6.** Myelin _____.
**A.** is a protective layer that coats some neurons; **B.** gives neurons a white appearance; **C.** prevents sideways transmission of nerve impulses, thereby increasing the rate of transmission; **D.** all of the above.

**7.** An action potential _____.
**A.** is a brief reversal of temperature in the neuronal membrane; **B.** is propagated from the terminal bouton toward the cell body; **C.** begins when sodium channels close in response to stimulation; **D.** is propagated as a wave of depolarization moves down the length of the neuron.

**8.** Neurotransmitters _____.
**A.** are electrical charges that move down myelinated axons; **B.** are released across the nodes of Ranvier to hasten nerve impulse transmission; **C.** are released when an electrical impulse arrives at the terminal bouton of the postsynaptic neuron; **D.** diffuse across a synapse and bind to receptors on the cell membrane of the postsynaptic neuron.

**9.** The effects of a neurotransmitter could be increased by _____.
**A.** increasing the number of receptors on the post-synaptic cell; **B.** preventing reuptake; **C.** providing more enzymes involved in synthesizing the neurotransmitter; **D.** inhibiting enzymes involved in breakdown of the neurotransmitter from the synapse; **E.** all of the above.

**10. True or False:** A reflex arc is generated after multiple exposures to a negative stimulus such as heat.

# Analyzing and Applying the Basics

**1.** Develop a neurobiological hypothesis for why ADD might be more severe in some people than others.

**2.** Nerve impulses travel more quickly at warmer temperatures and down axons that are larger in diameter. What would you predict about the axons of squid that make their home in the coldest depths of an ocean?

**3.** Describe how reflexes differ from other nervous system responses.

# Connecting the Science

**1.** Universities have started including statements about the unauthorized use of prescription medications to enhance academic performance in their policies on academic dishonesty. Do you think using these drugs is cheating? How does this differ from using steroids to gain an athletic advantage?

**2.** Should health-care practitioners at universities allow early refills to replace lost or stolen ADD medication? If a university clinician suspects a student is not taking the pills as prescribed should he or she be allowed to ask for a urine test before prescribing more medication?

Answers to **Stop & Stretch, Visualize This, Working with Data, Savvy Reader, Sounds Right, But Is It?,** and **Chapter Review** questions can be found in the **Answers** section at the back of the book.

# CHAPTER 24

# Feeding the World

**Agricultural practices in the Midwest were partially responsible for enormous dust storms in the 1930s.**

# Plant Structure and Growth

*And then the dispossessed were drawn west…. Car-loads, caravans, homeless and hungry; twenty thousand and fifty thousand and a hundred thousand and two hundred thousand. They streamed over the mountains, hungry and restless— restless as ants, scurrying to find work to do—to lift, to push, to pull, to pick, to cut—anything, any burden to bear, for food. The kids are hungry. We got no place to live.*

This passage from John Steinbeck's classic novel *The Grapes of Wrath* describes the movement of people from the American Midwest to California during the era of the Dust Bowl. From 1931 to 1936, a combination of intense drought, severe storms, and damaging farming practices in the short-grass prairies of the Great Plains led to crop failures and

**The damage to soil that occurred during this period caused many families to lose their farms and become homeless.**

**Similar dust storms have occurred more recently in China, carrying soil across the Pacific Ocean.**

widespread soil loss. Impoverished residents of the panhandles of Oklahoma and Texas faced a desperate choice between waiting out the bad times or pulling up stakes and leaving their homes. Nearly 25% of the population in these regions left over a period of just 4 years.

The farming practices that helped create the Dust Bowl were initially heralded as great improvements. Slow horse-drawn plows and harvesters were replaced with gasoline-powered tractors and combines, allowing more and more of the native prairie to be turned into wheat fields—but plowing prairie grasses destroyed the root systems that held the soil in place. Once these grasses were gone, soil could be picked up by fierce midwestern winds and blown thousands of miles in tremendous dust storms.

The environmental and human disaster during the Dust Bowl years was not a unique event. Today,

thousands of acres of formerly productive farmland in China are turning to shifting sand. As a result, hundreds of thousands of people are moving away from this Asian Dust Bowl to cities along China's Pacific Coast. Similar migrations have occurred in west-central Africa, parts of the Middle East, southern Russia, and elsewhere around the world as agriculture expands to ever-larger areas of Earth's surface. The mistakes of the Dust Bowl era are being repeated again and again.

Aggressive efforts to reverse soil erosion, combined with the return of normal rainfall levels, finally ended the Dust Bowl era in the United States. Thanks to the advances of modern agriculture, including the use of chemical fertilizers and irrigation from underground water stores, the Great

**The Dust Bowl is now one of the most productive regions on Earth, thanks to modern technology. But is this production sustainable?**

Plains is now part of the most productive corn- and wheat-producing region on Earth. Can newly destroyed regions recover in the way that areas depleted in the American Dust Bowl did? Or is the Dust Bowl recovery an illusion maintained by huge inputs of resources that will eventually run out? To answer these questions, we must first understand the objectives and methods of agricultural practice—beginning by understanding the source of our crops and what these plants need to produce abundant food.

# 24.1 **Plants as Food**

For the first 250,000 years of human existence, our ancestors survived by hunting game and gathering fruits, seeds, roots, and other edible plant parts. Beginning about 11,000 years ago, independent human groups in what are now modern Iraq, Mexico, Peru, China, New Guinea, and elsewhere began to practice **agriculture**—that is, to plant seeds, cultivate the soil, and harvest crops. What triggered the development of this radically different way of life?

## The Evolution of Agriculture: Food Plant Diversity

Nearly all agricultural plants belong to a single phylum—the **angiosperms,** (flowering plants), which produce energy-storing seeds within fruit. Around the time agriculture developed, certain angiosperms that were suitable for cultivation may have become more common due to climate change and increased human activity.

Clues about past climate conditions can be gathered from a variety of sources, including plant remains trapped in sediments, the size of annual growth rings in ancient trees and fossil wood, and air bubbles trapped in ice sheets. According to this record, Earth's climate changed dramatically beginning about 11,000 years ago from a cool and wet period to a warmer, drier condition. The new climate also included more dramatic seasonality—wet seasons alternating with dry.

**angio-** means vessel.

**sperm-** means seed.

**Annual plants,** angiosperms that complete their life cycle from seed to adult plant to seed in the course of a single season, were successful in these new climate conditions. Annuals are also quite suitable for agriculture, providing a rapidly produced, abundant, and easily transported food source. Nearly all major crops are annual plants.

The annuals that provide the majority of calories to most societies evolved from weedy ancestors. These ancestors were well suited to growing on open and disturbed land, which often surrounded human settlements. The hardiest weeds in these environments were usually from the same two plant families: Poaceae, or grasses, in the class known as **monocots,** and Fabaceae, or peas and beans, from the class commonly known as **dicots.** Monocots and dicots get their names from the number of **cotyledons,** or embryonic leaves, they produce in their seeds (**TABLE 24.1**). The seeds of grasses (known in agriculture as cereal grains) provide a rich source of carbohydrates and 80% of the calories in a typical diet. On the other hand, the seeds of peas and beans (commonly called legumes) are rich in proteins, especially in certain amino acids that are uncommon in cereal grains. Thus, these two groups of plants together provide a reasonably balanced diet. In the Middle East, the grain-legume combinations are wheat or barley with lentils or peas; in Asia, it is rice and soybeans; and in the Americas, corn and kidney beans.

Humans realized that nutritious and highly productive wild grasses and peas were colonizing "waste areas" and probably deliberately cleared additional

**TABLE 24.1** **Monocot and dicot seeds.** The two major classes of angiosperms are named for the number of cotyledons contained in the seed.

| Name | Number of Cotyledons (Embryonic Seed Leaves) | Example |
|---|---|---|
| Monocot | 1 cotyledon | Corn<br>Wheat<br>Rice<br>Lily |
| Dicot | 2 cotyledons | Pea<br>Tomato<br>Maple<br>Dandelion |

land and scattered seed to encourage the plants' growth. Over time, the wild weeds became **domesticated** so that their growth and reproduction came under human control (**FIGURE 24.1**). The process of artificial selection, by which farmers chose which plants would produce the next generation based on their growth rates and ease of harvesting, produced crop plants that cannot survive without human intervention—and human societies that cannot survive without crop plants.

Grains and peas are not the only domesticated plants, of course, but they are among the most important. The **staple crops**, or major sources of calories, in Europe, Asia, and Central America are wheat, rice, and corn, respectively. But in other places, people get most of their calories from other crops—for example, sweet potatoes in tropical Africa and Indonesia, white potatoes in South America, and manioc (also known as cassava or tapioca) in Central and South America and Western Africa. While potatoes and manioc are also angiosperms, their food value comes not from their fruit or seeds but from other plant organs.

**(a) Teosinte, ancestor of modern corn**

**(b) Modern corn**

**FIGURE 24.1 The effect of domestication.** (a) Teosinte, the ancestor of corn, produces a few small, loosely held kernels in a fragile "ear" that disintegrates when the kernels are ripe. (b) In contrast, corn's kernels are large, numerous, and do not detach from the cob.

◀ **Visualize This**

**Describe the actions of early farmers that would have caused teosinte to evolve via artificial selection into the familiar ear of corn.**

## Plant Structure

Angiosperms have a relatively simple structure, consisting of three nonreproductive **vegetative organs** and two **reproductive organs**. The three vegetative organs are the roots, stems, and leaves; the reproductive organs are flowers and fruit (**FIGURE 24.2**).

Plant organs are made up of only four types of tissue (**FIGURE 24.3**). The primary function of **meristematic tissue** is cell division, and it is most commonly found at the tips of roots and stems. Cells arising from meristematic tissue are undifferentiated, and their adult form depends on their location in a growing plant; thus, unlike animals, a plant does not have a completely specified final form. **Epidermal tissue** serves as an outer covering of the plant body, functioning in water absorption below ground and water retention above ground. **Vascular tissue** is made up of two subtypes of liquid-conducting tissue: **xylem**, which carries water and minerals from the roots to the leaves, and **phloem**, which carries food, in the form of sap, throughout the plant. **Ground tissue** makes up the remainder of the plant body and may consist of cells that perform photosynthesis or store starch, among other roles. The relative amounts of these types of tissue in a vegetative organ often determine its nutritional value; more vascular tissue means more fiber, but more ground tissue means a larger percentage of usable nutrients.

**Roots.** The primary function of a plant's root system is to absorb water and minerals from the soil. The roots of most monocots are fibrous and spread out broadly, giving these plants the ability to exploit water as soon as it becomes available in the soil. However, the large central taproot of most dicots permits these plants to access minerals and water found much deeper in the soil

**FIGURE 24.2 Plant structure.** Plants consist of five organs: the vegetative structures—roots, stems, and leaves—and the reproductive structures—flowers and fruit.

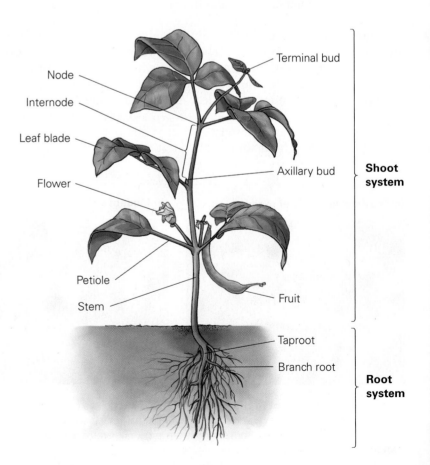

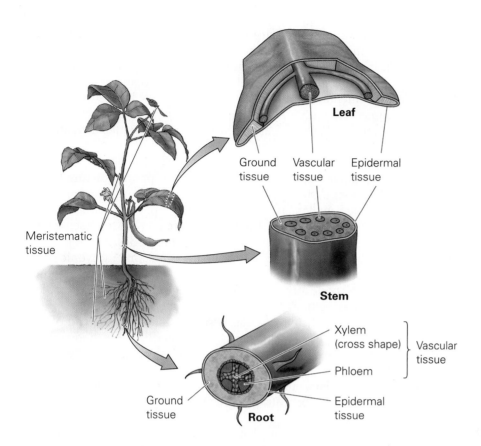

Leaf

Ground
tissue

Vascular
tissue

Epidermal
tissue

Meristematic
tissue

Stem

Xylem
(cross shape)
⎤
⎦ Vascular
tissue

Phloem

Ground
tissue

Root

Epidermal
tissue

◀ **Visualize This**

Based on this image, which of the
four tissue types do you think is
most common in the plant body?

**FIGURE 24.3 Plant tissue
systems.**  Plants possess four tissue
types—epidermal, ground, vascular, and
meristematic.

(**FIGURE 24.4**). Most of the water and minerals obtained by a plant are
absorbed across the surface of root hairs, tiny projections of the epidermal cells
(**FIGURE 24.5**). The meristematic tissue at the tips of roots allows the root
system to extend through the soil via growth.

**STOP & STRETCH**   If a gardener removes the stem and leaf of
a weedy plant without pulling out its roots, the weed will very often regrow. What does this
imply about the capabilities of plant cells?

**Visualize This** ▼

How do root hairs improve a root's
capacity to absorb water?

**(a) Typical monocot root system**

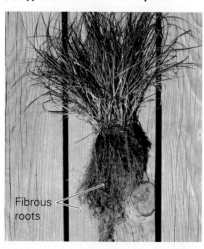

Fibrous
roots

**(b) Typical dicot root system**

Taproot

**FIGURE 24.4 Monocot and dicot root systems.**  (a) Most monocots produce a fibrous
root system that competes well for water but is not a strong support structure. (b) Most
dicots produce a much sturdier, deeper taproot.

**FIGURE 24.5 Root hairs.**  The growing
ends of a plant's roots are covered with tiny,
hairlike extensions of the epidermal cells.

**FIGURE 24.6 Manioc or cassava.** The storage roots of this plant are a staple in the diet of people throughout southeast Asia and elsewhere in the tropics. The roots contain large amounts of a cyanide-producing acid and must be extensively processed to make them edible.

---

**enni-** pertains to a year.

---

**rhizo-** means root.

---

**FIGURE 24.7 A modified stem.** A tuber is a modified stem, as the presence of sprouting buds on these potatoes indicates. These tubers are produced on another stem modified for underground life, a rhizome.

Roots may also serve as the primary storage organ of a plant, and some that store carbohydrates in their ground tissue, such as carrots, parsnips, and sugar beets, have been modified into agricultural crops. Many plants with storage roots are not annuals but **perennials** that live for many years. Perennials survive difficult environmental conditions by using food they have stored during favorable environmental conditions. Perennials grown as crops, such as sweet potatoes and manioc, are harvested after 1 year of growth to take advantage of this stored carbohydrate (**FIGURE 24.6**).

**Stems.** The principal role of the stem is to serve as a support structure for the other organs of the shoot system (see Figure 24.2). At the apex, or top, of the stem is the terminal bud, which encloses the growing tip, or **apical meristem**, of the plant. The nodes of the stem, the point where leaves are attached, are separated from each other by sections of stem called internodes. At the junction between the leaf and stem at each node is an axillary bud, an embryonic shoot containing meristematic tissue. Most axillary buds have the ability to grow into a new lateral node-internode structure (that is, a vegetative branch), although some will become flowering branches that cannot branch further.

Most stems contain significant vascular tissue, which is fibrous and of limited food value. However, like roots, stems can be modified for food storage, and some storage stems have become major crops. White potatoes and yams are tubers, enlarged structures found on underground stems called **rhizomes**. The "eyes" on a potato are axillary buds, each with the ability to sprout into a stem containing nodes and internodes (**FIGURE 24.7**).

**Leaves.** The "food factories" of a plant are its leaves, the primary site for photosynthesis. While no staple crops are leaves, leafy greens, including lettuce, spinach, chard, kale, and bok choy, are rich in essential vitamins and minerals and are important components of diets in many parts of the world. The leaf part that is more commonly consumed is the flat, photosynthetic blade, rather than the stiff and fibrous petiole, which consists primarily of vascular tissue. Humans have also domesticated plants that store food in modified leaves. Onions and other similar plants produce bulbs, underground storage structures that consist of short stems and fleshy, carbohydrate-filled leaves.

Humans have domesticated crops based on all imaginable plant parts (**FIGURE 24.8**). However, the vast majority of our foods are based on reproductive organs, primarily fruits and seeds.

## Plant Reproduction

We may not think of our breakfast toast and cornflakes as belonging to the same category of foods as our orange juice or bananas, but all of these foods are products of **fruit**, defined by botanists as mature flower ovaries containing seeds. Flowers and fruit are structures that are unique to angiosperms.

**Flowers** are the sexual organs of angiosperms (**FIGURE 24.9**). The typical flower produces both male and female structures. Male gametes are contained in pollen. The pollen is produced in structures called **anthers** on the male reproductive organ, called the **stamen**. Female gametes are contained in structures called **ovules** and produced in one or more chambers, called the **ovary**, of a flower's female reproductive organ, the **carpel**. The transfer of pollen to the **stigma**, a sticky pad on the top of a carpel, is called **pollination**. Pollination may occur via wind, as in corn and wheat, or via insects or other animals, as in soybeans. A flower's **petals** are adapted to facilitate pollination. After landing on the stigma, pollen grains germinate into tubes that extend down the length of the carpel to the ovules. A developing flower is enclosed in protective, modified leaves called **sepals** that in some plants also facilitate pollination.

Seed  Fruit  Leaf  Tuber (modified stem)

Flower  Root  Endosperm (of seed)  Bulb (modified stem with leaves)

**FIGURE 24.8 Plants as food.** Our diet includes a diversity of plant parts, including fruits, seeds, roots, and flowers.

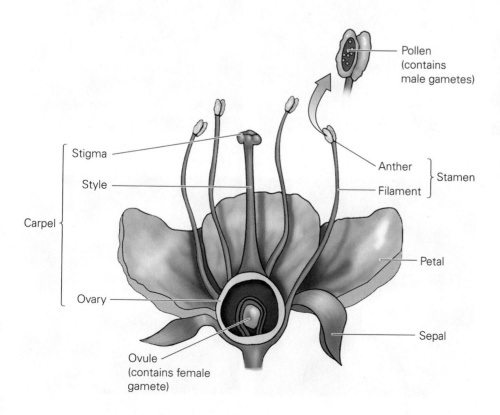

Pollen (contains male gametes)

Stigma

Style

Anther

Filament

Stamen

Carpel

Petal

Ovary

Sepal

Ovule (contains female gamete)

◀ **Visualize This**

**What are the advantages and disadvantages of producing both male and female parts in the same flower?**

**FIGURE 24.9 Flower structure.** The typical flower contains both male and female reproductive organs, although many angiosperm species produce two types of flowers: male only and female only.

In the case of wind-pollinated grass, petals are much reduced, whereas in insect-pollinated legumes, they are colorful and contain nectar to attract and reward their pollinators (**FIGURE 24.10a–b**). Many plants are pollinated only by a single species or group of species. These specific relationships evolved as pollinators selected plants that gave them the best rewards, and plants developed traits that were highly attractive to particular pollinators (**FIGURE 24.10c–e**).

Once in the ovules, the pollen tube releases two sperm. In a process known as **double fertilization,** one sperm fertilizes the egg to become the diploid embryo, and the other fuses with two nuclei within the ovule to produce a triploid cell, containing three copies of each chromosome. This cell divides and develops into **endosperm,** tissue that nourishes the developing embryo. The primary component of wheat flour is the starchy endosperm of wheat seeds; whole wheat flour also contains the embryo (the germ) and accessory tissues (the bran), whereas white flour is strictly the endosperm. In some species, such as in legumes, the embryo consumes the endosperm as it develops and the seed matures. The two halves of a bean seed or peanut are enlarged cotyledons produced by the digestion of the endosperm.

## Visualize This ▼

Some flowers have evolved to look like female wasps and even produce a scent that smells like the wasp's sex hormone. What animal likely pollinates these flowers?

**(a) Wind-pollinated flower**

**(b) Bee-pollinated flower**

**(c) Bird-pollinated flower**

**(d) Bat-pollinated flower**

**(e) Fly-pollinated flower**

**FIGURE 24.10 Petals and pollination.** A flower's petals are adapted to aid in the process of pollination. (a) The most successful wind-pollinated plants have reduced petals that do not interfere with pollen transfer. (b) Bee-pollinated flowers often provide a nectar reward at the base of a petal tube, while (c) bird-pollinated flowers typically have petals fused into a single deep, highly visible floral tube. (d) Flowers that are pollinated by bats often open their petals only at night. (e) The petals of fly-pollinated flowers may have the appearance—and odor—of rotting flesh.

**(a) Wind-dispersed fruit**   **(b) Water-dispersed fruit**   **(c) Animal-dispersed hitchhiker fruit**   **(d) Animal-dispersed edible fruit**

**FIGURE 24.11 Seed dispersal units.** Fruits are adapted to disperse seeds away from the parent plant. Many mechanisms are possible, including (a) wind, (b) water, (c) mammal fur, and (d) through an animal's digestive tract.

The embryo, endosperm, and remainder of the ovule tissues make up the **seed.** Triggered by successful fertilization, other parts of the flower—in nearly all cases, the ovary, but sometimes petals, sepals, or the flower stalk—develop into fruit. Most fruits are adapted to be vehicles for dispersing seeds far from the parent plant. Some fruits help seeds disperse on wind or water; other fruits help seeds disperse by using animals to carry the fruit away and drop its seeds elsewhere (**FIGURE 24.11**). Because humans have similar sensory systems and nutritional needs to those of other mammals, many of the crop plants that we rely on were selected from fruits adapted to attract mammal dispersers. Even grains are adapted to attract mammals, primarily rodents. **FIGURE 24.12** illustrates the relationship among flowers, seeds, and fruit in angiosperms.

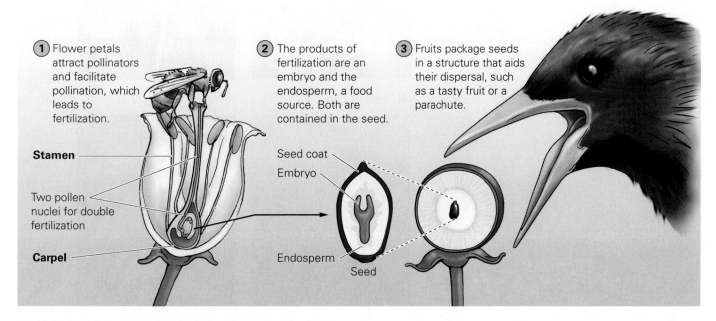

1 Flower petals attract pollinators and facilitate pollination, which leads to fertilization.

2 The products of fertilization are an embryo and the endosperm, a food source. Both are contained in the seed.

3 Fruits package seeds in a structure that aids their dispersal, such as a tasty fruit or a parachute.

**Stamen**

Two pollen nuclei for double fertilization

**Carpel**

Seed coat
Embryo

Endosperm
Seed

**FIGURE 24.12 A summary of angiosperm reproduction.**

**The Great Plains is now part of the most productive croplands on Earth.**

**STOP & STRETCH** If a parent plant is already successful, the conditions where it is growing must be ideal for that species. What is the advantage of dispersing seeds rather than keeping them near an obviously successful parent plant?

Producing abundant fruit for human consumption requires more than simply selecting plants for optimal fruit production. The agricultural practices of people around the world are designed to maximize the growth of these plants.

# 24.2 Plant Growth Requirements

Agricultural practices such as turning over soil, irrigating soil with water, and adding nutrients to soil developed independently everywhere that plants were domesticated by people. To understand why these techniques enhance agricultural production, we must understand first how plants grow.

## How Plants Grow

The first stage of plant growth is the emergence of the seedling, supported by the seed's internal food stores. Later plant growth can be divided into two types—growth in length, called **primary growth**, and growth in girth or circumference, called **secondary growth**.

**Germination.** The emergence of an embryo from a seed is called **germination**. The process begins when the seed begins to take up water, expands, and ruptures its confining seed coat (**FIGURE 24.13a**). As long as the seed coat is

**FIGURE 24.13 Germination.** (a) Seed germination begins when the seed takes up water. (b) The first organ to emerge is the root. (c) Water obtained by the root makes the cells in the embryonic shoot enlarge, allowing the shoot to emerge from the soil. (d) Germination is complete when the seedling is functioning independently of the stored food in the seed.

(a)

(b)

(c)

(d)

impermeable to water, the seed remains dormant, which means that it will not germinate. Dormancy allows seeds to act as "time travelers," passing through poor environmental conditions in a state of suspended animation. Many seeds require exposure to specific environmental conditions before they can escape dormancy.

> **STOP & STRETCH** In temperate regions with cold winters, some seeds need to be exposed to low temperatures for a certain amount of time before they can germinate. What is the advantage of this strategy in these environments?

The first structure to emerge from a germinating seed is the root, which enables the developing young plant, or **seedling**, to become anchored in the soil and absorb water (**FIGURE 24.13b**). Growth of the shoot soon follows as the seed's stored food is depleted, and photosynthesis becomes necessary (**FIGURE 24.13c**). The time between germination and when the seedling is established as an independent organism is the most vulnerable period in the life of a plant. Damage to the plant during this period is usually fatal. Once a seedling becomes established, however, damage often may be overcome by continued growth (**FIGURE 24.13d**).

### Primary Growth.

In plant stems, primary growth takes place through the production of additional nodes and internodes by growth at the shoot tips and by the growth of axillary buds. Growth in length occurs when cells at a meristem divide in a single plane, creating columns of daughter cells below them and pushing the apical meristem farther up into the air (**FIGURE 24.14**). A hormone produced by the terminal bud inhibits cell division in the axillary buds in many plants. Once the terminal apical meristem has grown away from a particular axillary bud, the bud's meristem begins to divide and produces a lateral branch, also containing axillary buds.

The node-internode system of the main stem and branches supports the **indeterminate growth** of a plant's shoot system—its ability to grow throughout its lifetime. This is in contrast to the **determinate growth** of most animals, which grow to a set size that they typically reach early in life. Not all plant growth is indeterminate—leaves and flowers tend to grow to a set size, for instance.

In roots, primary growth at the tip occurs due to cell division in the apical meristem. Cell division in the root occurs in two opposing directions, resulting in columns of cells that elongate and push the root tip forward through the soil, as well as the production of a loosely organized **root cap** that covers the tip of the root. As the root tip pushes through the soil, exterior cells of the root cap are worn off and replaced by younger cells more recently produced by the apical meristem. The primary meristematic tissue producing lateral roots is found near the center of the root, surrounding the vascular tissue. As a result, lateral roots can form anywhere along the length of a root (**FIGURE 24.15**).

Primary growth results in plants that are taller and bushier and have more extensive root systems. In longer-lived plants, continued primary growth presents a dilemma—a tall plant cannot support itself with a skinny stem. Many plants can increase their stability by making wider stems and roots via secondary growth.

## Visualize This ▼

Identify the nodes and internode visible in this micrograph.

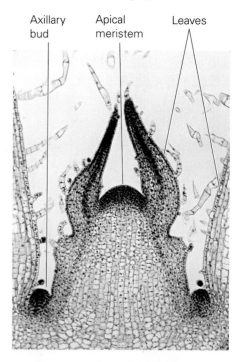

FIGURE 24.14 **Primary growth in stems.** This microscopic image of an apical meristem clearly shows the repeating node-internode system.

## Visualize This ▼

Why must the lateral root originate from the vascular cylinder in the center of the primary root?

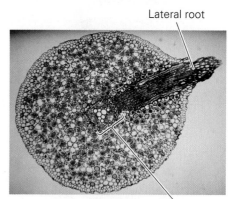

FIGURE 24.15 **Lateral root formation.** In this cross-section, meristematic tissue in the center of the root has produced a lateral root.

**Secondary Growth.** All of the flowering plants that undergo true secondary growth are dicots. In these plants, a ring of cells running through the bundles of vascular tissue in the stem becomes a meristem called the vascular cambium (**FIGURE 24.16**). This cambium produces phloem cells to the outside of the stem and xylem cells to the inside.

The additional xylem cells produced by the cambium make up the bulk of the material in these stems, and their stiffened cell walls produce the stem's **wood** (**FIGURE 24.17**). Epidermis on the surface of the stem breaks apart and is shed as the stem's surface area increases. Cells immediately under the epidermis become another meristem (called the **cork cambium**). Cell division in the cork cambium results in the production of **cork**, a water-resistant outer covering. Phloem, cork, and cork cambium make up the **bark** of a woody dicot stem. All the components of bark are continually shed and replenished by secondary phloem cells, which are produced by the vascular cambium as the tree increases in diameter.

> **STOP & STRETCH** Removing a small strip of bark around the complete circumference of a tree, a technique called girdling, always eventually kills the tree. In contrast, removing a large patch of bark from one side of a tree often does not kill the tree. What function of bark is totally disrupted by girdling but not totally disrupted by loss of a large patch?

Interestingly, although dicot trees are much more common than monocot trees, the only tree-based staple food in human societies is the monocot coconut, which is important to communities in Indochina and Central America. The thick stem of coconut palms consists of additional strands of vascular tissue, produced by another meristem that surrounds the apical meristem. The structure of this stem makes palm trees incredibly flexible (**FIGURE 24.18**).

Primary and secondary growth of plants can be maintained only by abundant energy. Plants obtain this energy by transforming light into chemical energy in the process of photosynthesis. Our agricultural practices are intended to maximize the photosynthesis of crop plants so that they will produce a surplus of carbohydrates that we can harvest.

**(a) Monocot stem**
Vascular tissue scattered

**(b) Dicot stem**
Vascular tissue in a ring

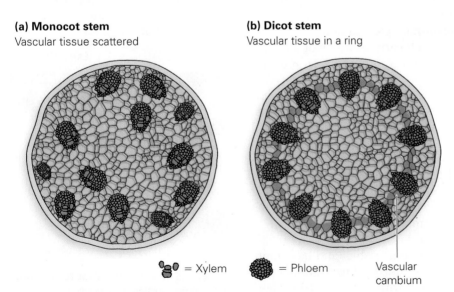

= Xylem    = Phloem    Vascular cambium

**FIGURE 24.16 Monocot and dicot stems.** Unlike monocot stems, which have scattered bundles of vascular tissue, dicot stems have vascular bundles arranged in a ring. This organization allows a single ring of meristematic tissue, the vascular cambium, to form between the xylem and phloem.

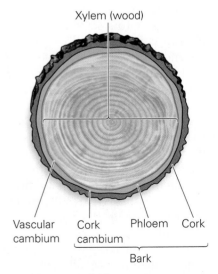

Xylem (wood)

Vascular cambium    Cork cambium    Phloem    Cork

Bark

**FIGURE 24.17 Dicot wood.** The wood of a dicot stem is made up of xylem cells, while the bark consists of phloem and cork.

**(a) Palm tree trunk cross-section**

**(b) Palm trees bending in hurricane-force winds**

**FIGURE 24.18** **Monocot "wood."** Secondary growth in monocots, such as this palm, produces much more fibrous wood (a), which is not nearly as strong as dicot wood but is very resilient (b).

## Maximizing Plant Growth: Water, Nutrients, and Pest Control

The basic equation of photosynthesis is as follows:

$$\text{Carbon Dioxide} + \text{Water} + \text{Light Energy} \rightarrow \text{Carbohydrate} + \text{Oxygen Gas}$$

In other words, photosynthesis in plants converts energy from sunlight, carbon dioxide from the atmosphere, and water from the soil into energy-rich carbohydrate molecules. Oxygen is essentially a "waste product" of this reaction (Chapter 5).

If we look only at this chemical reaction, it is clear that plants require carbon dioxide, water, and light to produce carbohydrates. However, the process also requires a few more components. For instance, to survive, plants require much more water than that needed simply for the chemical reactions of photosynthesis. Plants also require other chemical inputs in smaller amounts. For example, photosynthesis requires the pigment **chlorophyll** to proceed, and magnesium is an essential element in chlorophyll. Inputs such as magnesium and nitrogen, an essential element in proteins, are obtained from the soil. Pest organisms that damage photosynthetic tissue and fruits and seeds also affect plant production. Therefore, to increase crop production, modern farming practices attempt to maximize the amount of light, water, carbon dioxide, and nutrients available to crop plants as well as to minimize damage caused by plant predators (**FIGURE 24.19**).

**Plant growth requirements:**

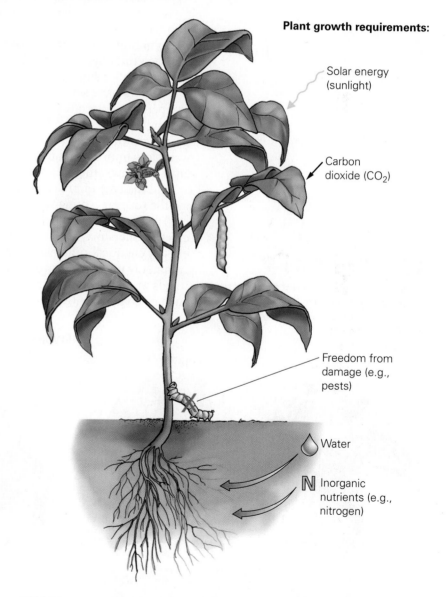

Solar energy (sunlight)

Carbon dioxide ($CO_2$)

Freedom from damage (e.g., pests)

Water

N Inorganic nutrients (e.g., nitrogen)

**FIGURE 24.19** **Five requirements for plant growth.** Plants require the basic ingredients for photosynthesis—water, carbon dioxide, and light—as well as nutrients from the soil and freedom from damage to produce excess carbohydrates for human consumption.

**Water, Sunlight, and Carbon Dioxide.** To hold its leaves perpendicular to the sun's rays, a plant's cells must be full of water; that is, they must be turgid (**FIGURE 24.20**). Recall that the membrane of a plant cell is surrounded by a stiff cell wall. The cell wall is tough, but it is also elastic, meaning that it can stretch and still hold its basic shape. As a plant cell's central vacuole fills with water, the cell wall balloons slightly, becoming turgid. It is the pressure of adjoining turgid cells that holds the leaves upright. When cells lose water, the pressure of cells pushing against each other decreases; the cells become flaccid, and the plant wilts. Thus, part of maximizing a plant's exposure to light is ensuring that it has plenty of water.

In many regions, it is necessary to prevent wilting of crop plants by adding water to the soil through **irrigation** (**FIGURE 24.21**). The enormous production of corn, wheat, and soybeans in the former Dust Bowl region is a result of an equally enormous amount of irrigation. The water for irrigation in this dry region comes from a huge underground reservoir called the Ogallala aquifer.

Another way to increase the amount of water that reaches preferred plants is to reduce the amount of competition for that water. Farmers do this by trying to minimize the number of nonpreferred plants, commonly called **weeds**, in agricultural fields. Controlling weeds has the additional benefit of reducing competition for sunlight.

Farmers have several techniques for controlling weeds. One is to remove competitors before the crop is planted. This process, called tilling, involves turning over the soil to kill weeds that have sprouted from seed. After a crop begins growing, it is possible to prevent the growth of competitors and reduce water loss from the soil by mulching the crop. To mulch, farmers spread a thick layer of straw, other dead plant material, sheets of newspaper, or dark plastic around the base of preferred plants. These materials block sunlight from reaching the soil, preventing weed seeds from germinating.

Tilling does not kill all weeds, and some spring back to life quickly. In these cases, farmers may use **herbicides,** chemicals that kill plants. Herbicides that exploit chemical differences between dicot and monocot plants are the most commonly used sprays in agriculture. These herbicides are used to control dicot weeds in monocot crops such as corn and wheat. Broad-spectrum herbicides, such as glyphosate (Roundup®), kill all plants; until recently, these herbicides

# Visualize This ▼

**What do you think would happen to the cells of a plant if it was watered with very salty water, and why?**

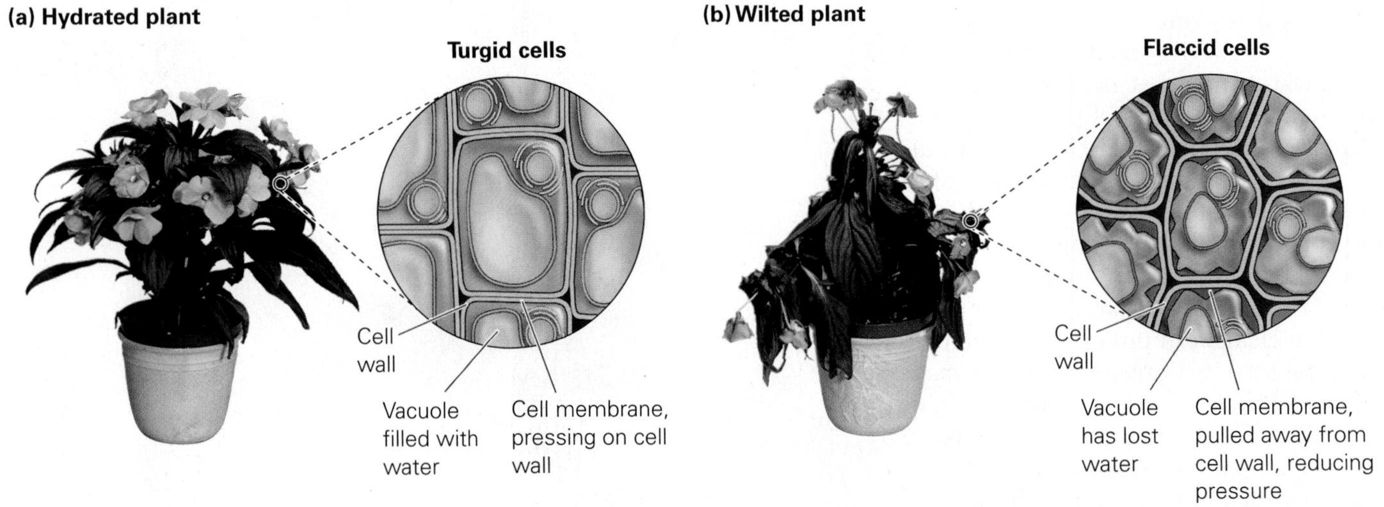

**(a) Hydrated plant**

Turgid cells

Cell wall

Vacuole filled with water

Cell membrane, pressing on cell wall

**(b) Wilted plant**

Flaccid cells

Cell wall

Vacuole has lost water

Cell membrane, pulled away from cell wall, reducing pressure

**FIGURE 24.20 Water in plant cells.** (a) When a plant is fully hydrated, the cell's vacuoles balloon with water, and the cells press against each other so that the plant remains "crisp." (b) As the plant dries out, the vacuoles lose water, and the cells no longer support each other, causing the plant to wilt.

have been useful only for controlling weeds before planting. Development of genetically modified organisms (GMOs) has increased the applicability of broad-spectrum herbicides to many more crop-and-weed combinations in recent years (Chapter 9).

Even when plants are fully hydrated and not wilting, their efforts to prevent water loss can interfere with growth. This is because plant strategies for conserving water can impose a cost to photosynthetic production.

The primary adaptation that plants have to reduce water loss is a waxy covering, called the cuticle, on the epidermis of the shoot system. The cuticle also restricts the diffusion of carbon dioxide into the plant. To enable plant cells to obtain adequate carbon dioxide for photosynthesis despite the cuticle, the shoot epidermis is pocked with tiny pores called **stomata** (singular: stoma). However, these pores come with a price—stomata also provide portals through which water can escape.

In most plants, each stoma is encircled by a pair of **guard cells** that regulate the size of the pore (**FIGURE 24.22**).

**FIGURE 24.21 Irrigation.** The circle pattern of lush plant growth derives from center-pivot irrigation sprayers, which spray water in an arc from the center point. The smaller circles in this image have a diameter of about 800 m.

## Visualize This ▼

Given the two competing roles of stomata, how do you expect the number of stomata to differ between plants in desert areas compared to those in tropical rain forests?

**(a) When water is abundant:**

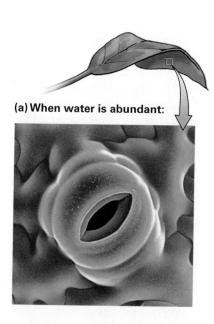

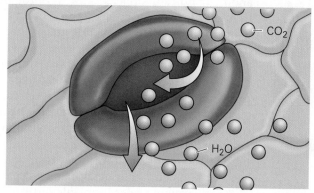

- Guard cells swell.
- Stoma (pore) is large.
- High carbon dioxide uptake
- High water loss

**(b) When water is scarce:**

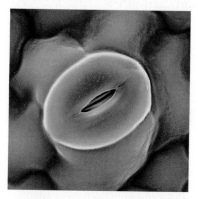

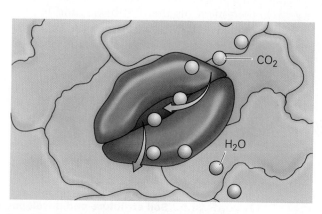

- Guard cells shrink.
- Stoma (pore) is small.
- Low carbon dioxide uptake
- Low water loss

**FIGURE 24.22 Guard cells.** Guard cells help to regulate the size of stomata on a plant's surface to minimize water loss or maximize carbon dioxide intake.

When the guard cells have abundant water, they increase in size, and their unique shape causes the pore to open. When they are water deprived, such as when the soil is dry, the skies are sunny, and the humidity is low, guard cells shrink and relax, causing the pore to be small. Thus, to maximize carbon dioxide uptake, plants need enough water to keep their stomata open.

Where precipitation or irrigation is sufficient to keep plants from wilting and the stomata open, the most important factor affecting plant growth is nutrient availability.

**Nutrients and Soil.** Elements needed by plants in large amounts, such as nitrogen and magnesium, are called macronutrients (**TABLE 24.2**). Nutrients needed in much smaller amounts—such as iron, which is required in the processes of chlorophyll synthesis and aerobic respiration—are known as micronutrients. Both categories of nutrients are available in the soil.

**Soil** is a unique material composed of both living and nonliving components. Healthy soil consists of a base layer of recently eroded rock; a top layer made up of dead and decaying remains of plants and animals, along with living organisms that eat them, and waste; and a middle layer where components of the base and top layers are mixed by earthworms and other soil organisms (**FIGURE 24.23**). The primary way to maximize a crop plant's access to soil nutrients is to reduce competition with weeds—using the same techniques used to reduce competition for water.

In natural systems, nutrients become available to plants through nutrient cycling (Chapter 15). In agricultural systems, the nutrient cycle is often broken. Instead of plants being eaten by animals, which are then eaten by other animals and decompose all in the same location, the plants from agricultural fields are removed and trucked to other locations where they will be consumed and their nutrients released. Most of the nutrients that are consumed end up in human or animal waste, which often flows into lakes or oceans. In other words, nutrients on croplands are not recycled back into the soil but mined from it (**FIGURE 24.24**).

Because nutrient cycles are broken in agricultural fields, farmers must add nutrients to soils in the form of **fertilizer.** Fertilizer that is applied to the soil can be **organic**—that is, it has carbon-containing molecules made up of the partially decomposed waste products of plants and animals. Fertilizer can also be **inorganic**—meaning having simple molecules that lack carbon, such as ammonia and nitrate, produced by an industrial process. Because plants require nutrients in inorganic form, soil organisms must first decompose organic fertilizer before its nutrients are available for plant growth.

Most farmers prefer inorganic fertilizer for two reasons: because the nutrients in it are more concentrated and much easier to transport, store, and apply than most organic fertilizers; and because plants can utilize these nutrients immediately. Most farmers in the United States fertilize with inorganic nitrogen, phosphorus, and potassium and occasionally will add other macronutrients such as calcium.

**STOP & STRETCH** Most people do not harvest their houseplants, yet many "feed" them inorganic fertilizer every few months. Why do houseplants need fertilizer?

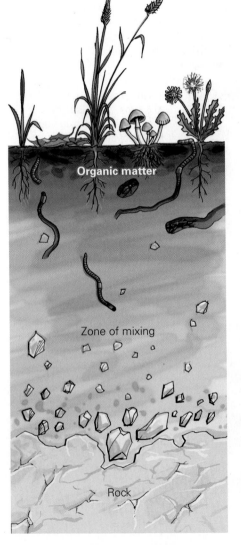

**FIGURE 24.23 Soil structure and development.** Soil forms from the breakdown of rock and the accumulation of organic matter from dead and decaying plants, fungi, animals, and animal wastes.

**TABLE 24.2 Plant nutrients and their functions.**

| Primary Functions | |
| --- | --- |
| **Macronutrient** | |
| **Nitrogen** | Component of proteins, DNA, and RNA |
| **Potassium** | Involved in opening and closing of stomata |
| **Calcium** | Component of cell walls; involved in cell membrane permeability |
| **Magnesium** | Component of chlorophyll |
| **Phosphorus** | Component of ATP, DNA, RNA, and cell membranes |
| **Sulfur** | Component of proteins |
| **Micronutrient** | |
| **Chlorine** | Involved in moving water into and out of cells |
| **Iron** | Required for chlorophyll synthesis and aerobic respiration |
| **Boron** | Influences calcium use; required for DNA and RNA synthesis |
| **Manganese** | Required for integrity of chloroplast membrane |
| **Zinc** | Component of many enzymes |
| **Copper** | Component of some enzymes |
| **Nickel** | Required for nitrogen metabolism |
| **Molybdenum** | Required for nitrogen metabolism |

# Visualize This ▼

How would the nitrogen flows in this image differ if
the system were a natural environment?

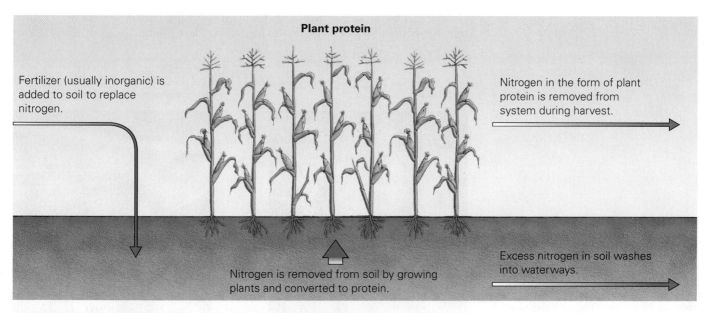

Plant protein

Fertilizer (usually inorganic) is added to soil to replace nitrogen.

Nitrogen in the form of plant protein is removed from system during harvest.

Nitrogen is removed from soil by growing plants and converted to protein.

Excess nitrogen in soil washes into waterways.

**FIGURE 24.24 Nutrient mining.** The movement of nitrogen in an agricultural system is illustrated here.

Farmers can also replace some nutrients in the soil through **crop rotation**—that is, by varying the crops that are planted on any one field over time. Surprisingly, some plants actually increase the levels of nutrients in the soil. These plants have a mutualistic relationship with **nitrogen-fixing bacteria** (**FIGURE 24.25**), which are microbes with the ability, unique among living organisms, to convert nitrogen gas from the atmosphere into a solid form—namely, ammonia.

Legumes develop structures called **nodules** on their roots in response to chemical signals from certain species of nitrogen-fixing bacteria. The nodules house the bacteria and supply them with carbohydrates. In return, the plants receive the excess nitrogen that the bacteria fix. When a legume is growing well, the bacteria fix so much nitrogen that the excess is released into the surrounding soil. If legumes are planted in alternating years with a crop that does not fix nitrogen (such as corn or wheat), the need for nitrogen fertilizer can be greatly reduced because of the release of this excess nitrogen and because the nitrogen-fixing crop remains can be plowed back into the soil. The benefits of crop rotation may be one reason that so many early agricultural societies relied on staples of grains and legumes.

Once a farmer has maximized the growth of his or her plants by providing abundant water and nutrients, these plants become attractive targets to other consumers—including other mammals, birds, insects, fungi, and bacteria. To ensure that most of the food produced by an agricultural crop goes to feed people, a farmer must try to reduce the damage these competitors do to the crop.

**Freedom from Pest Damage.** Living organisms that cause damage to agricultural products are generally referred to as pests. Up to 40% of the potential food production in major crops is lost to pests every year. The primary tools farmers use to reduce pest impact are **pesticides**, chemicals designed to kill or reduce the growth of a target pest. (Herbicides are often considered a class of pesticides. In this book, however, we use the word *pesticide* only when referring to chemicals that control insect, fungal, and bacterial pests.) Pesticide use became widespread in the years following World War II. Today, millions of tons of these chemicals are applied annually to crops all over the world.

While pesticide use has been reduced recently, thanks to integrated pest management, described in the next section, these chemicals are still in use in

**(a) Alfalfa**

**(b) Nodules on alfalfa roots**

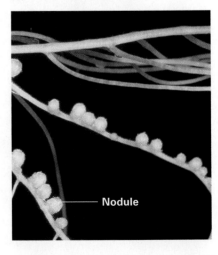

Nodule

**FIGURE 24.25 Nitrogen-fixing crop.** (a) Alfalfa can increase the nitrogen content of soil. (b) The nodules on the roots of alfalfa contain nitrogen-fixing bacteria.

modern agriculture. Surprisingly, pesticide use has not considerably reduced the overall loss of crops to pests. Instead, these chemicals have dramatically influenced farming methods—primarily by allowing farmers to plant a single crop over a wide acreage. This practice is called **monoculture.** Monocultural production of crops is very efficient. Farmers have one planting, fertilizing, and pest control schedule, and they require only the planting and harvesting equipment specific to their crop (**FIGURE 24.26a**). Monoculture and the accompanying mechanization of agriculture have greatly reduced the amount of labor required to produce crops. The percentage of the population working on farms in the United States has dropped from 25% to less than 2% in the past 60 years due to this change in farming practice.

On the other hand, monocultures are very susceptible to outbreaks of pests. If a small population of insects or disease infestation consumes or kills a host plant, another appropriate host for the pest is right next door. The easy availability of resources to pests in monocultural crops allows their populations to grow exponentially and can lead to the loss of an entire crop. Thus, while monoculture increases the efficiency of agricultural production, it results in an increased need for pesticides just to keep pest damage at historical levels.

Farmers can also use **cultural control** to minimize pest populations. Because most pests have evolved to attack a single crop, they will not cause damage to other, unrelated crops. Crop rotation, in addition to conserving soil nutrients, is a method of cultural control that moves plants away from their pests. After the growing season, a pest population spends the winter in the soil or on plant waste left in the field. If the same crop is planted the following year, the pest population has easy access to more resources and can continue to increase in population. However, if a different crop is planted, the pests must disperse in the spring in an attempt to find their required crop. Only some of these dispersers will find their host, and the population of pests in the new site will be relatively small.

Another form of cultural control is **polyculture**—planting many different crop plants over a single farm's acreage (**FIGURE 24.26b**). Unlike monocultures, polycultural planting keeps pest populations relatively small because, after

**(a) Monoculture**

**(b) Polyculture**

**FIGURE 24.26 Monoculture and polyculture.**  (a) Planting a single crop over many hundreds of acres (monoculture) allows for the efficient use of massive planting and harvesting equipment, as in this Alberta wheat field. (b) Planting many different crops together (polyculture) can reduce pest problems and buffer farmers from crop failures.

**From 1931 to 1936, intense drought and damaging farming practices in the Great Plains led to crop failures and population loss.**

the pests have consumed a patch of plants, they must disperse long distances to find another source of food. Polyculture also minimizes risk to farmers by ensuring that a pest outbreak on a single crop does not destroy all of the farm's production for that year.

Over the last 60 years, the use of pesticides, fertilizer, and large farm machinery has dramatically changed the way food is produced around the world. The revolution in biology—particularly the new science of genetics—has also played a role in this change.

## Designing Better Plants: Hybrids and Genetic Engineering

Traditionally, farmers used artificial selection to modify the characteristics of their crop plants and animals; that is, they selected for next year's seed stock or breeding animals those that performed best under the conditions on their farm. However, this process limits farmers to only the variation that is currently available. Two developments in plant science during the twentieth century reduced and finally eliminated this limitation.

Beginning in the 1920s, agricultural scientists began using plant breeding to produce better crops. Plant breeding generally consists of creating hybrids, the offspring of two different varieties of an agricultural crop. One of the first hybrids produced on a large scale was dwarf spring wheat. One parent variety was a nitrogen-loving, high-producing wheat that was difficult to harvest because it tended to fall over when the wheat grains were ripe. The other parent variety was a lower-producing dwarf wheat plant. The hybrid of these two varieties was high producing but short, so it responded well to fertilizer application and was easy to harvest with large machines. Because hybrids are the first-generation offspring of two distinct varieties, new hybrid seeds have to be produced every year from the parent populations. This is necessary because independent assortment occurs during meiosis (Chapter 7); thus, seeds produced by the hybrid plant will not contain the mix of chromosomes found in the first-generation plants.

In addition to the difficulty of producing hybrid seeds, plant breeding limits agricultural scientists to the genetic variation within a single species. More recently, DNA technology has permitted the creation of novel agricultural organisms containing genes from other species. These genetically modified organisms, or GMOs, include herbicide-resistant crop plants. In the United States, over 40% of farmland planted with corn, more than 70% of the cotton crop, and nearly 80% of soybean fields are planted with genetically modified varieties. However, many questions remain about GMOs, including the likelihood that their widespread use will cause the evolution of resistance in pests, their effects on nontarget organisms, and the consequences of the "escape" of engineered genes into noncrop plants (see Chapter 9).

The concerns related to GMOs should not be downplayed. However, in several ways, these concerns pale in comparison to those created by conventional agricultural practices, including environmental degradation and loss of soil.

# 24.3 The Future of Agriculture

The use of pesticides, herbicides, fertilizer, and irrigation over the past 60 years—a phenomenon known as the "Green Revolution"—has increased food production dramatically. However, many biologists argue that current

## Working with Data ▶

In what regions has the aquifer lost the most volume?

high levels of production are **unsustainable**—they compromise the ability of future generations to grow enough food to feed the human population. The future of agricultural production depends on finding ways to maximize food production while minimizing environmental damage.

## Modern Agriculture Causes Environmental Damage

What evidence supports the hypothesis that current agricultural practices are unsustainable? First, the production and application of agricultural chemicals such as pesticides and fertilizers requires large amounts of nonrenewable fossil fuels; second, the water sources used to provide irrigation water are limited and under increasing demand; third, the environmental costs of fertilizer and pesticide use may be greater than the human population is able to bear.

**Precious Liquids: Oil and Water.** Petroleum is required not only to run farm equipment, but also as a raw material for the production of fertilizer and pesticides. While scientists and oil industry experts disagree about if and when the world's oil supply will run out, it is clear that demand for oil continues to increase, causing prices to climb as a result. Dependence on the finite and ever more costly resource of oil makes current agricultural practices unsustainable.

Some environmental scientists argue that freshwater is the "new oil." Several statistics support this view. In the United States, 65% of total water withdrawals not used in power generation are used to irrigate crops. In western states, the portion of water use devoted to irrigation exceeds 90%. It is clear to geologists studying the Ogallala aquifer (**FIGURE 24.27**) that irrigation in many of the regions above the aquifer is using water much more quickly than it is being replaced by precipitation. Analysis of the rate of water loss indicates that some areas of the aquifer will dry up within decades.

In addition to its reliance on nonrenewable or limited resources, such as oil and freshwater, many of the processes and inputs of modern agriculture can cause serious environmental damage.

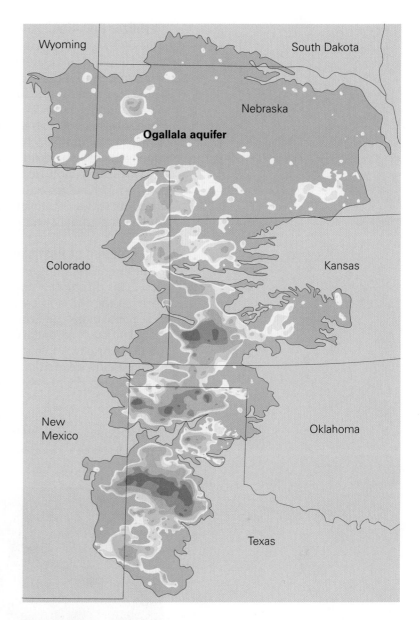

**Water-level change, in feet (decline)**

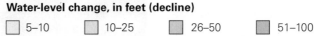

| | 5–10 | | 10–25 | | 26–50 | | 51–100 |
| 101–150 | | More than 150 |

**FIGURE 24.27 Unsustainable water use.** The Ogallala aquifer is the largest single pool of freshwater in the world; it underlies and supplies the "bread basket" region of the United States. However, water is being removed from the aquifer much more rapidly than it refills, leading to dramatic declines in some areas. This figure illustrates the drop in water level in areas of the aquifer through 2011.

**Fertilizer Pollution.** The minerals in inorganic fertilizer that are not taken up quickly by growing plants are carried away by water moving through the soil. This runoff accumulates in water reservoirs in lakes, streams, and oceans or deep underground in aquifers. When the nutrients from inorganic fertilizer reach high levels in the water, it becomes unsafe for human consumption. High levels of nitrate fertilizer in water wells in the American Midwest correlate with greater risks of miscarriage and bladder cancer in women who use this water. Some forms of nitrogen in drinking water can cause brain damage or death among infants who drink it in infant formula.

Fertilizer runoff from farms can also cause eutrophication (described in Chapter 15) and result in extensive fish kills (**FIGURE 24.28**). If the use of fertilizer increases as human populations grow, the loss of drinking water sources and fishing grounds will probably also increase.

**The Problem of Pesticides.** Over the past 50 years, the amount and toxicity of pesticides applied to crops in the United States has increased ten-fold. Pesticides directed toward one crop pest can cause another, secondary pest to thrive if the second pest is a competitor with the original target. In addition, pest populations can evolve resistance to pesticides that target them. To control populations of secondary pests or now-resistant target pests, higher levels of pesticide or new, more toxic pesticides must be applied. This situation is called the pesticide treadmill because farmers must continually expend energy and resources simply to stay in place and keep pest populations under reasonable control.

**STOP & STRETCH** Explain how natural selection causes a pest population to become resistant to a pesticide.

**FIGURE 24.28 Dead zone.** This satellite image shows the dispersal of fertilizer-laden sediment—the orange colors in the photo—flowing out of the mouth of the Mississippi River. Thousands of square miles of the Gulf of Mexico are deadly to fish during each summer as a result of the eutrophication that occurs because of this runoff.

One side effect of the increase in pesticide use has been an increase in the number of incidents of wildlife poisonings, especially of birds. Although trends can only be inferred from anecdotal reports, it appears that the number of wildlife poisonings has tripled since 1980. Even at low doses, many pesticides may have effects that do not cause death but may weaken individuals and make them more susceptible to predation or less able to compete—or may even affect their endocrine system (Chapter 22).

An additional problem associated with some types of pesticides is **biomagnification,** the concentration of persistent toxic chemicals at higher levels of a food web. Biomagnification is a product of the trophic pyramid, the bottom-heavy relationship between the biomass (total weight) of populations at each level of a food chain (Chapter 15). Thanks to the activism of ecologist Rachel Carson (**FIGURE 24.29**), the problem of biomagnification was recognized in the late 1960s—particularly the effects of magnified concentrations of the pesticide DDT in fish-eating birds such as bald eagles. This recognition led to a ban on DDT application. However, many persistent pesticides are still used in the United States and elsewhere, and these can accumulate in any organism that is high on a food chain—including humans.

**FIGURE 24.29 Rachel Carson.** *Silent Spring*, Rachel Carson's most influential book, made well known the hypothesis that pesticides like DDT bioaccumulate, and can kill birds and negatively affect people.

Long-term health effects of low-level exposure to pesticides are still mostly unknown, but some evidence suggests that these effects include increased risks of cancer, birth defects, and permanent nerve damage. The pesticide treadmill demands ever more powerful and toxic chemicals, and high pesticide levels that result from biomagnification are likely to negatively affect more and more of the human population.

Wise use of fertilizers and pesticides may help minimize the environmental effects of modern agriculture; these negative consequences fade when their use is reduced. However, one consequence of modern agriculture results in environmental damage that is not easily recovered—the loss of soil via erosion.

**Soil Erosion.**   Soil is held together by the organisms living and the roots of plants growing within it. Modern agricultural techniques destroy soil in two ways: by regularly removing the plants (through tilling or harvest) and by the application of inorganic fertilizer, which is in the form of salts and can be deadly to soil organisms. When soil loses its structure, it is easily picked up by strong winds and blown away—the Dust Bowl and the dust storms in China are dramatic examples of this type of soil erosion caused by agricultural practices.

Irrigation water running over the surface of disturbed soil can also result in erosion. In addition, irrigation can cause **salinization**—the "salting" of soil. Water from underground aquifers contains tiny amounts of mineral salts. When this water is applied to the soil, some evaporates, leaving salts behind. Eventually, so much salt accumulates that the soil becomes completely infertile. The production of soil from rock and organic material takes time, and soil lost through erosion and salinization is not easily replaced. Agricultural practices that rapidly degrade soil are unsustainable.

If we also consider the effects of rapid climate change on agriculture, it seems clear that current agricultural practices will not be sufficient to feed a growing human population indefinitely. Fortunately, some alternatives to these practices may help provide enough food more sustainably.

## How to Reduce the Damage

A more **sustainable** agricultural system is one that meets the needs of the current generation of humans without compromising the ability of future generations to meet their needs. One path to sustainability focuses on requiring

**Can newly destroyed regions recover in the way that the American Dust Bowl did?**

changes in agricultural practices. However, another path is to encourage consumers to change their eating and food-buying habits.

### Reducing Harmful Inputs.

We have already reviewed many of the farming techniques that can reduce the environmental costs of agriculture. Minimizing monocultural plantings may be the single most effective strategy. Not only does planting polycultures reduce the need for pesticides, it also helps minimize fertilizer inputs and preserves soil by increasing crop rotation. Additionally, planting polycultures helps to insulate farmers from environmental factors that destroy a single crop. Moving away from monocultural production will not be an easy task; it must be preceded by changes in government policies and agricultural economics. For instance, government subsidies that pay farmers to produce a few commodities, such as wheat, corn, cotton, and rice, encourage monocultural production.

Even without a switch to polyculture, farmers have good reason to reduce their use of fertilizer and pesticides—after all, these inputs cost money. One technique employs technology to closely monitor the production of agricultural fields so that fertilizer and pesticide applications are targeted more effectively.

Another strategy to reduce inputs has been adoption of GMOs that produce pesticides directly. In non-GMO crops, farmers can employ **integrated pest management (IPM),** which utilizes releases of predator insects, introduction of pest competitors, and changes in planting techniques to reduce pesticide use. Unfortunately, neither GMO nor IPM use has been a silver bullet. Since the introduction of both strategies in conventional agriculture, pesticide and fertilizer use has still continued to rise in the United States.

### The Role of the Consumer.

Farmers are only one-half of the food production equation. Consumers also determine how and what food is produced. If consumers consistently select certain types of foods, farmers will try to supply them. In many cases, consumer demand encourages some of the more damaging environmental consequences of modern agriculture.

Most of the pesticides sprayed on vegetable and fruit crops are applied to reduce crop "losses" that are caused when consumers refuse to purchase fruits and vegetables with superficial signs of pest damage. The only way to produce pristine products is to apply hundreds of pounds of pesticides to completely eliminate pest insects. If consumers were willing to accept a visible but small amount of pest damage to fresh vegetables, then the amount of pesticides used on these crops could drop tremendously (**FIGURE 24.30**).

Consumer demand for meat also fuels unsustainable farming practices. Most of the beef, pork, and poultry produced in the United States comes from animals that are fed field-grown grain in enormous feed lots, called factory farms (**FIGURE 24.31**). The same production process holds for hogs, chickens, and turkeys. In fact, 66% of the cereal grain consumed in the United States, including 80% of corn and 95% of oats, is used to feed these animals. Because much of the energy consumed by an animal is used for maintenance rather than growth, the grain used as animal feed yields much fewer calories in the form of meat. Put another way, a 10-acre field of corn can support 10 people if they eat the corn directly, but only 2 people if they feed the corn to cattle and then eat the beef.

When individuals consume many of their calories as conventionally produced meat and dairy products, they are forcing the high-intensity farming that causes the environmental problems discussed in this chapter. Reducing meat consumption would also reduce the number of acres farmed; as a result, some soils that are highly susceptible to degradation through erosion or salinization

**(a) Conventional apple**

**(b) Organic apple**

**FIGURE 24.30 Cosmetic pesticide use.** (a) This apple was produced on a conventional farm, with high levels of pesticides to control superficial damage. (b) This apple was produced on an organic farm, with no pesticides. The damage is only on the surface, but consumers often reject fruit that look like this.

**FIGURE 24.31 Modern meat production.** Most of the meat and dairy products available in modern supermarkets are produced in high-density "factory farms," such as the one pictured here.

could be left undisturbed. Less land needed for agricultural production also leaves more natural habitat for other species to survive.

Factory farming of cattle, hogs, and poultry carries environmental costs in addition to its prodigious use of crops. For example, the U.S. Department of Agriculture estimated that the meat industry produced 2 billion tons of animal waste in 2005; this is 100 times the amount of waste produced by the human population of the United States. Failures of manure storage ponds and inadequate disposal have led to spills, water contamination, and fish kills. In addition, the high-density populations of animals in factory farms provide prime conditions for the spread of disease. Many meat producers in the United States handle this risk by feeding large amounts of antibiotics to all of their animals, sick and healthy alike. The use of antibiotics in agricultural settings has led to the development of antibiotic resistance in several dangerous human pathogens, including *E. coli* O157:H7, a bacterium found in beef that sickens thousands of people—and kills over 100—every year.

**Organic Farming.** There is little doubt that consumer demand can change agricultural practices. In recent years, many consumers with concerns about the costs of modern agriculture have changed their buying habits (**FIGURE 24.32**). As a result, some farms have switched to organic farming, a practice that shuns the use of chemical fertilizers and pesticides as well as GMOs. The U.S. Department of Agriculture reported 20% growth in the retail sales of organically grown food in every year from 1990 to 2007. Organic farm acreage in the United States increased by 400% between 1992 and 2005. Organic farms are still very much the exception in American agriculture, representing less than 1% of the total land in agriculture, but consumer interest in organic food is clearly having an effect.

The focus of organic farming is the natural maintenance of soil fertility (using crop rotation and organic fertilizers) and the development of biological diversity on the farm (using polycultures) as a method to control pests and provide income stability. Organically grown food has lower levels of pesticide residues and is produced in a way that minimizes the environmental costs of

**FIGURE 24.32 A consumer movement.** Sales of organic foods have grown at a rate of 20% per year over the past 15 years, thanks to changes in the buying choices of millions of individuals.

agriculture (**TABLE 24.3**). Organic food does often cost more for consumers; grocery bills average 60% higher according to several analyses, although the cost differences decrease when consumers purchase seasonal foods and buy directly from the grower at farm markets. And because food is relatively inexpensive in the United States, a 60% increase in food cost for a family of four totals about $3500 a year—about what an average family spends annually on meals at restaurants. Organic food does not avoid other costs seen in the conventional agricultural system, including the transport of food over long distances, although purchasing from local organic producers can minimize that environmental impact.

**TABLE 24.3 Reducing the environmental impact of modern agriculture.**

This table summarizes the environmental costs of modern agriculture, along with the individual and collective actions we can employ to reduce these costs.

| Environmental Problem | Solution | Actions We Can Take |
|---|---|---|
| **Bioaccumulation of persistent pesticides**  | • Minimize monocultural plantings.<br>• Reduce use of pesticides.<br>• Increase use of integrated pest management. | • Accept produce that has superficial pest damage.<br>• Purchase organic foods.<br>• Add diversity of foods to our diets to promote polyculture.<br>• Reduce meat consumption and thus overall crop production.<br>• Support policies that reduce use of monocultural production (e.g., elimination of subsidies for crops).<br>• Support integrated pest management research. |
| **Eutrophication of surface water and nitrate pollution of groundwater**  | Reduce use of inorganic fertilizer. | • Purchase organic foods.<br>• Purchase foods grown in polyculture, where crop rotation is employed.<br>• Reduce meat consumption. |
| **Salinization of soil** 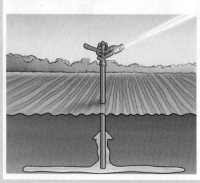 | • Reduce number of acres irrigated.<br>• Increase efficiency of irrigation. | • Reduce meat consumption.<br>• Support research and national agricultural policies to reduce agricultural water use. |

*continued*

**TABLE 24.3**  **Reducing the environmental impact of modern agriculture.** *continued*

| Environmental Problem | Solution | Actions We Can Take |
|---|---|---|
| **Soil erosion and desertification**<br /> | • Reduce use of tilling on marginal lands.<br />• Reduce amount of land used for crop production. | • Reduce meat consumption.<br />• Support research on effective nonchemical means of weed control.<br />• See also suggestions regarding population and aid to developing countries. |
| **Loss of habitat for nonhuman species**<br /> | Minimize the amount of land converted to agricultural production. | • Reduce meat consumption.<br />• Support policies that effectively lead to decreased human population growth rates.<br />• Support policies that provide developing countries with the tools to increase crop yield sustainably. |

Feeding the human population in a way that preserves environmental quality and leaves room for biodiversity will require thoughtful decision making. These decisions include those we make as individuals, such as how much meat to consume and what sort of produce we buy, and those we make collectively, such as how federal subsidies for crops are distributed. Better knowledge of plant and agricultural science and an understanding of the power and limitations of scientific knowledge can help you become an effective participant in this decision-making process.

# savvy reader

## Is There a Health Benefit to Organic Foods?

"There isn't much difference between organic and conventional foods, if you're an adult and making a decision based solely on your health," said Dena Bravata, MD, MS, the senior author of a paper comparing the nutrition of organic and non-organic foods, published in the Sept. 4 issue of *Annals of Internal Medicine.*

A team led by Bravata … did the most comprehensive meta-analysis to date of existing studies comparing organic and conventional foods.

…

For their study, the researchers sifted through thousands of papers and identified 237 of the most relevant to analyze. … [T]he duration of the studies involving human subjects ranged from two days to two years.

After analyzing the data, the researchers found little significant difference in health benefits between organic and conventional foods. No consistent differences were seen in the vitamin content of organic products. There was also no difference in protein or fat content between organic and conventional milk, though evidence from a limited number of studies suggested that organic milk may contain significantly higher levels of omega-3 fatty acids.

…

The review yielded scant evidence that conventional foods posed greater health risks than organic products. While researchers found that organic produce had a 30 percent lower risk of pesticide contamination than conventional fruits and vegetables, organic foods are not necessarily 100 percent free of pesticides. What's more, as the researchers noted, the pesticide levels of all foods generally fell within the allowable safety limits. Two studies of children consuming organic and conventional diets did find lower levels of pesticide residues in the urine of children on organic diets, though the significance of these findings on child health is unclear. Additionally, organic chicken and pork appeared to reduce exposure to antibiotic-resistant bacteria, but the clinical significance of this is also unclear.

"You have to remember that the conventionally produced veggies or crops have pesticide levels below the tolerance level regulated by government. They are there, but way below the tolerance level," he said.

1. What is the hypothesis the study reported here was testing?
2. What evidence from the summary sheds light on this hypothesis?
3. What evidence (of the costs or benefits of organic food) is missing from this story? What additional information would you need to determine if organic food is "worth" the cost?

http://med.stanford.edu/ism/2012/september/organic.html

# SOUNDS RIGHT BUT IS IT?

Nearly everyone knows that a plant that wilts needs water. But now you know that cell walls made of cellulose (a polysaccharide) also provide much of the "skeleton" of the plant. With this knowledge, you might think that watering your plant with a mixture of sugar and water (say, soda) will provide not only the water it needs, but the building blocks for making a stronger skeleton on cell walls. In other words:

**"Watering" a wilted plant with a solution of water and sugar will help the plant recover better than water alone.**

Sounds right, but it isn't.

1. Recall Chapter 4, where you learned about transport of materials and water across cell membranes. The water contained in plant cell vacuoles is nearly pure. Where would water flow in the case where you added sugar water to the soil—into or out of the cell? Explain why.

2. What does it mean that a plant cell wall is "elastic"?

3. What happens to the pressure inside plant cells that lose water? What do we call this condition?

4. How does the condition of cells relate to whether a plant is wilted or "crisp"?

5. Even if the sugar in soda did increase the thickness of plant cell walls, would that affect whether the plant was wilted or crisp under these conditions?

6. Given your answers to questions 1–5, explain why the statement bolded above sounds right, but isn't.

---

# Chapter Review MasteringBiology®

Go to the Study Area in MasteringBiology® for practice quizzes, myeBook, BioFlix™ 3-D animations, MP3Tutor sessions, videos, current events, and more.

## Summary

### Section 24.1

List the tissues and structures found on angiosperm plants, and describe their functions.

- Nearly all major crops belong to the angiosperm phylum, which are flower- and fruit-producing plants (p. 532).
- Roots absorb water and minerals from the soil but also may be modified to store carbohydrates (pp. 534–536).
- Stems serve as support for leaves and reproductive organs but may also be modified for carbohydrate storage (p. 536).
- Leaves are the photosynthetic organs of the plant and contain some essential nutrients for humans (p. 536).
- Flowers contain both male and female reproductive organs (pp. 536–537).

Describe the process of reproduction in angiosperms, including the role of double fertilization.

- When pollen is transferred from the male to the female angiosperm, two sperm are released. One sperm fertilizes the egg, and the other fuses two other nuclei inside the ovule, a process called double fertilization (p. 538).

- The resulting seed remains inside the flower structure, which becomes modified into a fruit that functions in seed dispersal (p. 539).

### Section 24.2

Compare and contrast primary and secondary growth in dicot plants.

- Dicot plants undergo primary growth, in the length of stems and roots, and secondary growth, in the width of these organs (pp. 541–542).
- Primary growth occurs because of cell division in the apical meristems. This meristem can produce additional meristems, leading to indeterminate growth patterns in plants (p. 541).
- Secondary growth in dicot stems occurs when the vascular cambium produces additional xylem, which becomes wood, and phloem, which becomes the inner part of the bark (p. 542).
- Seeds may survive difficult environmental conditions by becoming dormant. The emergence of a plant embryo from a seed is called germination (pp. 540–541).

Describe the role of water availability in maximizing a plant's exposure to sunlight and carbon dioxide.

- Plants require abundant water for photosynthesis, for effective interception of light, and to maximize the uptake of carbon dioxide. Farmers not only provide water through irrigation but also remove noncrop plants that compete for water (pp. 543–544).
- Plants regulate water loss but also cut off the uptake of carbon dioxide by closing the stomata on their leaf and stem surfaces (p. 545).

Provide examples of important nutrients for plant growth, and distinguish between how these nutrients occur in natural systems and how they are made available in agricultural systems.

- Soil is a matrix composed of minerals from eroded rock, dead and decaying material, and the living organisms that decompose and mix the components. Soil is the major source of nutrients for plant growth (p. 546).
- In agricultural systems, nutrients are mined from the soil. Farmers make up for the loss of nutrients by adding fertilizer and growing nitrogen-fixing crops (pp. 546, 548).

# Section 24.3

Describe the costs associated with modern agriculture, and list some possible ways these costs could be reduced.

- Widespread use of pesticides has led to monocultural production of crops. Before pesticides, polycultural plantings kept pest populations down and insulated farmers from pest outbreaks (pp. 548–550).
- Petroleum and water are essential to modern agriculture and are in limited or finite supply (p. 551).
- Fertilizer runs off into waterways, causing human health problems and damage to fish and other animal populations (p. 552).
- Farmers continually increase pesticide use as pests become resistant (p. 552).
- Pesticides can be persistent and accumulate in organisms at higher levels of the food web, including humans (p. 553).
- In many areas, soil is being lost through erosion and salinization faster than it is being replaced (p. 553).
- Fertilizer and pesticide use can be minimized using polyculture, by crop rotation, and by careful monitoring of the agricultural environment (pp. 549–550, 554).
- Consumers can help reduce the environmental costs of modern agriculture by accepting a small amount of pest damage on produce, eating less meat, and purchasing organically grown products (pp. 554–556).
- Organic agriculture may provide a sustainable model for future food production (pp. 555–556).

# Roots to Remember

**The following roots of words come mainly from Latin and Greek and will help you decipher terms:**

| | |
|---|---|
| **angio-** | means vessel. Chapter term: *angiosperm* |
| **enni-** | pertains to a year. Chapter term: *perennial* |
| **rhizo-** | means root. Chapter term: *rhizome* |
| **sperm-** | means seed. Chapter term: *angiosperm* |

# Learning the Basics

**1.** Sketch and label the parts of a plant and flower.

**2.** What does it mean to say a plant's growth is "indeterminate"?

**3.** All of the following plant organs have been modified in certain plants to store large amounts of carbohydrate, except _____.
A. fruit; B. stems; C. roots; D. leaves; E. apical meristems.

**4.** One difference between roots and leaves is _____.
A. roots have vascular tissue, but leaves do not; B. roots can store carbohydrates, but leaves cannot; C. leaves have only phloem, roots only xylem; D. root epidermis absorbs water, leaf epidermis conserves it; E. leaves are only produced by flowering plants, roots by all plants.

**5.** As a result of fertilization in flowering plants, _____.
A. pollen is picked up at the stigma by an animal for transfer to an anther; B. a single sperm is released by a pollen grain; C. a seed begins to absorb water and break free of its coat; D. a triploid tissue called endosperm is formed; E. a carpel is produced that later becomes a seed.

**6.** In dicot woody plants, secondary growth results in _____.
A. production of new xylem and phloem; B. increase in the girth or width of the stem; C. the production of additional nodes and internodes; D. A and B are correct; E. A, B, and C are correct.

**7.** The function of stomata is _____.
A. to prevent pests from damaging the photosynthetic surfaces of the plant; B. to allow carbon dioxide to enter the plant body; C. to sense water availability in the air; D. to provide stability for a growing plant; E. to increase nutrient availability in soil.

**8.** The first organ to emerge from a germinating seed is the _____.
A. axillary bud; B. root; C. cotyledon; D. endosperm; E. stem.

**9.** Soils in agricultural systems require fertilizer because

_____.

**A.** farming mines nutrients from the soil; **B.** weeds compete with crop plants for soil nutrients; **C.** most crop plants have nitrogen-fixing bacteria in their roots and need lots of nitrogen; **D.** most nutrients run off soils into waterways, causing eutrophication; **E.** pests remove most nutrients from crop plants.

**10.** When inorganic fertilizer ends up in waterways,

_____.

**A.** the waterways become eutrophic; **B.** algae populations explode in surface waters; **C.** fish kills result; **D.** drinking water may become contaminated; **E.** all of the above.

## Analyzing and Applying the Basics

**1.** Agricultural scientists have been successful at creating only a few types of seedless fruits through traditional breeding—notably bananas, apples, grapes, pineapples, navel oranges, and a few melons. Why might it be difficult to produce seedless fruits?

**2.** One way to make a houseplant look fuller is to cut off its primary growing tip. Based on what you've learned about primary growth, how does this action promote a bushy form?

**3.** Your friend has been planting the same variety of tomatoes in the same spot in his garden for the past 8 years. In recent years, he has noticed that his plants are less healthy, and that pests damage more of the tomato fruits. Apply what you have learned in this chapter to explain why his tomato crop is declining and suggest several ways that he can improve his tomato yield.

## Connecting the Science

**1.** The human population relies on approximately 6 staple crops and about 24 additional crops for most of our calories. In the United States, nearly 80% of the cropland is planted with three species: corn, wheat, and soybeans. What are the risks and benefits of relying on so few agricultural species? Are you concerned about the lack of diversity in agricultural production? Why or why not?

**2.** Have you ever considered the environmental impact of your diet? Are you willing to change your diet to reduce its environmental impact? Why or why not?

---

Answers to **Stop & Stretch, Visualize This, Working with Data, Savvy Reader, Sounds Right, But Is It?**, and **Chapter Review** questions can be found in the **Answers** section at the back of the book.

CHAPTER **25** | # Growing a Green Thumb

**What does it take to have a garden this beautiful?**

# Plant Physiology

Many of us know people who seem to have a "green thumb"—men and women whose homes are a jumble of thriving houseplants, who produce enough vegetables in their backyard gardens to saturate friends with excess tomatoes and zucchini, or who have flower gardens that make the neighbors, well, green with envy. How do they do it? Is it simply a gift that some are born with and some without?

A significant contributor to a gardener's success is time. Years of trial and error, weeks of preparation for a

**How can you make a backyard vegetable plot yield abundant produce ...**

**... or fill your home with thriving houseplants?**

growing season, and hours of work tending to gardens and indoor plants will create an experienced grower who can get the most out of his or her plants. Part of what a gardener gains through experience is a thorough understanding of how plants work. This knowledge helps a plant lover choose the right plants for the environment, design a landscape that is always in bloom, and even shape

the form of plants to fit his or her needs.

Fortunately for anyone who is an eager but inexperienced gardener, it is possible to start developing a green thumb by learning the basics of plant function. Our discussion of plant physiology fits within the context of planning a plant-filled space, such as a garden or houseplant collection. Let's get growing!

**It helps to have knowledge of plants, gained through experience and understanding.**

**Many of us know people who seem to have a "green thumb."**

trans- means across.

co- means with.

ad- means to.

# 25.1 The Right Plant for the Place: Water Relations

One of the first considerations of gardening is choosing plants that can thrive in the available environment. Many factors influence plant growth in a particular space, including the soil's chemical nature, light availability, and length of the **growing season,** which is the time from the last frost in the spring to the first frost in the fall. However, one of the most important factors influencing plant production is the amount of water available to the plant. How do plants obtain and manage water, and how can we use our understanding of this process to help plants thrive?

## Transpiration

Water and dissolved minerals obtained from the soil, together called **xylem sap,** travel up the xylem, the water-conducting cells inside a plant. This occurs primarily as a result of the loss of water vapor from the leaves through evaporation. Through this process, called **transpiration,** the plant essentially pulls a column of sap from the soil, through the roots, up tubes made of dead xylem cells in the stem, and into the leaves. Transpiration is effective in even the tallest trees (over 100 meters, or 330 feet) because of the hydrogen bonds formed by water molecules (Chapter 2). Transpiration evolved in plants after they colonized land 400 million years ago, and it has allowed plants to colonize nearly all land on Earth's surface.

The tendency for identical molecules to stick together, such as water molecules linked by hydrogen bonding, is referred to as **cohesion;** the tendency for unlike molecules to stick together, such as when water is hydrogen bonded to other polar molecules, is called **adhesion.** Water adheres to cellulose in plant cell walls. Cohesion and adhesion together maintain the continuity of a sap column during transpiration.

The pulling force on a column of xylem sap is generated by the evaporation of water from the leaves (**FIGURE 25.1**). Stomata, the pores on leaf surfaces, control the rate of gas exchange and water loss in land plants. When water evaporates out of stomata, the forces of cohesion and adhesion on the water molecules remaining inside the leaves create **tension,** or negative water pressure. Tension is the force that allows a column of water to be pulled into a syringe when the plunger is pulled up. In plants, tension causes water molecules from the xylem to be drawn out into the leaf to replace the water that has evaporated. This in turn increases tension on the water molecules immediately below them in the xylem, causing them to move toward the leaves, and so on, all the way down to the roots and the soil. In many plants, the surface area for water uptake is increased by the presence of symbiotic mycorrhizae, fungal strands that receive carbohydrate from the plant for their services (**FIGURE 25.2**). The presence of mycorrhizae is greatest in soil that contains abundant organic matter and is well aerated, which is why gardeners want you to stay out of their flowerbeds, so you won't compact the soil.

When a stem is injured, the continuity of the stream of water is disrupted and the tension is released. As a result, water pulls away from the cut surface. This leaves an air gap that remains even when the cut stem is placed in a water-filled vase. Because the stream of water in the xylem is no longer continuous, water from the vase cannot replace water evaporating from the xylem, and the stem will rapidly dry up and wilt. Gardeners who understand this know either to cut flowers in the early morning, when the plants are full of water and tension is lowest—producing little or no air gap—or to reestablish a continuous

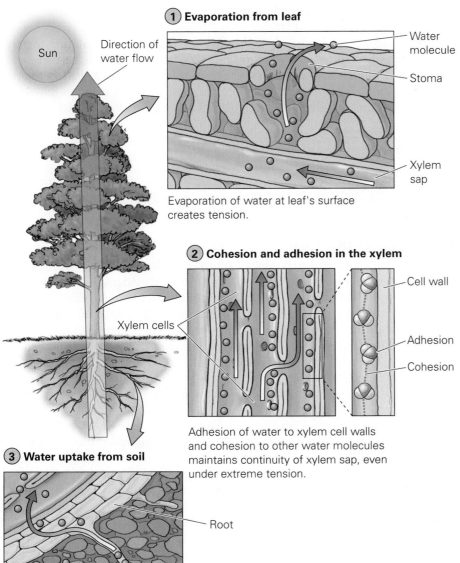

① **Evaporation from leaf**

Water molecule

Stoma

Xylem sap

Evaporation of water at leaf's surface creates tension.

② **Cohesion and adhesion in the xylem**

Xylem cells

Cell wall

Adhesion

Cohesion

Adhesion of water to xylem cell walls and cohesion to other water molecules maintains continuity of xylem sap, even under extreme tension.

③ **Water uptake from soil**

Root

Tension is transmitted to source of water, in this case the soil.

Sun

Direction of water flow

## ◀ Visualize This

**How are the conditions in the tree different at night? Does transpiration occur then?**

FIGURE 25.1 **Transpiration.** Evaporation of water from the leaves creates a tension that is transmitted to a column of water in the xylem. Water flows upward against the pull of gravity and maintains its continuity, thanks to the forces of cohesion and adhesion.

## Visualize This ▶

**If this is a symbiosis, then there must be a benefit to the fungus as well as the plant. How could a fungus benefit from being integrated with a plant's roots?**

FIGURE 25.2 **Mycorrhizae.** The symbiotic relationships between the roots of many species of plants and certain kinds of fungi greatly increase the surface area for nutrient and water absorption to support plant growth.

Root

Mycorrhizae

**FIGURE 25.3 A C$_4$ plant.** Crabgrass has the C$_4$ photosynthetic pathway, which gives it an advantage over Kentucky bluegrass in hot, dry midsummer and late summer.

stream of water by cutting the flower stem a second time, once it is placed in a water-filled vase.

Plants can actively modify the rate of transpiration by regulating the size of stomata (Chapter 24). In addition to these active movements, certain plants also possess adaptations to photosynthesis or leaf shape that can affect the rate of transpiration. The presence or absence of these adaptations helps determine which plants are best suited for particular climates and environments.

## Adaptations That Affect Transpiration

Plants have evolved several strategies for reducing transpiration water loss. These strategies fit into two general categories: (1) modifications to photosynthesis that make carbon dioxide acquisition more efficient, and (2) modifications to leaf shape and stomata placement that increase water conservation.

**Photosynthetic Adaptations.** Two adaptations to a plant's photosynthetic pathways help reduce water loss: C$_4$ and CAM photosynthesis (Chapter 5). Plants with C$_4$ photosynthesis have an additional chemical process that precedes the Calvin cycle, the step that converts carbon dioxide into sugar in all plants. The additional chemical cycle in a C$_4$ plant essentially pumps carbon dioxide from the air toward the chloroplasts, allowing photosynthesis to proceed even when the stomata are just barely open and carbon dioxide uptake is limited.

Many C$_4$ plants, including corn, sugarcane, and crabgrass, belong to the grass family. These plants have an advantage in hot, sunny, and dry conditions because they will continue photosynthesizing even when stomata are closed to reduce water loss. Homeowners across the United States can see the relative advantage of C$_4$ crabgrass over non-C$_4$ lawn grasses like Kentucky bluegrass during the warmest parts of summer. At these times, the yellowish, broad leaves of crabgrass will become dominant as growth of the less tolerant bluegrass slows (**FIGURE 25.3**). If the crabgrass spreads, patches of bluegrass may die off, leading to a patchy lawn when the temperature cools in late summer and the crabgrass dies.

CAM plants, including all cacti, pineapples, many orchids, aloe, and jade plants, accumulate carbon dioxide from the air during the night, when less water will evaporate through the stomata because of the cooler temperatures. The carbon dioxide is converted to a simple carbohydrate and is stored in the vacuoles of leaf cells as malic acid. At sunrise, the stomata close, and the accumulated acid is degraded, releasing carbon dioxide for photosynthesis. The growth of a CAM plant is greatly limited by the relatively small amount of carbon dioxide it can use in a day—only what can be chemically stored. These plants are most common in extremely dry environments, although they can be successful in other environments. One of the driest environments a plant can experience is a house in winter, and many popular houseplants use the CAM process (**FIGURE 25.4**).

So, what about the plants without either C$_4$ or CAM photosynthesis? Their strategies for reducing the rate of water loss through stomata are limited to the structural adaptations of their leaves.

**FIGURE 25.4 Jade, a CAM houseplant.** Plants with CAM photosynthesis are adapted to very dry conditions and thus are ideal houseplants to survive dry indoor air.

**Leaf Adaptations.** The amount of water lost by a plant via transpiration is generally correlated to its photosynthetic surface. In particular, a plant with many large leaves tends to transpire more water than does a plant with fewer or

**FIGURE 25.5 Small leaves limit transpiration loss.** In desert plants, small leaves help reduce the surface area for water evaporation. Some leaves may be so reduced that they are modified to spines.

smaller leaves. Gardeners use their understanding of this relationship between leaf surface area and transpiration to help transplanted plants survive. Because removing plants from soil (no matter how carefully) inevitably results in root damage, newly transplanted plants have a reduced capacity to acquire water. To compensate for this, gardeners remove leaves when transplanting to reduce water loss.

Of course, decreased leaf area also means less photosynthesis, so plants with smaller leaves tend to grow more slowly than plants with broader leaves in the same environment. This is the trade-off experienced by needle-leaved plants, such as pines. While their small leaves help conserve water, they cannot compete successfully in well-watered environments against broad-leaved trees. Plants that thrive in shady sites often have broad leaves because evaporation is lower and intercepting light is more crucial, while narrow-leaved plants tend to be more successful in bright sun. Leaf size is reduced to its extreme in cacti, which have photosynthetic stems and leaves modified into spines (**FIGURE 25.5**). Cacti grow quite slowly, but they make for another ideal houseplant that can survive even the most inattentive gardener.

Some broad-leaved plants have other adaptations to reduce water loss, such as the placement of stomata on the cooler, shaded undersurface of leaves or in hair-lined pits, which trap water molecules and thus reduce water loss even when the stomata are open (**FIGURE 25.6**). Where periods of drought occur reliably on an annual basis, many broad-leaved plants have evolved to become **deciduous,** that is, to actively drop their leaves to reduce the rate of transpiration. In Section 25.2, we describe this process in greater detail.

In environments where water is abundant, plants compete with each other for access to light. In these environments, faster-growing plants often have an advantage. Plants that have evolved strategies for maximizing transpiration, including changes in water-conducting tissues, can grow rapidly and quickly tower over their neighbors.

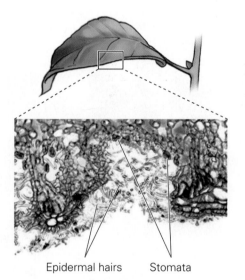

Epidermal hairs    Stomata

**FIGURE 25.6 Shaded stomata help conserve leaf water.** The stomata of this leaf are found on the shaded underside, where temperatures are lower and evaporation slower. The hair-lined pits trap evaporating water molecules and keep humidity high around them.

**Xylem Adaptations.** When a column of xylem is under extreme tension inside a plant, the forces of cohesion and adhesion may not be strong enough to prevent the water column from breaking. Breakage of the water column results in the formation of a bubble, called an embolism, which inhibits water flow within the xylem tubes. In general, larger-diameter xylem tubes are more likely to suffer embolisms than are smaller-diameter tubes. Many drought-adapted species have only very narrow or tapering xylem tube cells, called **tracheids** (**FIGURE 25.7a**).

However, the narrow passageway produced by a column of tracheids leads to increased friction, reducing the rate of water flow—a disadvantage in moist environments. In moist conditions, plants with wider, perforated **vessel elements** are favored because they move water to their leaves up to 10 times faster, resulting in a proportionally faster growth rate (**FIGURE 25.7b**).

The diameter of the xylem tubes in any given plant thus represents a trade-off between the risk of embolism and maximum growth rate. For example, in cold climates, where liquid water may be unavailable for a large part of the year, trees are at high risk of embolism and have much less porous-looking wood than do trees in warmer, wetter climates where embolism is rare but competition for light is intense.

The growth of a plant in a natural landscape is strongly influenced by the balance between water availability and transpiration rate. However, in gardens and houses, water is rarely limited as long as the gardener pays attention to the plant's needs. In fact, a common problem of houseplants in the hands of inexperienced gardeners is not excessive dryness but too much water.

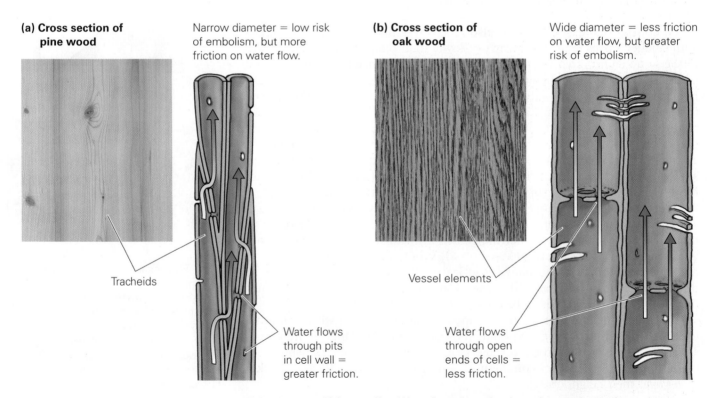

**(a) Cross section of pine wood**

Narrow diameter = low risk of embolism, but more friction on water flow.

Tracheids

Water flows through pits in cell wall = greater friction.

**(b) Cross section of oak wood**

Wide diameter = less friction on water flow, but greater risk of embolism.

Vessel elements

Water flows through open ends of cells = less friction.

**FIGURE 25.7 Xylem cells.** Water flows through xylem tubes made up of one or two cell types. (a) Plants in dry environments have a high proportion of long, narrow tracheids. (b) In moist environments, short, wide vessel elements are the most dominant xylem cell type.

**Overwatering.**  Excess water, by filling the air spaces in soil, prevents root cells from obtaining oxygen. As a result, the water-absorbing root hairs die, decreasing the transpiration rate and causing the plant to wilt. Inexperienced gardeners respond to an overwatered, wilting plant by adding more water, further destroying the roots and eventually killing it. As a rule of thumb, potting soil should be allowed to become completely dry down to 2 centimeters below the surface before more water is added.

Overwatering is a typical problem for houseplants, but garden and agricultural plants can also suffer from excess moisture. Soils that are high in clay, the tiniest mineral particles, are prone to becoming waterlogged because each clay particle can trap a shell of water around it, eventually filling up all of the soil's air spaces and leading to the same symptoms as those in overwatered houseplants. Gardeners with clay soils often need to amend the soil before planting to avoid this problem. The most common soil amendment is organic matter, such as compost or peat moss. In most areas of the country, plants require about 2.5 centimeters (1 inch) of water per week, through rainfall or supplemental watering, for optimal growth. Experienced gardeners know to look for signs of water stress (limp or wilting plants, a yellow or bluish cast to leaves) and to check how dry the soil is a few centimeters below the surface before watering their plants.

## An experienced grower can get the most out of his or her plants.

> **STOP & STRETCH**    If you have a plant that is suffering from overwatering, one solution is to remove a large number of leaves while waiting for the soil to dry out (a process that could take weeks if the soil is really soaked). Why does removing leaves help the plant survive this period?

A gardener's ability to supplement natural rainfall allows the growth of water-needy plants even in dry landscapes. However, it is far more difficult for gardeners to control the temperatures to which their landscape plants are exposed, so most garden plant choices are based on temperature requirements. Ability to tolerate freezing is one factor that determines whether a plant can survive in a particular temperature range.

## Water Inside Plant Cells

Plant cells contain a central vacuole that is mostly filled with water. As described in Chapter 24, water-filled vacuoles provide support to a plant. At low enough temperatures, this important fluid will turn to ice, with devastating consequences. Cold tolerance is primarily a function of a plant's ability to control the freezing of water within its cells.

Ice formation in plants is damaging in three ways: (1) When water outside plant cell walls turns into ice, liquid water from inside the cells will begin flowing out by osmosis—that is, down a concentration gradient across the cell membrane (**FIGURE 25.8**). As the cell loses increasing amounts of water to the ice crystal, normal cell processes become disrupted and may cease to function. (2) When water freezes inside the central vacuole of a plant cell, it produces an ice crystal. The jagged edges of this crystal can pierce the membrane, allowing cytosol (the fluid inside the cell)

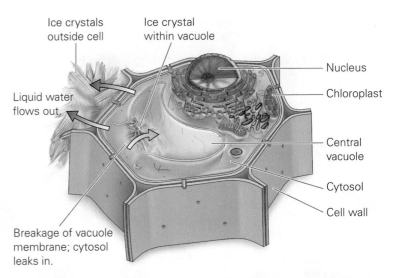

**FIGURE 25.8 Freezing inside plant cells.**  Ice outside plant cells can cause the cell to dehydrate, interfering with normal functions. An ice crystal inside the central vacuole can pierce its membrane, allowing the cytosol to leak into the vacuole.

to leak into the vacuole, thereby killing the cell. (3) When ice freezes within xylem tubes, air comes out of the water solution, forming embolisms and blocking future water uptake. All three of these processes can severely damage or kill plants.

A plant's tolerance of cold temperatures, known as its **hardiness**, depends on its ability to either prevent or manage freezing inside cells. Water freezes at temperatures lower than normal (0°C or 32°F) when it has solutes dissolved in it, in the same way that added antifreeze prevents water in a car's radiator from freezing. Tomato plants can survive temperatures down to −1°C (31°F), apple fruits −4°C (25°F), and cabbage as low as −9°C (16°F) due to their sugar-based cell antifreeze. At lower temperatures, certain adaptations can improve freezing tolerance. These adaptations include controlling ice formation within cells to prevent sharp edges and internal structures that prevent xylem embolisms.

Many cold-tolerant plants require both a gradual cooling period to become maximally hardy and uninterrupted cold temperatures to remain hardy. Cold snaps that occur in late summer and early fall may damage plants that can normally thrive in even colder temperatures because they have not become **hardened**, or fully able to resist the cold. Woody plants can become damaged by cold even after they are fully hardened when bright winter sunshine warms their stems, allowing water to transpire that cannot be replaced by frozen water in the soil. This is a common occurrence in evergreens in landscaping, which can suffer from winter "burn." Freezing temperatures are most likely to cause damage during the spring, however, when intolerant leaves and flowers are beginning to emerge from formerly cold-tolerant buds.

Because of the extreme damage that freezing temperatures can cause, gardeners must know two things before planning an outdoor garden: (1) the hardiness zone of their landscape, that is, how cold it gets (**FIGURE 25.9**), and (2) the rated hardiness—that is, the cold tolerance—of the plants they wish to grow. Experienced gardeners know, however, that with special care, plants that are less cold tolerant can survive freezing conditions. Plants that are grown outside their hardiness zone can be placed on the north sides of buildings, where they are shaded from the sun (preventing winter burn) and where cooler local temperatures retard budding and leafing until the threat of frost has decreased. Plants can also be covered with mulch such as straw or snow, and they can be protected from unusually early or late frosts by covering with plastic tents.

Less tolerant plants are also better able to survive cold winters if they have sufficient water and sugar reserves. These reserves allow the plants to produce adequate antifreeze at the appropriate times during the hardening process. Gardeners can ensure that their sensitive plants are ready for winter by giving them plenty of water during the fall and by preventing them from using up their stored starch and sugar reserves.

To best manage a plant's sugar supply, a gardener needs to understand how plants store and move sugar. This knowledge helps gardeners not only ensure that their favorite plants survive harsh conditions but also maximizes the beauty and productivity of these plants.

**FIGURE 25.9 Hardiness zones.** The U.S. Department of Agriculture generated this hardiness zone map based on average lowest temperatures experienced in thousands of sites across the country. The boundaries of these zones are changing rapidly as a result of climate change.

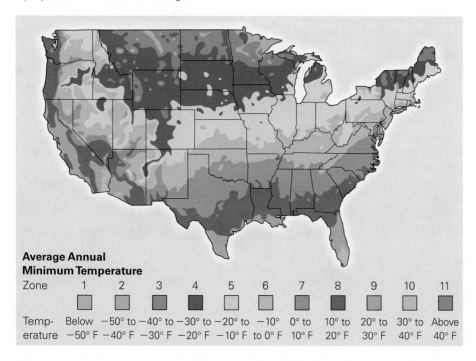

**Average Annual Minimum Temperature**

| Zone | 1 | 2 | 3 | 4 | 5 | 6 | 7 | 8 | 9 | 10 | 11 |
|---|---|---|---|---|---|---|---|---|---|---|---|
| Temperature | Below −50° F | −50° to −40° F | −40° to −30° F | −30° to −20° F | −20° to −10° F | −10° to 0° F | 0° to 10° F | 10° to 20° F | 20° to 30° F | 30° to 40° F | Above 40° F |

# 25.2 A Beautiful Garden: Sap Translocation, Photoperiodism, and Flower and Fruit Production

Each gardener has personal goals for a garden—from a summer-long supply of fresh vegetables, to a continually changing palette of showy flowers, to an aesthetically pleasing but low-maintenance decor. Because plants grow throughout their lives, each plant can allocate its resources in many possible directions at any time—to greater leaf production, flower production, fruit size, and so on. By understanding how plants allocate resources, gardeners can encourage plants to transfer resources to their different plant organs at appropriate times.

**What a gardener gains through experience is a thorough understanding of how plants work.**

## Translocation of Sugars and Nutrients

Sugars and other dissolved nutrients are moved around a plant's body in phloem tissue. The solution in the phloem, or **phloem sap**, always flows from a **source**, where sugar is in high concentration, to a **sink**, where nutrients are being used or accumulated. As a result, the movement of phloem sap, called **translocation**, can be in any direction within a plant. Different organs can act as sources and sinks throughout the year; in fact, separate phloem tubes in the same branch can be translocating sugars in opposite directions at the same time. For instance, some tubes in an apple tree branch may be sending sugar to the roots for storage, while others are sending sugar to the branch tips for fruit production.

Translocation of phloem sap results from pressure differences within a phloem tube between a source and a sink. **FIGURE 25.10** illustrates the **pressure flow mechanism** that causes phloem sap movement. In the first step of pressure flow, sugar is actively loaded into a phloem tube at the source. As sugar accumulates, water flows in passively via osmosis from nearby tissues and xylem tubes. At the same time, respiring cells at the sink are acquiring sugar from the phloem tube, either through active transport or passively, down a concentration gradient from the sugary phloem sap into a sugar-consuming cell. As a result of this movement into the sink cells, the sugar concentration in the nearby phloem tubes declines and water exits the phloem tube passively. As a result of the movements of sugar at the source and the sink and subsequent osmosis, water pressure in the phloem tube is high at the source and low at the sink. Just as in

## Visualize This ▼

How would sap flow in the phloem tubes be affected by drought?

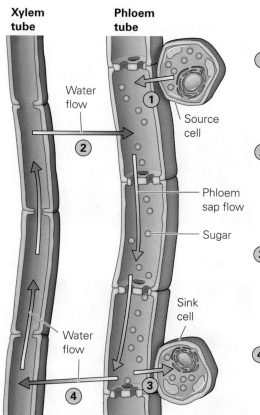

1. Sugar is actively loaded into phloem tubes at the source.

2. Water flows into phloem passively, down its concentration gradient, increasing the water pressure.

3. Sugar moves into sink cell (actively if it is a storage cell, passively if it is a growing or dividing cell).

4. Water flows out of the phloem tube passively, down its concentration gradient, decreasing the water pressure.

**FIGURE 25.10 The pressure flow mechanism of phloem translocation.** Phloem sap moves from source to sink because of pressure differentials in the phloem tube.

a garden hose, in which high water pressure at the faucet and low pressure in the open air cause water to flow out, this pressure differential causes the bulk movement of water from the high-pressure source end to the lower-pressure sink end of the phloem tubes in a plant.

> **STOP & STRETCH**  Maple syrup producers tap the phloem tubes of sugar maple trees in early spring to draw off the sugary phloem sap. Before they add the tap, what is the sugar source in the tree, and what is the sink? When a tap is inserted, how does the identity of the source or sink change?

## Managing Translocation

The key feature of translocation in plants is that materials always move from source to sink. Gardeners who recognize which plant organs are functioning as sources or sinks at any given time can modify a plant's structure to promote flowering, increase the quantity and quality of produce, and prepare plants for winter.

**Promoting Flowering.**  How a gardener manages translocation to increase flower production varies depending on the plant type. Plants that complete their life cycle within a single year and do not survive past that year are called annuals. Many common garden flowers, including petunia, pansy, morning glory, and cosmos, are annuals (**FIGURE 25.11a**).

The evolutionary strategy of annual plants is to produce as many seeds as possible in their single year of life; to do this, annuals produce many flowers

**(a) Annual plants**

**(b) Biennial plants**

**(c) Perennial plants**

**FIGURE 25.11 Plant seasonal growth cycles.** (a) Annual plants live only a single season or year, growing from a seed and producing a new seed crop at the end of their lives. Their evolutionary strategy is to put a significant amount of energy into flower production, making them ideal for garden displays. (b) Biennial plants typically make a low-profile set of leaves in the first year and send sugar into a storage root. In the second year, sugar from storage supplies a large floral display, after which the plant dies. (c) Perennials live for many years, putting energy into flower production and storage roots every year. Gardeners divide the roots of perennials after a few years of growth to promote a balance of sources and sinks.

to increase their chances of pollination and fertilization. The larger the sugar source available for flower production, the greater the number of flowers (sinks) that can be produced. To manage flower production in annuals, gardeners remove, or **prune,** many of the flowers on the young transplants they purchase to encourage the plant to send sugars to different sinks—meristems that are developing more leaves. Once the annual has reached sufficient size, the plant can be encouraged to produce flowers instead of more leaves; this can often be accomplished by restricting both water and fertilizer, which signals the plant that the growing season is coming to an end and thus flower production is a priority. If the plant is grown just for its flowers, individual blooms are removed once they fade (a process called *deadheading*) to eliminate the competitive sink of developing fruits and seeds.

Plants that live for 2 years—producing leaves the first year, surviving dormancy by storing nutrients in roots or underground stems, and producing flowers the second year—are called **biennials.** Some of the best-loved garden flowers, including hollyhock, forget-me-not, and delphinium, are biennial plants (**FIGURE 25.11b**). Gardeners who plant biennials from seed know that the role of the plant in its first year is to send sugars into the storage roots so that the floral display the following year is as large as possible. The most effective way to do this is to encourage leaf growth early in summer, thus maximizing the amount of sugar sent into storage in late summer. Typically, this means ensuring that the biennial leaves are not overly shaded by taller neighbors.

Plants that live for many years by storing nutrients in underground structures and producing flowers every year are called **perennials.** Favorite garden perennials include phlox, peony, daylilies, and tulips (**FIGURE 25.11c**). Perennials with stiffened aboveground stems, such as trees and shrubs, are called **woody plants.** Most garden perennials do not have woody stems and instead die back every year. Because of their long lives, the evolutionary strategy for perennials is more balanced—sinks are evenly divided between flower and fruit production and storage organ production. As with annuals, if perennials are grown for flowers, they are deadheaded to minimize the fruit and seed sink.

Many gardeners will remove perennial flowering stalks well before the last flower has bloomed so that adequate sugars go into the storage sink before **dormancy,** the plant's yearly rest period. In addition, because the underground storage organs continue to enlarge and become crowded over time, the stems and roots of older perennials may have difficulty accessing water and light. To deal with this problem, gardeners must regularly divide perennials by lifting the storage organs from the soil, removing old tissue that no longer produces buds, and replanting smaller, more vigorous portions.

### Increasing Produce Quality and Quantity.

When annuals, biennials, and perennials are grown for fruit or vegetable production, the goal is to increase the number and quality of these sinks. Many of the same techniques for flower production apply. For example, for annual crops, making sure that the plant puts energy into producing leaves and stems early ensures better production of fruit later. Other gardening strategies focus on different sinks; for instance, plants grown for leaves or stems instead of fruit (such as lettuce and asparagus) may be actively discouraged from flowering or setting seed because these activities will rob their edible parts of quality.

In many crop plants grown for fruit, gardeners must strike a balance between the number of fruits produced and their quality. If plants are allowed to continue making new fruit sinks, each fruit will be smaller and less flavorful because the sugar has to be divided among more sinks. Experienced gardeners know to prune additional tomato blossoms after a plant has a significant load of fruit and

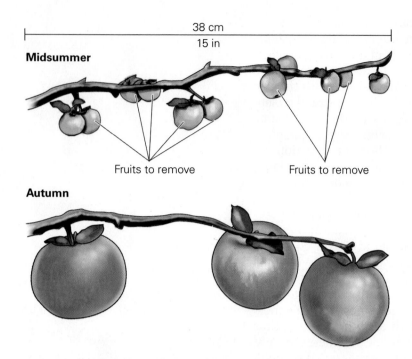

**Midsummer**

38 cm
15 in

Fruits to remove                    Fruits to remove

**Autumn**

**FIGURE 25.12 Maximizing fruit quality over quantity.** A flavorful apple requires about 40 leaves to support it. A good rule of thumb is to thin fruit until they are spaced about the width of your hand apart.

to pick off a proportion of the developing apples on a tree to produce larger, more flavorful fruit (**FIGURE 25.12**).

Although most fruits do not contain vascular tissue, the same source-sink principle can be applied when harvesting fruits to maximize flavor. Once a fruit is harvested, living cells within the fruit continue to act as a sink; however, because the fruit no longer has an external source, it begins to use its own stored sugars as an energy source. As these sugars are used up, the flavor of the fruit suffers. Plant cells use sugar at higher rates when they are warm, so keeping fruit as cool as possible during harvesting helps retain maximum sweetness. To keep vegetable quality high, gardeners try to harvest during the coolest parts of the day and load their vegetables in baskets that allow for air movement.

**STOP & STRETCH** The best time to prune woody plants is at the end of their dormant period, right before they open their leaf buds. According to your understanding of source-sink relationships, why is this the best time?

Modifications of plant sources and sinks work for gardeners because plants develop through indeterminate growth, in which their ultimate size and shape are subject to environmental conditions rather than a genetic program. However, other responses of a plant to its environment result from innate traits. In particular, the timing of plant flowering and the beginnings of dormancy are often controlled by genetic factors.

## Photoperiodism

In the early part of the twentieth century, scientists from the U.S. Department of Agriculture began studying a unique tobacco plant that had appeared in a Maryland farmer's field. This plant grew to twice the height of normal tobacco plants and continued producing leaves until the end of the growing season. Obviously, farmers were thrilled with this variety, except for one small problem—it rarely flowered in the field and thus produced little seed with which to sow an entire crop.

The scientists who worked on the strange tobacco suspected that the "Maryland mammoth" was not flowering because it was not responding to day length in the same way that other tobacco plants did. Through a process of trial and error, researchers discovered that these tobacco plants could be triggered to flower if the day length (that is, total sunlight) they experienced was shorter than a critical number of hours. Later investigations of other plants illustrated that many, like tobacco, exhibit **photoperiodism**—a biological response that is tied to a change in the proportion of light and dark within a 24-hour cycle.

Tobacco, like chrysanthemums, strawberries, primroses, and poinsettias, is considered a short-day plant because it will not flower if day length surpasses a critical value. Short-day plants tend to bloom when day length is short—like the poinsettias that flower in December. Other plants will flower only if day length is longer than a critical value. These long-day plants, like irises, spinach, and lettuce, are typically summer

**photo-** means light.

**phyto-** means plant.

**-chrome** means color.

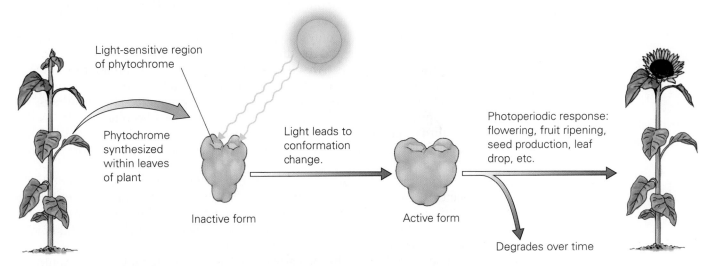

**FIGURE 25.13 How plants tell time.** The ratio between the active and inactive form of phytochrome allows plants to "sense" day length. A plant synthesizes phytochrome in an inactive form, while exposure to sunlight turns it into its active form. In the active form, phytochrome can have a number of effects, but it is slowly degraded over time.

flowers. Not all plants exhibit flowering photoperiodism. Those that do not are called day-*neutral* plants.

Long-day and short-day plants are actually responding to the concentrations of two forms of a protein called **phytochrome**. Phytochrome is produced by plants in one form that changes chemical shape, or conformation, when exposed to red light, a wavelength that is abundant in sunlight (**FIGURE 25.13**). In its light-activated conformation, phytochrome acts to trigger flowering in long-day plants but delays flowering in short-day plants.

Phytochrome is produced continuously within a photoperiodic plant. The switch between the inactive and active forms of phytochrome is nearly instantaneous on exposure to light, but because proteins are regularly recycled within cells, the active form disappears after a time when no light is present. The amount of time that active phytochrome is present provides the measure of day length. Because a few seconds of light will convert phytochrome to its active form, scientists more accurately measure a photoperiodic plant's **critical night length**, the minimum amount of darkness required to inhibit the effects of phytochrome. If night is shorter than this critical length, the active form of phytochrome can exert its effect (**FIGURE 25.14**). Indoor gardeners can manipulate a photoperiodic plant's

# Working with Data ▼

**Lettuce is typically planted so that it reaches edible size in early spring (April and May) and in fall (August to October). Given what the graph illustrates about the role of day length in determining flowering in lettuce, explain why is it less common to plant lettuce in late spring to harvest its leaves in midsummer.**

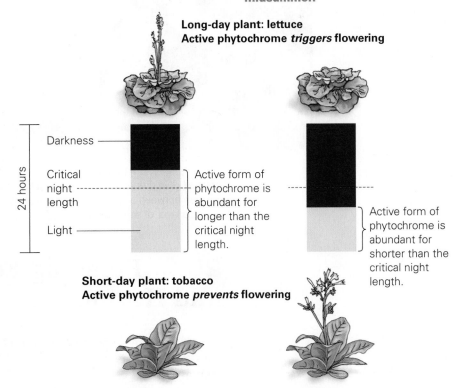

**FIGURE 25.14 Phytochrome, short-day plants, and long-day plants.** The presence of the active form of phytochrome has the opposite effect on short-day and long-day plants.

exposure to artificial light during the night, or shade it during the day, to promote or prevent flowering.

Flowering is not the only activity that is triggered by light cues. The opening of leaf buds in spring and the dropping of leaves in fall are also photoperiodic in many woody plants. **Abscission**, or leaf drop, is a complex process that is first triggered by increasing night length as winter approaches. In preparation for abscission, leaves cease to produce chlorophyll for photosynthesis. As chlorophyll levels decline, other pigments in the leaf become visible. As the leaf turns, cells at the boundary between leaf and stem break down, allowing the leaf to separate from the stem and leaving a protective cap over the junction (**FIGURE 25.15**).

Experienced gardeners know that they should match the photoperiod of a plant's native environment to their own environment. The same species of plant growing at different latitudes may have different critical night lengths and will thus respond differently to light conditions. Spinach, strawberries, and onions are all especially susceptible to being planted "out of place" and thus flower, fruit, and produce bulbs too early or too late in the growing season. Southern varieties of widespread deciduous tree species can suffer from damage if planted so far north that they freeze before dropping their leaves.

In addition to finding the right plant for the site and ensuring that their plants produce when and how they wish them to, gardeners are also concerned with the aesthetic form of their gardens. For many, this means actively managing the way that plants fill and share space with each other. To modify form, we must understand how plants grow in response to environmental cues.

**(a)**  **(b)**

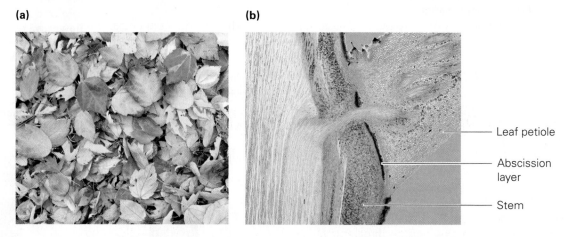

Leaf petiole

Abscission layer

Stem

**FIGURE 25.15 Abscission.** (a) Autumn colors appear when decidious plants reabsorb chlorophyll in preparation for dropping their leaves. (b) Leaf drop occurs because cells at the junction between a leaf and a stem break down, forming a weak layer that separates and leaves a layer of scar tissue.

# 25.3 Pleasing Forms: Tropisms and Hormones

On the surface, plants seem to be captives to their environment. They cannot run away when something tries to eat them. They cannot seek shade on hot days or find a sheltered spot from a freezing wind. However, plants are not as inert as they appear.

## Tropisms

Although a rooted plant cannot physically move itself from one environment to another, plants can grow toward the best possible environment. Directional growth in plants is called **tropism.** Tropic responses are visible in various plant organs in response to different environmental cues. Roots in germinating seeds show positive **gravitropism** (growing downward toward the source of gravity), while stems show negative gravitropism (**FIGURE 25.16a**). This pattern ensures that emerging plant roots maximize exposure to water in the soil and that young shoots maximize exposure to light. Houseplants often demonstrate **phototropism,** directional growth in response to light, when placed on a windowsill (**FIGURE 25.16b**). Other plants grow in response to touch, exhibiting **thigmotropism.** Vining plants show thigmotropism by twining around support stakes (**FIGURE 25.16c**).

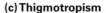

**trop-** means to turn.

Indoor gardeners can use their understanding of plant tropisms to maintain the beauty of their plants. Rotating houseplants regularly ensures that the photo- tropic growth of a plant is balanced around the whole pot. Handling plants occasionally by touching their leaves can promote a thigmotropic response of shorter, sturdier stems. It may be that the widespread belief that talking to plants is good for them comes from this response—they grow fuller not because of the kind words, but because of the increased amount of touch and movement that a well-tended plant experiences.

Plants bend toward light, or away from gravity, when cells on one side of a stem increase in length and cells on the other side do not, forcing the top of the plant to turn away from the expanding cells. This differential cell expansion is caused by different amounts of growth hormone in these cells.

**(a) Gravitropism**

**(b) Phototropism**

**(c) Thigmotropism**

**FIGURE 25.16 Plant tropisms.** (a) This plant stem is exhibiting negative gravitropism, orienting itself opposite the pull of gravity. The plant was allowed to reorient in the dark, indicating that the response is not to light. (b) This houseplant is growing toward a light source, exhibiting phototropism. (c) The tendril on this vine is curled around the support as a result of thigmotropism.

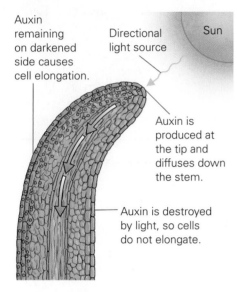

Auxin remaining on darkened side causes cell elongation.

Directional light source

Sun

Auxin is produced at the tip and diffuses down the stem.

Auxin is destroyed by light, so cells do not elongate.

**FIGURE 25.17 Directional growth.** Tropisms are caused when one side of a plant organ (in this case, a phototropic stem) grows more rapidly than the opposite side, causing the tip to bend.

# Hormones

**Plant hormones** regulate a plant's internal environment and control its responses to environmental conditions. Plant hormones are similar to animal hormones in several ways: (1) They are produced in tiny amounts but have large effects, (2) they are mobile so that they can be produced in one organ and have an effect on another, and (3) they have multiple and complex effects and interactions. **TABLE 25.1** lists the five major plant hormones, their effects, and their commercial applications. We will focus on the first plant hormone identified, and the one most commonly manipulated: auxin.

**Auxin** is the hormone that triggers the expansion of cells and is thus responsible for causing plants to increase in size. In tropic responses, auxin is destroyed in some cells in response to an environmental signal. As the cells that retain auxin grow in length while those without it remain small, the organ itself bends away from the auxin-containing side (**FIGURE 25.17**).

Like many hormones, auxin can have opposing effects on different plant organs. In most plants, auxin causes cells to expand in a main stem but inhibits the growth of axillary buds along the sides of the stem. Thus, auxin is responsible for **apical dominance**—the tendency for the plant apex, or uppermost bud, to grow more rapidly than the lower buds. Auxin produced at the shoot tip strongly inhibits branching right below the tip. Farther from the shoot tip, auxin concentrations are lower, and its inhibitory effects are less dramatic. Auxin-mediated apical dominance is responsible, for instance, for the "Christmas tree" shape of young conifers.

Apical dominance increases the height and leaf area of a plant but suppresses bushiness as well as flower and fruit production. For this reason, gardeners are often interested in reducing the amount of apical dominance that a plant exhibits. The primary method for reducing apical dominance is light pruning, also called pinching back (**FIGURE 25.18**). When the terminal bud is pruned, the plant's auxin levels are reduced, inhibition on lateral buds is lifted, and branches form. Eventually, these branches begin to exert apical dominance and must also be pinched back to maintain a plant's bushy form.

Auxin also interacts with other growth regulators. Abscission is promoted by the hormone **ethylene.** Recall that during abscission, cells at the boundary between the stem and a leaf, flower, or fruit break down so that

**TABLE 25.1 Plant hormones.**

| Hormone Type | Effects | Commercial Applications |
|---|---|---|
| **Auxin** | Promotes cell expansion, apical dominance, tropic responses, differentiation of vascular tissue; promotes roots on plant cuttings; inhibits fruit and leaf abscission; stimulates fruit development; stimulates ethylene production (in high doses) | Rooting powder for plant cuttings; prevents fruit and flower drop; promotes seedless fruit production; used as weed killer by triggering too-rapid growth |
| **Gibberellin** | Stimulates leaf and stem cell elongation and division; hastens seed germination, breaks seed dormancy; stimulates fruit development | Used to cause uniform germination of barley in brewing; prevents flower initiation; promotes seedless fruit production |
| **Cytokinin** | Promotes cell division; triggers lateral bud growth even in presence of auxin; with auxin, regulates root and shoot production; delays leaf aging | Used to trigger growth of whole plants from genetically engineered plant cells |
| **Ethylene** | Inhibits cell expansion; promotes fruit ripening; promotes abscission | Hastens ripening of tomatoes, walnuts, and grapes after harvest; used as fruit-thinning agent in commercial orchards; used to defoliate plants |
| **Abscisic acid** | Prevents seed germination; promotes dormancy | Inhibits growth and promotes flowering of potted plants for florists |

**Early spring**

Pruning cut

**Late summer**

Axillary buds, released from apical dominance, produce new branches.

**FIGURE 25.18 Overcoming apical dominance.** Apical dominance can be countered by light pruning, removing the terminal bud. Loss of the terminal bud reduces auxin concentration, lifting growth suppression on the lateral buds, which sprout.

the abscised organ can separate from the plant. Auxin reduces the sensitivity of boundary layer cells to the effects of ethylene and thus delays abscission. Levels of auxin production are positively correlated to a plant's overall health; as a result, stressed plants tend to abscise fruits and leaves earlier than do healthy plants. The antiabscission role of auxin makes it useful in commercial operations; for instance, citrus growers spray their orchards with auxin to prevent early fruit drop.

All of the plant hormones identified by scientists have been adopted for some commercial uses. Knowledge of the effects of these hormones, along with knowledge of transpiration, translocation, and photoperiodism, can contribute to the development of a gardener's "green thumb." But this knowledge has a deeper meaning as well. The discovery and human use of plant hormones reflects the way that evolution has created both highly complex biological processes and highly complex human brains. These brains have enabled us to modify organisms to meet human needs, and as a result, we have the opportunity to improve lives for ourselves, our descendants, and other species.

As this chapter has illustrated, an experienced gardener understands that he or she gets the best results by working with natural processes, not by ignoring them. And as many of the stories in this book have demonstrated, this basic gardening principle can be a guide for all of our interactions with the biological world (**FIGURE 25.19**).

**FIGURE 25.19 Earth as a garden.** The future of humanity depends on the wise use of our biological heritage. Learning about the natural world ideally inspires people to live within natural processes rather than trying to dominate them.

# savvy reader

## The Benefits of Gardening

A new study by the University of Exeter has revealed that houseplants at the workplace can enhance creativity and productivity of employees, which could eventually lead to promotion.

The researchers found that indoor plants improve staff well-being by 47 percent, creativity by 45 percent, and productivity by 38 percent.

For the study, the researchers measured the creativity, happiness and productivity of 350 visitors at the Chelsea Flower Show in a series of 90 experiments at four different designed work environments.

Psychologist Dr. Craig Knight from the University of Exeter said that the results from the horticultural show indicated that business profitability can be achieved by enhancing staff's psychological well-being and the design of the environment.

1. How could you find the research that this story is based on?

2. Describe how the results were obtained. How do the conditions and population tested in the experiments differ from the conditions and the population they are interested in? How might the difference affect the results?

3. Examine the claim made by the psychologist in the last paragraph of this article. Is there enough evidence presented in the article to support this claim? If not, what else would you need to know?

"Want a Promotion? Place a Potted Plant at Your Cubicle!" by Stephen Adkins, *University Herald*, December 7, 2013

# SOUNDS RIGHT BUT IS IT?

Generations of biology students have complained about having to learn about plants. "They're boring!" is a common refrain. We can summarize student reasoning about why plants are boring with the following phrase:

**Plants just sit there, doing nothing.**

Sounds right, but it isn't.

1. List three characteristics of living things.

2. Choose two of the items on the list and explain how plants demonstrate each of these characteristics.

3. Using your answers to questions 1 and 2, explain why the statement in bold sounds right, but isn't.

# Chapter Review MasteringBiology®

Go to the Study Area in MasteringBiology® for practice quizzes, myeBook, BioFlix™ 3-D animations, MP3Tutor sessions, videos, current events, and more.

## Summary

## Section 25.1

Describe how water moves up a plant stem.

- Water is drawn from the soil and into the stems and leaves of a plant because of evaporation at the leaf's surface (p. 564).
- Loss of water results in tension, which pulls on the water column. The column maintains its continuity via cohesion among water molecules and adhesion between water and the xylem walls (p. 564).

List the modifications of plant physiology and anatomy that reduce water loss or make plants more drought tolerant.

- Certain plants have modifications to photosynthesis that reduce water loss. These include $C_4$ photosynthesis, which pumps carbon dioxide to photosynthetic cells, and CAM photosynthesis, which allows carbon dioxide to be harvested from the air at night (p. 566).
- Smaller size, shaded stomata, and deciduousness are leaf adaptations that reduce transpiration (p. 567).
- Narrow tracheids in xylem prevent the occurrence of embolisms, which destroy a plant's capacity to take up water. However, narrow vessels deliver water more slowly, making this trait a disadvantage in moist environments (p. 568).
- Overwatered plants wilt because their roots die from lack of oxygen (p. 569).
- Cold-resistant plants control the rate of freezing inside cells to prevent cell damage and prevent or manage the ice-induced formation of embolisms in the xylem (pp. 569–570).

## Section 25.2

Describe the pressure flow mechanism of phloem transport.

- Sap moves in phloem from source to sink. Sugar is loaded into phloem tubes at the source, causing water to flow in as well. Sugar is unloaded at the sink, and water flows out. The difference in water pressure between the tube at the source and the tube at the sink causes sap to travel through the phloem tubes (pp. 571–572).
- Gardeners can modify relationships between sources and sinks to ensure that energy is directed to the gardener's preferred plant organ (pp. 572–574).

Define *photoperiodism,* and explain the mechanism by which plants respond to day length.

- Many plants exhibit photoperiodism—sensing the length of daylight and responding to it. One system for sensing light relies on a protein called phytochrome (pp. 574–575).
- Light-mediated responses in plants include the initiation of flower or fruit production, the emergence of leaves, and leaf abscission in deciduous trees (p. 576).

## Section 25.3

List the environmental factors that cause tropism in plants, and describe how differences in cell expansion lead to directional growth.

- Plants grow in response to environmental cues, such as the direction of a light source, gravity, and contact with other objects (p. 577).
- Directional growth is caused by differential cell expansion on opposite sides of an organ (pp. 577–578).

Define "plant hormone," and describe a number of ways in which one hormone, auxin, affects plant growth.

- Plant growth regulators, called hormones, tune a plant to its environment and like animal hormones are produced in one organ but have effects on different organs (p. 578).
- Auxin causes cell expansion in stem cells but inhibits the growth of lateral buds, leading to apical dominance. Light pruning can reduce apical dominance (pp. 578–579).
- Auxin interacts with the hormone ethylene to control abscission (pp. 578–579).

## Roots to Remember

**The following roots of words come mainly from Latin and Greek and will help you decipher terms.**

| | |
|---|---|
| **ad-** | means to. Chapter term: *adhesion* |
| **-chrome** | means color. Chapter term: *phytochrome* |
| **co-** | means with. Chapter term: *cohesion* |
| **photo-** | means light. Chapter terms: *photoperiodism, phototropism* |
| **phyto-** | means plant. Chapter term: *phytochrome* |
| **trans-** | means across. Chapter terms: *transpiration, translocation* |
| **trop-** | means to turn. Chapter terms: *tropism, gravitropism, phototropism, thigmotropism* |

## Learning the Basics

**1.** Why does a column of water in a xylem tube remain intact despite being under tremendous tension?

**2.** How do the $C_4$ and CAM adaptations to photosynthesis reduce water loss from plants?

**3.** How does light shining on a phototropic stem change conditions in the stem, and how does this change affect the stem's growth?

**4.** Water moves up a plant's stem as a result of _____.
A. the xylem pump; B. diffusion of water into roots; C. translocation of phloem sap; D. evaporation of water from the leaves; E. photosynthesis in xylem cells.

**5.** When water is under extremely high tension in a xylem tube, _____.
A. transpiration slows; B. water cannot be removed from the soil; C. the water column can "break," blocking further water flow; D. stomata open wide to permit greater water absorption; E. the plant is not photosynthesizing.

**6.** Which of the following adaptations provide an advantage to plants in warm, moist environments?
A. closing stomata in response to decreased water availability; B. large-diameter xylem vessels, permitting rapid water uptake; C. deciduousness, reducing water loss during dry periods; D. small leaf size, reducing sun exposure; E. vertically oriented photosynthetic stems, shading the stomata

**7.** Sap travels in phloem as a result of _____.
A. evaporation of sugar water from the leaves; B. the phloem pump; C. a countercurrent to xylem flow; D. differences in water pressure in a phloem tube between a sugar source and a sugar sink; E. cohesion and adhesion of water to the phloem walls.

**8.** Each of the following pairs represents a likely sugar source and its sink *except* _____.
A. an apple seed and the fruit in which it is contained; B. mature leaves of a tomato plant and a developing tomato; C. newly emerged leaves on a bean seedling and the seedling's developing roots; D. a sugar beet root in spring and its developing flower stalk; E. the roots of an oak tree and its leaf buds in spring.

**9.** A short-day plant can be prevented from flowering by _____.
A. keeping it constantly in the light; B. preventing phytochrome from converting into the active form; C. ensuring that day length does not exceed the plant's critical length; D. growing it in the same environment as a long-day plant; E. more than one of the above is correct.

**10.** Plant hormones are unlike animal hormones because _____.
A. they are produced only in small amounts; B. they have different effects on different organs; C. they are not produced in specialized glands; D. they are mobile and can move throughout the plant body; E. interactions among different plant hormones can determine their ultimate effect on an organ.

## Analyzing and Applying the Basics

**1.** You have decided to divide an overgrown daylily and move a portion of it to a new spot in your yard. Thinking about water relations in plants, how should you prepare the site and the lily for transplanting?

**2.** Poinsettias are short-day plants with a critical night length of about 14 hours. Describe how you could induce a poinsettia to flower during the long days of summer.

**3.** Christmas tree farmers want to encourage apical dominance so that their plants have the "correct" form. They encourage this when plants are a few years old by pruning. What should they prune to maintain a single tip to the tree? What should they prune to ensure bushy lower branches?

# Connecting the Science

1. In even the smallest apartments, there is usually enough window space and light to grow a few herbs or small lettuce plants, and people with larger yards can grow a significant amount of the fresh produce they consume during the growing season. Do you grow food to eat? Why or why not? Do you see a value in growing some of your own food?

2. In the last paragraph of the chapter, we suggest taking a gardener's approach—working with natural processes and having a desire to understand—in all of our interactions with the natural world. Describe how this approach could be applied to a current issue concerning human modification of nature. Do you agree that it might be helpful to "think like a gardener," or do you see disadvantages of this attitude as well?

Answers to **Stop & Stretch, Visualize This, Working with Data, Savvy Reader, Sounds Right, But Is It?**, and **Chapter Review** questions can be found in the **Answers** section at the back of the book.

# Appendix
## Metric System Conversions

| To Convert Metric Units: | Multiply by: | To Get English Equivalent: |
|---|---|---|
| **Length** | | |
| Centimeters (cm) | 0.3937 | Inches (in) |
| Meters (m) | 3.2808 | Feet (ft) |
| Meters (m) | 1.0936 | Yards (yd) |
| Kilometers (km) | 0.6214 | Miles (mi) |
| **Area** | | |
| Square centimeters (cm$^2$) | 0.155 | Square inches (in$^2$) |
| Square meters (m$^2$) | 10.7639 | Square feet (ft$^2$) |
| Square meters (m$^2$) | 1.196 | Square yards (yd$^2$) |
| Square kilometers (km$^2$) | 0.3831 | Square miles (mi$^2$) |
| Hectare (ha) (10,000 m$^2$) | 2.471 | Acres (a) |
| **Volume** | | |
| Cubic centimeters (cm$^3$) | 0.06 | Cubic inches (in$^3$) |
| Cubic meters (m$^3$) | 35.30 | Cubic feet (ft$^3$) |
| Cubic meters (m$^3$) | 1.3079 | Cubic yards (yd$^3$) |
| Cubic kilometers (km$^3$) | 0.24 | Cubic miles (mi$^3$) |
| Liters (L) | 1.0567 | Quarts (qt), U.S. |
| Liters (L) | 0.26 | Gallons (gal), U.S. |
| **Mass** | | |
| Grams (g) | 0.03527 | Ounces (oz) |
| Kilograms (kg) | 2.2046 | Pounds (lb) |
| Metric ton (tonne) (t) | 1.10 | Ton (tn), U.S. |
| **Speed** | | |
| Meters/second (mps) | 2.24 | Miles/hour (mph) |
| Kilometers/hour (kmph) | 0.62 | Miles/hour (mph) |

| To Convert English Units: | Multiply by: | To Get Metric Equivalent: |
|---|---|---|
| **Length** | | |
| Inches (in) | 2.54 | Centimeters (cm) |
| Feet (ft) | 0.3048 | Meters (m) |
| Yards (yd) | 0.9144 | Meters (m) |
| Miles (mi) | 1.6094 | Kilometers (km) |
| **Area** | | |
| Square inches (in$^2$) | 6.45 | Square centimeters (cm$^2$) |
| Square feet (ft$^2$) | 0.0929 | Square meters (m$^2$) |
| Square yards (yd$^2$) | 0.8361 | Square meters (m$^2$) |
| Square miles (mi$^2$) | 2.5900 | Square kilometers (km$^2$) |
| Acres (a) | 0.4047 | Hectare (ha) (10,000 m$^2$) |
| **Volume** | | |
| Cubic inches (in$^3$) | 16.39 | Cubic centimeters (cm$^3$) |
| Cubic feet (ft$^3$) | 0.028 | Cubic meters (m$^3$) |
| Cubic yards (yd$^3$) | 0.765 | Cubic meters (m$^3$) |
| Cubic miles (mi$^3$) | 4.17 | Cubic kilometers (km$^3$) |
| Quarts (qt), U.S. | 0.9463 | Liters (L) |
| Gallons (gal), U.S. | 3.8 | Liters (L) |
| **Mass** | | |
| Ounces (oz) | 28.3495 | Grams (g) |
| Pounds (lb) | 0.4536 | Kilograms (kg) |
| Ton (tn), U.S. | 0.91 | Metric ton (tonne) (t) |
| **Speed** | | |
| Miles/hour (mph) | 0.448 | Meters/second (mps) |
| Miles/hour (mph) | 1.6094 | Kilometers/hour (kmph) |

| Metric Prefixes | | |
|---|---|---|
| **Prefix** | | **Meaning** |
| giga- | G | $10^9 = 1,000,000,000$ |
| mega- | M | $10^6 = 1,000,000$ |
| kilo- | k | $10^3 = 1,000$ |
| hecto- | h | $10^2 = 100$ |
| deka- | da | $10^1 = 10$ |
| | | $10^0 = 1$ |
| deci- | d | $10^{-1} = 0.1$ |
| centi- | c | $10^{-2} = 0.01$ |
| milli- | m | $10^{-3} = 0.001$ |
| micro- | µ | $10^{-6} = 0.000001$ |

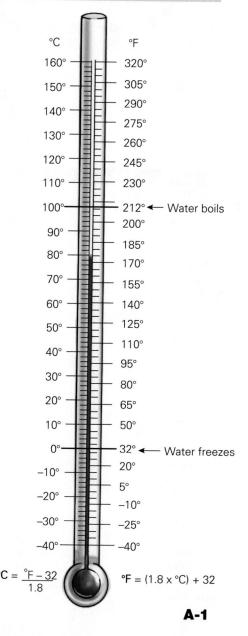

$$°C = \frac{°F - 32}{1.8}$$

$$°F = (1.8 \times °C) + 32$$

# Periodic Table of the Elements

| Metals | Metalloids | Nonmetals |

| 1 | | | | | | | | | | | | | | | | | 8 |
|---|---|---|---|---|---|---|---|---|---|---|---|---|---|---|---|---|---|
| **1**<br>**H**<br>1.008 | 2 | | | | | | | | | | | 3 | 4 | 5 | 6 | 7 | **2**<br>**He**<br>4.003 |
| **3**<br>**Li**<br>6.941 | **4**<br>**Be**<br>9.012 | | | | | | | | | | | **5**<br>**B**<br>10.81 | **6**<br>**C**<br>12.01 | **7**<br>**N**<br>14.01 | **8**<br>**O**<br>16.00 | **9**<br>**F**<br>19.00 | **10**<br>**Ne**<br>20.18 |
| **11**<br>**Na**<br>22.99 | **12**<br>**Mg**<br>24.31 | | | | | | | | | | | **13**<br>**Al**<br>26.98 | **14**<br>**Si**<br>28.09 | **15**<br>**P**<br>30.97 | **16**<br>**S**<br>32.07 | **17**<br>**Cl**<br>35.45 | **18**<br>**Ar**<br>39.95 |
| **19**<br>**K**<br>39.10 | **20**<br>**Ca**<br>40.08 | **21**<br>**Sc**<br>44.96 | **22**<br>**Ti**<br>47.87 | **23**<br>**V**<br>50.94 | **24**<br>**Cr**<br>52.00 | **25**<br>**Mn**<br>54.94 | **26**<br>**Fe**<br>55.85 | **27**<br>**Co**<br>58.93 | **28**<br>**Ni**<br>58.69 | **29**<br>**Cu**<br>63.55 | **30**<br>**Zn**<br>65.41 | **31**<br>**Ga**<br>69.72 | **32**<br>**Ge**<br>72.64 | **33**<br>**As**<br>74.92 | **34**<br>**Se**<br>78.96 | **35**<br>**Br**<br>79.90 | **36**<br>**Kr**<br>83.80 |
| **37**<br>**Rb**<br>85.47 | **38**<br>**Sr**<br>87.62 | **39**<br>**Y**<br>88.91 | **40**<br>**Zr**<br>91.22 | **41**<br>**Nb**<br>92.91 | **42**<br>**Mo**<br>95.94 | **43**<br>**Tc**<br>(98) | **44**<br>**Ru**<br>101.1 | **45**<br>**Rh**<br>102.9 | **46**<br>**Pd**<br>106.4 | **47**<br>**Ag**<br>107.9 | **48**<br>**Cd**<br>112.4 | **49**<br>**In**<br>114.8 | **50**<br>**Sn**<br>118.7 | **51**<br>**Sb**<br>121.8 | **52**<br>**Te**<br>127.6 | **53**<br>**I**<br>126.9 | **54**<br>**Xe**<br>131.3 |
| **55**<br>**Cs**<br>132.9 | **56**<br>**Ba**<br>137.3 | **57**<br>**La**<br>138.9 | **72**<br>**Hf**<br>178.5 | **73**<br>**Ta**<br>180.9 | **74**<br>**W**<br>183.8 | **75**<br>**Re**<br>186.2 | **76**<br>**Os**<br>190.2 | **77**<br>**Ir**<br>192.2 | **78**<br>**Pt**<br>195.1 | **79**<br>**Au**<br>197.0 | **80**<br>**Hg**<br>200.6 | **81**<br>**Tl**<br>204.4 | **82**<br>**Pb**<br>207.2 | **83**<br>**Bi**<br>209.0 | **84**<br>**Po**<br>(209) | **85**<br>**At**<br>(210) | **86**<br>**Rn**<br>(222) |
| **87**<br>**Fr**<br>(223) | **88**<br>**Ra**<br>(226) | **89**<br>**Ac**<br>(227) | **104**<br>**Rf**<br>(261) | **105**<br>**Db**<br>(262) | **106**<br>**Sg**<br>(266) | **107**<br>**Bh**<br>(264) | **108**<br>**Hs**<br>(269) | **109**<br>**Mt**<br>(268) | **110**<br>**Ds**<br>(271) | **111**<br>—<br>(272) | **112**<br>—<br>(285) | **113**<br>—<br>(284) | **114**<br>—<br>(289) | **115**<br>—<br>(288) | | | |

Group number

Atomic number

Transition elements

**Lanthanides**

| **58**<br>**Ce**<br>140.1 | **59**<br>**Pr**<br>140.9 | **60**<br>**Nd**<br>144.2 | **61**<br>**Pm**<br>(145) | **62**<br>**Sm**<br>150.4 | **63**<br>**Eu**<br>152.0 | **64**<br>**Gd**<br>157.3 | **65**<br>**Tb**<br>158.9 | **66**<br>**Dy**<br>162.5 | **67**<br>**Ho**<br>164.9 | **68**<br>**Er**<br>167.3 | **69**<br>**Tm**<br>168.9 | **70**<br>**Yb**<br>173.0 | **71**<br>**Lu**<br>175.0 |
|---|---|---|---|---|---|---|---|---|---|---|---|---|---|

**Actinides**

| **90**<br>**Th**<br>232.0 | **91**<br>**Pa**<br>231.0 | **92**<br>**U**<br>238.0 | **93**<br>**Np**<br>(237) | **94**<br>**Pu**<br>(244) | **95**<br>**Am**<br>(243) | **96**<br>**Cm**<br>(247) | **97**<br>**Bk**<br>(247) | **98**<br>**Cf**<br>(251) | **99**<br>**Es**<br>252 | **100**<br>**Fm**<br>257 | **101**<br>**Md**<br>258 | **102**<br>**No**<br>259 | **103**<br>**Lr**<br>260 |
|---|---|---|---|---|---|---|---|---|---|---|---|---|---|

| Name | Symbol | Name | Symbol | Name | Symbol | Name | Symbol |
|---|---|---|---|---|---|---|---|
| Actinium | Ac | Einsteinium | Es | Mendelevium | Md | Samarium | Sm |
| Aluminum | Al | Erbium | Er | Mercury | Hg | Scandium | Sc |
| Americium | Am | Europium | Eu | Molybdenum | Mo | Seaborgium | Sg |
| Antimony | Sb | Fermium | Fm | Neodymium | Nd | Selenium | Se |
| Argon | Ar | Fluorine | F | Neon | Ne | Silicon | Si |
| Arsenic | As | Francium | Fr | Neptunium | Np | Silver | Ag |
| Astatine | At | Gadolinium | Gd | Nickel | Ni | Sodium | Na |
| Barium | Ba | Gallium | Ga | Niobium | Nb | Strontium | Sr |
| Berkelium | Bk | Germanium | Ge | Nitrogen | N | Sulfur | S |
| Beryllium | Be | Gold | Au | Nobelium | No | Tantalum | Ta |
| Bismuth | Bi | Hafnium | Hf | Osmium | Os | Technetium | Tc |
| Bohrium | Bh | Hassium | Hs | Oxygen | O | Tellurium | Te |
| Boron | B | Helium | He | Palladium | Pd | Terbium | Tb |
| Bromine | Br | Holmium | Ho | Phosphorus | P | Thallium | Tl |
| Cadmium | Cd | Hydrogen | H | Platinum | Pt | Thorium | Th |
| Calcium | Ca | Indium | In | Plutonium | Pu | Thulium | Tm |
| Californium | Cf | Iodine | I | Polonium | Po | Tin | Sn |
| Carbon | C | Iridium | Ir | Potassium | K | Titanium | Ti |
| Cerium | Ce | Iron | Fe | Praseodymium | Pr | Tungsten | W |
| Cesium | Cs | Krypton | Kr | Promethium | Pm | Uranium | U |
| Chlorine | Cl | Lanthanum | La | Protactinium | Pa | Vanadium | V |
| Chromium | Cr | Lawrencium | Lr | Radium | Ra | Xenon | Xe |
| Cobalt | Co | Lead | Pb | Radon | Rn | Ytterbium | Yb |
| Copper | Cu | Lithium | Li | Rhenium | Re | Yttrium | Y |
| Curium | Cm | Lutetium | Lu | Rhodium | Rh | Zinc | Zn |
| Darmstadtium | Ds | Magnesium | Mg | Rubidium | Rb | Zirconium | Zr |
| Dubnium | Db | Manganese | Mn | Ruthenium | Ru | | |
| Dysprosium | Dy | Meitnerium | Mt | Rutherfordium | Rf | | |

# Answers

## Chapter 1

STOP & STRETCH

**p. 6:** Given the information here, it is most likely that someone brought a box of doughnuts to share, and that Homer ate a large number of them. This hypothesis is based on inductive reasoning.

**p. 7:** No, because even though Homer is a likely culprit, there are probably other individuals in your workplace who may have eaten some or all of the doughnuts.

**p. 10:** The independent variable is the presence of *H. pylori* infection, the dependent variable is number of ulcers per animal.

**p. 11:** Participants could have been exposed to different cold viruses; early arrivals may have been infected by a severe virus that was spreading in the hospital, while later arrivals may have been infected by a milder virus.

**p. 13:** The independent variable is stress, and the dependent variable is susceptibility to developing a cold when exposed to a virus.

**p. 17:** No, if the result is not statistically significant, we cannot conclude that hair styles are longer in 2015.

**p. 22:** It is much easier to relate to individual experiences than to a more abstract piece of information. Anecdotes are the way humans have learned from each other throughout history; polling or statistical data are much more recent.

**p. 24:** You could use the databases available through an academic library search service. Make sure the database you choose leads you to peer-reviewed research.

VISUALIZE THIS

**Figure 1.1:** Good scientists generate good hypotheses. This requires imagination, experience, and insight, as well as an understanding of science. These traits can be nurtured by taking a wide range of non-science courses.

**Figure 1.3:** Even with a well-designed experiment, it is possible that an alternative factor can explain the results, and researchers need to consider if that might be the case in their experiment.

**Figure 1.4:** Preventing virus attachment and invasion of cells. Preventing activation of the immune system response. Other answers may be acceptable.

**Figure 1.11:** Exercise may reduce feelings of stress and may independently strengthen the immune system (perhaps by improving blood circulation). Thus low levels of exercise could contribute to both high stress and high cold susceptibility.

**Figure 1.12:** A lack of racial and ethnic diversity in an experimental sample can affect the results and interpretation of a study because it means the sample may differ from the population in systematic ways. In this case, the average hair length in the class may not reflect the entire population because different subpopulations are not represented.

**Figure 1.14:** Outliers (unusually high or unusually low data points) have less impact if there are more data points in the sample.

**Figure 1.15:** Hypothesis: true. Difference between control and experimental groups: large. Sample size: large. This result is very likely to be both statistically and practically significant.

**Figure 1.16:** The reviewers do not have any stake in the outcome of the experiment and can catch errors in interpretation or experimental design that the original researchers may have missed.

WORKING WITH DATA

**Figure 1.7:** Students should notice that this change in the graph makes the Echinacea tea look much more effective.

**Figure 1.10:** There may not have been enough participants at a particular stress level to allow effective analysis of the correlation. If individuals at stress level 3 have the same cold susceptibility as those at stress level 4, that should cause us to question the correlation because we'd expect those at stress level 4 to have greater susceptibility.

**Figure 1.13:** The two data points are not "connected" to each other—that is, one does not follow the other, they are simply the same measurement in two different groups.

SAVVY READER

Red flags for questions 1, 2, 9, 10, and 11. Most other questions from the checklist do not pertain to this particular excerpt.

## SOUNDS RIGHT, BUT IS IT?

1. The experiment would be a double-blind, random assignment, placebo-controlled study. In this case, patients with acne would be randomly assigned use the cream or a placebo-alternative for enough time to possible see results in acne levels. Patients and evaluators would not know whether the cream used contained the active ingredient.

2. We don't know. Statistical significance does not equal practical significance—without information on how much difference there was between treatment and placebo samples, we don't know how large the effect of the cream was.

3. It is highly prone to bias, either by the consumer or by the clinician recording the results. Without a placebo control, these types of experiments also fail to account for other differences between non-treated and treated individuals, including the other aspects of skin care that might come along with using the cream (e.g., how many times per day the face is washed, etc.).

4. No, a single experiment could give a result that is statistically significant purely as a result of chance sampling error. In addition, even with a well-controlled experiment, there may be other explanations for a difference in acne levels between treatment and placebo groups besides the effectiveness of the active ingredient—for instance, perhaps one of the ingredients in the placebo cream actually causes additional acne. It's only after multiple tests of a hypothesis that we can come close to "proving" it.

5. Not necessarily. It means that the average amount of acne was less in the experimental group. If there was significant variability in results, it still could be that some individuals using the "real" cream had more acne or no change in acne after treatment.

6. We don't know from the statement "clinically proven" by itself if the experiment was designed to eliminate bias and/or showed a practically significant result. We also can't be confident that the hypothesis is proven without multiple tests that allow us to reject alternative hypotheses. Finally, one can't tell from the statement whether there was significant variability in the results, such that some individuals were benefitted a lot by the product and others not at all.

## LEARNING THE BASICS

1. A placebo allows members of the control group to be treated exactly like the members of the experimental group *except* for their exposure to the independent variable.

**2.** B; **3.** B; **4.** D; **5.** A; **6.** D; **7.** E; **8.** C; **9.** B; **10.** C; **11.** B; **12.** E

## ANALYZING AND APPLYING THE BASICS

1. No, another factor that causes type 2 diabetes could lead to obesity as well, or it is possible that type 2 diabetes causes obesity, not vice versa.

2. Neither the participants nor the researchers were blind, meaning that both subject expectation and observer bias might influence the results. If participants felt that vitamin C was likely to be effective, they may underreport their cold symptoms. If the clinic workers thought one or the other treatment was less effective, they might let that bias show when they were talking to the participants, influencing the results.

3. See the answer to number 1; BDNF may increase as a result of exercise, or excess BDNF might cause an increased interest in exercise.

# Chapter 2

## STOP & STRETCH

**p. 32:** Zombies require humans to prey upon. If the Zombie apocalypse occurs, zombies will die out for lack of human prey.

**p. 35:** Students that care more about exam scores might be more likely to think ahead and bring water. Likewise, students that are better prepared may have more time to focus on such details versus spending all their time studying.

**p. 37:** The carbons should each be bonded to two hydrogens, not one, to complete their valence shells.

**p. 42:** Kids are given snacks and drinks at parties or other exciting celebrations. An alternative hypothesis could be that attending a party makes children hyperactive, whether or not sugary drinks and snacks are served.

**p. 45:** You and your first cousin share more recent common ancestors in your grandparents. More distant relations, for example, second cousins, share less recent common ancestors and will therefore have fewer genetic similarities. The common ancestors to second cousins are the great-grandparents.

## VISUALIZE THIS

**Figure 2.4:** Oxygen.

**Figure 2.9:** The other locations should have similar rates of "disappearance" as the Bermuda triangle.

**Figure 2.12:** 100 times.

**Figure 2.16:** The red spheres represent sugars and the purple represent phosphates. The two colors of and sizes of nitrogenous bases represent the double ring (blue) purines and the single ring (yellow) pyrimidines.

## WORKING WITH DATA

**Figure 2.2:** The dependent variable is average exam score.

## SAVVY READER

1. Not really; how would a negative charge produce hydroxyl ions? While it is true that hydroxyl ions are negatively charged, if you remove the hydroxyl ion from water you would be left with the positively charged hydrogen ion. Hydroxyl ions and the hydrogen ions would bind to each other to produce water again.

2. Red flags—untested assertions made by those who stand to benefit financially. .

3. Amazing, hugely improving, etc.

4. No, the graph is not measuring cancer rates.

5. Error bars.

## SOUNDS RIGHT, BUT IS IT?

1. Computers cannot grow, reproduce or move. Neither can they metabolize or undergo homeostasis.

2. Maybe, but it seems unlikely.

3. Memorization does not allow easy extrapolation to novel questions.

4. Understanding usually beats memorization when applying knowledge to new situations.

5. Until computers can learn, they are going to require human input.

## LEARNING THE BASICS

1. Carbohydrates, lipids, proteins, nucleic acids.

2. Genetic material, cell wall, cell membrane, ribosomes.

3. Living organisms share a common ancestor. They differ due to changes that have occurred over time.

4. B; **5.** A; **6.** A; **7.** D; **8.** D; **9.** D; **10.** D; **11.** B; **12.** C

## ANALYZING AND APPLYING THE BASICS

1. No. Living organisms can reproduce—viruses can't reproduce on their own.

2. Because water is a good solvent and facilitates chemical reactions.

3. Only nucleic acids and proteins.

# Chapter 3

## STOP & STRETCH

**p. 53:** These food pairings, when eaten together, provide all the essential amino acids. Therefore, the meal is complete in protein.

**p. 62:** Enzymes, being proteins, would be digested by the stomach into amino acids that would be absorbed into the bloodstream and would not have catalytic effects on other tissues.

**p. 63:** Electrolytes and glucose need transporters.

## VISUALIZE THIS

**Figure 3.3:** The top tail in the fat shown in part b is polyunsaturated.

**Figure 3.4:** Plant cells have a cell wall, chloroplasts and large central vacuole and animal cells do not.

**Figure 3.18:** The hydrophobic portion of the cholesterol molecule interacts with the phospholipid head groups and the hydrophobic portion of the cholesterol molecule interacts with the fatty acid tails of the bilayer.

## WORKING WITH DATA

**Figure 3.17:** Ingesting chlorophyll tablets is not the same as applying a topical solution to a wound. Ingested chlorophyll could be digested by the stomach before it made its way to her blood stream. In addition, there is no indication of concentration of chlorophyll to be used. She should also consult her physician for possible contraindications before ingesting chlorophyll.

## SAVVY READER

1. Use of peer-reviewed studies, independent scientists evaluating data.

2. It makes sense to be less skeptical of a University putting together a database for the public than of a private company. The university will not gain more for arriving at a particular conclusion. A private company that manufactures the supplement will only benefit from the conclusion that the supplement is helpful.

3. Again, it makes sense to be more skeptical when there is something to be gained.

## SOUNDS RIGHT, BUT IS IT?

1. Fertilizer has a high concentration of dissolved solute.

2. Adding fertilizer would increase the concentration of solute outside the plant cell.

3. Water would move out of the plant cell.

4. The plant will wilt.

5. A wilting plant should be watered, not fertilized.

LEARNING THE BASICS

1. Nutrients serve as energy stores and as building blocks for cellular structures.

2. Carbon dioxide, oxygen, and water.

3. D; **4.** B; **5.** B; **6.** B; **7.** D; **8.** D; **9.** E; **10.** B

ANALYZING AND APPLYING THE BASICS

1. She could eat lots of protein-rich foods like beans.

2. With heart disease plotted on the x axis and trans fat consumption on the axis, the line showing the relationship should be a diagonal line.

3. Other vitamins, minerals, antioxidants and fiber.

# Chapter 4

STOP & STRETCH

**p. 74:** Body weight and calories burned during exercise are positively correlated—the more you weigh, the more calories you burn during exercise.

**p. 80:** More oxygen is available for the conversion of lactic acid to pyruvic acid.

**p. 82:** Grade 1 obesity is not associated with poor health outcomes and grade 2 is so it does not make sense to lump these two categories together.

VISUALIZE THIS

**Figure 4.2:** No.

**Figure 4.3:** Three negative charges would cause forceful repulsion between magnets.

**Figure 4.10:** 2 electrons.

**Figure 4.12:** It would take 2 turns of the cycle to release 4 of the 6 carbons in glucose. Each glucose molecule is converted into two three carbon pyruvates and each of those pyruvates is decarboxylated (losing one carbon) before entering the cycle. So, each turn of the cycle releases two carbons.

**Figure 4.13:** Oxygen exists in cells as molecular oxygen ($O_2$). Because only one oxygen atom is required to make water ($H_2O$), the convention is to place the fraction ½ before $O_2$.

WORKING WITH DATA

**Table 4.1:** 28.

**Figure 4.16:** Severe obesity—underweight—moderate obesity—overweight—normal weight.

SAVVY READER

1. No, pathogenic bacteria, plant toxins, snake venom, hurricanes and ice storms are all natural but can be very bad for humans.

2. No. Consult with a physician.

3. No. There needs to be evidence presented from well-controlled scientific studies.

SOUNDS RIGHT, BUT IS IT?

1. No.

2. No.

3. Carbon dioxide.

4. Exercise increases breathing rate. It is actually the loss of the carbon and oxygen atoms that were once part of fats that causes weight loss.

5. Weight loss occurs when fat is metabolized into carbon dioxide and released from the body. Laxatives work only on materials in the large intestine that will never become part of the body.

LEARNING THE BASICS

1. All of the building up and breaking down chemical reactions that occur in a cell.

2. When a substrate binds to the active site of an enzyme, the enzyme clamps down on the substrate, causing increased stress on the bonds holding the substrate together.

3. Reactants of respiration are oxygen and glucose. Products of respiration are carbon dioxide and water.

4. C; **5.** B; **6.** D; **7.** D; **8.** E; **9.** B; **10.** C

ANALYZING AND APPLYING THE BASICS

1. No. Fats and proteins also feed into cellular respiration at various points in the cycle and fats actually store more energy than carbohydrates.

2. The carbon atom would either become part of a cellular structure inside your body or be released as $CO_2$.

3. Keep doing what you are doing—your body weight is a healthful one and your BMI classification may be misleading if you are muscular.

# Chapter 5

## STOP & STRETCH

**p. 89:** Because the moon has no or very little greenhouse gas, heat from the sun is not retained to be reradiated overnight, which is why the nighttime temperatures are cold. The warm temperatures during the day occur because water is not present to absorb some of the heat energy without changing temperature.

**p. 92:** The carbon dioxide released by the burning of logs or sawdust is essentially recycled when replacements for plants that produced these logs or sawdust are re-grown. This leads to no net increase in carbon dioxide. Fossil fuel burning releases carbon that was removed from the atmosphere millions of years ago so is causing a current net increase.

**p. 94:** The chlorophyll is reabsorbed because it is so valuable to the plant—its components can be stored and the chlorophyll can be reproduced in the spring.

**p. 97:** Possible answers include Nitrogen and Sulphur (important components of proteins), Phosphorous (important in nucleic acids), and Iron (in photosynthetic pigments). Potassium is also important, but it is less structural.

**p. 98:** It may be that any mutation that reduces rubisco's ability to bind oxygen also reduces its ability to bind carbon dioxide. Or that these mutations affect the protein's ability to function within the Calvin cycle. In other words, the benefit of changing rubisco must be lower than the cost of photorespiration.

**p. 100:** No, since Swedes have a high standard of living with one-fourth of the carbon dioxide emissions per person.

## VISUALIZE THIS

**Figure 5.1:** The red "greenhouse gas" portion of the figure will be thicker and thus there will be an increase in the amount of heat (illustrated by more heat arrows) that is reradiated back to Earth.

**Figure 5.2:** Water levels would rise and shorelines would erode.

**Figure 5.3:** Water molecules should appear in the "air" portion of the center box. Evaporation occurs more rapidly when the water is warming because the lack of hydrogen bonds can allow the molecules to escape the liquid form.

**Figure 5.4:** Human activity.

**Figure 5.11:** Carbon dioxide, water, and light energy are obtained from the environment, and oxygen and carbohydrate are released. NADP+ and ATP/ADP are contained within the chloroplast.

**Figure 5.12:** The reactions are similar in that energized electrons are used to create a gradient of hydrogen ions, which is then used to generate ATP. However, the energized electrons in photosynthesis come from chlorophyll that has been energized by the sun; in respiration the electrons come from NADP, which was produced by the breakdown of sugar. At the end of the chain, electrons in photosynthesis are used to produce sugar; in respiration, the electrons are picked up by oxygen, which reacts with hydrogen to produce water.

**Figure 5.13:** Without sunlight, ATP and NADPH inputs from the light reactions would be absent and the cycle could not function.

**Figure 5.15:** Stomata are more likely to be open during the day to allow in $CO_2$, because light is required for photosynthesis to proceed.

## WORKING WITH DATA

**Figure 5.6:** The increase in concentration is steeper during the period 2000–2005 when compared to 1960–1970.

**Figure 5.8:** The left axis is for the blue line and the right for the red line. They are on the same graph to better illustrate the relationship between carbon dioxide level and temperature over time.

**Figure 5.10:** The absorption of light would be small in the lower wavelengths (red) and thus the line would be closer to the x axis on the left of the graph. The absorption of light should be higher in the middle and perhaps highest wavelengths and thus the line would be far from the x axis in the middle and right side of the graph.

## SAVVY READER

1. Climate change could affect economic systems, ecological systems, financial systems, and humanitarian concerns. The research primarily dealt with economic systems.

2. Consensus means general agreement. The author evaluated 14 studies on the question of the economic costs and benefits of climate change and had his analysis peer reviewed by "leading economists." This does not appear to represent a consensus of the world's economists.

3. Several possible. We don't know the funding source for the scientist, the reporter did not put the results into a larger context or include comments from other scientists.

## SOUNDS RIGHT, BUT IS IT?

1. To grow, reproduce, maintain homeostasis.

2. ATP

3. $6CO_2 + 6H_2O + \text{Light energy} \rightarrow C_6H_{12}O_6 + 6O_2$

4. No, ATP produced during photosynthesis is used to produce sugars.

5. Aerobic respiration

6. $C_6H_{12}O_6 + 6O_2 \rightarrow 6CO_2 + 6H_2O + \text{Energy (ATP)}$

7. Even though, unlike animals, plants consume carbon dioxide and release oxygen in the process of photosynthesis, because plants need to perform aerobic respiration to produce ATP for cellular activities, they also consume oxygen and release carbon dioxide, just like animals. So it is not correct to say that plants perform a sort of "opposite respiration" that the statement implies.

## LEARNING THE BASICS

1. Reactants of photosynthesis are carbon dioxide and water. Products of photosynthesis are oxygen and glucose.

2. C; **3.** D; **4.** E; **5.** D; **6.** A; **7.** C; **8.** A; **9.** D; **10.** D

## ANALYZING AND APPLYING THE BASICS

1. Because colder water has the capacity to absorb more heat before water molecules can evaporate into the atmosphere, colder temperatures mean that less water circulates in the water cycle.

2. The water cycle would remain basically unchanged. However, the carbon cycle would be quite different. The only way carbon dioxide would move from the atmosphere to Earth's surface would be by absorption into water and perhaps incorporation into rocks. It would return to the atmosphere via diffusion from water and volcanic activity.

3. High rates of photosynthesis are continually pumping oxygen into the air. So even though the molecule is being "used up" through reactions with other compounds, its continual production ensures high levels in the atmosphere.

# Chapter 6

## STOP & STRETCH

**p. 109:** 120; 870

**p. 111:** ~ 23,000

**p. 114:** 8pg at the end of S

**p. 118:** Multiple hit means that several mutations (hits) to genes must occur for a cell to escape all regulatory controls and lead to cancer. One mutation alone is not enough to cause cancer.

**p. 119:** If radiation treatment damages cell cycle control genes of normal cells, they could become cancerous.

**p. 123:** Both meiosis I and mitosis produce two daughter cells. Meiosis I differs from mitosis in that meiosis I separates homologous pairs of chromosomes from each other. Mitosis does not do this.

## VISUALIZE THIS

**Figure 6.4:** (a) 2; (b) asexual reproduction would result in two stems with leaves, not one larger plant.

**Figure 6.6:** 2 molecules of DNA—one 100% purple, one 50% purple.

**Figure 6.12:** At the metaphase checkpoint.

**Figure 6.17:** The X chromosome.

**Figure 6.20:** No. No DNA duplication occurs between Meiosis I and Meiosis II.

**Figure 6.22:** 8 different alignments are possible at metaphase I.

## WORKING WITH DATA

**Figure 6.2:** Multiplicative. Smoking only increases risk 10%; Drinking only increases risk 2%. Smoking and drinking together increases risk 20%. 10 × 2 = 20.

## SAVVY READER

1. Ask your physician or do your own research using Google Scholar or other trustworthy resource.

2. Non-peer-reviewed journals are less credible than peer-reviewed journals.

3. What percent of cancer patients who did not take this supplement were still alive one year later.

4. Yes and Yes.

5. No.

6. Desperation may be one reason.

## SOUNDS RIGHT, BUT IS IT?

1. No. You would need to know the cancer risk of being in the sun for a few hours before you could decide whether that behavior was safe.

2. It will take longer to get a tan, so people will have to pay for more or longer tanning sessions.

3. No.

4. Eat fruits and vegetables rich in Vitamin D. Take a vitamin D supplement.

5. The use of tanning beds is not safe. It injures skin and increases cancer risk. Benefits if increased vitamin D are more safely obtained through healthy eating.

## LEARNING THE BASICS

**1.** Mitosis occurs in somatic cells, meiosis in cells that give rise to gametes. Mitosis produces daughter cells that are genetically identical to parent cells, while meiosis produces daughter cells with novel combinations of chromosomes and half as many chromosomes as compared to parent cells.

**2.** Rapid cell division.

**3.** D; **4.** A; **5.** C; **6.** D; **7.** B; **8.** C; **9.** D

**10.** Random alignment of homologues and crossing over.

**11.** Proto-oncogenes are genes that control the cell cycle. Oncogenes are mutated versions of proto-oncogenes.

**12.** Plant and animal cells both have interphases followed by M phases. Cytokinesis differs in plant and animal cells.

**13.** Cancer cells divide when they should not, invade surrounding tissues, and move to other locations in the body.

## ANALYZING AND APPLYING THE BASICS

**1.** No. Only mutations in cells that give rise to gametes are passed on. The mutation in skin cancer was acquired and will only be found in the affected skin cells.

**2.** No. Some tumors remain benign.

**3.** Radiation can only be used on tumors in particular locations and its effects are more localized. Chemotherapy moves throughout the body attacking rapidly dividing cells.

# Chapter 7

## STOP & STRETCH

**p. 137:** A mutation in a promoter portion of the DNA may affect whether the gene is copied or translated. A mutation in a structural portion of a chromosome may weaken the chromosome or change its ability to be transcribed. A mutation in "nonsense" DNA is likely to have no effect.

**p. 138:** Eight different gametes. The relationship is $2^n$, where n is the number of chromosome pairs.

**p. 143:** Its effects are not seen until later adulthood, after an individual is likely to have had children.

**p. 145:** Eight possible gamete types (e.g., T Y R; T y R; t Y R, etc.) and 64 possible cells inside the Punnett square.

**p. 148:** Immigration from countries where average height is shorter may be changing the genetics of height in the United States such that there are a larger number of "small" alleles today than in the past. Dietary changes may also be leading to reduced growth or early puberty, which stunts growth in height.

**p. 153:** Even a trait that is highly heritable can be strongly influenced by the environment. If the current environment results in some individuals having very low IQ scores relative to others, it is certainly possible that a changed environment would lead to smaller differences among individuals. See the case of the "maze-dull" and "maze-bright" rats.

## VISUALIZE THIS

**Figure 7.1:** the most dramatic effects of a mutagen would happen if exposure occurred at the zygote stage, since all cells would inherit any mutations. The least effect would occur at the adult stage, when affected cells would leave few or no offspring.

**Figure 7.3:** Many possibilities.

**Figure 7.4:** A nonsense mutation (c) is the most likely type of mutation, because most random errors would result in a misspelling that is not a "word."

**Figure 7.6:** Blood group gene from his dad and eye color gene from his dad; Blood group gene from his mom and eye color gene from his mom.

**Figure 7.7:** Monozygotic twins are more likely to be conjoined than dizygotic twins, because the genetic similarity of their cells can promote fusion of the embryos.

**Figure 7.9:** No, they are non-identical siblings because the genes in one egg are different from the genes in the other, and each is fertilized by a genetically unique pollen grain.

**Figure 7.13:** It should decrease the likelihood of having a child who carried two mutant alleles.

**Figure 7.22:** They should revert to the "normal" size seen at the top of the figure.

## WORKING WITH DATA

**Figure 7.14:** 50%

**Figure 7.15:** 9/16 or 56%

**Figure 7.16:** No, the majority of men were not 5 feet 10 inches. There is a large range of heights in this population.

**Figure 7.19:** Points would be more widely scattered, and the correlation line would be flatter.

## SAVVY READER

**1.** No, these were identical twins and thus had identical genes.

**2.** Gene methylation affects the expression of gene, that is, how much protein is produced by the gene. If two individuals with the same gene differ in gene methylation, the amount of protein product they produce is different, which means that their cells will not be identical.

3. Two reasons: 1) Increased SERT methylation and increased risk of bullying may be correlated to another factor that could differ between twins (for example, birth weight differences caused by chance differences in uterine development), or in fact, increased SERT methylation may cause increased risk of bullying because bullies target kids with particular personalities. 2) The changes described in SERT have not been definitely connected to differences in behavior, at least according to this article.

SOUNDS RIGHT, BUT IS IT?

1. More likely to be determined by many genes.
2. Yes.
3. Yes.
4. It is likely that genes from both parents influence eye color and shape.
5. Harry's eyes are the product of genetic information from both his mother and his father as expressed during his development. It is nearly impossible that he has the same genetic information as his mother for this trait—so while his eyes may be a similar color, they cannot be identical. (And if you are a Harry Potter fan, you know this is true—Lily Potter did not wear glasses, but Harry is dependent on his!)

LEARNING THE BASICS

1. Genotype refers to the alleles you possess for a particular gene or set of genes. Phenotype is the physical trait itself, which may be influenced by genotype and environmental factors.
2. Multiple genes or multiple alleles influencing a trait, environmental effects that affect the expression of a trait, or a mixture of both factors.
3. B; 4. A; 5. B; 6. A; 7. B; 8. C; 9. E; 10. D

GENETICS PROBLEMS

1. All have genotype Tt and are tall.
2. One-quarter are TT and are tall; one-half are Tt and are tall; and one-quarter are tt and are short.
3. Both are alleles for the same trait; P is dominant, and p is recessive.
4. Both parents must be heterozygous, Aa.
5. 50%
6. a. yellow; b.YY and yy

7. Yellow, round parent: Yy Rr; green, wrinkled parent: yy rr.
8. a. 50%; b. 50%; c. 25%; d. yes
9. a. dominant; b. The phenotype does not inevitably surface when the allele is present.

ANALYZING AND APPLYING THE BASICS

1. The trait could be recessive, and both parents could be heterozygotes. If the trait is coded for by a single gene with two alleles, one would expect 25% of the children (1 of 4) to have blue eyes. That 50% (2 of 4) have blue eyes does not refute Mendel because the probability of any genotype is independent for each child and is not dependent on the genotype of his or her siblings. (Consider that some families may have 4 boys and no girls; this does not refute the idea that the Y chromosome is carried by only 50% of a man's sperm cells.)
2. No, most quantitative traits are influenced by the environment. Even if differences in genes account for most of the differences among individuals in a trait, if the environment changed, the trait would very likely change as well.
3. No, heritability doesn't explain why any two individuals differ. For instance, John and Jerry could be identical twins, but Jerry might suffered a minor accidental head injury that reduced his IQ.

# Chapter 8

STOP & STRETCH

**p. 162:** $I^C I^C$ type C; $I^A I^C$ type AC; $I^B I^C$ type BC; $I^C I$ type C; along with previously described AB types

**p. 164:** Her father must have the disease, and her mother is a carrier or has the disease.

**p. 166:** It is far more likely that the mutation originated in the ovary of Queen Victoria's mother. When Queen Victoria's mother produced egg cells containing the mutation, all cells that arise from that egg cell during development would have that mutation. Therefore Queen Victoria would have would have a 50% chance of passing the hemophilai allele to each of her children. It is very unlikely that the mutation would occurred three separate times in Queen Victoria's ovary.

**p. 170:** Siblings share 50% of their DNA so they should share 4 bands. Identical twins would share all 8 bands.

VISUALIZE THIS

**Figure 8.1:** Red

**Figure 8.3:** 3.2

**Figure 8.4:** One X chromosome would be genetically identical to the X chromosome the woman in the figure received in the egg cell her mother ovulated. The woman in the figure would have inherited the other X chromosome from her father when his sperm fertilized her mother's egg cell.

**Figure 8.5:** A cross involving a hemophilic female.

**Figure 8.6:** The first generation sb a square connected to a circle connected to square. Below the line connecting the first square and circle is a circle. Below the line connecting the circle and second square is a circle then 2 squares.

**Figure 8.13:** Every band in the child's profile is present in one of the parent's profiles.

**Figure 8.16:** Prince George is not at risk for hemophilia because no one on his father and grandfather's side of the family, going all the way back to the tsarina's sister, has had the disease. The tsarina's sister did not inherit the allele and none of her heirs appears to carry it.

WORKING WITH DATA

**Figure 8.9:** To calculate this probability, multiply ten times itself 13 times i.e. $10^{13}$ Therefore, the odds that any two non-identical twins would have the same DNA profile is 1 ten trillion, which is very unlikely since there are 7 billion people on earth!

SAVVY READER

1. Odds of having a child of a particular biological sex are 50%. Product is 60% so claim is 10% increase.
2. No, we would need statistical analysis to see if the data the manufacturer is supplying is a real difference and not just a sampling error, etc.
3. No.
4. No.
5. Inability of males to find partners with whom to reproduce.

SOUNDS RIGHT, BUT IS IT?

1. ½ × ½ × ½ = 1/8th
2. ½ × ½ × ½ = 1/8th
3. All three girls, two girls and one boy, or two boys and one girl.
4. The couple already has two boys. This eliminates the probability of them having all three girls or two girls and one boy.

5. The probability of the couple having a girl is 50% no matter how many boys they already have.

LEARNING THE BASICS

1. When both alleles are equally expressed in the phenotype codominance is occurring. When the dominant allele fails to completely mask the effects of the recessive allele in a heterozygote incomplete dominance is occurring.
2. A trait with multiple alleles has more than two alleles for that gene. An individual can only carry up to 2 different alleles of a gene.
3. Females and males inherit one X chromosome from their mom in the egg cell. Females inherit another X chromosome from their dad in the sperm cell while males inherit a Y chromosome from their dad.
4. DNA is isolated, then amplified using PCR. DNA is cut with restriction enzymes, loaded on gel and subject to electrophoresis to separate fragments by size.
5. $I^A i$ (dad), $I^B i$ (mom)
6. 2/8 A+; 2/8 A−; 1/8 AB+; 1/8 AB−; 1/8 B+; 1/8B−
7. C; **8.** B; **9.** E; **10.** D; **11.** B

ANALYZING AND APPLYING THE BASICS

1. DNA profiling can unambiguously identify an individual. Blood typing can only exclude individuals from consideration.
2. Banding patterns of parents should mostly differ from each other. Every band in a child should be present in one or the other parent and roughly 50% of the bands in the sisters profiles should be the same.
3. Pedigrees can vary, but the first cousins must share great grandparents. If one of the great grandparents carries a rare recessive allele, it can be passed down both sides of the pedigree and each of the first cousins can carry that recessive allele. If the first cousins mate, their child can have the disease encoded by the rare recessive alleles.

# Chapter 9

STOP & STRETCH

**p. 183:** Pro-ASN. Yes, Pro is also coded for by codons CCC;CCA and CCG. ASP is also coded for by codon AAC.

**p. 189:** Answers will vary but most people would assume that hormone or rBGH free was safer.

**p. 192:** Larger fish will likely outcompete smaller fish for resources.

**p. 193:** Some people believe that embryos are humans or could become human and should not be used for scientific research. Others believe that embryos are just a collection of cells and not yet human and thus acceptable to use in scientific research.

## VISUALIZE THIS

**Figure 9.1:** The second carbon of the sugar, ribose, is bonded to -OH while the same carbon in deoxyribose is attached to hydrogen. There is a CH3 group in thymine that is not present in uracil.

**Figure 9.2:** 12/4 = 3

**Figure 9.3:** Sample answer starting after the fork in the DNA: AGCTGGGCAGGTAC. From this particular DNA, the mRNA would be UCGACCCGUCCAUG.

**Figure 9.5:** Adenine paired with Uracil and Cytosine paired with Guanine.

**Figure 9.22:** The cloned baby would be identical to the man and to his twin brother.

## WORKING WITH DATA

**Figure 9.16:** Use increased by ~20 percent. Initially pests were killed by the pesticide so herbicide used dropped. Then, as herbs evolved resistance, herbicide use increased again.

## SAVVY READER

1. It may be because people tend to be more frightened of bacteria ad viruses than other organisms used in cloning, for example plants.

2. The DNA would be digested in the stomach—not used to make viral or bacterial proteins.

3. A more reliable article will tend to at least point out counter arguments.

4. Hopefully you would prefer a more balanced, data-driven approach.

## SOUNDS RIGHT, BUT IS IT?

1. It has to develop inside a woman's uterus.

2. This process would initially produce one fertilized cell which, if it develops inside a woman's uterus and is born, could grow to be an adult like any other child.

3. Probably not, some would be shorter and some taller.

4. Socio-economic status, family experience, exposure and access to different hobbies, etc.

5. Cloned humans would be the same genetically but all traits that are determined by environmental factors or a combination of genetics and environment could differ.

## LEARNING THE BASICS

1. GCUAAUGAAU

2. gln, arg, ile, leu

3. A nucleus is transferred in to an enucleated egg cell.

4. B; **5.** C; **6.** B; **7.** C; **8.** C; **9.** B; **10.** D ; **11.** T; **12.** F

## ANALYZING AND APPLYING THE BASICS

1. The same amino acid is often coded for by codons that differ in the third position. For instance, UUU and UUC both code for phenylalanine.

2. Transcription would be slowed or stopped in that cell.

3. GAATTC; CTTAAG

# Chapter 10

## STOP & STRETCH

**p. 202:** This is very unlikely to be an evolutionary change, which results from genetic changes in a population that are inherited. The teenagers in 2008 are unlikely to have a significant genetic difference from those in 1982, instead access to (and interest in) orthodontic treatment has increased.

**p. 207:** Faith is the belief that something is true despite having no physical evidence for that belief; science is based on observations of the natural world. Because statements of faith are based on nonrepeatable, mostly unobserved phenomenon, they cannot be falsified. Thus, they are not subject to scientific hypothesis testing.

**p. 213:** It may be that the human diet was harder on teeth in the past, so that teeth were lost due to breakage, leaving "room" for the wisdom teeth to erupt. These teeth may have been beneficial, in fact, replacing the role of missing molars.

**p. 220:** They are very close. The oldest human ancestor is slightly younger than the estimated date of divergence between chimpanzees and humans, as is expected.

**p. 226:** Nucleus, ribosomes, cytoskeleton, plasma membrane, endoplasmic reticulum, Golgi apparatus, glycolysis, aerobic respiration, mitotic division (others possible).

VISUALIZE THIS

**Figure 10.1:** All lice would have been eliminated by the pesticide, and the population would not have evolved.

**Figure 10.2:** These branches represent organisms that are extinct.

**Figure 10.7:** Perhaps the presence of plant-eating tortoises on the islands made those cacti with longer stems (thus out of reach of the tortoises) more likely to survive, leading to the evolution of longer stems. There also may be a lack of other trees on the islands, meaning that taller cacti were able to capture more sunlight and thus reproduce more than smaller plants.

**Figure 10.10:** Hair-covered bodies, birth to live young, other characteristics of all mammals (also forward-facing eyes and opposable thumbs).

**Figure 10.11:** Squirrel would be the farthest left branch, originating above the mammal ancestor. The shared ancestor is the junction point.

**Figure 10.12:** The green bones are longer in the bat and sea lion and function to make the limb paddle shaped. In lions the green bones are very short and function mostly as a stable foot pad. In chimpanzees and humans, the bones are medium length and allow for fine movements of individual digits.

**Figure 10.19:** The minerals cannot accumulate fast enough to fill in the spaces left by the decaying tissue, so the fossil loses structure.

WORKING WITH DATA:

**Figure 10.16:** The black turnstone shares a more recent common ancestor with the red knot than the Caspian tern does, as evidenced by greater similarity in morphology, but more convincingly, DNA sequence.

**Figure 10. 17:** The percentages indicate the differences between humans and these two species. An analogy might help: coffee might be 99.01% similar to cola and 89.90% similar to tea, but this does not mean cola and tea are more similar to each other than either is to coffee.

**Figure 10.21:** About 2 million years ago.

**Figure 10.22:** 3 million years old.

SAVVY READER

1. The argument is essentially that the theory of evolution is suspect because several hundred scientists are skeptical of the theory.

2. Science is not a democratic process, but is based on evidence. The theory of evolution has support from many different subfields of biology, as well as chemistry and geology. A scientific theory that fits the evidence more effectively has not been proposed by these skeptics. In addition, the article provides little evidence that the skeptics are experts in evolutionary theory. For example, the scientist quoted is an expert in human medicine.

3. Dr. Egnor is a professor of neurosurgery and pediatrics—he does not appear to be an expert on evolution, so the article is unconvincing that the debate is among those who study evolution.

4. Look at the "About" page of an Internet site to learn about the mission of the host organization. An organization with a particular agenda is more likely to cherry pick data that supports their agenda than to present a balanced view of an issue.

SOUNDS, RIGHT, BUT IS IT?

1. No. The observations of the natural world related to the origin of biological diversity have led to scientists convincingly rejecting the other possible theories.

2. No.

3. No. Supernatural causes for natural phenomenon cannot be falsified—there is no observation of the natural world that could cause us to reject the existence of an entity that is omnipotent and not bound by natural laws. Therefore, supernatural explanations are not within the realm of science.

4. Not clear. It seems that teachers who "teach the controversy" would either include evidence for theories that have been falsified, misleading "evidence" that the theory of evolution should be falsified, or non-scientific alternative theories.

5. Presenting an issue as a scientific controversy by bringing in an alternative non-scientific explanation or rejecting established scientific understanding only serves to confuse students about the nature of science and whether scientific investigation can ever answer questions.

LEARNING THE BASICS

1. The theory of common descent describes that the similarities among living species can be explained as a result of their descent from a common ancestor.

2. B; **3.** D; **4.** D; **5.** B; **6.** E; **7.** B; **8.** A; **9.** D; **10.** D

ANALYZING AND APPLYING THE BASICS

1. Parentheses surround groups that share a common ancestor (((pasture rose and multiflora rose) spring avens) live forever) spring vetch).

**2.**

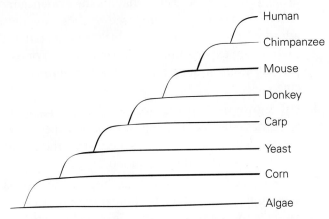

- Human
- Chimpanzee
- Mouse
- Donkey
- Carp
- Yeast
- Corn
- Algae

**3.** The comparison is only between humans and these organisms in genetic sequence. There is no comparison on the degree of relatedness among the organisms. The differences between humans and yeast may be in totally different segments of DNA than the differences between humans and algae.

# Chapter 11

## STOP & STRETCH

**p. 236:** The TB bacteria can continue to multiply, doing damage to the lungs and resulting in a larger population. More cells mean more likelihood of resistant variants, as well as a more difficult time clearing out the infection.

**p. 240:** Humans have invented technologies and disease cures that decrease the differences in fitness among individuals with different genetics (for example, corrective lenses have increased the survival of individuals with poor eyesight). We are still subject to natural selection for traits that are less amenable to technological fixes and even in our ability to respond to technology.

**p. 244:** The cystic fibrosis allele may provide protection from diseases that are common in more urban environments. Individuals carrying the allele in these environments were more likely to survive, thus causing the allele to become more common in those populations.

**p. 249:** Mutations appear by chance and may not have much affect on fitness unless the environment changes. The resistance allele appeared randomly.

## VISUALIZE THIS

**Figure 11.6:** If other flowers farther away are blooming, it may be an advantage to be able to pick up more distantly related pollen so that the plant's offspring carry novel alleles that might be advantageous.

**Figure 11.8:** Because the elephants are using most of their resources in Generation 3, a percentage of Generation 4 will die from starvation.

**Figure 11.11:** The population would change in the direction of demonstrating that behavior under the right conditions.

**Figure 11.13:** The individual with the fast metabolism of alcohol would not have proportionally more offspring than those without the allele, so the ratio of fast to slow metabolizers would stay the same on each "shelf."

## WORKING WITH DATA

**Figure 11.9:** It decreased by a large amount.

**Figure 11.12:** The x-axis would have a scale of "generation number" from 1–60. The y-axis would have the same scale as at present. The percentage of flies with the fast metabolizing allele would be noted at generation 1 and generation 57 for both populations and the two points for each population would be connected. A legend would be required to distinguish the two lines from each other.

**Figure 11.16:** The fitness of the red flowered plants would be lower, so the population as a whole would move toward the white direction as a result of directional selection.

## SAVVY READER

**1.** Airline air is polluted and leads to virus spreading. Andrew Speaker knowingly put his bride and others at risk of contracting a dangerous form of TB. There is no evidence provided for either statement.

**2.** Other postings by this author; evidence that the author linked to credible sites to back up statements of fact.

**3.** Students may provide various acceptable answers.

## SOUNDS RIGHT, BUT IS IT?

**1.** Genetic mutation.

**2.** $10000 \times 3 = 30000$.

**3.** No, natural selection sorts out variations—those changes that lead to greater survival and reproduction are passed on and become more common, those that reduce survival and reproduction are lost.

**4.** The thumb evolved from a wrist bone.

**5.** No.

**6.** Yes.

**7.** The number of heritable variations that can occur in a population is large. If any of these variants lead to greater survival and reproduction, they will become common in the population. Over time, structures can become more and more modified and complex as an accumulation of changes occurs, even taking on new structures and functions.

## LEARNING THE BASICS

1. Antibiotics have helped reduce deaths due to tuberculosis, but they have become less effective as strains of antibiotic-resistant *M. tuberculosis* have appeared in human populations.

2. Artificial selection is a process of selection of plants and animals by humans who control the survival and reproduction of members of a population to increase the frequency of human-preferred traits. Artificial selection is like natural selection in that it causes evolution; however, it differs because humans are directly choosing which organisms reproduce. In natural selection, environmental conditions cause one variant to have higher fitness than other variants.

3. Within a single patient, the *Mycobacterium tuberculosis* consists of a large number of genetic variants. These variants differ in a number of traits, including their susceptibility to a particular antibiotic. When the patient takes an antibiotic, the individual bacteria that are more resistant to the drug have higher fitness, and natural selection leads to the evolution of a bacteria population that is resistant to the drug.

4. C; 5. D; 6. E; 7. E; 8. B; 9. C; 10. D

## ANALYZING AND APPLYING THE BASICS

1. Farmers who desired larger strawberries must have saved seeds from the wild plants that produced the largest berries. Over the course of many generations of the same type of selection in domesticated strawberries, average berry size increased dramatically. Tradeoffs include: domestic strawberries have less energy to put into roots and leaves than do wild strawberries; domesticated strawberries require a larger supply of water and nutrients, and perhaps domesticated strawberries have fewer chemical defenses against insects and other predators.

2. Individual zebras with some sort of striping must have had greater survival than those who were not striped. Over the course of many generations of this type of selection, individuals in the population evolved dramatic stripes.

3. No, not all features are adaptations; some might arise by chance. A trait is an adaptation if it increases fitness relative to others without the trait. One can look for ways that the trait increases fitness or examine individuals who lack or have a less dramatic version of the trait to measure its fitness effects.

# Chapter 12

## STOP & STRETCH

**p. 258:** Lions live on the open plain made up of dry grass—it may be that a solid light brown color blends in better in the grass.

**p. 260:** No, because the varieties are the same species, so the number of chromosomes should be the same in both parents.

**p. 266:** Morphological concept, because they would be identifying birds by sight (and sound) without information about mating habits or genetic relatedness.

**p. 273:** For example: average height varies but is not typically used to identify similar groups. Adding this category might lead to identifying a larger number of races.

**p. 278:** "Ma" or "mmmm" is one of the first sounds an infant can make. Since the mother is so associated with infant care, this sound became associated with her. ("Da" or "Pa" is a very early sound as well.)

**p. 281:** Females put more investment (energy) in offspring compared to males, so have a higher incentive for making sure that the offspring has the greatest chance of survival and later reproduction.

## VISUALIZE THIS

**Figure 12.2:** No, because the chromosomes from the two different parents are not homologous and thus still could not pair properly.

**Figure 12.3:** No, natural selection or other mechanisms of evolution could be creating differences in traits other than those related to appearance.

**Figure 12.14:** African Americans with ancestry from sub-Saharan Africa as well as Italian-, Greek-, Turkish-, Indian- and certain Arab-American groups, and people whose ancestors came from the Balkans.

**Figure 12.16:** Torpedo-shaped, highly maneuverable thanks to their flippers, fast swimmers, light belly with a darker back.

**Figure 12.18:** An individual with dark skin in a low-UV environment could take vitamin D to ensure proper bone growth, while an individual with light skin in a high-UV environment could avoid sun exposure, use sunscreen, and/or take folate supplements.

## WORKING WITH DATA:

**Figure 12.5:** The Honeycrisp might slow divergence, because newly emerged hawthorne flies may be able to mix with later-emerging apple flies on this later-flowering tree.

**Figure 12.9:** 6 out of 20 or 0.3.

**Figure 12.10:** Heterozygotes can form via (p) males × (q) females and by (p) females × (q) males.

**Figure 12.13:** The frequency of type O blood would likely increase with distance from Asia.

**Figure 12.17:** Various possibilities as visualized on the graph (for the first question, dots of the same color that are distant from each other on the y-axis; for the second question, dots of different color that are in close proximity on the y-axis).

SAVVY READER

1. The primary evidence presented is that African Americans have higher rates of hypertension than U.S. whites; research shows that the age-adjusted prevalence in blacks is 41.8 percent, versus 29.8 percent for whites.

2. U.S. whites have a higher prevalence of hypertension than Nigerians, while African Americans have a lower prevalence than Germans and Finns, indicating that the difference may not be due to an underlying biological similarity among blacks and consistent difference from whites.

3. Is hypertension in African Americans associated with a higher risk of heart disease? (If not, the difference in blood pressure may not be practically significant.) Does a reduction in salt intake have a significant effect on hypertension in African Americans? (If not, the dietary guidelines may simply impose an extra burden—the article noted that few adults have a sodium intake as low as the recommendation—rather than provide a health benefit.) Note that the answer to both of these questions may be "yes," but this article excerpt does not provide those answers.

SOUNDS RIGHT, BUT IS IT?

1. Yes, as with all physical traits, it is likely variation existed for this trait.

2. Yes, as with most behavioral traits, it is likely variation existed in foraging behavior.

3. No, only those who had the alleles that coded for different bill shapes or different behaviors.

4. Yes, these traits could improve survival and reproduction of the individuals who possess them.

5. Individuals with higher evolutionary fitness leave more offspring. Thus, the traits they possess become more common in the population over time.

6. "In order" implies some sort of striving toward a goal. Instead, the evolution of nectar-feeding honeycreepers

occurred as a result of natural selection acting on the variations present in the population. Those individuals with longer, curved bills—traits that happened to increase fitness—left more offspring, thus a portion of the population changed over time to consist of more individuals with longer, curved bills who fed on particular flowers.

LEARNING THE BASICS

1. The gene pools of populations must become isolated from each other, they must diverge (as a result of natural selection or genetic drift) while they are isolated, and reproductive incompatibility must evolve.

2. Because it takes time for mutations to accumulate in a population and/or species, the relative lack of mutations is evidence of recent origin.

3. Genetic drift can occur as a result of founder effect (a small subpopulation moves away and contains an unrepresentative sample of the original gene pool), the bottleneck effect (the few survivors of a disaster have an unrepresentative sample of the original gene pool), or by the chance loss of alleles (death or lack of reproduction not having to do with the traits coded by the gene) from small, isolated gene pools. By changing allele frequencies, all of these events result in the evolution of a population.

**4.** D; **5.** B; **6.** A; **7.** C; **8.** C; **9.** D; **10.** C

ANALYZING AND APPLYING THE BASICS

1. It depends on the species definition one uses. The two populations of wolves are not different biological species. They are likely different genealogical species—the definition that corresponds to "race." To determine if they are different races, one would have to look for the standard evidence: unique alleles or unique allele frequencies in each population.

2. Ireland, 0.01; Britain 0.007; Scandinavia 0.005. The allele frequency could be different as a result of genetic drift, particularly the founder effect, in these different populations. It also could be different because the allele has some yet unknown fitness advantage in the environment of Ireland relative to the environment of Britain or Scandinavia.

3. If the early mutation becomes widespread, the timing difference between the early and normal populations should cause reproductive isolation. The same is possible with the expanded mutation if certain pollinators become associated with the individuals that bloom earlier than expected.

# Chapter 13

## STOP & STRETCH

**p. 293:** Grouping species by similarity in diets could help biologists understand the common needs for species of interest and perhaps simplify natural resource management. Grouping species by similarity in habitat need can help identify whether species with similar habitat needs may be at risk of extinction due to habitat loss.

**p. 297:** Mitochondria in plant and animal cells are identical, indicating that they evolved before plants and animals diverged.

**p. 311:** A beak that can tear meat, long wings, large size, and the ability to soar.

## VISUALIZE THIS

**Figure 13.7:** The second membrane resulted from the infolding of the host cell membrane when it engulfed the bacterium.

**Figure 13.12:** To allow dispersal of spores away from the food source that is already in use.

**Figure 13.16:** The embryo is like an astronaut in that it is protected within a tough outer coating that contains all the resources it requires for at least short term survival in a harsh environment. It's different in many ways, including that it can still exchange materials (for example, gases) with the environment, while an astronaut cannot get any materials required from life from space.

## WORKING WITH DATA

**Figure 13.2:** About 8400 genera have been described of the 8750 or so predicted to exist.

**Figure 13.3:** We'd have to draw an extra branch on the tree that shows a common ancestor between animals and this protistan group. The "phylum" of Protista would also be broken up, potentially.

**Figure 13.4:** They use bases that are common to all groups as a framework and then infer that older species are more likely to possess the ancestral bases than younger species.

**Figure 13.19:** It probably had little head striping, like the junco, but a sparrow-colored body (brown-striped) like all of the sparrows.

## SAVVY READER

1. That middle-aged women carry extra fat, which would allow them to survive for periods without eating and leave extra food for the young. In addition, middle-aged people "run the world."

2. Women who stored extra fat were able to survive longer during hard times and provide support for younger people in their family group while not competing as much for food as those who did not store fat. Families of these women survived periods of low food more successfully than families of women who did not store fat as effectively, so the trait became more common in the population.

3. There is no evidence presented in the article that storing more fat had a selective advantage.

4. Middle-aged humans are more likely to be living with serious problems—including back problems, loss of hearing and eyesight—that could possibly reduce the fitness of their group (i.e., they are less able to provide for others, but still require resources). The stories in this chapter provide examples of other species that are arguably better suited for their environment.

## SOUNDS RIGHT, BUT IS IT?

1. Approximately, yes.

2. Flight, metamorphosis, ability to sense different chemicals or light wavelengths than we can, etc.

3. Sea star

4. No, the tree would still indicate the same evolutionary relationship among its members—humans and starfish share a recent common ancestor, humans and insects a less recent ancestor.

5. Humans are the product of millions of years of evolution from the first animal—as are all modern animals. While humans have many complex structures and traits, other animals have other, different complex structures and traits. An evolutionary tree can be drawn in many different ways to indicate the same relationships—in this tree, humans do not have to be on the very top.

## LEARNING THE BASICS

1. Scientists have identified over 1.5 million different species, but at least 10 million, and possibly over 100 million, different species might exist on Earth today.

2. After flowers evolved, the number of different kinds of plants with flowers increased rapidly. One hypothesis for why this occurred is that flowering plants were better adapted to dry conditions, another is that the relationship between plants and animals (as pollinators and dispersers) contributed to the diversity of forms through co-evolution.

**3.** By comparing the DNA sequences of organisms to see if the pattern of similarity and difference matches what is predicted, and by examination of the fossil record, looking for the record of evolutionary change in the group of organisms

**4.** B; **5.** E; **6.** E; **7.** D; **8.** C; **9.** D; **10.** B

ANALYZING AND APPLYING THE BASICS

**1.** An oxygen-sensitive cell that kept an oxygen-consuming prokaryote alive inside of it would be protected from the damaging effects of oxygen, as any nearby oxygen would quickly be "used up" by the prokaryote. This would be an advantage over cells that did not have this prokaryote.

**2.** Although it is not very motile, this is probably an animal since it consumes other organisms (is not photosynthetic) by ingesting.

**3.** These chloroplasts likely evolved through a process of secondary endosymbiosis, in which the ancestral Euglena cell engulfed but did not consume a eukaryotic green algal cell. In this case the original bacterial membrane is surrounded by the secondary membrane from the engulfed cell and both of these are surrounded by the vacuole membrane from the ancestral Euglenid.

# Chapter 14

STOP & STRETCH

**p. 319:** 10,000 (solve for x: 1/100 = 100/x)

**p. 321:** Average life span equals the sum of the ages at death divided by the number of deaths. Deaths of infants thus greatly bring down the average life span, and reducing these deaths will raise life expectancy dramatically, even if the otherwise population health has not changed.

**p. 324:** Infectious diseases require a population of susceptible individuals to survive and spread. If the population is small, the infectious organism cannot maintain a population.

**p. 327:** A large population with a high birth rate produces a large number of offspring, as these offspring have their own offspring, they can potentially overshoot the carrying capacity. A large population with a low birth rate produces many fewer young, so even when the offspring reproduce, they don't add huge numbers to the population, thus allowing a gradual approach to the carrying capacity.

VISUALIZE THIS

**Figure 14.1:** Trap-happiness would cause the population estimate to be too small because a greater proportion of returnees will be marked. Trap-shyness causes the opposite effect because fewer returnees will be marked.

**Figure 14.14:** Several reasons: larger immigration rate, younger population on average, religious traditions which encourage larger families.

WORKING WITH DATA

**Figure 14.3:** Yes, growth was still in relation to the total, but the rate of growth was small.

**Figure 14.4:** Approximately 23 years (half the time it takes if population is growing 1.5%—also 70 divided by 3).

**Figure 14.5:** Where birth rates and death rates are closest to equal; that is, before the transition and after (Areas 1 and 3 on the graph).

**Figure 14.6:** Where the line is most vertical; that is, at the point where the "S" turns from rising to flattening out. Here at a population size of around 450.

**Figure 14.8:** In the low-growth scenario, where the graph of population size begins to level out.

**Figure 14.11:** The same basic pattern of population fluctuation would occur, just at the higher carrying capacity.

**Figure 14.13:** The United States has about one-quarter the population of India.

SAVVY READER

**1.** The Christmas Bird Count and Breeding Bird Surveys, "citizen science" programs (i.e., birdwatchers participate in data collection).

**2.** Many answers possible, such as evidence that the causes Audubon identifies are truly responsible, the types of actions that are needed, how effective these actions are in preserving bird populations, etc.

SOUNDS RIGHT, BUT IS IT?

**1.** Growth that occurs in proportion to the current total.

**2.** (b), the proposal that starts at 1 cent per day.

**3.** $30,000

**4.** $167,772.16

**5.** Because the wage growth is always in relation to the previous day's wage, the number of pennies earned rises very quickly, eventually far exceeding the constant daily wage. Exponential growth causes both populations, and earnings, to rise much more rapidly than simple geometric growth. This principle can be applied to many situations, including the exponential growth of an algae bloom in a swimming pool, an invading species in a new area, or the growth of a retirement savings account.

LEARNING THE BASICS

**1.** A decrease in death rate, especially a decrease in infant mortality, has led to increasing growth rates. In addition, the process of exponential growth lends itself to these population explosions as the number of people added in any year is a function of the population of the previous year.

**2.** In most populations, growth rate declines because death rate increases (or birth rate decreases) as resources are "used up" by the population.

**3.** B; **4.** C; **5.** B; **6.** A; **7.** B; **8.** D; **9.** D; **10.** D

ANALYZING AND APPLYING THE BASICS

**1.** 2500. Solve for x: 1/50 = 50/x.

**2.** −0.002 (r = Birth rate − Death rate = 25/1000 − 27/1000). This population is likely near carrying capacity because it is not growing, but in fact declining slightly, as would be expected if resources are limited.

**3.** The carrying capacity is expected to rise since more resources are available. However, because the flies are in a restricted environment (a culture bottle), the accumulation of waste may become a problem, and a large population of flies may not be able to survive this accumulation. Depending on fly behavior, there may also be an increase in aggressive interactions or a decrease in mating in more dense situations.

# Chapter 15

STOP & STRETCH

**p. 343:** Habitat destruction, pollution, introduction of exotics, even over-exploitation of colorful or dramatic species. It would help to know the environmental situation of many of the species at risk to determine if they are threatened by a common cause. It may be just that these animals are especially vulnerable to all threats because of their unique biology.

**p. 348:** All species pairs are competitors; the organisms they compete for are: a) fish, b) krill, c) penguins, d) squid, e) herbivorous zooplankton. The most straightforward way to test if competition is occurring is to remove one of the competing species and see if the remaining species increases in population size.

**p. 351:** The cycle is changed because more nitrogen is made available for cycling by pulling it out of the atmosphere and applying it to farm fields. In addition, the nitrogen does not stay in the fields where it is spread, but is released to the water in human waste (or buried in cemeteries), so the cycle is more global, with waterways becoming an important sink for nitrogen.

**p. 357:** An organism that is highly social and will not mate, hunt, or successfully produce young without the presence of a large group may fail to reproduce if there is not a critical mass of other individuals. Organisms that are highly choosy of mates may also fail to reproduce in small groups.

VISUALIZE THIS

**Figure 15.9:** Answer will vary depending on location of student.

**Figure 15.12:** If baleen whales disappeared, krill populations would probably rise, leading to higher populations of fish and other carnivorous zooplankton that compete with whales for krill. However, it might lead to a decrease in smaller toothed whales, which prey on the baleen whales.

**Figure 15.16:** The total number of bacterial cells reaches the carrying capacity of the environment (that is, the intestines) in both cases—it is just the makeup of the bacterial community that changes.

**Figure 15.18:** In man-made systems, nutrients typically flow in one direction (from crops to humans to waste water and cemeteries), so we can no longer refer to the process as a nutrient cycle.

**Figure 15.21:** More are closer to the equator.

**Figure 15.22:** Fewer offspring per pair per year, longer development time with parental care required.

**Figure 15.24:** Because the normal allele does not cause disease, homozygotes for the normal allele are healthy. This is not the case for individuals who are homozygous for the mutant allele, who are ill.

WORKING WITH DATA

**Figure 15.2:** 10 to 50 million years

**Figure 15.3:** 1 million years

**Figure 15.4:** 1600–1849: 21; 1850–2000: 50

**Figure 15.7:** If the island lost 67% of its habitat (10,000 of 15,000 square kilometers), you'd expect about 25% of the species to become extinct.

**Figure 15.8:** About 1000 mountain lions.

**Figure 15.25:** About 50% is lost.

SAVVY READER

1. The representative cites unnamed "scientists and wolf recovery advocates" as experts who agree that the wolf is "back." Without listing who they are and their qualifications for making these statements, the politician's statement is not convincing.

2. It doesn't seem likely the wolf's return will wipe out elk and moose. There is always greater biomass of prey than of predator species, and the experience in Yellowstone indicates that wolves don't cause loss of prey, but may change prey's behavior.

SOUNDS RIGHT, BUT IS IT?

1. a) Mosquitos are predators of microbes; b) dragonflies are predators of mosquitos; c) mosquitos and the plants are mutualists; and d) mosquitos are predators (parasites) of humans.

2. Bacteria, algae, and other microbes in water would increase, dragonflies, some fish, and crustaceans would decrease, plants pollinated by mosquitos would decrease, and humans may increase.

3. Many answers possible. Examples: more microbes in water may lead to a population increase of another predator, or may lead to increased risk of disease of organisms that are hosts to these bacteria; decrease in dragonflies could negatively affect bird species that eat them, etc.

4. A keystone species is one that has a disproportionately large effect on other species in its food web. In other words, removing a keystone species can cause the rest of the system to radically change.

5. We do not know. It can be impossible to determine until a species is removed from a system how important it is to that system.

6. Mosquitos are connected to many other species and their loss would affect these species in positive or negative ways. Whether these changes would in turn negatively affect humans is nearly impossible to predict given how complicated food webs are.

LEARNING THE BASICS

1. Habitat fragmentation endangers species because it interferes with their ability to disperse from unsuitable to suitable habitat and because it increases their exposure to humans and human-modified environments. Species that have very specialized habitat requirements, those that cannot move rapidly, and those that are very susceptible to human disturbance are most negatively affected by fragmentation.

2. Mutualism is a relationship among species in which all partners benefit. Predation is a relationship among species in which one benefits and others are consumed. Competition is a relationship among species in which all partners are harmed by the presence of the others.

3. D; 4. D; 5. A; 6. A; 7. B; 8. C; 9. D; 10. A 11. D; 12. D

ANALYZING AND APPLYING THE BASICS

1. Kelp → sea urchins → sea otters. Kelp also provides habitat for fishes, clams, and abalones. When sea otters are removed, sea urchins devastate the kelp forest, leading to the loss of these associated species. Sea otter thus serves as a keystone.

2. Even though there are more individuals in situation A, with so few fathers contributing genes to the next generation, the likelihood of large changes in allele frequency due to drift (i.e., the founder effect) is much higher in situation A than in situation B.

3. How uncertain are the environmental conditions—that is, do density-independent factors cause the death of large numbers of individuals when they occur? What is the mating system—are a few males mating with the majority of the females, or is the species more monogamous? What is the likely carrying capacity for piping plovers in this environment?

# Chapter 16

STOP & STRETCH

**p. 370:** A reduction in tilt would reduce seasonality because areas would be receiving the same solar irradiance throughout the year. An increase in tilt would increase the yearly range of temperatures in seasonal areas.

**p. 378:** Chlorophyll may be "expensive" for the trees to make in some way. The most likely expense is nitrogen, which is limited in the soil and may not be easy to replace if simply dropped to the soil.

**p. 384:** Wetlands slow water flow and thus can absorb extra nutrients before they accumulate in the open water body.

**p. 387:** The population is expected to be 9 billion, if two-thirds is 6 billion.

**p. 388:** Fossil fuels were easy to access just by digging. The technology to convert the fuels to electricity (burning and generating steam) or to power an engine (controlled explosions) were a lot simpler than the technology needed to convert sunlight into electricity. Wind and solar power is not as reliable as the power generated from fossil fuels.

## VISUALIZE THIS

**Figure 16.2:** Solar irradiance is higher in Florida compared to Vermont because it is closer to the equator. Thus Florida is warmer than Vermont.

**Figure 16.3:** Because when the Northern Hemisphere is tilted away from the sun, the Southern Hemisphere (where Australia is located) is tilted toward it.

**Figure 16.5:** Near 30° north and south and near the equator, where air is more likely to be moving vertically (up or down) rather than horizontally, thus not producing a directional breeze.

**Figure 16.7:** This period is the hottest part of the summer, and areas near the sea will be cooler than areas inland and also should experience a cooling breeze.

**Figure 16.15:** Cattle are not nearly as large or strong as elephants and cannot uproot these large trees.

## WORKING WITH DATA

**Figure 16.9:** Cold desert, temperate deciduous forest, temperate rainforest, woodland/shrubland/or grassland. Depends on precipitation.

**Figure 16.22:** Grassland and temperate forest.

**Figure 16.25:** 97.9 million tons, 175.1 million tons, and 163.5 million tons, respectively.

## SAVVY READER

1. None of the statements include objective support, but some imply that the company has evidence to support their claim.

2. On the website where the claims are made. The evidence to support each claim should be obviously and easily located, for instance one hyperlink click away or on a clearly labeled drop down menu item.

3. Answers will vary, but should focus on objective evidence.

## SOUNDS RIGHT, BUT IS IT?

1. Midwest: January is winter. Australia: January is summer.

2. The amount of energy per square meter falling on a surface.

3. Solar irradiance is higher in July in the Midwest because this is the time of year when the North Pole is pointing toward the sun—in January, the pole points away, so solar irradiance is lower. In Australia during July, the nearest pole (the South Pole) is facing away from the sun, so solar irradiance is lower during July compared to January in the same location.

4. No.

5. Australian summer is our winter—if summer seasonality was determined by distance to the sun, both the Midwest and Australia would have summer and winter at the same time. Instead, the average temperature in an area depends on solar irradiance. The fact that Earth's axis is tilted is more important than distance to the sun in determining seasonal differences in temperatures.

## LEARNING THE BASICS

1. The tilt of Earth's axis means that as the planet revolves around the sun, the Northern Hemisphere is tilted toward the sun during part of the year and away during part of the year. The temperature at a given point on Earth's surface is determined in large part by the solar irradiance (i.e., the strength of sunlight) striking the surface. Because the Northern Hemisphere is tilted away in the winter, solar irradiance is low, as are temperatures. The opposite occurs in the summer.

2. Water is slow to gain and lose heat, so when surrounding land areas cool as a result of declining solar irradiance, warmer air from over the water keeps the temperature higher. The opposite is true as the surrounding land warms as a result of increasing solar irradiance.

3. B; **4.** C; **5.** E; **6.** E; **7.** C; **8.** A; **9.** B; **10.** D

## ANALYZING AND APPLYING THE BASICS

1. Semitropical with seasonal wetter and drier periods as a result of latitude. Relatively rainy due to location on windward side of mountain range.

2. The temperature will become less moderate because warm water will not be carried to the coasts of Europe, warming nearby landmasses. Seasonality will be more pronounced and winters colder and longer.

3. Bend, Yakima, and perhaps Walla Walla are on the leeward side of the Cascades, while the other cities are on the windward side, since more rain is likely on the leeward side.

# Chapter 17

## STOP & STRETCH

**p. 400:** The lack of blood vessels in cartilage makes it difficult for transplanted tissues to receive the nutrients they need to survive.

**p. 404:** Organism–vehicle; organ system–fuel injection system, organ–engine, tissue, valves and pistons.

**p. 408:** Negative.

## VISUALIZE THIS

**Figure 17.3:** The cellular component of both is fibroblasts. The matrices of both contain collagen fibers. Fibrous connective tissue contains more collagen fibers and they are arranged in parallel. Loose connective tissue contains elastin fibers in the matrix also.

**Figure 17.4:** Actin and myosin.

**Figure 17.6:** Cardiovascular and urinary are two examples.

**Figure 17.8:** Damage to the liver or kidney.

## WORKING WITH DATA

**Figure 17.10:** Y-axis is number of people waiting; x-axis is particular organ.

## SAVVY READER

1. Both scientists are affiliated with universities—Goyal with Geisinger Health System in State College, Pennsylvania and Sehgal with the CWRU School of Medicine.

2. They reported the data even though it was not what they expected.

3. Less credible than the JAMA article because the organ-trafficking agency stands to benefit financially from organ trade.

## SOUNDS RIGHT, BUT IS IT?

1. Yes.

2. No.

3. In a coma.

4. No.

5. In a coma.

6. No.

7. A brain-dead person will not recover the ability to survive off life support.

## LEARNING THE BASICS

1. Epithelia are tightly packed sheets of cells that cover and line organs, vessels, and body cavities. Connective tissues are loosely organized tissues that bind organs and tissues to each other. Muscle tissue consists of bundles of long, thin, cylindrical muscle fibers that are able to contract. Nervous tissue is composed mainly of neurons that can conduct and transmit electrical impulses.

2. Loose connective tissue, adipose tissue, blood, fibrous connective tissue, cartilage, and bone.

3. Muscle tissues differ based on the presence or absence of striations and whether the muscle is voluntary or involuntary. Cardiac muscle is striated and involuntary; skeletal is striated and voluntary; smooth muscle is not striated and involuntary.

4. C; **5.** E; **6.** A; **7.** A; **8.** B; **9.** C; **10.** D; **11.** False

## ANALYZING AND APPLYING THE BASICS

1. When muscles contract and expand during shivering, heat energy is produced.

2. It secretes insulin.

3. Answers will vary. An example of negative feedback is regulating home temperature. An example of positive feedback is getting a good grade makes a student work harder.

# Chapter 18

## STOP & STRETCH

**p. 420:** An increase in HCl would cause gastrin release to slow, resulting in a decrease in the secretion of other digestive enzymes. This prevents the pH of the stomach from getting so low that the stomach lining is damaged.

**p. 422:** It would contain larger materials such as whole proteins and perhaps some blood cells.

## VISUALIZE THIS

**Figure 18.1:** 28. Third molars are found on both sides of the upper and lower jaw.

**Figure 18.5:** Bacteria can move from bladder through renal vein to kidney.

**Figure 18.6:** To the bladder.

## WORKING WITH DATA

**Table 18.2:** A 120 lb woman could only have 2 drinks in 2 hours because 3 drinks in 2 hours puts her right at 0.8, which is legally drunk. Answers for the second question will vary for each student.

## SAVVY READER

1. This could be the case. Most colleges and universities are very concerned about reporting rape statistics because of concerns about scaring away potential students.

2. Answers will vary.

**3.** Answers will vary.

**4.** Answers will vary.

SOUNDS RIGHT, BUT IS IT?

**1.** At least two hours.

**2.** No.

**3.** Slapping, pinching, and yelling.

**4.** Call 911.

**5.** Because his BAC has not peaked, your friend needs careful attention for at least two hours.

LEARNING THE BASICS

**1.** Mouth, pharynx, esophagus, stomach, small intestine, large intestine.

**2.** Liver, pancreas, gall bladder.

**3.** The kidneys remove waste and excess water from the blood, producing urine.

**4.** D; **5.** D; **6.** C; **7.** F; **8.** A; **9.** C; **10.** T

ANALYZING AND APPLYING THE BASICS

**1.** Taking aspirin before drinking might make the symptoms experienced after drinking worse, not better.

**2.** Food moved from the pharynx to the trachea instead of the esophagus.

**3.** No—food is just moving through a tube inside the body.

# Chapter 19

STOP & STRETCH

**p. 432:** By forcing the diaphragm upward violently, the volume of the lungs is quickly reduced, resulting in a large exhalation of air.

**p. 435:** Less surfactant means that alveolar walls are sticking together, requiring greater force to allow air to gain access to the respiratory surface.

**p. 437:** The only "cure" is lung transplant. However, increasing respiratory capacity (i.e., strength of the respiratory muscles) can help by drawing more air into the lungs.

**p. 440:** The brain.

**p. 442:** The valves that prevent backflow.

VISUALIZE THIS

**Figure 19.2:** It causes a rapid intake of air.

**Figure 19.6:** Less oxygen crosses the respiratory membrane, meaning that individuals must breathe more rapidly to obtain adequate oxygen.

**Figure 19.8:** The disadvantages would be greater likelihood that the alveoli walls will stick together, the creation of more friction impeding the exchange of gases with the outside air, and by devoting more energy to build cells in the lung there is less energy for other organs.

**Figure 19.11:** An abundance of one or more blood cell type, typically ones that are not fully differentiated.

**Figure 19.13:** More force is required to move blood to the body extremities than to the lungs, and too much force sending blood to the lungs can cause fluid to leak from capillaries into alveoli, interfering with breathing.

**Figure 19.16:** By preventing blood from pooling in the veins of the lower leg, compression socks can increase blood circulation and thus the rate of oxygen delivery.

**Figure 19.17:** Deoxygenated blood from the right atrium can flow into the left atrium, meaning that the blood pumped to the body has an overall lower oxygen content.

**Figure 19.18:** It crushes the plaque while increasing the diameter of the artery.

WORKING WITH DATA

**Figure 19.3:** About 500 ml is exchanged at rest.

SAVVY READER

**1.** Most New Yorkers appear to oppose an e-cigarette ban; while in Oregon the majority want to ban their sales to minors; and fewer than half the people in Syracuse are sure that e-cigarette are safe.

**2.** Only those individuals who feel strongly about the question are likely to participate in these surveys. This means that e-cigarette users may be overrepresented or underrepresented in the sample.

**3.** Answers may vary, but should focus on employing surveys that take randomized samples of the population.

SOUNDS RIGHT, BUT IS IT?

**1.** Numerous carcinogens, including radioactive compounds, in the plant itself.

**2.** It increases heart rate, blood pressure, and levels of "bad" cholesterol, all of which increase the risk of heart attack and stroke.

**3.** Absorbed across the cells lining these organs.

**4.** Smokeless tobacco contains many of the compounds known to result in negative health outcomes as found in tobacco smoke. These compounds are still absorbed into the bloodstream and can have the same effect, regardless of the delivery system.

## LEARNING THE BASICS

**1.** Contraction of the diaphragm flattens out this dome-shaped muscle at the base of the chest cavity. This has the effect of increasing the size of the lung cavity, lowering air pressure and allowing outside, oxygen-rich air to flow through the upper respiratory system and into the lungs.

**2.** Blood flowing from body tissues enters the right atrium of the heart and proceeds through a valve to the right ventricle. From the right ventricle, it is pumped out through the pulmonary arteries to the lungs. Oxygen-rich blood from the lungs enters the left atrium and proceeds through a valve to the left ventricle. From the left ventricle, it is pumped out through the aorta to the body.

**3.** E; **4.** D; **5.** A; **6.** D; **7.** B; **8.** D; **9.** D; **10.** E

## ANALYZING AND APPLYING THE BASICS

**1.** The blood after doping has greater oxygen-carrying capacity and can supply muscles with oxygen more effectively. The danger is the risk of clots forming from excess blood cells.

**2.** Oxygen delivery to tissues will fail to meet the body's demand, resulting in fatigue and shortness of breath (breathing rate will be triggered to increase in response to low $O_2$ levels).

**3.** By crossing into the bloodstream at the respiratory membrane and then being carried throughout the body by the circulatory system.

# Chapter 20

## STOP & STRETCH

**p. 452:** Infectious but not contagious.

**p. 458:** They could be infected and become ill or even die.

**p. 464:** It's possible that an individual who can fight off the infection has a slightly different set of genes than those who can't.

## VISUALIZE THIS

**Figure 20.1:** Yes, because ribosomes are not bounded by membranes.

**Figure 20.4:** Surface proteins.

**Figure 20.6:** The odds of an epidemic increase as the numbers of vaccinated people decreases.

**Figure 20.10:** White blood cells.

## WORKING WITH DATA

**Figure 20.3:** The 2 bacteria present at noon would double to 4 at 12:20, double again to 8 bacteria by 12:40 and double again by 1 o'clock giving 16 bacteria. Exponential growth curve is linear at first, then shows a rapid increase (concave line) as cells multiply. The line showing growth at cooler temperatures would have this same concave shape, but would grow much less rapidly.

## SAVVY READER

**1.** Sometimes there are subtle differences in interpretation of data. For example, in the case of AIDS, there might be some debate about the rate at which mutations occur in HIV. However, there is no debate about the larger issue—that HIV causes AIDS.

**2.** Answers will vary.

**3.** Answers will vary.

## SOUNDS RIGHT, BUT IS IT?

**1.** The immune system is fully formed at birth.

**2.** The immune system gets stronger with exposure to antigens.

**3** Modern vaccines contain small parts of viruses.

**4.** Whole viruses would cause the generation of more kinds of antibodies.

**5.** It is not true that modern children are exposed to more antigen in vaccines because the whole virus vaccines of old would have presented more of an immune challenge than modern vaccines.

## LEARNING THE BASICS

**1.** Bacteria are single-celled organisms that have no nucleus or membrane-bounded organelles. They house their DNA in the nucleoid region and are surrounded by a cell wall. Viruses are composed of genetic material surrounded by a protein coat called the capsid. Some viruses are also surrounded by a membranous envelope.

**2.** They can remove the infectious organism from the tissue it infects.

**3.** B cells produce receptors that are found on cell surfaces or secreted as antibodies and destroy invaders. T cells also produce receptors but not ones that are secreted.

**4.** D; **5.** B; **6.** C; **7.** D; **8.** B; **9.** C; **10.** C ; **11.** E; **12.** F

ANALYZING AND APPLYING THE BASICS

1. They cannot replicate themselves without the aid of a host cell.

2. Because the flu virus is a rapidly mutating virus.

3. Because the presence of memory cells allows a stronger and more rapid response.

# Chapter 21

STOP & STRETCH

**p. 474:** The gene that encodes the protein receptor for testosterone undergoes a mutation so that it makes no receptor or makes a receptor than cannot bind testosterone.

**p. 476:** Estrogen inhibits bone resorption activities by osteoclasts.

**p. 481:** No, not if arm length and muscle mass are equal.

**p. 486:** At menopause, when estrogen declines, women begin to store fat like men, in abdomen.

VISUALIZE THIS

**Figure 21.2:** The breast tissue would develop less and the larynx tissue more.

**Figure 21.8:** The femur (thigh) and patella (kneecap) and the shin bone.

WORKING WITH DATA

**Figure 21.12:** (a) Within; (b) No.

SAVVY READER

1. Random selection of participants; collecting data from same set of colleges over different years.

2. Symptoms of steroid abuse are hard to accept. May lead to escapism via drug and alcohol use, or students who are likely to abuse alcohol and drugs may be more likely to use steroids also.

3. Answers will vary. It may be competition for scholarships increased or steroid availability increased.

SOUNDS RIGHT, BUT IS IT?

1. 25 days
2. 25 mg
3. 24 mg
4. 1.75 mg
5. The birth control pill can be manufactured to deliver the same monthly amount of estrogen as the normal menstrual cycle delivers.

LEARNING THE BASICS

1. The hypothalamus regulates body temperature and affects behaviors such as hunger, thirst, and reproduction. The pituitary gland secretes two hormones involved with producing sex differences: follicle-stimulating hormones (FSH) for sperm production and egg cell development, and luteinizing hormones (LH) for testosterone production and the release of an egg cell during ovulation. Adrenal glands secrete the hormone adrenaline (epinephrine) in response to stress or excitement and androgens, including testosterone and estrogen. Testes secrete testosterone, the hormone that aids in sperm production, hair thickness and distribution, increased muscle mass, and voice deepening. Ovaries produce and secrete estrogen, which regulates menstruation, the maturation of egg cells, breast development, pregnancy, and menopause.

2. Bone tissue can be tightly or loosely packed. Compact bone forms the hard outer shell of bones, and spongy bone is the porous honeycomb-like inner bone. Blood vessels run throughout bone. Bone marrow inside bones helps to produce blood cells.

3. A muscle contains bundles of parallel muscle fibers. Within each muscle fiber is a parallel array of threadlike filaments called myofibrils. Each myofibril is a linear arrangement of sarcomeres. A sarcomere is composed of thin filaments of the protein actin and thick filaments of the protein myosin.

**4.** C; **5.** D; **6.** A; **7.** B; **8.** B; **9.** B; **10.** B

ANALYZING AND APPLYING THE BASICS

1. Differences between members of the same sex prevent generalizations from having much meaning.

2. Sperm production is highest at temperatures slightly lower than body temperature. Looser clothing keeps the testes cooler.

3. Lowering testosterone levels may allow the estrogen present in males to have a larger effect, leading to development of male breast tissue.

# Chapter 22

STOP & STRETCH

**p. 489:** Using the same amount of estrogen found in a typical waterway, comparing numbers of feminized fish before and after treatment in the same lake, and comparing numbers against fish in untreated lakes. From the results we can conclude only that the level of estrogen in the treated lake does cause physical malformations in the fish studied. We cannot conclude that estrogen from the birth control would have the same affect without further study.

**p. 498:** The experiment could simply track start dates of menstrual cycles for the women over a year or so to see whether their cycles converge.

**p. 502:** An example of an alternate hypothesis would be that men in rural areas are more likely to have jobs with a greater component of manual labor. It may be that the consistently higher body temperatures experienced by manual laborers interfere with sperm production, and that atrazine exposure has little effect.

## VISUALIZE THIS

**Figure 22.1:** Each cell will, potentially at least, land on the ocean floor and begin dividing to produce another sponge.

**Figure 22.3:** Bacteria can move from vulva to vagina to uterus to oviducts. An opening between each oviduct and ovary allows bacteria entrance to abdomen.

**Figure 22.5:** Development occurs in one location. It is depicted as moving around the ovary just to show the changes that occur.

**Figure 22.8:** Binding of the sperm head to the zona pellucida is the species-specific step.

**Figure 22.12:** The drug would cause the uterus to contract, which would stimulate the release of oxytocin via positive feedback.

## WORKING WITH DATA

**Figure 22.6:** Ages 20–24 are most susceptible.

## SAVVY READER

1. No data outlining the rate of precocious puberty in other years is presented. Based on the data from one year presented in this article, it is not possible to determine whether rates are increasing, decreasing, or staying the same.

2. The rate of plastic usage should increase as well.

3. Check reputable medical and government websites. Speak with your physician.

4. There is no way to know for certain from the website. She could be a very committed physician, dedicated to helping as many people as possible, or she could be an unethical physician trying to scare people to make money off of their fears.

## SOUNDS RIGHT, BUT IS IT?

1. Day 12
2. Day 12

3. Day 14
4. Yes.
5. While the egg is only viable for 24 hours, sperm already in the reproductive tract are able to fertilize the egg cell when it is ovulated. Therefore, a woman can get pregnant if unprotected intercourse occurs several days before ovulation, as well as the 24 hours after ovulation.

## LEARNING THE BASICS

1. Asexual reproduction occurs when one parent produces offspring that are genetically identical to the parent and to each other. Sexual reproduction requires genetic input from two parents.

2. See Figures 22.2 and 22.3.

3. See Figures 22.4 and 22.5.

4. A; **5.** A; **6.** C; **7.** E; **8.** C; **9.** A; **10.** E

## ANALYZING AND APPLYING THE BASICS

1. Asexual

2. Vulva

3. Progesterone in pills would keep uterine lining thickened instead of allowing it to slough off.

# Chapter 23

## STOP & STRETCH

**p. 516:** The distance to the brain is longer than the distance to the spinal cord. Therefore, by the time the pain message reaches your brain, your spinal cord has already sent a message to your hand to remove it from the hot surface.

**p. 519:** A less well developed frontal lobe can lead to less planning and impulse control.

**p. 521:** Blocking sodium channels would prevent action potentials from being generated.

## VISUALIZE THIS

**Figure 23.5:** See part (a) of figure.

**Figure 23.6:** Cerebral cortex because altered functioning in this structure could lead to some of the symptoms of ADD.

## WORKING WITH DATA

**Figure 23.9:** a) If an impulse travels 100m/sec, it would travel 2 meters in 1/50 of a second. b) 1/500 second.

SAVVY READER

1. Random selection of students, 100 different colleges.

2. Without the use of statistics, we cannot know whether any measured differences were real.

3. No. They only comment on information for which they have data.

SOUNDS RIGHT, BUT IS IT?

1. No.

2. Through the corpus callosum, a band of nerve fibers that connects the two hemispheres of the brain.

3. No. If math skills are coordinated out of one side of the brain, you would expect only one side of the brain to be active when solving math problems.

4. The two hemispheres complement each other.

5. Not really.

6. The two hemispheres of the brain contain the same structures, communicate with each other and perform functions that complement each other.

LEARNING THE BASICS

1. The CNS is composed of the brain and spinal cord and is responsible for integrating, processing, and coordinating information taken in by the senses. It is also the seat of functions such as intelligence, learning, memory, and emotion. The PNS includes the network of nerves outside the brain and spinal cord and functions to link the CNS with senses.

2. Sensory neurons carry information to the brain. They are connected to the motor neurons, which carry information away from the brain, by interneurons.

3. General senses do not have special organs, but are found all over the body. These include sense of touch, pain and temperature.

4. A; 5. D; 6. D; 7. D; 8. D; 9. E; 10. F

ANALYZING AND APPLYING THE BASICS

1. A more strongly affected individual may have less dopamine or norepinephrine (or both) or they may have fewer dopamine or norepinephrine receptors (or both).

2. They could have axons that are larger in diameter.

3. Reflexes are prewired and automatic responses, whereas other nervous system responses, require sensory activity followed by integration and motor output.

# Chapter 24

STOP & STRETCH

**p. 535:** Plant cells can transform from one type (e.g., root) into others (e.g., stem).

**p. 540:** The successful adult is probably using all of the resources available to it quite well. The young plants would generally have a difficult time establishing under these conditions.

**p. 541:** The requirement of an extended freezing spell ensures that the seeds will not germinate in fall, which would doom the seedling to death during winter conditions.

**p. 541:** Transfer of nutrients from the leaves to the roots of the plant. If only a patch is removed, phloem still remains intact between the majority of leaves and roots.

**p. 546:** They do get larger and flower and thus use up nutrients. They also do not get additional nutrients from a natural cycle.

**p. 552:** If some individuals in a population are resistant to the pesticide, they survive and produce the next generation, passing the "resistance" trait on to the next generation.

VISUALIZE THIS

**Figure 24.1:** By saving seeds from the plants that produced the largest ears and those ears that kept their seeds, farmers drove the evolution of the crop toward larger and larger, tightly bound fruits.

**Figure 24.3:** Ground tissue.

**Figure 24.5:** By increasing the surface area for absorption many-fold.

**Figure 24.9:** Advantages are that a pollinator can both pick up pollen for dispersal and facilitate fertilization of eggs in a single visit. Disadvantages include increased risk of self-pollination (and thus inbreeding).

**Figure 24.10:** Male wasps are the pollinators because they will attempt to mate with the flower! However, you could also hypothesize that predators of the wasp are pollinators.

**Figure 24.14:** Needs to be indicated on the photo. They are the darker, domed regions.

**Figure 24.15:** So that the vascular tissue is continuous from the branch root to the stem.

**Figure 24.20:** The plant would wilt because water would be drawn osmotically from the cells.

**Figure 24.22:** There should be fewer stomata on the leaves of plants in desert areas compared to tropical areas because plants in the desert need to conserve water.

**Figure 24.24:** Nutrients would return to the soil after consumption by organisms present on the site.

WORKING WITH DATA

**Figure 24.27:** The panhandles of Texas and Oklahoma.

SAVVY READER

1. Organic food is measurably better for consumers than conventionally produced food.

2. As asserted by a scientist at the University of Minnesota, organic food does contain less pesticide residue than conventionally produced food, but there is no evidence that organic food is more nutritious.

3. What are the health costs associated with conventional agriculture that are less direct than what you eat (e.g., air, soil, water pollution)? How much more expensive is organic produce than conventionally grown produce?

SOUNDS RIGHT, BUT IS IT?

1. Water will flow out of cell, from a region where it is high concentration to where it is in lower concentration.

2. It will stretch and hold its shape, putting pressure on the cell contents.

3. It goes down. Cells become flaccid.

4. Flaccid cells do not support each other within an organ, so the plant will wilt.

5. No, because the elasticity is the important part, not the overall thickness.

6. The thickness of the cell walls is not much of a factor in determining whether a plant will wilt—instead it is the water in the cells that matters. Adding a sugary solution to the soil will not strengthen the plant, instead it will draw water out of the plant, causing it to wilt further.

LEARNING THE BASICS

1. See Figures 24.2, 24.3, and 24.9.

2. Plants with indeterminate growth do not have a final defined size they will reach and always have the capability to grow under the right conditions.

3. E; **4.** D; **5.** D; **6.** D; **7.** B; **8.** B; **9.** A; **10.** E

ANALYZING AND APPLYING THE BASICS

1. Most plants will only produce fruit if the ovules are fertilized. It is difficult to trick a plant into thinking it has been fertilized when it has not.

2. New growth will have to occur at the axillary buds, causing the formation of side branches.

3. The soil may be depleted of nutrients that are especially helpful to tomatoes, and pests specific to tomatoes may have established populations in that area of his garden so they are ready to attack the plants as soon as the plants appear. The easiest solution is to move the tomatoes to a new area of the yard or garden. He could also add organic fertilizer to the soil to replenish it.

# Chapter 25

STOP & STRETCH

**p. 569:** If many of the roots have died as a result of oxygen deprivation, the rate of transpiration may be too high for them to adequately replace water lost from the plant. Removing leaves reduces this rate.

**p. 572:** Before the tap, the sugar source is roots and the sink is tree leaf buds; after the tap, the sugar source is still the roots, but some of the sink is the taps.

**p. 574:** So that when the tree "wakes up" and the roots (the source) begin sending sugars to the buds (the sinks), the only buds that remain are that the gardener wants to grow.

VISUALIZE THIS

**Figure 25.1:** Evaporation is much reduced and the stomata are most likely closed (since the light reactions of photosynthesis cannot proceed without sunlight). Transpiration is unlikely.

**Figure 25.2:** The fungus benefits by having a reliable source of carbohydrate produced by the photosynthetic activities of the plant.

**Figure 25.10:** Sap flow would be reduced because water would be less "available" to move into the phloem tubes.

WORKING WITH DATA

**Figure 25.14:** Because days are long in midsummer, the lettuce does not experience the night length required to lose enough active phytochrome. Since active phytochrome promotes flowering in lettuce, plants sown in summer will flower before they are large enough to harvest.

SAVVY READER

1. Use a library database to search for articles by Craig Knight, the psychologist quoted in the story.

2. Experiments were performed on visitors to the Chelsea flower show. Presumably, visitors to the show have a positive view of plants, and we're not sure how that compares to the general "working" population. It may be that these subjects have a stronger positive response to the presence of plants than someone who did not visit the show. In addition, the experiments took place at the show and were presumably short term, so it's difficult to determine if the benefits last over the course of days or weeks.

3. There is not enough evidence to support the claim that profitability will increase as the result of plants. We would need to know if the positive effects are long-lasting, and that the "costs" of having plants in the environment were balanced by the increase in creative output.

SOUNDS RIGHT, BUT IS IT?

1. Answers are numerous. They exchange materials with the environment, transform energy, evolve, respond to environmental conditions, etc.

2. Again, variable answers. We are looking for students to connect what you've learned about plants in Chapters 24 and 25 to the basic characteristics of life.

3. Another variable answer. Here we are asking you simply to state how plants are more than … potted plants.

LEARNING THE BASICS

1. Hydrogen bonding among water molecules (cohesion) and between water molecules and xylem cell walls (adhesion) provides a force that counters the strong negative pressure and keeps the water column together.

2. The $C_4$ adaptation pumps carbon dioxide toward actively photosynthetic cells, allowing stomata to remain only partially open. The CAM adaptation allows plants to accumulate carbon dioxide during the cooler night when less water will evaporate so that they may keep their stomata closed during of side the hottest parts of the day.

3. Auxin is destroyed on the illuminated part of the stem, preventing cell elongation. As the cells on the darker side continue to elongate, the stem tip is pushed toward the light source.

**4.** D; **5.** C; **6.** B; **7.** D; **8.** A; **9.** A; **10.** C

ANALYZING AND APPLYING THE BASICS

1. Make sure that the soil in the site is well watered and mulched. The lily should have some of its leaves removed to reduce transpiration.

2. Keep it in a dark closet for a good portion of each day until it produces flower buds.

3. Prune competing apical meristems so that only one remains to promote a single tip. Prune apical meristems of side branches to encourage these side branches to branch more to improve bushiness.

# Glossary

**ABO blood system** A system for categorizing human blood based on the presence or absence of carbohydrates on the surface of red blood cells. (Chapter 8)

**abscisic acid** A hormone in plants associated with dormancy in seeds and buds. (Chapter 25)

**abscission** The dropping off of leaves, flowers, fruits, or other plant organs. (Chapter 25)

**accessory organs** Organs including the pancreas, liver, and gallbladder, which aid the digestive system. (Chapter 18)

**acetylcholine** A neurotransmitter with many functions, including facilitating muscle movements, and thought to be involved in the development of Alzheimer's disease. (Chapter 23)

**acid** A substance that increases the concentration of hydrogen ions in a solution. (Chapter 2)

**acid rain** Rain (or other precipitation) that is unusually acidic; caused by air pollution in the form of sulfur and nitrogen dioxides. (Chapter 16)

**acquired immune deficiency syndrome (AIDS)** Syndrome characterized by severely reduced immune system function and numerous opportunistic infections. Results from infection with HIV. (Chapter 20)

**acrosome** An organelle at the tip of the sperm cell containing enzymes that help the sperm penetrate the egg cell. (Chapter 22)

**actin** A protein found in muscle tissue that, together with myosin, facilitates contraction. (Chapter 21)

**action potential** Wave of depolarization in a neuron propagated to the end of the axon—also called a nerve impulse. (Chapter 23)

**activation energy** The amount of energy that reactants in a chemical reaction must absorb before the reaction can start. (Chapter 4)

**activator** A protein that serves to enhance the transcription of a gene. (Chapter 9)

**active immunity** Immunity that results from the production of antibodies, as differentiated from passive immunity. (Chapter 20)

**active site** Substrate-binding region of an enzyme. (Chapter 4)

**active smoker** An individual who smokes tobacco. (Chapter 19)

**active transport** The ATP-requiring movement of substances across a membrane against their concentration gradient. (Chapter 3)

**adaptation** Trait that is favored by natural selection and increases an individual's fitness in a particular environment. (Chapters 11, 12, 13)

**adaptive radiation** Diversification of one or a few species into large and very diverse groups of descendant species. (Chapter 13)

**ADD** See *attention deficit disorder*. (Chapter 23)

**adenine** Nitrogenous base in DNA, a purine. (Chapters 2, 9)

**adenosine diphosphate (ADP)** A nucleotide composed of adenine, a sugar, and two phosphate groups. Produced by the hydrolysis of the terminal phosphate bond of ATP. (Chapter 4)

**adenosine triphosphate (ATP)** A nucleotide composed of adenine, the sugar ribose, and three phosphate groups that can be hydrolyzed to release energy. Form of energy that cells can use. (Chapter 4)

**adhesion** The sticking together of unlike materials—often water and a particular surface. (Chapter 25)

**adipose tissue** Fat-storing connective tissue. (Chapter 17)

**adrenal gland** Either of two endocrine glands, one located atop each kidney, that secrete adrenaline in response to stress or excitement, help maintain water and salt balance, and secrete small amounts of sex hormones. (Chapter 21)

**adrenaline** A hormone secreted by the adrenal glands in response to stress or excitement. (Chapter 20)

**aerobic** An organism, environment, or cellular process that requires oxygen. (Chapter 4)

**aerobic respiration** Cellular respiration that uses oxygen as the electron acceptor. (Chapter 4)

**agarose gel** A jelly-like slab used to separate molecules on the basis of molecular weight. (Chapter 8)

**agriculture** Cultivation of crops, raising of livestock; farming. (Chapter 24)

**algae** Photosynthetic protists. (Chapter 13)

**alimentary canal** The tube inside the body, from the mouth to the anus, through which food passes during digestion. (Chapter 18)

**allele** Alternate versions of the same gene, produced by mutations. (Chapters 6, 7, 9, 11, 15)

**allele frequency** The percentage of the gene copies in a population that are of a particular form, or allele. (Chapters 12, 15)

**allergy** An abnormally high sensitivity to allergens such as pollen or microorganisms. Can cause sneezing, itching, runny nose, and watery eyes. (Chapter 20)

**allopatric** Geographic separation of a population of organisms from others of the same species. Usually in reference to speciation. (Chapter 12)

**alternative hypothesis** Factor other than the tested hypothesis that may explain observations. (Chapter 1)

**alveoli** (singular: alveolus) Sacs inside lungs, making up the respiratory surface in land vertebrates and some fish. (Chapter 19)

**Alzheimer's disease** Progressive mental deterioration in which there is memory loss along with the loss of control of bodily functions, ultimately resulting in death. (Chapter 23)

**amenorrhea** Abnormal cessation of menstrual cycle. (Chapters 4, 22)

**amino acid** Monomer subunit of a protein. Contains an amino, a carboxyl, and a unique side group. (Chapters 2, 3, 9)

**amnion** The fluid-filled sac in which a developing embryo is suspended. (Chapter 22)

**anabolic steroid** A derivative, usually synthetic, of the steroid hormone testosterone. (Chapter 21)

**anaerobic** An organism, environment, or cellular process that does not require oxygen. (Chapter 4)

**anaerobic respiration** A process of energy generation that uses molecules other than oxygen as electron acceptors. (Chapter 4)

**anaphase** Stage of mitosis during which microtubules contract and separate sister chromatids. (Chapter 6)

**androgen** A masculinizing hormone, such as testosterone, secreted by the adrenal glands. (Chapters 21, 22)

**anecdotal evidence** Information based on an individual's personal experience. (Chapter 1)

**angiosperm** Plant in the phyla Angiospermae, which produce seeds borne within fruit. (Chapters 13, 24, 25)

**animal** An organism that obtains energy and carbon by ingesting other organisms and is typically motile for part of its life cycle (Chapter 13)

**Animalia** Kingdom of Eukarya containing organisms that ingest others and are typically motile for at least part of their life cycle. (Chapters 10, 13)

**annual plant** Plant that completes its life cycle in a single growing season. (Chapters 16, 24, 25)

**annual growth rate** Proportional change in population size over a single year. Growth rate is a function of the birth rate minus the death rate of the population. (Chapter 14)

**anorexia** Self-starvation. (Chapter 4)

**antagonistic muscle pair** A set of muscles whose actions oppose each other. (Chapter 21)

**anther** The pollen-containing structure on the stamen of a flower. (Chapter 24)

**antibiotic** A chemical that kills or disables bacteria. (Chapter 11, 13, 20)

**antibiotic resistant** Characteristic of certain bacteria; a physiological characteristic that permits them to survive in the presence of particular antibiotics. (Chapter 11)

**antibody** Protein made by the immune system in response to the presence of foreign substances or antigens. Can serve as a receptor on a B cell or be secreted by plasma cells. (Chapter 20)

**antibody-mediated immunity** Immunity that occurs via secreted antibodies, versus immunity mediated by T cells. (Chapter 20)

**anticodon** Region of tRNA that binds to an mRNA codon. (Chapter 9)

**antigen** Short for antibody-generating substances; an antigen is a molecule that is foreign to the host and stimulates the immune system to react. (Chapter 20)

**antigen receptor** Protein in B- and T-cell membrane that bind to specific antigens. (Chapter 20)

**antioxidant** Certain vitamins and other substances that protect the body from the damaging effects of free radicals. (Chapter 3)

**antiparallel** Feature of DNA double helix in which nucleotides face *up* on one side of the helix and *down* on the other. (Chapter 2)

**apical dominance** The influence exerted by the terminal bud in suppressing the growth of axillary or lateral buds. Mediated by the hormone, auxin. (Chapter 25)

**apical meristem** The actively dividing cells at the tip of stems and roots in plants. (Chapter 24)

**appendicular skeleton** The part of the skeleton composed of the bones of the hip, shoulder, and limbs. (Chapter 21)

**aquaporin** A transport protein in the membrane of a plant or animal cell that facilitates the diffusion of water across the membrane by osmosis. (Chapter 3)

**aquatic** Of, or relating to, water. (Chapter 16)

**Archaea** Domain of prokaryotic organisms made up of species known from extreme environments. (Chapter 13)

**artery** Blood vessel that carries oxygenated blood from the heart to body tissues. (Chapter 19)

**artificial selection** Selective breeding of domesticated animals and plants to increase the frequency of desirable traits. (Chapter 11)

**asexual reproduction** A type of reproduction in which one parent gives rise to genetically identical offspring. (Chapters 6, 12, 22)

**assortative mating** Tendency for individuals to mate with other individuals who are like themselves. (Chapter 12)

**asthma** A respiratory disease characterized by spasmodic inflammation of the air passages in the lungs and overproduction of mucus. Often triggered by air contaminants. (Chapter 19, 20)

**asymptomatic** Stage in an infection that is characterized by relatively unnoticeable, or absent, symptoms of illness. (Chapter 20)

**atherosclerosis** Accumulation of fatty deposits within blood vessels; hardening of the arteries. (Chapter 19)

**atom** The smallest unit of matter that retains the properties of an element. (Chapter 2)

**atomic number** The number of protons in the nucleus of an atom. Unique to each element, this number is designated by a subscript to the left of the symbol for the element. (Chapter 2)

**ATP synthase** Enzyme found in the mitochondrial membrane that helps synthesize ATP. (Chapter 4)

**atrium** An upper chamber of the heart that receives blood from the body or lungs and pumps it to a ventricle. (Chapter 19)

**attention deficit disorder (ADD)** Syndrome characterized by forgetfulness; distractibility, fidgeting; restlessness; impatience; difficulty sustaining attention in work, play, or conversation; or difficulty following instructions and completing tasks in more than one setting. (Chapter 23)

**autoimmune disease** Any of the diseases that result from an attack by the immune system on normal body cells. (Chapter 20)

**autosome** Non-sex chromosome, of which there are 22 pairs in humans. (Chapter 6)

**auxin** A class of plant hormones that control cell elongation, among other effects. (Chapter 25)

**AV valves** Heart valves between the atria and the ventricles. (Chapter 19)

**axial skeleton** Part of the skeleton that supports the trunk of the body and consists largely of the bones making up the vertebral column or spine and much of the skull. (Chapter 21)

**axillary bud** A bud located at the junction between a plant stem and leaf. (Chapters 24, 25)

**axon** Long, wire-like portion of the neuron that ends in a terminal bouton. (Chapter 23)

**B lymphocyte (B cell)** The type of white blood cell responsible for antibody-mediated immunity. (Chapter 20)

**background extinction rate** The rate of extinction resulting from the normal process of species turnover. (Chapter 15)

**bacteria** Domain of prokaryotic organisms. (Chapters 2, 9, 13, 15, 20)

**ball and socket joint** Type of joint in the hips and shoulders that enables arms and legs to move in three dimensions. (Chapter 21)

**bark** The outer layer of a woody stem, consisting of cork, cork cambium, and phloem. (Chapter 24)

**basal metabolic rate** Resting energy use of an awake, alert person. (Chapter 4)

**base** A substance that reduces the concentration of hydrogen ions in a solution. (Chapter 2)

**base-pairing rule** A rule governing the pairing of nitrogenous bases. In DNA, adenine pairs with thymine and cytosine with guanine. (Chapters 2, 9)

**behavioral isolation** Prevention of mating between individuals in two different populations based on differences in behavior. (Chapter 11)

**benign** Tumor that stays in one place and does not affect surrounding tissues. (Chapter 6)

**bias** Influence of research participants' opinions on experimental results. (Chapter 1)

**biennial** Plant that completes its life cycle over the course of two growing seasons. (Chapter 24)

**bile** Mixture of substances produced in the liver that aids in digestion by emulsifying fats. (Chapter 18)

**binary fission** An asexual form of bacterial reproduction. (Chapter 20)

**binge drinking** The consumption of more than four drinks in a two-hour time period. (Chapter 18)

**bioaccumulation** A phenomenon that results in the concentration of persistent (i.e., slow to degrade) pollutants in the bodies of animals at high levels of a food chain. (Chapter 16)

**biodiversity** Variety within and among living organisms. (Chapters 13, 15)

**biogeography** The study of the geographic distribution of organisms. (Chapter 10)

**biological classification** Field of science attempting to organize biodiversity into discrete, logical categories. (Chapters 10, 13)

**biological diversity** Entire variety of living organisms. (Chapter 13)

**biological evolution** See *evolution*. (Chapter 10)

**biological population** Individuals of the same species that live and breed in the same geographic area. (Chapters 10, 12)

**biological race** Populations of a single species that have diverged from each other. Biologists do not agree on a definition of *race*. See also *subspecies*. (Chapter 12)

**biological species concept** Definition of a species as a group of individuals that can interbreed and produce fertile offspring but typically cannot breed with members of another species. (Chapter 12)

**biology** The study of living organisms. (Chapter 1)

**biomagnification** Concentration of toxic chemicals in the higher levels of a food web. (Chapter 15, 24)

**biomass** The mass of all individuals of a species, or of all individuals on a level of a food web, within an ecosystem. (Chapter 15)

**biome** A broad ecological community defined by a particular vegetation type (e.g., temperate forest, prairie), which is typically determined by climate factors. (Chapter 16)

**biophilia** Humans' innate desire to be surrounded by natural landscapes and objects. (Chapter 15)

**biopsy** Surgical removal of some cells, tissue, or fluid to determine whether cells are cancerous. (Chapter 6)

**bipedal** Walking upright on two limbs. (Chapter 10)

**blade** The broad part of a leaf. (Chapter 24)

**blastocyst** An embryonic stage consisting of a hollow ball of cells. (Chapter 22)

**blind experiment** Test in which subjects are not aware of exactly what they are predicted to experience. (Chapter 1)

**blood** The combination of cells and liquid that flows through blood vessels in the cardiovascular system; made up of red blood cells, plasma, white blood cells, and platelets. (Chapters 17, 19)

**blood clotting** The process by which a blood clot, a mass of the protein fibrin and dead blood cells, forms in the region of blood vessel damage. (Chapter 19)

**blood pressure** The force of the blood as it travels through the arteries; partially determined by artery diameter and elasticity. (Chapter 19)

**blood vessel** The structure that carries blood via the circulatory system throughout the body; arteries, capillaries, and veins. (Chapter 19)

**body mass index (BMI)** Calculation using height and weight to determine a number that correlates to an estimate of a person's amount of body fat with health risks. (Chapter 4)

**bolus** In digestion, a soft ball of chewed food. (Chapter 17)

**bone** A type of connective tissue consisting of living cells in a matrix rich in collagen and calcium. (Chapter 17)

**bone marrow** Network of soft connective tissue that fills bones and is involved in the production of red blood cells. (Chapters 19, 21)

**boreal forest** A biome type found in regions with long, cold winters and short, cool summers. Characterized by coniferous trees. (Chapter 16)

**botanist** Plant biologist. (Chapter 13)

**brain stem** Region of the brain that lies below the thalamus and hypothalamus; it governs reflexes and some involuntary functions such as breathing and swallowing. (Chapter 23)

**bronchi** The large air passageways from the trachea into the lungs. (Chapter 19)

**bronchiole** The branching air passageway inside the lungs. (Chapter 19)

**bronchitis** Inflammation of the bronchi and bronchioles in the lungs. (Chapter 19)

**budding** An asexual mechanism of propagation in which outgrowths of the parent form and pinch off, producing new individuals. (Chapter 22)

**bulbourethral gland** Either of the two glands at the base of the penis that secrete acid-neutralizing fluids into semen. (Chapter 22)

**bulimia** Binge eating followed by purging. (Chapter 3)

**$C_3$ plant** Plant that uses the Calvin cycle of photosynthesis to incorporate carbon dioxide into a 3-carbon compound. (Chapter 5)

**$C_4$ plant** Plant that performs reactions incorporating carbon dioxide into a 4-carbon compound that ultimately provides carbon dioxide for the Calvin cycle. (Chapter 5)

**calcium** Nutrient required in plant cells for the production of cell walls and in humans for bone strength and blood clotting. (Chapter 3)

**Calorie** A kilocalorie or 1000 calories. (Chapter 4)

**calorie** Amount of energy required to raise the temperature of one gram of water by 1°C. (Chapter 4)

**Calvin cycle** A series of reactions that occur in the chloroplast during photosynthesis and that utilize NADPH and ATP to reduce carbon dioxide and produce sugars. (Chapter 5)

**CAM plant** A plant that uses Crassulacean acid metabolism, a variant of photosynthesis during which carbon dioxide is stored in sugars at night and released during the day to prevent water loss. (Chapter 5)

**Cambrian explosion** Relatively rapid evolution of the modern forms of multicellular life that occurred approximately 550 million years ago. (Chapter 13)

**cancer** A disease that occurs when cell division escapes regulatory controls. (Chapter 6)

**capillary** The smallest blood vessel of the cardiovascular system, connecting arteries to veins and allowing material exchange across their thin walls. (Chapter 19)

**capillary bed** A branching network of capillaries supplying a particular organ or region of the body. (Chapter 19)

**capsid** Protein coat that surrounds a virus. (Chapter 20)

**capsule** Gelatinous outer covering of bacterial cells that aids in attachment to host cells during an infection. (Chapter 20)

**carbohydrate** Energy-rich molecule that is the major source of energy for the cell. Consists of carbon, hydrogen, and oxygen in the ratio $CH_2O$. (Chapters 2, 3)

**carcinogen** Substance that causes cancer or increases the rate of its development. (Chapters 6, 19)

**cardiac cycle** The cycle of contraction and relaxation that a normally functioning heart undergoes over the course of a single heart beat. (Chapter 19)

**cardiac muscle** Muscle that forms the contractile wall of the heart. (Chapter 17)

**cardiovascular disease** Malfunction of the cardiovascular system, including, but not limited to, heart attack, stroke, and hypertension. (Chapter 19)

**cardiovascular system** The organ system made up of the heart and circulatory system, including arteries, capillaries, and veins. (Chapters 17, 19)

**carpel** The female structure of a flower, containing the ovary, style, and stigma. (Chapter 24)

**carrier** Individual who is heterozygous for a recessive allele. (Chapters 7, 8)

**carrying capacity** Maximum population that the environment can support. (Chapter 14)

**cartilage** Connective tissue found in the skeletal system that is rich in collagen fibers. (Chapter 17)

**catalyst** A substance that lowers the activation energy of a chemical reaction, thereby speeding up the reaction. (Chapter 4)

**catalyze** To speed up the rate of a chemical reaction. Enzymes are biological catalysts. (Chapter 4)

**cell** Basic unit of life; an organism's fundamental building-block units. (Chapters 2, 3)

**cell body** Portion of the neuron that houses the nucleus and organelles. (Chapter 23)

**cell cycle** An ordered sequence of events in the life cycle of a eukaryotic cell from its origin until its division to produce daughter cells. Consists of M, $G_1$, S, and $G_2$ phases. (Chapter 6)

**cell division** Process a cell undergoes when it makes copies of itself. Production of daughter cells from an original parent cell. (Chapter 6)

**cell plate** A double layer of new cell membrane that appears in the middle of a dividing plant cell and divides the cytoplasm of the dividing cell. (Chapter 6)

**cell wall** Tough but elastic structure surrounding plant and bacterial cell membranes. (Chapters 2, 3, 6, 24)

**cell-mediated immunity** Type of specific immune response carried out by T cells. (Chapter 20)

**cellular respiration** Metabolic reactions occurring in cells that result in the oxidation of macromolecules to produce ATP. (Chapter 4)

**cellulose** A structural polysaccharide found in cell walls and composed of glucose molecules. (Chapters 2, 3, 5)

**central nervous system (CNS)** Includes brain and spinal cord and is responsible for integrating, processing, and coordinating information taken in by the senses. It is the seat of functions such as intelligence, learning, memory, and emotion. (Chapter 23)

**central vacuole** A membrane-enclosed sac in a plant cell that functions to store many different substances. (Chapters 3, 25)

**centriole** A structure in animal cells that helps anchor for microtubules during cell division. (Chapters 3, 6)

**centromere** Region of a chromosome where sister chromatids are attached and to which microtubules bind. (Chapter 6)

**cereal grains** Staples of the human diet that are from the grass family, namely, wheat, corn, oats, rye, and barley. (Chapter 24)

**cerebellum** Region of the brain that controls balance, muscle movement, and coordination. (Chapter 23)

**cerebral cortex** Deeply wrinkled outer surface of the cerebrum where conscious activity and higher thought originate. (Chapter 23)

**cerebrospinal fluid** Protective liquid bath that surrounds the brain within the skull. (Chapter 23)

**cerebrum** Portion of the brain in which language, memory, sensations, and decision making are controlled. The cerebrum has two hemispheres, each of which has four lobes. (Chapter 23)

**cervix** The lower narrow portion of the uterus at the top end of the vagina. (Chapter 22)

**chaparral** A biome characteristic of climates with hot, dry summers and mild, wet winters and a dominant vegetation of aromatic shrubs. (Chapter 16)

**checkpoint** Stoppage during cell division that occurs to verify that division is proceeding correctly. (Chapter 6)

**chemical reaction** A process by which one or more chemical substances are transformed into one or more different chemical substances. (Chapter 2)

**chemotherapy** Using chemicals to try to kill rapidly dividing (cancerous) cells. (Chapter 6)

**chlorophyll** Green pigment found in the chloroplast of plant cells. (Chapter 5)

**chloroplast** An organelle found in plant cells that absorbs sunlight and uses the energy derived to produce sugars. (Chapters 5, 13)

**cholesterol** A steroid found in animal cell membranes that affects membrane fluidity. Serves as the precursor to estrogen and testosterone. (Chapters 2, 3, 19)

**chondrocyte** A type of cartilage cell that produces collagen and other substances. (Chapter 17)

**chromosome** Subcellular structure composed of a long single molecule of DNA and associated proteins, housed inside the nucleus. (Chapters 6, 7, 8, 9)

**chyme** Partially digested food and enzymes that are passed from the stomach to the intestine. (Chapter 17)

**circulatory system** The vessels that transport blood, nutrients, and waste around the body. (Chapters 17, 19)

**citric acid cycle** A chemical cycle occurring in the matrix of the mitochondria that breaks the remains of sugars down to produce carbon dioxide. (Chapter 4)

**cladistic analysis** A technique for determining the evolutionary relationships among organisms that relies on identification and comparison of newly evolved traits. (Chapter 13)

**classification system** Method for organizing biological diversity. (Chapters 10, 13)

**cleavage** In embryology, the period of rapid cell division that occurs during animal development. (Chapter 22)

**climate** The average temperature and precipitation as well as seasonality. (Chapter 16)

**climax community** The group of species that is stable over time in a particular set of environmental conditions. (Chapter 16)

**clitoris** Sensitive erectile tissue found in the external genitalia of females that functions in sexual arousal. (Chapter 22)

**clonal population** Population of identical cells copied from the immune cell that first encounters an antigen. The entire clonal population has the same DNA arrangement, and all cells in a clonal population carry the same receptor on their membrane. (Chapter 20)

**cloning** Producing copies of a gene or an organism that are genetically identical. (Chapter 9)

**clumped distribution** A spatial arrangement of individuals in a population where large numbers are concentrated in patches with intervening, sparsely populated areas separating them. (Chapter 14)

**codominant** Two different alleles of a gene that are equally expressed in the heterozygote. (Chapters 7, 8)

**codon** A triplet of mRNA nucleotides. Transfer RNA molecules bind to codons during protein synthesis. (Chapter 9)

**coenzyme** (or cofactor) Substances such as vitamins that help enzymes catalyze chemical reactions. (Chapter 3)

**coevolution** Occurs when change in one biological species triggers a change in an associated species. Coevolution commonly occurs between predator and prey or parasite and host. (Chapter 13)

**cohesion** The tendency for molecules of the same material to stick together. (Chapter 2)

**collecting duct** A structure in the kidney that accepts filtrate from multiple nephrons and transmits it to the renal pelvis. Each kidney contains hundreds of collecting ducts. The amount of water retained by the kidneys is largely controlled at the collecting ducts. (Chapter 18)

**colon** Tube between the small intestine and anus of the digestive system. (Chapter 17)

**combination drug therapy** The use of more than one drug simultaneously to treat a disease. Often used for disease organisms that mutate quickly or are difficult to control to combat the problem of drug resistance. (Chapter 11)

**commensalism** In ecology, a relationship between two species in which one is benefitted and the other is neither harmed nor benefitted. (Chapter 15)

**common descent** The theory that all living organisms on Earth descended from a single common ancestor that appeared in the distant past. (Chapter 10)

**community** A group of interacting species in the same geographic area. (Chapter 15)

**compact bone** The hard, outer shell of bones. (Chapter 21)

**competition** Interaction that occurs when two species of organisms both require the same resources within a habitat; competition tends to limit the size of populations. (Chapter 15)

**competitive exclusion** Reduction or elimination of one species in an environment resulting from the presence of another species that requires the same or similar resources. (Chapter 15)

**complementary** Complementary bases pair with each other by hydrogen bonding across the DNA helix. Adenine is complementary to thymine and to guanine. (Chapters 2, 9)

**complement protein** Type of blood protein with which an antibody-antigen complex can combine in order to kill bacterial cells. Enhances the immune response on many levels. (Chapter 22)

**complementary base pair** Nitrogenous bases that hydrogen bond to each other. In DNA, adenine is complementary to thymine, and cytosine is complementary to guanine. In RNA, adenine is complementary to uracil and guanine to cytosine. (Chapters 2, 9)

**complete protein** Dietary protein that contains all the essential amino acids. (Chapter 3)

**complex carbohydrate** Carbohydrate consisting of two or more monosaccharides. (Chapter 3)

**compound** A substance consisting of two or more elements in a fixed ratio. (Chapter 2)

**connective tissue** Animal tissue that functions to bind and support other tissues. Composed of a small number of cells embedded in a matrix. (Chapter 17)

**consilience** The unity of knowledge. Used to describe a scientific theory that has multiple lines of evidence to support it. (Chapter 10)

**contagious** Spreading from one organism to another. (Chapter 20)

**continuous variation** A range of slightly different values for a trait in a population. (Chapter 7)

**control** Subject for an experiment who is similar to experimental subject except is not exposed to the experimental treatment. Used as baseline values for comparison. (Chapter 1)

**convergent evolution** Evolution of the same trait or set of traits in different populations as a result of shared environmental conditions rather than shared ancestry. (Chapters 10, 12)

**copulation** Sexual intercourse. (Chapter 22)

**coral reef** Highly diverse biome found in warm, shallow salt water, dominated by the limestone structures created by coral animals. (Chapters 13, 16)

**cork** A tissue produced by the cork cambium that provides protection to a woody stem by forming outer bark. (Chapter 24)

**cork cambium** A meristematic tissue in bark that arises from phloem cells and creates cork cells before being shed. (Chapter 24)

**corpus callosum** Bundle of nerve fibers at the base of the cerebral fissure that provides a communication link between the cerebral hemispheres. (Chapter 23)

**corpus luteum** Hormone-producing tissue (the ovarian follicle after ovulation) that makes progesterone and estrogen and degenerates about 12 days after ovulation if fertilization does not occur. (Chapter 22)

**correlation** Describes a relationship between two factors. (Chapters 1, 7, 12)

**cotyledon** The first leaves produced by an embryonic seed plant. (Chapter 24)

**covalent bond** A type of strong chemical bond in which two atoms share electrons. (Chapter 2)

**critical night length** The night length that a plant must experience in order to trigger a photoperiodic response. (Chapter 25)

**crop rotation** The practice of growing different crops at different times on the same field in order to maintain soil fertility and decrease pest damage. (Chapter 24)

**cross** In genetics, the mating of two organisms. (Chapter 7)

**crossing over** Gene for gene exchange of genetic information between members of a homologous pair of chromosomes. (Chapter 6)

**cryptorchidism** A developmental defect in which the testes fail to descend to the scrotum. (Chapter 22)

**cultural control** Control of agricultural pests through environmental techniques, such as crop rotation and maintenance of native vegetation. (Chapter 24)

**cuticle** The waxy layer on the outer surface of plant epidermal cells. (Chapter 24)

**cyst** Noncancerous, fluid-filled growth. (Chapter 6)

**cytokinesis** Part of the cell cycle during which two daughter cells are formed by the cytoplasm splitting. (Chapter 6)

**cytoplasm** The entire contents of the cell (except the nucleus) surrounded by the plasma membrane. (Chapter 3)

**cytosine** Nitrogenous base, a pyrimidine. (Chapters 2, 9)

**cytoskeleton** A network of tubules and fibers that branch throughout the cytoplasm. (Chapter 3)

**cytosol** The semifluid portion of the cytoplasm. (Chapters 3, 4)

**cytotoxic T cell** Immune-system cell that attacks and kills body cells that have become infected with a virus before the virus has had time to replicate. (Chapter 20)

**data** Information collected by scientists during hypothesis testing. (Chapter 1)

**daughter cells** The offspring cells that are produced by the process of cell division. (Chapter 6)

**death rate** Number of deaths averaged over the population as a whole. (Chapter 14)

**deciduous** Pertaining to woody plants that drop their leaves at the end of a growing season. (Chapters 16, 25)

**decomposer** An organism, typically bacteria and fungi in the soil, whose action breaks down complex molecules into simpler ones. (Chapter 15)

**decomposition** The breakdown of organic material into smaller molecules. (Chapter 16)

**deductive reasoning** Making a prediction about the outcome of a test; *if/then* statements. (Chapter 1)

**defensive protein** Nonspecific protein of the immune system including interferons and complement proteins. (Chapter 20)

**deforestation** The removal of forest lands, often to enable the development of agriculture. (Chapters 5, 15)

**degenerative disease** Disease characterized by progressive deterioration. (Chapter 9)

**dehydration** Loss of water. (Chapter 3)

**deleterious** In genetics, said of a mutation that reduces an individual's fitness. (Chapter 15)

**demographic momentum** Lag between the time that humans reduce birth rates and the time that population numbers respond. (Chapter 14)

**demographic transition** The period of time between when death rates in a human population fall (as a result of improved technology) and when birth rates fall (as a result of voluntary limitation of pregnancy). (Chapter 14)

**denature** (1) In proteins, the process where proteins unravel and change their native shape, thus losing their biological activity. (2) For DNA, the breaking of hydrogen bonds between the two strands of the double-stranded DNA helix, resulting in single-stranded DNA. (Chapter 8)

**dendrite** Short extension of the neuron that receives signals from other cells. (Chapter 23)

**density-dependent factor** Any of the factors related to a population's size that influence the current growth rate of a population—for example, communicable disease or starvation. (Chapter 14)

**density-independent factor** Any of the factors unrelated to a population's size that influence the current growth rate of a population—for example, natural disasters or poor weather conditions. (Chapter 14)

**deoxyribonucleic acid (DNA)** Molecule of heredity that stores the information required for making all of the proteins required by the cell. (Chapters 2, 3, 6, 8, 9)

**deoxyribose** The five-carbon sugar in DNA. (Chapters 2, 6, 8, 9)

**dependent variable** The variable in a study that is expected to change in response to changes in the independent variable. (Chapter 1)

**depolarization** Reduction in the charge difference across the neuronal membrane. (Chapter 23)

**depression** Disease that involves feelings of helplessness and despair, and sometimes thoughts of suicide. (Chapter 23)

**desert** Biome found in areas of minimal rainfall. Characterized by sparse vegetation. (Chapter 16)

**desertification** The process by which formerly productive land is converted to unproductive land, typically by overgrazing of cattle or unsustainable agricultural practices, but also increasingly by the effects of climate change. (Chapter 16)

**determinate growth** Growth of limited duration. (Chapter 24)

**developed country** A term used by the United Nations to describe a country with high per-capita income and significant industrial development. (Chapter 14)

**development** All of the progressive changes that produce an organism's body. (Chapters 13, 21)

**diabetes** Disorder of carbohydrate metabolism characterized by impaired ability to produce or respond to the hormone insulin. (Chapter 4)

**diaphragm** (1) Dome-shaped muscle at the base of the chest cavity. Contraction of this muscle helps draw air into the lungs. (Chapter 19) (2) A birth control device consisting of a flexible contraceptive disk that covers the cervix to prevent the entry of sperm. (Chapter 22)

**diastole** The stage of the cardiac cycle when the heart relaxes and fills with blood. (Chapter 19)

**dicot** The class of angiosperms characterized by having two cotyledons. (Chapter 24)

**differentiation** Structural and functional divergence of cells as they become specialized. (Chapter 22)

**diffusion** The spontaneous movement of substances from a region of their own high concentration to a region of their own low concentration. (Chapter 3)

**dihybrid cross** A genetic cross involving the alleles of two different genes. For example, AaBb x AaBb. (Chapter 7)

**diploid cell** A cell containing homologous pairs of chromosomes (2n). (Chapters 6, 22)

**directional selection** Natural selection for individuals at one end of a range of phenotypes. (Chapter 11)

**disaccharide** A double sugar consisting of two monosaccharides joined together by a glycosidic linkage. (Chapter 2)

**diverge** See *divergence*. (Chapter 12)

**divergence** Occurs when gene flow is eliminated between two populations. Over time, traits found in one population begin to differ from traits found in the other population. (Chapters 10, 12, 13)

**diversifying selection** Natural selection for individuals at both ends of a range of phenotypes but against the *average* phenotype. (Chapter 11)

**dizygotic twins** Fraternal twins (nonidentical) that develop when two different sperm fertilize two different egg cells. (Chapter 7)

**DNA** See *deoxyribonucleic acid*. (Chapters 2, 3, 6, 7, 8, 9)

**DNA profiling** A technique used to identify individuals on the basis of DNA sequences. (Chapter 8)

**DNA polymerase** Enzyme that facilitates base pairing during DNA synthesis. (Chapter 6)

**DNA replication** The synthesis of two daughter DNA molecules from one original parent molecule. Takes place during the S phase of interphase. (Chapter 6)

**domain** Most inclusive biological category. Biologists group life into three major domains. (Chapters 10, 13)

**domesticated** Referring to animals or plants that are now largely dependent on humans for reproduction and survival as a result of artificial selection. (Chapter 24)

**dominant** Applies to an allele with an effect that is visible in a heterozygote. (Chapter 7)

**dopamine** Neurotransmitter active in pathways that control emotions and complex movements. (Chapter 23)

**dormancy** A condition of arrested growth and development in plants. (Chapter 25)

**dormant** In a state of dormancy, that is, alive, but not growing or developing. (Chapter 24)

**double blind** Experimental design protocol when both research subjects and scientists performing the measurements are unaware of either the experimental hypothesis or who is in the control or experimental group. (Chapter 1)

**double fertilization** The fusion of egg and sperm and the simultaneous fusion of a sperm with nuclei in the ovule of angiosperms. (Chapters 13, 24)

**ecological footprint** A measure of the natural resources used by a human population or society. (Chapter 16)

**ecological niche** The functional role of a species within a community or ecosystem, including its resource use and interactions with other species. (Chapter 15)

**ecology** Field of biology that focuses on the interactions between organisms and their environment. (Chapters 14, 15)

**ecosystem** All of the organisms and natural features in a given area. (Chapter 15)

**ecotourism** The visitation of specific geographical sites by tourists interested in natural attractions, especially animals and plants. (Chapter 15)

**ectoderm** The outermost of the three germ layers that arise during animal development. (Chapter 22)

**ectotherm** An animal that must use energy from the sun and other environmental sources to regulate its body temperature. (Chapter 17)

**effector** Muscle, gland, or organ stimulated by a nerve. (Chapter 23)

**egg cell** Gamete produced by a female organism. (Chapters 6, 22)

**electron** A negatively charged subatomic particle. (Chapters 2, 4)

**electron shell** An energy level representing the distance of an electron from the nucleus of an atom. (Chapter 2)

**electron transport chain** A series of proteins in the mitochondrial and chloroplast membranes that move electrons during the redox reactions that release energy to produce ATP. (Chapter 4)

**electronegative** The tendency to attract electrons to form a chemical bond. (Chapter 2)

**element** A substance that cannot be broken down into any other substance. (Chapter 2)

**embryo** The developmental stage commencing after the first mitotic divisions of the zygote and ending when body structures begin to appear; from about the second week after fertilization to about the ninth week. (Chapters 7, 10, 22)

**emphysema** A lung disease caused by the breakdown of alveoli walls; characterized by shortness of breath and an expanded chest cavity. (Chapter 19)

**encephalitis** Pathology, or disease, of the brain. (Chapter 20)

**Endangered Species Act (ESA)** U.S. law intended to protect and encourage the population growth of threatened and endangered species, enacted in 1973. (Chapter 15)

**endocrine disrupter** Chemical that disrupts the functions of the hormone-producing endocrine system. (Chapter 22)

**endocrine gland** Any of the glands that secrete hormones into the bloodstream. (Chapter 21)

**endocrine system** The internal system of chemical signals involving hormones, the organs and glands that produce and secrete them, and the target cells that respond to them. (Chapter 21)

**endocytosis** The uptake of substances into cells by a pinching inward of the plasma membrane. (Chapter 3)

**endoderm** The innermost of the three germ layers that arise during animal development. (Chapter 22)

**endometriosis** The abnormal occurrence of functional endometrial tissue outside the uterus, resulting in painful menstrual cycles. (Chapter 22)

**endometrium** Lining of the uterus, shed during menstruation. (Chapter 22)

**endoplasmic reticulum (ER)** A network of membranes in eukaryotic cells. When rough, or studded with ribosomes, it functions as a workbench for protein synthesis. When devoid of ribosomes, or smooth, it functions in phospholipid and steroid synthesis and detoxification. (Chapter 3)

**endosperm** The triploid structure formed by the fusion of a sperm and two nuclei in the ovule of an angiosperm plant. (Chapter 24)

**endospore** Resistant form of certain bacterial species. Produced as a resting stage to bring cell through poor conditions. (Chapter 13)

**endosymbiotic theory** Theory that organelles such as mitochondria and chloroplasts in eukaryotic cells evolved from prokaryotic cells that took up residence inside ancestral eukaryotes. (Chapter 13)

**endotherm** An animal that uses its own metabolic energy to maintain a constant body temperature. (Chapter 17)

**enveloped viruses** A virus that is surrounded by a membrane. (Chapter 20)

**environmental tobacco smoke (ETS)** The tobacco smoke in the air that results from smoldering tobacco on the lit ends of cigarettes and pipes as well as the smoke exhaled by active smokers. (Chapter 19)

**enzyme** Protein that catalyzes and regulates the rate of metabolic reactions. (Chapters 2, 3, 4)

**epidemic** Contagious disease that spreads rapidly and extensively among many individuals. (Chapter 20)

**epidermal tissue** The outermost layer of cells of the leaf and of young stems and roots. (Chapter 24)

**epididymis** A coiled tube located adjacent to the testes where sperm are stored. (Chapter 22)

**epiglottis** Flap that blocks the windpipe so food goes down the pharynx, not into the lungs. (Chapter 17)

**epithelial tissue (epithelia)** Tightly packed sheets of cells that line organs and body cavities. (Chapter 17)

**equator** The circle around Earth that is equidistant to both poles. (Chapter 16)

**esophagus** Tube that conducts food from the pharynx to the stomach. (Chapter 17)

**essential amino acid** Any of the amino acids that humans cannot synthesize and thus must be obtained from the diet. (Chapter 3)

**essential fatty acid** Any of the fatty acids that animals cannot synthesize and must be obtained from the diet. (Chapter 3)

**estrogen** Any of the feminizing hormones secreted by the ovary in females and adrenal glands in both sexes. (Chapters 21, 22)

**estuary** An aquatic biome that forms at the outlet of a river into a larger body of water such as a lake or ocean. (Chapter 16)

**ethylene** A plant hormone involved in fruit ripening and abscission. (Chapter 25)

**eukaryote** Cell that has a nucleus and membrane-bounded organelles. (Chapters 2, 13)

**eutrophication** Process resulting in periods of dangerously low oxygen levels in water, sometimes caused by high levels of nitrogen and phosphorus from fertilizer runoff, that result in increased growth of algae in waterways. (Chapters 15, 24)

**evolution** Changes in the features (traits) of individuals in a biological population that occur over the course of generations. See also *theory of evolution*. (Chapters 1, 2, 10, 11)

**evolutionary classification** System of organizing biodiversity according to the evolutionary relationships among living organisms. (Chapter 13)

**excretion** The release of urine from the kidneys into the storage organ of the bladder. (Chapter 18)

**exocytosis** The secretion of molecules from a cell via fusion of membrane-bounded vesicles with the plasma membrane. (Chapter 3)

**experiment** Contrived situation designed to test specific hypotheses. (Chapter 1)

**exponential growth** Growth that occurs in proportion to the current total. (Chapter 14)

**external fertilization** Fertilization that does not require that the parents ever have physical contact with each other. Seen in many aquatic invertebrates and most fish and amphibians. (Chapter 22)

**extinction** Complete loss of a species. (Chapter 15)

**extinction vortex** A process by which an endangered population is driven toward extinction via loss of genetic diversity, increased vulnerability to the effects of density-independent factors, and inbreeding depression. (Chapter 15)

**facilitated diffusion** The spontaneous passage of molecules, through membrane proteins, down their concentration gradient. (Chapter 3)

**falsifiable** Able to be proved false. (Chapter 1)

**fat** Energy-rich, hydrophobic lipid molecule composed of a three-carbon glycerol skeleton bonded to three fatty acids. (Chapters 2, 3, 4)

**fatty acid** A long acidic chain of hydrocarbons bonded to glycerol. Fatty acids vary on the basis of their length and on the number and placement of double bonds. (Chapters 2, 3)

**femur** Bone extending from the pelvis to the knee; also called thighbone. (Chapter 21)

**fermentation** A process that makes a small amount of ATP from glucose without using an electron transport chain. Ethyl alcohol and lactic acid are produced by this process. (Chapters 4, 13)

**fertility** Ability to produce viable gametes. (Chapter 22)

**fertilization** The fusion of haploid gametes (in humans, egg and sperm) to produce a diploid zygote. (Chapters 6, 7, 22)

**fertilizer** Any of a variety of growth medium amendments that increase the level of plant nutrients. (Chapter 24)

**fetus** The term used to describe a developing human from the ninth week of development until birth. (Chapter 22)

**fever** Abnormally high body temperature. (Chapter 20)

**fibrin** A protein produced during the process of blood clotting that makes up a net to trap and block blood flow from a damaged blood vessel. (Chapter 19)

**fibroblast** A protein-secreting cell found in loose connective tissue. (Chapter 17)

**fibrous connective tissue** A dense tissue found in tendons and ligaments composed largely of collagen fibers. (Chapter 17)

**filtration** In the kidneys, the removal of plasma from the bloodstream through capillaries surrounding a nephron. (Chapter 18)

**fitness** Relative survival and reproduction of one variant compared to others in the same population. (Chapters 11, 15)

**flagellum** (*plural:* flagella) A long cellular projection that aids in motility. (Chapter 20)

**flower** Reproductive structure of a flowering plant. (Chapters 24, 25)

**flowering plant** Member of the kingdom Plantae, which produce flowers and fruit. See also *angiosperm*. (Chapter 13)

**fluid mosaic model** The accepted model for how membranes are structured with proteins bobbing in a sea of phospholipids. (Chapter 3)

**follicle** Fluid filled sac in the ovary that contains the developing egg cell and secretes estrogen. (Chapter 22)

**follicle-stimulating hormone (FSH)** Hormone secreted by the pituitary gland involved in sperm production, regulation of ovulation, and regulation of menstruation. (Chapters 21, 22)

**food chain** The linear relationship between trophic levels from producers to primary consumers, and so on. (Chapter 15)

**food web** The feeding connections between and among organisms in an environment. (Chapter 15)

**foramen magnum** Hole in the skull that allows passage of the spinal cord. (Chapter 10)

**forest** Terrestrial community characterized by the presence of trees. (Chapter 16)

**fossil** Remains of plants or animals that once existed, left in soil or rock. (Chapters 10, 13, 15)

**fossil fuel** Nonrenewable resource consisting of the buried remains of ancient plants that have been transformed by heat and pressure into coal and oil. (Chapters 5, 16)

**fossil record** Physical evidence left by organisms that existed in the past. (Chapter 10)

**founder effect** Type of sampling error that occurs when a small subset of individuals emigrates from the main population and begins a new population, leading to differences in the gene pools of both. (Chapter 12)

**founder hypothesis** The hypothesis that the diversity of unique species in isolated habitats results from divergence from a single founding population. (Chapter 12)

**frameshift mutation** A mutation that occurs when the number of nucleotides inserted or deleted from a DNA sequence is not a multiple of three. (Chapter 9)

**free radical** A substance containing an unpaired electron that is therefore unstable and highly reactive, causing damage to cells. (Chapter 3)

**frontal bone** Upper front portion of the cranium, or the forehead. (Chapter 21)

**frontal lobe** The largest and most anterior portion of each cerebral hemisphere. (Chapter 21)

**fruit** A mature ovary of an angiosperm, containing the seeds. (Chapters 24, 25)

**Fungi** Kingdom of eukaryotes made up of members that are immobile, rely on other organisms as their food source, and are made up of hyphae that secrete digestive enzymes into the environment and that absorb the digested materials. (Chapters 10, 13, 15)

**gall bladder** Organ that stores bile and empties into the small intestine. (Chapter 17)

**gamete** Specialized sex cell (sperm and egg in humans) that contains half as many chromosomes as other body cells and is therefore haploid. (Chapters 6, 7, 22)

**gamete incompatibility** An isolating mechanism between species in which sperm from one cannot fertilize eggs from another. (Chapter 12)

**gametogenesis** The production of gametes. (Chapter 22)

**gas exchange** The passage of gases, as a result of diffusion, from one compartment to another. (Chapter 19)

**gastrin** Hormone that stimulates the secretion of stomach acid. (Chapter 18)

**gastrula** The two-layered, cup-shaped stage of embryonic development. (Chapter 22)

**gel electrophoresis** The separation of biological molecules on the basis of their size and charge by measuring their rate of movement through an electric field. (Chapter 8)

**gene** Discrete unit of heritable information about genetic traits. Consists of a sequence of DNA that codes for a specific polypeptide—a protein or part of a protein. (Chapters 6, 7)

**gene expression** Turning a gene on or off. A gene is expressed when the protein it encodes is synthesized. (Chapter 9)

**gene flow** Spread of an allele throughout a species' gene pool. (Chapter 12)

**gene gun** Device used to shoot DNA-coated pellets into plant cells. (Chapter 9)

**gene pool** All of the alleles found in all of the individuals of a species. (Chapter 12)

**gene therapy** Replacing defective genes (or their protein products) with functional ones. (Chapter 9)

**genealogical species concept** A scheme that identifies as separate species all populations with a unique lineage. (Chapter 12)

**general sense** Also called proprioception. Any of the senses including temperature, pain, touch, pressure, and body position, with sensory receptors scattered throughout the body. (Chapter 23)

**genetic code** Table showing which mRNA codons code for which amino acids. (Chapters 9, 10)

**genetic drift** Change in allele frequency that occurs as a result of chance. (Chapters 12, 15)

**genetic variability** All of the forms of genes, and the distribution of these forms, found within a species. (Chapter 15)

**genetic variation** Differences in alleles that exist among individuals in a population. (Chapter 7)

**genetically modified organisms (GMOs)** Organisms whose genome incorporates genes from another organism; also called transgenic or genetically engineered organisms. (Chapter 9)

**genome** Entire suite of genes present in an organism. (Chapter 9)

**genotype** Genetic composition of an individual. (Chapters 7, 8)

**genus** Broader biological category to which several similar species may belong. (Chapters 10, 12)

**geologic period** A unit of time defined according to the rocks and fossils characteristic of that period. (Chapter 13)

**germ theory** The scientific theory that all infectious diseases are caused by microorganisms. (Chapter 1)

**germination** The beginning of growth by a seed. (Chapter 24)

**gibberellin** A plant hormone that causes the elongation of stem cells. (Chapter 25)

**glans penis** The head of the penis. (Chapter 22)

**glial cell** A type of cells within the brain that does not carry messages but rather supports neurons by supplying nutrients, repairing the brain after injury, and attacking invading bacteria. (Chapter 23)

**global warming** Increase in average global temperatures as a result of the release of increased amounts of carbon dioxide and other greenhouse gases into the atmosphere. (Chapters 5, 15)

**global climate change** Changes in regional patterns of temperature and precipitation that are occurring around Earth as a result of increasing greenhouse gases in the atmosphere. (Chapter 5)

**glycolysis** The splitting of glucose into pyruvate, which helps drive the synthesis of a small amount of ATP. (Chapter 4)

**Golgi apparatus** An organelle in eukaryotic cells consisting of flattened membranous sacs that modify and sort proteins and other substances. (Chapter 3)

**gonads** The male and female sex organs; testicles in human males or ovaries in human females. (Chapters 20, 21, 22)

**gonadotropin-releasing hormone (GnRH)** Hormone produced by the hypothalamus that stimulates the pituitary gland to release FSH and LH, thereby stimulating the activities of the gonads. (Chapters 21, 22)

**gradualism** The hypothesis that evolutionary change occurs in tiny increments over long periods of time. (Chapter 12)

**grana** Stacks of thylakoids in the chloroplast. (Chapters 3, 5)

**grassland** Biome characterized by the dominance of grasses, usually found in regions of lower precipitation. (Chapter 16)

**gravitropism** Directional plant growth in response to gravity. (Chapter 25)

**gray matter** Unmyelinated axons, combined with dendrites and cell bodies of other neurons that appear gray in cross section. (Chapter 23)

**greenhouse effect** The retention of heat in the atmosphere by carbon dioxide and other greenhouse gases. (Chapter 5)

**ground tissue** The plant tissues other than the vascular tissue, epidermis, and meristem. (Chapter 24)

**growing season** The length of time from last freeze in spring to first freeze in autumn. (Chapter 25)

**growth rate** Annual death rate in a population subtracted from the annual birth rate. (Chapter 14)

**guanine** Nitrogenous base in DNA, a purine. (Chapters 2, 9)

**guard cell** Either of the paired cells encircling stomata that serve to regulate the size of the stomatal pore in leaves. (Chapters 5, 24)

**gymnosperm** Member of a group of plants made up of several phyla that produce seeds, but not true flowers. (Chapters 24, 25)

**habitat** Place where an organism lives. (Chapter 15)

**habitat destruction** Modification and degradation of natural forests, grasslands, wetlands, and waterways by people; primary cause of species loss. (Chapter 15)

**habitat fragmentation** Threat to biodiversity caused by humans that occurs when large areas of intact natural habitat are subdivided by human activities. (Chapter 15)

**half-life** Amount of time required for one-half the amount of a radioactive element that is originally present to decay into the daughter product. (Chapter 10)

**haploid** Describes cells containing only one member of each homologous pair of chromosomes; in humans, these cells are eggs and sperm. (Chapters 6, 22)

**hardened** Relating to plants that have undergone the appropriate physiological changes that allow them to survive freezing temperatures. (Chapter 25)

**hardiness** In plants, the ability to survive freezing temperatures. (Chapter 25)

**Hardy Weinberg theorem** Theorem that holds that allele frequencies remain stable in populations that are large in size, randomly mating, and experiencing no migration or natural selection. Used as a baseline to predict how allele frequencies would change if any of its assumptions were violated. (Chapter 12)

**heart** The muscular organ that pumps blood via the circulatory system to the lungs and body. (Chapters 17, 19)

**heart attack** An acute condition, during which blood flow is blocked to a portion of the heart muscle, causing part of the muscle to be damaged or die. (Chapter 19)

**heat** The total amount of energy associated with the movement of atoms and molecules in a substance. (Chapter 5)

**helper T cell** A type of immune-system cell that enhances cell-mediated immunity and humoral immunity by secreting a substance that increases the strength of the immune response. See also T4 cell. (Chapter 20)

**hemoglobin** An iron-containing protein that carries oxygen in red blood cells. (Chapter 19)

**hemophilia** Rare genetic disorder caused by a sex-linked recessive allele that prevents normal blood clotting. (Chapter 8)

**hepatocyte** Liver cell. (Chapter 17)

**herbicide** A chemical that kills plants. (Chapter 24)

**heritability** The amount of variation for a trait in a population that can be explained by differences in genes among individuals. (Chapter 7)

**hermaphrodite** An individual with both male and female reproductive organs. (Chapter 22)

**heterozygote** Individual carrying two different alleles of a particular gene. (Chapters 7, 8, 12)

**heterozygote advantage** The tendency for individuals that are heterozygous for a larger number of genes to have higher fitness. (Chapter 15)

**heterozygous** Said of a genotype containing two different alleles of a gene. (Chapter 7)

**high-density lipoprotein (HDL)** A cholesterol-carrying particle in the blood that is high in protein and low in cholesterol. (Chapter 19)

**hinge joint** A joint that allows back and forth movement. (Chapter 21)

**histamine** Chemical released from mast cells during allergic reactions. Causes blood vessels to dilate and become more permeable and lowers blood pressure. (Chapter 20)

**HIV** See *human immunodeficiency virus*. (Chapters 11, 20)

**homeostasis** The steady-state condition an organism works to maintain. (Chapters 2, 17)

**hominin** Referring to humans and human ancestors. (Chapters 10, 12)

**homologous pair** Set of two chromosomes of the same size and shape with centromeres in the same position. Homologous pairs of chromosomes carry the same genes in the same locations but may carry different alleles. (Chapters 6, 7)

**homology** Similarity in characteristics as a result of common ancestry. (Chapter 10)

**homozygous** Having two copies of the same allele of a gene. (Chapters 7, 15)

**hormones** A protein or steroid produced in one tissue that travels through the circulatory system to act on another tissue to produce some physiological effect. (Chapter 21)

**human immunodeficiency virus (HIV)** Agent identified as causing the transmission and symptoms of AIDS. (Chapter 20)

**humoral immunity** B-cell-mediated immunity that occurs when a B-cell receptor binds to an antigen. The B cell divides to produce a clonal population of memory cells; the cell also produces plasma cells. (Chapter 21)

**hybrid** Offspring of two different strains of an agricultural crop. See also *interspecies hybrid*. (Chapters 12, 24)

**hydrocarbon** A compound consisting of carbons and hydrogens. (Chapter 2)

**hydrogen atom** One negatively charged electron and one positively charged proton. (Chapters 2, 4, 5)

**hydrogen bond** A type of weak chemical bond in which a hydrogen atom of one molecule is attracted to an electronegative atom of another molecule. (Chapters 2, 16)

**hydrogen ion** The positively charged ion of hydrogen (H+) formed by removal of the electron from a hydrogen atom. (Chapters 2, 4, 5)

**hydrogenation** Adding hydrogen gas under pressure to make liquid oils more solid. (Chapter 3)

**hydrophilic** Readily dissolving in water. (Chapter 2)

**hydrophobic** Not able to dissolve in water. (Chapter 2)

**hypertension** High blood pressure. (Chapter 19)

**hyphae** Thin, stringy fungal material that grows over and within a food source. (Chapter 13)

**hypothalamus** Gland that helps regulate body temperature; influences behaviors such as hunger, thirst, and reproduction; and secretes a hormone (GnRH) that stimulates the activities of the gonads. (Chapters 21, 22)

**hypothesis** Tentative explanation for an observation that requires testing to validate. (Chapters 1, 10, 12)

**immune response** Ability of the immune system to respond to an infection, resulting from increased production of B cells and T cells. (Chapter 20)

**immune system** The organ system that produces cells and cell products, such as antibodies, that help remove pathogenic organisms. (Chapter 20)

**in vitro fertilization** Fertilization that takes place when sperm and egg are combined in glass or a test tube. (Chapter 9)

**inbreeding** Mating between related individuals. (Chapter 15)

**inbreeding depression** Negative effect of homozygosity on the fitness of members of a population. (Chapter 15)

**incomplete dominance** A type of inheritance where the heterozygote has a phenotype intermediate to both homozygotes. (Chapter 8)

**independent assortment** The separation of homologous pairs of chromosomes into gametes independently of one another during meiosis. (Chapter 7)

**independent variable** A factor whose value influences the value of the dependent variable, but is not influenced by it. In experiments, the variable that is manipulated. (Chapter 1)

**indeterminate growth** Unrestricted or unlimited growth. (Chapter 24)

**indirect effect** In ecology, a condition where one species affects another indirectly through intervening species. (Chapter 15)

**induced fit** A change in shape of the active site of an enzyme so that it binds tightly to a substrate. (Chapter 4)

**inductive reasoning** A logical process that argues from specific instances to a general conclusion. (Chapter 1)

**infant mortality** Death rate of infants and children under the age of five. (Chapter 14)

**infectious** Applies to a pathogen that finds a tissue inside the body that will support its growth. (Chapter 20)

**infertility** The inability to conceive after one year of unprotected intercourse. (Chapter 22)

**inflammatory response** A line of defense triggered by a pathogen penetrating the skin or mucus membranes. (Chapter 20)

**inner cell mass** A cluster of cells in the blastocyst that eventually develops into the embryo. (Chapter 22)

**inorganic** Said of chemical compounds that do not contain carbon. (Chapter 24)

**insulin** A hormone secreted by the pancreas that lowers blood glucose levels by promoting the uptake of glucose by cells and the storage of glucose as glycogen in the liver. (Chapter 17)

**integrated pest management (IPM)** A strategy for pest control that relies on a mix of pesticides, cultural control, and better knowledge of pest populations. (Chapter 24)

**interferon** A chemical messenger produced by virus-infected cells that helps other cells resist infection. (Chapter 20)

**intermembrane space** The space between two membranes; for example, the space between the inner and outer mitochondrial membrane. (Chapters 3, 4)

**internal fertilization** The union of gametes inside the body of the female. (Chapter 22)

**interneuron** A neuron located between sensory and motor neurons that functions to integrate sensory input and motor output. (Chapter 23)

**internode** The region of a plant stem between nodes. (Chapter 24)

**interphase** Part of the cell cycle when a cell is preparing for division and the DNA is duplicated. Consists of $G_1$, S, and $G_2$. (Chapter 6)

**interspecies hybrid** Organism with parents from two different species. (Chapter 12)

**intertidal zone** The biome that forms on ocean shorelines between the high tide elevation and the low tide elevation. (Chapter 16)

**introduced species** A nonnative species that was intentionally or unintentionally brought to a new environment by humans. (Chapter 15)

**invertebrate** Animal without backbone. (Chapter 13)

**involuntary muscle** Muscle tissue whose action requires no conscious thought. (Chapter 17)

**ion** Electrically charged atom. (Chapter 2)

**ionic bond** A chemical bond resulting from the attraction of oppositely charged ions. (Chapter 2)

**irrigation** The technique of supplying additional water to crop plants, typically via flooding or spraying. (Chapter 24)

**karyotype** The chromosomes of a cell, displayed with chromosomes arranged in homologous pairs and according to size. (Chapter 6)

**keystone species** A species that has an unusually strong effect on the structure of the community it inhabits. (Chapter 15)

**kidney** Major organ of the excretory system, responsible for filtering liquid waste from the blood. (Chapters 17, 21)

**kingdom** In some classifications, the most inclusive group of organisms; life is typically categorized into five or six. In other classification systems, the level below domain on the hierarchy. (Chapters 10, 13)

**labia majora** Paired thick folds of skin that enclose and protect the labia minor of the vulva. (Chapter 22)

**labia minora** Paired thin folds of the vulva; enclose the urinary and vaginal openings and clitoris. (Chapter 22)

**labor** Strong rhythmic contractions that force a developing baby from the uterus through the vagina during childbirth. (Chapter 22)

**lactation** Production of milk to nurse offspring. (Chapter 22)

**lake** An aquatic biome that is completely landlocked. (Chapter 16)

**large intestine** Colon, portion of the digestive system located between the small intestine and the anus that absorbs water and forms feces. (Chapter 17)

**larynx** A portion of the upper respiratory tract made up primarily of stiff cartilage. Also known as the voice box. (Chapter 19)

**latent virus** Dormant virus. (Chapter 20)

**leaf** The primary photosynthetic organ in plants. (Chapters 24, 25)

**legumes** A family of plants that produces seeds in pods, like peas and beans. (Chapter 24)

**leptin** A hormone by fat cells that may be involved in the regulation of appetite. (Chapter 4)

**less developed country** A term used by the United Nations to describe a country with low per-capita income and often poor population health and economic prospects. (Chapter 14)

**Leydig cell** Cells scattered between the seminiferous tubules of the testicles that produce testosterone and other androgens. (Chapter 22)

**life cycle** Description of the growth and reproduction of an individual. (Chapter 7)

**ligament** A band of fibrous connective tissue joining bones. (Chapter 21)

**light reactions** A series of reactions that occur in chloroplasts during photosynthesis and serve to convert energy from the sun into the energy stored in ATP. These reactions also produce oxygen gas. (Chapter 5)

**linked gene** Genes located on the same chromosome. (Chapter 6)

**lipid** Hydrophobic molecule, including fats, phospholipids, and steroids. (Chapters 2, 3)

**liver** Organ with many functions, including the production of bile to aid in the absorption of fats. (Chapter 17)

**lobule** Subdivision of the lobes of the liver. (Chapter 17)

**logistic growth** Pattern of growth seen in populations that are limited by resources available in the environment. A graph of logistic growth over time typically takes the form of an S-shaped curve. (Chapter 14)

**loose connective tissue** Connective tissue that serves to bind epithelia to underlying tissues and to hold organs in place. (Chapter 17)

**low-density lipoprotein (LDL)** Cholesterol-carrying substance in the blood that is high in cholesterol and low in protein. (Chapter 19)

**lung** The primary organ of the respiratory system; the site where gas exchange occurs. (Chapter 19)

**luteinizing hormone (LH)** Hormone involved in sperm production, regulation of ovulation, and regulation of menstruation. (Chapters 21, 22)

**lymph node** Organ located along lymph vessels that filter lymph and help defend against bacteria and viruses. (Chapter 20)

**lymphatic system** A system of vessels and nodes that return fluid and protein to the blood. (Chapter 20)

**lymphocyte** White blood cells that make up part of the immune system. (Chapter 20)

**lysosome** A membrane-bounded sac of hydrolytic enzymes found in the cytoplasm of many cells. (Chapter 3)

**macroevolution** Large-scale evolutionary change, usually referring to the origin of new species. (Chapter 10)

**macromolecule** Any of the large molecules including polysaccharides, proteins, and nucleic acids, composed of subunits joined by dehydration synthesis. (Chapter 2)

**macronutrient** Nutrient required in large quantities. (Chapter 3)

**macrophage** Phagocytic white blood cell that swells and releases toxins to kill bacteria. (Chapter 20)

**malignant** Describes a tumor that is cancerous, whether it is invasive or metastatic. (Chapter 6)

**mandible** Bone of the lower jaw. (Chapter 21)

**marine** Of, or pertaining to, salt water. (Chapter 16)

**mark-recapture method** A technique for estimating population size, consisting of capturing and marking a number of individuals, releasing them, and recapturing more individuals to determine what proportion are marked. (Chapter 14)

**mass extinction** Loss of species that is rapid and global in scale, and affects a wide variety of organisms. (Chapter 15)

**matrix** (1) In a mitochondrion, the semifluid substance inside the inner mitochondrial membrane, which houses the enzymes of the citric acid cycle. (Chapters 2, 4)

(2) In connective tissue, a nonliving substance between cells, ranging from fluid blood plasma to fibrous matrix in tendons to solid bone matrix. (Chapter 17)

**mean** Average value of a group of measurements. (Chapter 1)

**mechanical isolation** A form of reproductive isolation between species that depends on the incompatibility of the genitalia of individuals of different species. (Chapter 12)

**medulla** The center of an organ or gland, such as the kidney or adrenal gland. (Chapter 17)

**medulla oblongata** Region of the brain stem that is a continuation of the spinal cord and conveys information between the spinal cord and other parts of the brain. (Chapter 23)

**meiosis** Process that diploid sex cells undergo in order to produce haploid daughter cells. Occurs during gametogenesis. (Chapters 6, 7, 22)

**memory cell** Cell that is part of a clonal population, programmed to respond to a specific antigen, that helps the body respond quickly if the infectious agent is encountered again. (Chapter 20)

**menopause** Cessation of menstruation. (Chapter 22)

**menstrual cycle** Changes that occur in the uterus and depend on intricate interrelationships among the brain, ovaries, and lining of the uterus. (Chapter 22)

**menstruation** The shedding of the lining of the uterus during the menstrual cycle. (Chapter 22)

**meristematic tissue** Undifferentiated plant tissue from which new plant cells arise. (Chapter 24)

**mesoderm** The middle of three germ layers that arise during animal development. (Chapter 22)

**messenger RNA (mRNA)** Complementary RNA copy of a DNA gene, produced during transcription. The mRNA undergoes translation to synthesize a protein. (Chapter 9)

**metabolic rate** Measure of an individual's energy use. (Chapter 4)

**metabolism** All of the physical and chemical reactions that produce and use energy. (Chapters 2, 4, 5)

**metaphase** Stage of mitosis during which duplicated chromosomes align across the middle of the cell. (Chapter 6)

**metastasis** When cells from a tumor break away and start new cancers at distant locations. (Chapter 6)

**microbe** Microscopic organism, especially Bacteria and Archaea. (Chapter 13)

**microbiologists** Scientists who study microscopic organisms, especially referring to those who study prokaryotes. (Chapter 13)

**microevolution** Changes that occur in the characteristics of a population. (Chapter 10)

**micronutrient** Nutrient needed in small quantities. (Chapter 3)

**microorganism** See *microbe*. (Chapter 13)

**microtubule** Protein structure that moves chromosomes around during mitosis and meiosis. (Chapters 2, 6)

**microvillus** Fine fingerlike projection composed of epithelial cells that function in absorption. (Chapter 17)

**micturation** Release of urine from the bladder. Also known as urination. (Chapter 18)

**midbrain** Uppermost region of the brain stem, which adjusts the sensitivity of the eyes to light and of the ears to sound. (Chapter 23)

**mineral** Inorganic nutrient essential to many cell functions. (Chapter 3)

**mitochondria** Organelles in which products of the digestive system are converted to ATP. (Chapters 3, 4)

**mitosis** The division of the nucleus that produces daughter cells that are genetically identical to the parent cell. Also, portion of the cell cycle in which DNA is apportioned into two daughter cells. (Chapter 6)

**model organism** Any nonhuman organism used in genetic studies to help scientists understand human genes because they share genes with humans. (Chapter 1)

**model systems** See *model organisms* (Chapter 1)

**mold** A fungal form characterized by rapid, asexual reproduction. (Chapter 13)

**molecular clock** Principle that DNA mutations accumulate in the genome of a species at a constant rate, permitting estimates of when the common ancestor of two species existed. (Chapter 10)

**molecule** Two or more atoms held together by covalent bonds. (Chapter 2)

**monocot** One of the two classes of flowering plants, also called narrow-leaved plants. Monocot seeds contain one leaf (cotyledon). (Chapter 24)

**monoculture** Practice of planting a single crop over a wide acreage. (Chapter 24)

**monomer** An individual molecule that binds to other molecules to form a polymer. (Chapter 2)

**monosaccharide** Simple sugar. (Chapter 2)

**monozygotic twins** Identical twins that developed from one zygote. (Chapter 7)

**morphological species concept** Definition of species that relies on differences in physical characteristics among them. (Chapter 12)

**morphology** Appearance or outward physical characteristics. (Chapter 12)

**motor neuron** Neuron that carries information away from the brain or spinal cord to muscles or glands. (Chapter 23)

**mulching** Covering the ground around favored plants with light-blocking material to prevent weed growth. (Chapter 24)

**multicellular** The condition of being composed of many coordinated cells. (Chapter 13)

**multiple allelism** A gene for which there are more than two alleles in the population. (Chapter 8)

**multiple hit model** The notion that many different genetic mutations are required for a cancer to develop. (Chapter 6)

**multiple sclerosis** Chronic, degenerative nervous system disease caused by breakdown of myelin. (Chapter 23)

**muscle fiber** Single cell that aligns with others in parallel bundles to form muscles. (Chapter 21)

**muscle tissue** Specialized contractile tissue. (Chapter 17)

**mutation** Change to a DNA sequence that may result in the production of altered proteins. (Chapters 6, 7, 9)

**mutualism** Interaction between two species that provides benefits to both species. (Chapter 15)

**mycologist** Scientist who specializes in the study of fungi. (Chapter 13)

**myelin sheath** Protective layer that coats many axons, formed by supporting cells such as Schwann cells. The myelin sheath increases the speed at which the electrochemical impulse travels down the axon. (Chapter 23)

**myofibril** Structure found in muscle cells, composed of thin filaments of actin and thick filaments of myosin. (Chapter 21)

**myosin** A type of protein filament that, along with actin, causes muscle cells to contract. (Chapters 17, 21)

**natural experiment** Situation where unique circumstances allow a hypothesis test without prior intervention by researchers. (Chapter 7)

**natural killer cell** A cell that attacks virus-infected cells or tumor cells without being activated by an immune system cell or antibody. (Chapter 20)

**natural selection** Process by which individuals with certain traits have greater survival and reproduction than individuals who lack these traits, resulting in an increase in the frequency of successful alleles and a decrease in the frequency of unsuccessful ones. (Chapters 10, 11, 12, 21)

**negative feedback** A mechanism of maintaining homeostasis in which the product of the process inhibits the process. (Chapter 17)

**nephron** The functional structure within a kidney where waste filtration and urine concentration occurs. (Chapter 18)

**nerve** Bundle of neurons; nerves branch out from the brain and spinal cord to eyes, ears, internal organs, skin, and bones. (Chapter 23)

**nerve impulse** Electrochemical signal that controls the activities of muscles, glands, organs, and organ systems. (Chapter 23)

**nervous system** Brain, spinal cord, sense organs, and nerves that connect organs and link this system with other organ systems. (Chapter 23)

**nervous tissue** Tissue composed of neurons and associated cells. (Chapters 17, 23)

**net primary production (NPP)** Amount of solar energy converted to chemical energy by plants, minus the amount of this chemical energy plants need to support themselves. A measure of plant growth, typically over the course of a single year. (Chapter 14)

**neuron** Specialized message-carrying cell of the nervous system. (Chapters 17, 23)

**neurotransmitter** One of many chemicals released by the presynaptic neuron into the synapse, which then diffuse across the synapse and bind to receptors on the membrane of the postsynaptic neuron. (Chapter 23)

**neutral mutation** A genetic mutation that confers no selective advantage or disadvantage. (Chapters 9, 12)

**neutron** An electrically neutral particle found in the nucleus of an atom. (Chapter 2)

**nutrient cycling** A process by which inorganic nutrients are passed between living and non-living entities in an ecosystem. (Chapter 15)

**nutrients** Atoms other than carbon, hydrogen, and oxygen that must be obtained from an organisms environment. (Chapters 3, 15, 24)

**nicotinamide adenine dinucleotide** Intracellular electron carrier. Oxidized form is NAD+; reduced form is NADH. (Chapter 4)

**nicotine** The active drug in tobacco that stimulates dopamine receptors in the brain. (Chapter 19)

**nitrogen-fixing bacteria** Organisms that convert nitrogen gas from the atmosphere into a form that can be taken up by plant roots; some species live in the root nodules of legumes. (Chapters 15, 24)

**nitrogenous base** Nitrogen-containing base found in DNA: A, C, G, and T and in RNA: U. (Chapters 2, 8, 9)

**node** Point on a plant stem where a leaf and axillary bud arise. (Chapter 24)

**node of Ranvier** Small indentation separating segments of the myelin sheath. Nerve impulses *jump* successively from one node of Ranvier to the next. (Chapter 23)

**nodule** Compartment housing nitrogen-fixing bacteria, produced on the roots of legume plants such as beans and alfalfa. (Chapter 24)

**nonpolar** Won't dissolve in water. Hydrophobic. (Chapter 2)

**nonrenewable resource** Resource that is a one-time supply and cannot be easily replaced. (Chapter 14)

**nonspecific defenses** Defense system against infection that does not distinguish one pathogen from another. Includes the skin, secretions, and mucous membranes. (Chapter 20)

**normal distribution** Bell-shaped curve, as for the distribution of quantitative traits in a population. (Chapter 7)

**notochord** A long, flexible rod that runs through the axis of the vertebral body in the future position of the spinal cord. (Chapter 23)

**nuclear envelope** The double membrane enclosing the nucleus in eukaryotes. (Chapters 3, 6)

**nuclear transfer** Transfer of a nucleus from one cell to another cell that has had its nucleus removed. (Chapter 9)

**nucleic acids** Polymers of nucleotides that comprise DNA and RNA. (Chapters 2, 6, 8, 9)

**nucleoid region** The region of a prokaryotic cell where the DNA is located. (Chapter 2, 20)

**nucleotides** Building blocks of nucleic acids that include a sugar, a phosphate, and a nitrogenous base. (Chapters 2, 4, 6, 8, 9)

**nucleus** Cell structure that houses DNA; found in eukaryotes. (Chapters 2, 3, 6, 8, 9)

**nutrient cycling** Process by which nutrients become available to plants. Nutrient cycling in a natural environment relies on a healthy community of decomposers within the soil. (Chapter 15)

**nutrients** Substances that provide nourishment. (Chapter 3)

**obesity** Condition of having a BMI of 30 or greater. (Chapter 4)

**objective** Without bias. (Chapter 1)

**occipital lobe** The posterior lobe of each cerebral hemisphere, containing the visual center of the brain. (Chapter 23)

**ocean** A biome consisting of open stretches of salt water. (Chapter 16)

**oncogene** Mutant version of a cell cycle controlling proto-oncogene. (Chapter 6)

**oogenesis** Formation and development of female gametes, which occurs in the ovaries and results in the production of egg cells. (Chapter 22)

**organ** A specialized structure composed of several different types of tissues. (Chapter 17)

**organ systems** Suites of organs working together to perform a function or functions. (Chapter 17)

**organelle** Subcellular structure found in the cytoplasm of eukaryotic cells that performs a specific job. (Chapter 3)

**organic** When pertaining to agriculture, refers to products grown without the use of manufactured pesticides and inorganic fertilizer. (Chapter 24)

**organic chemistry** The chemistry of carbon-containing substances. (Chapter 2)

**osmosis** The diffusion of water across a selectively permeable membrane. (Chapter 3)

**ossa coxae** The paired bones that form the bony pelvis. (Chapter 21)

**osteoblast** Bone-forming cell responsible for the deposition of collagen. (Chapter 21)

**osteoclast** Bone-reabsorbing cell that liberates calcium. (Chapter 21)

**osteocyte** Highly branched cell found in bone. (Chapter 17, 21)

**osteoporosis** A condition resulting in an elevated risk of bone breakage from weakened bones. (Chapter 3, 21)

**ovary** (1) In animals, the paired abdominal structures that produce egg cells and secrete female hormones. (Chapters 6, 22) (2) In plants, a chamber of the carpel containing the ovules. (Chapter 24)

**overexploitation** Threat to biodiversity caused by humans that encompasses overhunting and overharvesting. (Chapter 15)

**oviduct** Egg-carrying duct that brings egg cells from ovaries to uterus. (Chapter 22)

**ovulation** Release of an egg cell from the ovary. (Chapter 22)

**ovule** Structure in flowering plants consisting of eggs and accessory tissues. After fertilization, will develop into a seed. (Chapter 24)

**ovum** An egg; the female gamete. Cell produced during oogenesis that receives the majority of the cytoplasmic nutrients and organelles. (Chapter 22)

**oxytocin** Pituitary hormone that stimulates the contraction of smooth muscle of the uterus during labor and facilitates secretion of milk from the breast during nursing. (Chapter 22)

**pacemaker** A patch of heart tissue or an implanted device that produces a regular electrical signal, setting the rhythm for the cardiac cycle. (Chapter 19)

**paleontologist** Scientist who searches for, describes, and studies ancient organisms. (Chapter 11)

**pancreas** Gland that secretes digestive enzymes and insulin. (Chapters 17, 21)

**parasite** An organism that benefits from an association with another organism that is harmed by the association. (Chapter 20)

**parathyroid gland** One of four endocrine glands located on the thyroid that secrete parathyroid hormone in order to regulate blood calcium levels. (Chapter 21)

**parietal lobe** Part of the brain that processes information about touch and is involved in self-awareness. (Chapter 23)

**Parkinson's disease** Disease that results in tremors, rigidity, and slowed movements. May be due to faulty dopamine production. (Chapter 23)

**particulate** Tiny airborne particle found in smoke and other pollutants. (Chapter 19)

**passive immunity** Immunity acquired when antibodies are passed from one individual to another, as from mother to child during breast-feeding. (Chapter 20)

**passive smoker** An individual who inhales environmental tobacco smoke but who does not actively consume tobacco. (Chapter 19)

**passive transport** The diffusion of substances across a membrane with their concentration gradient and not requiring an input of ATP. (Chapter 3)

**pathogen** Disease-causing organism. (Chapters 11, 20)

**pedigree** Family tree that follows the inheritance of a genetic trait for many generations. (Chapter 8)

**peer review** The process by which reports of scientific research are examined and critiqued by other researchers before they are published in scholarly journals. (Chapter 1)

**penis** The copulatory structure in males. (Chapter 22)

**peptide bond** Covalent bond that joins the amino group and carboxyl group of adjacent amino acids. (Chapter 2)

**perennial** See *perennial plant*. (Chapter 25)

**perennial plant** Plant that lives for many years. (Chapters 16, 24, 25)

**peripheral nervous system (PNS)** Network of nerves outside the brain and spinal cord that links the CNS with sense organs. (Chapter 23)

**peristalsis** Rhythmic muscle contractions that move food through the digestive system. (Chapter 17)

**permafrost** Permanently frozen soil. (Chapter 16)

**pest** Any organism that competes with humans for agricultural production or other resources. (Chapter 24)

**pesticide** Chemical that kills or disables agricultural pests. (Chapter 24)

**pesticide treadmill** The tendency for pesticide toxicity and total applications to increase once a farmer begins to employ pesticides. (Chapter 24)

**petal** A flower part that typically functions to attract pollinators. (Chapter 24)

**petiole** The stalk of a leaf. (Chapter 24)

**pH** A logarithmic measure of the hydrogen ion concentration ranging from 0 to 14. Lower numbers indicate higher hydrogen ion concentrations. (Chapter 2)

**phagocytosis** Ingestion of food or pathogens by cells. (Chapter 3)

**pharming** Genetic engineering to make pharmaceuticals. (Chapter 9)

**pharynx** Tube and muscles connecting the mouth to the esophagus; throat. (Chapter 17)

**phenotype** Physical and physiological traits of an individual. (Chapters 6, 7, 8)

**phenotypic ratio** Proportion of individuals produced by a genetic cross who possess each of the various phenotypes that cross can generate. (Chapter 7)

**phloem** The portion of a plant's vascular tissue modified for multidirectional transport of nutrients throughout the body of the plant. (Chapter 24)

**phloem sap** The liquid, often containing dissolved carbohydrates, carried in the phloem tubes in a plant. (Chapter 25)

**phospholipid** One of three types of lipids; phospholipids are components of cell membranes. (Chapters 2, 3)

**phospholipid bilayer** The membrane that surrounds cells and organelles and is composed of two layers of phospholipids. (Chapters 2, 3)

**phosphorylation** To introduce a phosphoryl group into an organic compound. (Chapter 4)

**photoperiodism** Response to the duration and timing of day and night length in plants. (Chapter 25)

**photorespiration** A series of reactions triggered by the closing of stomatal openings to prevent water loss. (Chapter 5)

**photosynthesis** Process by which plants, along with algae and some bacteria, transform light energy to chemical energy. (Chapter 5)

**phototropism** Growth in response to directional light. (Chapter 25)

**phyla** (singular: phylum) The taxonomic category below kingdom and above class. (Chapters 10, 13)

**phylogeny** Evolutionary history of a group of organisms. (Chapter 13)

**phytochrome** The light-sensitive chemical in plants responsible for photoperiodic responses. (Chapter 25)

**pituitary gland** Small gland attached by a stalk to the base of the brain that secretes growth hormone, reproductive hormones, and other hormones. (Chapters 21, 22)

**pivot joint** A type of joint that allows freedom of movement. (Chapter 21)

**placebo** Sham treatments in experiments. (Chapter 1)

**placenta** Membrane produced by a developing fetus that releases a hormone to extend the life of the corpus luteum. (Chapter 22)

**plant hormone** Substance in plants that helps tune the organisms' response to the environment. (Chapter 25)

**Plantae** Multicellular photosynthetic eukaryotes, excluding algae. (Chapter 13)

**plasma** The liquid portion of blood. (Chapter 19)

**plasma cell** Cell produced by a clonal population that secretes antibodies specific to an antigen. (Chapter 20)

**plasma membrane** Structure that encloses a cell, defining the cell's outer boundary. (Chapters 2, 3)

**plasmid** Circular piece of bacterial DNA that normally exists separate from the bacterial chromosome and can make copies of itself. (Chapter 9)

**platelet** Cell in the blood that carries constituents required for the clotting response. (Chapter 19)

**pleiotropy** The ability of one gene to affect many different functions. (Chapters 7, 8)

**polar** Describes a molecule with regions having different charges; goes into solution in water. (Chapter 2)

**polarization** The difference in charge between the inside and outside of the resting neuron cell. (Chapter 23)

**poles** Opposite ends of a sphere, such as of a cell (Chapter 6) or of a planet such as Earth. (Chapter 16)

**pollen** The male gametophyte of seed plants. (Chapters 12, 13, 15, 24)

**pollination** The process by which pollen is transferred to the female structures of a plant. (Chapters 13, 24)

**pollinator** An organism that transfers sperm (pollen grains) from one flower to the female reproductive structures of another flower. (Chapter 15)

**pollution** Human-caused threat to biodiversity involving the release of poisons, excess nutrients, and other wastes into the environment. (Chapter 15)

**polyculture** Practice of planting many different crop plants over a single farm's acreage. (Chapter 24)

**polygenic trait** A trait influenced by many genes. (Chapters 7, 8)

**polymer** General term for a macromolecule composed of many chemically bonded monomers. (Chapter 2)

**polymerase** An enzyme that catalyzes phosphodiester bond formation between nucleotides. (Chapters 6, 8)

**polymerase chain reaction (PCR)** A laboratory technique that allows the production of many identical DNA molecules. (Chapter 8)

**polyploidy** A chromosomal condition involving more than two sets of chromosomes. (Chapter 12)

**polysaccharide** A carbohydrate composed of three or more monosaccharides. (Chapter 2)

**polyunsaturated** Relating to fats consisting of carbon chains with many double bonds unsaturated by hydrogen atoms. (Chapter 3)

**pond** An aquatic biome that is completely landlocked. (Chapter 16)

**pons** A structure located on the brain stem, between the brain and spinal cord. (Chapter 23)

**population** Subgroup of a species that is somewhat independent from other groups. (Chapters 12, 14)

**population bottleneck** Dramatic but short-lived reduction in population size followed by an increase in population. (Chapter 12)

**population crash** Steep decline in number that may occur when a population grows larger than the carrying capacity of its environment. (Chapter 14)

**population cycle** In some populations, the tendency to increase in number above the environment's carrying capacity, resulting in a crash, following by an overshoot of the carrying capacity and another crash, continuing indefinitely. (Chapter 14)

**population genetics** Study of the factors in a population that determine allele frequencies and their change over time. (Chapter 12)

**population pyramid** A visual representation of the number of individuals in different age categories in a population. (Chapter 14)

**positive feedback** A relatively uncommon homeostatic mechanism in which the product of a process intensifies the process, thereby promoting change. (Chapter 17)

**postsynaptic neuron** The neuron that responds to neurotransmitter released from the presynaptic neuron. (Chapter 23)

**prairie** A grassland biome. (Chapter 16)

**precipitation** When water vapor in the atmosphere turns to liquid or solid form and falls to Earth's surface. (Chapter 16)

**predation** Act of capturing and consuming an individual of another species. (Chapter 15)

**predator** Organism that eats other organisms. (Chapter 15)

**prediction** Result expected from a particular test of a hypothesis if the hypothesis were true. (Chapter 1)

**pressure flow mechanism** The process by which phloem sap moves from a sugar source to a sugar sink within a plant. (Chapter 25)

**presynaptic** The neuron that secretes neurotransmitter into a synapse, transmitting a signal. (Chapter 23)

**primary consumer** Organism that eats plants. (Chapter 15)

**primary growth** Growth occurring at the tips of a plant, originating in the apical meristems. (Chapter 24)

**primary source** Article reporting research results, written by researchers, and reviewed by the scientific community. (Chapter 1)

**primers** Short nucleotide sequences used to help initiate replication of nucleic acids. (Chapter 8)

**probability** Likelihood that something is the case or will happen. (Chapter 1)

**processed food** Food that has been modified from its original form to increase shelf life, transportability, or the like. (Chapter 3)

**producer** Organism that produces carbohydrates from inorganic carbon, typically via photosynthesis. (Chapter 15)

**product** The modified chemical that results from a chemical or enzymatic reaction. (Chapter 2)

**progesterone** Ovarian hormone. High levels have a negative feedback effect on the hypothalamus, causing GnRH secretion to decrease. (Chapter 22)

**prokaryote** Type of cell that does not have a nucleus or membrane-bounded organelles. (Chapters 2, 13)

**prolactin** A hormone produced by the pituitary gland that stimulates the development of mammary glands. (Chapter 22)

**promoter** Sequence of nucleotides to which the polymerase binds to start transcription. (Chapter 9)

**prophase** Stage of mitosis during which duplicated chromosomes condense. (Chapter 6)

**prostate** A gland that secretes an acid-neutralizing fluid into semen. (Chapter 22)

**protein** Cellular constituent made of amino acids coded for by genes. Proteins can have structural, transport, or enzymatic roles. (Chapters 2, 3, 8, 9)

**protein synthesis** Joining amino acids together, in an order dictated by a gene, to produce a protein. (Chapter 9)

**Protista** Kingdom in the domain Eukarya containing a diversity of eukaryotic organisms, most of which are unicellular. (Chapter 13)

**proton** A positively charged subatomic particle. (Chapters 2, 4, 5)

**proto-oncogenes** Genes that encode proteins that regulate the cell cycle. Mutated proto-oncogenes (oncogenes) can lead to cancer. (Chapter 6)

**prune** To trim back the growing tips of a plant. (Chapter 25)

**pseudoscience** Information presented as scientific but does not hold up under scientific scrutiny. (Chapter 2)

**psychosis** Abnormal condition of the mind. (Chapter 23)

**puberty** The point in human development when male and female hormones are triggered in the body. Males produce sperm at this time, and females begin the menstrual cycle. (Chapter 21)

**pulmonary circuit** The path of blood through vessels from the heart, through the lungs, and back to the heart. (Chapter 19)

**pulse** The volume of blood that passes into the arteries as a result of the heart's contraction. (Chapter 19)

**punctuated equilibrium** The hypothesis that evolutionary changes occur rapidly and in short bursts, followed by long periods of little change. (Chapter 12)

**Punnett square** Table that lists the different kinds of sperm or eggs parents can produce relative to the gene or genes in question and predicts the possible outcomes of a cross between these parents. (Chapter 7)

**purine** Nitrogenous base (A or G) with a two-ring structure. (Chapter 2)

**pus** Fluid, dead cells, and microorganisms that accumulate at the site of the infection. (Chapter 20)

**pyrimidine** Nitrogenous base (C, T, or U) with a single-ring structure. (Chapter 2)

**pyruvic acid** The three-carbon molecule produced by glycolysis. (Chapter 4)

**Q angle** Angle of the femur in relation to a horizontal line drawn through the kneecap. (Chapter 20)

**quantitative trait** Trait that has many possible values. (Chapter 7)

**race** See *biological race*. (Chapter 12)

**racism** Idea that some groups of people are naturally superior to others. (Chapter 12)

**radiation therapy** Focusing beams of reactive particles at a tumor to kill the dividing cells. (Chapter 6)

**radioactive decay** Natural, spontaneous breakdown of radioactive elements into different elements, or *daughter products*. (Chapter 10)

**radiometric dating** Technique that relies on radioactive decay to estimate a fossil's age. (Chapters 10, 12)

**random alignment** When members of a homologous pair line up randomly with respect to maternal or paternal origin during metaphase I of meiosis, thus increasing the genetic diversity of offspring. (Chapters 6, 7)

**random assignment** Placing individuals into experimental and control groups randomly to eliminate systematic differences between the groups. (Chapter 1)

**random distribution** The dispersion of individuals in a population without pattern. (Chapter 14)

**random fertilization** The unpredictability of exactly which gametes will fuse during the process of sexual reproduction. (Chapter 7)

**reabsorption** Reuptake of water and other essential substances by a nephron from the filtrate initially squeezed into the kidney. (Chapter 18)

**reactant** Any starting material in a chemical reaction. (Chapter 2)

**reading frame** The grouping of mRNAs into three base codons for translation. (Chapter 9)

**receptor** (1) Protein on the surface of a cell that recognizes and binds to a specific chemical signal. (Chapter 21) (2) See *sensory receptor*. (Chapter 23)

**recessive** Applies to an allele with an effect that is not visible in a heterozygote. (Chapter 7)

**recombinant** Produced by manipulating a DNA sequence. (Chapter 9)

**recombinant bovine growth hormone (rBGH)** Growth hormone produced in a laboratory and injected into cows to increase their size and ability to produce milk. (Chapter 9)

**red blood cell** Primary cellular component of blood, responsible for ferrying oxygen throughout the body. (Chapter 19)

**reflex** Automatic response to a stimulus. (Chapter 23)

**reflex arc** Nerve pathway followed during a reflex consisting of a sensory receptor, a sensory neuron, an interneuron, a motor neuron, and an effector. (Chapter 23)

**renal artery** The vessel that carries blood to the kidney for filtering. (Chapter 18)

**renal vein** The vessel that carries filtered blood from the kidney. (Chapter 18)

**repolarization** The restoration of a charge difference across a membrane. (Chapter 23)

**repressor** A protein that suppresses the expression of a gene. (Chapter 9)

**reproductive cloning** Transferring the nucleus from a donor adult cell to an egg cell without a nucleus in order to clone the adult. (Chapter 9)

**reproductive isolation** Prevention of gene flow between different biological species due to failure to produce fertile offspring; can include premating barriers and postmating barriers. (Chapter 12)

**reproductive organ** Internal and external genitalia involved in production and delivery of gametes. (Chapters 21, 22)

**respiratory surface** Body surface across which gas exchange occurs. (Chapter 19)

**respiratory system** The organ system involved in gas exchange between an animal and its environment. In humans, the lungs and air passages. (Chapter 19)

**restriction enzyme** An enzyme that cleaves DNA at specific nucleotide sequences. (Chapters 9, 13)

**retrovirus** RNA virus that synthesizes DNA using reverse transcriptase. (Chapter 20)

**reuptake** In neurons, the process by which neurotransmitters are reabsorbed by the neuron that secreted them. (Chapter 23)

**reverse transcriptase** Enzyme in RNA viruses that produces DNA by transcription of viral RNA. (Chapter 20)

**Rh factor** Surface molecule found on some red blood cells. (Chapter 8)

**rhizomes** Underground stems in plants. (Chapter 24)

**ribose** The five-carbon sugar in RNA. (Chapter 9)

**ribosome** Subcellular structure that helps translate genetic material into proteins by anchoring and exposing small sequences of mRNA. (Chapters 3, 9, 13)

**risk factor** Any exposure or behavior that increases the likelihood of disease. (Chapter 6)

**Ritalin** Stimulant used to treat ADD. (Chapter 23)

**river** Aquatic biome characterized by flowing water. (Chapter 16)

**RNA (ribonucleic acid)** Information-carrying molecule composed of nucleotides. (Chapters 2, 9)

**RNA polymerase** Enzyme that synthesizes mRNA from a DNA template during transcription. (Chapter 9)

**root cap** Loosely organized cells that cover the growing tip of a root and are continually shed as the root pushes through the soil. (Chapter 24)

**root hair** Extension of an epidermal cell on young roots, which maximize the surface area for water and nutrient uptake. (Chapter 25)

**root system** The plant organ system responsible for water and nutrient uptake and anchorage. (Chapter 24)

**rough endoplasmic reticulum** Ribosome-studded subcellular membranes found in the cytoplasm and responsible for some protein synthesis. (Chapter 3)

**rubisco** Abbreviation for ribulose bisphosphate carboxylase oxygenase, the enzyme that catalyzes the first step in the Calvin cycle of photosynthesis. (Chapter 5)

**runoff** Water moving across a land surface, picking up contaminants, and eventually flowing into lakes, streams, or groundwater. (Chapter 24)

**salinization** Degradation of soil by mineral salts deposited as a result of irrigation. (Chapter 24)

**salivary amylase** An enzyme secreted by salivary glands of the mouth to break down starch. (Chapter 17)

**salt** A charged substance that ionizes in solution. (Chapter 2)

**sample** Small subgroup of a population used in an experimental test. (Chapter 1)

**sample size** Number of individuals in both the experimental and control groups. (Chapter 1)

**sampling error** Effect of chance on experimental results. (Chapter 1)

**sarcomere** Any of the repeating units in a myofibril of striated muscle, bounded by Z discs. (Chapter 21)

**saturated fat** Type of lipid rich in single bonds. Found in butter and other fats that are solids at room temperature. This type of fat is associated with higher blood cholesterol levels. (Chapter 3)

**savanna** Grassland biome containing scattered trees. (Chapter 16)

**Schwann cell** Cells that form the myelin sheath along the axons of nerve cells in the peripheral nervous system. (Chapter 23)

**scientific method** A systematic method of research consisting of putting a hypothesis to a test designed to disprove it, if it is in fact false. (Chapter 1)

**scientific theory** Body of scientifically accepted general principles that explain natural phenomena. (Chapters 1, 10)

**scrotum** The pouch of skin that houses the testes. (Chapters 21, 22)

**secondary compounds** Chemicals produced by plants and some other organisms as side reactions to normal metabolic pathways and that typically have an antipredator or antibiotic function. (Chapter 13)

**secondary consumers** Animals that eat primary consumers; predators. (Chapter 15)

**secondary growth** Growth in girth of plants, as a result of cell division in lateral meristems. (Chapter 24)

**secondary sources** Books, news media, and advertisements as sources of scientific information. (Chapter 1)

**secretion** A step in waste removal in the kidney, in which contaminants in low concentration in the blood are actively absorbed into the urine. (Chapter 18)

**seed** A plant embryo packaged with a food source and surrounded by a seed coat. (Chapters 13, 24)

**seedling** A young plant that develops from a germinating seed. (Chapter 24)

**segregation** Separation of pairs of alleles during the production of gametes. Results in a 50% probability that a given gamete contains one allele rather than the other. (Chapter 7)

**semen** Sperm and energy-rich associated fluids. (Chapter 22)

**semiconservative replication** DNA replication results in the production of two daughter DNA molecules, each with one conserved and one new strand of DNA. (Chapter 6)

**semilunar valve** Heart valve controlling blood flow from the ventricles into blood vessels leading away from the heart. (Chapter 19)

**seminal vesicle** Either of two pouchlike glands located on both sides of the bladder that add a fructose-rich fluid to semen prior to ejaculation. (Chapter 22)

**seminiferous tubule** Highly coiled tube in the testicles where sperm are formed. (Chapter 22)

**semipermeable** In biological membranes, a membrane that allows some substances to pass but prohibits the passage of others. (Chapter 3)

**sensory neuron** A neuron that conducts impulses from a sense organ to the central nervous system. (Chapter 23)

**sensory receptor** Any of the cellular systems that collect information about the environment inside or outside the body and transmit that information to the brain. (Chapter 23)

**sepal** The outermost floral structure, usually enclosing the other flower parts in a bud. (Chapter 24)

**Sertoli cell** Testicular cell that secretes substances that aid in the development of mature sperm. (Chapter 22)

**sex chromosome** Any of the sex-determining chromosomes (X and Y in humans). (Chapters 6, 7)

**sex determination** Determining the biological sex of an offspring. Humans have a chromosomal mechanism of sex determination in which two X chromosomes produce a female and an X and a Y chromosome produce a male. (Chapter 8)

**sex hormone** Any of the steroid hormones that affect development and functions of reproductive structures and secondary sex characteristics. (Chapters 21, 22)

**sex-linked gene** Any of the genes found on the X or Y sex chromosomes. (Chapter 8)

**sexual reproduction** Reproduction involving two parents that gives rise to offspring that have unique combinations of genes. (Chapters 6, 22)

**sexual selection** Form of natural selection that occurs when a trait influences the likelihood of mating. (Chapter 12)

**shoot system** The organs of a plant that are modified for photosynthesis: stem and leaf. (Chapter 24)

**short tandem repeat** Small repeat sequences of DNA used in profiling. (Chapter 8)

**signal transduction** When a change in a cell or its environment is relayed through various molecules and results in a cellular response. (Chapter 20)

**single nucleotide polymorphism (SNP)** A DNA sequence variation that occurs when members of species differ from each other at a single nucleotide (A, T, C, or G) locus. (Chapter 12)

**sink** A plant organ that is using carbohydrate, either for growth and metabolism or for conversion into starch. (Chapter 25)

**sinoatrial (SA) node** Region of the heart muscle that generates an electrical signal that controls heart rate. See *pacemaker*. (Chapter 19)

**sister chromatid** Either of the two duplicated, identical copies of a chromosome formed after DNA synthesis. (Chapter 6)

**skeletal muscle** Striated muscle involved with voluntary movements. (Chapter 17)

**sliding filament model** The theory that muscles contract when actin filaments slide across myosin filaments, shortening the sarcomere. (Chapter 21)

**small intestine** The narrow, twisting, upper part of the intestine where nutrients are absorbed into the blood. (Chapter 17)

**smog** Products of fossil fuel combustion in combination with sunlight, producing a brownish haze in still air. (Chapter 16)

**smooth endoplasmic reticulum** The subcellular, cytoplasmic membrane system responsible for lipid and steroid biosynthesis. (Chapter 3)

**smooth muscle** Nonstriated, spindle-shaped muscle cells that line organs and blood vessels. (Chapter 17)

**sodium potassium pump** A protein pump in a cell membrane that moves sodium out of the cell and potassium into the cell, both against their concentration gradients. (Chapter 23)

**soil erosion** Loss of topsoil. (Chapter 24)

**soil** Medium for plant growth made up of mineral particles, partially decayed organic matter, and living organisms. (Chapter 24)

**solar irradiance** The amount of solar energy hitting Earth's surface at any given point. (Chapter 16)

**solid waste** Garbage. (Chapter 16)

**solstice** When the sun reaches its maximum and minimum elevation in the sky. (Chapter 16)

**solute** The substance that is dissolved in a solution. (Chapter 2)

**solution** A mixture of two or more substances. (Chapter 2)

**solvent** A substance, such as water, that a solute is dissolved in to make a solution. (Chapter 2)

**somatic cell** Any of the body cells in an organism. Any cell that is not a gamete. (Chapter 6)

**somatic cell gene therapy** Changes to malfunctioning genes in somatic or body cells. These changes will not be passed to offspring. (Chapter 9)

**somatomammotropin** A hormone produced by the placenta that plays a role in stimulating milk production. (Chapter 22)

**source** Plant organs that are actively generating sugar. (Chapter 25)

**spatial isolation** A mechanism for reproductive isolation that depends on the geographic separation of populations. (Chapter 12)

**special creation** The hypothesis that all organisms on Earth arose as a result of the actions of a supernatural creator. (Chapter 10)

**special senses** Senses that have specialized organs. These include sight, hearing, equilibrium, taste, touch, and smell. (Chapter 23)

**speciation** Evolution of one or more species from an ancestral form; macroevolution. (Chapter 12)

**species** A group of individuals that regularly breed together and are generally distinct from other species in appearance or behavior. (Chapters 10, 12, 13, 15)

**species-area curve** Graph describing the relationship between the size of a natural landscape and the relative number of species it contains. (Chapter 15)

**specific defense** Defense against pathogens that utilizes white blood cells of the immune system. (Chapter 20)

**specificity** Phenomenon of enzyme shape determining the reaction the enzyme catalyzes. (Chapter 4)

**sperm** Gametes produced by males. (Chapters 6, 7, 22)

**spermatogenesis** Production of sperm. (Chapter 22)

**spermatozoa** Mature sperm composed of a small head containing DNA, a midpiece that contains mitochondria, and a tail (flagellum). (Chapter 22)

**spinal cord** Thick cord of nervous tissue that extends from the base of the brain through the spinal column. (Chapters 17, 23)

**spongy bone** The porous, honeycomblike material of the inner bone. (Chapter 21)

**spore** Reproductive cell in plants and fungi that is capable of developing into an adult without fusing with another cell. (Chapters 13, 24)

**stabilizing selection** Natural selection that favors the average phenotype and selects against the extremes in the population. (Chapter 11)

**stamen** Male flower part, producing the pollen. (Chapter 24)

**standard error** A measure of the variance of a sample; essentially the average distance a single data point is from the mean value for the sample. (Chapter 1)

**staple crop** One of a number of agricultural crops that make up the majority of calories in human societies. (Chapter 24)

**statistical test** Mathematical formulation that helps scientists evaluate whether the results of a single experiment demonstrate the effect of treatment. (Chapter 1)

**statistically significant** Said of results for which there is a low probability that experimental groups differ simply by chance. (Chapter 1)

**statistics** Specialized branch of mathematics used in the evaluation of experimental data. (Chapter 1)

**stem** The part of vascular plants that is above ground, as well as similar structures found underground. (Chapter 24)

**stem cell** Cells that can divide indefinitely and can differentiate into other cell types. (Chapter 9)

**steppe** Biome characterized by short grasses; found in regions with relatively little annual precipitation. (Chapter 16)

**steroid** Naturally occurring or synthetic organic fat-soluble substance that produces physiologic effects. (Chapters 2, 21)

**stigma** Sticky pad on the tip of a flower's carpel where pollen is deposited. (Chapter 24)

**stomach** A sac-like enlargement of the alimentary canal involved in digestion. (Chapter 17)

**stomata** Pores on the photosynthetic surfaces of plants that allow air into the internal structure of leaves and green stems. Stomata also provide portals through which water can escape. (Chapters 5, 24)

**stop codon** An mRNA codon that does not code for an amino acid and causes the amino acid chain to be released into the cytoplasm. (Chapter 9)

**stream** Biome characterized by flowing water, sometimes seasonal. Typically smaller than rivers. (Chapter 16)

**striated muscle** A voluntary muscle made up of elongated, multinucleated fibers. Typically skeletal and cardiac muscle that is distinguished from smooth muscle by the in-register banding patterns of actin and myosin filaments. (Chapter 17)

**stroke** Acute condition caused by a blood clot that blocks blood flow to an organ or other region of the body. (Chapter 19)

**stroma** The semi-fluid matrix inside a chloroplast where the Calvin cycle of photosynthesis occurs. (Chapters 3, 5)

**style** Stalk connecting stigma to ovary in a flower's carpel. (Chapter 24)

**subspecies** Subdivision of a species that is not reproductively isolated but represents a population or set of populations with a unique evolutionary history. See also *biological race*. (Chapter 12)

**substrate** The substance upon which an enzyme reacts. (Chapter 4)

**succession** Replacement of ecological communities over time since a disturbance, until finally reaching a stable state. (Chapter 16)

**sugar-phosphate backbone** Series of alternating sugars and phosphates along the length of the DNA helix. (Chapter 2)

**supernatural** Not constrained by the laws of nature. (Chapter 1)

**sustainable** Referring to human activities that are able to continue indefinitely with no degradation of the resources on which they depend. (Chapter 24)

**symbiosis** A relationship between two species. (Chapter 13)

**sympatric** In the same geographic region. (Chapter 12)

**synapse** Gap between neurons consisting of the terminal boutons of the presynaptic neuron, the space between the two adjacent neurons, and the membrane of the postsynaptic neuron. (Chapter 23)

**synergistic** The creation of a whole that is greater than the sum of the parts. (Chapter 6)

**systematist** Biologist who specializes in describing and categorizing a particular group of organisms. (Chapter 13)

**systemic circuit** Flow of blood from the heart to body capillaries and back to the heart. (Chapter 19)

**systole** Portion of the cardiac cycle when the heart is contracting, forcing blood into arteries. (Chapter 19)

**T lymphocyte (T cell)** Immune-system cell that develops in the thymus gland; it facilitates cell-mediated immunity. (Chapter 20)

**Taq polymerase** An enzyme that can withstand high temperatures, which is used during polymerase chain reactions. (Chapter 8)

**target cell** A cell that responds to regulatory signals such as hormones. (Chapter 21)

**telophase** Stage of mitosis during which the nuclear envelope forms around the newly produced daughter nucleus, and chromosomes decondense. (Chapter 6)

**temperate forest** Biome dominated by deciduous trees. (Chapter 16)

**temperature** A measure of the intensity of heat or kinetic energy. (Chapters 5, 16)

**temporal bone** Either of the paired bones forming the sides and base of the cranium. (Chapter 21)

**temporal isolation** Reproductive isolation between populations maintained by differences in the timing of mating or emergence. (Chapter 12)

**temporal lobe** Part of the cerebral hemisphere that processes auditory and visual information, memory, and emotion. Located in front of the occipital lobe. (Chapter 23)

**tension** In plants, negative water pressure. (Chapter 25)

**terminal bouton** Knoblike structure at the end of an axon. (Chapter 23)

**terminal bud** Apical meristem on the tip of a stem or branch. (Chapter 24)

**testable** Possible to evaluate through observations of the measurable universe. (Chapter 1)

**testicle** (*plural: testes*) The paired male gonads, involved in gametogenesis and secretion of reproductive hormones. (Chapters 21, 22)

**testosterone** Masculinizing hormone secreted by the testes. (Chapters 21, 22)

**thalamus** Main relay center between the spinal cord and the cerebrum. (Chapter 23)

**theory** See *scientific theory*. (Chapters 1, 10)

**theory of evolution** Theory that all organisms on Earth today are descendants of a single ancestor that arose in the distant past. See also *evolution*. (Chapters 1, 2, 10)

**therapeutic cloning** Using early embryos as donors of stem cells for the replacement of damaged tissues and organs in another individual. (Chapter 9)

**thermoregulation** The ability to maintain a body temperatures within a narrow range. (Chapter 17)

**thigmotropism** Growth in response to contact with a solid object. (Chapter 25)

**thylakoid** Flattened membranous sac located in the chloroplast stroma. Function in photosynthesis. (Chapters 3, 5)

**thymine** Nitrogenous base in DNA, a pyrimidine. (Chapters 2, 9)

**thymus** An endocrine gland located in the neck region that helps establish the immune system. (Chapter 21)

**thyroid gland** An endocrine gland in the neck region that stimulates metabolism and regulates blood calcium levels. (Chapter 21)

**tilling** Turning over the soil to kill weed seedlings, with the goal of removing competitors before a crop is planted. (Chapter 24)

**tissue** A group of cells with a common function. (Chapter 17)

**tongue** A muscular structure in the mouth that has taste buds that help you taste food. It aids in breaking down food for the digestive process. (Chapter 17)

**totipotent** Describes a cell able to specialize into any cell type of its species, including embryonic membrane. Compare with multipotent, pluripotent. (Chapter 9)

**toxin** poisonous substance produced by bacteria or other cell. (Chapter 22)

**trachea** Air passage from upper respiratory system into lower respiratory system. Also called windpipe. (Chapter 17)

**tracheid** Narrow xylem cell with pitted walls. (Chapter 25)

**trans fat** Contains unsaturated fatty acids that have been hydrogenated, which changes the fat from a liquid to a solid at room temperature. (Chapter 3)

**transcription** Production of an RNA copy of the protein coding DNA gene sequence. (Chapters 7, 9)

**transcription regulators** Proteins that determine whether and how frequently a gene will be transcribed into messenger RNA. (Chapter 7)

**transfer RNA (tRNA)** Amino-acid-carrying RNA structure with an anti-codon that binds to an mRNA codon. (Chapter 9)

**transgenic organism** Organism whose genome incorporates genes from another organism; also called genetically modified organism (GMO). (Chapter 9)

**translation** Process by which an mRNA sequence is used to produce a protein. (Chapter 9)

**translocation** Movement of phloem sap around the body of a plant. (Chapter 25)

**transpiration** Movement of water from the roots to the leaves of a plant, powered by evaporation of water at the leaves and the cohesive and adhesive properties of water. (Chapters 5, 25)

**trophic level** Feeding level or position on a food chain; for example, producers and primary consumers. (Chapter 15)

**trophic pyramid** Relationship among the mass of populations at each level of a food web. (Chapter 15)

**trophoblast** The outer layer of a developing blastocyst that supplies nutrition to the embryo. (Chapter 22)

**tropical forest** Biome dominated by broad-leaved, evergreen trees; found in areas where temperatures never drop below the freezing point of water. (Chapter 16)

**tropism** In plants, directional growth. (Chapter 25)

**tuberculosis (TB)** Degenerative lung disease caused by infection with the bacterium *Mycobacterium tuberculosis*. (Chapter 11)

**tumor** Mass of tissue that has no apparent function in the body. (Chapter 6)

**tumor suppressor** Cellular protein that stops tumor formation by suppressing cell division. When mutated leads to increased likelihood of cancer. (Chapter 6)

**tundra** Biome that forms under very low temperature conditions. Characterized by low-growing plants. (Chapter 16)

**turgid** In plants, cells that are filled with enough water so that the cell walls are deformed. (Chapter 24)

**understory** The level of a forest below the canopy trees and shrub cover, typically consisting of herbaceous perennial plants and tree and shrub seedlings. (Chapter 16)

**undifferentiated** A cell that is not specialized. (Chapter 9)

**unicellular** Made up of a single cell. (Chapter 13)

**uniform distribution** Occurs when individuals in a population are disbursed in a uniform manner across a habitat. (Chapter 14)

**unsaturated fat** Type of lipid containing many carbon-to-carbon double bonds; liquid at room temperature. (Chapter 3)

**unsustainable** Relating to practices that may compromise the ability of future generations to support a large human population with an adequate quality of life. (Chapter 24)

**uracil** Nitrogenous base in RNA, a pyrimidine. (Chapters 2, 9)

**urban sprawl** The tendency for the boundaries of urban areas to grow over time as people build housing and commercial districts farther and farther from an urban core. (Chapter 16)

**ureter** Tube that delivers urine from a kidney to the bladder. (Chapter 18)

**urethra** Urine-carrying duct that also carries sperm in males. (Chapters 18, 20)

**urinary bladder** An organ of the excretory system that stores urine after it is excreted from the kidneys. (Chapter 18)

**urinary system** The organ system responsible for the filtering, collection, and excretion of liquid waste; consisting of the kidneys, ureters, bladder, and urethra. (Chapter 18)

**urine** Liquid expressed by the kidneys and expelled from the bladder in mammals, containing the soluble waste products of metabolism. (Chapter 18)

**uterus** Pear-shaped muscular organ in females that can support pregnancy and that undergoes menstruation when its lining is shed. (Chapter 22)

**vaccination** A preparation of a weakened or killed pathogen, or portion of a pathogen, that will stimulate the immune system of a recipient to prepare a long-term defense (memory cells) against that pathogen. (Chapter 20)

**vagina** Muscular canal in females leading from the cervix to the vulva. (Chapter 22)

**valence shell** The outermost energy shell of an atom containing the valence electrons which are most involved in the chemical reactions of the atom. (Chapter 2)

**variable** A factor that varies in a population or over time. (Chapter 1)

**variance** Mathematical term for the amount of variation in a population. (Chapter 7)

**variant** An individual in a population that differs genetically from other individuals in the population. (Chapter 11)

**vas deferens** Either of the two ducts in males that carry sperm from the epididymis to the urethra. (Chapter 22)

**vascular cambium** Meristematic tissue that forms in vascular bundles of dicot roots and stems and permits secondary growth. (Chapter 24)

**vascular system** Plant tissue system responsible for the delivery of water and dissolved solutes throughout the plant body. (Chapter 24)

**vascular tissue** Cells that transport water and other materials within a plant. (Chapters 13, 24)

**vector** An organism that carries a pathogen from one host to another, such as a mosquito carrying West Nile virus. (Chapter 20)

**vegetative organ** Plant organ that is not involved in sexual reproduction. (Chapter 24)

**vegetative reproduction** Asexual cloning of plants. (Chapter 22)

**vein** Vessel that carries blood from the body tissues back to the heart. (Chapter 19)

**ventricle** Chamber of the heart that pumps blood from the heart to the lungs or systemic circulation. (Chapter 19)

**vertebra** (*plural*: vertebrae) Bone of the spinal column through which the spinal cord passes. (Chapter 22)

**vertebral column** Also called the spine; the series of vertebrae and cartilaginous disks extending from the brain to the pelvis. (Chapter 22)

**vertebrate** Animal with a backbone. (Chapter 13)

**vesicle** Membrane-bounded sac-like structure. In neurons, these structures are found in the terminal bouton and store neurotransmitters. (Chapter 23)

**vessel element** Wide xylem cell with perforated ends, found only in angiosperms. (Chapter 25)

**vestigial trait** Modified with no or relatively minor function compared to the function in other descendants of the same ancestor. (Chapter 10)

**villus** (*plural:* villi) Small fingerlike projections on the inside of the small intestine that function in nutrient absorption. (Chapter 18)

**viral envelope** Layer formed around some virus protein coats (capsids) that is derived from the cell membrane of the host cell and may also contain some proteins encoded by the viral genome. (Chapter 20)

**virus** Infectious intracellular parasite with its own genetic material that can only reproduce by forcing its host to make copies of it. (Chapters 1, 9)

**vitamin** Organic nutrient needed in small amounts. Most vitamins function as coenzymes. (Chapter 3)

**voluntary** Muscle normally under conscious control. Mainly skeletal muscle. (Chapter 17)

**vulva** The outer portion of the female external genitalia including the labia majora, labia minora, clitoris, and vaginal and urethral openings. (Chapter 22)

**wastewater** Liquid waste produced by residential, commercial, or industrial activities. (Chapter 16)

**water** One molecule of water consists of one oxygen and two hydrogen atoms. (Chapter 2)

**weather** Current temperature and precipitation conditions. (Chapter 16)

**weed** Common term for nonpreferred plant. (Chapter 24)

**wetland** Biome characterized by standing water, shallow enough to permit plant rooting. (Chapter 16)

**white blood cell** General term for cell of the immune system. (Chapter 19)

**white matter** Nervous system tissue, especially in the brain and spinal cord, made of myelinated cells. (Chapter 23)

**whole food** Any food that has not undergone processing. Includes grains, beans, nuts, seeds, fruits, and vegetables. (Chapter 3)

**wood** Xylem cells produced by secondary growth of stems and roots. (Chapter 24)

**woody plant** Plant that produces stiffened stems via secondary production of xylem. (Chapter 25)

**x-axis** The horizontal axis of a graph. Typically describes the independent variable. (Chapter 1)

**X-linked gene** Any of the genes located on the X chromosome. (Chapter 8)

**xylem** Plant vascular tissue, dead at maturity, that carries water and dissolved minerals in a one-way flow from the roots to the shoots of a plant. (Chapters 24, 25)

**xylem sap** Water and dissolved minerals flowing in the xylem vessels. (Chapter 25)

**y-axis** The vertical axis of a graph. Typically describes the dependent variable. (Chapter 1)

**yeast** Single-celled eukaryotic organisms found in bread dough. Often used as model organisms and in genetic engineering. (Chapters 10, 13)

**Y-linked gene** Any of the genes located on the Y chromosome. (Chapter 8)

**Z disc** The border of a sarcomere in muscle. (Chapter 23)

**zona pellucida** The substance surrounding an egg cell that a sperm must penetrate for fertilization to occur. (Chapter 22)

**zoologist** Scientist who specializes in the study of animals. (Chapter 13)

**zygote** Single cell resulting from the fusion of gametes (egg and sperm). (Chapters 6, 22)

# Credits

## Text and Art Credits

CHAPTER 1 **Savvy Reader.** Source: Gupta, Terry, "Get back your health with vitamin C, tea," The Telegraph Nashua, NH, Oct. 26, 2007. Used with permission.

**Figure 1-07.** Source: Data from: Lindenmuth, G. Frank, Ph.D. and Elise B. Lindenmuth, Ph.D., "The efficacy of echinacea compound herbal tea preparation on the severity and duration of upper respiratory and flu symptoms: A randomized, double-blind placebo-controlled study," *The Journal of Alternative and Complementary Medicine,* Vol. 6: 4, pp. 327–334, 2000, Mary Ann Liebert, Inc.

**Figure 1-10.** Source: Based on Cohen, Sheldon, Ph.D. et al., "Psychological Stress and Susceptibility to the Common Cold," *New England Journal of Medicine,* Vol. 325: 9, pp. 606–612, fig. 1, Aug. 29, 1991. Massachusetts Medical Society.

**Figure 1-13.** Source: Data from: Mossad, Sherif B., MD et al., "Zinc Gluconate Lozenges for Treating the Common Cold, A Randomized, Double-Blind, Placebo-Controlled Study," *Annals of Internal Medicine,* Vol. 125: 2, July 15, 1996.

CHAPTER 2 **Savvy Reader.** Source: © Pearson Education, Inc.

**Figure 2-09.** Source: © Pearson Education, Inc.

CHAPTER 3 **Savvy Reader.** Source: University of California at Berkeley Wellness Letter. http://www.berkeleywellness.com/article/probiotics-pros-and-cons

CHAPTER 5 **Savvy Reader.** Excerpted from: Ridley, Matt. Why Climate Change is Good for the World. The Spectator, 19 October 2013 - http://www.spectator.co.uk/features/9057151/carry-on-warming/

**Figure 5-06.** Source: Data from: "Trends in Atmospheric Carbon Dioxide—Mauna Loa," NOAA/ESRL Global Monitoring Division,www.esrl.noaa.gov/gmd/ccgg/trends/

**Figure 5-08.** Source: Based on Macmillan Publishers Ltd., from J.R. Petit et al., "Climate and atmospheric history of the past 420,000 years from the Vostok ice core, Antarctica," *Nature,* 399 (3):429– 436, June 1999. Nature Publishing, Inc.

CHAPTER 6 **Savvy Reader.** Source: © Pearson Education, Inc.

CHAPTER 7 **Savvy Reader.** Source: Livescience.com. http://www.livescience.com/25840-bullying-may-alter-genes.html

CHAPTER 8 **Savvy Reader.** Source: © Pearson Education, Inc.

CHAPTER 9 **Savvy Reader.** Source: GMO Free NY. http://gmo-freeny.net/thecaseforgmolabeling.html

**Figure 9-19.** Source: © Pearson Education, Inc.

CHAPTER 10 **Savvy Reader.** Source: "Ranks of Scientists Doubting Darwin's Theory on the Rise," Discovery Institute, Feb. 8, 2007, Discovery Institute. Used with permission.

CHAPTER 11 **Savvy Reader.** Source: Tandy, Florence, "What Gall!" http://fwtandy.wordpress.com/2007/06/08/what-gall/. Used with permission.

CHAPTER 12 **Savvy Reader.** Source: From *Slate,* April 18 © 2011 The Slate Group. All rights reserved. Used by permission and protected by the Copyright Laws of the United States. The printing, copying, redistribution, or retransmission of this Content without express written permission is prohibited.

**p.257.** Source: Elizabeth Alexander, "Praise Song for the Day," Graywolf Press, 2009.

CHAPTER 13 **Savvy Reader.** Source: *The Telegraph* - http://www.telegraph.co.uk/science/science-news/9129205/Middle-aged-people-the-pinnacle-of-evolution.html?fbA

CHAPTER 14 **Savvy Reader.** Source: "Common Birds in Decline: What's happening to birds we know and love?" State of the Birds, National Audubon Society, Summer 2007 update, http://stateofthebirds.audubon.org/cbid/. Used with permission.

CHAPTER 16 **Savvy Reader.** Source: © Pearson Education, Inc.

CHAPTER 17 **Savvy Reader.** Source: "Trafficking in Kidneys," "Study reveals consequences for people who sell kidneys," Stamatis, George, Reprinted with permission of Case Western Reserve University

CHAPTER 18 **Savvy Reader.** Source: © Pearson Education, Inc.

CHAPTER 19 **Savvy Reader.** Source: Porter, Sabrina. "Students speak out in smoking ban survey," *The Tartan,* Feb. 26, 2007. Used with permission. https://thetartan.org/2007/2/26/news/smoking

CHAPTER 20 **Savvy Reader.** Source: © Pearson Education, Inc.

CHAPTER 21 **Savvy Reader.** Source: Republished with permission of Elsevier Science Ltd., from *Drug and Alcohol Dependence,* vol. 90, nos. 2–3, 2007; permission conveyed through Copyright Clearance Center, Inc.

CHAPTER 22 **Savvy Reader.** Source: "What's a mom to do? Preventing early puberty and hormone problems in our daughters-here's the why and how." Feb. 12, 2014. Aviva Romm, MD. http://avivaromm.com/preventing-early-puberty-and-hormone-problems-in-our-daughters-heres-the-why-and-how

CHAPTER 23 **Savvy Reader.** Source: McCabe SE, Knight JR, Teter CJ, Wechsler H. *Addiction* 2005, Jan; 100(1): 96–106.

CHAPTER 24 **Savvy Reader.** Source: "Little evidence of health benefits from organic foods, study finds," by Michelle Brandt. Sept. 3, 2012. Reprinted with permission from the Stanford School of Medicine's Office of Communication and Public Affairs.

**p.531.** Source: John Steinbeck (1939). *The Grapes of Wrath,* Viking Press.

CHAPTER 25  **Savvy Reader.** Source: *University Herald*, "Want A Promotion? Place a Potted Plant at Your Cubicle!", by Stephan Adkins, Dec. 7, 2013. http://www.universityherald.com/articles/6056/20131207/promotion-potted-plant-workplace-decorative-purposes-positive-psychological-effects-health-benefits-university-of-exeter.htm

# Photo Credits

Icon (lime trees): Klaus Leidorf/Corbis

CHAPTER 1  Opener, Paul Bradbury/OJO Images/Getty Images; opener (L) Apollofoto/Shutterstock; opener (C) Edyta Pawlowska/Shutterstock; opener (R) michaeljung/Shutterstock; 1.2a Eye of Science/Science Source; 1.2b Anders Wiklund/Reuters/Landov; 1.5 The Natural History Museum/Alamy; 1.6 Zina Seletskaya/Shutterstock; 1.9a Heiti Paves/Alamy; 1.9b Stefan Klein/Getty Images; 1.9c Thomas Deerinck, NCMIR/Science Source; 1.17 GVictoria/Shutterstock

CHAPTER 2  Opener, Global Cinema Distribution/Everett Collection; opener (L) Amanda Nicholls/Shutterstock; opener (C) Suzanne Tucker/Shutterstock; opener (R) txking/Shutterstock; 2.1 Photos.com; 2.18a Juergen Berger/Science Source; 2.19a Dr. Gary Gaugler/Science Source; 2.19b Eye of Science/Science Source; 2.19c, Christopher Meder/Fotolia; 2.19d Alan49/Shutterstock; 2.19e joefotofl/Fotolia; 2.19f M. I. Walker/Science Source; 2.19g bc, Ivonne Wierink/Fotolia; 2.19h, Claude Huot/Shutterstock

CHAPTER 3  Opener, Hans Neleman/Getty Images; opener (L) Jupiterimages/Getty Images; opener (C) Tony Gentile/Reuters/Corbis; opener (R) Fuse/Getty Images; 3.1a AidanStock/Alamy; 3.1b Seralex/Fotolia; 3.7 Dr. David Furness, Keele University/Science Source; 3.8 CNRI/Science Source; 3.9 Callista Images/Cultura RM/Alamy; 3.10 Don W. Fawcett/Science Source; 3.11 Science Source; 3.12 Don W. Fawcett/Science Source; 3.13 SPL/Science Source; 3.14 David M. Phillips/Science Source; 3.15 Don W. Fawcett/Science Source; 3.16 Ron Boardman/Life Science Image/FLPA/Science Source; 3.17 BSIP SA/Alamy

CHAPTER 4  Opener, Henry Arden/Corbis; opener (L) Digital Vision/Photodisc/Getty Images; opener (C) Robert McKean/KRT/Newscom; opener (R) Christopher LaMarca/Redux; 4.9a Professors Pietro M. Motta & Tomonori Naguro/Science Source; 4.15a Lorimer Images/Shutterstock; 4.15b SCIMAT/Science Source; 4.17 Peter Anderson/DK Images

CHAPTER 5  Opener, Keith Bedford/Reuters/Corbis; opener (L) STR New/Reuters; opener (C) Ashley Cooper/Alamy; opener (R) Richard Green/Alamy; 5.5 Reuters; 5.7 British Antarctic Survey/Science Source; 5.9a Dr. David Furness/Keele University/Science Source; 5.14 br, Winton Patnode/Science Source; Table 5.1a FLariviere/Shutterstock; Table 5.1b GJones Creative/Shutterstock; Table 5.1c James Forte/National Geographic/Getty Images

CHAPTER 6  Opener, StClair/Massie/Splash News/Newscom; opener (L) Felix Vogel/imageBroker/Alamy; opener (C) Humannet/Shutterstock; opener (R) Phanie/Science Source; 6.2 Jim Arbogast/Digital Vision/Getty Images; 6.3a Cortier/BSIP/Alamy; 6.3b Christian MartA-nez Kempin/Getty Images; 6.4a Lebendkulturen.de/Shutterstock; 6.4b Rossco/Shutterstock; 6.5a Biophoto Associates/Science Source; 6.5b Science Source; 6.10a Dr. Gopal Murti/Science Source; 6.10b Kent Wood/Science Source; 6.15 PCN Photography/Alamy;

6.17a Mediscan/Medical-on-Line/Alamy; 6.17b Mediscan/Medical-on-Line/Alamy; 6.17c Mediscan/Medical-on-Line/Alamy

CHAPTER 7  Opener, Tony Brain/SPL/Science Source; opener (L) Andresr/Shutterstock; opener (C) Oksana Kuzmina/Fotolia; opener (R) Purestock/Alamy; 7.2a Gerald McCormack; 7.8 Pictorial Press Ltd/Alamy; 7.10 Linda Gordon; 7.11 Simon Fraser/Royal Victoria Infirmary, Newcastle upon Tyne/Science Source; 7.12 Dr. Kathryn Lovell, Ph.D; 7.17a Darrick E. Antell, M.D., F.A.C.S; 7.17b Darrick E. Antell, M.D., F.A.C.S; 7.18a Peter Bernik/123RF; 7.18b Peace!/A. collection/amana images/Getty Images; 7.21 Angel Franco/The New York Times/Redux Pictures; 7.23 John Sunderland/Getty Images; 7.24 Juice Images/Alamy

CHAPTER 8  Opener, Fine Art Images/Heritage Image Partnership Ltd/Alamy; opener (L) Anatoly Maltsev/EPA/Newscom; opener (C) Dr. Sergey Nikitin; opener (R) Adam Gault/OJO Images Ltd/Alamy; 8.1a Gtstudio/Shutterstock; 8.1b Jlarrumbe/Shutterstock; 8.1c Maksim Shebeko/Fotolia; 8.2a Andre@cf/Fotolia; 8.2b Luri/Shutterstock; 8.2c Sam Wirzba/Alamy; Table 8.3a Johan Pienaar/Shutterstock; Table 8.3b Shutterstock; Table 8.3c Shutterstock; Table 8.3d Arnon Ayal/Shutterstock; Table 8.3e Henrik Larsson/Shutterstock; 8.7a Biophoto Associates/Science Source; 8.7b Piotr Marcinski/Shutterstock; 8.7c Christina Kennedy/Alamy; 8.9 Zmeel Photography/E+/Getty Images; 8.10 Science and Society/SuperStock; 8.12 BSIP/Newscom; 8.13 Pearson Education; 8.15a Public domain; 8.15b Pictorial Press Ltd/Alamy

CHAPTER 9  Opener, B Christopher/Alamy; opener (L) CB2/ZOB/WENN/Newscom; opener (C) Handout/Getty Images; opener (R) Irina Rogova/Shutterstock; 9.11a Biology Pics/Science Source; 9.11b David McCarthy/Science Source; 9.14 Professor John Doebley/University of Wisconsin; 9.16a Fotokostic/Getty Images; 9.17a Pearson Education; 9.17b Freer/Shutterstock; 9.18 MCT/Landov; 9.20 Shutterstock

CHAPTER 10  Opener, Digital Vision/Getty Images; opener (L) Scala/Art Resource, NY; opener (C) The Print Collector/Fine Art/Corbis; opener (R) William Campbell/Sygma/Corbis; 10.1 Carolina K Smith MD/Fotolia; 10.3 D. Bayes/Lebrecht Music and Arts Photo Library/Alamy; 10.4a MLE/ZOJ WENN Photos/Newscom; 10.4b Interfoto/Alamy; 10.6a Daniel Padavona/Shutterstock; 10.6b Van der Meer Marica/Arterra Picture Library/Alamy; 10.7a Jeffrey M. Frank/Shutterstock; 10.7b Paul Sterry/Worldwide Picture Library/Alamy; 10.9 Fuse/Getty Images; 10.13a Karloss/Shutterstock; 10.13b Jerome Wexler/Science Source; 10.13c Martin Shields/Alamy; 10.14b1 Tom Brakefield/Stockbyte/Getty Images; 10.14b2 Bele Olmez/ImageBroker/Alamy; 10.15 M. K. Richardson; 10.16a Lori Skelton/Shutterstock; 10.16b Michael Stubblefield/Alamy; 10.16c Zoology/Inter Foto/Alamy; 10.16d Dennis Donohue/Shutterstock; 10.19d Javier Trueba/Madrid Scientific Films/Science Source; 10.24 SPL/Science Source; 10.26a Thorsten Rust/Shutterstock; 10.26b Dario Sabljak/Shutterstock; 10.26c Denis Tabler/Fotolia; 10.26d Ross Petukhov/Fotolia

CHAPTER 11  Opener, Larry W. Smith/EPA/Newscom; opener (L) Mike Cassese/Reuters; opener (C) Mediacolor's/Alamy; opener (R) Michael Coddington/Shutterstock; 11.1 Kwangshin Kim/Science Source; 11.2 Stockdevil/Fotolia; 11.3 Jenny Matthews/Alamy; 11.4 Nick Gregory/Alamy; 11.5 Courtesy of Caufield & Shook Collection/Special Collections/University of Louisville; 11.6a Dennis Donohue/Shutterstock; 11.6b Sil63/Shutterstock; 11.7 PetStockBoys/Alamy;

11.10 Archana bhartia/Shutterstock; 11.14 Heather Angel/Natural Visions/Alamy; 11.15a Chas53/Fotolia; 11.15b Nigel Cattlin/Alamy; 11.19 BSIP SA/Alamy

CHAPTER 12    Opener, Ron Edmonds/AP Images; opener (L) Sofiaworld/Shutterstock; opener (C) Demis Maryannakis/Splash News/Newscom; opener (R) Design Pics Inc./Alamy; 12.1a Keith Levit/Shutterstock; 12.1b Keith Levit/Shutterstock; 12.2a1 Eric Isselee/Shutterstock; 12.2a2 Eric Isselee/Shutterstock; 12.2a3 DK Images; Table 12.1b Mariko Yuki/Shutterstock; Table 12.1e Image Quest Marine; 12.6a Suzifoo/E+/Getty Images; 12.6b 68/Ocean/Corbis; 12.6c Peter Grosch/Shutterstock; 12.7a Professor Forrest L. Mitchell; 12.7b David Allan Brandt/Iconica/Getty Images; 12.15a Buena Vista Images/Digital Vision/Getty Images; 12.15b Ariadne Van Zandbergen/Alamy; 12.16a Dex Image/Collage/Corbis; 12.16b Hiroshi Sato/Shutterstock; 12.20a Derek Hall/DK Images; 12.20b Ivanna Reznichenko/Getty Images; 12.21a Dave Logan/Getty Images; 12.21b Bobkeenan Photography/Shutterstock; 12.21c Kdmot/Getty Images; 12.22 Eric Gay/AP Images

CHAPTER 13    Opener, Jgz/Fotolia; opener (L) Sheri Armstrong/Fotolia; opener (C) Michael Abbey/Science Source; opener (R) Imago Sportfotodienst/Newscom; 13.1a ImageBroker/Alamy; 13.1b Mauro Fermariello/Science Source; 13.5 J. William Schopf, University of California at Los Angeles; 13.6a Custom Medical Stock Photo/Alamy; 13.6b NASA; 13.6c David Wall/Alamy; 13.8a Mike Peres/RBP SPAS/CMSP Biology/Newscom; 13.8b Alex Wild Photography; 13.9 Chase Studio/Science Source; Table 13.3a Jubal Harshaw/Shutterstock; Table 13.3b MedicalRF.com/Alamy; Table 13.3c Lebendkulturen.de/Shutterstock; Table 13.3d Torsten Lorenz/Shutterstock; Table 13.3e Dee Breger/Science Source; Table 13.3f Jeffrey Waibel/Getty Images; Table 13.3g Lebendkulturen.de/Shutterstock; 13.10a PhotosByNancy/Shutterstock; 13.10b Jean-Paul Ferrero/Mary Evans Picture Library Ltd/AGE Fotostock; 13.19c Crisod/Fotolia; 13.10d Steffen & Alexandra/Mary Evans Picture Library Ltd/AGE Fotostock; Table 13.4a Jolanta Wojcicka/Shutterstock; Table 13.4b Mark Aplet/Shutterstock; Table 13.4c Dickson Despommier/Science Source; Table 13.4d Thierry Duran/Shutterstock.com; Table 13.4e D. Kucharski K. Kucharska/Shutterstock; Table 13.4f Carsten Medom Madsen/Shutterstock; Table 13.4g paytai/Shutterstock; Table 13.4h NatalieJean/Shutterstock; Table 13.4i Tom McHugh/Science Source; 13.11a Thomas J. Peterson/Alamy; 13.11b Images & Stories/Alamy; 13.11c Ekkapan Poddamrong8/Shutterstock; 13.13 Patrick Landmann/Science Source; Table 13.5a Biophoto Associates/Science Source; Table 13.5b Robin Treadwell/Science Source; Table 13.5c Nmelnychuk/Fotolia; Table 13.5d Henk Bentlage/Shutterstock; 13.14 Robin Treadwell/Science Source; Table 13.6a Lucie Zapletalova/Shutterstock; Table 13.6b Chursina Viktoriia/Shutterstock; Table 13.6c Debbie Aird Photography/Shutterstock; Table 13.6d Digital Vision/Photodisc/Getty Images; 13.15a Christian Ziegler/Newscom; 13.15b konzeptm/Fotolia; 13.15c Photo by Peter S. Goltra for the National Tropical Botanical Garden; 13.17 Dr. Linda M. Stannard/University of Cape Town/Science Source; 13.18a Bruce MacQueen/Shutterstock; 13.18b Goran Bogicevic/Getty Images; 13.18c Steven Cooper/Getty Images; 13.18d iphoto5151/Fotolia

CHAPTER 14    Opener, Reto Stockli/Goddard Laboratory for Atmospheres/NASA; opener (L), Raga Jose Fuste/Prisma Bildagentur AG/Alamy; opener (C) Neil Cooper/Alamy; opener (R) Monkey Business Images/Shutterstock; 14.2a Onepony/Fotolia; 14.2b

Kevin Schafer/Alamy; 14.2c Matauw/Getty Images; 14.4a Danita Delimont/DanitaDelimont.com/Pete Oxford/Alamy; 14.7a Martin Shields/Alamy; 14.7b Robert Pickett/Papilio/Alamy; 14.7c Wojciech Nowak/Fotolia; 14.7d Bartosz Hadyniak/Getty Images; 14.10a Hungry Planet/Peter Menzel; 14.10b Hungry Planet/Peter Menzel; 14.12 Alexander Chaikin/Shutterstock; 14.15 John Van Hasselt/Sygma/Corbis

CHAPTER 15    Opener, lrh847/Getty Images; opener (L) EPA/Landov; opener (C) Jana Shea/Shutterstock; opener (R) Michael Neelon/Alamy; 15.1 Laura Romin & Larry Dalton/Alamy; 15.5a David Ponton/Design Pics/Stockbyte/Getty Images; 15.5b Tailored Photography/Shutterstock; 15.5c Adrian Arbib/Alamy; 15.5d Thomas Pickard/E+/Getty Images; 15.6 Petr Malek/Getty Images; 15.9 Chuck Pratt/Bruce Coleman Inc./Alamy; 15.10 Steve Helber/AP Images; 15.11a Norman Chan/Shutterstock; 15.11b John Doebley/University of Wisconsin; 15.11c Juan A. Morales-Ramos; 15.13 Peter Malsbury/E+/Getty Images; 15.14 Esterio/Shutterstock; 15.15a G. Ronald Austing/Science Source; 15.15b Tracy Ferrero/Alamy; 15.17a h46it/Getty Images; 15.17b Jupiterimages/Getty Images; 15.19a Dave Hansen/University of Minnesota Agricultural Experiment Station; 15.19b Dave Hansen/University of Minnesota Agricultural Experiment Station; 15.20 Ostill/Shutterstock; 15.27 Petty Officer 2nd Class Molly A. Burgess, USN/U.S. Department of Defense Visual Information Center

CHAPTER 16    Opener, Davor Lovincic/Getty Images; opener (L) Seread/Fotolia; opener (C) Jim Kruger/E+/Getty Images; opener (R) Photodisc/Houghton Mifflin Harcourt/Getty Images; 16.4 Frank Zullo/Getty Images; 16.8 Dhoxax/Shutterstock; Table 16.1a Stillfx/Getty Images; Table 16.1b R. S. Ryan/Shutterstock; Table 16.1c TT/Getty Images; Table 16.1d Mguntow/123RF; Table 16.1e Mac99/Getty Images; Table 16.1f JosjeN/Shutterstock; Table 16.1g Charles Krebs/Getty Images; Table 16.1h Jennifer Stone/Shutterstock; 16.10 Thomas R. Fletcher/Alamy; 16.11 Eco Images/Universal Images Group/Getty Images; 16.12 Gucio_55/Shutterstock; 16.13a BARNpix/Alamy; 16.13b Natali26/Shutterstock; 16.13c Mares Lucian/Shutterstock; 16.14 Shutterstock; 16.15 AfriPics.com/Alamy; 16.16 Liz Every/AGE Fotostock; 16.18a Aurora Photos/Alamy; 16.18b Alexandra Jones/Ecoscene; Table 16.2a Robert Pickett/Papilio/Alamy; Table 16.2b Ernest Manewal/Photolibrary/Getty Images; Table 16.2c Lynda Lehmann/Shutterstock; Table 16.2d1 Shime/Fotolia; Table 16.2d2 Ralph White/Corbis; Table 16.2d3 Bernard Castelein/Nature Picture Library; Table 16.2e Fotolia; Table 16.2f Martin H Smith/FLPA/AGE Fotostock; 16.19a Przemyslaw Wasilewski/Shutterstock; 16.19b Roberto Tetsuo Okamura/Shutterstock; 16.20 D P Wilson/FLPA/AGE Fotostock; 16.21a Borisoff/Fotolia; 16.21b Tubuceo/Shutterstock; 16.23 Alan Gignoux/Alamy; 16.24 Billy Wellborn/Fotolia; 16.26 Urbanhearts/Fotolia; 16.27 Kansas City Star/Getty Images

CHAPTER 17    Opener, St Petersburg Times/ZUMA Press/Newscom; opener (L) l, Superstock; opener (C) Lord_Ghost/Fotolia; opener (R) Monkey Business Images/Shutterstock; 17.1b Ed Reschke/Getty Images; 17.1c Nina Zanetti/Pearson Education; 17.2 AFP/Getty Images/Newscom; 17.4a Robert Tallitsch, Pearson Education; 17.4b Robert Tallitsch, Pearson Education; 17.4c Robert Tallitsch, Pearson Education; 17.9 Brian Walker/AP Images

CHAPTER 18    Opener, Amana Images Inc./Alamy; opener (L) A J James/Getty Images; opener (C) Getty Images/Comstock/Thinkstock; opener (R) Facai/Shutterstock; 18.3 CNRI/Science Source

# Index

Page references that refer to figures are in **bold**. Page references that refer to tables are in *italic*.